Mathematical Modeling of Fluid Flow and Heat Transfer in Petroleum Industries and Geothermal Applications

Mathematical Modeling of Fluid Flow and Heat Transfer in Petroleum Industries and Geothermal Applications

Special Issue Editor
Mehrdad Massoudi

MDPI • Basel • Beijing • Wuhan • Barcelona • Belgrade • Manchester • Tokyo • Cluj • Tianjin

Special Issue Editor
Mehrdad Massoudi
Carnegie Mellon University
USA

Editorial Office
MDPI
St. Alban-Anlage 66
4052 Basel, Switzerland

This is a reprint of articles from the Special Issue published online in the open access journal *Energies* (ISSN 1996-1073) (available at: https://www.mdpi.com/journal/energies/special_issues/fluid_heat_petroleum_geothermal).

For citation purposes, cite each article independently as indicated on the article page online and as indicated below:

LastName, A.A.; LastName, B.B.; LastName, C.C. Article Title. *Journal Name* **Year**, *Article Number*, Page Range.

ISBN 978-3-03928-720-8 (Pbk)
ISBN 978-3-03928-721-5 (PDF)

© 2020 by the authors. Articles in this book are Open Access and distributed under the Creative Commons Attribution (CC BY) license, which allows users to download, copy and build upon published articles, as long as the author and publisher are properly credited, which ensures maximum dissemination and a wider impact of our publications.

The book as a whole is distributed by MDPI under the terms and conditions of the Creative Commons license CC BY-NC-ND.

Contents

About the Special Issue Editor . . . . **ix**

Mehrdad Massoudi
Mathematical Modeling of Fluid Flow and Heat Transfer in Petroleum Industries and Geothermal Applications
Reprinted from: *Energies* **2020**, *13*, 1344, doi:10.3390/en13061344 . . . . **1**

Jingnan Dong, Mian Chen, Yuwei Li, Shiyong Wang, Chao Zeng and Musharraf Zaman
Experimental and Theoretical Study on Dynamic Hydraulic Fracture
Reprinted from: *Energies* **2019**, *12*, 397, doi:10.3390/en12030397 . . . . **5**

Hui Liu, Shanjun Mao and Mei Li
A Case Study of an Optimized Intermittent Ventilation Strategy Based on CFD Modeling and the Concept of FCT
Reprinted from: *Energies* **2019**, *12*, 721, doi:10.3390/en12040721 . . . . **27**

Daigang Wang, Yong Li, Jing Zhang, Chenji Wei, Yuwei Jiao and Qi Wang
Improved CRM Model for Inter-Well Connectivity Estimation and Production Optimization: Case Study for Karst Reservoirs
Reprinted from: *Energies* **2019**, *12*, 816, doi:10.3390/en12050816 . . . . **43**

Jie Bai, Huiqing Liu, Jing Wang, Genbao Qian, Yongcan Peng, Yang Gao, Lin Yan and Fulin Chen
CO_2, Water and N_2 Injection for Enhanced Oil Recovery with Spatial Arrangement of Fractures in Tight-Oil Reservoirs Using Huff-'n-puff
Reprinted from: *Energies* **2019**, *12*, 823, doi:10.3390/en12050823 . . . . **59**

He Meng, Qiang Shi, Tangyan Liu, FengXin Liu and Peng Chen
The Percolation Properties of Electrical Conductivity and Permeability for Fractal Porous Media
Reprinted from: *Energies* **2019**, *12*, 1085, doi:10.3390/en12061085 . . . . **91**

Kai Liao, Shicheng Zhang, Xinfang Ma and Yushi Zou
Numerical Investigation of Fracture Compressibility and Uncertainty on Water-Loss and Production Performance in Tight Oil Reservoirs
Reprinted from: *Energies* **2019**, *12*, 1189, doi:10.3390/en12071189 . . . . **107**

Dongdong Liu and Yanyong Xiang
A Semi-Analytical Method for Three-Dimensional Heat Transfer in Multi-Fracture Enhanced Geothermal Systems
Reprinted from: *Energies* **2019**, *12*, 1211, doi:10.3390/en12071211 . . . . **127**

Vasyl Zapukhliak, Lyubomyr Poberezhny, Pavlo Maruschak, Volodymyr Grudz Jr., Roman Stasiuk, Janette Brezinová and Anna Guzanová
Mathematical Modeling of Unsteady Gas Transmission System Operating Conditions under Insufficient Loading
Reprinted from: *Energies* **2019**, *12*, 1325, doi:10.3390/en12071325 . . . . **139**

Penghui Su, Zhaohui Xia, Ping Wang, Wei Ding, Yunpeng Hu, Wenqi Zhang and Yujie Peng
Fractal and Multifractal Analysis of Pore Size Distribution in Low Permeability Reservoirs Based on Mercury Intrusion Porosimetry
Reprinted from: *Energies* **2019**, *12*, 1337, doi:10.3390/en12071337 . . . . **153**

Peng Wang, Shangmin Ao, Bo Yu, Dongxu Han and Yue Xiang
An Efficiently Decoupled Implicit Method for Complex Natural Gas Pipeline Network Simulation
Reprinted from: *Energies* **2019**, *12*, 1516, doi:10.3390/en12081516 169

Muhammad Kabir Abba, Athari Al-Otaibi, Abubakar Jibrin Abbas, Ghasem Ghavami Nasr and Martin Burby
Influence of Permeability and Injection Orientation Variations on Dispersion Coefficient during Enhanced Gas Recovery by CO_2 Injection
Reprinted from: *Energies* **2019**, *12*, 2328, doi:10.3390/en12122328 197

Pengfei Shan, Xingping Lai and Xiaoming Liu
Correlational Analytical Characterization of Energy Dissipation-Liberation and Acoustic Emission during Coal and Rock Fracture Inducing by Underground Coal Excavation
Reprinted from: *Energies* **2019**, *12*, 2382, doi:10.3390/en12122382 213

Chengcheng Tao, Barbara G. Kutchko, Eilis Rosenbaum, Wei-Tao Wu and Mehrdad Massoudi
Steady Flow of a Cement Slurry
Reprinted from: *Energies* **2019**, *12*, 2604, doi:10.3390/en12132604 229

Natalie Nakaten and Thomas Kempka
Techno-Economic Comparison of Onshore and Offshore Underground Coal Gasification End-Product Competitiveness
Reprinted from: *Energies* **2019**, *12*, 3252, doi:10.3390/en12173252 255

Qiang Wang, Yongquan Hu, Jinzhou Zhao, Lan Ren, Chaoneng Zhao and Jin Zhao
Multiscale Apparent Permeability Model of Shale Nanopores Based on Fractal Theory
Reprinted from: *Energies* **2019**, *12*, 3381, doi:10.3390/en12173381 283

Ruiyao Zhang, Jun Li, Gonghui Liu, Hongwei Yang and Hailong Jiang
Analysis of Coupled Wellbore Temperature and Pressure Calculation Model and Influence Factors under Multi-Pressure System in Deep-Water Drilling
Reprinted from: *Energies* **2019**, *12*, 3533, doi:10.3390/en12183533 301

Xinbo Lei, Xiuhua Zheng, Chenyang Duan, Jianhong Ye and Kang Liu
Three-Dimensional Numerical Simulation of Geothermal Field of Buried Pipe Group Coupled with Heat and Permeable Groundwater
Reprinted from: *Energies* **2019**, *12*, 3698, doi:10.3390/en12193698 329

Shuo Mi, Zongliu Huang, Xin Jin, Mahdi Tabatabaei Malazi and Mingming Liu
Numerical Study of Highly Viscous Fluid Sloshing in the Real-Scale Membrane-Type Tank
Reprinted from: *Energies* **2019**, *12*, 4244, doi:10.3390/en12224244 345

Yuzhe Cai and Arash Dahi Taleghani
Semi-Analytical Model for Two-Phase Flowback in Complex Fracture Networks in Shale Oil Reservoirs
Reprinted from: *Energies* **2019**, *12*, 4746, doi:10.3390/en12244746 359

Yongquan Hu, Qiang Wang, Jinzhou Zhao, Shouchang Xie and Hong Jiang
A Novel Porous Media Permeability Model Based on Fractal Theory and Ideal Particle Pore-Space Geometry Assumption
Reprinted from: *Energies* **2020**, *13*, 510, doi:10.3390/en13030510 385

Chengcheng Tao, Barbara G. Kutchko, Eilis Rosenbaum and Mehrdad Massoudi
A Review of Rheological Modeling of Cement Slurry in Oil Well Applications
Reprinted from: *Energies* **2020**, *13*, 570, doi:10.3390/en13030570 . . . . **403**

About the Special Issue Editor

Mehrdad Massoudi, PhD. Adjunct Professor. Department of Biomedical Engineering, Carnegie Mellon University. Pittsburgh, PA 15213. Dr. Massoudi received his BS, MS, and PhD in 1979, 1982, and 1986 from the University of Pittsburgh. His research interests are in the areas of mathematical modeling of non-linear materials, non-Newtonian fluid mechanics, multiphase flows, and granular materials. He is an ASME Fellow and the Editor-in-Chief of Fluids.

Editorial

Mathematical Modeling of Fluid Flow and Heat Transfer in Petroleum Industries and Geothermal Applications

Mehrdad Massoudi

U. S. Department of Energy, National Energy Technology Laboratory (NETL), Pittsburgh, PA 15236-0940, USA; Mehrdad.Massoudi@NETL.DOE.GOV

Received: 12 February 2020; Accepted: 12 March 2020; Published: 13 March 2020

Keywords: mathematical modeling; computational fluid dynamics (CFD); drilling; porous media; multiphase flow; hydraulic fracturing; geothermal; enhanced oil recovery

---

Introduction

This Special Issue of Energies is dedicated to all aspects of fluid flow and heat transfer in geothermal applications, including the ground heat exchanger, conduction, and convection in porous media. The emphasis is on mathematical and computational aspects of fluid flow in conventional and unconventional reservoirs, geothermal engineering, fluid flow, and heat transfer in drilling engineering and enhanced oil recovery applications. I would like to thank all the authors who contributed to this special issue. A brief outline of each paper is given below.

For the first time, the direct observations and theoretical analyses of the relationship between the crack tip and the fluid front in a dynamic hydraulic fracture are presented. Dong et al. [1] present their observations and theoretical analyses of the relationship between the crack tip and the fluid front in a dynamic hydraulic fracture. Liu et al. [2] use the concept of frequency conversion technology (FCT) with an optimized intermittent ventilation strategy to model a coal mine by using computational fluid dynamic (CFD) approaches to study the spatiotemporal characteristics of airflow behavior and methane distribution. Wang et al. [3] use the reservoir flow dynamics approach based on the traditional capacitance–resistance (CRM) models and Darcy's percolation theory to study injector–producer-pair-based CRM models with the newly-developed Stochastic Simplex Approximate Gradient (StoSAG) optimization algorithm to estimate a waterflood operation. Bai et al. [4] use the natural tight cores from J field in China to conduct experimental studies on different fluid huff-'n-puff processes; they proposed a new core-scale fracture lab-simulation method, and their results showed that, regardless of the arrangement of fractures, CO_2 has mostly obvious advantages over water and N_2 in tight reservoir development in huff-'n-puff modes. Meng at al. [5] develop analytical percolation expressions for conductivity and permeability using fractal theory by introducing the critical porosity. Their simulations indicate that the critical porosity could lead to the non-Archie phenomenon and that increasing critical porosity could significantly affect the permeability and the conductivity.

Liao et al. [6] look at a series of mechanistic models by considering stress-dependent porosity and permeability; they study the impacts of fracture uncertainties, such as natural fracture density, proppant distribution, and natural fracture heterogeneity, etc., and notice that fracture closure during the flowback can promote water imbibition into the matrix and delay the oil breakthrough time. Liu and Xiang [7] present a temporal semi-analytical method to study an enhanced geothermal system (EGS), where the finite-scale fractures and three-dimensional conduction in the rock matrix are considered. Their results indicate that enlarging the spacing between the fractures and increasing the number of fractures can improve the heat extraction. Zapukhliak et al. [8] propose a mathematical model to estimate the amplitude of the pressure fluctuations in a gas pipeline; their study indicates that the shutdown of compressor stations along the pipeline route can significantly impact the unsteady transient operating conditions. In order to understand the nature of the pore structures in sandstone reservoirs, Su et al. [9] apply single fractal theory and multifractal theory to study the characteristics of pore size distributions based on mercury intrusion porosimetry. Their results indicate that multifractal theory can better quantitatively characterize the heterogeneity of pore structures. Wang et al. [10] extend their Decoupled Implicit Method for Efficient Network Simulation (DIMENS), which was based on the 'Divide-and-Conquer Approach' ideal, to the continuity/momentum and energy equations coupled with the complex pipeline network. Their results indicate that the accuracy of the proposed method is equivalent to that of the Stoner Pipeline Simulator (SPS), used in commercially available simulation core codes, while the new method is more efficient than that of the SPS. Abba et al. [11] study the influence of permeability of the porous media with respect to the injection orientations during enhanced gas recovery (EGR) by CO2 injection using different core samples with different petrophysical properties. Their study suggests a revision of the CO2 plume propagation at reservoir conditions during injection.

Shan et al. [12] use an acoustic emission (AE) test to study the energy dissipation of coal and rock fracture due to underground coal excavation. They also devise a testing method to look at the energy release mechanism from damage to fracture of the unloading coal and rock under uniaxial compressive loading. Tao et al. [13] study the fully developed flow of a cement slurry. The slurry in the wellbore is modeled as a modified form of the second grade (Rivlin–Ericksen) fluid; they also suggest using a diffusion flux vector for the concentration of particles. They perform a parametric study and look at the effect of various dimensionless numbers on the velocity and the volume fraction profiles. Nakaten and Kempka [14] attempt to identify economically-competitive, site-specific end-use options for onshore- and offshore-produced underground coal gasification (UCG) synthesis gas, where the capture and storage (CCS) and/or utilization (CCU) of produced CO2 is considered. Their study shows that air-blown gasification scenarios are the most cost-effective ones. Wang et al. [15] develop a mathematical model of gas flow in nano-pores of shale and obtain a new shale apparent permeability model. Their study shows the influences of pressure, temperature, shape of the pore surface, and the tortuous fractal dimension on the apparent permeability, slip flow, Knudsen diffusion, and surface diffusion of the shale gas transport mechanism. Zhang et al. [16] discuss the variation of wellbore temperature and bottom-hole pressure during deep-water drilling circulation. They develop a simple model which is discretized and solved using the finite difference method and Gauss Seidel iteration. Their results indicate that the temperate variation in the annulus is sensitive to the location of the multi-pressure system.

To study the flow and heat transfer in non-isothermal conditions in porous media, where the effects of permeable groundwater and tube group are included, Lei et al. [17] develop a model and test its accuracy via the sandbox test and on-site thermal response test. Their results indicate that the tube group and permeable groundwater can affect the heat transfer and the ground temperature field of a buried pipe. Mi et al. [18] look at the highly viscous fluid (glycerin) sloshing. They use the full-scale membrane-type tank and numerically study the two-phase flow based on the spatially averaged Navier–Stokes equations. The pressures and the slamming effects are analyzed, and the frequency response is identified by the fast Fourier transformation technology. Cai and Dahi Taleghani [19] propose a semi-analytical method which can be used to characterize complex fracture networks generated during hydraulic fracturing. Their two-phase oil–water flowback model with a matrix oil influx for wells with bi-wing planar fractures can provide a simple way of studying early flowback in complex fracture networks. Hu et al. [20] propose a new porous media permeability model by using particle models, capillary bundle models, and fractal theory. Their results indicate that the tortuosity fractal dimension is negatively correlated with porosity, whereas the pore area fractal dimension is positively correlated with porosity. Also, they mention that as the particle radius increases, the greater the permeability difference coefficient will become. Tao et al. [21] provide a comprehensive review of the rheological modeling of cement slurries, which in general, behave as complex non-linear fluids. They also discuss the impact of the concentration of cement particles, water-to-cement ratio, additives/admixtures, shear rate, temperature and pressure, mixing methods, and the thixotropic behavior of cement on the stress tensor.

Acknowledgments: Without the help of qualified reviewers, it would not have been possible to organize this Special Issue. I am grateful to all the anonymous reviewers for their help. A personal note of appreciation and gratitude to Howland (Hao) Wu and the editorial staff at Energies; without their help and assistance, Energies could not publish high quality papers in a short period of time.

Conflicts of Interest: The author declares no conflict of interest.

References

1. Dong, J.; Chen, M.; Li, Y.; Wang, S.; Zeng, C.; Zaman, M. Experimental and Theoretical Study on Dynamic Hydraulic Fracture. *Energies* **2019**, *12*, 397. [CrossRef]
2. Liu, H.; Mao, S.; Li, M. A Case Study of an Optimized Intermittent Ventilation Strategy Based on CFD Modeling and the Concept of FCT. *Energies* **2019**, *12*, 721. [CrossRef]
3. Wang, D.; Li, Y.; Zhang, J.; Wei, C.; Jiao, Y.; Wang, Q. Improved CRM Model for Inter-Well Connectivity Estimation and Production Optimization: Case Study for Karst Reservoirs. *Energies* **2019**, *12*, 816. [CrossRef]
4. Bai, J.; Liu, H.; Wang, J.; Qian, G.; Peng, Y.; Gao, Y.; Yan, L.; Chen, F. CO2, Water and N2 Injection for Enhanced Oil Recovery with Spatial Arrangement of Fractures in Tight-Oil Reservoirs Using Huff-'n-puff. *Energies* **2019**, *12*, 823. [CrossRef]
5. Meng, H.; Shi, Q.; Liu, T.; Liu, F.; Chen, P. The Percolation Properties of Electrical Conductivity and Permeability for Fractal Porous Media. *Energies* **2019**, *12*, 1085. [CrossRef]
6. Liao, K.; Zhang, S.; Ma, X.; Zou, Y. Numerical Investigation of Fracture Compressibility and Uncertainty on Water-Loss and Production Performance in Tight Oil Reservoirs. *Energies* **2019**, *12*, 1189. [CrossRef]
7. Liu, D.; Xiang, Y. A Semi-Analytical Method for Three-Dimensional Heat Transfer in Multi-Fracture Enhanced Geothermal Systems. *Energies* **2019**, *12*, 1211. [CrossRef]
8. Zapukhliak, V.; Poberezhny, L.; Maruschak, P.; Grudz Jr., V.; Stasiuk, R.; Brezinová, J.; Guzanová, A. Mathematical Modeling of Unsteady Gas Transmission System Operating Conditions under Insufficient Loading. *Energies* **2019**, *12*, 1325. [CrossRef]
9. Su, P.; Xia, Z.; Wang, P.; Ding, W.; Hu, Y.; Zhang, W.; Peng, Y. Fractal and Multifractal Analysis of Pore Size Distribution in Low Permeability Reservoirs Based on Mercury Intrusion Porosimetry. *Energies* **2019**, *12*, 1337. [CrossRef]
10. Wang, P.; Ao, S.; Yu, B.; Han, D.; Xiang, Y. An Efficiently Decoupled Implicit Method for Complex Natural Gas Pipeline Network Simulation. *Energies* **2019**, *12*, 1516. [CrossRef]
11. Abba, M.K.; Al-Otaibi, A.; Abbas, A.J.; Nasr, G.G.; Burby, M. Influence of Permeability and Injection Orientation Variations on Dispersion Coefficient during Enhanced Gas Recovery by CO2 Injection. *Energies* **2019**, *12*, 2328. [CrossRef]
12. Shan, P.; Lai, X.; Liu, X. Correlational Analytical Characterization of Energy Dissipation-Liberation and Acoustic Emission during Coal and Rock Fracture Inducing by Underground Coal Excavation. *Energies* **2019**, *12*, 2382. [CrossRef]
13. Tao, C.; Kutchko, B.G.; Rosenbaum, E.; Wu, W.-T.; Massoudi, M. Steady Flow of a Cement Slurry. *Energies* **2019**, *12*, 2604. [CrossRef]
14. Nakaten, N.; Kempka, T. Techno Economic Comparison of Onshore and Offshore Underground Coal Gasification End-Product Competitiveness. *Energies* **2019**, *12*, 3252. [CrossRef]
15. Wang, Q.; Hu, Y.; Zhao, J.; Ren, L.; Zhao, C.; Zhao, J. Multiscale Apparent Permeability Model of Shale Nanopores Based on Fractal Theory. *Energies* **2019**, *12*, 3381. [CrossRef]
16. Zhang, R.; Li, J.; Liu, G.; Yang, H.; Jiang, H. Analysis of Coupled Wellbore Temperature and Pressure Calculation Model and Influence Factors under Multi-Pressure System in Deep-Water Drilling. *Energies* **2019**, *12*, 3533. [CrossRef]
17. Lei, X.; Zheng, X.; Duan, C.; Ye, J.; Liu, K. Three-Dimensional Numerical Simulation of Geothermal Field of Buried Pipe Group Coupled with Heat and Permeable Groundwater. *Energies* **2019**, *12*, 3698. [CrossRef]
18. Mi, S.; Huang, Z.; Jin, X.; Tabatabaei Malazi, M.; Liu, M. Numerical Study of Highly Viscous Fluid Sloshing in the Real-Scale Membrane-Type Tank. *Energies* **2019**, *12*, 4244. [CrossRef]

19. Cai, Y.; Dahi Taleghani, A. Semi-Analytical Model for Two-Phase Flowback in Complex Fracture Networks in Shale Oil Reservoirs. *Energies* **2019**, *12*, 4746. [CrossRef]
20. Hu, Y.; Wang, Q.; Zhao, J.; Xie, S.; Jiang, H. A Novel Porous Media Permeability Model Based on Fractal Theory and Ideal Particle Pore-Space Geometry Assumption. *Energies* **2020**, *13*, 510. [CrossRef]
21. Tao, C.; Kutchko, B.G.; Rosenbaum, E.; Massoudi, M. A Review of Rheological Modeling of Cement Slurry in Oil Well Applications. *Energies* **2020**, *13*, 570. [CrossRef]

© 2020 by the author. Licensee MDPI, Basel, Switzerland. This article is an open access article distributed under the terms and conditions of the Creative Commons Attribution (CC BY) license (http://creativecommons.org/licenses/by/4.0/).

Article

Experimental and Theoretical Study on Dynamic Hydraulic Fracture

Jingnan Dong [1,2,3], Mian Chen [1,2,4,*], Yuwei Li [4], Shiyong Wang [1,2], Chao Zeng [5] and Musharraf Zaman [3]

1. State Key Laboratory of Petroleum Resources and Prospecting, Beijing 102249, China; 15210877909@163.com (J.D.); shiyong_ponder@outlook.com (S.W.)
2. College of Petroleum Engineering, China University of Petroleum, Beijing 102249, China
3. Mewbourne School of Petroleum and Geological Engineering, University of Oklahoma, Norman, OK 73019, USA; zaman@ou.edu
4. Institute of Unconventional Oil & Gas, Northeast Petroleum University, Daqing 163318, China; liyuweibox@126.com
5. Department of Civil, Architectural and Environmental Engineering, Missouri University of Sciences and Technology, Rolla, MO 65409, USA; zc727@mst.edu
* Correspondence: chenchinacup@163.com; Tel.: +86-138-0103-9464

Received: 7 January 2019; Accepted: 24 January 2019; Published: 27 January 2019

Abstract: Hydraulic fracturing is vital in the stimulation of oil and gas reservoirs, whereas the dynamic process during hydraulic fracturing is still unclear due to the difficulty in capturing the behavior of both fluid and fracture in the transient process. For the first time, the direct observations and theoretical analyses of the relationship between the crack tip and the fluid front in a dynamic hydraulic fracture are presented. A laboratory-scale hydraulic fracturing device is built. The momentum-balance equation of the fracturing fluid is established and numerically solved. The theoretical predictions conform well to the directly observed relationship between the crack tip and the fluid front. The kinetic energy of the fluid occupies over half of the total input energy. Using dimensionless analyses, the existence of equilibrium state of the driving fluid in this dynamic system is theoretically established and experimentally verified. The dimensionless separation criterion of the crack tip and the fluid front in the dynamic situation is established and conforms well to the experimental data. The dynamic analyses show that the separation of crack tip and fluid front is dominated by the crack profile and the equilibrium fluid velocity. This study provides a better understanding of the dynamic hydraulic fracture.

Keywords: dynamic hydraulic-fracturing experiments; dynamic crack tip; fluid front kinetics; energy conservation analysis

1. Introduction

Hydraulic fracturing is an inherently unstable process. Even the controlled and overall stable hydraulic fracture propagation is unstable at the small scale [1]. Most previous studies on hydraulic fracture are based on the quasi-static treatment of the fracturing fluid and the inertial force of the fluid is neglected [2–4]. The dynamic situation is fundamentally different from the quasi-static simulation. For example, it has been shown that in the quasi-static fluid-driven crack, the fluid front has a relationship of $x \sim t^{1/2}$ with time t in a quasi-static situation where the inertial force is neglected and viscosity dominates the fluid behavior [4]. According to the dynamic simulation considered herein, the velocity of fluid front advancement is approximately constant in a propagating crack traveling at a constant speed. A recent numerical study [5] on a dynamic hydraulic fracture taking the inertia effect into account has shown that the hydraulic fracture propagates in a stepwise manner with fluid pressure oscillations. The stepwise propagation of the crack and the pressure oscillations indicate that the fluid inertia plays an important role in the dynamic fracturing, especially within the short time period right after the crack initiation.

The most widely known phenomenon of the relationship between crack tip and fluid front is the "lag region (dry zone)". The lag region is the space between the crack tip and fluid front. Under some specific conditions [4], the lag region could be considered infinitely small, exerting no influence on the fracturing behavior. There are also studies [6–9] showing that although the length of lag zone is indeed very small, its effect on fracture geometry cannot be ignored. For example, Bunger [10] considered the lag region to be filled by vapor from the fluid at a pressure that is negligibly small compared to the characteristic pressure in the fracturing process. The existence of the lag region induces another characteristic length besides the crack length. Previous studies related to the lag region have all been based on static or quasi-static assumptions of fluid behavior. The present study provides both experimental and theoretical insights on the lag region in dynamic situation.

Recently, the dynamic hydraulic fracturing was numerically investigated by Khoei et al. [11] and Kostov et al. [12] using Extended Finite Element Method (XFEM). To the authors' knowledge, no previous studies have reported direct observations of the real dynamic process of hydraulic fracture. In terms of experimental observations, we have limited knowledge on the effect of inertia after the initiation of the hydraulic fracture. In most previous experimental studies [10,13–22], the time spans of the experiments were on the scale of seconds and the data acquisition interval was too large to capture the dynamic response within seconds. Moreover, the data from the oil field [23] were on a larger time scale making dynamic analysis intractable. In this paper it demonstrates that the inertial force of the fluid plays an important role in dynamic hydraulic fracture. This is achieved by direct observations and theoretical analyses of the relationship between the crack tip and the fluid front in dynamic hydraulic fracture. The dimensionless separation criterion of the crack tip and the fluid front is also established to provide a guideline of distinguishing dynamic hydraulic fracture from the quasi-static fluid-driven fracture in numerical simulations.

In order to describe the transient motion of the fluid in the fracture, the momentum balance equation regarding the fracturing fluid is established. The treatment generally follows the approach by Bosanquet [24]. A proper treatment of the crack profile is important to the dynamic analyses. The most well-known crack tip profile is the parabola (Linear Elastic Fracture Mechanics) described as $D \sim X^{1/2}$, where D is the crack opening and X is the distance from the crack tip. Recent studies also showed that the hydraulic fracture had a profile of $D \sim X^{3/2}$ (zero-toughness situation) in the near tip region [25], and it transitioned to the form of $D \sim X^{2/3}$ away from the near-tip region [26–28] and finally it became $D \sim X^{1/3}$ after the crack attained blade-like geometry (Perkins–Kern–Nordgren model [29,30]). Thus, without loss of generality, the crack profile of $D \sim X^n$ ($0 < n < 2$) is assumed in this study.

In this paper, a laboratory-scale hydraulic fracturing device is built to create dynamic Mode I hydraulic fracture (HF) in a double-cantilever-beam (DCB) shaped Polymethyl methacrylate (PMMA) specimens. The analytic solution of the beam deflection and stress intensity factor (SIF) are derived. To validate the calculated SIF, the analytic SIF is compared with the existing theoretical study. Then, the relationship between crack tip and fluid front in dynamic HF is investigated. The momentum balance equation of dynamic flow is established. Numerical results by solving this equation are compared with the pertinent experimental results. The corresponding energy form of equations are considered. The existence of equilibrium state is predicted theoretically and experimentally verified. The influence of crack profile is taken into consideration. A dimensionless factor Π_2 is proposed as the dimensionless separation criterion to study the separation of the crack tip and the fluid front, as well as the existence of equilibrium state. The validity of Π_2 is verified by the experimental results.

2. Experimental Setup and Verification

The purpose of this section is to provide an overview of the experimental setup, the theoretical and experimental verification of this device and the tests performed. Two types of tests are performed in this study. The first focuses on the experimental verification of SIF using a combination of the Digital Image Correlation (DIC) method and the Williams' series. An optical microscope is used to obtain the image. The second type captures the relationship between crack tip and fluid front in the dynamically propagating crack. A high-speed camera is used for this purpose.

2.1. Experimental Setup

In this study, a laboratory-scale hydraulic fracture device, based on the DCB test, is designed and built. In order to acquire photographs using both optical microscope and high-speed camera, two transparent and thick PMMA boards are machined into certain shape and used as the frame structure. As a result of the difficulty in sealing the top of the specimen against fluid leaks, a relatively soft silicone ring is used as the sealing material. To seal the specimen surface, two U-shape silicone strips are placed on both sides of the specimen surface. As shown in Figure 1a, this design creates two narrow spaces filled with pressurized fracturing fluid on both sides of the specimen. Silicone seals offer great performance according to our experimental observations. In order to obtain a straight crack path, a small force parallel to the crack direction is applied to the DCB specimen. This small force not only ensures a straight crack path, but also makes the specimen squeeze the silicone ring against any fluid leaks. Care was taken to avoid applying a relatively large force that can lead to fluid leaks. The fluid is injected into the void between the two beams of the specimen. To capture the fluid front during the high-speed dynamic crack propagation, distilled water with a small amount of black ink is used as fracturing fluid. For tests conducted to verify the SIF value using an optical microscope, only distilled water is used as the fracturing fluid.

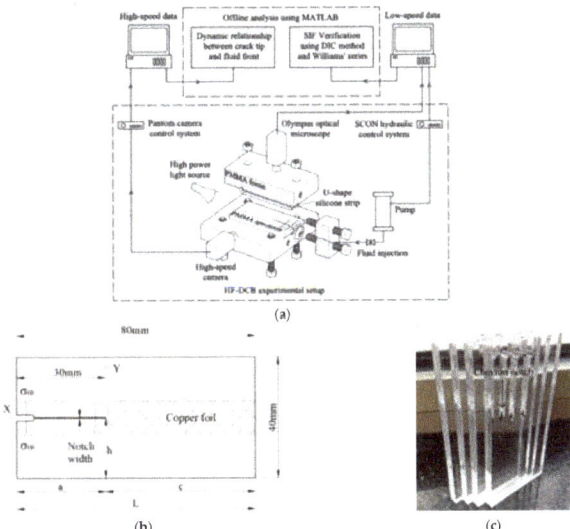

Figure 1. The experimental setup and specimen configuration. (a) Laboratory-scale hydraulic fracturing device. The high-speed camera direction is parallel to the specimen surface and the optical microscope direction is perpendicular to the specimen surface. Both the specimen and frame are transparent; (b) Configuration of the hydraulic fracture in double-cantilever-beam (HF-DCB) specimen. The to-be-fractured area is pasted with 0.05 mm thin copper foil to avoid side flow; (c) HF-DCB with intrinsically sharp crack tip induced by Chevron notch.

The PMMA specimens are commercial products having an elastic modulus of 2.6 GPa and a Poisson's ratio of 0.33, evaluating using uniaxial compression test in a RTR-1500 tri-axial machine manufactured by GCTS. The details of the geometrical parameters are shown in Figure 1b. The thickness of the specimen is maintained at b = 5.0 mm ± 0.1 mm. Crack notch of different widths is studied. The width varied from intrinsically sharp to 2.0 mm. The intrinsically sharp crack is induced by Chevron notch, as shown in Figure 1c, with a length a = 35 mm ± 2 mm. The other notch widths (0.5 mm and 1.0 mm and 2.0 mm) with length a = 30 mm, are produced by laser cutting.

In order to create an ideal two-dimensional flow, a copper foil of 0.05 mm thickness (1% of the sample thickness) is pasted on each side of the specimen to avoid side flow of fracturing fluid to the crack tip. This experimental treatment ensures that the fracturing fluid is only supplied by the rear end of the fracture. According to our observations, the copper foil offers great performance in resisting the side flow from the specimen surface and it does not exert any influence on the fracturing behavior of the specimen.

In the tests herein, a RTR-1500 hydraulic system and a SCON controlling system from GCTS are used as the hydraulic system and the controlling system, respectively. A high-speed camera is used to capture the dynamic relationship of crack tip and fluid front. A typical data acquisition interval of 5 microseconds is used. A 2000 watt high-power light source and several high-power flashlights are used together to provide sufficient light for the camera for capturing images with a very small exposure time (about 4 microseconds). In this paper, the dynamic relationship between fluid front and crack tip is directly captured within 1 millisecond using a high-speed camera with a sampling rate of 200,000 Hz (sampling interval: 5 μs).

Tests are performed in a displacement (fluid volume) control mode. The injection rate is kept at 17.4 mL/min. The following data are collected: transient fluid pressure as a function of time (using the GCTS system), dynamic relationship of crack tip and fluid front (using high-speed camera) and full-field displacement (using optical microscope (OM)).

2.2. The Theoretical and Experimental Verification of Stress Intensity Factor (SIF)

2.2.1. Theoretical Verification

The deflection and SIF of HF-DCB specimen under hydraulic pressure are derived. Based on the Euler-Bernoulli beam theory and the Winkler foundation, Kanninen [31] derived the following closed-form analytic solution of the beam deflection of the DCB specimen under point load.

$$w(x,a) = \frac{6P}{Ebh^3\lambda^3}\left[-\frac{\lambda^3}{3}x^3 + a\lambda^3 x^2 + (D + 2a\lambda H)\lambda x + F + a\lambda D\right] \quad (1)$$

where,

$$D = \frac{\sinh^2(\lambda c) + \sin^2(\lambda c)}{\sinh^2(\lambda c) - \sin^2(\lambda c)}$$
$$H = \frac{\sinh(\lambda c)\cosh(\lambda c) + \sin(\lambda c)\cos(\lambda c)}{\sinh^2(\lambda c) - \sin^2(\lambda c)}$$
$$F = \frac{\sinh(\lambda c)\cosh(\lambda c) - \sin(\lambda c)\cos(\lambda c)}{\sinh^2(\lambda c) - \sin^2(\lambda c)}$$
$$\lambda = 1.565/h$$

All other parameters are shown in Figure 1b. Based on the point load solution shown above, the solution of the DCB specimen under uniformly distributed load, namely the solution of the HF-DCB specimen is derived. The analytic solution of beam deflection, the total strain energy stored in the HF-DCB specimen, the energy release rate and the SIF of the HF-DCB specimen are derived as follows:

$$w_{HF}(x) = \frac{\sigma}{Eh^3\lambda^3}\varphi(x) \quad (2)$$

$$U(a) = \sigma^2\Phi(a) \quad (3)$$

$$G = \frac{1}{b}\sigma^2 \frac{d\Phi(a)}{da} \tag{4}$$

$$K = \frac{\sigma}{h^{3/2}} \sqrt{\left[3(1 - \frac{1}{\lambda}\frac{\partial H}{\partial c})a^4 + \frac{1}{\lambda}(12H - \frac{7}{2}\frac{1}{\lambda}\frac{\partial D}{\partial c})a^3 + (\frac{21}{2}\frac{1}{\lambda^2}D - \frac{1}{\lambda^3}\frac{\partial F}{\partial c})a^2 + \frac{2}{\lambda^3}Fa\right]} \tag{5}$$

where,

$$\varphi(x) = \frac{1}{2}\lambda^3 x^4 - 2\lambda^3 ax^3 + 3\lambda^3 a^2 x^2 + 6(\lambda^2 Ha^2 + \lambda Da)x + aF + \frac{1}{2}a^2\lambda D \tag{6}$$

$$\Phi(a) = \frac{b}{Eh^3\lambda^3}(\frac{3}{5}\lambda^3 a^5 + 3\lambda^2 Ha^4 + \frac{7}{2}\lambda Da^3 + Fa^2) \tag{7}$$

The details of the derivation are shown in Appendix A. The fluid volume between two beams is expressed as follows:

$$V_{HF}(a) = 2\sigma\Phi(a) \tag{8}$$

According to Equations (3) and (8), the total strain energy has the following relationship with volume: $U(a) = \frac{1}{2}\sigma V_{HF}(a)$.

As commonly seen in DCB analyses, Equation (5) is very insensitive to c when $c/h > 2$, in which situation it reduces to the following:

$$K = \sigma\sqrt{a} \cdot \sqrt{3(\frac{a}{h})^3 + 7.7(\frac{a}{h})^2 + 4.3\frac{a}{h} + 0.52} \tag{9}$$

The SIF is compared with the results by Fett [32] who derived the SIF under a distributed load using the boundary collocation method and the weight functions. According to his study, K can be expressed as follows:

$$K = \int_0^a \chi(x) \cdot \sigma(x) dx \tag{10}$$

where $\sigma(x)$ is a constant and the weight function $\chi(x)$ is given by

$$\chi = \sqrt{\frac{12}{h}}(\frac{a}{h} + 0.68) - \sqrt{\frac{12}{h}}\frac{x}{h} + \sqrt{\frac{2}{\Pi(a-x)}}\exp(-\sqrt{\frac{12(a-x)}{h}}) \tag{11}$$

Thus, K can be written as follows

$$K = \sigma\sqrt{a}\left[\sqrt{\frac{12a}{h}}(\frac{1}{2}\frac{a}{h} + 0.68) + \frac{2}{\sqrt{6\Pi}}\sqrt{\frac{h}{a}}\left(1 - \exp(-\sqrt{\frac{12a}{h}})\right)\right] \tag{12}$$

A comparison between the Fett's results (Equation (12)) and the analytic solution derived in this study (Equation (5)) is shown in Figure 2. It can be seen that the analytic results are in good agreement with the Fett's results. It should be noted that the Fett's results present a numerical fitting and are only applicable in the condition of $c/h > 2$. The analytic solution in Equation (5) provides a generalized analytic solution without this constraint of c/h.

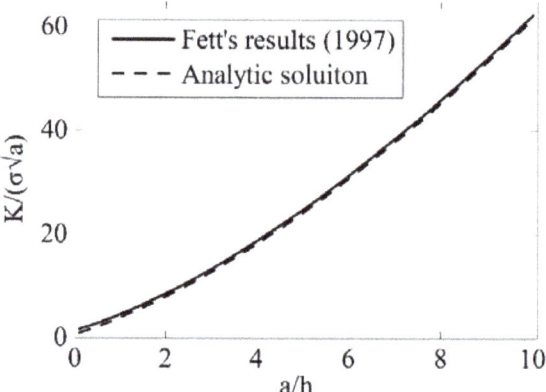

Figure 2. Comparison between stress intensity factor (SIF) between the analytic solution (Equation (9)) using beam theory and the Fett's result (Equation (12)) using boundary collocation method and the weight functions. for $c/h = 3$. The two results show good agreement with each other in a wide range. The variables on the horizontal and the vertical coordinates are dimensionless numbers.

2.2.2. Experimental Verification

The theoretical SIF expressed in Equation (9) is experimentally verified using a combined method of DIC full-field displacement extraction and the fitting of Williams' series. The crack tip position, Mode I and Mode II SIFs are unknown parameters in the fitting. This non-linear fitting is performed in the least square sense using a derivative-free Levenberg-Marquardt algorithm [33,34]. A comparison between theoretical SIFs and experimental SIFs results under different values of a is shown in Figure 3. The images for DIC analyses are taken by an Olympus optical microscope with a resolution of 1624 × 1236 pixels, and a pixel length of 8.09 μm. For better DIC identification, the specimen surface around crack tip is abraded by 800 grit sandpaper. All the fitting process generally follows the method by Yates [35]. The details of the fitting process are not shown here.

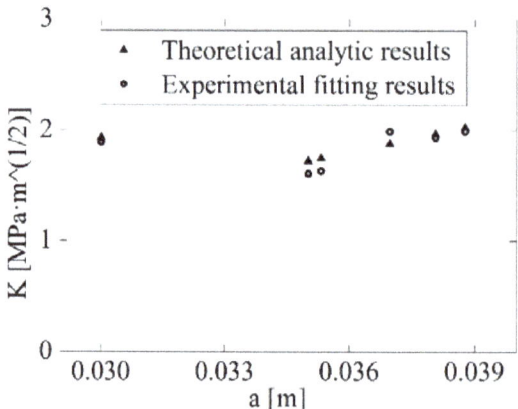

Figure 3. Comparison between theoretical SIF results and experimental SIF results. The theoretical SIFs are derived using Equation (9). The experimental SIFs are derived using a combined method of Digital Image Correlation (DIC) full-field displacement extraction and the fitting of DIC full-field displacement extraction and the fitting of Williams' series. Williams' series. The specimens are loaded to the SIF that is slightly lower than the fracture toughness.

3. The Relationship Between the Crack Tip and the Fluid Front in Dynamic Hydraulic Fracture (HF)

3.1. Governing Equations

The two-dimensional dynamic flow in hydraulic fracture studied here can be approximated as dynamic flow along slowly-varying channels. The momentum-balance equation of the fracturing fluid between the slowly-varying plates under constant fracturing pressure is as follows [36,37]:

$$\frac{d(M(x,t))}{dt} + F_V(x,t) + F_C(x,t) = F_{Pf}(t) \tag{13}$$

In this equation, the coordinate system is established as show in Figure 4. x is the position of the fluid front (not the same x in beam deflection analyses), the first term $d(M(x,t))/dt$ on the left side of the equation is the inertial resistance, $M(x,t)$ is the total momentum of the fluid, the second term $F_V(x,t)$ stands for the drag force induced by viscosity, the third term $F_C(x,t)$ is the resistance induced by the slowly-varying flanks of the crack, and $F_{Pf}(t)$ is the total driving force. It may be worth noting that x is also a function of t. This momentum balance equation has the following expanded form:

$$\frac{d}{dt}\left(\rho b D(x,t) x \frac{dx}{dt}\right) + 12b\eta D(x,t)\frac{dx}{dt}\int_0^x \frac{1}{D(x_0,t)^2}dx_0 + \int_0^x bP(x_0,t)\left(-\frac{\partial D(x_0,t)}{\partial x_0}\right)dx_0 = P_f b D(0,t) \tag{14}$$

where $D(x,t)$ is the crack opening width as a function of time t and fluid front position x.

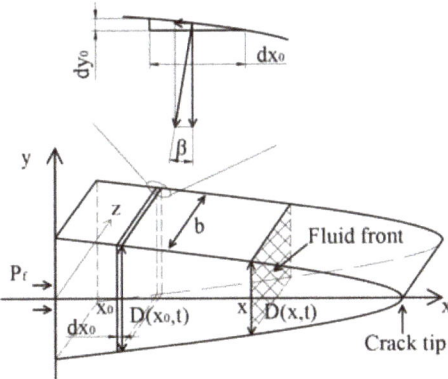

Figure 4. Illustration of dynamic fluid analyses in the crack. The coordinate origin is where the crack starts to propagate. The crack profile is assumed to stay the same while propagating. x is the position of the fluid front. x_0 is the point for the integral analyses in the fluid. D is the crack opening width. The upper part of the figure is the illustration of the analyses of the crack flank resistance.

The first term on the left side of Equation (14) is the time derivative of the total momentum of the fluid in the propagating crack:

$$\frac{d(M(x,t))}{dt} = \frac{d(\int V(x_0,t)dm)}{dt} \tag{15}$$

where $dm = \rho b D(x_0,t)dx_0$ and $V(x_0,t) = D(x,t)/D(x_0,t)\frac{dx}{dt}$ (volume conservation) as shown in Figure 4. The second term on the left side of Equation (14) is the viscous force:

$$F_V(x,t) = \int dF_V = 2\int_0^{x_0} \tau(x_0,t)b dx_0 \tag{16}$$

where $\tau(x_0,t) = 6\eta V(x_0,t)/D(x_0,t) = 6\eta D(x,t)/D(x_0,t)^2 \frac{dx}{dt}$.

The third term is the horizontal component of the resistance applied by the crack flanks:

$$F_C(x,t) = \int bP(x_0,t)dy_0 = \int_0^x bP(x_0,t)\left(-\frac{\partial D(x_0,t)}{\partial x_0}\right)dx_0 \tag{17}$$

Strictly speaking, the derivation of the third term requires an explicit $P(x_0,t)$, which can only be obtained by solving Equation (14). This makes Equation (14) implicit. Here the pressure distribution for the calculation of the third term is assumed to have the following linear form:

$$P(x_0,t) = -\frac{P_f}{x}(x_0 - x) \tag{18}$$

This first order assumption ensures the pressure at the fluid front is zero and is P_f at the injection point. The influence of the crack-flank resistance term is discussed later in this paper.

The derivation of Equation (14) contains the following assumptions. Firstly, the fluid is assumed to fully fill the crack in the rear of fluid front. Secondly, the fluid front is assumed to have a flat profile. The velocity of the fluid at point x_0 is considered as the mean velocity $V(x_0,t) = \frac{dx_0}{dt}$ at any cross section.

It can be inferred that the theoretical fluid front as a function of time is strongly influenced by the time-dependent crack profile, $D(x,t)$. The determination of $D(x,t)$ is also where the difficulty resides in. According to our experimental results of crack tip velocity, for some of the tests, before first arrest of the crack, the hydraulic crack in DCB specimen roughly propagates at a constant velocity. Thus, the constant-velocity approximations are made in sample #15 and sample #16. As for sample #12 and sample #14, nine-order polynomial fitting is used to fit the crack tip position as a function of time. It should be noted that if the details of the curve are not omitted, five-order polynomial fitting is suggested as the fitting becomes sensitive to the error when the order increases. In the case we investigate here, the crack profile in the DCB specimen is approximated by linear approximation. The comparison between approximation of the crack profile and the theoretical quasi-static crack profile derived by Equation (A7) is shown in Figure 5. As shown, the linear crack profile provides a good approximation in the range of crack length we investigated here.

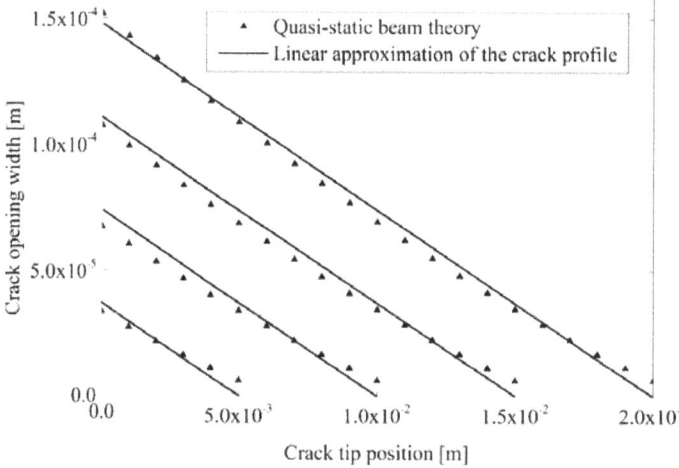

Figure 5. Comparison between quasi-static beam deflection and the approximation of linear crack profile. The quasi-static beam deflection is derived using beam theory. The linear crack profile provides a good approximation of the crack profile in the considered area.

According to the approximations of linear crack profile, the $D(x,t)$ can be expressed as:

$$D(x,t) = a_1[C(t) - x] \tag{19}$$

a_1 is a constant in all the tests and has a non-dimensional value of 1.43×10^{-2}. a_1 means the slope of the linear crack profile and is derived by fitting the theoretical quasi-static crack profile as shown in Figure 5. The theoretical quasi-static crack profile is obtained by substituting Equation (5) into Equation (A7) and eliminating σ. And the value of K is set as the arrest toughness K_a (1.0 $MPa\sqrt{m}$, regular dry DCB tests). In this way a_1 is obtained.

$C(t)$ has the following form under the approximation of constant-velocity crack.

$$C(t) = a_1 V_C t \tag{20}$$

In the nine-order polynomial fitting of the crack tip position, $C(t)$ is expressed as:

$$C(t) = \sum_{n=0}^{9} A_n t^n \tag{21}$$

Substituting Equation (19) into Equation (14) we have the following differential equation:

$$x'' - \frac{P_f C(t)}{\rho x(C(t) - x)} + \frac{P_f}{2\rho(C(t) - x)} + \frac{x'^2}{x} + x' \frac{C'(t) - x'}{C(t) - x} + \frac{12\eta x'}{a_1^2 \rho(C(t) - x)C(t)} = 0 \tag{22}$$

Equation (22) is a general equation for linear-profile crack. After substituting Equations (20) and (21) into Equation (22) separately, Equation (22) is numerically solved on MATLAB using "ode45". The initial values of t, and x and are assigned a small positive value (typically 10^{-9}) because Equation (22) is singular at $x = 0$ and $t = 0$. All the experimental and numerical results are shown in Figures 6 and 7.

3.2. Energy Conservation in the Dynamic Process

The energy conservation of the fluid within the crack is investigated here. The major energy terms taking part in the energy conservation regarding the dynamic fluid flow include the input energy W_I, the kinetic energy of the fluid W_K, the energy dissipated by viscosity W_V, and the work done by the fluid on the crack flanks W_C. These terms as a function of time can be derived via the time integral of the rate of these energy terms, which are shown as follows:

$$\dot{W}_I = Q(t) \cdot P_f = x' D(x,t) b P_f \tag{23}$$

$$\dot{W}_K = \frac{d\left(\int \frac{1}{2} V(x_0,t)^2 dm\right)}{dt} = \frac{1}{2} b\rho \frac{d\left(x'^2 D(x,t)^2 \int_0^x \frac{1}{D(x_0,t)} dx_0\right)}{dt} \tag{24}$$

$$\dot{W}_V = \int V(x_0,t) dF_V(x_0,t) = 12 b\eta x'^2 D(x,t)^2 \int_0^x \frac{1}{D(x_0,t)^3} dx_0 \tag{25}$$

$$\dot{W}_C = \int C'(t) dF_C(x_0,t) = \int_0^x C'(t) b P(x_0,t) \left(-\frac{\partial D(x_0,t)}{\partial x_0}\right) dx_0 \tag{26}$$

The rate of energy dissipated by viscosity $\dot{W}_V$ is calculated by visco-friction between the crack and fluid. The rate of work done on the crack flank $\dot{W}_C$ involves the calculation of the force component on the direction of crack propagation.

The conservation of energy should be satisfied:

$$W_I = W_K + W_V + W_C \tag{27}$$

which can be written,

$$\int_0^t \dot{W}_I = \int_0^t \dot{W}_K + \int_0^t \dot{W}_V + \int_0^t \dot{W}_C \tag{28}$$

In the following section, the energy conservation is verified. After solving Equation (22), the total input energy W_I as a function of time is compared with the total energy consumed by the fluid $W_K + W_V + W_C$.

3.3. Comparison between Experimental and Theoretical Results

All the detailed data of the samples are shown in Table 1. The fracturing fluid is water with black ink. Because the viscosity of black ink is only slightly larger than the viscosity of water, the viscosity η for calculation is set as 0.9×10^{-3} Pa·s, which is the viscosity of water at 25 degrees centigrade. Figures 6 and 7 show the typical experimental and theoretical results of the dynamic process. The positions of the crack tip and fluid front are recorded frame by frame. As it can be seen in Figure 6a,b and Figure 7a,b, the theoretical dynamic fluid fronts are in good agreement with the experimental results. Both experimental results and numerical results show that during the dynamic propagation of the hydraulic fracture, the fluid front separates with the crack tip and has a different velocity compared to the crack. Before the fluid front approaches the crack tip, the results of the calculation with and without the crack-flank resistance term provide similar results. However, when the fluid front approaches the crack tip, the calculation with that term provides a much better result. It is because when approaching the crack tip, the fluid occupies most of the crack, and the horizontal component of the force applied by the crack flanks increases to an important part of the total resistance. The results indicate that when the fluid occupies most of the crack, the crack-flank resistant force is not negligible in a dynamic calculation.

Table 1. Sample data.

NO.	Crack Notch Width (mm)	Initial Crack Length a (mm)	Fracturing Pressure P_f (MPa)	Initiation Toughness * (MPa$\sqrt{m}$)
#5	Intrinsically sharp crack	35	1.95	2.52
#6	Intrinsically sharp crack	35	1.72	2.22
#7	Intrinsically sharp crack	37	1.52	2.15
#11	0.5	30	3.56	3.62
#12	0.5	30	2.74	2.78
#13	0.5	30	2.26	2.30
#14	2.0	30	4.06	4.13
#15	2.0	30	2.64	2.68
#16	2.0	30	2.53	2.57

* According to our preliminary study, the initiation toughness is related to both the crack notch width and the pressure-hold time before initiation which results in craze and blunting in the near-tip region.

As for the energy terms, in most of the time during propagation, the total input energy is almost the same as the total energy consumed by the fluid as shown in Figure 6c,d and Figure 7c,d. This is a good verification of the first-order approximation of pressure distribution. As can be seen, the viscosity energy is only a very small part of the total input energy. The kinetic energy of the fluid and the work done on the crack flank takes up most of the energy during the dynamic process. The results show that the value of W_C/W_I has a positive relationship with the velocity of the crack tip. When the crack arrests, the W_C stops to increase and the W_C/W_I starts to decrease. This result is predicted by Equation (8), which involves the crack tip velocity. Overall, the kinetic energy of the fluid takes up over half of the total input energy during this process. When the crack slows down or is arrested, this ratio further increases. The W_C and the W_V share the rest of the input energy.

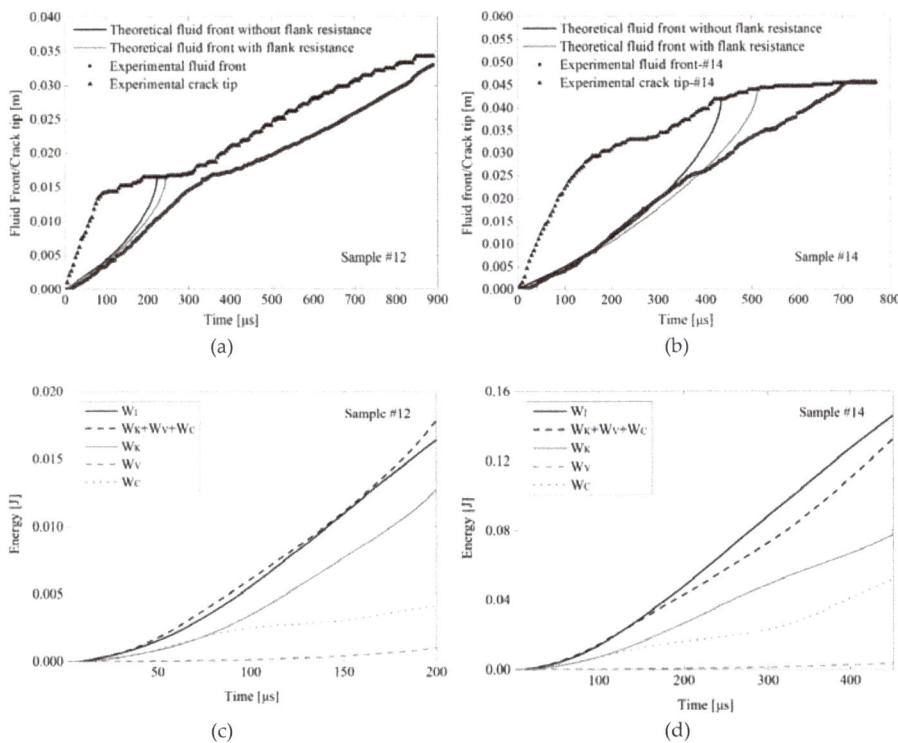

Figure 6. (**a**,**b**) are the comparison between experimental results and theoretical results of fluid front for sample #12 (fracturing pressure: 2.74 MPa) and sample #14 (fracturing pressure: 4.06 MPa) respectively. The fluid front positions are the numerical results by solving Equation (22) with and without the term of the crack flank resistance. The calculations with the crack flank resistance term provide better results of the fluid front position; (**c**,**d**) are corresponding energy as a function of time. The energy terms are calculated using the equations from Equation (23) to Equation (26). The kinetic energy W_K of the fluid occupies around half of the total energy. The work done by the fluid on the crack flank is a non-negligible part of the total energy. The input energy and the consumed energy conform well to each other.

Recent numerical simulation [5] showed that the dynamic fracturing (in 1 millisecond) is not smooth or continuous but was a distinct stepwise process concomitant with fluid pressure oscillations. The experimental results here provide a direct evidence for the stepwise propagation of the crack in the dynamic situation. As it can be seen, the crack is prone to re-initiation when the fluid front approaches the crack tip. The crack speed of the re-initiation is higher than the fluid velocity, and the crack arrests before the fluid front approaches again. The experimental and theoretical results of crack tip and fluid front also agree with the existing studies [38–40].

Despite that the numerical results provided by solving Equation (22) are in good agreement with the experimental results, there are still problems that need further investigation. According to our numerical results, the fluid front allowing for the first-order approximation of the crack-flank resistance predict an extremely high fluid velocity when the fluid front further approaches the crack tip. It is not clear whether the behavior at such high speed is real or is simply an artifact of the model. Interestingly, the experimental results seem to provide some evidence of this high-speed phenomenon. Figure 8 shows the dynamic images of sample #12, which shares the same time coordinate with the data in Figure 6a. As can be seen, the fluid front becomes unstable after 0.235 milliseconds. This may result from the high speed of the fluid, which leads to turbulence. The data of the fluid front position in Figure 6a is not sampled at the very front of the unstable fluid after 0.235 milliseconds, as a result of which the velocity seems to be moderate when approaching the crack tip. If the very front of fluid is regarded as the fluid front, there will be the high speed of the fluid when approaching the crack tip. It should be noted that when the fluid front gets close to the crack tip, which is beyond the scope of this study, all the kinetic energy will turn into the work done on the crack flanks in the way of possible high fluid pressure. This may provide an explanation for the phenomenon that every time the fluid front approaches the crack tip, the crack starts to propagate. This phenomenon is seen in all the tests. Previous research by Rubin [41] showed that in the case of a saturated permeable media, the cavity between crack tip and fluid front was either filled with the vapor of the fracturing fluid, with a nearly constant vapor pressure in the case of an impermeable medium, or the pore-fluid infiltrated from the surrounding domain.

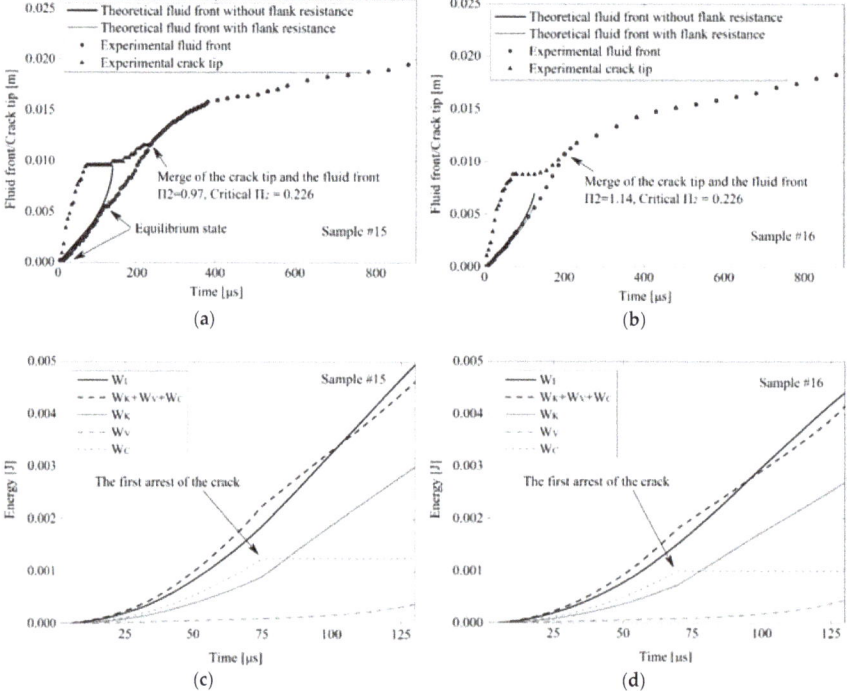

Figure 7. (**a**,**b**) are comparison between experimental results and theoretical results of fluid front for sample #15 (fracturing pressure: 2.64 MPa) and sample #16 (fracturing pressure: 2.53 MPa) respectively. The fluid front positions are the numerical results by solving Equation (22) with and without the term of the crack flank resistance. The calculations with the crack flank resistance term provide much more realistic results of the fluid front position; (**c**,**d**) are corresponding energy as a function of time. The kinetic energy W_K of the fluid occupies over half of the total energy. W_C stops increasing after the arrest of the crack as expected. The input energy and the consumed energy conform well to each other.

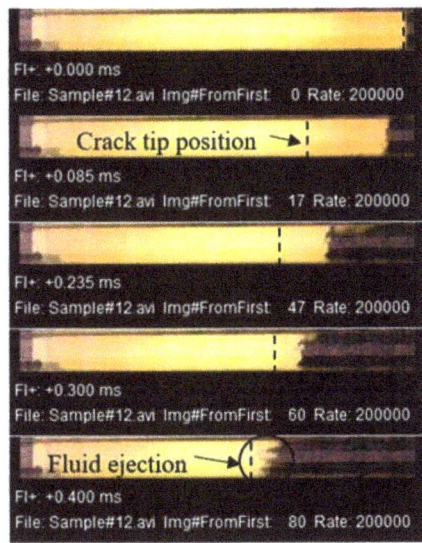
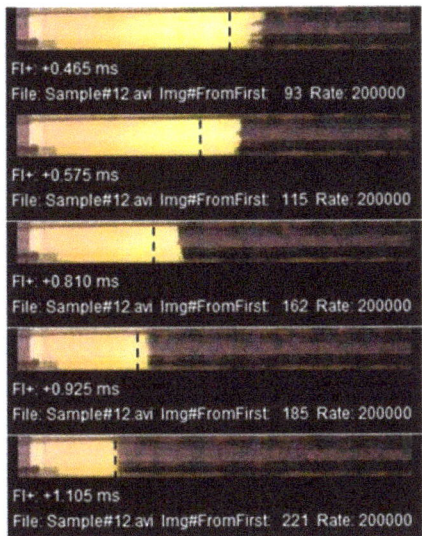

Figure 8. Dynamic images of the relationship between crack tip and fluid front of sample #12. The direction of the high-speed camera is shown in Figure 1a. The crack tip and the fluid front separate at the beginning of the propagation, and the fluid front catches up with the crack tip when it slows down. The phenomenon of fluid ejection presents when the fluid front approaches the crack tip, and meanwhile the fluid velocity is theoretically high but actually low in the real case.

As can be seen, the crack tip shown in Figure 8 is vague. However, it is clear and readily identifiable in the video version in the Supplementary Materials.

3.4. Will the Fluid Front Catch Up with a Propagating Crack?

Now let us turn our attention to the separation of the fluid front and the crack tip during the dynamic process, namely the existence of the lag region. The main purpose of this section is to determine whether the fluid front and the crack tip share the same front or not after initiation, and to investigate the existence of the "equilibrium state" which indicates that the fluid front will travel at a constant velocity shortly after the initiation of the crack. Earlier discussion has already shown that the momentum-balance equation of the fracturing fluid between the linearly-slowly-varying plates can be expressed by Equation (14). It should be noted that the linear profile assumption is only valid in the crack of HF-DCB specimen. In the following analyses, a more generalized-form crack profile is investigated:

$$D(x,t) = A(V_c t - x)^n \tag{29}$$

where n is a positive real number. The coordinate is the same as the one shown in Figure 4. Equation (29) describes a more generalized profile of the crack propagating at a constant velocity. Using the first order approximation of crack-flank resistance, the governing equation can be written in the following form by substituting Equations (29) and (18) into Equation (14):

$$\frac{d}{dt}\left(\rho b A(V_c t - x)^n x \frac{dx}{dt}\right) + 12\eta b [A(V_c t - x)^n] \frac{dx}{dt} \int_0^x \frac{1}{[A(V_c t - x_0)^n]^2} dx_0 = \\ P_f b A(V_c t)^n - \int_0^x b(-\frac{P_f}{x}(x_0 - x))(-\frac{\partial [A(V_c t - x_0)^n]}{\partial x_0}) dx_0 \tag{30}$$

After the integral, Equation (30) becomes

$$\frac{d}{dt}\left((V_c t)^n (1-\frac{x}{V_c t})^n x \frac{dx}{dt}\right) + (V_c t)^n \frac{12\eta(1-\frac{x}{V_c t})^n x'}{\rho A^2 (2n-1)}\left[(V_c t - x)^{-2n+1} - (V_c t)^{-2n+1}\right] = \\ (V_c t)^n \frac{P_f}{\rho} - (V_c t)^n \frac{P_f V_c t}{\rho(n+1)x}\left[(n+1)\frac{x}{V_c t} + (1-\frac{x}{V_c t})^{n+1} - 1\right] \quad (31)$$

Equation (31) can be written in the following dimensionless form:

$$\frac{d}{dt}\left(V_c (V_c t)^{n+1}(1-\Pi_1)^n \Pi_1 P_1\right) = \\ (V_c t)^n V_c^2 \left[2\Pi_2 - \frac{\Pi_3 (1-\Pi_1)^n P_1}{(2n-1)}\left[(1-\Pi_1)^{-2n+1} - 1\right] - \frac{2\Pi_2}{(n+1)\Pi_1}\left[(n+1)\Pi_1 + (1-\Pi_1)^{n+1} - 1\right]\right] \quad (32)$$

where $P_1 = \frac{x'}{V_c}$, the dimensionless instantaneous velocity of the fluid front; $\Pi_1 = \frac{x}{tV_c}$, the dimensionless secant velocity of the fluid front; $\Pi_2 = \frac{P_f}{2\rho V_c^2}$ and $\Pi_3 = \frac{12\eta}{A^2 \rho V_c^{2n} t^{2n-1}}$. Π_1, Π_2 and Π_3 are the composed criteria and P_1 is the simple criterion.

This governing equation mathematically predicts a dimensionless velocity: "equilibrium fluid velocity P_E". It indicates an equilibrium state where after a long period of time ($t \to \infty$) the fluid front travels at a constant velocity $P_E V_C$, and will not catch up with the running crack tip. The equilibrium state can only be approached. After a long period of time ($t \to \infty$), for there to be an equilibrium state, the instantaneous velocity of the fluid front equals the secant velocity and both are time independent, which leads to $P_1 = \Pi_1 = P_E = constant$. Also notice that $\lim_{t \to \infty} \Pi_3 = 0$ for $n > \frac{1}{2}$. This makes the terms on the right side of Equation (14) readily integrable. After the time integral of Equation (32) over the range from 0 to t and by taking $P_1 = \Pi_1 = P_E$, we have the following dimensionless equation:

$$\frac{(n+1)^2 (1-P_E)^n P_E^3}{2\left[1-(1-P_E)^{n+1}\right]} = \Pi_2 \quad (33)$$

When $n > 1/2$, for there to be an equilibrium state, Equation (33) has to be satisfied. This equation can be utilized to examine the existence of the equilibrium state. It is easy to find that for an arbitrary n, Equation (33) first goes up then goes down when P_E increases from 0 to 1. Thus, for Equation (33) to have a solution within the range of $0 < P_E < 1$, there exists a critical Π_2, which can be expressed as $\Pi_{2c}(n)$. When $\Pi_2 < \Pi_{2c}(n)$, the solution exists and the equilibrium state will finally be reached. After a long period of time, the fluid front will travel at a constant velocity $P_E V_c$ that is slower than the crack tip and will not meet the crack tip. The equilibrium fluid velocity can be obtained by substituting a given set of Π_2 and n into Equation (33) and solving P_E. In the case of $\Pi_2 > \Pi_{2c}(n)$, the solution does not exist and there is not an equilibrium state for the system. The fluid front and the crack tip either keep sharing the same front right (low-viscosity situation) after the initiation of the crack or separate after initiation (viscosity-dominating situation). It should be noted that the relationship between $\Pi_{2c}(n)$ and n can only be numerically obtained as there is no explicit solution. It is worth noting that there is a maximum equilibrium fluid velocity P_{Emax} when an equilibrium state exists. The value of P_{Emax} can be solved by substituting a given n and corresponding $\Pi_{2c}(n)$ into Equation (33). The value of $\Pi_{2c}(n)$ and maximum equilibrium fluid velocity P_{Emax} as a function of n are shown in Figure 9.

As can be seen in Figure 9, the maximum equilibrium fluid velocity cannot reach 1.0 over the range investigated. This means that when approaching the equilibrium state, despite the fluid length keeps occupying an invariable portion (P_E) of the crack length as time goes by, the crack tip and fluid front keep getting away from each other. It is also worth noting that the parameter A, which characterizes the opening level of the crack, does not enter Equation (33). This means that the existence of equilibrium states is not related to the degree of the crack opening.

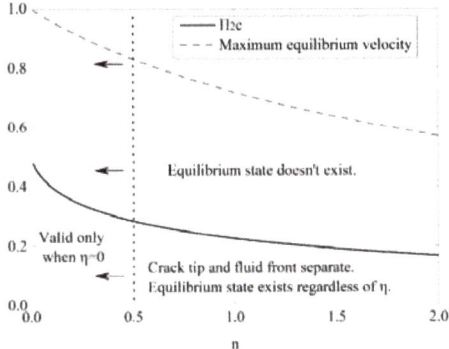

Figure 9. Π_{2c} and P_{Emax} as a function of n. The solid line is the criterion of the existence of the equilibrium state. It is valid regardless of η on the right of the dot line ($n = 0.5$), and is valid only when $\eta = 0$ on the left of the dot line. The maximum equilibrium velocity indicates the equilibrium velocity when $\Pi_2 = \Pi_{2c}$.

Figure 10 shows the numerical results of the dimensionless fluid velocity as a function of time under different Π_2 and viscosity when $n = 1$. The numerical results verify the theoretical prediction of the existence of equilibrium state. Fluid fronts accelerate after the initiation of crack and gradually approach the equilibrium fluid velocity as long as $\Pi_2 < \Pi_{2c}$. Once the Π_2 becomes slightly larger than Π_{2c}, there is no equilibrium state and the fluid front quickly rushes to the crack tip after a certain period of time t_c. Provided that the Π_2 stays the same and is smaller than Π_{2c}, the fluid front finally travels at the same constant velocity after reaching the equilibrium state despite the difference of viscosity. High viscosity leads to a larger t_c. It should be noted that the time t_c needed for the fluid to approach the equilibrium state is actually related to V_c. Higher V_c leads to smaller t_c. As can be seen in the earlier discussion, the lack of characteristic time indicates that t_c can be regarded as the dimensional value carrying the dimension of time and enters the governing equation. The existence of t_c is also seen in the experiments as shown in Figure 6a,b and Figure 7a,b: the fluid is relatively slow or even static in the first 5 microseconds to 25 microseconds. The fluid tends to accelerate from zero velocity to an approximately constant velocity after the initiation of the crack. The t_c observed in the experiments varies from 5 microseconds to 25 microseconds.

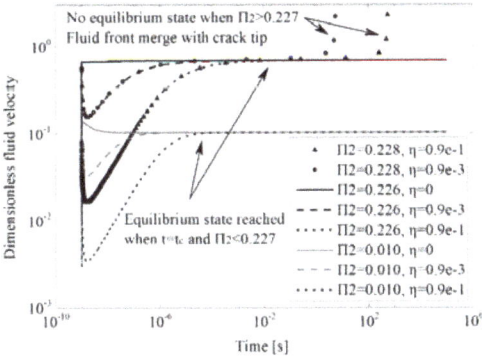

Figure 10. Numerical results of dimensionless fluid velocity V_f/V_C as a function of time under different Π_2 and viscosity η. These numerical results are obtained using liner profile, $n = 1$, $\Pi_{2c} = 0.227$, and $V_C = 1,000 \ m/s$. There is no equilibrium state when $\Pi_2 > \Pi_{2c}$. The initial "slow down" at the time of 10^{-9} s is an artificial result of the initial value of the numerical calculation. The numerical results conform well to the theoretical prediction of Π_{2c}.

Figure 11a shows the results of numerical (by solving Equation (30) using MATLAB "ode45") and theoretical (by solving Equation (33)) equilibrium fluid velocity as a function of Π_2. As can be seen, the numerical solution and theoretical prediction are in consistent with each other. The numerical equilibrium fluid velocities are precisely predicted by Equation (33). The curves are much the same and are nearly linear in the range of critical Π_2. For $n > 1/2$, the equilibrium velocities are not influenced by viscosity. Figure 11b shows the comparison between experimental and theoretical equilibrium fluid velocity as a function of Π_2 for linear crack profile. As it can be seen, the experimental results are in good agreement with the theoretical prediction of the equilibrium fluid velocity. The experimental results of whether the fluid front and the crack tip separate or not are shown in Figure 12. As it can be seen, the curve of Π_{2c} provides a very good prediction of the separation of the two fronts.

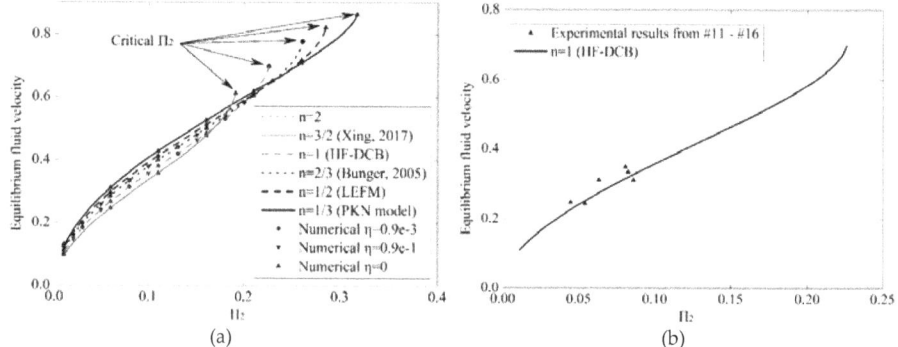

Figure 11. (a) Comparison between numerical and theoretical equilibrium fluid velocity as a function of Π_2 for different crack profiles. The numerical results are derived by solving Equation (22). The theoretical results are from Equation (33). Some typical profiles are investigated; (b) Comparison between experimental and theoretical equilibrium fluid velocity as a function of Π_2 for $n = 1$. The experimental results of equilibrium fluid velocity are extracted from the latter half of the equilibrium state as shown in Figure 7a.

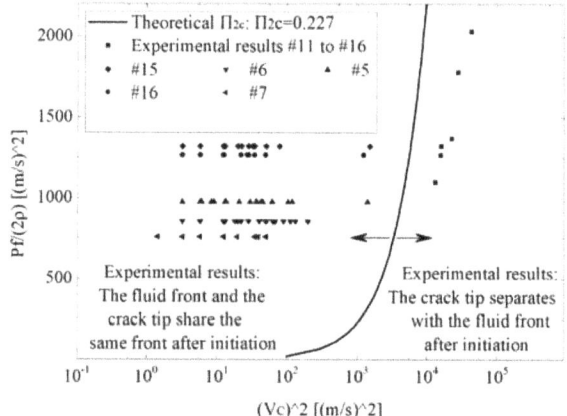

Figure 12. Verification of Π_{2c} and Π_{2s} compared with experimental results. The massive data on the left are extracted from varies experiments in which cracks propagate in a stepwise manner and never separate with fluid fronts. Two of the data points close to the curve of Π_{2s} are shown in Figure 7a,b. The sporadic data on the right are the initiation parts of the cracks in the experiments, which are prone to the separation due to the high speed of crack tip. Both Π_{2c} and Π_{2s} provide good predictions of the experimental results.

4. Discussion

The momentum balance of Equation (14) involves the determination of the fluid pressure distribution within the crack, which leads to the determination of crack flank resistance F_C. This resistance is only a small part of the driving force if the fluid has not occupied most of the crack length, as a result of which the linear approximation of pressure distribution in this paper is reasonable and provides good results before the fluid front approaches the crack tip. The F_C starts to neutralize most of the driving force when the fluid runs close to the crack tip and the results are sensitive to the accuracy of the approximation of the pressure distribution. Strictly speaking, Equation (14) is only a boundary condition of the following momentum balance equation:

$$\frac{d(M(x_0,t))}{dt} + F_V(x_0,t) + F_C(x_0,t) + F_P(x_0,t) = F_{Pf}(t) \tag{34}$$

where $F_P(x_0,t)$ is the resistance from the fluid pressure. F_P is expressed as

$$F_P(x_0,t) = bP(x_0,t)D(x_0,t) \tag{35}$$

This momentum-balance equation is established between the origin and x_0 ($x_0 < x$). The boundary condition is: $F_P(x,t) = 0$ ($t \neq t_c$), which leads to Equation (14). The boundary condition only implies that the pressure at the fluid front is 0 before reaching the crack tip, it reveals no details of the pressure distribution of the fluid. For further studies to improve the accuracy dealing with the fluid approaching the crack tip, Equation (34) shall be further investigated. Another limitation regarding the model here is that the crack profile is assumed to be the quasi-static one. The fully dynamic model should consider the inertia and stress wave of the material that forms the crack. The model will be extremely complicated considering these two factors. Further study should take these into consideration.

The experiments here have some drawbacks that should be improved in the future. The study presented here used PMMA as specimen for the convenience of observation. In terms of rock, as we know, the fracture toughness is not a constant and does not translate across scales. Rocks are well known to have cracks and defects at all scales [42–48]. In some case rocks are even elastic-anisotropic [49,50]. Another problem regarding performing the actual field-scale experiment is that the pore pressure was not considered in this study. The actual stratum is penetrable, which may lead to a fundamental difference compared to the results in this study [51]. Last but not least, the confining pressure is not considered in this study. In the further study, the confining pressure and the saturation of rock materials are to be considered. The experiments considering these two factors are to be performed in the further study.

As we know, the proppant plays an important role in the real hydraulic fracturing process. Considering proppant transport would make a big difference compared to the experiments performed in this study. The proppant will slow down the hydraulic fluid when the crack initiates, because the proppants are subjected to a larger resistance from the crack flank compared to the hydraulic fluid. In dynamic situation, the proppants also enhance the pressure drop of the hydraulic fluid in the crack as it introduces another resistance term in Equation (34). The influence of proppants is well studied in the previous studies by Siddhamshetty et al. [52,53] These research works may provide guidance for further study.

5. Conclusions

A laboratory-scale hydraulic fracturing device is built to create dynamic hydraulic fracture and investigate the dynamic relationship between the crack tip and the fluid front. The momentum-balance equation of the fracturing fluid under constant fracturing pressure is established. The numerical solutions of the momentum balance equation conform well to the experimental results. Theoretical analyses show that the horizontal component of crack-flank resistance starts to exert influence when the fluid front approaches the crack tip. In the energy analyses, results show that the kinetic energy of the fluid occupies over half of the total input energy during the whole process, and when the crack slows down or arrests, this ratio further increases.

Then the existence of equilibrium state of the dynamic system is investigated. The equilibrium state indicates that the hydraulic fracture and the fluid front travel at different but constant after a certain period of time. Moreover, the fluid front will not catch up with the running crack tip. The existence of equilibrium state is theoretically found and experimentally verified. The theoretical results of equilibrium velocity P_E are in good agreement with the experimental observations. Furthermore, the dynamic analyses show that the separation of crack tip and fluid front is dominated by both the crack profile and the equilibrium fluid velocity. The dimensionless separation criterion of the crack tip and the fluid front in the dynamic situation is established and is found to conform well to the experimental data. The criterion provides a guideline to distinguish a dynamic hydraulic fracture from a quasi-static hydraulic fracture in numerical simulations.

Supplementary Materials: The following are available online at http://www.mdpi.com/1996-1073/12/3/397/s1, Data S1: Data.zip, Video S1: Video-Craze process of intrinsically sharp crack tip Duration 15 min, Video S2: Video-Sample#12-Dynamic propagation of hydraulic fracture.

Author Contributions: Conceptualization, J.D.; Funding acquisition, M.C.; Methodology, C.Z. and S.W.; Data curation, Y.L.; Supervision, M.Z.; Writing J.D.

Funding: This research was funded by National Natural Science Foundation of China (NSFC) for Major Projects, grant number 51490651.

Conflicts of Interest: The authors declare no conflict of interest.

Nomenclature

a, a_1	Initial crack length, a non-dimensional value (1.43×10^{-2} in this paper) characterizing the slope of the linear crack profile
A	A factor controlling the opening of the crack width in the general $D(x,t)$ with a dimension of L^{1-n} (L is the length dimension)
b	Thickness of the specimen
c	$L - a$
$C(t), C\prime(t)$	The position of crack tip, the velocity of the crack tip
$D(x,t)$	The crack opening width as a function of time t and fluid front position x
E	Elastic modulus of the specimen
$F_V(x,t)$	The drag force induced by viscosity
$F_C(x,t)$	The resistance induced by the slowly-varying flanks of the crack
$F_{Pf}(t)$	The total driving force
$F_P(x_0,t)$	The resistance from the fluid pressure
f_1, f_2	Point loads in beam analysis
G	Energy release rate
K	Stress intensity factor
h	Half width of the specimen
L	Length of the specimen
$M(x,t)$	The total momentum of the fluid
P	Point load in beam analysis
P_1	The dimensionless instantaneous velocity of the fluid front $x\prime/V_c$
P_i, P_f	Input pressure, fracturing pressure
P_E	Equilibrium fluid velocity
$P(x_0,t)$	Stress distribution of the fracturing fluid
$Q(t)$	Flow rate
r	The distance to crack tip
t	Time

t_c	The time needed for the fluid to approach the equilibrium state
U	Total strain energy stored in the specimen
$V(x,t)$, dx/dt	Velocity of the fluid front
V_{HF}	Volume between the two beams
V_C	The constant velocity of the crack tip
W_I, $\dot{W}_I$	The (rate of the) input energy
W_K, $\dot{W}_K$	The (rate of the) kinetic energy of the fluid
W_V, $\dot{W}_V$	The (rate of the) energy dissipated by viscosity
W_C, $\dot{W}_C$	The (rate of the) work done by the fluid on the crack flanks
w	Beam deflection in double-cantilever-beam (DCB) specimen
w_{HF}	Beam deflection in hydraulic fracture in double-cantilever-beam (HF-DCB) specimen
x, $x\prime$, $x\prime\prime$ x_0	The position of fluid front, the velocity of the fluid front, the acceleration of the fluid front, a point in the fluid
Π_1	The dimensionless secant velocity of the fluid front x/tV_c
Π_2	$P_f/2\rho V_c^2$
Π_3	$12\eta/A^2\rho V_c^{2n} t^{2n-1}$
$\Pi_{2c}(n)$	Critical Π_2 as a function of n controlling the existence of equilibrium state
$\Pi_{2s}(n)$	Critical Π_2 as a function of n controlling the separation between the fluid front and the crack tip
ρ	Density of the fluid
σ, P_f	Hydraulic pressure
η	Dynamic viscosity of the fracturing fluid (water with black ink): $0.9\cdot10^{-3}$ Pa·s
θ	Beam deflection angle in beam analysis and a polar-coordinate parameter in Williams' fitting
μ	Shear modulus of the specimen
ν	Poisson's ratio
τ	Shear stress of the fluid

Appendix A. Hydraulic Fracture in Double-Cantilever-Beam (HF-DCB) Solutions

The distributed load can be regarded as infinite numbers of small point loads df distributed along the beam. As it can be seen in Figure A1, the deflection at x_0 is contributed by the point loads along the beam, and is separately considered, namely $0 < x_1 < x_0$, and $x_0 < x_2 < a$. It should be noted that the x_0 is not the same x_0 as in dynamic fluid analyses. The deflection at x_1 induced by a small point load at x_1 ($0 < x < x_1$) is be expressed as:

$$w_m(x_1, x_1) = \frac{6}{Ebh^3\lambda^3}\left[\frac{2}{3}\lambda^3 x_1^3 + 2\lambda^2 H x_1^2 + 2\lambda D x_1 + F\right]df_1 \tag{A1}$$

where $df_1 = \sigma b dx_1$. The angle of deflection at x_1 can be expressed as

$$\tan(\theta(x_1)) = \left.\frac{\partial w(x, x_1)}{\partial x}\right|_{x=x_1} \tag{A2}$$

Taking both the deflection and angle of deflection into consideration, the deflection contributed by a small point load at x_1 to the point at x_0 is expressed as:

$$\begin{aligned}w_1(x_1, x_0) &= w_m(x_1, x_1) + \tan(\theta(x_1))(x_0 - x_1) \\ &= \frac{6}{Ebh^3\lambda^3}\left[-\frac{1}{3}\lambda^3 x_1^3 + \lambda^3 x_0 x_1^2 + (2\lambda^2 H x_0 + \lambda D)x_1 + \lambda D x_0 + F\right]df_1\end{aligned} \tag{A3}$$

The sum of the deflection contributed by distributed load along x_1 ($0 < x_1 < x_0$) to the point at x_0 can be derived as:

$$w_1(x_0) = \int_0^{x_0}[w_m(x_1, x_1) + \tan(\theta(x_1))(x_0 - x_1)] \tag{A4}$$

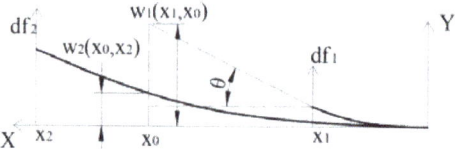

Figure A1. Schematics of beam deflection. The total deflection is the sum of w_1 and w_2. Both w_1 and w_2 are calculated by regarding the distributed load as infinite numbers of small point loads. The coordinate is not the same as the coordinate used in Figure 4.

Now consider the deflection contributed by the distributed load at x_2 where $x_0 < x_2 < a$. The contribution by a small point load at x_2 is expressed as:

$$w_2(x_0, x_2) = \frac{6}{Ebh^3\lambda^3}\left[-\frac{1}{3}\lambda^3 x_0^3 + \lambda^3 x_2 x_0^2 + (D + 2x_2\lambda H)\lambda x_0 + F + x_2\lambda D\right]df_2 \tag{A5}$$

where $df_2 = \sigma b dx_2$. The total deflection contributed by distributed load at x_2 to the point at x_0 can be derived as:

$$w_2(x_0) = \int_{x_0}^{a} w(x_0, x_2) \tag{A6}$$

Thus, the total deflection can be derived as the sum of the Equations (A4) and (A6)

$$w_{HF}(x) = w_1(x) + w_2(x) = \frac{\sigma}{Eh^3\lambda^3}\varphi(x) \tag{A7}$$

where $\varphi(x)$ is expressed in Equation (6).

The strain energy stored in the beam can be calculated by the work done by the distributed load

$$U(a) = 2\int_0^a \frac{1}{2}\sigma b w_{HF}(x)dx = \sigma^2 \Phi(a) \tag{A8}$$

The energy release rate is expressed as

$$G = \frac{1}{b}\frac{dW}{da} = \frac{1}{b}\sigma^2 \Phi\prime(a) \tag{A9}$$

where

$$\Phi\prime(a) = \frac{\partial \Phi}{\partial a} - \frac{\partial \Phi}{\partial c} \tag{A10}$$

According to the well known relationship between the energy release rate G and the stress intensity factor K, $K = \sqrt{EG}$, the stress intensity factor K can be expressed as Equation (5).

References

1. Germanovich, L.N.; Astakhov, D.K.; Mayerhofer, M.J.; Shlyapobersky, J.; Ring, L.M. Hydraulic fracture with multiple segments I. Observations and model formulation. *Int. J. Rock Mech. Min. Sci.* **1997**, *34*, 97.e1–97.e19. [CrossRef]
2. Garagash, D.I.; Detournay, E.; Adachi, J.I. Multiscale tip asymptotics in hydraulic fracture with leak-off. *J. Fluid Mech.* **2011**, *669*, 260–297. [CrossRef]
3. Sachau, T.; Bons, P.D.; Gomez-Rivas, E. Transport efficiency and dynamics of hydraulic fracture networks. *Front. Phys.* **2015**, *3*, 63. [CrossRef]
4. Yew, C.H.; Weng, X. Mechanics of hydraulic fracturing. *Gulf Prof. Publ.* **2014**, *23*, 164–166.
5. Cao, T.D.; Hussain, F.; Schrefler, B.A. Porous media fracturing dynamics: Stepwise crack advancement and fluid pressure oscillations. *J. Mech. Phys. Solids* **2018**, *111*, 113–133. [CrossRef]
6. Abe, H.; Keer, L.M.; Mura, T. Growth rate of a penny-shaped crack in hydraulic fracturing of rocks, 2. *J. Geophys. Res.* **1976**, *81*, 6292–6298. [CrossRef]
7. Garagash, D.; Detournay, E. The tip region of a fluid-driven fracture in an elastic medium. *J. Appl. Mech.* **2000**, *67*, 183–192. [CrossRef]

8. Jeffrey, R.G. The combined effect of fluid lag and fracture toughness on hydraulic fracture propagation. In Low Permeability Reservoirs Symposium. *Soc. Petroleum Eng.* **1989**. [CrossRef]
9. Yew, C.H.; Liu, G.F. The fracture tip and KIc of a hydraulically induced fracture. *Spe Prod. Eng. J.* **1991**, 171–177.
10. Bunger, A.P.; Gordeliy, E.; Detournay, E. Comparison between laboratory experiments and coupled simulations of saucer-shaped hydraulic fractures in homogeneous brittle-elastic solids. *J. Mech. Phys. Solids* **2013**, *61*, 1636–1654. [CrossRef]
11. Khoei, A.R.; Vahab, M.; Haghighat, E.; Moallemi, S. A mesh-independent finite element formulation for modeling crack growth in saturated porous media based on an enriched-FEM technique. *Int. J. Fract.* **2014**, *188*, 79–108. [CrossRef]
12. Kostov, N.M.; Ning, J.; Gosavi, S.V.; Gupta, P.; Kulkarni, K.; Sanz, P.F. Dynamic hydraulic fracture modeling for wellbore integrity prediction in a porous medium. In Proceedings of the 2015 SIMULIA Community Conference, Berlin, Germany, 18–21 May 2015.
13. Chang, H. Hydraulic Fracturing in Particulate Materials. Ph.D. Thesis, Georgia Institute of Technology, Atlanta, Georgia, 2004.
14. Frash, L.P. Laboratory-Scale Study of Hydraulic Fracturing in Heterogeneous Media for Enhanced Geothermal Systems and General Well Stimulation. Ph.D. Thesis, Colorado School of Mines, Golden, CO, USA, 2014.
15. Lhomme, T.P.; De Pater, C.J.; Helfferich, P.H. Experimental study of hydraulic fracture initiation in Colton sandstone. In Proceedings of the SPE/ISRM Rock Mechanics Conference, Irving, TX, USA, 20–23 October 2002.
16. Lhomme, T.P.Y. Initiation of Hydraulic Fractures in Natural Sandstones. Ph.D. Thesis, Delft University of Technology, Delft, The Netherlands, 2005.
17. Pizzocolo, F.; Huyghe, J.M.; Ito, K. Mode I crack propagation in hydrogels is step wise. *Eng. Fract. Mech.* **2013**, *97*, 72–79. [CrossRef]
18. Tan, P.; Jin, Y.; Han, K.; Hou, B.; Chen, M.; Guo, X.; Gao, J. Analysis of hydraulic fracture initiation and vertical propagation behavior in laminated shale formation. *Fuel* **2017**, *206*, 482–493. [CrossRef]
19. Wu, R. Some Fundamental Mechanisms of Hydraulic Fracturing. Ph.D. Thesis, Georgia Institute of Technology, Atlanta, Georgia, 2006.
20. Wu, R.; Bunger, A.P.; Jeffrey, R.G.; Siebrits, E. A comparison of numerical and experimental results of hydraulic fracture growth into a zone of lower confining stress. In Proceedings of the 42nd US Rock Mechanics Symposium (USRMS), San Francisco, CA, USA, 29 June–2 July 2008.
21. Zhang, G.Q.; Chen, M. Dynamic fracture propagation in hydraulic re-fracturing. *J. Pet. Sci. Eng.* **2010**, *70*, 266–272. [CrossRef]
22. Zhou, J.; Chen, M.; Jin, Y.; Zhang, G.Q. Analysis of fracture propagation behavior and fracture geometry using a tri-axial fracturing system in naturally fractured reservoirs. *Int. J. Rock Mech. Min. Sci.* **2008**, *45*, 1143–1152. [CrossRef]
23. Fisher, M.K.; Warpinski, N.R. Hydraulic-fracture-height growth: Real data. *Spe Prod. Oper.* **2012**, *27*, 8–19. [CrossRef]
24. Bosanquet, C.H. LV. on the flow of liquids into capillary tubes. *Lond. Edinb. Dublin Philos. Mag. J. Sci.* **1923**, *45*, 525–531. [CrossRef]
25. Xing, P.; Yoshioka, K.; Adachi, J.; El-Fayoumi, A.; Bunger, A.P. Laboratory measurement of tip and global behavior for zero-toughness hydraulic fractures with circular and blade-shaped (PKN) geometry. *J. Mech. Phys. Solids* **2017**, *104*, 172–186. [CrossRef]
26. Bunger, A.P. Near-Surface Hydraulic Fracture. Ph.D. Thesis, University of Minnesota, Minneapolis, MN, USA, 2005.
27. Bunger, A.P.; Detournay, E. Experimental validation of the tip asymptotics for a fluid-driven crack. *J. Mech. Phys. Solids* **2008**, *56*, 3101–3115. [CrossRef]
28. Desroches, J.; Detournay, E.; Lenoach, B.; Papanastasiou, P.; Pearson, J.R.A.; Thiercelin, M.; Cheng, A. The crack tip region in hydraulic fracturing. *Proc. R. Soc. Lond. A* **1994**, *447*, 39–48. [CrossRef]
29. Nordgren, R.P. Propagation of a vertical hydraulic fracture. *Soc. Pet. Eng. J.* **1972**, *12*, 306–314. [CrossRef]
30. Perkins, T.K.; Kern, L.R. Widths of Hydraulic Fractures. *J. Pet. Technol.* **1961**, *13*, 937–949. [CrossRef]

31. Kanninen, M.F. An augmented double cantilever beam model for studying crack propagation and arrest. *Int. J. Fract.* **1973**, *9*, 83–92.
32. Fett, T.; Munz, D. Stress intensity factors and weight functions. *Comput. Mech.* **1997**, *1*, 112–115.
33. Levenberg, K. A Method for the Solution of Certain Problems in Least-Squares. *Q. Appl. Math.* **1944**, *2*, 164–168. [CrossRef]
34. Marquardt, D. An Algorithm for Least-squares Estimation of Nonlinear Parameters. *Siam J. Appl. Math.* **1963**, *11*, 431–441. [CrossRef]
35. Yates, J.R.; Zanganeh, M.; Tai, Y.H. Quantifying crack tip displacement fields with DIC. *Eng. Fract. Mech.* **2010**, *77*, 2063–2076. [CrossRef]
36. Schechter, R.S.; Wade, W.H.; Wingrave, J.A. Sorption isotherm hysteresis and turbidity phenomena in mesoporous media. *J. Colloid Interface Sci.* **1977**, *59*, 7–23. [CrossRef]
37. Ridgway, C.J.; Gane, P.A.; Schoelkopf, J. Effect of capillary element aspect ratio on the dynamic imbibition within porous networks. *J. Colloid Interface Sci.* **2002**, *252*, 373–382. [CrossRef]
38. Garagash, D.I. Propagation of a plane-strain hydraulic fracture with a fluid lag: Early-time solution. *Int. J. Solids Struct.* **2006**, *43*, 5811–5835. [CrossRef]
39. Hunsweck, M.J.; Shen, Y.; Lew, A.J. A finite element approach to the simulation of hydraulic fractures with lag. *Int. J. Numer. Anal. Methods Geomech.* **2013**, *37*, 993–1015. [CrossRef]
40. Vahab, M.; Khalili, N. Computational Algorithm for the Anticipation of the Fluid-Lag Zone in Hydraulic Fracturing Treatments. *Int. J. Geomech.* **2018**, *18*, 04018139. [CrossRef]
41. Rubin, A.M. Propagation of magma-filled cracks. *Annu. Rev. Earth Planet. Sci.* **1995**, *23*, 287–336. [CrossRef]
42. Xia, Y.; Jin, Y.; Chen, M.; Chen, K.P. An Enriched Approach for Modeling Multiscale Discrete-Fracture/Matrix Interaction for Unconventional-Reservoir Simulations. *SPE J.* **2018**. [CrossRef]
43. Zeng, C.; Deng, W. The Effect of Radial Cracking on the Integrity of Asperity Under Thermal Cooling Process. In Proceedings of the 52nd US Rock Mechanics/Geomechanics Symposium, Seattle, WA, USA, 17–20 June 2008.
44. Dong, J.; Chen, M.; Jin, Y.; Hong, G.; Zaman, M.; Li, Y. Study on micro-scale properties of cohesive zone in shale. *Int. J. Solids Struct.* **2019**, in press. [CrossRef]
45. Li, Y.; Jia, D.; Rui, Z.; Peng, J.; Fu, C.; Zhang, J. Evaluation method of rock brittleness based on statistical constitutive relations for rock damage. *J. Pet. Sci. Eng.* **2017**, *153*, 123–132. [CrossRef]
46. Li, Y.; Zuo, L.; Yu, W.; Chen, Y. A Fully Three Dimensional Semianalytical Model for Shale Gas Reservoirs with Hydraulic Fractures. *Energies* **2018**, *11*, 436. [CrossRef]
47. Li, Y.; Rui, Z.; Zhao, W.; Bo, Y.; Fu, C.; Chen, G.; Patil, S. Study on the mechanism of rupture and propagation of T-type fractures in coal fracturing. *J. Nat. Gas Sci. Eng.* **2018**, *52*, 379–389. [CrossRef]
48. Li, Y.W.; Zhang, J.; Liu, Y. Effects of loading direction on failure load test results for Brazilian tests on coal rock. *Rock Mech. Rock Eng.* **2016**, *49*, 2173–2180. [CrossRef]
49. Geng, Z.; Bonnelye, A.; Chen, M.; Jin, Y.; Dick, P.; David, C.; Fang, X.; Schubnel, A. Time and Temperature Dependent Creep in Tournemire Shale. *J. Geophys. Res. Solid Earth* **2018**, *123*, 9658–9675. [CrossRef]
50. Geng, Z.; Bonnelye, A.; Chen, M.; Jin, Y.; Dick, P.; David, C.; Fang, X.; Schubnel, A. Elastic anisotropy reversal during brittle creep in shale. *Geophys. Res. Lett.* **2017**, *44*, 21. [CrossRef]
51. Jin, Y.; Chen, K.P.; Chen, M. An asymptotic solution for fluid production from an elliptical hydraulic fracture at early-times. *Mech. Res. Commun.* **2015**, *63*, 48–53. [CrossRef]
52. Siddhamshetty, P.; Kwon, J.S.; Liu, S.; Valkó, P.P. Feedback control of proppant bank heights during hydraulic fracturing for enhanced productivity in shale formations. *Aiche J.* **2018**, *64*, 1638–1650. [CrossRef]
53. Siddhamshetty, P.; Yang, S.; Kwon, J.S. Modeling of hydraulic fracturing and designing of online pumping schedules to achieve uniform proppant concentration in conventional oil reservoirs. *Comput. Chem. Eng.* **2018**, *114*, 306–317. [CrossRef]

© 2019 by the authors. Licensee MDPI, Basel, Switzerland. This article is an open access article distributed under the terms and conditions of the Creative Commons Attribution (CC BY) license (http://creativecommons.org/licenses/by/4.0/).

Article

A Case Study of an Optimized Intermittent Ventilation Strategy Based on CFD Modeling and the Concept of FCT

Hui Liu, Shanjun Mao * and Mei Li

Institute of Remote Sensing and Geographic Information System, Peking University, Beijing 100871, China; huil@pku.edu.cn (H.L.); meili@pku.edu.cn (M.L.)
* Correspondence: sjmao_@pku.edu.cn; Tel.: +86-10-6275-5420

Received: 15 January 2019; Accepted: 4 February 2019; Published: 22 February 2019

Abstract: With the increasing operation costs and implementation of carbon tax in the underground coal mining systems, a cost-effective ventilation system with well methane removal efficiency becomes highly required. Since the intermittent ventilation provides a novel approach in energy saving, there still exists some scientific issues. This paper adopts the design concept of frequency conversion technology (FCT) to the ventilation pattern design for the first time, and an optimized intermittent ventilation strategy is proposed. Specifically, a real excavation laneway of a coal mine in China is established as the physical model, and computational fluid dynamic (CFD) approaches are utilized to investigate the spatiotemporal characteristics of airflow behavior and methane distribution. Plus, the period of intermittency and appropriate air velocity are scientifically defined based on the conception of FCT by conducting the parametric studies. Therefore, an optimized case is brought out with well methane removal efficiency and remarkable energy reduction of 39.2% in a ventilation period. Furthermore, the simulation result is verified to be reliable by comparing with field measurements. The result demonstrates that a balance of significant energy saving and the methane removal requirement based on the concept of FTC is possible, which could in turn provide good operational support for FTC.

Keywords: cost-effective; frequency conversion technology (FCT); ventilation; methane removal; computational fluid dynamic (CFD); spatiotemporal characteristics

1. Introduction

In the underground coal mining operation environment, methane gas is one of the most hazardous emitted gases for it is highly explosive under certain conditions inside a laneway of the coal mine, which also contributes to global warming [1–3]. Numerous reported methane-related incidents and accidents with fatalities show that appropriate methane gas control and the related progressing in mine safety still remains a longstanding challenge and an emerging issue [4,5]. To avoid the methane accumulation in the mining face area, and therefore eliminate potential incidents and accidents, a good ventilation system is mandatory and highly desirable. In accordance with coal mining regulations in many countries, methane concentrations should be maintained below 3%v/v in US, 2.5% in Spain, 2% in France, and even lower in other countries: 1.25% in UK, and 1% in Germany and China [6,7].

Generally, large ventilation system is installed in underground coal mines to supply fresh air and to dilute the methane gas below the maximum allowable level. However, it is showed that, nowadays, most ventilation systems generally supply excess fresh air to remove methane gas and ensure a safe working environment as required by local codes, and the specific energy consumption value differs from different site conditions. For example, Jeswiet stated that at least 25% of electric energy consumption in mining is costed by ventilation purposes, Mielli and Bongiovanni indicated that approximately 15%–40% of the operating costs can be attributed to mine ventilation systems, and up to 60% of the total mining operation costs are connected with mine ventilation as revealed by Reddy [8–11]. With the increase of energy and operation costs and implementation of a carbon tax, a cost-effective ventilation and methane control strategy is strongly required for the mining industry, which means that the ventilation system should be designed to meet the regulatory limits and ensure a safe and productive working environment whilst keeping energy usage and operating costs to a minimum. As such, safety and cost are the two main competing requirements that have to be balanced, for which the careful ventilation design is mandatory and a concept of ventilation-on-demand has received considerable attention [12–15].

In this sense, a cost-effective ventilation and methane removal strategy is expected to be a thorough solution to save operating costs as well as maintain the safety standards for underground coal mining systems. A fundamental challenge to effectively design the ventilation strategy is to understand the physical mechanism of methane gas distribution as well as its coupling with ventilation airflow. Numerous simulation experiments based on the principles of computational fluid dynamics (CFD) have been widely applied to evaluate and investigate the airflow behavior, methane dispersion regularity, and control strategies in underground coal mines, which allows innovation at very low costs. Heerden and Sullivan (1993) [16] were among the first researchers by utilizing the CFD modeling to investigate airflow behavior, methane emission, and dust dispersion in the underground coal mine, while validation of the model against field measured data was not conducted till 1997 by Uchino and Inoue [17]. Nakayama et al. (1999) conducted similar studies by utilizing CFD software, LASAR95/98, and pointed out that transversely in and along the corner space and the area underneath the ventilation duct are the locations where higher methane gas generally exists [18]. Since in the case of extraction of flammable minerals, the forced ventilation system is address to be more effective and safe compared with exhaust only ventilation system by several studies [19–21], numerous methane removal strategies have been studied based on the forced ventilation system. For example, Wala et al. numerically simulated methane dispersion in room and pillar mines and verified the CFD model with the experimental data, and later extended their experiment by combining the scrubber operation on the face ventilation [22]. Besides, Torano et al. (2011) compared various turbulence models to evaluate methane distribution, and later added dust to their model on the basis of the validation with experimental data [23]. Szlązak N. and Zhong M. studied the optimization of air flow and concerned the methane hazard in earlier research [24,25]. However, these studies still remain very limited due to the neglect of dynamic regularity of methane distribution during ventilation design process, which renders the principle and technology of methane removal a key scientific issue.

Recently, Sasmito and Kurnia et al. (2013, 2017) carried out a series of simulations of different ventilation scenarios by applying auxiliary equipment to improve ventilation as well as the control of methane gas whilst keeping low energy usage consumption [26–29]. And Pach et al. presented some latest research of optimization of forced air distribution by comparing positive and negative regulations, and also showed the airflow optimization due to methane hazards and low-cost ventilation [30,31]. Among these studies, it suggests that the additional ventilation equipment may cause the complexity and severe energy usage to the system, such as the combination of air curtains together with a physical brattice, and brattice ventilation with no additional power source may even limit the flexibility and movement of miners and mining fleets. One of the most significant research results is the novel intermittent ventilation strategy, which is implemented by alternating the airflow velocity between a predetermined high and low value [32]. However, this method is just being proposed, some scientific issues have to be carefully solved. And there mainly exists two major issues, as presented by Guang Xu [33]: one is the control of alternating flow is challenging; another is that this study assumes the methane concentration is still below the limit value during the period of low flow, which shows that the increase of flow velocity is probably not necessary since keeping the low flow rate would save more power costs. Where the first issue of implementation is feasible with the application of frequency conversion technology (FCT) in many underground coal mines [34,35], and the second issue is primarily due to the predetermined intermittency for high and low flow rate, which ignored the dynamic regularity of methane accumulation in an intermittency of low flow rate. Besides, although the CFD modeling provides practical results in present studies, validation is needed by comparison with field measured data.

Based on the above issues, this study applies the design concept of FCT for the first time to investigate an optimized intermittent ventilation strategy with the goal of reducing energy costs whilst maintaining the methane concentration at a low level. Firstly, a geometric laneway model is established and discretized with different mesh generations, and the appropriate mesh size is chosen by testing and comparing each other in terms of elapsed time and computational efficiency. Secondly, computational fluid dynamic (CFD) models are applied to study the spatiotemporal characteristics of airflow behavior and methane distribution inside a laneway of coal mine, and intermittent ventilation patterns with different speed functions are tested and compared with respect to the potential energy saving and main impact factors. Finally, in accordance with the design concept of FCT, the key time point is defined by analyzing the dynamic regularity of methane concentration, thus the optimal intermittent ventilation pattern is scientifically determined, which provides excellent performance in balancing both methane control efficiency and energy costs reduction. Furthermore, the simulation results are validated with field measured data in Sijiazhuang coal mine. It is noted that the simpler design of this intermittent ventilation pattern can be suitable and practical in underground mine applications, and can provide decision supports for FCT when utilized in local fan of ventilation system.

2. Model Development

2.1. Geometric Model and Research Background

In this study, a fully mechanized excavation laneway of 1-5205 (identification) in Sijiazhuang coal mine is selected as the physical laneway prototype, which is located in China, Shanxi province. The main reason for choosing this laneway is the accessibility of continuously and precisely monitored mine data. The latitude and longitude extensions of this coal mine range from 101,962.58 to 106,993.34 m and from 72,077.19 to 66,008.85 m, respectively. Figure 1 shows the geometric model of the laneway, which is established to perform the case study. The excavation laneway is 32 m long with rectangular cross-section of 4.4 m wide and 4 m high, and the ventilation duct with a diameter of 0.8 m is hung at 3 m height from the laneway floor and 0.7 m from the laneway wall on the access road. The distance from its outlet to the driving face is 11 m.

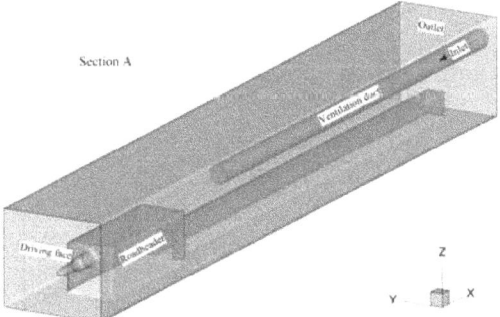

Figure 1. Schematic views of the laneway model.

The safety and operation regulations in Shijiazhuang coal mine shows that the mine has a primary structure coal, and adopts the comprehensive mechanized driving method in the construction of the laneway. Besides, what we care about is the airflow quantity, which is calculated based on the basic requirement of the evaluated methane gas or carbon dioxide emission quantity, the number of miners in the laneway, the actual wind suction of local fan, and the maximum and minimum air velocity during the process of laneway excavation. The safety and operation regulations indicate that the required airflow quantity in the excavation laneway of 1-5205 is 7.53 m^3/s (0.428 m/s), so the velocity of airflow in ventilation duct is 15 m/s, the evaluated volume flow rate of methane gas around the driving face is approximately 0.031 m^3/s, and the maximum number of miners allowed in the 1-5205 laneway is 9. In this study, a representative longitudinal section is chosen to investigate the spatiotemporal distribution of airflow behavior and methane distributions in the numerical simulation. As can be seen in Figure 1, where section A is specified in the outtake side of the laneway with 0.8 m away from the nearest wall of the laneway, which is also the region where miners typically walk.

2.2. Mathematical Formulation

The essential regularity of methane dispersion and its coupling with the air flow is a key issue for the methane gas removal problem, which belongs to viscous Newtonian fluid, and is governed by the Navier-Stokes (N-S) equations. During the process of coal mining, a large amount of methane gas generates from the driving face and disperses with impact of airflow from the ventilation duct, which is defined as species transport motion. We assume that the airflow and gases are the continuous medium gas, and heat and mass transfer are ignored during the process of dispersing.

2.2.1. Governing Conservation Equations

In the laneway flow, methane is released and dispersed with ventilation airflow from the specified sources in the driving face, and turbulent mass, momentum, energy, and species transport occur [7,32]. The conservation equations of which need to be considered can be expressed as:

$$\frac{\partial \rho}{\partial t} = -\nabla \cdot \rho \mathbf{U} \tag{1}$$

$$\frac{\partial}{\partial t}(\rho \mathbf{U}) = -\nabla \cdot \rho \mathbf{U}\mathbf{U} - \nabla p + \rho g + \nabla \cdot \boldsymbol{\tau} \tag{2}$$

$$\frac{\partial}{\partial t}(\rho c_p T) = \nabla \cdot (k_{eff} + \frac{c_p \mu_t}{\sigma_{rt}}) \nabla T - \nabla \cdot (\rho c_p \mathbf{U} T) \tag{3}$$

$$\frac{\partial}{\partial t}(\rho \omega_i) = \nabla \cdot (\rho D_{i,eff} + \frac{\mu_t}{sc_t}) \nabla T - \nabla \cdot (\rho \omega_i \mathbf{U}) \tag{4}$$

where ρ is the fluid density, $\mathbf{U}$ is the fluid velocity, and p is the pressure; $\boldsymbol{\tau}$ is the viscous stress tensor, c_p is the specific heat of the fluid, g is the gravity acceleration, k_{eff} is the effective fluid thermal conductivity, and T is the temperature. ω_i, $D_{i,eff}$ are the mass fraction and the effective diffusivity of species i (O_2, CH_4 and N_2), respectively, μ_t is the turbulent viscosity, sc_t is the turbulent Schmidt number, and σ_{rt} is the turbulent Prandtl number.

2.2.2. Constitutive Relations

The viscous stress tensor for fluid can be described as [7,32]:

$$\boldsymbol{\tau} + 2/3[(\mu + \mu_t)(\nabla \cdot \mathbf{U})\mathbf{I} + \rho k \mathbf{I}] = (\mu + \mu_t)(\nabla \cdot \mathbf{U} + (\nabla \cdot \mathbf{U})^T) \tag{5}$$

where μ is the dynamic viscosity, $\mathbf{I}$ is the identity or second order unit tensor, and k is the turbulent kinetic energy.

It is noted that in the mining laneway, the interaction between the species follows the incompressible ideal gas law, which is captured in the mixture density, thus a ternary species mixture consisting of methane gas, oxygen, and water vapour is solved, and the ideal gas law is described as:

$$\rho = \frac{pM}{RT} \tag{6}$$

where R is the universal gas constant, M denotes the mixture molecular weight, which is given by:

$$M = [\frac{\omega_{CH_4}}{M_{CH_4}} + \frac{\omega_{O_2}}{M_{O_2}} + \frac{\omega_{H_2O}}{M_{H_2O}} + \frac{\omega_{N_2}}{M_{N_2}}]^{-1} \tag{7}$$

ω_i and M_i are the mass fraction and the molar mass of species i (O_2, CH_4 and N_2), respectively, where the mass fraction of nitrogen is calculated as:

$$\omega_{N_2} = 1 - (\omega_{CH_4} + \omega_{O_4} + \omega_{H_2O}) \tag{8}$$

The molar fractions are related to the mass fraction, calculated by:

$$x_i = \frac{\omega_i M}{M_i} \tag{9}$$

The fluid mixture viscosity is given as:

$$\mu = \sum_i [x_i u_i / (\sum_j x_i \Phi_{i,j})] \text{ with } i, j = CH_4, O_2, H_2O, N_2 \tag{10}$$

where $x_{i,j}$ are the mole fraction of species i and j, and $\Phi_{i,j}$ is defined as:

$$\Phi_{i,j} = \frac{1}{\sqrt{8}}(1 + \frac{M_i}{M_j})^{-1/2}[1 + (\frac{\mu_i}{\mu_j})^{1/2}(\frac{M_i}{M_j})^{1/4}]^2 \tag{11}$$

In consideration of the practical purpose, the unit of methane concentration commonly used in the regulation/law is presented in %v/v, which is defined as:

$$CH_4 = \omega_{CH_4} \times 100\% \tag{12}$$

In addition, the fan power is calculated as:

$$P_{fan} = \Delta P_{fan} \dot{Q}_{fan} \tag{13}$$

where ΔP_{fan} stands for the pressure difference from the inlet, $\dot{Q}_{fan}$ is the volumetric flow rate of the fan. It is noted that depending on the actual fan efficiency, the fan power would be different.

2.2.3. Turbulence Model

The turbulence model is the key foundation of the species transport motion in representing flow behaviour, because it can directly affect the simulation accuracy for dynamic characteristics of methane dispersion. Thus, it is of great importance to select the appropriate turbulence model. As for the most widely used models, such as K-epsilon, K-omega, RSM and Spalart-Allmaras. The comparison experiment demonstrated by Kurnia et al. and our previous work shows that the K-epsilon model gives a reasonably good prediction [7], so K-epsilon model is selected in the current study. In short, the model considers a two-equation model and solves for turbulent kinetic energy, k, and its rate of dissipation, ε, which is coupled with turbulent viscosity [26,27]. The equations are given as:

$$\nabla \cdot [(\mu + \frac{\mu_t}{\sigma_k})\nabla k] + G = \frac{\partial}{\partial t}(\rho k) + \nabla \cdot (\rho \mathbf{U} k) \tag{14}$$

$$\nabla \cdot [(\mu + \frac{\mu_t}{\sigma_\varepsilon})\nabla \varepsilon] + C = \frac{\partial}{\partial t}(\rho \varepsilon) + \nabla \cdot (\rho \mathbf{U} \varepsilon) \tag{15}$$

where $G = G_k - \rho\varepsilon$, $C = C_{1\varepsilon}\frac{\varepsilon G_k}{k} + C_{2\varepsilon}\rho\frac{\varepsilon^2}{k}$, G_k represents the generation of turbulence kinetic energy due to the mean velocity gradients. σ_k and σ_ε are the turbulent Prandtl numbers for k and ε respectively. The turbulent viscosity is computed by combining k and ε as follows:

$$\mu_t = \rho C_\mu \frac{k^2}{\varepsilon} \qquad (16)$$

where $C_{1\varepsilon}$, $C_{2\varepsilon}$, C_μ, σ_k and σ_ε are constants, the values are 1.44, 1.92, 0.09, 1, and 1.3, respectively.

2.2.4. Boundary Conditions

The boundary conditions of the model are summarized as follows:

(i) Walls: standard wall function is defined;
(ii) Inlet: the inlet of ventilation duct is prescribed as the velocity-inlet, the basic 15 m/s and various time varying air velocity are specified according to different ventilation patterns;
(iii) Driving face: methane gas is released evenly at a volume flow rate of 0.031 m^3/s;
(iv) Outlet: the outlet of the laneway is set to the pressure-outlet boundary condition with standard atmospheric pressure (101.325 kPa), and with the average temperature in the Sijiazhuang coal mine of 300 K.

3. Numerical Methodology

The three-dimensional computational domain of the laneway was established, meshed and labelled with boundary conditions using the pre-processor Gambit 2.3.16. In order to ensure a mesh independent solution, four different amount of mesh 3.94×10^5, 5.11×10^5, 1.15×10^6 and 4.04×10^6 were tested and compared in terms of airflow velocity and methane concentration, and detailed comparisons were presented in Figures 2 and 3. The results indicated that mesh amount of 1.15 million elements was sufficient for the numerical simulation purposes: a fine structure near the ventilation duct and coarser mesh in other areas to reduce the computational cost, see Table 1 for detailed information. It shows that the mesh amount of 1.15×10^6 gives about 4.2% deviation compared with 4.04×10^6, while it requires significantly shorter mesh generation time and less RAM of about 4.8 GB; whereas, the mesh amount of 3.94×10^5 and 5.11×10^5 gives up to 14.6% and 10.4% deviation compared to the number of 4.04×10^6, although they consume shorter time to generate mesh. Thus, the mesh amount of 1.15 million elements was chosen to study the following research. The meshed laneway is shown in Figure 4.

Table 1. Details for the mesh generation of computational domains.

Mesh Amount	Interval of Laneway (m)	Interval of Windpipes (m)	Elapsed Time (s)	RAM Required (GB)	Deviation
3.94×10^5	0.2	0.2	11	3.4	14.6%
5.11×10^5	0.2	0.1	18	3.9	10.4%
1.15×10^6	0.15	0.1	35	4.8	4.2%
4.04×10^6	0.1	0.05	114	7.1	-

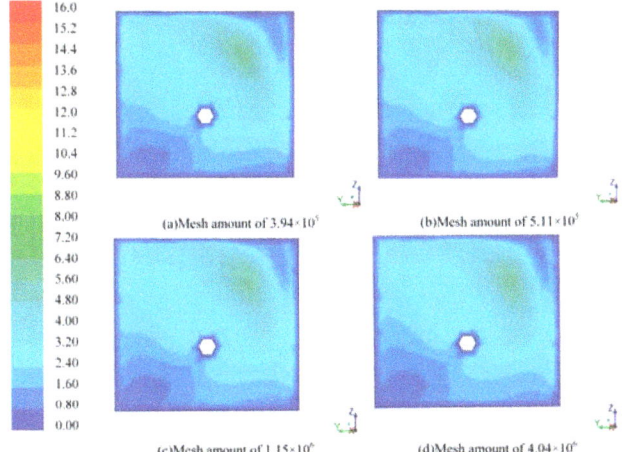

Figure 2. Velocity field of the airflow (m/s) under different mesh generation.

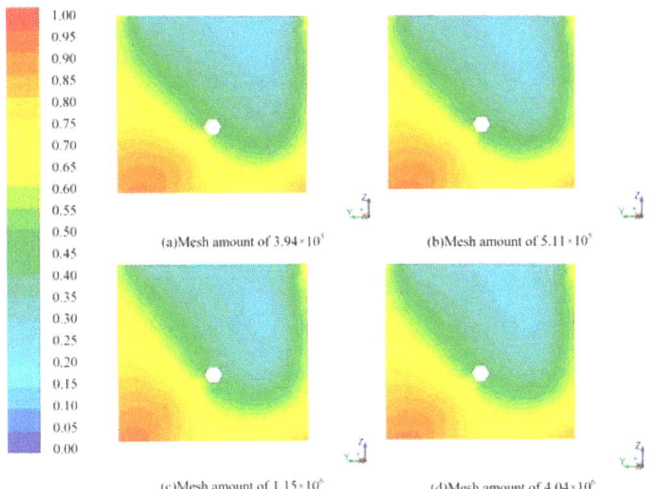

Figure 3. Methane concentration (%v/v) under different mesh generation.

Figure 4. Mesh generation.

The established model was numerically investigated utilising a Computational Fluid Dynamics platform (FLUENT Inc., New York, NY, USA) [36]. All terms of the equations were solved with the semi-implicit pressure-linked equation (SIMPLE) algorithm, second-order upwind scheme. The turbulence model and boundary conditions were solved using finite volume solver. On average, each simulation required 3600 iterations with convergence tolerance of 1×10^{-6} for all variables, and other detailed parameters are presented in Table 2. It takes around 11–15 h on computers with four core processors and 8 GB RAM.

Table 2. Parameters for the turbulence models.

Parameters	Setting
Air density (kg/m^3)	1.225
Turbulent viscosity (m^2/s)	1.7894×10^{-5}
Turbulent kinetic energy	1.3
The relative intensity of turbulence	5.1
Volume flow rate of methane gas (m^3/s)	0.031
Convergence criteria	1×10^{-6}
Initialization method	Hybrid Initialization
Number of time steps for transient simulation	1800
Time step size (s)	1
Max iterations/time step	15

4. Results and Discussion

4.1. Intermittent Ventilation Strategies and Parametric Studies

Here, the spatiotemporal characteristics of air flow behavior and methane distribution with different configurations of ventilation strategies are investigated, and parametric studies are conducted and discussed to compare the effects of various factors.

It is showed that to supply sufficient fresh air for coal miners and to keep methane gas concentrations below the threshold value, most underground coal mines usually drive excessive amounts of airflow to various locations [23], and large main fans are utilized to provide airflow from the surface and smaller branches with auxiliary ventilation system are further distributed to ventilate airflow. The majority of coal mining operators employed this kind of ventilation system with fans working continuously during coal mining operation, which requires massive amounts of electricity to power the fan. Similarly, Sijiazhuang coal mine adopts the typical ventilation system. In order to save energy whilst ensuring the methane removal efficiency, here, we introduce the design and implementation concept of FTC [34,35]: methane control is the main purpose in normal period of ventilation, when the methane concentration is beyond the specified value close to the alarm value of the methane sensor (0.7%v/v in this study), the velocity should be increased to reduce the concentration; when the methane concentration is under the specified value (0.4%v/v in this study), the velocity should be decreased to ensure methane gas to be recovered for operation of ventilation air methane energy extraction. It is noted that in China, the maximum allowable value of methane concentration is 1%v/v in general, and according to the safety and operation regulations in Sijiazhuang coal mine, the specified alarm value is 0.8%v/v. The upper limit of methane sensor is usually set as 0.7%v/v in actual site applications and there leaves a 0.1% interval between the alarm value, and the lower limit is usually set as 0.4%v/v. The ventilation system using FTC is composed of frequency conversion controller, methane sensor, automatic air distribution device, ventilation machine, silencer etc. The special-use switch of the frequency conversion controller has a programmable logic controller called PLC system, which can analyze and determine whether the methane concentration exceeds the upper limit or below the lower limit of the set methane concentration, and then decide to automatically adjust the rotate speed of motor to change the output power of the local fan as well as the airflow velocity. Thus, we have to scientifically define the appropriate air velocity value and the intermittency period of step changes for velocity. Based on the essential requirements of ventilation velocity, we proposed six possible intermittent ventilation patterns and employ parametric studies to define the key parameters.

Here, the performance of various ventilation patterns is investigated and compared with the typical steady flow pattern (15 m/s, defined as case 0) in terms of the potential energy saving and methane removal efficiency, and we mainly focus on the dynamic changes of methane concentration about the time it requires to be stable when the velocity is changed over time. Specifically, the inlet velocity for ventilation duct are defined with different dynamic patterns, which are presented in Figure 5: (case 1) 300 s high velocity (15 m/s) and 300 s low velocity (12 m/s); (case 2) 300 s high velocity (15 m/s) and 300 s low velocity (9 m/s); (case 3) sinusoidal flow, the velocity changing curve is defined as the sine law with period of 600 s, the peak and valley values are 15 m/s and 12 m/s respectively; (case 4) 300 s high velocity (15 m/s) and 600 s low velocity (12 m/s); (case 5) 300 s high velocity (15 m/s) and 600 s low velocity (9 m/s), and (case 6) 300 s high velocity (15 m/s) and 300 s low velocity (6 m/s). It should be noted that the intermittent flow control can be implemented by using FCT, which can be done by installing a frequency converter for the local fan in the circuit without affecting main ventilation fans and other apparatus in the ventilation system.

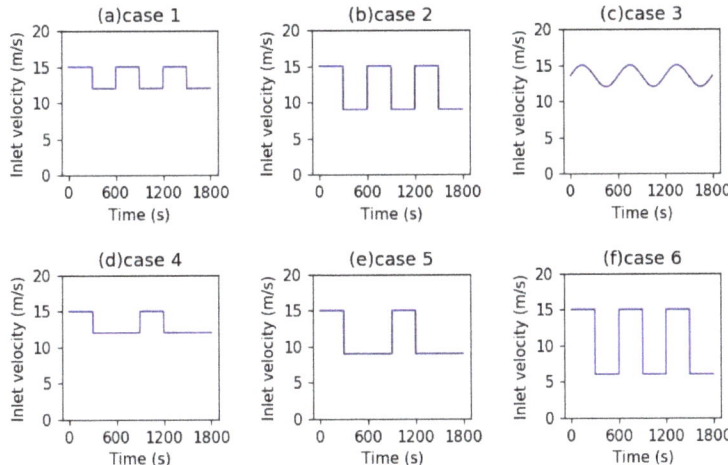

Figure 5. Six different ventilation patterns.

Therefore, the typical steady flow with six dynamic ventilation patterns were simulated utilizing the computational fluid dynamic platforms. Firstly, a line in section A with a height of 3 m was selected to observe the methane concentration for various ventilation patterns along the excavation laneway at different time points, and the variation trend of methane concentration among the seven ventilation patterns was presented in Figure 6 with time 300 s, 600 s, 900 s, 1200 s. Several features here are apparent: methane gas is generated at the driving face and then pushed into the whole laneway, which is reflected by high methane concentration on the first few meters and the lower trend with the increasing laneway length. On average, methane concentration keeps under the maximum allowable value of 1%v/v as well as the alarm value of 0.8%v/v in the areas where miners work, which is typically farther than 5 m away from the driving face. Specifically, it can be found that at 300 s, the methane concentration under all the cases maintained at a low level. At 600 s, the methane concentrations for case 1–6 all increased due to the changing low velocity in this period with the computed average value of 0.51%v/v, 0.65%v/v, 0.45%v/v, 0.51%v/v, 0.69%v/v, 0.90%v/v, increased up to 32.26%, 67.07%, 17.01%, 31.52%, 78.45%, 130.76%, respectively, compared with base case 0 of 0.39%v/v. The same figure at 900 s for six dynamic cases are 0.45%v/v, 0.60%v/v, 0.48%v/v,

0.55%v/v, 0.68%v/v, 0.60%v/v, which increased up to 8.31%, 44.00%, 16.81%, 32.80%, 64.29% and 45.52%, respectively, compared with typical steady flow with an average methane concentration of 0.41%. Besides, the methane concentrations at 1200s for case 0–6 are 0.42%v/v, 0.57%v/v, 0.61%v/v, 0.56%v/v, 0.51%v/v, 0.66%v/v, 0.89%v/v, and the dynamic patterns increased 36.09%, 44.00%, 34.62%, 20.32% 57.66%, and 110.61%, respectively. It can be concluded that case 5 performs worst due to the long period of low velocity and followed by case 2 with the same low velocity of different period.

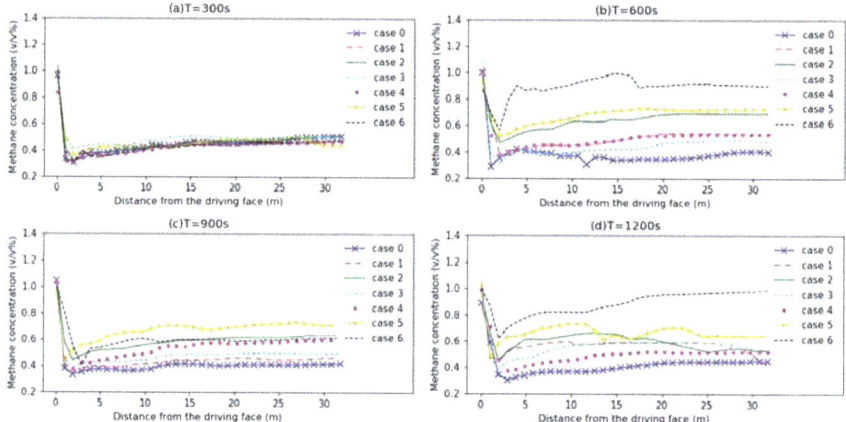

Figure 6. Average methane concentration along the selected line.

Furthermore, we look into the dynamic regularity of methane concentration at the selected point, which is located on section A with 3 m high and 10 m away from the driving face, as presented in Figure 7. It demonstrated that intermittency based on frequency conversion of inlet velocity leads to dynamic behaviour of methane concentration as relatively high methane concentration develops when low velocity is applied, and all the cases can control the methane concentration below the alarm value of 0.8%v/v except for case 6 due to the low velocity. Specifically, by comparing the different velocities of 6 m/s, 9 m/s, 12 m/s, and 15 m/s, it shows that the steady flow of base case 0 with 15 m/s maintains methane concentration around 0.4–0.45%v/v; case 1, case 3 and case 4 with the low velocity of 12 m/s increase the methane concentration up to about 0.5%v/v; case 2 and case 5 with intermittency low velocity of 9 m/s leads the methane concentration rise to 0.65%v/v to 0.7%v/v; while case 6 with lower velocity of 6 m/s increases the methane concentration to the critical explosive level of 1%v/v, which cannot satisfy the requirement of ventilation. Thus, we conclude that 9 m/s is the minimum inlet velocity in this case study to ensure the methane concentration below the alarm value. Besides, one of our research goals is to scientifically define the frequency conversion period of inlet velocity in order to save energy whilst keeping the methane concentration below the maximum allowable value. By comparing of case 2 and case 5, the results suggest that the intermittency duration plays an important role as the short period of the low velocity leads to a short-term and lesser accumulation of methane concentration, while the long period of the low velocity induces a long-term and more accumulation of methane concentration, which will also lead to slightly increase of methane concentration in the next ventilation period. In addition, an important finding is that the methane concentration takes about 3 min to be steady with the step changes of velocity.

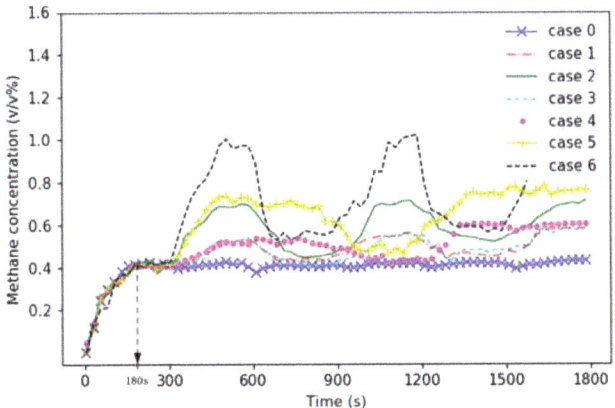

Figure 7. Average methane concentration in a selected point at different times.

Although the intermittent ventilation patterns based on frequency conversion of velocity cannot apparently improve methane removal efficiency compared with normal steady flow, it offers potential in energy saving and still guarantees that the average methane concentration under the limited level with velocity over 9 m/s, which confirms that continuously ventilated high-speed velocity typically used in Sijiazhuang coal mine is really not that necessary, and proper velocity decrease based on the dynamic characteristics of methane concentration can reduce energy consumption. The significant amount of energy saving and other parameters on average are presented in Table 3, which are calculated based on Bernoulli equation within 1800 s. It indicates that case 5 performs best in energy saving while it is dangerous since the methane concentration close to alarm value with long period of low ventilation velocity; followed by case 6, but the methane concentration is beyond the alarm value in this case as we analyzed previously; cases 1, 2, 3, 4 can effectively save energy usage whilst maintain the methane concentration under the limited value, among which case 2 offers balanced performance with energy saving up to 39.20% and controls methane concentration at safe level. The result shows that the potential energy saving based on the intermittent ventilation patterns is essentially significant for economic benefits.

Table 3. Energy saving for various ventilation patterns in 1800 s.

Cases	Average Pressure Difference (Pa)	Average Volumetric Flow Rate (m³/s)	Average Fan Power (Watt)	Energy Saving (%)
Case 0	137.81	7.54	1038.56	0
Case 1	113.01	6.78	785.15	24.40
Case 2	93.71	6.03	631.44	39.20
Case 3	113.01	6.78	785.15	24.40
Case 4	104.74	6.53	700.68	32.53
Case 5	79.01	5.53	495.74	52.27
Case 6	79.93	5.28	552.51	46.80

4.2. The Spatiotemporal Characteristics of the Airflow and Methane Distribution

The ventilation airflow is the main factor which directly influence the methane dispersion and distribution, here case 2 is chosen to investigate the spatiotemporal characteristics of airflow behaviour and the methane distribution. Figure 8 presents the predicted velocity profiles at different cross-sections for four time points, and Figure 9 shows the methane distribution at 1 m high from the laneway floor with the same time points. It is clear that the pressing air discharges from the ventilation duct, flows through the driving face to the outtake side of laneway with gradually decreased velocity; with the impact of the pressing air, large amount of methane gas disperses and is driven to the outtake side of laneway with decreasing concentration. The result also suggests that the frequency conversion of inlet velocity changes the overall behaviour of velocity and methane concentration inside the whole laneway: with high inlet velocity of 15 m/s (Figure 8a,c), methane gas is dispersed and forced to leave the laneway with a relatively low concentration (Figure 9a,c); conversely, with the action of a low inlet velocity of 9 m/s, the airflow velocity throughout the whole laneway reduces significantly (Figure 8b,d), with that the methane concentration rises from about 0.45%v/v to around 0.7%v/v (Figure 9b,d) and then reduces back to low concentration as the high velocity is periodically applied.

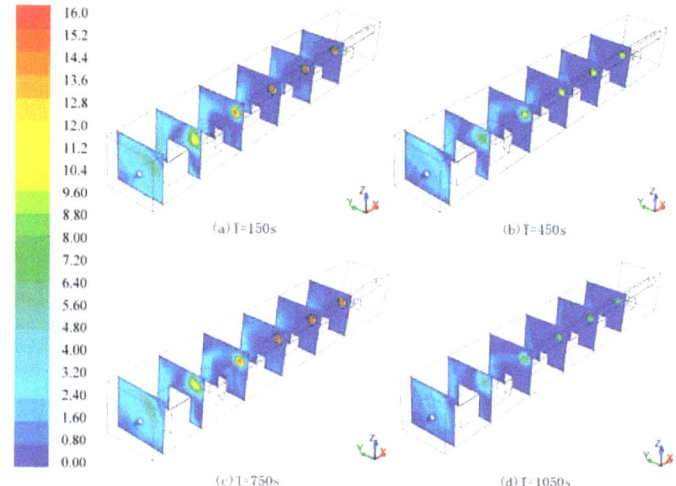

Figure 8. Airflow velocity (m/s) profiles at different cross sections (with a distance of 1 m, 5 m, 10 m, 15 m, 20 m to the driving face).

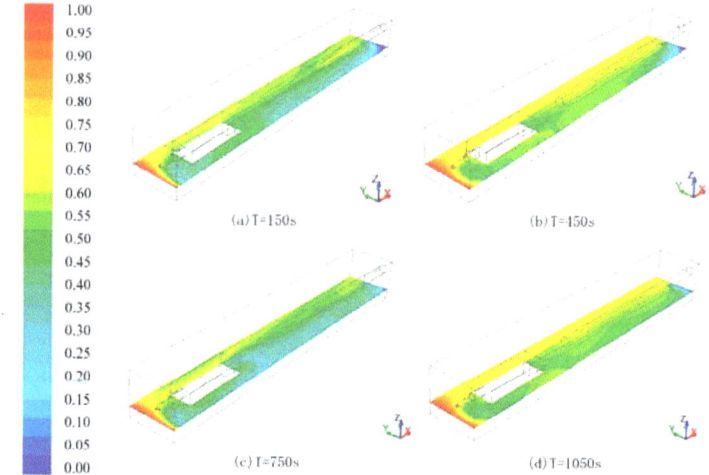

Figure 9. Methane distribution (%v/v) at height of 1 m above the laneway floor.

4.3. The Optimal Intermittent Ventilation Pattern and Data Validation

Based on the previous parametric studies and the spatiotemporal characteristics of airflow behavior and methane distribution, we have concluded that the intermittent ventilation patterns are potential in energy saving. And an important finding is that the inlet velocity value and its period of step changes plays an important role in the changing regularity of methane concentration, where the inlet velocity of 9 m/s is determined as the minimum value and it takes about 3 min for the methane concentration to be steady at one period of velocity step changes. Therefore, we investigated case 7 in this section, which is defined as 180 s high velocity (15 m/s) and 180 s low velocity (9 m/s), as shown in Figure 10. Two points A and B in section A with 3 m high and 10 m and 20 m respectively away from the driving face are monitored, as can be seen in Figure 11. The methane concentration is changed with this idea: since the methane concentration maintains at a stable level after about 180 s, which illustrates that the methane removal efficiency with continuously high speed velocity is not significant anymore, so the inlet velocity is decreased slightly to save energy consumption; as the methane concentration increases to around 0.7%v/v in a period of low airflow velocity, the inlet velocity is increased to control methane concentration to a low level. It is noteworthy that numerical result can provide operation supports for FTC with the consistent fundamental design conception. Besides, it is necessary to compare the methane concentration of all these cases. As point A is also chosen as the observation point in Figure 7, the average methane concentration is calculated at this point for case 0–7, which is 0.45%v/v, 0.49%v/v, 0.59%v/v, 0.50%v/v, 0.53%v/v, 0.63%v/v, 0.74%v/v, 0.52%v/v respectively. Furthermore, as can be seen, case 7 saves energy consumption saving of up to 39.20% as presented in Table 4, from which we concluded that case 7 performs best as it saves energy usage whilst maintaining methane concentration at the relatively low level.

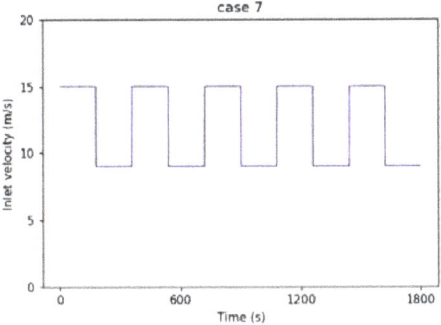

Figure 10. Optimized velocity pattern for the ventilation and methane removal strategy.

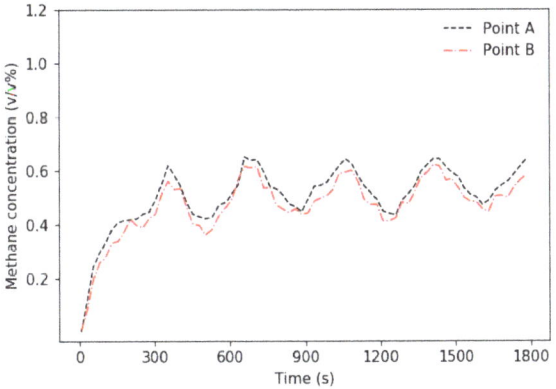

Figure 11. Methane concentration changed over time at two specified points.

Table 4. Energy saving for case 7 in 1800 s.

Average Pressure Difference (Pa)	Average Volumetric Flow Rate (m³/s)	Average Fan Power (Watt)	Energy Saving (%)
93.71	6.03	631.44	39.20

Before accepting the research results as decision supports for practical application, the ventilation pattern of base case 0 must be verified and validated with monitored data in Sijiazhuang coal mine. In this experiment, the methane sensor of type KJ9701A was used in 1-5205 excavation laneway, which is 3.7 m high from the laneway floor, 5 m away from the driving face and 0.4 m away from the laneway wall. Specifications of the methane sensor with type KJ9701A is shown in Table 5, and data comparison among field monitored data and simulated data of base case 0 is illustrated in Figure 12. The validation result indicates that the two sets of data agrees well with each other and exhibits a stable low level of methane concentration, which is within the alarm value of 0.8%v/v, indicating that the methane quality has remained the standard. Thus, the numerical results can be used to guide field measurements and operation to a certain degree.

Table 5. Specifications of the methane sensor (type KJ9701A).

Specifications	Range
Measurement range	0–4%v/v CH_4
Relative error	0–1% $CH_4 \leq \pm 0.1\%$ CH_4 1–3% $CH_4 \leq 10\%$ CH_4 3–4% $CH_4 \leq \pm 0.3\%$ CH_4
Working voltage	9VDC–24VDC

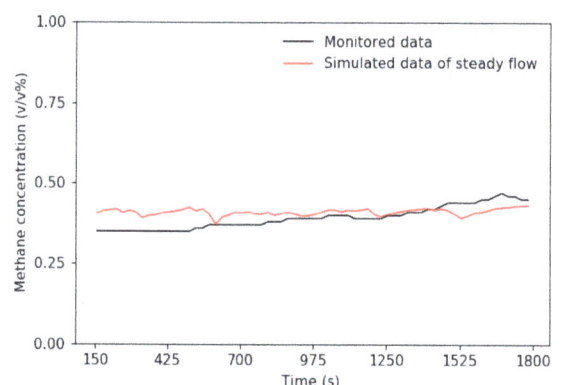

Figure 12. Comparison of the simulated data with field monitored data.

5. Conclusions

In this study, based on the conception of FTC, an optimized intermittent ventilation and methane removal strategy is proposed and evaluated with the goal of saving energy usage whilst keeping methane level below the alarm value inside a laneway of coal mine. An important finding is that by conducting parametric studies from the perspective of time, factors, such as the inlet velocity value and the period of step changes, can be scientifically defined to obtain a balanced result between energy savings and methane removal efficiency.

Specifically, by investigating the spatiotemporal characteristics of methane concentration based on the parametric studies, the minimum inlet velocity value is determined as 9 m/s and the period of intermittency is defined as 180 s in this specific underground mine size, which can be utilized to FTC design and applied to optimize the intermittent ventilation pattern. Therefore, the cost-effective ventilation and methane removal strategy which performs best is defined as case 7, providing remarkable energy reduction of 39.2% compared with steady flow and controls methane concentrations at the lower level compared with case 2.

In conclusion, this study provides a possibility to design an optimized intermittent ventilation pattern with the same conception of FTC based on the spatiotemporal characteristics of airflow behavior and methane distribution; however, the physical model in this study remains quite simple. Maybe more complex ventilation approach with a brattice installation or air curtain based on parametric studies are required to investigate in the future.

Author Contributions: During this study, H.L. performed the numerical simulation and wrote the paper, S.M. and M.L. contributed the analysis tools and materials.

Funding: This work is financially supported by Mega-projects of Science Research for the 13th Five-year Plan: [Grant Number 2017YFC0804303].

Acknowledgments: The authors would like to acknowledge the commercial Fluent (ANSYS) software and appreciate the anonymous reviewers who contributed to the quality of this article by providing helpful suggestions.

Conflicts of Interest: No potential conflict of interest was reported by the authors.

References

1. Baris, K. Assessing ventilation air methane (VAM) mitigation and utilization opportunities: A case study at Kozlu Mine, Turkey. *Energy Sustain.* **2013**, *17*, 13–23. [CrossRef]
2. Karakurt, I.; Aydin, G.; Aydiner, K. Mine ventilation air methane as a sustainable energy source. *Renew. Sustain. Energy Rev.* **2011**, *15*, 1042–1049. [CrossRef]
3. Somers, J.; Burklin, C. A 2102 update on the world VAM oxidizer technology market. In Proceedings of the 14th U.S./North American Mine Ventilation Symposium, Salt Lake City, UT, USA, 17–20 June 2012; pp. 259–263.
4. Torano, J.; Torno, S.; Menendez, M.; Gent, M.; Velasco, J. Models of methane behaviour in auxiliary ventilation of underground coal mining. *Int. J. Coal Geol.* **2009**, *80*, 35–43. [CrossRef]
5. Zheng, Y.; Tien, J.C. DPM dispersion study using CFD for underground metal/nonmetal mines. In Proceedings of the 12th North American Mine Ventilation Symposium, University of Nevada, Reno, NV, USA, 9–11 June 2008; pp. 281–286.
6. Noack, K. Control of gas emissions in underground coal mines. *Intern. J. Coal Geol.* **1998**, *35*, 57–82. [CrossRef]
7. Sasmito, A.P.; Birgersson, E.; Ly, H.; Mujumdar, A.S. Some approaches to improve ventilation system in underground coal mines environment-A computational fluid dynamic study. *Tunn. Undergr. Space Technol.* **2013**, *34*, 82–95. [CrossRef]
8. Reddy, A.C. Development of a Coal Reserve GIS Model and Estimation of the Recoverability and Extraction Costs. Master's Thesis, West Virginia University, Morgantown, WV, USA, 2009.
9. Jeswiet, J.; Szekeres, A. Energy Consumption in Mining Comminution. *Proced. CIRP* **2016**, *48*, 140–145. [CrossRef]
10. Jeswiet, J.; Archibald, J.; Thorley, U.; De Souza, E. Energy Use in Premanufacture (Mining). *Proced. CIRP* **2015**, *29*, 816–821. [CrossRef]
11. Mielli, F.; Bongiovanni, M. Production energy optimization in mining. In Proceedings of the World Mining Congress, Montreal, QC, Canada, 11–15 August 2013.
12. Tuck, M.A.; Finch, C.; Holden, J. Ventilation on demand: A preliminary study for Ballarat Goldfields NL. In Proceedings of the 11th North American Mine Ventilation Symposium, University Park, PA, USA, 5–7 June 2006; pp. 11–14.
13. Hu, S.; Wang, Z. Temporal and Spatial Distribution of Respirable Dust After Blasting of Coal Roadway Driving Faces: A Case Study. *Minerals* **2015**, *5*, 679–692. [CrossRef]
14. Allen, C.; Keen, B. Ventilation on demand (VOD) project—Vale Inco Ltd., Coleman Mine. In Proceedings of the 12th North American Mine Ventilation Symposium, Reno, NV, USA, 9–11 June 2008; pp. 45–49.
15. Liu, H.; Wu, X.; Mao, S. A Time Varying Ventilation and Dust Control Strategy Based on the Temporospatial Characteristics of Dust Dispersion. *Minerals* **2017**, *7*, 59. [CrossRef]

16. Heerden, J.; Sullivan, P. The application of CFD for evaluation of dust suppression and auxiliary ventilation systems used with continuous miners. In Proceedings of the 6th US Mine Ventilation Symposium, Littleton, CO, USA, 21–23 June 1993; pp. 293–297.
17. Uchino, K.; Inoue, M. Auxiliary ventilation at a heading of a face by a fan. In Proceedings of the 6th International Mine Ventilation Congress, Pittsburgh, PA, USA, 17–22 May 1997; pp. 493–496.
18. Nakayama, S.; Kim, Y.K.; Jo, Y.D. Simulation of methane gas distribution by computational fluid dynamics. In *Mining and Science Technology*; Xie, H.P., Golosinski, T.S., Eds.; Balkema Publisher: Brookfield, WI, USA, 1999; pp. 259–262.
19. Niu, W.; Jiang, Z.G.; Tian, D.M. Numerical simulation of the factors influencing dust in drilling tunnels: It's application. *Min. Sci. Technol.* **2011**, *21*, 11–15.
20. Parra, M.T.; Villafruela, J.M.; Castro, F.; Mendez, C. Numerical and experimental analysis of different ventilation systems in deep mines. *Build. Environ.* **2006**, *41*, 87–93. [CrossRef]
21. De Souza, E. Application of ventilation management programs for improved mine Safety. *Intern. J. Min. Sci. Technol.* **2017**, *27*, 647–650. [CrossRef]
22. Wala, A.M.; Vytla, S.; Taylor, C.D.; Huang, G. Mine face ventilation: A comparison of CFD results against benchmark experiments for the CFD code validation. *Min. Eng.* **2007**, *59*, 1–7.
23. Torano, J.; Torno, S.; Mendez, M.; Gent, M. Auxiliary ventilation in mining roadways driven with road headers: Validated CFD modeling of dust behavior. *Tunn. Undergr. Space Technol.* **2011**, *26*, 201–210. [CrossRef]
24. Szlązak, N.; Obracaj, D.; Borowski, M.; Swolkień, J.; Korzec, M. Monitoring and controlling methane hazard in excavation in hard coal mines. *AGH J. Min. Geoeng.* **2013**, *37*, 105–116. [CrossRef]
25. Zhong, M.; Xing, W.; Weicheng, F.; Peide, L.; Baozhi, C. Airflow optimizing control research based on genetic algorithm during mine fire period. *J. Fire Sci.* **2003**, *21*, 131–153. [CrossRef]
26. Kurnia, J.C.; Sasmito, A.P.; Mujumdar, A.S. CFD simulation of methane dispersion and innovative methane management in underground mining faces. *Appl. Math Model* **2014**, *38*, 3467–3484. [CrossRef]
27. Kurnia, J.C.; Sasmito, A.P.; Mujumdar, A.S. Prediction and innovative control strategies for oxygen and hazardous gases from diesel emission in underground mines. *Sci. Total Environ.* **2014**, *481*, 317–334. [CrossRef]
28. Kurnia, J.C.; Sasmito, A.P.; Hassani, F.P. Introduction and evaluation of a novel hybrid brattice for improved dust control in underground mining faces: A computational study. *Int. J. Min. Sci. Technol.* **2015**, *24*, 537–543. [CrossRef]
29. Kurnia, J.C.; Peng, X.; Sasmito, A.P. A novel concept of enhanced gas recovery strategy from ventilation air methane in underground coal mines—A computational investigation. *J. Nat. Gas Sci. Eng.* **2016**, *35*, 661–672. [CrossRef]
30. Pach, G. Optimization of Forced Air Flow by the Comparison of Positive and Negative Regulations in Mine Ventilation Network. *Arch. Min. Sci.* **2018**, *63*, 853–870.
31. Pach, G.; Sułkowski, J.; Różański, Z.; Wrona, P. Costs Reduction of Main Fans Operation According to Safety Ventilation in Mines—A Case Study. *Arch. Min. Sci.* **2018**, *63*, 43–60.
32. Kurnia, J.C.; Sasmito, A.P.; Mujumdar, A.S. Simulation of a novel intermittent ventilation system for underground mines. *Tunn. Undergr. Space Technol.* **2014**, *42*, 206–215. [CrossRef]
33. Xu, G.; Kray, D.; Luxbacher, S.R.; Xu, J.L.; Ding, X.H. Computational fluid dynamics applied to mining engineering: A review. *Intern. J. Min. Reclam. Environ.* **2017**, *31*, 251–275. [CrossRef]
34. Wu, J.H. Research on Mine ventilator frequency conversion energy saving technology ventilator. *Opencast Min. Technol.* **2010**, *4*, 68–72.
35. Yang, Z.; Yu, Z.Z. Research on frequency conversion technology of metro station's ventilation and air-conditioning system. *Appl. Therm. Eng.* **2017**, *69*, 123–129. [CrossRef]
36. Fluent 6.3 documentations. Available online: https://www.ansys.com/ (accessed on 31 January 2019).

© 2019 by the authors. Licensee MDPI, Basel, Switzerland. This article is an open access article distributed under the terms and conditions of the Creative Commons Attribution (CC BY) license (http://creativecommons.org/licenses/by/4.0/).

Article

Improved CRM Model for Inter-Well Connectivity Estimation and Production Optimization: Case Study for Karst Reservoirs

Daigang Wang [1,*], Yong Li [2,*], Jing Zhang [2], Chenji Wei [2], Yuwei Jiao [2] and Qi Wang [2]

1. Beijing International Center for Gas Hydrate, Peking University, Beijing 100871, China
2. Research Institute of Petroleum Exploration and Development, PetroChina, Beijing 100083, China; zing@petrochina.com.cn (J.Z.); weichenji@petrochina.com.cn (C.W.); jiaoyuwei@petrochina.com.cn (Y.J.); wangqi.riped@petrochina.com.cn (Q.W.)
* Correspondence: dgwang@pku.edu.cn (D.W.); liyongph@petrochina.com.cn (Y.L.); Tel.: +86-18310865935 (D.W.); +86-13811134328 (Y.L.)

Received: 22 January 2019; Accepted: 25 February 2019; Published: 1 March 2019

Abstract: Due to the coexistence of multiple types of reservoir bodies and widely distributed aquifer support in karst carbonate reservoirs, it remains a great challenge to understand the reservoir flow dynamics based on traditional capacitance–resistance (CRM) models and Darcy's percolation theory. To solve this issue, an improved injector–producer-pair-based CRM model coupling the effect of active aquifer support was first developed and combined with the newly-developed Stochastic Simplex Approximate Gradient (StoSAG) optimization algorithm for accurate inter-well connectivity estimation in a waterflood operation. The improved CRM–StoSAG workflow was further applied for real-time production optimization to find the optimal water injection rate at each control step by maximizing the net present value of production. The case study conducted for a typical karst reservoir indicated that the proposed workflow can provide good insight into complex multi-phase flow behaviors in karst carbonate reservoirs. Low connectivity coefficient and time delay constant most likely refer to active aquifer support through a high-permeable flow channel. Moreover, the injector–producer pair may be interconnected by complex fissure zones when both the connectivity coefficient and time delay constant are relatively large.

Keywords: capacitance-resistance model; aquifer support; inter-well connectivity; production optimization; karst carbonate reservoir

1. Introduction

The hydrocarbon resources stored in carbonate reservoirs play an important role in new proven reserves worldwide over the last decade. Based on the type of reservoir bodies and geological origin and production response, carbonate reservoirs can be divided into three major categories, i.e., porous type, fractured-porous type, and karst type. So far, many karst carbonate reservoirs have been found in the northern Tarim basin, China. Compared with porous or fractured-porous carbonate reservoirs which are mainly distributed in the Middle East, Central Asia, and North America, domestic karst carbonate reservoirs are usually subjected to multiscale geological structures and strong heterogeneity. The coexistence of matrix, fracture, and karst caves usually leads to extremely complicated multiphase fluid dynamics in this type of reservoir. On the other hand, the lack of powerful techniques to accurately evaluate reservoir dynamic connectivity while considering the impact of aquifer support, further increases the uncertainty for finding an optimal waterflood development scheme. Therefore, it is urgent to establish a practical methodology for inter-well connectivity estimation while coupling the effect of aquifer support and subsequent real-time injection production optimization in order to improve waterflooding recovery as much as possible.

A typical workflow for analyzing waterflood performance involves a combination of reservoir characterization, geological models, and numerical simulation engines for modeling multiphase flow in porous media. This ab initio integrated type of model is useful for evaluating strategies at the larger scale, but is not suited to discrete, localized waterflood operations in karst carbonate reservoirs. Over the last decades, data-driven physics-based models have emerged as an attempt to provide a simple picture of reservoir fluid dynamics, providing some vital information, e.g., inter-well connectivity and advanced decision-making processes by scarifying part of the exact description (e.g., evolution of the saturation field over time). Production history, well logs, and interventions are common examples of hard data that are used to train a data-driven reservoir model to answer specific questions. The data-driven models for inter-well connectivity estimation mainly include capacitance-resistance (CRM) models [1–4], flow-network models [5], and inter-well numerical simulation models [6–8].

As a powerful approach for effective analogy between source/sink terms in hydrocarbon reservoirs and electrical conductors, the CRM only requires the most readily available production history and producers' bottom hole pressure (BHP) when available. In addition, the CRM can easily handle dynamic boundary conditions such as new wells or shut-ins. Full-field history matching and channeling detection can be carried out in minutes making it a suitable tool for real-time monitoring. The CRM concept was initiated by the work of Albertoni and Lake [9]. Yousef et al. [1] advanced the model by incorporating the tank material balance concept and derived the model to multiple producers and injectors by applying superposition in space. Sayarpour et al. [2] further developed the CRM equations to several types of reservoir control volumes: tank model (CRMT), producer-based model (CRMP), and injector-producer pair based model (CRMIP). The CRM has also been integrated with other analytical tools (e.g., rate-transient analysis, (RTA)) for screening or monitoring the performance of various enhanced oil recovery (EOR) processes [10–15]. Recently, Mamghaderi and Pourafshary [16] established an improved CRM to investigate the effect of layers using data from production logging tools (PLTs). Zhang et al. [17,18] presented a system of multi-layer CRM models for layered reservoirs by considering both BHP and crossflow. Holanda et al. [19] further introduced the state-space (SS) theory to describe the dynamic behavior of CRMs as a multi-input/multi-output system. Capacitance-resistance models were also incorporated with fractional flow models [20,21] to allow the prediction of oil rates. It indicated that the Koval model proposed by Cao et al. [21] may not be a good choice for mature waterfloods, but it is effective enough to revisit abrupt breakthroughs caused by active aquifer support, which widely exist in karst carbonate reservoirs. During production of karst reservoirs, large-scale fractured-vuggy units are usually treated as isolated targets for continuous waterflood, thereby the previous CRM models should be modified to supply the aquifer influx rate as needed with all contributing injection rates.

Owing to the fact that understanding reservoir fluid dynamics to achieve optimal decision-making by grid-based reservoir models is computationally expensive and frequently require large volumes of uncertain data to estimate petrophysical properties, full field-scale models are not easy for rapid reservoir analysis to make reliable decisions. As a newly developed technique for optimal decision, production optimization where a grid-based model is substituted by a useful yet tractable data-driven model for waterflood systems was developed to find the optimal well controls that maximize cumulative oil production or net present value (NPV). The constrained optimization we consider here is essentially a non-linear, least-squares problem. The efficient optimization mostly depends on the computations of gradients of objective function to control variables. Theoretically, gradient of objective function can be accurately obtained using the adjoint method that requires only two simulations for this calculation, no matter how many variables [22,23]. However, calculation of the adjoint gradient by traditional reservoir simulation is inconvenient and code intrusive. Besides, it is not feasible to compute finite difference gradient when the number of control variables is large.

A viable alternative is approximation of the gradient by an ensemble optimization technique known as EnOpt, which has received appealing attention over the past years by reservoir engineers after the pioneering work of Chen et al. [24,25]. Using the deterministic or standard EnOpt, hybrid constrained optimization problems were resolved by generalized non-linear programming algorithms [7,26–28]. Do and Reynolds [29] analyzed the deterministic EnOpt and demonstrated its close connection with other stochastic gradient algorithms. Stordal et al. [30] confirmed that deterministic EnOpt can be treated as a special example of Gaussian mutation. Jafroodi and Zhang [31] utilized the CRM model as an underlying reservoir proxy for ensemble-based production optimization and used a power law relationship to predict oil productivity. In their work, The CRM equation was only used to detect faults or low-permeability areas between wells, rather than acting as a proxy for the simulator, with the advantage of using actual production and injection data. Hong et al. [32] presented a proxy-model workflow where a grid-based model was supplemented by the useful yet tractable CRM model as a proxy for waterflood operations. The results indicated that CRM models have high potential to serve as a cogent proxy model for waterflooding related decision-making and obtain robust results that result in a near-optimal solution. Recently, a novel ensemble-based technique, stochastic-simplex-approximate-gradient (StoSAG), was developed by Fonseca et al. [33,34]. The StoSAG deals with reservoir simulator as a black box and approximates gradient through the inputs and outputs of all the ensemble runs. Various theoretical analyses [35–39] showed that StoSAG can yield a significantly higher NPV than that obtained with the standard EnOpt. However, there are few proposals using the newly developed StoSAG for estimation of inter-well connectivity. We will explore the possibility of using StoSAG for the case where two objectives are to assimilate the production history data to infer inter-well connectivity when active aquifer support exists and to perform real-time production optimization by maximizing the NPV or cumulative oil production under hybrid non-linear constraints, respectively.

In this paper, an improved CRMIP model by coupling the effect of active aquifer support is first proposed and integrated with the newly developed StoSAG optimization algorithm for better understanding of inter-well connectivity in a waterflood operation. The improved CRM–StoSAG workflow is further employed for real-time production optimization to find the optimal injection rate at each control step by maximizing the objective function, i.e., net present value (NPV) of waterfloods with regard to typical karst carbonate reservoirs. Finally, the conclusions of this work will be provided.

2. Methodology

This section discusses in detail the integrated workflow for inter-well connectivity estimation and production optimization in a waterflood reservoir.

2.1. The Improved CRM–Koval Model

With regard to different CRM representations (e.g., CRMT, CRMP, and CRMIP), the two main parameter of a CRM are connectivity coefficient and time delay constant if BHP data of producers are not available. For an oil–water system, the time delay constant denotes how long a pressure wave from injectors takes to reach a producer. Due to reservoir heterogeneity, time delays of displacing fluid from adjacent injectors to a certain producer may differ from each other significantly. So, assuming only a time delay constant for each producer, as in CRMP, may be unreliable. In such a case, it is necessary to develop one continuity equation for every injector–producer pair, i.e., the CRMIP model. However, the traditional CRMIP model is not suited to investigate the impact of aquifer support on inter-well connectivity estimates, which are widely distributed in karst carbonate reservoirs. As shown in Figure 1, when aquifer support is inevitable, the governing differential equation should be modified, as follows

$$\frac{dq_{ij}(t)}{dt} + \frac{1}{\tau_{ij}} q_{ij}(t) = \frac{1}{\tau_{ij}} \left[f_{ij} I_i(t) + e_{wij} \right] - J_{ij} \frac{dP_{wf,j}}{dt} \qquad (1)$$

where $q_{ij}(t)$ is the liquid production rate of an injector–producer pair at time t, m^3/d; τ_{ij} is the time delay constant, d; e_{wij} is the water influx rate, m^3/d; $I_i(t)$ is the injection rate of injector i, m^3/d; J_{ij} is the liquid production index of an injector–producer pair, m^3/(MPa·d); $P_{wf,j}$ is the bottom hole pressure of producer j at time t, MPa; f_{ij} is the inter-well connectivity coefficient between injector i and producer j, $\in [0,1]$.

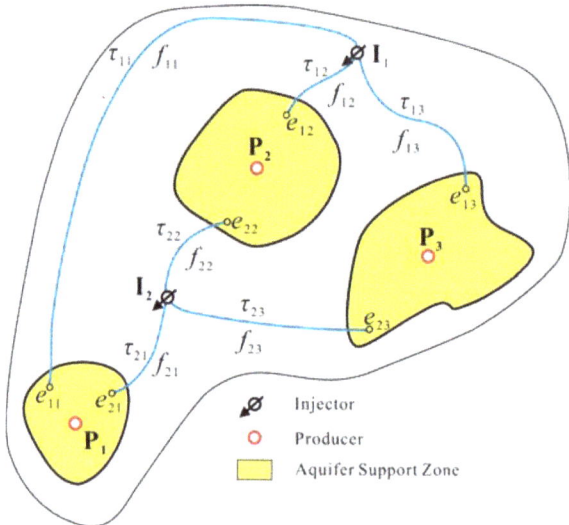

Figure 1. Schematic diagram of an injector–producer-pair-based control volume that is widely distributed in karst waterflood carbonate reservoirs.

The semi-analytical solution of Equation (1) will be further derived using superposition in space, which is stated as

$$q_{lij}^{cal}(t_k) = q_{lij}^{cal}(t_{k-1})e^{-(\frac{\Delta t_k}{\tau'_{ij}/M_{ij}^k})} + \left(1 - e^{-\frac{\Delta t_k}{\tau'_{ij}/M_{ij}^k}}\right)\left[e_{wij} + f_{ij} \cdot I_i^{(k)} - J'_{ij} \cdot \tau'_{ij}\frac{\Delta p_{wf,j}^{(k)}}{\Delta t_k}\right] \qquad (2)$$

According to superposition in time, Equation (2) has the following form

$$q_{ij}(t_k) = q_{ij}(t_0)e^{-(\frac{t_k-t_0}{\tau_{ij}})} \\ + \sum_{s=1}^{k}\left[\left(1 - e^{-\frac{\Delta t_s}{\tau_{ij}}}\right)\left(e_{wij} + f_{ij}I_i^{(s)} - J_{ij}\tau_{ij}\frac{\Delta p_{wf,j}^{(s)}}{\Delta t_s}\right)e^{-(\frac{t_k-t_s}{\tau_{ij}})}\right] \qquad (3)$$

For injector–producer pairs, the liquid production rate $q_j(t_k)$ of producer j at time t_k can be given by

$$q_j(t_k) = \sum_{i=1}^{N_{inj}} q_{ij}(t_k) \qquad (4)$$

When aquifer support is active and BHP change of producer with time is negligible, there are four unknown parameters for each injector–producer-pair based control volume, i.e., f_{ij}, $q_{ij}(t_0)$, τ_{ij} and e_{wij}. The total number of unknown parameters for waterflood reservoir is equal to $4 \times N_{pro} \times N_{inj}$. Obviously, when the waterflood reservoir is not affected by aquifer support, the semi-analytical solution Equation (3) can be simplified as the traditional CRMIP model. In order to guarantee a reasonable balance between injection and production, the inter-well connectivity coefficient f_{ij} should satisfy inequality constraints. In many cases, the liquid production rate $q_{ij}(t_0)$ must be subjected to some equality constraints as follows

$$\sum_{i}^{N_{inj}} f_{ij} \leq 1, j = 1, 2, \cdots, N_{pro} \qquad (5)$$

$$\sum_{i}^{N_{inj}} q_{ij}(t_0) = q_j(t_0) \qquad (6)$$

If the two-phase flow in each control volume can be regarded as steady-state flow, the Koval model proposed by Cao et al. [21] will be used to compute the fractional flow of water in porous media by considering the effect of local heterogeneity and viscosity ratio, which is given by

$$f_w = \begin{cases} 0 & t_D < \frac{1}{K_{val}} \\ \frac{K_{val} - \sqrt{\frac{K_{val}}{t_D}}}{K_{val} - 1} & \frac{1}{K_{val}} < t_D < K_{val} \\ 1 & t_D \geq K_{val} \end{cases} \qquad (7)$$

where K_{val} is the Koval factor, reflecting both reservoir heterogeneity and fluid–viscosity contrast. A large Koval factor (greater than unity) usually implies either a high degree of reservoir heterogeneity or a large oil-water viscosity ratio. t_D is a dimensionless time denoting the cumulative water injection in control volume.

$$t_D = \frac{\sum_k \sum_i f_{ij} I_i}{V_{pj}} \qquad (8)$$

where f_{ij} is the interwell connectivity coefficient, which can be estimated using the improved CRMIP model to assimilate the production history; V_{pj} is the drainage volume of a injector–producer pair, m³; I_i is the injection contribution to the producer at timestep t_k, m³/d.

For oil–water two-phase system, the oil or water production rate of jth producer at timestep t_k can be easily obtained based on the physical meaning of fraction-flow equation, which can be written as [21]

$$q_{wj}(t_k) = q_j(t_k) f_{wj}(t_k) \qquad (9)$$

$$q_{wj}(t_k) = q_j(t_k) [1 - f_{wj}(t_k)] \qquad (10)$$

Using the improved CRM–Koval model for inter-well connectivity estimation, the six unknown parameters for each injector–producer pair, i.e., connectivity coefficient, time delay constant, water influx rate, liquid production rate at timestep t_0, Koval factor, and drainage volume were estimated with non-linear multivariate regression, the required least-squares objective function were finally described as

$$\min_{u \in R^{N_u}} J(u) = \sum_{k=1}^{N_t} \sum_{j=1}^{N_{pro}} \left\{ \left[q_j^{cal}(t_k) - q_j^{obs}(t_k) \right]^2 + \left[q_{oj}^{cal}(t_k) - q_{oj}^{obs}(t_k) \right]^2 \right\} \qquad (11)$$

Except for Equations (5) and (6), the objective function was also constrained by

$$\tau_{ij} \geq 0, e_{wij} \geq 0, K_{val,j} \geq 1 \qquad (12)$$

$$\sum_{j}^{N_{pro}} V_{pj} \leq V_{pField} \qquad (13)$$

where the superscript ^{obs} and ^{cal} denote the observed and predicted production data, respectively; V_{pField} denotes the total pore (drainage) volume of a reservoir or block, m³.

2.2. Waterflood Production Optimization

When the improved CRM–Koval model was used for a better understanding of reservoir fluid dynamics including inter-well connectivity, aquifer influx rate, etc., the NPV of production was defined as the objective function by having the proxy model serve as a precursor of a grid-based reservoir model and finally maximized in order to find the optimal well controls (rate or pressure) under hybrid non-linear constraints, which can be given as follows [22]

$$N(\boldsymbol{u}) = \sum_{n=1}^{N_t} \left\{ \frac{\Delta t_n}{(1+b)^{\frac{t_n}{365}}} \left[\sum_{j=1}^{P} \left(r_o \cdot \overline{q_{o,j}^n} - c_w \cdot \overline{q_{w,j}^n} \right) - \sum_{k=1}^{I} \left(c_{wi} \cdot \overline{q_{wi,k}^n} \right) \right] \right\} \tag{14}$$

where u is a N_u-dimensional column vector containing all the well controls over the production lifetime. For the problem of interest here, we treated the injection rate of each injector as the control variable; r_o is the oil revenue, USD/STB; c_w and c_{wi}, respectively, are the disposal cost of produced water and the cost of water injection, USD/STB; b is annual discount rate; t_n denotes the time at end of the nth time step of the proxy model; Δt_n is the size of the nth time step; N_t is the total number of time steps; P and I denote the number of producers and injectors, respectively; $\overline{q_{o,j}^n}$ and $\overline{q_{w,j}^n}$ respectively, denote the average oil and water production rate at the jth producer, STB/day; $\overline{q_{wi,k}^n}$ is the average water injection rate at the kth injector, STB/day.

The generalized waterflood production optimization problem can be written as

$$\max_{u \in R^{N_u}} mize N(\boldsymbol{u}) \tag{15}$$

Satisfying the following constraints

$$u_i^{low} \le u_i \le u_i^{up}, i = 1, 2, \cdots, N_u \tag{16}$$

$$c_i(\boldsymbol{u}) \le 0, i = 1, 2, \cdots, n_i \tag{17}$$

$$e_i(\boldsymbol{u}) = 0, i = 1, 2, \cdots, n_e \tag{18}$$

where $N(\boldsymbol{u})$ is the objective function for waterflood optimization; Equations (16)–(18) denote the boundary constraint, inequality, and equality constraints, respectively; u_i^{low} and u_i^{up} are the lower and upper limits of ith control variable u_i, respectively; n_i and n_e denote the number of inequality and equality constraints, respectively.

2.3. Ensemble-Based Optimization Method

The key to sequential data assimilation or robust optimization in large-scale non-linear dynamics is to obtain the gradients of objective function and handle the hybrid inequality and equality constraints. Due to the complexity of fluid flow in heterogeneous porous media, accurate computation of gradient with the adjoint or finite difference method in a traditional reservoir simulation code is usually time consuming and code intrusive. To overcome the drawbacks, an ensemble-based method, StoSAG algorithm, was used to obtain the approximate gradient of least-squares objective function. The StoSAG algorithm can treat the reservoir simulator or proxy model as a black box and approximates the gradient of objective function through the inputs and outputs of all the ensemble runs efficiently. Moreover, the augmented Lagrange method and log-transformation method will be integrated to handle the hybrid nonlinear constraints.

2.3.1. Augmented Lagrange Objective Function

For the kth control variable u_k, an unbounded optimization problem can be obtained by applying a log transformation method [23] to transform the original bound-constrained optimization problem, which takes the form of

$$v_k = \ln\left(\frac{u_k - u_k^{low}}{u_k^{up} - u_k}\right) \tag{19}$$

where u_k^{low} and u_k^{up} are the lower and upper bounds of the kth control variable u_k, respectively; v_k is the log-domain kth control variable varying from $-\infty$ to ∞.

The robust optimization is performed in log-domain, but the control variable u_k can be obtained with the inverse log-transformation after each iteration, which can be illustrated as

$$u_k = \frac{\exp(v_k)u_k^{up} + u_k^{low}}{1 + \exp(v_k)} \tag{20}$$

Using the log-transformation to eliminate the bound constraints, the following augmented Lagrange objective function [15] will be ultimately established

$$\begin{aligned}L_a(u,\lambda,\mu) = & N(u) - \sum_{j=1}^{n_e}\lambda_{e,j}[s_{e,j}\cdot e_j(u)] + \frac{1}{2\mu}\sum_{j=1}^{n_e}[s_{e,j}\cdot e_j(u)]^2 \\ & - \sum_{i=1}^{n_i}\lambda_{c,i}\max[s_{c,i}\cdot c_i(u), -\mu\cdot\lambda_{c,i}] \\ & + \frac{1}{2\mu}\sum_{i=1}^{n_i}\{\max[s_{c,i}c_i(u), -\mu\lambda_{c,i}]\}^2\end{aligned} \tag{21}$$

where $\lambda_{e,j}$ and $\lambda_{c,i}$, respectively, denote the Lagrangian multiplier of equality and inequality constraints; μ denotes the penalty parameter; $s_{e,j}$ and $s_{c,i}$, respectively, denote the scaling factors for equality and inequality constraints. Note that the term $N(u)$ in Equation (14) is substituted with $L_a(u,\lambda,\mu)$ for a hybrid constrained optimization problem.

2.3.2. StoSAG Gradient Computation

To address the data assimilation problem of Equation (11) and waterflood optimization problem of Equation (15), we used the steepest ascent algorithm [24] to compute the estimate of the optimal control vector at the $(k+1)$th iteration, which is given by

$$u^{k+1} = u^k + a_k\left[\frac{d_k}{\|d_k\|_\infty}\right] \tag{22}$$

where u^0 is the initial guess; u^k is the optimal control vector at the kth iteration. a_k is the step size; d_k denotes the search direction vector. For our study, it was a stochastic search direction.

The StoSAG algorithm proposed by Fonseca et al. [33] with the stochastic search direction, described as following, was used to maximize the augmented Lagrange objective function $L_a(u,\lambda,\mu)$:

$$\begin{aligned}d_{k,sto} &= \frac{1}{N_e}\sum_{j=1}^{N_e}\left(\delta\widetilde{u}_j^k\left(\delta\widetilde{u}_j^k\right)^T\right)^+\delta\widetilde{u}_j^k\cdot\delta L_a\big|_j^k \\ &= \frac{1}{N_e}\sum_{j=1}^{N_e}\frac{\delta\widetilde{u}_j^k}{\|\delta\widetilde{u}_j^k\|_2^2}\cdot\delta L_a\big|_j^k\end{aligned} \tag{23}$$

where $\delta\widetilde{u}_j^k = \widetilde{u}_j^k - u^k$, and $\delta L_a\big|_j^k = L_a\left(\widetilde{u}_j^k\right) - L_a\left(u^k\right)$. The superscript "+" on a matrix denotes the Moore–Penrose pseudo-inverse. the superscript "T" denotes the transpose process for a vector or matrix. Note that, $\widetilde{u}_j^k$ denotes N_e ensembles of Gaussian random vector, where $X \sim N\left(u^k, I\right)$ (i.e., the mean of $\widetilde{u}_j^k$ equals to u^k and its covariance matrix I), and can be written as

$$\widetilde{u}_j^k = u^k + L \cdot Z_j, j = 1, 2, \cdots N_e \qquad (24)$$

where k is the iteration number of inner loop; u^k is the optimal control vector at the kth iteration; L is a lower triangular matrix obtained by Cholesky decomposition of the covariance matrix C_U; Z_j is the vector satisfying a Gaussian distribution $N(0, I)$, and I is unit matrix with dimension of $N_u \times N_u$. Therefore, $L \cdot Z_j$ denotes a $(N_u \times 1)$ Gaussian random vector, namely, $L \cdot Z_j \sim N(0, C_U)$. The following spherical covariance function is further adopted to compute the (i, j) entry of covariance matrix C_U by $C_{i,j}$,

$$C_{i,j} = \begin{cases} \sigma^2 \left[1 - \frac{3}{2}\left(\frac{|i-j|}{N_s}\right) + \frac{1}{2}\left(\frac{|i-j|}{N_s}\right)^3 \right] & if \; |i-j| < N_s \\ 0 & otherwise \end{cases} \qquad (25)$$

Figure 2 displays the integrated CRMIP–StoSAG workflow for inter-well connectivity estimation and waterflood optimization. This workflow was divided into two parts: data assimilation by the CRMIP–Koval model (shaded in orange) and real-time waterflood optimization based on good understandings of inter-well connectivity relationships (shaded in blue).

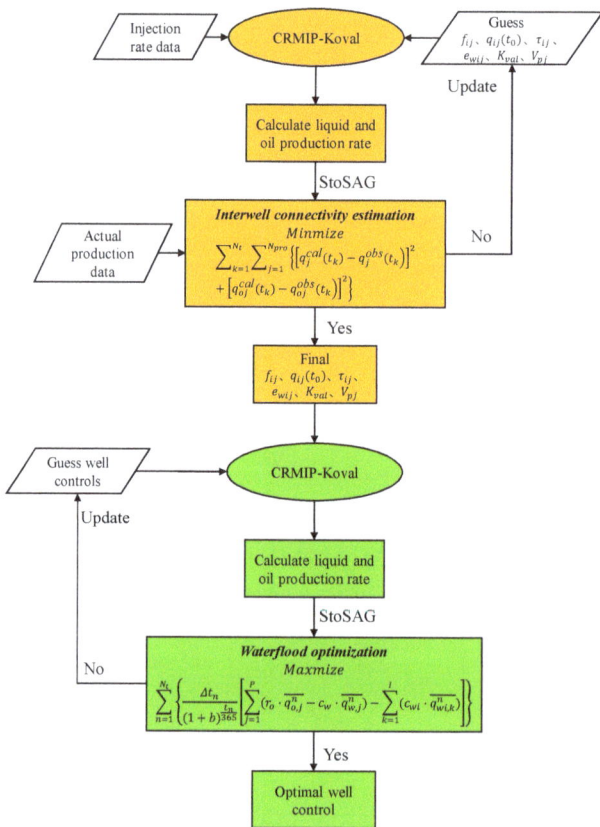

Figure 2. Workflow for inter-well connectivity estimation and waterflood optimization.

3. Case Study for Karst Carbonate Reservoir

As previously mentioned, when the impact of aquifer support is negligible, the improved CRMIP model will be simplified to the traditional CRM representation as described in Sayarpour et al. [2]. Accuracy of the improved CRM–StoSAG workflow and its feasibility for strongly-heterogeneous waterflood reservoir was validated in our previous work [15], which will not be revisited here. The following focuses on using the improved CRM–StoSAG workflow to infer the inter-well connectivity relationship of karst carbonate reservoirs and subsequent waterflood optimization.

The typical karst carbonate reservoir we consider here is located in northern Tarim Basin, which is subject to strong heterogeneity and contains multiple types of reservoir bodies in many cases, mainly including karst caves, fractured-vuggy bodies, etc. [40]. As a matter of fact, karst caves are usually regarded as the major reservoir spaces. Rapid drilling time drops, drilling rigs ventilation, overflow and large mud loss often occurs. In addition, strong bead reflection, as shown in Figure 3a, can be utilized to determine the karst cave systems. As for the fractured-vuggy reservoir bodies, the main components are composed of dissolved pores and small caves. In addition to communicating the dissolved pores and caves, a relatively high hydrocarbon storage is founded in high-angle tectonic fractures with varying conductivities. The main seismic amplitude for this type of reservoir body is strong flake and weak messy reflection, see Figure 3b.

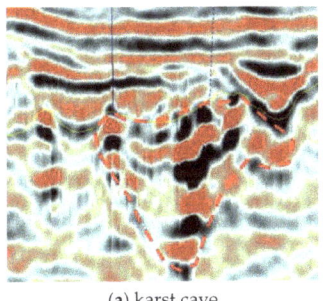
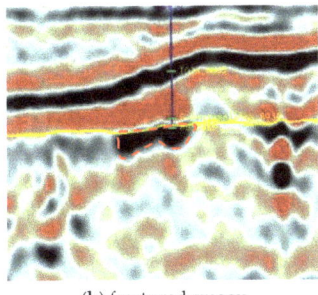

(a) karst cave (b) fractured-vuggy

Figure 3. Seismic amplitude of different reservoir bodies in the Tarim Basin: (**a**) strong bead-like reflection for karst caves; (**b**) strong flake-like reflection for fractured-vuggy bodies.

In addition, due to the complex distribution patterns of active aquifer support in karst carbonate reservoir, water influx behaviors observed mainly conform to abrupt watered-out and fluctuation or staircase rise characteristics, as shown in Figure 4. Moreover, the lack of efficient techniques to replenish formation void age may further result in a relatively low recovery efficiency and high natural decline rate of oil production. On the other hand, the majority of previously-used mathematical models are based on Darcy's flow, and the coexistence of porous and free-flow domains over a wide range of scales usually lead to sophisticated fluid flow dynamics in karst carbonate reservoirs. The traditional percolation theory and reservoir simulators are not adapted to represent the coupled flow dynamics in this type of waterflood reservoir.

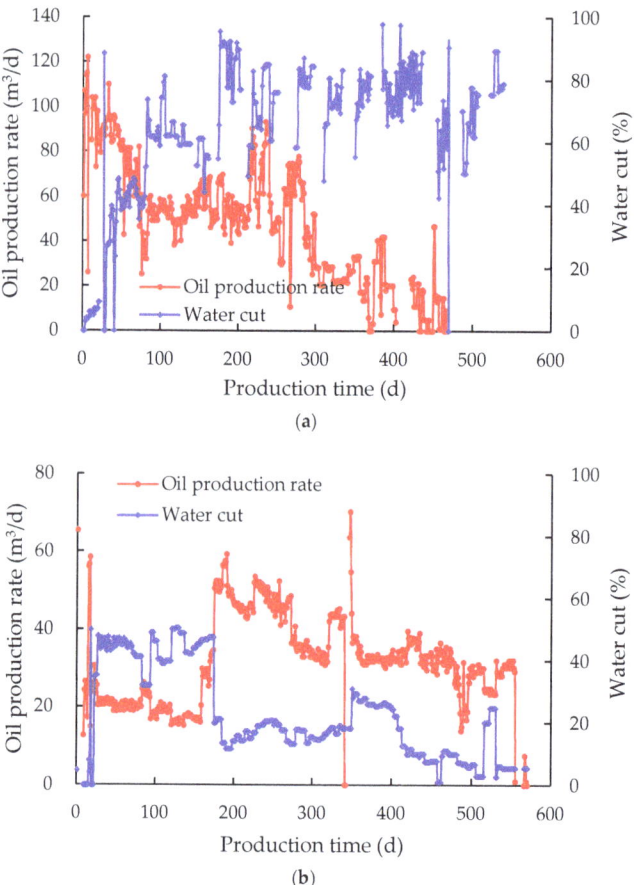

Figure 4. Schematic diagram of typical water influx behavior patterns: (**a**) abrupt watered-out; (**b**) fluctuation or staircase rise in water cut.

According to the problems of rapid production decline and low recovery rate caused by weak natural energy and rapid rise of water cut, pilot tests of sequential waterflooding have been carried out for some potentially interconnected injector–producer pairs, which are sorted out according to tracer surveillance and interference well testing. But with the amount of water injection increasing, oil yield effect gradually becomes worse, and much of the remaining oil is still unexploited around producers. For this study, typical injector–producer-pairs, Pro_1, Inj_1, and Pro_2, corresponding to the same large-scale fractured-vuggy unit, was screened for application for the improved CRM–StoSAG workflow. Using the liquid and oil production data of Pro_1, Inj_1, and Pro_2 with a relatively stable production scheme for history matching, the control variables for inter-well connectivity including connectivity coefficient, time delay constant, water influx rate, Koval factor, and drainage volume of each injector–producer pair were ultimately estimated and summarized in Table 1. The CRMIP oil production match results of Pro_1 and Pro_2 are shown in Figures 5 and 6, respectively. The iteration procedure of the least-squares objective function using the improved CRM–StoSAG workflow for inter-well connectivity estimation is illustrated in Figure 7.

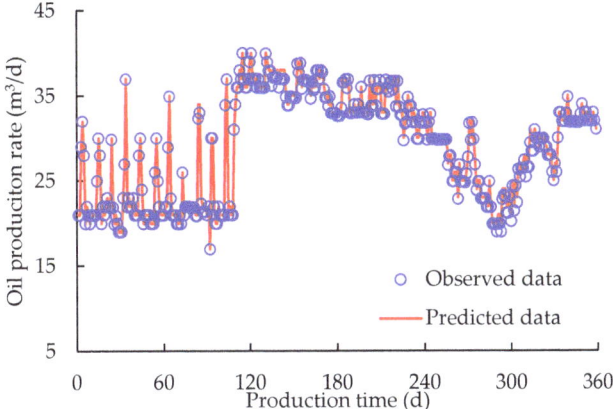

Figure 5. CRMIP oil production match for Pro_1.

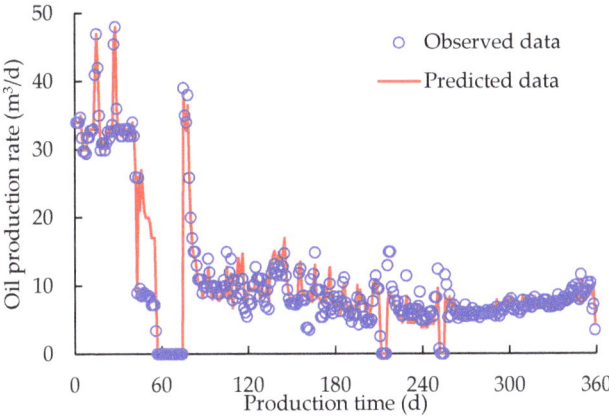

Figure 6. CRMIP oil production match for Pro_2.

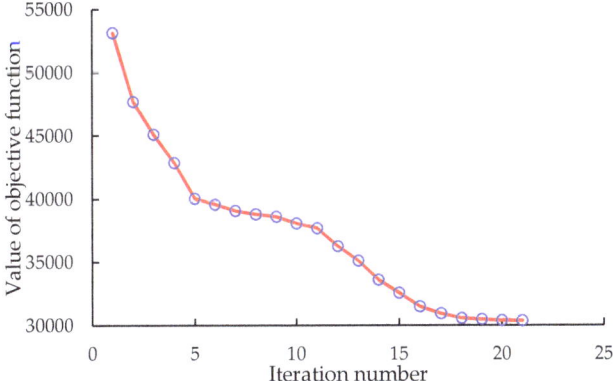

Figure 7. Iteration procedure of objective function based on the CRM–Koval model.

Table 1. The estimated control variables of CRMIP–Koval model.

The Estimated Control Variables	Inj_1				
	Connectivity Coefficient	Time Delay Constant (d)	Water Influx Rate (m^3/d)	Koval Factor	Drainage Volume (m^3)
Pro_1	0.433	230.34	0.36	4.22	1.54×10^5
Pro_2	0.037	39.40	9.59	4.45	1537.6

It demonstrates that, the estimated results of connectivity coefficient and time delay constant between Pro_2 and Inj_1 was far less than those between Pro_1 and Inj_1, suggesting that the injected water makes little contribution to oil production, which is mainly influenced by active aquifer support, and the most likely geological structure between Pro_2 and Inj_1 was high-permeable flow channel. Moreover, the estimated water influx rate of Pro_1 was lower than 1.0 m^3/d, indicating no active aquifer support from Inj_1. The remarkable oil yielding effect was mainly dependent on water injection. It is inferred that Pro_1 and Inj_1 are interconnected with a complex fissure zone, which is in good agreement with the understandings obtained from interference well testing. When performing production optimization for Pro_1, Inj_1, and Pro_2 pairs, water injection for Pro_1 needs to be properly increased along with real-time adjustment of operation parameters, e.g., nozzle size, shut-off, etc., in order to prohibit too-early water breakthroughs from nearby injectors.

Based on the accurate estimation of inter-well connectivity relationships, water injection rates of Inj_1 at each control step were regarded as unknown control variables, and the NPV of oil production with regard to this large-scale fractured-vuggy unit was maximized using the StoSAG optimization method in order to find the optimal well control variables. Figure 8 displays the optimized distribution of water injection rate at each timestep. Comparison of well group cumulative oil production prior to and after optimization is shown in Figure 9. The result indicates that, when cumulative volume of water injection is assumed constant, the cumulative oil increase of this well group after real-time injection-production optimization in karst carbonate reservoir was equal to 1290.2 m^3, indicating a remarkable oil yielding effect. Obviously, the proposed workflow can provide good insights into accurate estimation of dynamic connectivity and subsequent waterflood optimization in karst carbonate reservoirs with aquifer support widely existing.

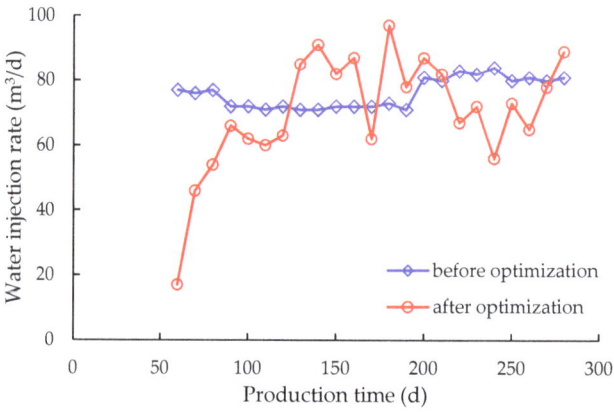

Figure 8. Distribution of water injection rate obtained by waterflood optimization.

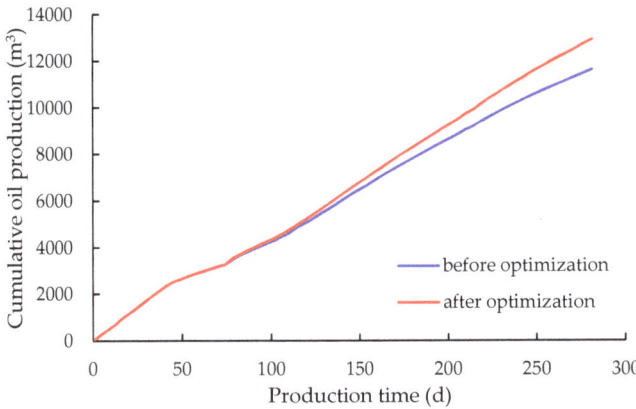

Figure 9. Comparison of cumulative oil production before and after optimization.

4. Conclusions

We develop an improved CRMIP model by coupling the effect of active aquifer support and integrated it with the newly developed StoSAG optimization algorithm for accurate evaluation of reservoir dynamic connectivity in a waterflood operation. Then, the improved CRM–StoSAG workflow was employed for real-time waterflood production optimization in order to find the optimal water injection rate at each control step by maximizing the objective function, i.e., net present value (NPV) of production.

Case studies showed that the proposed workflow can provide good insights into accurate estimation of inter-well connectivity and subsequent waterflood optimization. With regard to typical karst carbonate reservoirs, low connectivity coefficient and time delay constant most likely refer to active aquifer support through a high-permeable flow channel. The injector–producer pair may also be interconnected by complex fissure zones when both the estimated connectivity coefficient and time delay constant are relatively large.

Author Contributions: Conceptualization, D.W. and Y.L.; methodology, D.W.; validation, J.Z.; formal analysis, D.W.; investigation, C.W., Y.J.; resources, Q.W.; data curation, D.W.; writing—original draft preparation, D.W.; writing—review and editing, Y.L.; visualization, Q.W.; supervision, Y.L.; project administration, Y.L.; funding acquisition, D.W., Y.L.

Funding: This research was funded by the National Natural Science Foundation of China (Grant No. 41806070 and 51874346), the PetroChina Innovation Foundation (Grant No. 2018D-5007-0201), and the funding provided by the China Postdoctoral Science Foundation (Grant No. 2018M641069).

Acknowledgments: The authors sincerely acknowledge the Research Institute of Petroleum Exploration and Development, PetroChina for permission to publish this paper. We are also grateful to all the anonymous reviewers for their constructive comments.

Conflicts of Interest: The authors declare no conflict of interest.

References

1. Yousef, A.A.; Gentil, P.H.; Jensen, J.L.; Lake, L.W. A capacitance model to infer interwell connectivity from production and injection rate fluctuations. *SPE Reserv. Eval. Eng.* **2006**, *9*, 630–646. [CrossRef]
2. Sayarpour, M.; Zuluaga, E.; Kabir, C.S.; Lake, L.W. The use of capacitance-resistance models for rapid estimation of waterflood performance and optimization. *J. Pet. Sci. Eng.* **2009**, *69*, 227–238. [CrossRef]
3. Weber, D. The Use of Capacitance-resistance Models to Optimize Injection Allocation and Well Location in Water Floods. Ph.D. Thesis, University of Texas, Austin, TX, USA, 2009.

4. De Holanda, R.W.; Gildin, E.; Jensen, J.L.; Lake, L.W.; Kabir, C.S. A State-of-the-Art Literature Review on Capacitance Resistance Models for Reservoir Characterization and Performance Forecasting. *Energies* **2018**, *11*, 3368. [CrossRef]
5. Lerlertpakdee, P.; Jafarpour, B.; Gildin, E. Efficient production optimization with flow-network models. *SPE J.* **2014**, *19*, 1083–1095. [CrossRef]
6. Zhao, H.; Li, Y.; Cui, S.; Shang, G.; Reynolds, A.C.; Guo, Z.; Li, H.A. History matching and production optimization of water flooding based on a data-driven interwell numerical simulation model. *J. Nat. Gas Sci. Eng.* **2016**, *31*, 48–66. [CrossRef]
7. Guo, Z.; Reynolds, A.C.; Zhao, H. Waterflooding optimization with the INSIM-FT data-driven model. *Comput. Geosci.* **2018**, *22*, 745–761. [CrossRef]
8. Guo, Z.; Reynolds, A.C. INSIM-FT in three-dimensions with gravity. *J. Comput. Phys.* **2019**, *380*, 143–169. [CrossRef]
9. Albertoni, A.; Lake, L. Inferring interwell connectivity only from well-rate fluctuations in water floods. *SPE Reserv. Eval. Eng.* **2003**, *6*, 6–16. [CrossRef]
10. Parekh, B.; Kabir, C.S. A case study of improved understanding of reservoir connectivity in an evolving water flood with surveillance data. *J. Pet. Sci. Eng.* **2013**, *102*, 1–9. [CrossRef]
11. Moreno, G.A.; Lake, L.W. On the uncertainty of interwell connectivity estimations from the capacitance-resistance model. *Pet. Sci.* **2014**, *11*, 265–271. [CrossRef]
12. Laochamroonvorapongse, R.; Kabir, C.S.; Lake, L.W. Performance assessment of miscible and immiscible water-alternating gas floods with simple tools. *J. Pet. Sci. Eng.* **2014**, *122*, 18–30. [CrossRef]
13. Olsen, C.; Kabir, C.S. Waterflood performance evaluation in a chalk reservoir with an ensemble of tools. *J. Pet. Sci. Eng.* **2014**, *124*, 60–71. [CrossRef]
14. Tao, Q.; Bryant, S.L. Optimizing carbon sequestration with the capacitance/resistance model. *SPE J.* **2015**, *20*, 1094–1102. [CrossRef]
15. Wang, D.; Li, Y.; Chen, B.; Hu, Y.; Li, B.; Gao, D.; Fu, B. Ensemble-based optimization of interwell connectivity in heterogeneous waterflooding reservoirs. *J. Nat. Gas Sci. Eng.* **2017**, *38*, 245–256. [CrossRef]
16. Mamghaderi, A.; Pourafshary, P. Water flooding performance prediction in layered reservoirs using improved capacitance-resistive model. *J. Pet. Sci. Eng.* **2013**, *108*, 107–117. [CrossRef]
17. Zhang, Z.; Li, H.; Zhang, D. Water flooding performance prediction by multi-layer capacitance-resistive models combined with the ensemble Kalman filter. *J. Pet. Sci. Eng.* **2015**, *127*, 1–19. [CrossRef]
18. Zhang, Z.; Li, H.; Zhang, D. Reservoir characterization and production optimization using the ensemble-based optimization method and multi-layer capacitance-resistive models. *J. Pet. Sci. Eng.* **2017**, *156*, 633–653. [CrossRef]
19. Holanda, R.W.; Gildin, E.; Jensen, J.L. A generalized framework for Capacitance Resistance Models and a comparison with streamline allocation factors. *J. Pet. Sci. Eng.* **2018**, *162*, 260–282. [CrossRef]
20. Gentil, P.H. The Use of Multi-Linear Regression Models in Patterned Waterfloods: Physical Meaning of the Regression Coefficients. Ph.D. Thesis, The University of Texas, Austin, TX, USA, 2005.
21. Cao, F.; Luo, H.S.; Lake, L.W. Oil-rate forecast by inferring fractional-flow models from field data with Koval method combined with the capacitance-resistance model. *SPE Reserv. Eval. Eng.* **2015**, *18*, 534–553. [CrossRef]
22. Jansen, J.D. Adjoint-based optimization of multiphase flow through porous media-a review. *Comput. Fluids* **2011**, *46*, 40–51. [CrossRef]
23. Brouwer, D.R.; Jansen, J.D. Dynamic optimization of water flooding with smart wells using optimal control theory. *SPE J.* **2004**, *9*, 391–402. [CrossRef]
24. Chen, Y.; Oliver, D.S.; Zhang, D. Efficient ensemble-based closed-loop production optimization. *SPE J.* **2009**, *14*, 634–645. [CrossRef]
25. Chen, Y.; Oliver, D.S. Ensemble-based closed-loop optimization applied to Brugge field. *SPE Reserv. Eval. Eng.* **2010**, *13*, 56–71. [CrossRef]
26. Su, H.J.; Oliver, D.S. Smart well production optimization using an ensemble-based method. *SPE Reserv. Eval. Eng.* **2010**, *13*, 884–892. [CrossRef]
27. Dehdari, V.; Oliver, D.S. Sequential quadratic programming for solving constrained production optimization: Case study from Brugge field. *SPE J.* **2012**, *17*, 874–884. [CrossRef]

28. Sarma, P.; Chen, W. Improved estimation of the stochastic gradient with Quasi-Monte Carlo methods. In Proceedings of the 14th European Conference on the Mathematics of Oil Recovery (ECMOR XIV), Catania, Italy, 8–11 September 2014; pp. 8–11.
29. Do, S.T.; Reynolds, A.C. Theoretical connections between optimization algorithms based on an approximate gradient. *Comput. Geosci.* **2013**, *17*, 959–973. [CrossRef]
30. Stordal, A.S.; Szklarz, S.P.; Leeuwenburgh, O. A theoretical look at ensemble-based optimization in reservoir management. *Math. Geosci.* **2016**, *48*, 399–417. [CrossRef]
31. Jafroodi, N.; Zhang, D. New method for reservoir characterization and optimization using CRM-EnOpt approach. *J. Pet. Sci. Eng.* **2011**, *77*, 155–171. [CrossRef]
32. Hong, A.J.; Bratvold, R.B.; Nævdal, G. Robust production optimization with capacitance-resistance model as precursor. *Comput. Geosci.* **2017**, *21*, 1423–1442. [CrossRef]
33. Fonseca, R.M.; Chen, B.; Jansen, J.D.; Reynolds, A.C. A stochastic simplex approximate gradient (StoSAG) for optimization under uncertainty. *Int. J. Numer. Methods Eng.* **2016**, *109*, 1756–1776. [CrossRef]
34. Fonseca, R.M.; Kahrobaei, S.S.; van Gastel, L.J.T.; Leeuwenburgh, O.; Jansen, J.D. Quantification of the impact of ensemble size on the quality of an ensemble gradient using principles of hypothesis testing. In Proceedings of the SPE Reservoir Simulation Symposium, Houston, TX, USA, 23–25 February 2015.
35. Chen, B.; Reynolds, A.C. Optimal control of ICV's and well operating conditions for the water-alternating-gas injection process. *J. Pet. Sci. Eng.* **2017**, *149*, 623–640. [CrossRef]
36. Chen, B.; Fonseca, R.M.; Leeuwenburgh, O.; Reynolds, A.C. Minimizing the Risk in the robust life-cycle production optimization using stochastic simplex approximate gradient. *J. Pet. Sci. Eng.* **2017**, *153*, 331–344. [CrossRef]
37. Ding, S.W.; Lu, R.R.; Xi, Y.; Wang, S.L.; Wu, Y.P. Well placement optimization using direct mapping of productivity potential and threshold value of productivity potential management strategy. *Comput. Chem. Eng.* **2019**, *121*, 327–337. [CrossRef]
38. Guo, Z.; Reynolds, A.C. Robust Life-Cycle Production Optimization with a Support-Vector-Regression Proxy. *SPE J.* **2018**, *23*, 2409–2427. [CrossRef]
39. Moraes, R.J.; Fonseca, R.M.; Helici, M.A.; Heemink, A.W.; Jansen, J.D. An Efficient Robust Optimization Workflow using Multiscale Simulation and Stochastic Gradients. *J. Pet. Sci. Eng.* **2019**, *172*, 247–258. [CrossRef]
40. Wang, D.; Li, Y.; Hu, Y.; Li, B.; Deng, X.; Liu, Z. Integrated dynamic evaluation of depletion-drive performance in naturally fractured-vuggy carbonate reservoirs using DPSO–FCM clustering. *Fuel* **2016**, *181*, 996–1010. [CrossRef]

© 2019 by the authors. Licensee MDPI, Basel, Switzerland. This article is an open access article distributed under the terms and conditions of the Creative Commons Attribution (CC BY) license (http://creativecommons.org/licenses/by/4.0/).

Article

CO₂, Water and N₂ Injection for Enhanced Oil Recovery with Spatial Arrangement of Fractures in Tight-Oil Reservoirs Using Huff-'n-puff

Jie Bai [1,2], Huiqing Liu [1,2,*], Jing Wang [1,2], Genbao Qian [3], Yongcan Peng [3], Yang Gao [3], Lin Yan [4] and Fulin Chen [4]

1. State Key Laboratory of Petroleum Resources and Engineering, China University of Petroleum-Beijing, Beijing 102249, China; jaybye@163.com (J.B.); wangjing8510@163.com (J.W.)
2. MOE Key Laboratory of Petroleum Engineering in China University of Petroleum, Beijing 102249, China
3. PetroChina Xinjiang Oilfield, Karamay 834000, China; qgenbao@petrochina.com.cn (G.Q.); pyongcan@petrochina.com.cn (Y.P.); gaoyang@petrochina.com.cn (Y.G.)
4. PetroChina Research Institute of Petroleum Exploration & Development, Beijing 100083, China; linyan@petrochina.com.cn (L.Y.); Chenfulin@petrochina.com.cn (F.C.)
* Correspondence: liuhq@cup.edu.cn

Received: 2 February 2019; Accepted: 24 February 2019; Published: 1 March 2019

Abstract: Tight oil has been effectively developed thanks to artificial fracture technology. The basic mechanism of effective production through fractures lies in the contact between the fractures (both natural and artificial) and the matrix. In this paper, the natural tight cores from J field in China are used to conduct experimental studies on the different fluid huff-'n-puff process. A new core-scale fracture lab-simulation method is proposed. Woven metallic wires were attached to the outer surface of the core to create a space between the core holder and core as a high permeable zone, an equivalent fracture. Three different injecting fluids are used, including CO_2, N_2 and water. The equivalent core scale reservoir numerical models in depletion and huff-n-puff mode are then restored by numerical simulation with the Computer Modeling Group—Compositional & Unconventional Reservoir Simulator (CMG GEM). Simulation cases with eight different fracture patterns are used in the study to understand how fracture mechanistically impact Enhanced Oil Recovery (EOR) in huff n puff mode for the different injected fluids. The results showed: Firstly, regardless of the arrangement of fractures, CO_2 has mostly obvious advantages over water and N_2 in tight reservoir development in huff-'n-puff mode. Through EOR mechanism analysis, CO_2 is the only fluid that is miscible with oil (even 90% mole fraction CO_2 is dissolved in the oil phase), which results in the lowest oil phase viscosity. The CO_2 diffusion mechanism is also pronounced in the huff-'n-puff process. Water may impact on the oil recovery through gravity and the capillary force imbibition effect. N_2, cannot recover more crude oil only by elasticity and swelling effects. Secondly, the fracture arrangement in space has the most impact on CO_2 huff-'n-puff, followed by water and finally N_2. The fractures primarily supply more efficient and convenient channels and contact relationships. The spatial arrangement of fractures mainly impacts the performance of CO_2 through viscosity reduction in the contact between CO_2 and crude oil. Similarly, the contact between water in fractures and crude oil in the matrix is also the key to imbibition. In the process of N_2 huff-'n-puff, the elasticity energy is dominant and fracture arrangement in space hardly to improve oil recovery. In addition, when considering anisotropy, water huff-'n-puff is more sensitive to it, while N_2 and CO_2 are not. Finally, comparing the relationship between fracture contact area and oil recovery, oil production is insensitive to contact area between fracture and matrix for water and N_2 cases.

Keywords: tight reservoir; huff-'n-puff; fracture simulation; enhanced oil recovery; CO_2 diffusion

1. Introduction

The key dominant feature difference between a tight reservoir and a traditional reservoir lies in the tight pore. The pore throat of the porous medium of the tight reservoir is very narrow, and its movable fluid saturation is very low. The fluid flow requires a higher-pressure gradient through the pore throat (when <50 nm), and the mobility of fluids is poor [1–6], so in the development of tight reservoirs, fracturing is a key technology with an indelible contribution to oil production. The main objective of fracturing is to reconstruct the reservoir seepage formation and to achieve the purpose of the effective expansion and development of the fracture through the mechanical mechanism and provide a more efficient and convenient flow channel for the crude oil [2,7]. However, according to the evaluation of the oil production of tight reservoirs both in China and America, the production rate of tight reservoirs decreases rapidly, without a long and stable production period [2,6,8–11]. Thus, it is difficult to obtain satisfactory oil recovery only relying on the depressurization process even with fracturing.

The tight porous reservoir structure is dense, the pore throat is narrow, and at the nanometer scale, and the fluid flow is also affected by the boundary layer effect. This results in higher-pressure gradients in the flooding process in the matrix that restricts traditional flooding application in tight reservoirs [12–14], while with the existence of fractures, the breakthrough of gas/water usually occurs during the tight reservoir flooding process, so the huff-'n-puff method is considered a promising energy increasing method in tight reservoirs over the depressurization mode.

Many scholars have developed studies on effective displacement of different fluids in tight reservoirs, mainly including gas injection and chemical injection in Bakken tight oil formations [15,16]. Due to the narrow pore throat in tight reservoirs, gas injection, with its lower viscosity and larger injectivity, is considered to be much easier than water injection. CO_2 is one of the most effective injection gases for enhanced oil recovery, and the use of CO_2 to improve the recovery also serves a role of greenhouse gas storage, so it is attracting increasing attention. In the study of the enhanced oil recovery (EOR) by fluid injection, most of the studies focus on the comparison of the miscible process and immiscible process mechanisms [13,14,17–22].

In the CO_2 huff-'n-puff process, CO_2 is first injected into a producer well, then the well is shut off for soaking, finally, the well is reopened for fluids to be produced [23]. Its application to unconventional reservoirs has been investigated by lots of researchers recently [18,24–27]. The traditional EOR mechanisms of CO_2 are: (1) viscosity reduction, (2) oil swelling, (3) solution gas drive, (4) hydrocarbon extraction by CO_2. The most important operating parameters in CO_2 huff-'n-puff include injection rate and time, number of cycles, soaking time, and pressure.

Hawthorne et al. [28] explained CO_2 EOR process in tight reservoirs relatively clearly: (1) CO_2 is preferentially injected into the fractures, (2) CO_2 contacts the matrix rock through fractures, (3) CO_2 will further seep into the matrix, dominated by pressure gradients. When CO_2 dissolves in crude oil, the oil swells and its viscosity decreases, (4) oil drains into the fractures under the effect of CO_2, and (5) when the pressure gradient drops, oil transport is dominated by the concentration gradient across the matrix and the fractures.

A considerable number of scholars have also focused on the study of CO_2 in porous media and crude oil mass transfer mechanisms [15,29]. The mechanisms of CO_2 EOR in tight oil reservoirs is considered to be significantly different from those in conventional reservoirs, due to the special petrophysical properties, reservoir fluid thermodynamics, and mass transport mechanisms. Diffusion is considered to be an especially important mechanism affecting gas transportat under tight reservoir conditions, and its effect on enhanced oil recovery of tight oil cannot be neglected based on the core scale experiments.

In addition, experimental results from both 2D core slice and 3D homogeneous core scale models show that diffusion plays a significant role [30]. However, their field scale huff-'n-puff simulation model showed that convection is considered the dominating mechanism instead of diffusion. Vega carried out CO_2 injection experiments in miscible conditions [31]. The results showed that both diffusive and convective transfer mechanisms were significant.

In addition to CO_2, N_2 was also used to explore the EOR potential in tight reservoirs [32]. Cores from Eagle Ford were used in this research. Similar experiments can also be found in Li and Sheng's research [33]. The results indicated that oil recovery was improved when the soaking pressure increased. Extending soaking time can also significantly improve oil recovery. In addition, increasing huff-'n-puff cycles improved oil recovery. After six cycles, the recovery reached 30.99%.

Different from traditional steam huff-'n-and puff, cold water injection relies on the mechanism of imbibition for oil displacement in water-wet reservoirs in the presence of fractures [34,35]. This requires the rock to be water-wet. To alter the rock wettability, surfactants can be used [36–38].

In the extensive study of tight reservoirs, it is also found that the fracture formation process in tight reservoirs is complex and diverse. The fracture form (such as the Figure 1) is mainly influenced by the natural fractures and properties of the reservoir (the parameters of brittleness, modulus of elasticity, etc.) and the distribution characteristics of natural fractures [2,3,7,10,39,40]. The expansion and connection of the fractures occurs both in the plane and three-dimensional direction. The diversity in the fracture expansion and connection leads to the fact that the contact between fractures and matrix is a multi-contact relationship [39]. This undoubtedly increases the difficulty of evaluating the development effect of tight reservoirs, because the fractures are often simplified to simple uniform and single contact surfaces. In the fluid injection process, especially in huff-'n-puff mode, the mass exchange between crude oil and injected fluid is unavoidably partially ignored due to the lack of multi-contact relationships between fractures and the matrix. Research on the effect of processing under complex fracture conditions has also progressed. In Molero's research, different types of complex fractures were simulated and the effects of CO_2 huff-'n-puff evaluated. The results confirmed that the appropriate modeling of fracture geometry plays a critical role in the estimation of the incremental oil recovery. In Meng's research about imbibition effects, for fractured reservoirs the area of the water-covered-face between the fractures and matrix also had a significant effect on the oil production process. Therefore, the simulation of fractures is essential when evaluating the oil recovery in both depletion and huff-'n-puff processes [38].

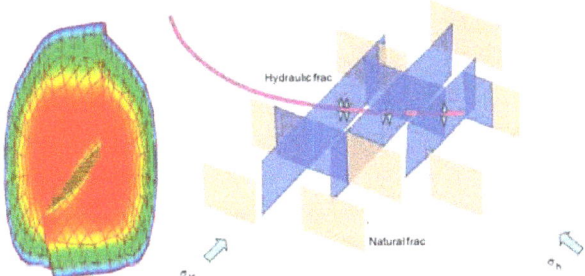

Figure 1. Nonplanar fracture model in a naturally fractured formation.

In terms of core scale experiments, cutting cores and then re-splicing them is the most common method for fracturing simulation. By cutting a complete vertical core with a saw or cutter, the core is separated in half. The fracture is then simulated by splicing the two core parts. The disadvantages of this method are usually a large fracture width (2 mm) and a loss of core and single shape of fractures. The cores also cannot be reused. Tovar et al. [30] simulated fractures by surrounding tight cores with glass beads. To prevent the loss of the glass beads, two sandstone plugs were mounted at both ends as filters. This method doesn't break the cores and keep them reusable. However, the shape of the fracture is still too single, and the whole surrounding space of the core is treated as fracture, so the rationality is questionable. In addition, the spatial arrangement of fractures still cannot be easily simulated in these methods.

On the basis of the established importance of complex fractures for tight oil huff-'n-puff processes, this research aims to mechanistically understand how complex fractures impact EOR in huff-'n-puff for different injected fluids. First, indoor core experiments were carried out, and a new fracture simulation method using woven metallic wires was proposed. Then a numerical simulation method was used to perform historical fitting of the experimental results with GMG GEM, which fully considers unconventional fluid flow, composition interaction and the mass transfer mechanism mentioned in fractured tight reservoirs. Three different injecting fluids are used, including CO_2, N_2 and water. Finally, core scale simulation cases with eight different fracture patterns are used in the study to understand how fractures mechanistically impact EOR in huff-'n-puff for different injecting fluids with a numerical method. Through this research, a more regular understanding of the influence of fracture and matrix distribution relationship on EOR process can be clear. It is of great significance for guiding fracturing construction. When selecting the optimal injected fluids corresponding to spatial arrangement fracture, theoretical support can also be obtained from this study.

2. Methodology

2.1. Experiments

The core plug used in this experiment was from the J Oilfield with a water-wet wettability. Three cores were selected to perform the huff-'n-puff process with CO_2, water and N_2. For the experimental system, a degassed crude oil is selected, so the weighing method is adopted to measure the amount of oil produced. Then the oil recovery of each round is calculated. The length is 5 cm with a diameter of 2.5 cm. The permeability is about 0.0375 mD, and porosity is 17.5%. Three cores are selected and their properties are listed in Table 1.

Table 1. Experimental core data.

Core Number	Porosity, %	Gas Permeability, mD	Liquid Permeability, mD
1	0.177513	0.0356	0.02835
2	0.175812	0.03691	0.0304
3	0.173267	0.03994	0.0339
Average	0.175531	0.037483	0.030883

In this paper, a new method for the simulation of core-scale fractures was designed. As shown in Figure 2, woven metallic wires attached to the outer surface of the core create a space between the core holder and core as a high permeable zone, equivalent to a fracture. The shape of the metallic cloths can be adjusted to mimic different fracture geometries and relative positions in space. In order to ensure the same fracture area, the same metal meshes were used and the cropped different shaped metal meshes kept at the same weight. The flow chart is shown in Figure 3.

In a specific experimental process, first, we wash, dry and weigh the tight core. The core is saturated with a vacuum saturate. The weight of the core before and after oil saturation is measured separately to determine the weight of oil. Finally, 3.5 mL oil is saturated in the tight core. Then, the core is placed into the core holder. A confining pressure of 30 MPa was applied to the core holder by the hand pump. Then it was saturated with an extra 1.2 mL of oil when initial pressure reached 30 MPa. The viscosity of the oil is 6 mPa·s at 60 °C. The entire system was placed in an incubator at 60 °C. Thus, the core scale tight reservoir system has been simulated using a lab method.

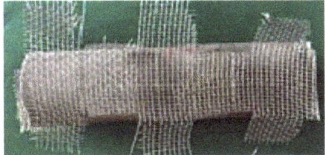 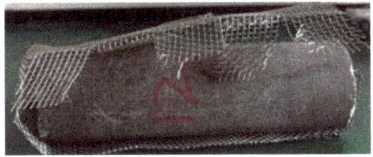

(**a**) fracture arrangement in case 2 (ij profile)   (**b**) fracture arrangement in case 2 (ik profile)

(**c**) fracture arrangement in case 4 (ij profile)   (**d**) fracture arrangement in case 4 (ik profile)

Figure 2. Basic fracture-matrix core.

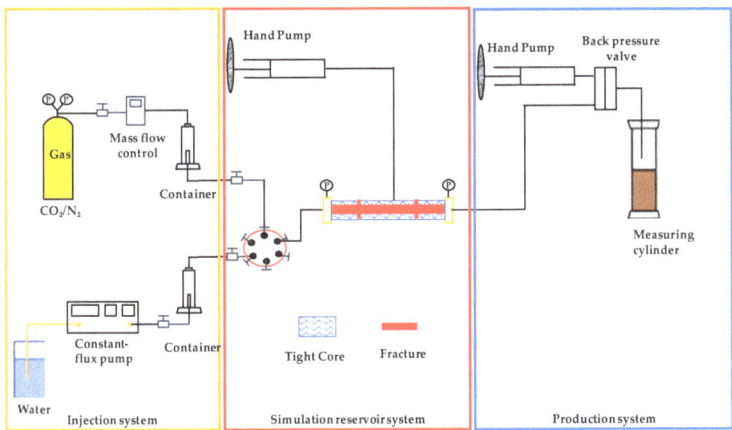

Figure 3. Flowchart of the huff-'n-puff experiments.

The injection scheme for the huff-'n-puff process is as follows: first, reduce the core pressure from an initial pressure of 30 MPa to 5 MPa. Then, inject CO_2/N_2/Water until the BHP reaches 40 MPa. The huff-'n-puff is repeated 12 times. The oil produced in the depletion process and CO_2/N_2 huff-'n-puff processes can be directly weighed, while in water huff-'n-puff, the water should be dried before weighing the produced oil.

The lab method can only provide practical oil recovery data. Subjected to laboratory conditions, the internal pressure and saturation, as well as other properties of fluids, are difficult to measure during a core-scale huff-'n-puff process. Then based on the experimental results, the parameters of relative permeability and fracture permeability are determined by fitting historical experimental data (shown in Figure 4). Finally, the numerical simulation method is used for more detailed analysis and research in the tight core production process.

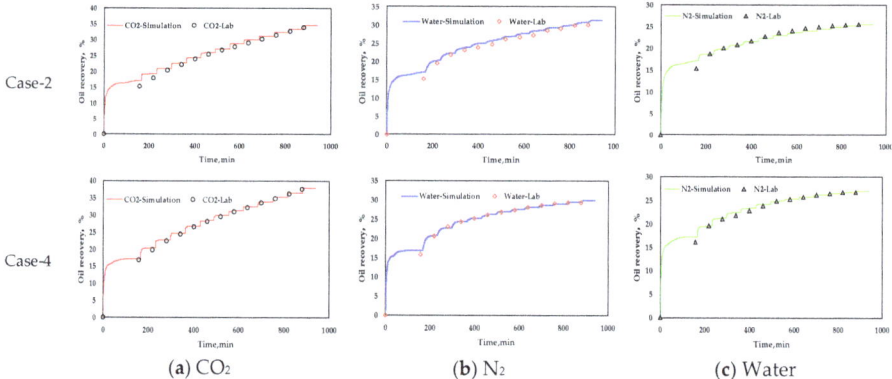

(a) CO₂   (b) N₂   (c) Water

Figure 4. Comparisons of oil recovery between lab data and huff-'n-puff simulation.

2.2. Numerical Simulation

2.2.1. Reservoir Simulation Model

To investigate the several mechanism effect on oil recovery, a Cartesian grid model is built in CMG GEM [41] based on the core experiments. The governing equation describing the mass balance of component i in the oil/gas phases is given by the following expression [15], including the cumulative term of component i in the rock and fluid phases, and the convection, dispersion and molecular diffusion term of component i in phase j:

$$\frac{\partial}{\partial t}[(1-\phi)\rho_s w_{is} + \phi \sum_{j=1}^{n} \rho_j S_j w_{ij}] + \vec{\nabla} \cdot [\sum_{j=1}^{N_p} \rho_j w_{ij} u_j - \phi \rho_j S_j K_{ij} \nabla w_{ij}] - r_i = 0, \quad (1)$$
$$i = 1, 2, \ldots N_c,$$
$$j = 1, 2, \ldots N_p,$$

where ϕ is the porosity of the tight matrix, ρ_s is the density of the tight matrix, ρ_j is the density the of j phase, S_j is the saturation of phase j, w_{is} is the fraction of component i that precipitates in the tight matrix rock, w_{ij} is the fraction of component i in the phase j. r_i is the injection or production rate as a source or sink term, N_p is the total number of phases, N_c is the total number of components, u_j is the flow velocity, which is defined based on Darcy's law as:

$$u_j = \frac{K k_{rj}}{\mu_j}(\nabla p_j - \rho_j g) \quad (2)$$

In particular when both oil and water phases exist, the capillary pressure also has effect on the flow velocity, and the equation is described as:

$$u_w = \frac{K k_{rw}}{\mu_w}(\nabla p_w + \nabla p_{\text{cow}}(S_w) - \rho_w g) \quad (3)$$

where K is the formation permeability, k_{rj} is the relative permeability of phase j, p_j is the pressure of phase j, and μ_j is the viscosity of phase j. K_{ij} is the dispersivity coefficient of component i in the phase j, which is defined as:

$$K_{ij} = \frac{D_{ij}}{\tau} + \frac{\alpha_l |u_l|}{\phi S_l} \quad (4)$$

where τ is the tortuosity of the tight matrix, D_{ij} is the diffusion coefficient of component i in phase j, and α_l is the dispersity coefficient of fluid j in different directions.

D_{ij} (its units are cm^2/s) can be calculated based on the Sigmund correlation [42]. The diffusion in water phase is ignored. The diffusion coefficient D_{ik} is calculated by:

$$D_{ik} = \frac{\rho_j^0 D_{ik}^0}{\rho_j}(0.99589 + 0.096016\rho_{jr} - 0.22035\rho_{jr}^2 + 0.032674\rho_{jr}^3) \tag{5}$$

where $\rho_j^0 D_{ik}^0$ is the zero-pressure limit of the product of density and diffusivity, which can be calculated by:

$$\rho_j^0 D_{ik}^0 = \frac{0.001853 T^{0.5}}{\sigma_{ik}^0 \Omega_{ik} R}\left(\frac{1}{M_i} + \frac{1}{M_k}\right)^{0.5} \tag{6}$$

where T is the absolute temperature (K), σ_{ik} is the collision diameter (Å), Ω_{ik} is the collision integral of the Lennard-Jones Potential, R is the universal gas constant, and M_i is the molecular weight of component i. These parameters can be calculated by the following equations [43]:

$$\sigma_i = (2.3551 - 0.087\omega_i) \times \left(\frac{T_{ci}}{P_{ci}}\right)^{1/3} \tag{7}$$

$$\varepsilon_i = \kappa_B(0.7915 - 0.1963\omega_i)T_{ci} \tag{8}$$

$$\sigma_{ij} = \frac{\sigma_i + \sigma_j}{2} \tag{9}$$

$$\varepsilon_{ij} = \sqrt{\varepsilon_i \varepsilon_j} \tag{10}$$

$$T_{ij}^* = \frac{\kappa_B T}{\varepsilon_{ij}} \tag{11}$$

$$\Omega_{ij} = \frac{1.06036}{(T_{ij}^*)^{-0.15610}} + \frac{0.19300}{\exp(-0.47635 T_{ij}^*)} + \frac{1.03587}{\exp(-1.52996 T_{ij}^*)} + \frac{1.76464}{\exp(-3.89411 T_{ij}^*)} \tag{12}$$

where ω is the acentric factor, P_c is the critical pressure (atm), T_c is the critical temperature (K), ε is the characteristic Lennard-Jones energy and κ_B is the Boltzmann constant:

$$\kappa_B = 1.3805 \times 10^{-16} ergs/K \tag{13}$$

ρ_{jr} is the reduced density, which can be calculated by:

$$\rho_{jr} = \rho_j\left(\frac{\sum\limits_{i}^{N_c} y_{ij} V_{ci}^{5/3}}{\sum\limits_{i}^{N_c} y_{ij} V_{ci}^{2/3}}\right) \tag{14}$$

V_{ci} is the critical volume of component i, y_{ij} is the mole fraction of component i in phase j. Nc is the total number of components. The diffusion coefficient of component i in the mixture D_{ij} is defined as:

$$D_{ij} = \frac{1 - y_{ij}}{\sum\limits_{i \neq k} \frac{y_{ij}}{D_{ik}}} \tag{15}$$

In this paper, the core scale reservoir depletion and huff-'n-puff process were simulated with CMG GEM based on the partial differential equation (PDE) of Equation (1).

2.2.2. Reservoir Model Description

The matrix is rectangular with dimensions of 5 × 2.5 × 2.5 cm (Figure 5). This core is divided into 10 × 5 × 5 gridblocks with dimensions of 0.5 cm × 0.5 cm × 0.5 cm. The matrix is assumed to be homogeneous and isotropic with a porosity of 0.175 and permeability of 0.03 mD. The relative permeability curve inside the matrix is given in Figure 6. Two layers of grid in each direction with dimensions of 0.1 cm × 0.5 cm × 0.5 cm, 0.5 cm × 0.1 cm × 0.5 cm and 0.5 cm × 0.5 cm × 0.1 cm are added to the periphery of the matrix model to serve as the grid space of the fracture, and thus the total number of model grids is 12 × 7 × 7.

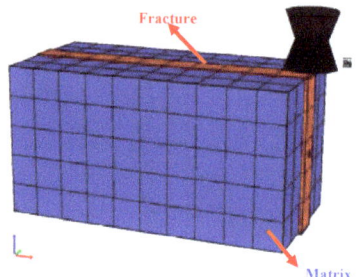

Figure 5. Basic Fracture-Matrix Model in CMG.

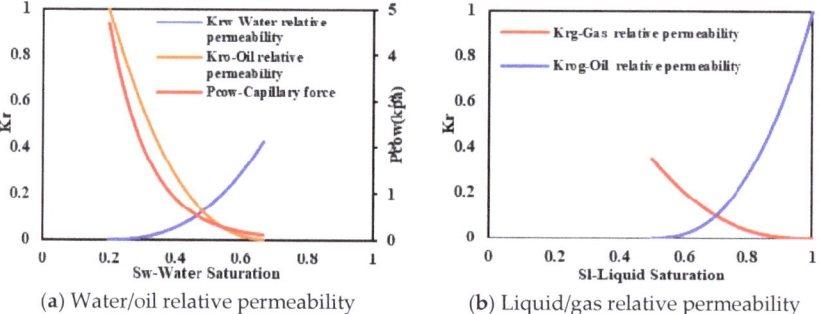

Figure 6. Relative permeability curve of the matrix.

The fractures are assumed to be thin highly permeable passages covering the surface of the core sample matrix. These assumptions are experimentally possible as shown in Figure 5. Woven metallic wires attached to the outer surface of the core create space between the core holder and core as a high permeability zone, an equivalent fracture. The shape of the metallic cloths can be adjusted to mimic different fracture geometries and relative positions in space.

Validating the model with experimental data is critical to ensure that the simulation results are correct. In the history matching work, the main adjustment parameters are the relative permeability curves and the fracture permeability. The porosity of the fractures is assumed to be 0.5 and permeability 50 mD through the history fitting of the experimental data. The relative permeability inside the matrix and the fractures are shown in Figures 6 and 7.

The injected fluids for huff-'n-puff in this paper are CO_2, N_2 and water. The initial oil and water saturations are 0.8/0.2. The initial pressure is 30MPa. The oil viscosity is 6 mPa·s. The CO_2 diffusion coefficient is set based on published simulated and laboratory measurements [15,30,44,45]. Other initial reservoir properties can be found in Table 2. The phase behavior of water/oil, water/oil/CO_2, and water/oil/N_2 are modeled with the Peng-Robinson equation of state. Inputs for the EOS model is given in Tables 3 and 4. The initial oil components are provided by the J field, and shown in Table 5. The well is located at (12,4,4) as indicated in Figure 5. The injection scheme for the huff-n-puff process are as follows: first, depletion and pressure to 5 MPa. Then, inject CO_2/N_2/Water until BHP reaching 40 MPa. Huff-'n-puff is repeated 12 times.

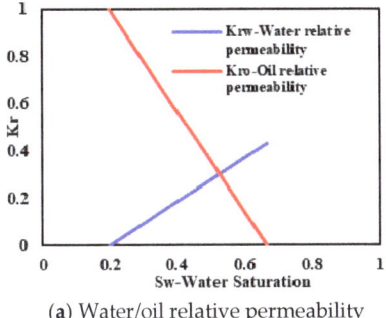
(a) Water/oil relative permeability

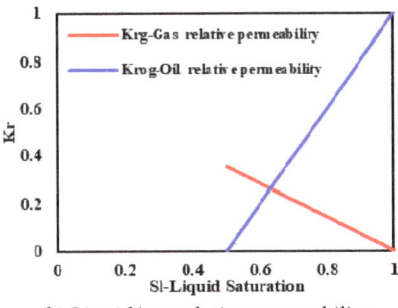
(b) Liquid/gas relative permeability.

Figure 7. Relative permeability curve of the fracture.

Table 2. Basic reservoir properties in the numerical model.

Parameter	Value	Unit
Initial reservoir pressure	30	MPa
Porosity	0.175	-
Oil viscosity	6	mPa·s
Soak pressure	40	MPa
Soak time	45	min
Production time	6	min
Cycle number	12	-
Temperature	60	°C
Initial water saturation	0.2	-
Matrix permeability	0.03	mD
Fracture permeability	50	mD
CO_2 diffusion coefficient	0.0005	cm^2/s

Table 3. Binary interaction parameters.

Component	CO_2	C1	C4	C7	C12	C19	C30	N_2
CO_2								
C1	0.103							
C4	0.1317	0.013						
C7	0.1421	0.0358	0.0059					
C12	0.1501	0.0561	0.016	0.0025				
C19	0.1502	0.0976	0.0424	0.0172	0.0067			
C30	0.1503	0.1449	0.0779	0.0427	0.0251	0.0061		
N_2	−0.02	0.031	0	0	0	0	0	

The relative arrangement of matrix and several fractures are modeled on a core scale with CMG GEM. Table 6 lists the core-scale tight oil-reservoirs with the same contact area between fractures and cores, but with different fracture spatial arrangements.

Table 4. Component properties and composition.

Component	Pc/atm (atm)	Tc (K)	Acentric Factor	Mol. Weight (g/gmole)	Volume Shift	Crit. Volume (m³/kgmole)	Omega A	Omega B	Specific Gravity (SG)	Tb (°C)	Parachor
CO_2	72.80	304.20	0.23	44.01	−0.09	0.10	0.46	0.08	0.60	−102.51	126.49
C1	45.24	189.67	0.01	16.21	−0.15	0.09	0.46	0.08	0.35	−161.15	40.38
C4	43.49	412.47	0.15	44.79	−0.09	0.23	0.46	0.08	0.80	−17.09	128.86
C7	37.69	556.92	0.25	83.46	−0.01	0.35	0.46	0.08	0.95	86.79	242.85
C12	31.04	667.52	0.33	120.52	0.06	0.50	0.46	0.08	1.05	176.75	345.92
C19	19.29	673.76	0.57	220.34	0.16	0.89	0.46	0.08	1.30	217.98	593.49
C30	15.38	792.40	0.94	321.52	0.21	1.06	0.46	0.08	1.50	331.76	799.68
N_2	33.50	126.20	0.04	28.01	0.00	0.09	0.00	0.00	0.81	−195.75	41.00

Table 5. Oil composition in the simulation model.

Competent	Percent/%
ZGLOBALC 'N$_2$'	1
ZGLOBALC 'CO$_2$'	0.01
ZGLOBALC 'C1'	23.79
ZGLOBALC 'C4'	4.38
ZGLOBALC 'C7'	11.26
ZGLOBALC 'C12'	36.18
ZGLOBALC 'C19'	16.51
ZGLOBALC 'C30'	7.87
Sum	100

Table 6. Settings of fracture contact relationships.

Fracture Pattern		I = 1	I = 10	Remarks
0				Baseline
1				Scattered fracture-1
2				Scattered fracture-2
3				Scattered fracture-3
4				Scattered fracture-4
5				Near well

Table 6. Cont.

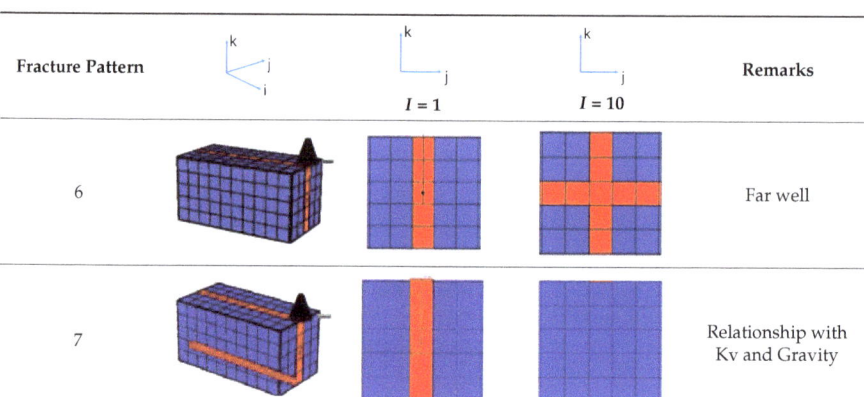

3. Results

3.1. Depletion and Huff-'n-puff with Different Fluids

3.1.1. Depletion Mode

By comparing the simulation results of the different fracture mode cases (Figure 8), it is found that in the depressurization period, the existence of fractures is significant for the enhanced oil process. With the same contact area between fracture and matrix, the cases with fractures have approximately 18% higher oil recovery than the case without any fracture. Figure 9 shows the pressure distribution in several cases. Without fractures, the pressure drop is not obvious. The pressure change area in the cases with fracture is wider and the pressure drop is more significant, especially in Case-4. This shows that the fractures supply the more efficient and convenient channels and contact relationships.

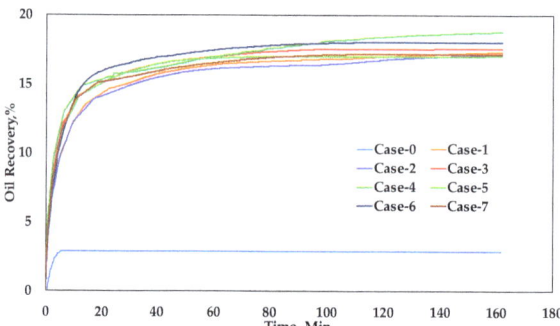

Figure 8. The comparison of oil recovery in fracture and no-fracture conditions.

In the depletion mode, crude oil flows mainly by the pressure drop caused by elastic energy, and there is no contact between other fluids and crude oil in the matrix through fractures. Therefore, in the case of depletion mode, when fractures have the same contact area with spatial arrangement, the obvious difference in oil recovery cannot be observed.

Case-4 represents the most typical contact relationship between horizontal well fracturing artificial fracture and tight matrix without natural fracture. Case-4 is consistent with the experimental scheme. Figure 9 shows that the fracture type in Case-4 has the best pressure spread effect. In the next study, Case-4 is used as the basis for comparing different fluid using huff-'n-puff.

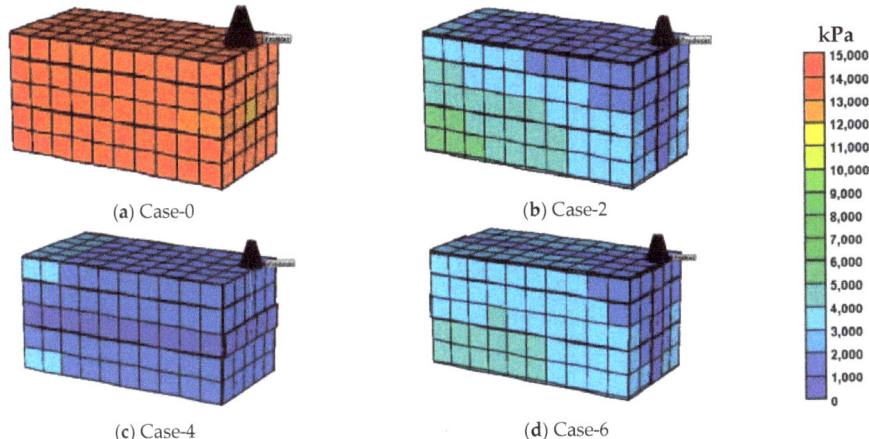

(a) Case-0   (b) Case-2   (c) Case-4   (d) Case-6

Figure 9. Pressure comparison between different scattered fractures in the depletion process.

3.1.2. Comparison of CO_2, N_2 and Water as Injecting Fluids for Huff-n-puff

- Oil recovery comparison

In this part, Case-4 is used to illustrate the mechanisms of using CO_2, N_2 and water for the huff-'n-puff process. Figure 10 shows the oil recovery factor on surface conditions by using the three different injecting fluids. CO_2 was able to enhance oil recovery by 21% after depletion, water by 13%, and N_2 by 10% so CO_2 has obvious advantages over water and N_2.

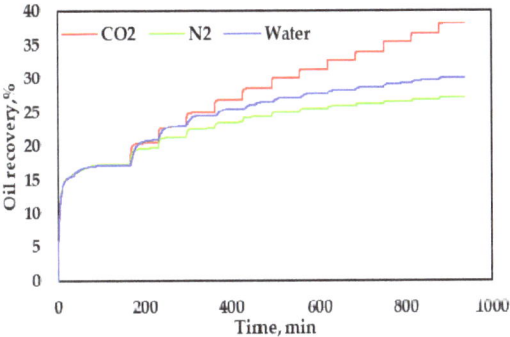

Figure 10. Oil recovery comparison of different fluids (Case-4).

- Gas/liquid injection volume

Firstly, injection fluid volume under the reservoir conditions was selected as a comparison index for the injectivity (Figure 11) to evaluate the EOR effect of different fluids.

From the comparison it can be found that the injectivity of N_2 is the highest among the three fluids, but its impact on EOR is the least. CO_2 has the secondary injectivity with the best EOR effect. The injected volume of water is the least, but with a moderate oil recovery.

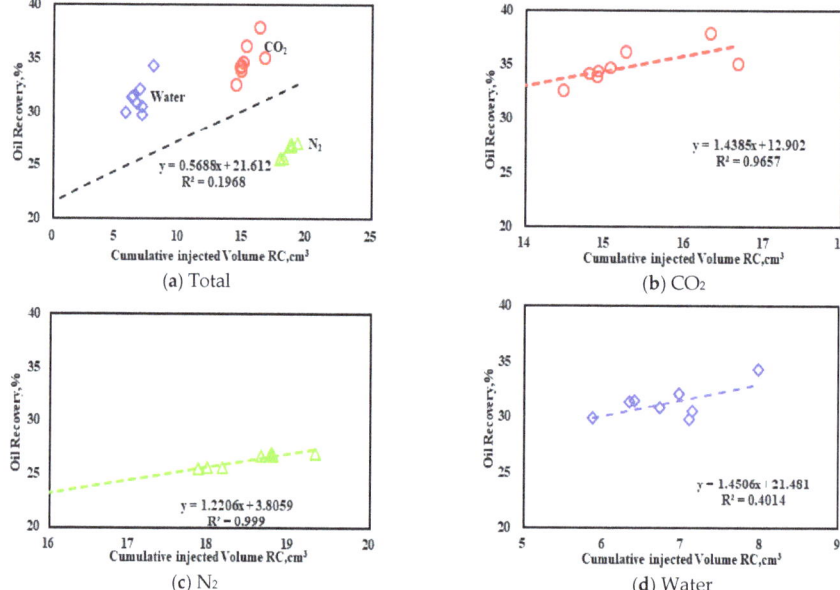

Figure 11. Relationship between injection volume RC and oil recovery in huff-'n-puff mode.

Then, we also found the relationship between the volume of the accumulative injection of fluid and the oil recovery under the reservoir conditions of different fluids (CO_2, N_2 and water) cannot be satisfactorily correlated (shown in Figure 11a). It shows that that the injectivity is only part of the basis for the EOR effect of fluids. Because of the obvious difference of EOR mechanism among the three fluids, the oil recovery only shows a good linear relationship with cumulative injected volume under reservoir conditions in same fluid cases (shown in Figure 11b–d).

The following is a detailed comparison of EOR processes of different fluids, to analyze how distributions of the fractures mechanistically affect the performance of the three different injectants:

- Viscosity profile and CO_2 concentration in oil phase

To further understand the EOR mechanisms of the three different injecting fluids, the influential factors in oil recovery are examined. One of the most important factors is the viscosity of the oil phase, which is affected by the thermodynamic conditions and oil composition. Figure 12 shows the viscosity of oil phase during the soaking process in the 12th cycle by using the three different fluids. It can be found that the effect of CO_2 on crude oil viscosity reduction is very significant, especially the viscosity of crude oil near the well and near the interface of fractures has the most significant change from the initial 6 mPa·s to about 1 mPa·s. However, N_2 is also mixed as a gas during the contact with crude oil, but there is no obvious viscosity reduction effect. After the water injection, the viscosity of crude oil may even increase to 7.5 mPa·s.

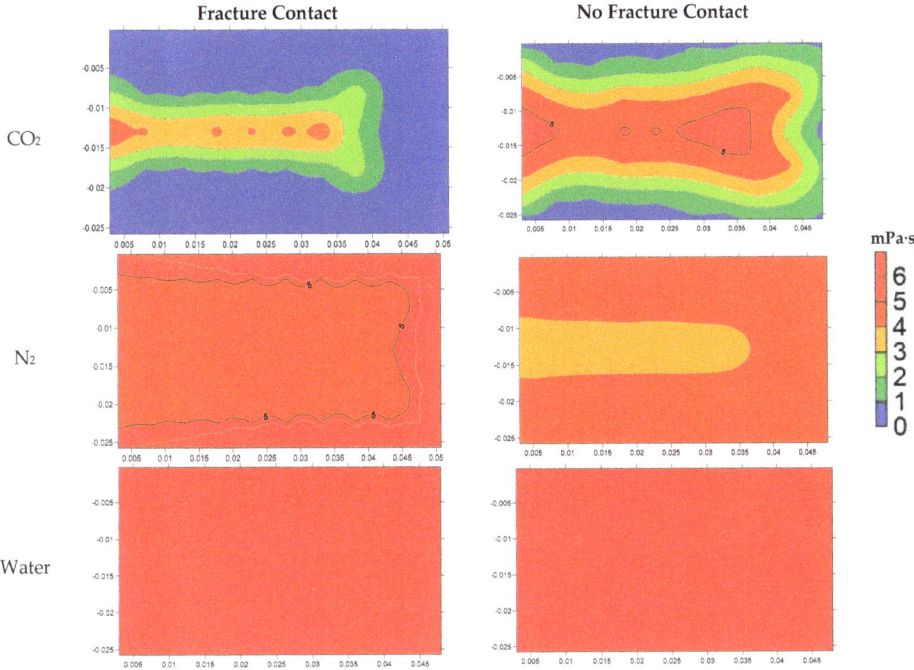

Figure 12. Oil viscosity of different fluids in Case-4 in soaking process (ij profile).

CO$_2$ is only fluid that is miscible with oil (even at 90% mole fraction with CO$_2$ in the oil phase as shown in Figures 13–15). The solubility of CO$_2$ results in the lowest oil phase viscosity of the three cases. In contrast, N$_2$ mainly exists in the gas phase and is much less soluble in the oil phase compared to CO$_2$. Water does not dissolve in the oil and water mainly enhances oil recovery through capillary force, gravity and elasticity displacement.

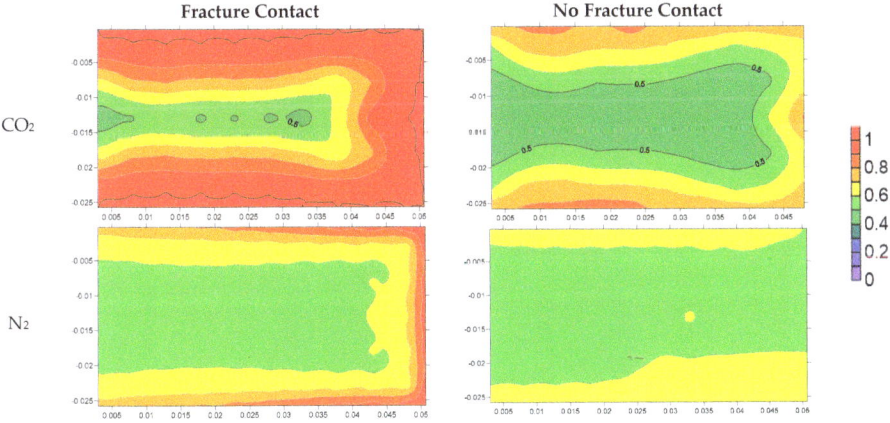

Figure 13. Mole fraction of injection fluid in oil in Case-4 in soaking process (ij profile).

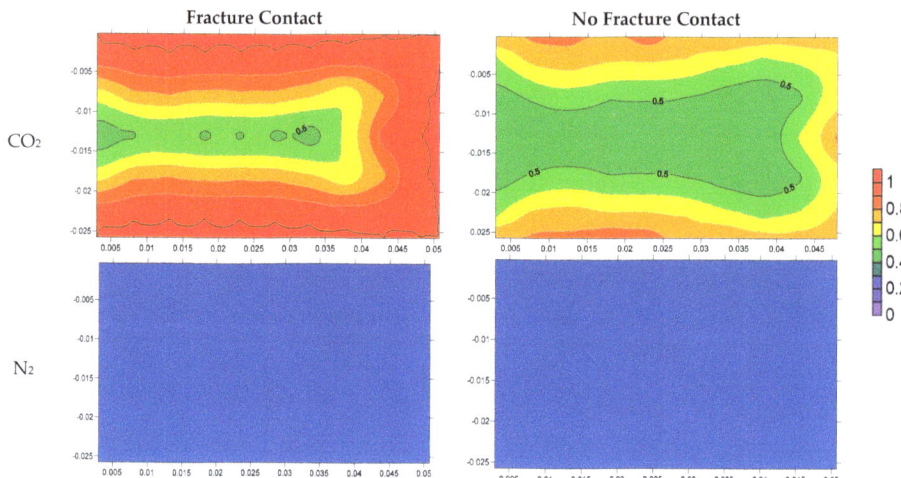

Figure 14. Mole fraction of injection fluid in oil in Case-4 in soaking process (ij profile).

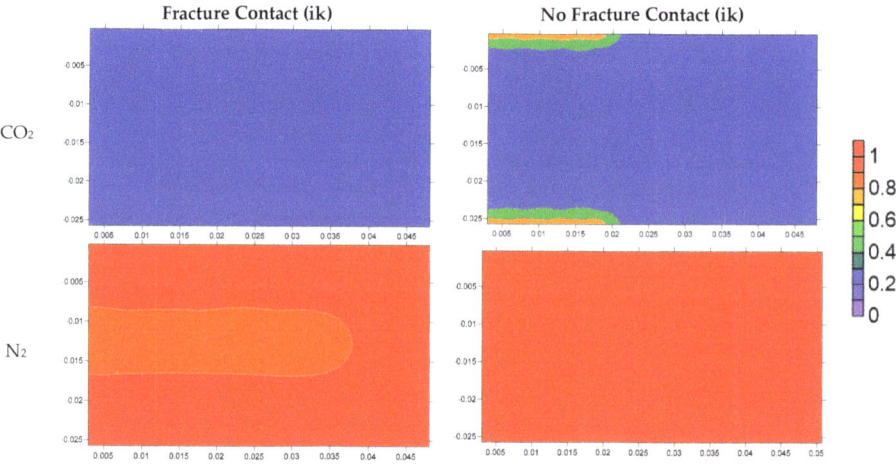

Figure 15. Mole fraction of injection fluid in gas in Case-4 in soaking process (ij profile).

Figure 16 shows the produced components of all three cases. CO_2 gives the most incremental C19 and C30 compared to depletion. However, in the water case, more C7 are produced, leaving a deprived heavy mass that causes an increase in the viscosity of the crude oil. N_2 has the least incremental effect on oil components in all carbon number ranges.

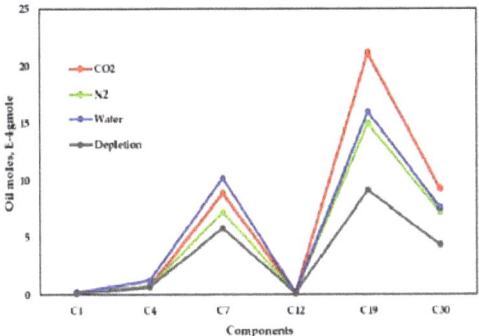

Figure 16. Production component comparison in Case-4.

- Oil swelling

The mixture of gas and crude oil will cause a volume expansion of crude oil. Figure 17 shows that the volume expansion of crude oil under reservoir conditions is most obvious after CO_2 injection, and N_2 has only a slight expansion effect on oil volume. When injecting water, the oil is compressed and its volume in reservoir condition decreased. Based on the expansion mechanism, CO_2 showed much better displacement performance than N_2 and water.

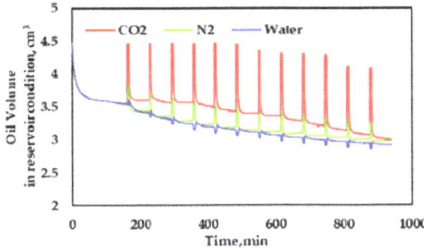
(**a**) oil volume in reservoir condition

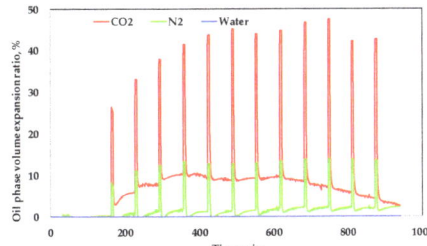
(**b**) oil volume expansion ratio in reservoir condition

Figure 17. Oil volume in reservoir conditions comparison in Case-4.

- Saturation profile and relative permeability profile

Another influential factor is the relative permeability of the oil phase. Figure 18 shows the relative permeability of the oil phase in the same production process in the 12th cycle. Figure 19 shows the oil, water and gas saturations at the same time. It can be seen with similar oil saturation, the relative oil relative permeability in the CO_2 and water case is much higher than the N_2 case. This is a consequence of the much lower relative permeability to liquid with the existence of gas phase. Based on the gas/liquid relative permeability curve, when two-phase flow occurs, the gas flows more easily than the liquid. Because the gas-liquid two-phase flow area is obviously narrower than the oil-water two-phase flow area (Figures 6 and 7).

From the saturation distribution of production stage in Figure 19, it can be seen that for gas huff-'n-puff, oil saturation in some areas is even higher than that in water huff-'n-puff. In water huff-'n-puff, because of the wider two-phase flow area of water and oil, the overall saturation decline area is larger. However, influenced by comprehensive mechanisms, water's EOR effect is still not as good as CO_2 with viscosity reduction mechanism, but better than N_2.

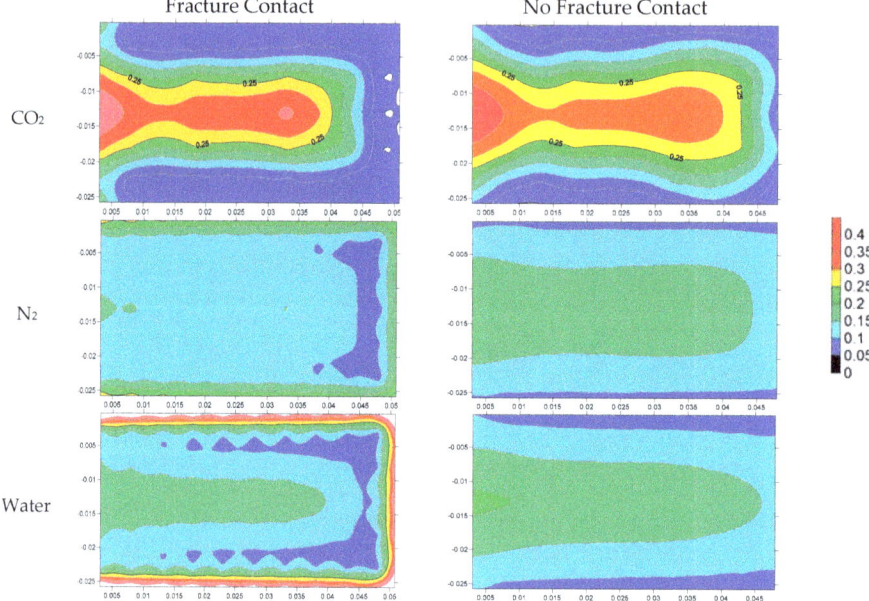

Figure 18. Oil relative permeability of different fluids in Case-4 in the production process (ij profile).

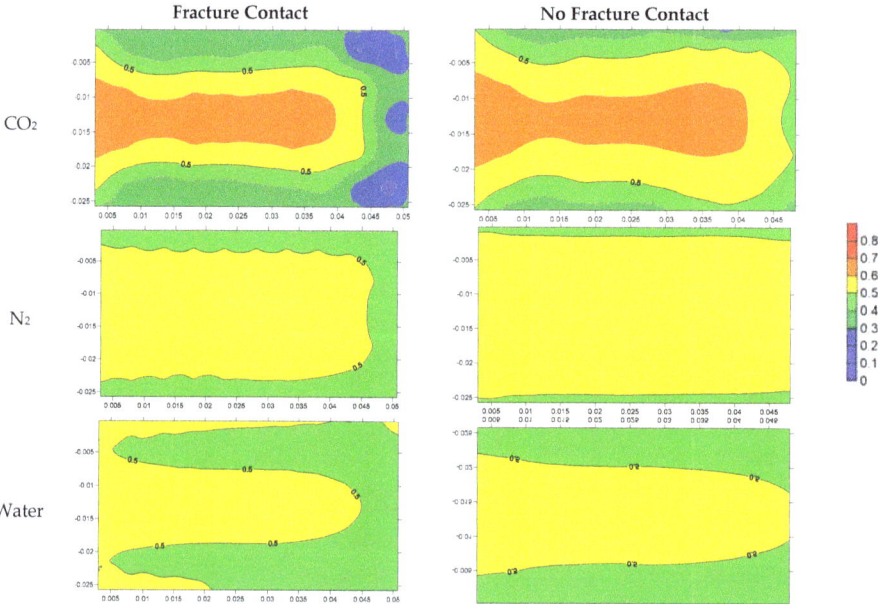

Figure 19. Oil saturation of different fluids in Case-4 in production process (ij profile).

- CO_2 diffusion

Figure 20 shows that the oil recovery considering the CO_2 diffusion effect at 0.005 cm^2/s is 1.42% higher than the cases without considering it. Figure 21 compares the CO_2 global mole fraction distribution profile with and without CO_2 diffusion. From the profile, the matrix grids have higher

CO_2 mole fraction with consideration of CO_2 diffusion, which indicates that more CO_2 may enter into the tight matrix from fractures to mix with the crude oil. Therefore, in order to accurately enhance the oil recovery effect of CO_2 in the tight reservoir simulation model, CO_2 diffusion effects cannot be ignored. On the basis of CO_2 diffusion to improve CO_2 EOR, a reasonable value of diffusion coefficient is critical for the reasonable evaluation of oil recovery in CO_2 huff-'n-puff.

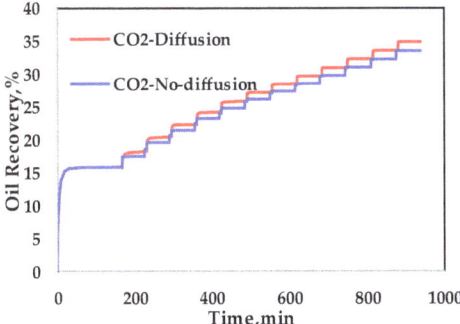

Figure 20. Oil recovery comparison of CO_2 huff-'n-puff with and without diffusion (Case-4).

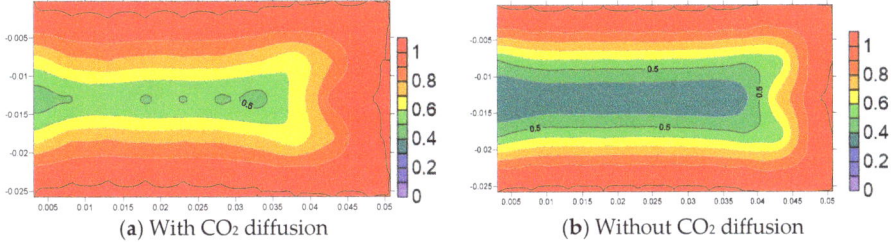

(**a**) With CO_2 diffusion   (**b**) Without CO_2 diffusion

Figure 21. CO_2 global mole fraction of CO_2 huff-'n-puff in Case-4 in soaking process (ij profile).

- Capillary imbibition

According to Meng's research, in the presence of fractures the capillary force can play a role of imbibition, which can also promote the drainage process [39]. In comparison (Figure 22), oil recovery considering the imbibition effect is 1.3% higher than cases without considering it. From the water saturation profile (Figure 23) the matrix grids have higher water saturation with consideration of imbibition, which indicates that more water in the fracture may enter the matrix, and then impact the oil drainage.

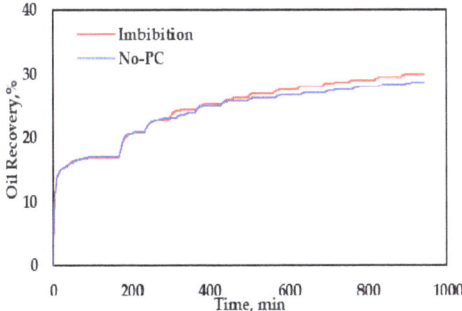

Figure 22. Oil recovery comparison of water huff-'n-puff with and without imbibition (Case-4).

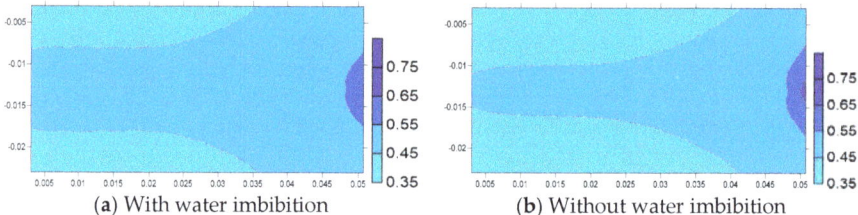

(a) With water imbibition   (b) Without water imbibition

Figure 23. Water saturation of water huff-'n-puff in Case-4 in soaking process (ij profile).

- Gravity

Another observation of the water saturation profiles is that water can be imbibed and is subject to density differences in the water case, which can displace more oil towards the fracture. Water is denser than oil, and N_2 is less dense than oil. Aggregation of different phase during the soaking period may affect the performance of huff-'n-puff as well. By setting the $Kv = 1$, the permeability anisotropy is eliminated, and the difference between Case-4 and Case-7 is dominated by gravity effects. It can be seen from Figure 24 that the gravity effect is almost negligible in the depletion mode and the gas huff-'n-puff mode. In water huff-'n-puff, the oil recovery finally has a difference of 0.95% in Case-4 and Case-7. By comparing the oil saturation profiles of different layers (Figure 25), there are slight differences between top and bottom layers in CO_2 and N_2 huff-'n-puff, but the differences are not significant. The difference of saturation between top and bottom layers is more obvious in water huff-'n-puff. Thus, the gravity effect can be neglected in core scale model. Whether there is a significant impact on large size will be verified and discussed in the authors' later studies.

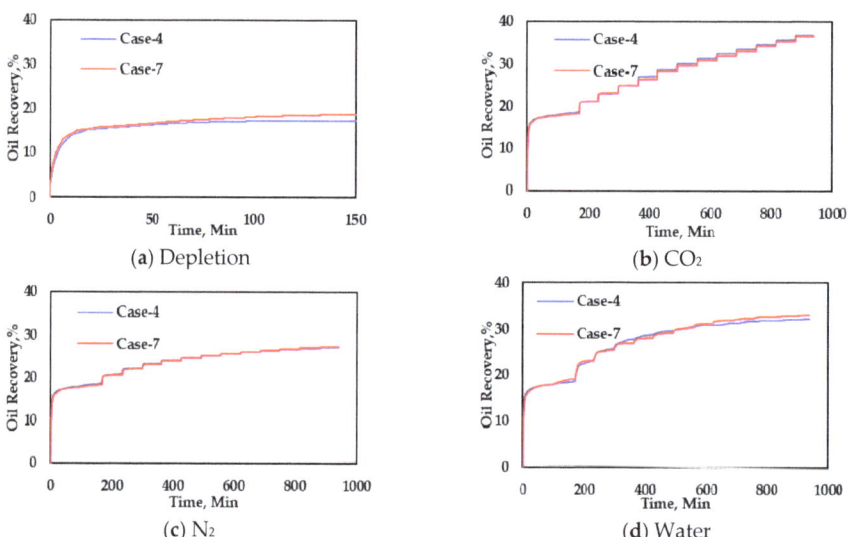

(a) Depletion   (b) CO_2
(c) N_2 (d) Water

Figure 24. Oil recovery comparison with gravity effect.

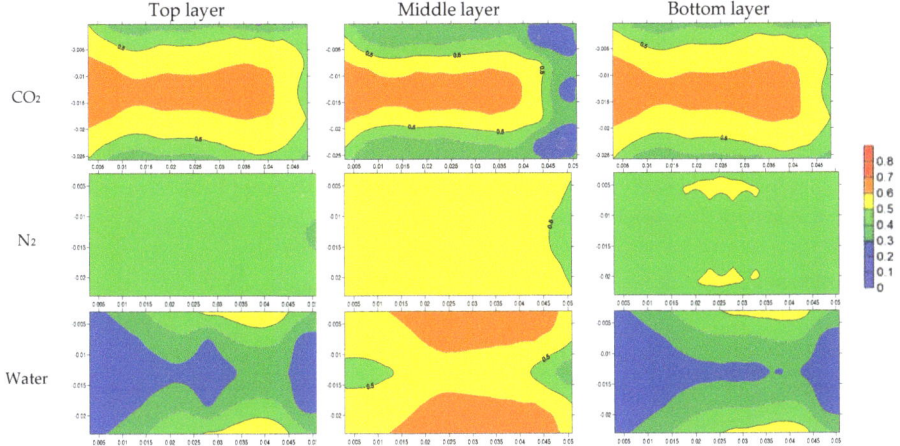

Figure 25. Oil saturation comparison with gravity effect (ij profile).

3.2. Fracture Characteristics

Now that the fundamental mechanisms of huff-'n-puff by using CO_2, N_2 and water are clear, it is of interest to understand how the distribution of fractures affects their performance in huff-'n-puff. For different fracture characteristics, the direction of scattered fracture, fracture in end face or lateral face, Scattered pattern of fractures, fracture relative location with well and fracture area are designed to mechanistically understand the impact of fracture characteristics on huff-'n-puff.

3.2.1. Scattered Pattern of Fractures

It is found that the dispersion characteristics of fractures has little effect on the recovery in N_2 and water huff-'n-puff. However, it has a great influence on the EOR effect in CO_2 case (Figure 26). When designing a fracture contact relationship, we compared four scattered fracture (Cases-1–4).

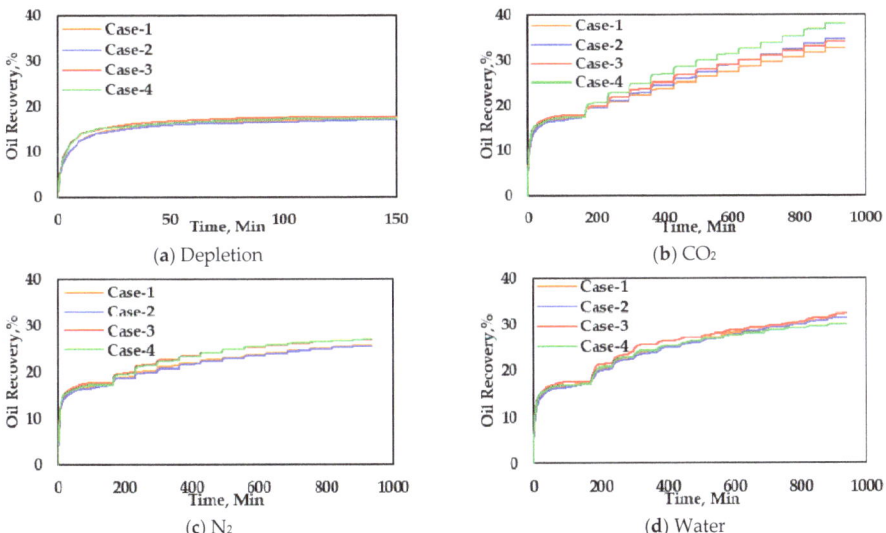

Figure 26. Oil recovery comparison between different scattered fractures.

In the depletion development mode, the final oil recoveries are not very different. In CO_2 huff-'n-puff, the oil recovery in Case-4 is the highest. The EOR effect in Case-1 is the worst. It can be seen from the comparison of CO_2 that the more scattered and farther in distance, the better the recovery effect of CO_2 huff-'n-puff. In N_2 huff-'n-puff cases, Case-3 and Case-4 are the best. In water huff-'n-puff cases, Case-4 is the best, followed by Case-3.

By comparing the oil viscosity distribution in different fractures (Figure 27), the fracture type in Case-4 can result in much less oil viscosity in the soaking process, and hence more oil can be produced. Therefore, the oil saturation is much lower in the Case-4 profile (Figure 28).

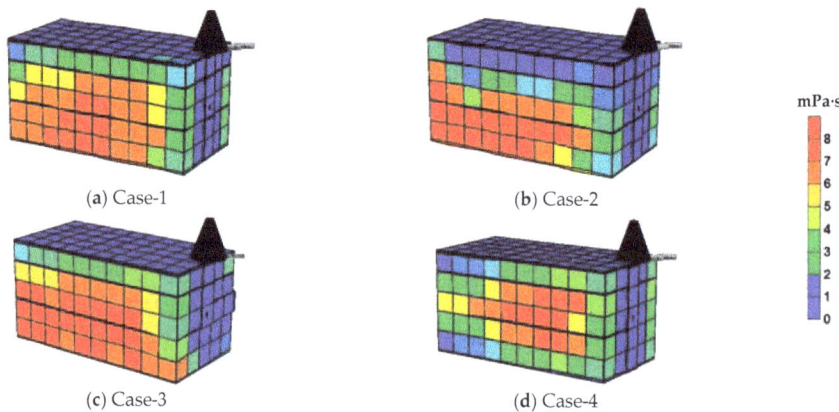

Figure 27. Oil viscosity comparison between different scattered fractures in the CO_2 soaking process.

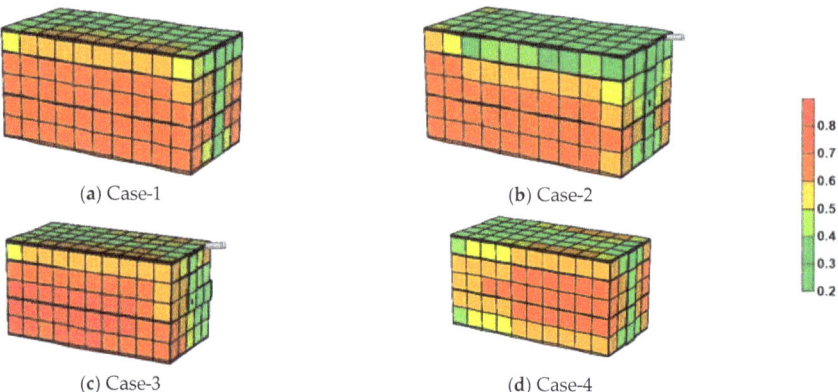

Figure 28. Oil Saturation comparison between different scattered fractures in CO_2 production process.

3.2.2. Direction of Scattered Fracture (with $Kv = 0.2$)

In this section, Case-4 and Case-7 are used to illustrate the mechanisms of permeability anisotropy by setting fractures in different directions. Figure 29 shows the oil recovery factor effect on surface conditions by using three different injecting fluids with fractures in different directions. The conclusion of this comparison is that permeability anisotropy plays a significant role in the water case (Case-7 is 4.38% higher than Case-4; but gravity has a negligible impact on CO_2 and N_2). The cross section of Cases-4 and -7 in k-j direction is specifically selected for analysis (Figures 30 and 31). From the viscosity comparison (Figure 30), the viscosity reduction area in Case-4 is larger than the area in Case-7.

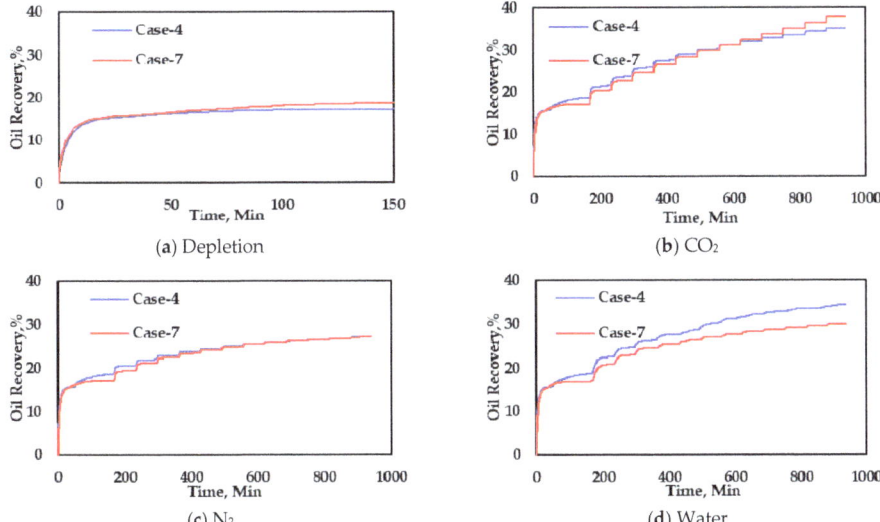

Figure 29. Oil recovery comparison between different directions of scattered fracture.

From the saturation profiles (Figure 31), it can be seen that in Case-4, more water can enter the matrix grid, but in Case-7 less oil can be displaced by water in this type fracture. As in the Kv and gravity effect, in Case-4, the oil recovery is significantly better than in Case-7. The fracture patterns in Cases-4 and -7 represent the typical horizontal well fracture and vertical well fracture. When it is extended to engineering applications, it can be considered that the horizontal well fractures are more effective than the vertical well fracture, and the horizontal fracture well is more suitable for water huff-n-puff. CO_2 huff-n-puff can be used both in horizontal and vertical fracture wells.

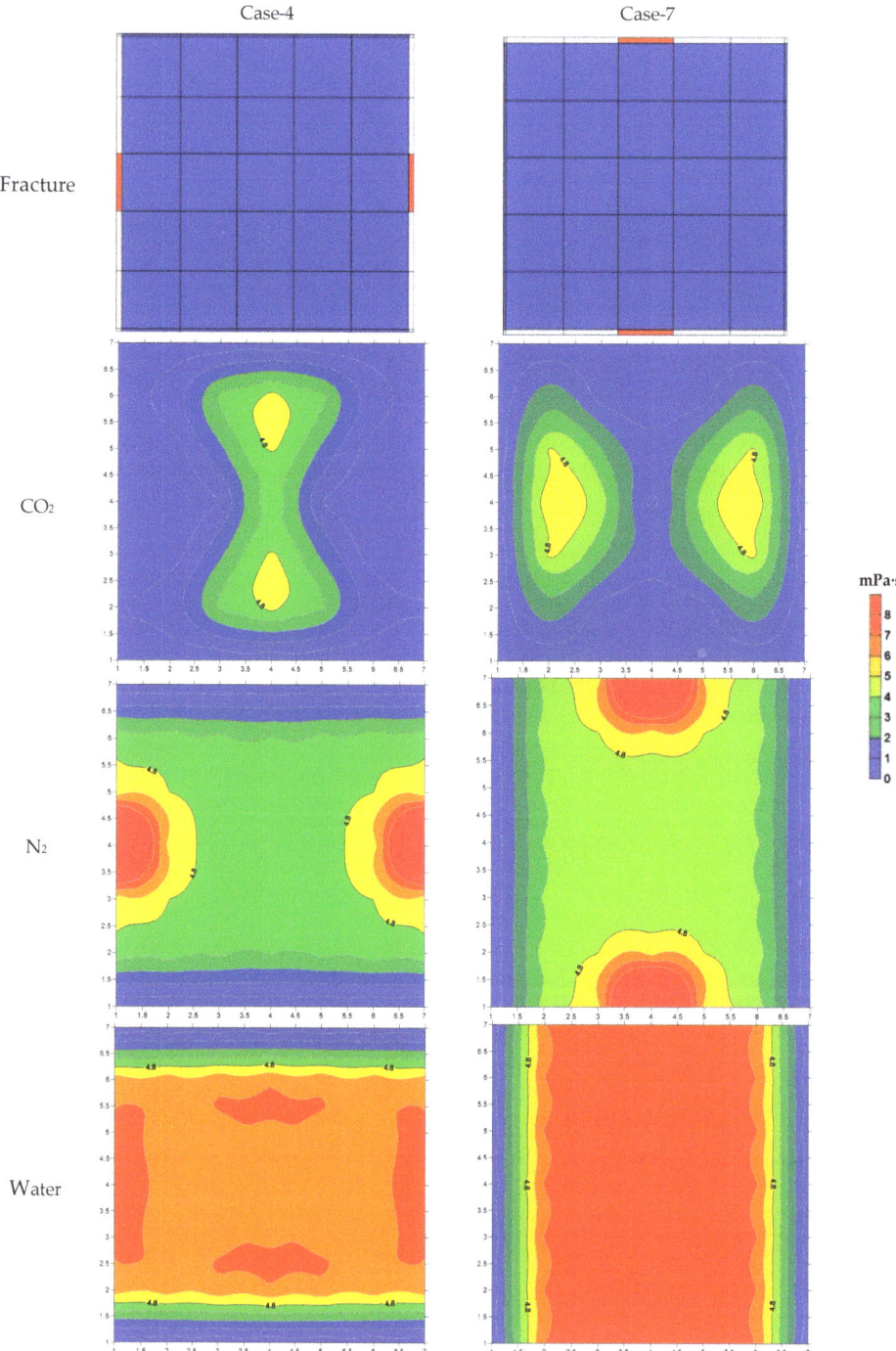

Figure 30. Oil viscosity of different fluids in soaking process (*jk* profile).

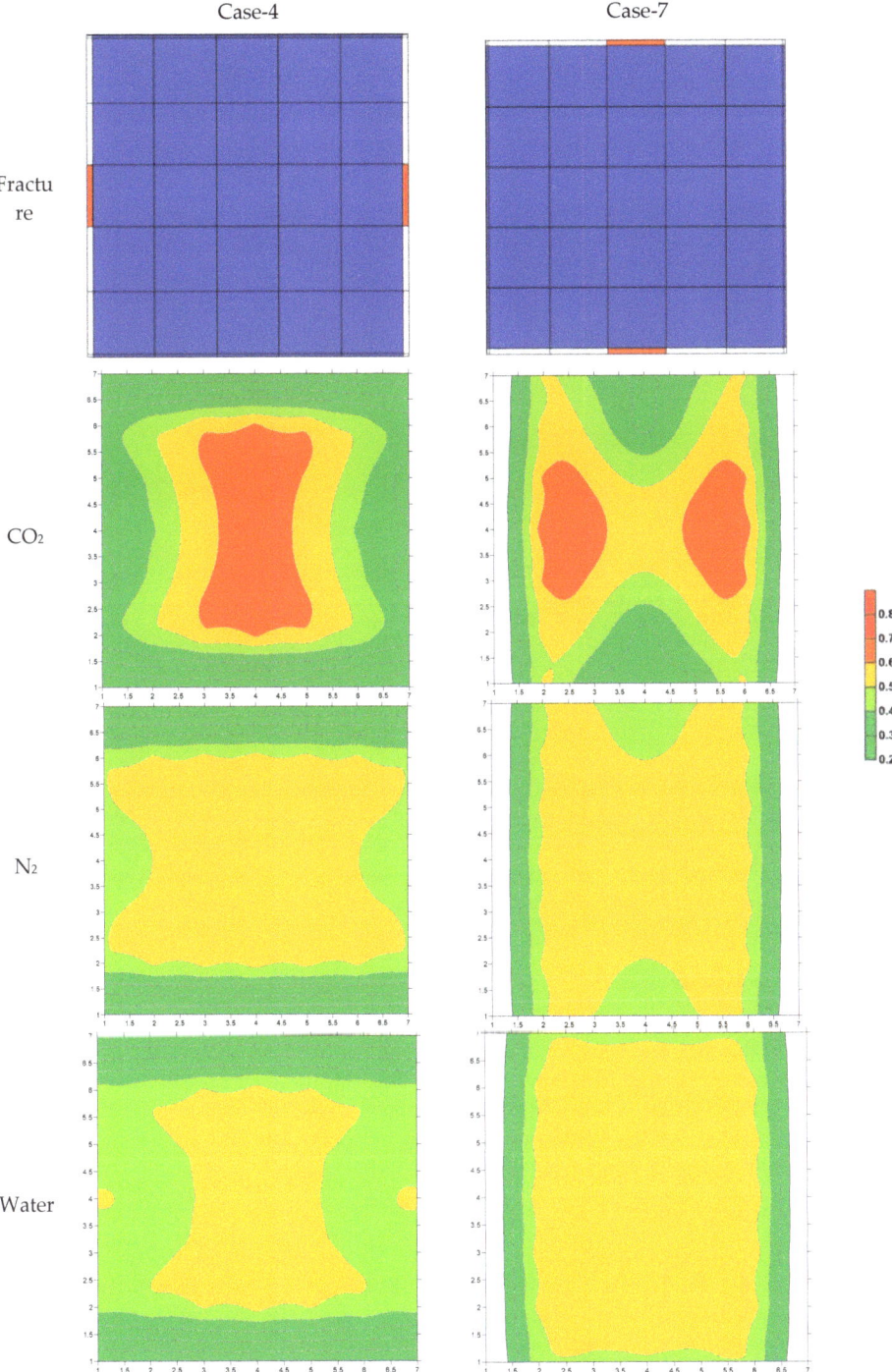

Figure 31. Oil saturation of different fluids in production process.

3.2.3. Fracture Near Well and Far Well

Due to the high conductivity of the fracture, the crude oil can flow into the well through the fractures, and the difference of oil recovery between the near-well and the far-well cases is small (Figure 32).

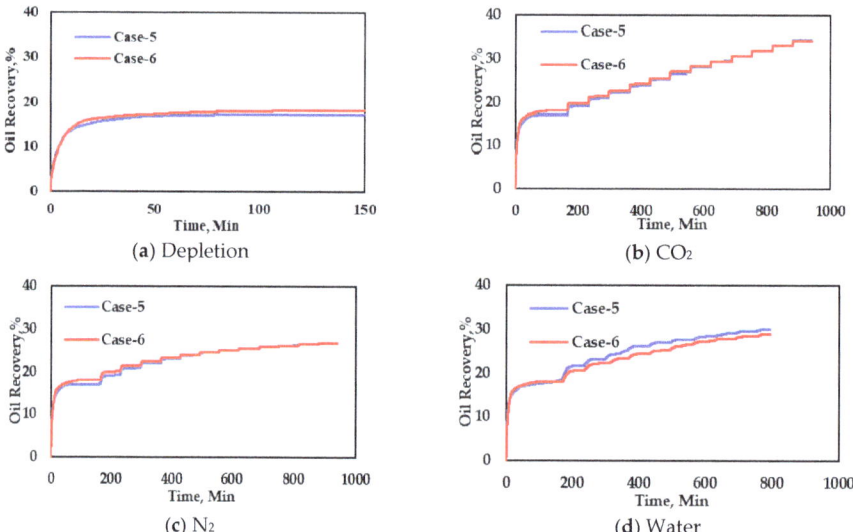

Figure 32. Oil recovery comparison between fractures near well and far well.

Comparing the EOR effects of near-well and far-well fracture in different development modes (Cases-5,6), it can be found that the results are not consistent either. For the depletion period, the far-well fracture (in Case-6) has an advantage over the near-well case (Case-5) by 1% higher oil recovery.

In the CO_2 and N_2 huff-''n-puff periods, the oil recovery of the two fracture types are close to each other. In water huff-'n-puff, the fracture in Case-5 has a higher incremental oil content. In the fourth round, the difference in oil recovery can even reach 3.1%, but in the later rounds, the difference is less than 0.74%.

3.2.4. Fracture Contact Area

Comparing the relationship between fracture contact area and oil recovery (Figure 33), it can be found that in the depletion period, the larger the contact area of the fracture, the higher the oil recovery, but when the contact area is greater than 25, the increase of the oil recovery is obviously slowed down. In the CO_2 huff-'n-puff, oil recovery is the highest when the contact area is 25. In N_2 and water huff-'n-puff, there is still not a clear trend between the contact area and oil recovery. Therefore, the contact area is not sensitive for the huff-'n-puff process based on this research.

3.2.5. Components Analysis

Through the previous comparison of oil recovery, viscosity, phase permeability and mobility, the EOR effect of different fluids in different fracture can be examined and compared macroscopically. In order to analyze the microscopic differences in more detail, we also selected Cases-4, 5 and 8 as the comparison experiments, adding the analysis of EOR produced components of different fluids in different fracture modes.

As can be seen from the figure (Figure 34), the mainly produced crude oil components include C7, C19 and C30. The proportion of C1 and C4 components in crude oil is originally small, and their production amount did not differ significantly between different fractures. First, for C7 analysis, the most component C7 can be produced in water huff-'n-puff, followed by CO_2, with the least amount of C7 produced during N_2 huff-'n-puff. For C19 and C30, CO_2 has a better drainage effect on them, followed by water, and N_2 is still the worst. This shows that different fluids have different drainage effects on different oil components. More interestingly, the fracture condition also has an effect on the composition of the components produced. Taking C19 as an example, C19 can be produced much easier in Case-5 in CO_2 huff-'n-puff. For water, the fracture in Case-8 is more conducive to the production of C19. This indicates that the fracture distribution pattern also has a non-negligible effect on the EOR effect of different fluids.

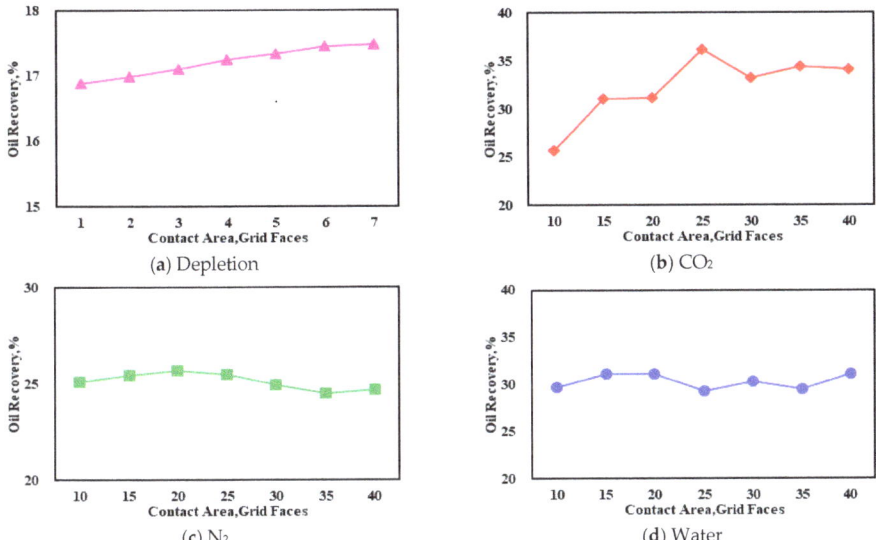

Figure 33. Relationship between contact area and oil recovery in different modes.

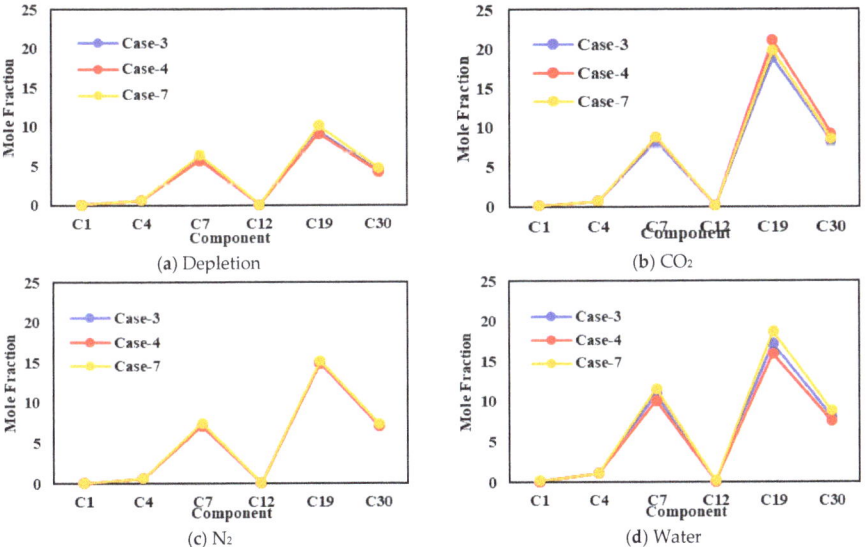

Figure 34. Production component comparison in different cases.

Finally, it should be noted that the main EOR effect is still due to the fluid. Fractures only provide a more efficient and convenient channel and contact relationship for fluids to exert their EOR effects. Finally, in the fluid selection of tight reservoir EOR huff-'n-puff, it is necessary to accurately and reasonably evaluate the distribution characteristics of fractures, and select the optimal injected fluid for specific fracture morphology.

4. Conclusions

This paper takes a fractured tight reservoir as the basic research object. Natural tight cores from the J field in China are used to conduct experimental studies on different fluid huff-'n-puff processes. A new experimental core-level fracture simulation method that is easy to operate and reusable is proposed. Three different injecting fluids are used, including CO_2, N_2 and water. The equivalent core scale reservoir numerical models in depletion and huff-'n-puff mode are then restored by numerical simulation with CMG GEM. Simulation cases with eight different fracture patterns were used in the detail study to understand how fracture mechanistically impact EOR in huff and puff for different injecting fluids including CO_2, N_2 and water. Finally, the key findings can be summarized as follows:

(1) In core-scale fractures, a new experimental core-level fracture simulation method that is easy to operate and reusable is proposed. In this method, woven metallic wires are attached to the outer surface of the core to create a space between the core holder and the core as a high permeability zone, equivalent to a fracture. This avoids the defects of single fracture shape, sanding, and difficulty in reusing the core. Different spatial arrangement of fractures can be set up in core-scale experiments.

(2) Based on the core-scale experiments and numerical simulation analyses, the existence of fractures is significant for the enhanced oil process in tight reservoirs. The presence of fractures not only has a significant effect on the depletion mode, but also has an important effect on the enhanced oil recovery (EOR) performance of the huff-'n-puff method. After comparison, fracture arrangement in space has most impact on CO_2 huff-'n-puff, followed by water and finally N_2.

(3) In the depletion mode, crude oil flows mainly due to the pressure drop caused by elastic energy, and there is no contact between other fluids and crude oil in the matrix through fractures. Therefore, in the case of depletion mode, when fractures have the same contact area with spatial arrangement, the obvious difference in oil recovery cannot be observed.

(4) CO_2 has more advantages over water and N_2 in tight reservoir with huff-'n-puff. Through the EOR mechanism analysis, CO_2 is the only fluid that is miscible with oil (even a 90% mole fraction with CO_2 in the oil phase is possible). This solubility of CO_2 results in the lowest oil phase viscosity in the three cases. The CO_2 diffusion mechanism is pronounced. In contrast, N_2 mainly exists in the gas phase and is much less soluble in the oil phase compared to CO_2, however, it still has a certain swelling effect on crude oil. Water does not dissolve in the oil and water mainly enhances oil recovery through capillary force, gravity and elasticity displacement.

(5) When considering the anisotropy, the direction of fracture will have a greater impact on oil recovery for the water huff-'n-puff. The conclusion from this comparison is that permeability anisotropy plays a significant role in the water case (Case-7 is 4.38% higher than Case-4; but anisotropy has a negligible impact on CO_2 and N_2).

(6) When it comes to the relative location between well and fracture, for the depletion development mode, the far-well fracture has an advantage over the near well case by 1% higher oil recovery. In the CO_2 and N_2 huff-'n-puff modes, the oil recovery of the two fracture types are not much different. In water huff-'n-puff, the fractures near the well have a better effect on the EOR process.

(7) The fracture contact area is not sensitive for the huff-'n-puff process for N_2 and water cases based on this research.

(8) From analysis of the produced components, CO_2 gives the most incremental C19 and C30 compared to depletion. However, in the water case, more C7 are produced, leaving a depleted

heavy mass that causes an increase in the viscosity of the crude oil. N_2 has the least incremental effect oil components for all carbon number ranges.

(9) Finally, the main EOR effect is mainly applied by the fluid. Fractures only provide a more efficient and convenient channel, and offer a better contact relationship for fluids to exert their EOR effects. In the fluid selection of tight reservoir EOR huff-n-puff, it is necessary to accurately and reasonably evaluate the distribution characteristics of fractures, and select the optimal injected fluid for specific fracture morphology.

Author Contributions: Conceptualization, J.B. and H.L.; Methodology, J.B.; Software, J.B.; Validation, L.Y. and F.C.; Formal Analysis, H.L.; Investigation, J.B.; Resources, G.Q. and Y.P.; Data Curation, Y.G.; Writing—Original Draft Preparation, J.B.; Writing—Review & Editing, H.L.; Visualization, J.B.; Supervision, J.W.; Project Administration, G.Q.; Funding Acquisition, G.Q.

Funding: This research was funded by National Program on Key Basic Research Project (No. 2015CB250906) and Innovation Fund of CNPC (No. 2015D-5006-0204).

Acknowledgments: We thank the teams at PetroChina Xinjiang Oilfield and Beijing RIPED.

Conflicts of Interest: The authors declare no conflict of interest.

References

1. Pollastro, R.M. Total petroleum system assessment of undiscovered resources in the giant Barnett Shale continuous (unconventional) gas accumulation, Fort Worth Basin, Texas. *AAPG Bull.* **2007**, *91*, 551–578. [CrossRef]
2. Miller, B.; Paneitz, J.; Mullen, M.; Meijs, R.; Tunstall, K.; Garcia, M. The Successful Application of a Compartmental Completion Technique Used to Isolate Multiple Hydraulic-Fracture Treatments in Horizontal Bakken Shale Wells in North Dakota. In Proceedings of the SPE Annual Technical Conference and Exhibition, Denver, CO, USA, 21–24 September 2008.
3. Loucks, R.G.; Reed, R.M.; Ruppel, S.C.; Jarvie, D.M. Morphology, Genesis, and Distribution of Nanometer-Scale Pores in Siliceous Mudstones of the Mississippian Barnett Shale. *J. Sediment. Res.* **2009**, *79*, 848–861. [CrossRef]
4. Jia, C.; Caineng, Z.; Jianzhong, L.; Denghua, L.; Min, Z. Evaluation criteria, main types, basic characteristics and resource prospects of China's tight oil. *Acta Pet. Sin.* **2012**, *33*, 343–350.
5. Hawthorne, S.B.; Gorecki, C.D.; Sorensen, J.A.; Miller, D.J.; Harju, J.A.; Melzer, L.S. Hydrocarbon Mobilization Mechanisms from Upper, Middle, and Lower Bakken Reservoir Rocks Exposed to CO. In Proceedings of the SPE Unconventional Resources Conference Canada, Calgary, AB, Canada, 5–7 November 2013.
6. Jinhu, D.; Haiqing, H.; Jianzhong, L.; Fuji, H.; Bingcheng, G.; Weipeng, Y. Progress in China's Tight Oil Exploration and Challenges. *China Pet. Explor.* **2014**, *19*, 1–9.
7. Rungamornrat, J.; Wheeler, M.F.; Mear, M.E. Coupling of Fracture/Non-Newtonian Flow for Simulating Nonplanar Evolution of Hydraulic Fractures. In Proceedings of the SPE Annual Technical Conference and Exhibition, Dallas, TX, USA, 9–12 October 2005.
8. Kabir, S.; Rasdi, F.; Igboalisi, B.O. Analyzing Production Data from Tight Oil Wells. *J. Can. Pet. Technol.* **2011**, *50*, 48–58. [CrossRef]
9. West, D.R.; Harkrider, J.D.; Barham, M.; Besler, M.R.; Mahrer, K.D. Optimized Production in the Bakken Shale: South Antelope Case Study. In Proceedings of the SPE Unconventional Resources Conference Canada, Calgary, AB, Canada, 5–7 November 2013.
10. Wang, H.; Liao, X.; Lu, N.; Cai, Z.; Liao, C.; Dou, X. A study on development effect of horizontal well with SRV in unconventional tight oil reservoir. *J. Energy Inst.* **2014**, *87*, 114–120. [CrossRef]
11. Wei, Y.; Ran, Q.; Ran, L.I.; Yuan, J.; Dong, J. Determination of dynamic reserves of fractured horizontal wells in tight oil reservoirs by multi-region material balance method. *Pet. Explor. Dev.* **2016**, *43*, 490–498. [CrossRef]
12. Sohrabi, M.; Tehrani, D.H.; Danesh, A.; Henderson, G.D. Visualisation of Oil Recovery by Water Alternating Gas (WAG) Injection Using High Pressure Micromodels—Oil-Wet & Mixed-Wet Systems. In Proceedings of the SPE Annual Technical Conference and Exhibition, New Orleans, LA, USA, 30 September–3 October 2001.

13. Ghaderi, S.M.; Clarkson, C.R.; Kaviani, D. Evaluation of Recovery Performance of Miscible Displacement and WAG Processes in Tight Oil Formations. In Proceedings of the SPE/EAGE European Unconventional Resources Conference and Exhibition, Society of Petroleum Engineers, Vienna, Austria, 20–22 March 2012.
14. Holtz, M.H. Immiscible Water Alternating Gas (IWAG) EOR: Current State of the Art. In Proceedings of the SPE Improved Oil Recovery Conference, Tulsa, OK, USA, 11–13 April 2016.
15. Yu, W.; Lashgari, H.R.; Wu, K.; Sepehrnoori, K. CO_2 injection for enhanced oil recovery in Bakken tight oil reservoirs. *Fuel* **2015**, *159*, 354–363. [CrossRef]
16. Hoffman, B.T. Comparison of Various Gases for Enhanced Recovery from Shale Oil Reservoirs. In Proceedings of the SPE Improved Oil Recovery Symposium, Society of Petroleum Engineers, Tulsa, OK, USA, 14–18 April 2012.
17. Arshad, A.; Al-Majed, A.A.; Menouar, H.; Muhammadain, A.M.; Mtawaa, B. Carbon Dioxide (CO_2) Miscible Flooding in Tight Oil Reservoirs: A Case Study. In Proceedings of the Kuwait International Petroleum Conference and Exhibition, Kuwait City, Kuwait, 14–16 December 2009.
18. Song, C.; Yang, D. Performance Evaluation of CO_2 Huff-n-Puff Processes in Tight Oil Formations. In Proceedings of the SPE Unconventional Resources Conference Canada, Calgary, AB, Canada, 5–7 November 2013.
19. Wang, H.; Liao, X.; Lu, N.; Cai, Z.; Liao, C.; Dou, X. The Study of CO_2 Flooding of Horizontal Well with SRV in Tight Oil Reservoir. In Proceedings of the SPE Energy Resources Conference, Port of Spain, Trinidad and Tobago, 9–11 June 2014.
20. Ren, B.; Ren, S.; Zhang, L.; Huang, H.; Chen, G.; Zhang, H. Monitoring on CO_2 Migration in a Tight Oil Reservoir during CO_2-EOR Process. In Proceedings of the Carbon Management Technology Conference, Carbon Management Technology Conference: Sugar Land, TX, USA, 17–19 November 2015.
21. Sheng, J.J. Enhanced oil recovery in shale reservoirs by gas injection. *J. Nat. Gas Sci. Eng.* **2015**, *22*, 252–259. [CrossRef]
22. Yang, Z.; Liu, X.; Zhang, Z.H.; Zhou, T.; Zhao, S.X. Physical simulation of CO_2 injection in fracturing horizontal wells in tight oil reservoirss. *Acta Pet. Sin.* **2015**, *36*, 724–729.
23. Torabi, F.; Asghari, K. Performance of CO_2 Huff-and-Puff Process in Fractured Media (Experimental Results). In Proceedings of the Canadian International Petroleum Conference, Petroleum Society of Canada: Calgary, AB, USA, 12–14 June 2007; p. 11.
24. Gamadi, T.D.; Sheng, J.J.; Soliman, M.Y. An Experimental Study of Cyclic CO_2 Injection to Improve Shale Oil Recovery. In Proceedings of the SPE Improved Oil Recovery Symposium, Society of Petroleum Engineers: Tulsa, OK, USA, 12–16 April 2014; p. 9.
25. Ma, J.; Wang, X.Z.; Gao, R.M.; Zeng, F.H.; Huang, C.X.; Tontiwachwuthikul, P.; Liang, Z.W. Enhanced light oil recovery from tight formations through CO_2 huff 'n' puff processes. *Fuel* **2015**, *154*, 35–44. [CrossRef]
26. Todd, H.B.; Evans, J.G. Improved Oil Recovery IOR Pilot Projects in the Bakken Formation. In Proceedings of the SPE Low Perm Symposium, Society of Petroleum Engineers: Denver, CO, USA, 5–6 May 2016; p. 22.
27. Wang, L.; Tian, Y.; Yu, X.Y.; Wang, C.; Yao, B.W.; Wang, S.H.; Winterfeld, P.H.; Wang, X.; Yang, Z.Z.; Wang, Y.H. Advances in improved/enhanced oil recovery technologies for tight and shale reservoirs. *Fuel* **2017**, *210*, 425–445. [CrossRef]
28. Hawthorne, S.B.; Gorecki, C.D.; Sorensen, J.A.; Steadman, E.N.; Harju, J.A.; Melzer, S. Hydrocarbon Mobilization Mechanisms Using CO_2 in an Unconventional Oil Play. *Energy Procedia* **2014**, *63*, 7717–7723. [CrossRef]
29. Yu, W.; Lashgari, H.; Sepehrnoori, K. Simulation Study of CO_2 Huff-n-Puff Process in Bakken Tight Oil Reservoirs. In Proceedings of the SPE Western North American and Rocky Mountain Joint Meeting, Denver, CO, USA, 17–18 April 2014.
30. Tovar, F.D.; Eide, O.; Graue, A.; Schechter, D. Experimental Investigation of Enhanced Recovery in Unconventional Liquid Reservoirs using CO_2: A Look Ahead to the Future of Unconventional EOR. In Proceedings of the SPE Unconventional Resources Conference, Society of Petroleum Engineers: The Woodlands, TX, USA, 1–3 April 2014; p. 9.
31. Vega, B.; O'Brien, W.J.; Kovscek, A.R. Experimental Investigation of Oil Recovery from Siliceous Shale by Miscible CO_2 Injection. In Proceedings of the SPE Annual Technical Conference and Exhibition, Society of Petroleum Engineers: Florence, Italy, 19–22 September 2010; p. 21.

32. Gamadi, T.D.; Sheng, J.J.; Soliman, M.Y. An Experimental Study of Cyclic Gas Injection to Improve Shale Oil Recovery. In Proceedings of the SPE Annual Technical Conference and Exhibition, New Orleans, LA, USA, 30 September–2 October 2013; p. 9.
33. Lei, L.; Sheng, J.J. Upscale Methodology for Gas Huff-n-Puff Process in Shale Oil Reservoirs. *J. Pet. Sci. Eng.* **2017**, *153*, 36–46. [CrossRef]
34. Jing, W. Investigations on spontaneous imbibition and the influencing factors in tight oil reservoirs. *Fuel* **2019**, *236*, 755–768. [CrossRef]
35. Jing, W.; Huiqing, L.; Genbao, Q.; Yongcan, P.; Yang, G. Experimental investigation on water flooding and continued EOR techniques in buried-hill metamorphic fractured reservoirs. *J. Pet. Sci. Eng.* **2018**, *171*, 529–541.
36. Kurtoglu, B.; Salman, A. How to Utilize Hydraulic Fracture Interference to Improve Unconventional Development. In Proceedings of the Abu Dhabi International Petroleum Exhibition and Conference, Society of Petroleum Engineers: Abu Dhabi, The United Arab Emirates, 9–12 November 2015; p. 18.
37. Wei, B.; Li, Q.; Jin, F.; Li, H.; Wang, C. The Potential of a Novel Nanofluid in Enhancing Oil Recovery. *Energy Fuels* **2016**, *30*, 2882–2891. [CrossRef]
38. Meng, Q.; Cai, Z.X.; Cai, J.C.; Yang, F. Oil recovery by spontaneous imbibition from partially water-covered matrix blocks with different boundary conditions. *J. Pet. Sci. Eng.* **2019**, *172*, 454–464. [CrossRef]
39. Weng, X.W.; Kresse, O.; Cohen, C.E.; Wu, R.T.; Gu, H.R. Modeling of Hydraulic-Fracture-Network Propagation in a Naturally Fractured Formation. *SPE Prod. Oper.* **2011**, *26*, 368–380.
40. Zuloaga-Molero, P.; Yu, W.; Xu, Y.; Sepehrnoori, K.; Li, B. Simulation Study of CO_2-EOR in Tight Oil Reservoirs with Complex Fracture Geometries. *Sci. Rep.* **2016**, *6*, 33445. [CrossRef] [PubMed]
41. CMG. *CMG: User's Guide GEM*; Computer Modeling Group, Ltd.: Calgary, AB, Canada, 2015.
42. Sigmund, P.M. Prediction of Molecular Diffusion at Reservoir Conditions. Part II—Estimating the Effects of Molecular Diffusion and Convective Mixing in Multicomponent Systems. *J. Can. Pet. Technol.* **1976**, *15*, 11. [CrossRef]
43. Reid, R.C.; Sherwood, T.K. The properties of gases and liquids: Their estimation and correlation. *Phys. Today* **1959**, *12*, 38–40. [CrossRef]
44. Zhang, Y.; Yu, W.; Li, Z.P.; Sepehrnoori, K. Simulation study of factors affecting CO_2 Huff-n-Puff process in tight oil reservoirs. *J. Pet. Sci. Eng.* **2018**, *163*, 264–269. [CrossRef]
45. Li, S.; Li, Z.; Dong, Q. Diffusion coefficients of supercritical CO_2 in oil-saturated cores under low permeability reservoir conditions. *J. CO_2 Util.* **2016**, *14*, 47–60. [CrossRef]

© 2019 by the authors. Licensee MDPI, Basel, Switzerland. This article is an open access article distributed under the terms and conditions of the Creative Commons Attribution (CC BY) license (http://creativecommons.org/licenses/by/4.0/).

Article

The Percolation Properties of Electrical Conductivity and Permeability for Fractal Porous Media

He Meng [1,2], Qiang Shi [3], Tangyan Liu [1,*], FengXin Liu [3] and Peng Chen [3]

1. State Key Lab of Marine Geology, Tongji University, Shanghai 200092, China; 1610887@tongji.edu.cn
2. College of Earth and Mineral Sciences, The Pennsylvania State University, University Park, PA 16802, USA
3. Research Institute of Petroleum Exploration and Development, PetroChina, Langfang 065007, China; shiqiang69@petrochina.com.cn (Q.S.); lfx69@petrochina.com.cn (F.X.L.); chenpeng52169@petrochina.com.cn (P.C.)
* Correspondence: tyliu05169@tongji.edu.cn; Tel.: +86-1376-451-9890

Received: 13 February 2019; Accepted: 15 March 2019; Published: 21 March 2019

Abstract: Many cases have indicated that the conductivity and permeability of porous media may decrease to zero at a nonzero percolation porosity instead of zero porosity. However, there is still a lack of a theoretical basis for the percolation mechanisms of the conductivity and permeability. In this paper, the analytical percolation expressions of both conductivity and permeability are derived based on fractal theory by introducing the critical porosity. The percolation models of the conductivity and permeability were found to be closely related to the critical porosity and microstructural parameters. The simulation results demonstrated that the existence of the critical could lead to the non-Archie phenomenon. Meanwhile, the increasing critical porosity could significantly decrease the permeability and the conductivity at low porosity. Besides, the complex microstructure could result in more stagnant pores and a higher critical porosity. This study proves the importance of the critical porosity in accurately evaluating the conductivity and permeability, and reveals the percolation mechanisms of the conductivity and permeability in complex reservoirs. By comparing the predicted conductivity and permeability with the available experimental data, the validity of the proposed percolation models is verified.

Keywords: percolation model; fractal theory; microstructure; critical porosity; conductivity; permeability

---

1. Introduction

Electrical conductivity and permeability are the crucial macroscopic parameters in characterizing current and fluid flow behaviors in porous media, and have numerous applications in the geology, petroleum, and chemical engineering, hydrology, and soil science. Although electrical conductivity and permeability have been studied over the past decades, modeling and predicting conductivity and permeability are still great challenges. The inherent complexity of both conductivity and permeability is due to its dependence on microscopic, pore-scale properties such as connectivity, tortuosity, and pore sizes. Because the conductivity and permeability of a porous medium are strongly affected by the pore structure, the reliable theoretical assessment of the conductivity and permeability based on the medium structural characteristics is important.

Due to the complex geometric microstructure and multiscale pore structure [1–3], it is difficult to use conventional geometric methods to accurately describe the conductivity and permeability of porous media. Since fractal theory was introduced by Mandelbrot [4], it has contributed significantly to the research on the rock-electric and seepage characteristics in complex reservoirs [5–13]. Direct experimental measurements have been applied to the study of fractal characterizations [14], and fractal analysis has become a powerful tool for quantifying the irregularity and complexity of porous media [15–21]. In the application, there are numerous examples that present the application of fractal theory to analyze porous media. Roy and Tarafdar [22] simulated a realistic Archie exponent by using a three-dimensional fractal model, and used a random walk parameter to show the fractal nature of the pore space. Coleman and Vassilicos [11,12] further proved the validity of that method. Nigmatullin et al. [23] assumed that the minimum pore size involved in electrical conduction through saturated rocks is the same as the minimum pore size in the fractal pore structure, and presented a fractal model that led to the Archie's equation. By means of the fractal geometry method, Wei et al. [24] derived an analytical model for electrical conductivity that has a direct physical basis. In addition, Pitchumani and Ramakrishnan [25] gave the permeability in terms of pore fractal dimension and tortuosity fractal dimension, which respectively describe the size distribution and the tortuous degree of the capillaries. Yu and Li [21] deduced a unified model to study the fractal character of porous media, and proposed a criterion to determine whether a porous medium can be characterized by fractal theory. Xu and Yu [26] developed a new form of permeability and Kozeny–Carman constant by means of fractal geometry. However, several observations and studies have suggested that there exists a percolation porosity below which the remaining porosity is disconnected and does not contribute to current and fluid flow [27–29]. Many traditional and recently presented fractal models have been unable to explain this phenomenon. Many experimental studies have shown that the percolation behavior of pore space is one of the most important characteristics, having a huge influence on the conductivity and permeability. Several empirical percolation models were proposed by the analysis of the experimental data to model and study the electrical and fluid transport properties in porous media [30–38], but a theoretical basis for the percolation mechanisms of the conductivity and permeability is still lacking.

In order to develop a good understanding of the percolation mechanisms of the conductivity and permeability in porous media (especially in tight sandstones and microporous carbonates), in this paper, the analytical expressions are developed using the fractal geometry theory to link the conductivity and permeability to the critical porosity and microstructural parameters. The objectives of this work were to provide the theoretical basis for the percolation mechanisms of the conductivity and permeability, study the significant effect of the critical porosity and microstructure on the rock-electric and seepage characteristics, and estimate the conductivity and permeability more accurately. Both the conductivity and permeability percolation models are experimentally verified.

2. Theoretical Formulation

2.1. Fractal Characteristics of Porous Media

In porous media, studies have shown that pore microstructure (e.g., the pore size distribution) displays fractal characteristics [39–42]. It is assumed that porous media consist of tortuous capillaries with variable lengths and diameters. According to fractal theory, the number of cumulative capillaries has been proven to obey the fractal scaling law [43]:

$$N(Diameter \geq r) = \left(\frac{r_{max}}{r}\right)^{D_f}, \quad (1)$$

where N is the number of capillaries whose diameters are greater than or equal to r, D_f is the fractal dimension for pore space, the typical range of D_f is $1 < D_f < 2$ in two dimensions or $2 < D_f < 3$ in three dimensions, and r_{max} is the maximum diameter of capillaries.

Replacing r with r_{min} in Equation (1) yields the total number of pores/capillaries:

$$N_t(Diameter \geq r_{min}) = \left(\frac{r_{max}}{r_{min}}\right)^{D_f}, \quad (2)$$

where N_t is the total number of capillaries in porous media, and r_{min} is the minimum diameter of capillaries.

For fractal porous media, pore size distribution is considered as a continuous and differentiable function. Through differentiating Equation (1) with respect to r, one can obtain the number of capillaries/pores whose sizes are within the infinitesimal range r to $r + dr$:

$$-dN = D_f r_{max}^{D_f} r^{-D_f-1} dr. \qquad (3)$$

The negative sign on the left side means that the number of pores decreases with increasing pore diameter. Dividing Equation (3) by Equation (2) yields the probability density function of pore size distribution [44]:

$$f(r) = -dN/N_t = D_f r_{min}^{D_f} r^{-D_f-1}. \qquad (4)$$

According to probability theory, the integration result of $f(r)$ should satisfy the normalization relationship:

$$\int_{-\infty}^{+\infty} f(r)dr = \int_{r_{min}}^{r_{max}} f(r)dr = 1 - \left(\frac{r_{min}}{r_{max}}\right)^{D_f} = 1. \qquad (5)$$

Thus, one can obtain the following equation from Equation (5):

$$\left(\frac{r_{min}}{r_{max}}\right)^{D_f} = 0. \qquad (6)$$

Generally, r_{min} is assumed to be much smaller than r_{max}, thereby Equation (6) holds for fractal porous media.

As for the complex pore space, suppose that there exist tortuous capillaries rather than straight capillaries. Tortuous capillaries in porous media have been proven to show self-similar and fractal behaviors, and thus the tortuous length of capillaries also follows the fractal scaling law [45]:

$$L(r) = r^{1-D_t} L_0^{D_t}, \qquad (7)$$

where D_t is the tortuosity fractal dimension of pore space, with $1 \leq D_t \leq 2$ and $1 \leq D_t \leq 3$ for two- and three-dimensional spaces, respectively. A value of $D_t = 1$ represents a straight capillary, while a greater value of $D_t > 1$ corresponds to a tortuous capillary.

Then, the tortuosity is defined as the ratio of the tortuous length L to the straight-line length L_0 [46], and the tortuosity is also scale dependent. Therefore, the fractal tortuosity can be derived from Equation (7):

$$\tau(r) = \frac{L(r)}{L_0} = r^{1-D_t} L_0^{D_t-1}, \qquad (8)$$

where τ is the fractal tortuosity.

For the case of tortuous capillaries, the total cumulative volume of pores can be obtained by:

$$V = -\int_{r_{min}}^{r_{max}} \frac{\pi}{4} r^2 L(r) dN = \frac{\pi}{4} \frac{D_f}{3 - D_t - D_f} L_0^{D_t} r_{max}^{D_f} \left(r_{max}^{3-D_t-D_f} - r_{min}^{3-D_t-D_f}\right). \qquad (9)$$

Meanwhile, the pore volume with pore diameter less than r_c can be expressed as follows:

$$V(< r_c) = -\int_{r_{min}}^{r_c} \frac{\pi}{4} r^2 L(r) dN = \frac{\pi}{4} \frac{D_f}{3 - D_t - D_f} L_0^{D_t} r_{max}^{D_f} \left(r_c^{3-D_t-D_f} - r_{min}^{3-D_t-D_f}\right), \qquad (10)$$

where r_c denotes the critical pore diameter, which controls the conductive or seepage properties, and below the critical pore diameter, it is considered that there is no current and fluid flowing through a porous medium (i.e., the conductivity and permeability become zero).

It is possible to relate the critical porosity to the corresponding critical pore diameter using Equations (9) and (10):

$$\frac{\phi_c}{\phi} = \frac{V(<r_c)}{V} = \frac{r_c^{3-D_t-D_f} - r_{min}^{3-D_t-D_f}}{r_{max}^{3-D_t-D_f} - r_{min}^{3-D_t-D_f}}, \quad (11)$$

where ϕ_c represents the critical porosity or stagnant porosity, which corresponds to the critical pore diameter and reflects the pore connectivity, and the greater the critical porosity is, the worse the connectivity is.

The porosity, pore sizes, and the fractal dimensions are related by:

$$\phi = \left(\frac{r_{min}}{r_{max}}\right)^{D_e - D_f}, \quad (12)$$

where D_e is the Euclidean dimension, with $D_e = 2$ and 3 for two- and three-dimensional spaces. In the following analysis and discussion, the taken value of D_e is 3.

Assuming $r_{min} \ll r_{max}$, and using Equation (12) can simplify Equation (11) as:

$$\frac{\phi_c}{\phi} = \frac{r_c^{3-D_t-D_f}}{r_{max}^{3-D_t-D_f}} - \phi^{\frac{3-D_t-D_f}{D_e-D_f}}. \quad (13)$$

2.2. Fractal Analysis of the Rock-Electric Property

As for the tortuous pore space fully saturated with water, Archie's equation can be expressed in terms of the tortuosity factor [47]:

$$\sigma = \sigma_w \frac{\phi}{\tau}, \quad (14)$$

where σ_w and σ are the electrical conductivity of water and porous media, respectively, and ϕ and τ are the porosity and tortuosity, respectively.

Considering that only effective connected pores contribute to the conductivity, by taking the critical porosity into account, Equation (14) is further modified as follows:

$$\sigma = \sigma_w C \frac{\phi - \phi_c}{\tau(\phi - \phi_c)}, \quad (15)$$

where C is an undetermined coefficient, and $\tau(\phi - \phi_c)$ represents the effective tortuosity of the electrically connected pores. Moreover, Equation (15) should satisfy the following boundary condition: $\phi = 1$, $\sigma = \sigma_w$, and taking it into Equation (15) gives $C = \frac{\tau(1-\phi_c)}{1-\phi_c}$. Then, Equation (15) becomes:

$$\sigma = \sigma_w \frac{\phi - \phi_c}{1 - \phi_c} \frac{\tau(1 - \phi_c)}{\tau(\phi - \phi_c)}. \quad (16)$$

As mentioned above, only the pore space above the critical porosity forms the effective conductive pathway, and it is appropriate to use the effective tortuosity to describe the effective conductive pathway. Therefore, averaging the tortuosity over the pores with pore diameter larger than r_c yields the effective tortuosity:

$$\tau(\phi - \phi_c) = \int_{r_c}^{r_{max}} \tau_p(r) * f(r) dr = \frac{D_f}{1 - D_f - D_t} L_0^{D_t-1} r_{max}^{1-D_t} \left(\frac{r_{min}^{D_f}}{r_{max}^{D_f}} - \frac{r_{min}^{D_f} r_c^{1-D_t-D_f}}{r_{max}^{1-D_t}}\right). \quad (17)$$

Using Equation (6) further simplifies Equation (17) as:

$$\tau(\phi - \phi_c) = \frac{D_f}{D_f + D_t - 1} L_0^{D_t-1} r_{max}^{1-D_t} \frac{r_c^{1-D_t-D_f}}{r_{max}^{1-D_t-D_f}} \frac{r_{max}^{1-D_t-D_f}}{r_{min}^{1-D_t-D_f}} \frac{r_{min}^{1-D_t}}{r_{max}^{1-D_t}}. \tag{18}$$

Substituting Equations (12) and (13) into Equation (18) expresses the effective tortuosity as:

$$\tau(\phi - \phi_c) = \frac{D_f}{D_f + D_t - 1} L_0^{D_t-1} r_{max}^{1-D_t} \phi^{\frac{D_f}{D_e-D_f}} \left(\frac{\phi_c}{\phi} + \phi^{\frac{3-D_t-D_f}{D_e-D_f}}\right)^{\frac{1-D_t-D_f}{3-D_t-D_f}}. \tag{19}$$

Then, we can obtain the following expression:

$$\tau(1 - \phi_c) = \frac{D_f}{D_f + D_t - 1} L_0^{D_t-1} r_{max}^{1-D_t} (\phi_c + 1)^{\frac{1-D_t-D_f}{3-D_t-D_f}}. \tag{20}$$

Furthermore, substituting Equations (19) and (20) into Equation (16) yields the conductivity:

$$\sigma = \sigma_w \frac{\phi - \phi_c}{1 - \phi_c} \frac{(\phi_c + 1)^{\frac{1-D_t-D_f}{3-D_t-D_f}}}{\phi^{\frac{D_f}{D_e-D_f}} \left(\frac{\phi_c}{\phi} + \phi^{\frac{3-D_t-D_f}{D_e-D_f}}\right)^{\frac{1-D_t-D_f}{3-D_t-D_f}}}. \tag{21}$$

According to Archie's law [48], the formation factor F can be expressed as:

$$F = \frac{\sigma_w}{\sigma} = \frac{1 - \phi_c}{\phi - \phi_c} \phi^{\frac{D_f}{D_e-D_f}} \left(\frac{\frac{\phi_c}{\phi} + \phi^{\frac{3-D_t-D_f}{D_e-D_f}}}{\phi_c + 1}\right)^{\frac{1-D_t-D_f}{3-D_t-D_f}}. \tag{22}$$

Equations (21) and (22) give the analytical percolation expression of the conductivity, which reveals that Archie's equation is closely related to the critical porosity and microstructure.

2.3. Fractal Analysis of the Permeability Property

The flow in a single tortuous capillary is assumed to obey the Hagen–Poiseuille equation:

$$q = \frac{\pi r^4}{128\mu} \frac{\Delta P}{L(r)}, \tag{23}$$

where q is the flow rate of a single capillary, μ is the viscosity, and ΔP is the pressure drop.

For a porous medium, the total volumetric flow rate is the sum of the flow rates through all the capillaries. Considering that the pore sizes below the critical pore size cannot contribute to the flow rate, integrating the individual flow rate over the pores with pore diameter larger than r_c gives the total flow rate Q as:

$$Q = -\int_{r_c}^{r_{max}} q dN(r) = \frac{\pi}{128\mu} \frac{\Delta P}{L_0^{D_t}} \frac{D_f}{3 + D_t - D_f} r_{max}^{3+D_t} \left(1 - \frac{r_c^{3+D_t-D_f}}{r_{max}^{3+D_t-D_f}}\right). \tag{24}$$

Moreover, the total flow rate Q can also be given by Darcy's law:

$$Q = \frac{KA_0}{\mu} \frac{\Delta P}{L_0}. \tag{25}$$

Then, combining Equations (24) and (25) can obtain the absolute permeability as:

$$K = \frac{\pi}{128 A_0} \frac{1}{L_0^{D_t-1}} \frac{D_f}{3+D_t-D_f} r_{max}^{3+D_t} \left(1 - \frac{r_c^{3+D_t-D_f}}{r_{max}^{3+D_t-D_f}}\right), \quad (26)$$

where K denotes the absolute permeability, and A_0 represents the cross-sectional area of a porous medium, which is written as follows:

$$A_0 = \frac{V}{\phi L_0}. \quad (27)$$

Substituting Equation (9) into Equation (27) obtains:

$$A_0 = \frac{\pi}{4} \frac{D_f}{3-D_t-D_f} L_0^{D_t-1} \frac{r_{max}^{3-D_t}}{\phi}\left(1 - \phi^{\frac{3-D_t-D_f}{D_e-D_f}}\right). \quad (28)$$

Then, combining Equations (11), (26), and (28) yields the analytical percolation expression of the permeability:

$$K = \frac{1}{32} \frac{3-D_t-D_f}{3+D_t-D_f} L_0^{2-2D_t} r_{max}^{2D_t} \phi \frac{1 - \left(\frac{\phi_c}{\phi} + \left(1 - \frac{\phi_c}{\phi}\right)\phi^{\frac{3-D_t-D_f}{D_e-D_f}}\right)^{\frac{3+D_t-D_f}{3-D_t-D_f}}}{1 - \phi^{\frac{3-D_t-D_f}{D_e-D_f}}}. \quad (29)$$

Without involving the empirical constants, Equation (29) gives the permeability in terms of the porosity, critical porosity, fractal dimension D_f, tortuosity fractal dimension D_t, and structural parameters L_0 and r_{max}, which can better describe the seepage property in porous media with complex pore structure. Equation (29) also indicates that the permeability is very sensitive to the maximum pore diameter r_{max}: the larger the maximum pore diameter is, the greater the permeability becomes, which is consistent with the practical situation. Note that the critical pore diameter or critical porosity is not necessarily the same for the electrical conductivity and permeability in a porous medium.

3. Results and Discussion

In porous media, fluid transport is dominated by the amount of pore space available for flow. Usually, fluid and electrical current flow through the effective pore space, while the stagnant pore space plays a passive role in fluid transport. Therefore, using total porosity values rather than effective porosity in the conductivity and permeability modeling will lead to erroneous results. In this work, the critical pore diameter or critical porosity—which reflects the effective pore space available for flow—is introduced into the fractal geometry to describe the percolation behaviors of the conductivity and permeability. In the following discussion, the percolation models of the conductivity and permeability are utilized to model the rock-electric and seepage characteristics at different critical porosity, study the effect of microstructure on the critical porosity, and compare with the experimental data. The analytical percolation expressions on the one hand provide insights into the significant influence of the critical porosity and microstructural parameters (e.g., pore fractal dimension, tortuosity fractal dimension) on the conductivity and permeability. On the other hand, they also reveal the correlation between the critical porosity and microstructural parameters. Besides, the percolation models could be used to fit and interpret the experimental data well.

3.1. The Rock-Electric Characteristics

To start with, the pore microstructural parameters D_f and D_t remained constant, the conductivity–porosity relationships at different critical porosity are depicted in Figure 1. It shows that the conductivities at the critical porosity decreased to zero, which proved that the electrical current could not flow through the stagnant pore space. In addition, it also demonstrates that the conductivities at high porosity were nearly the same, which means that the critical porosity had little effect on the conductive property at high porosity, while the conductivities at low porosity changed greatly. In short, the critical porosity can be an indicator of the amount of effective and stagnant pores, and as the critical porosity increased, there were more stagnant pores and fewer effective pores, which led to the decrease of the conductive ability. Therefore, ignoring the critical porosity may lead to the inaccuracy of the predicted conductivity. Furthermore, the critical porosity was set to 0.05, and the conductive properties at different microstructural parameters are plotted in Figures 2 and 3. As an indicator of pore microstructure, the high values of D_f and D_t represent the strong heterogeneity and complexity of the pore structure. In Figures 2 and 3, as D_f and D_t increased and the conductivity decreased, demonstrating that the complex pore structure is an important reason for the high resistivity of water-saturated reservoirs. On the other hand, it was found that when D_f and D_t increased, the conductivity tended to zero at higher porosity, meaning that high D_f and D_t can lead to increases of the critical porosity and complex structure can increase the stagnant porosity, which indicates that the critical porosity is a function of the pore geometry. In fact, the percolation behaviors have been found in some Archie's data, and the deviation between Archie's data and the simulation results could not be explained by the classic Archie's law. Archie's data [48] in Figures 4 and 5 illustrate two important phenomena. One phenomenon is that the measured conductivities in Figure 4 decreased to zero at certain porosity greater than zero. Another is that Archie's data in Figure 5 deviate from the fitting results obtained by the Archie equation and exhibits the non-Archie phenomenon on log–log scale. Apparently, the classic Archie's law could not explain these phenomena well. By considering the critical porosity, Archie's data in Figures 4 and 5 could be fitted and explained well using Equation (22), verifying the validity of the conductivity percolation model.

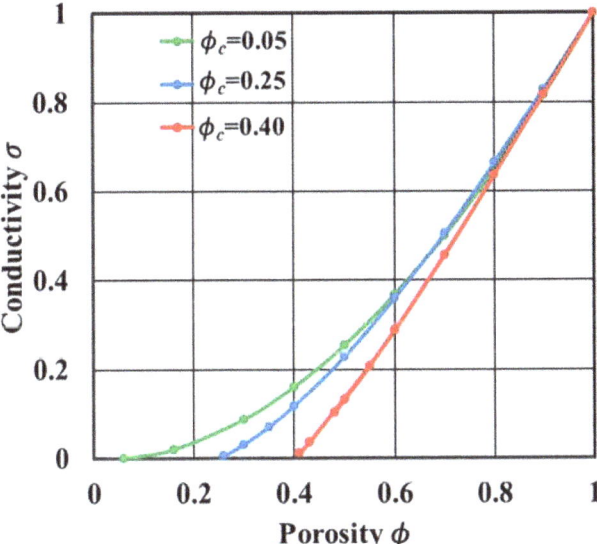

Figure 1. The conductivity σ versus porosity ϕ at σ_w = 1.0, D_f = 2.5, D_t = 1.5, and the critical porosity ϕ_c of 0.05, 0.25, and 0.40, respectively.

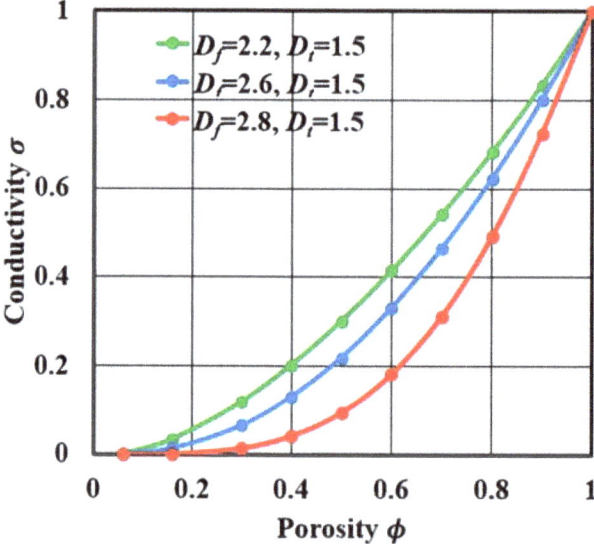

Figure 2. The conductivity σ versus porosity ϕ at different D_f at $\sigma_w = 1.0$, $\phi_c = 0.05$.

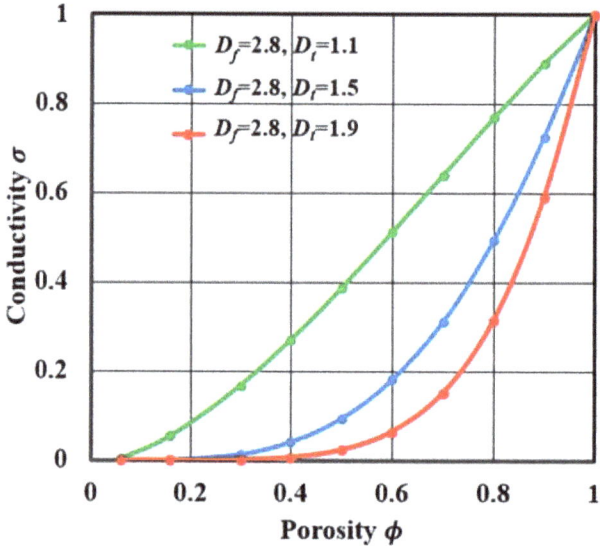

Figure 3. The conductivity σ versus porosity ϕ at different D_t at $\sigma_w = 1.0$, $\phi_c = 0.05$.

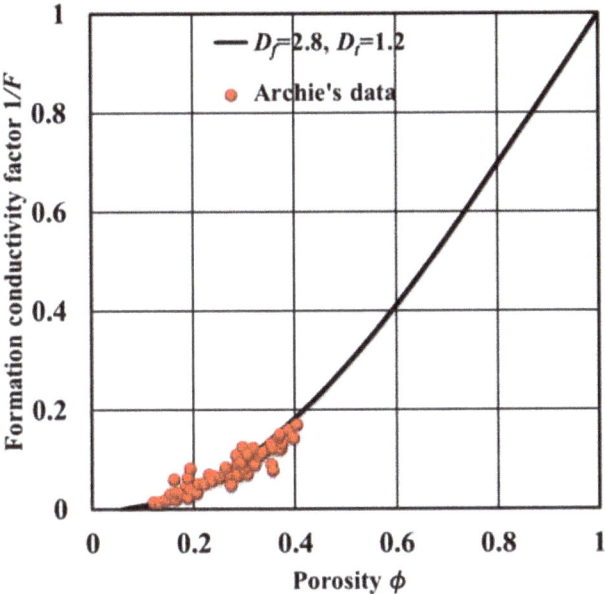

Figure 4. The comparison of Archie's data [48] and the conductivity factor $1/F$ calculated by Equation (22) at $\phi_c = 0.05$. Reproduced from [48], Society of Petroleum Engineers: 1942.

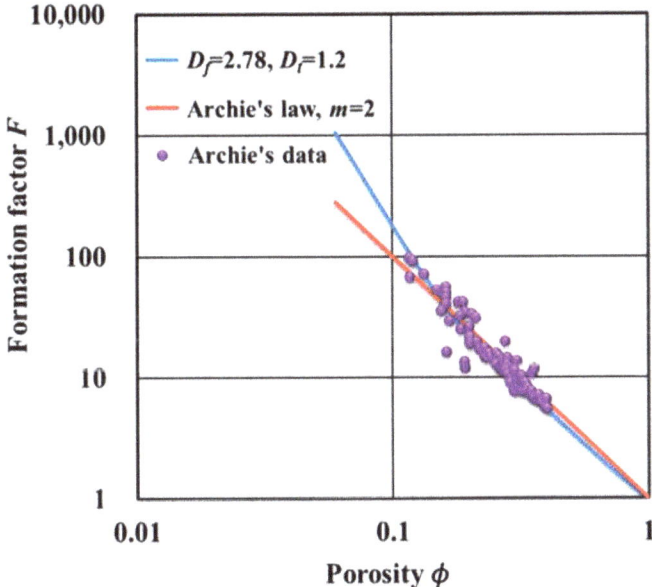

Figure 5. The comparison of Archie's data [48] and the formation factor F calculated by Archie's law and Equation (22) at $\phi_c = 0.05$. Reproduced from [48], Society of Petroleum Engineers: 1942.

3.2. The Permeability Characteristics

In this section, the percolation model of the permeability (Equation (29)) is applied to calculate the permeability at different critical porosity and microstructural parameters depicted in Figures 6–8, and interpret the experimental data displayed in Figures 9 and 10. It was assumed that the maximum capillary diameter of 2 μm and the capillary straight-line length of 200 μm in the present calculation were taken in Figures 6–8. In the case of constant parameters D_f and D_t, Figure 6 shows that the effective permeability was greater than zero at the porosity above the critical porosity, indicating that the fluid only flowed through the effective pore space rather than all the pore space. Moreover, it shows that the increase of the critical porosity could greatly decrease the permeability, which means that the critical porosity is a crucial factor affecting the seepage ability. Besides, with the change of D_f and D_t in Figures 7 and 8, it indicates that the complex pore structure (e.g., high D_f or D_t) could result in the decrease of the permeability and the increase of the critical porosity. The increasing critical porosity indicates more stagnant pore space, which is considered as an important reason for the low permeability. Next, the experimental data [49,50] in Figures 9 and 10 were used to test the percolation model of the permeability, r_{max} and L_0 were taken as 190 μm and 100 mm in Figure 9, and as 1.3 mm and 100 mm in Figure 10, respectively. Before calculating the permeability, we first determined the microstructural parameters D_f and D_t. Although some methods have been developed to obtain the fractal dimensions [19,51], the precision of the results is still not proven. Alternatively, here we can determine the fractal dimension D_f using Equation (12), and Feng et al. [52] indicated that D_f can be best fitted by $r_{max}/r_{min} = 1000$ for natural and artificial porous media, and the tortuosity fractal D_t can be estimated via D_f using $D_t = 2 \times (3 - D_f) + 1$. Then, the permeability can be directly calculated by the analytical expressions (Equation (29)). By the comparison, the calculation results of the permeability obtained by Equation (29) were not only in good agreement with the experimental data, but were also basically identical to the fitting trendlines ($R^2 = 0.9223$ and $R^2 = 0.9136$ for Figures 9 and 10) of the experimental data, which verifies the effectiveness of the proposed percolation model. In addition to critical porosity and microstructure, the permeability may be affected by other factors (e.g., the irreducible water or multi-scale pores), and the fitting error can be eliminated to a greater extent by reasonably introducing multiple influencing factors into the model in future study.

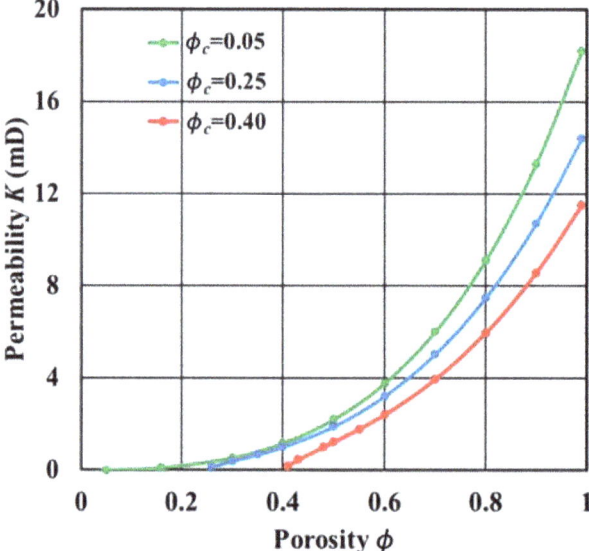

Figure 6. The permeability K versus the porosity ϕ at $D_f = 2.5$, $D_t = 1.2$, and the critical porosity ϕ_c of 0.05, 0.25, and 0.40, respectively.

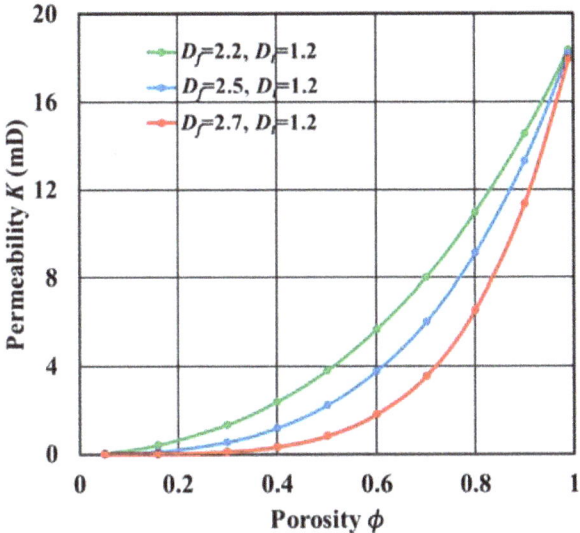

Figure 7. The permeability K versus the porosity ϕ at different D_f at $\phi_c = 0.05$.

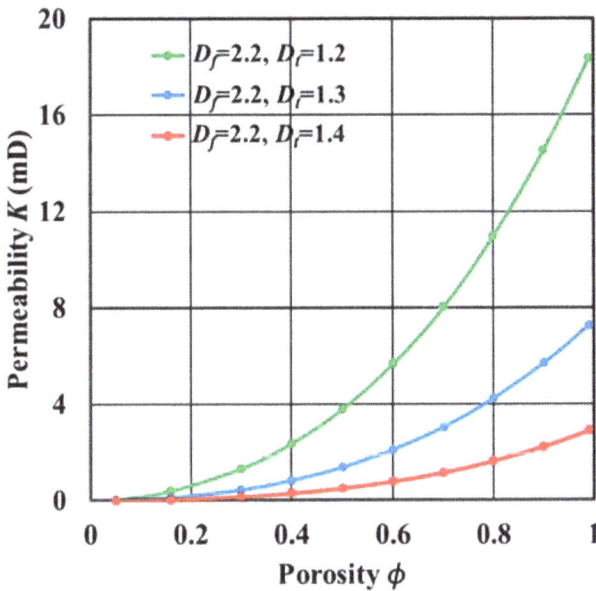

Figure 8. The permeability K versus the porosity ϕ at different D_t at $\phi_c = 0.05$.

In summary, the critical porosity is an important factor for the accurate interpretation of rock-electric and seepage characteristics in complex porous media, and ignoring the effect of critical porosity may cause the overestimation of the conductivity and permeability, which eventually influence the formation evaluation. Moreover, the critical porosity is also closely related to the pore microstructure. A complex pore structure can increase the critical porosity, resulting in more stagnant pores, which is the fundamental reason for the decrease of the conductive and seepage abilities in complex reservoirs.

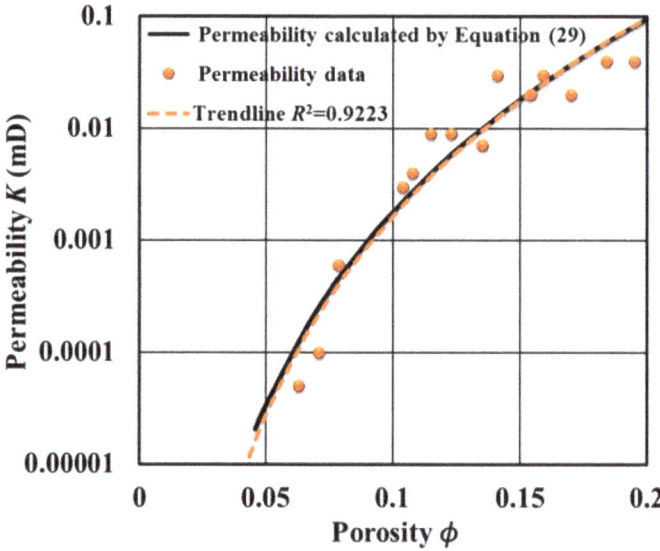

Figure 9. The comparison of permeability data [49] and the permeability K calculated by Equation (29) at ϕ_c = 0.045. D_f and D_t were estimated by Equation (12) and $D_t = 2 \times (3 - D_f) + 1$. Reproduced from [49], North-Holland Publishing Company: 1982.

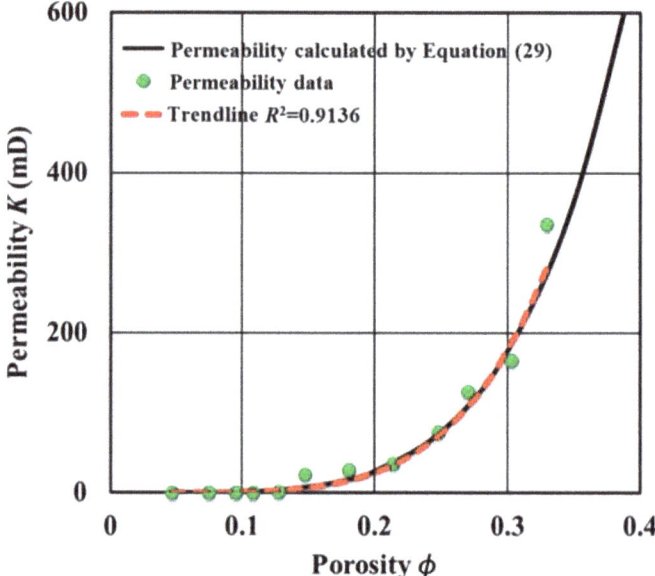

Figure 10. The comparison of permeability data [50] and the permeability K calculated by Equation (29) at ϕ_c = 0.05. D_f and D_t were estimated by Equation (12) and $D_t = 2 \times (3 - D_f) + 1$. Reproduced from [50], Nedra: 1965.

4. Conclusions

In this study, the analytical expressions to connect the critical porosity with the conductivity and permeability were derived without involving any empirical constants using the fractal geometry theory, which provides the theoretical basis of the percolation mechanisms of the conductivity and permeability. The critical porosity and microstructural parameters (e.g., pore fractal dimension and tortuosity fractal dimension) were considered as major determining factors in the proposed percolation models, which contributes to predicting the conductivity and permeability in complex reservoirs more accurately. The simulation results revealed that increasing the critical porosity could reduce the conductive and the seepage abilities of porous media, and the critical porosity is a crucial factor resulting in non-Archie phenomenon. The results also demonstrated that the complex pore structure could decrease the effective porosity and increase the critical porosity, which is an important reason for low conductivity and permeability in complex media. Due to the incorporation of the critical porosity, the percolation models could be used to better interpret the high-resistivity water layer in complex reservoirs. The validity of the percolation models was confirmed after its application to the available data.

Given that porous media consist of various types of pores, each type of pore should have independent critical porosity and fractal parameters, and multi-fractal analysis may better characterize the complex pore structure and petrophysical properties. Therefore, future research will extend the present work to the multi-fractal case.

Author Contributions: Conceptualization, T.L.; Methodology, T.L. and H.M.; Validation, H.M. and Q.S.; Formal Analysis, H.M.; Investigation, Q.S., F.X.L., and P.C.; Data Curation, Q.S., F.X.L., and P.C.; Writing—Original Draft Preparation, H.M.; Writing—Review & Editing, T.L.; Supervision, T.L.; Funding Acquisition, T.L.

Funding: This research was funded by National 863 Program of China (No. 2006AA06Z214), National Natural Science Foundation of China (No. 41476027) and National Key Technology Program (No. 2017ZX05035003-002).

Conflicts of Interest: The authors declare no conflict of interest.

Nomenclature

r	pore diameter
r_{max}	maximum pore diameter
r_{min}	minimum pore diameter
r_c	critical pore diameter
N	number of capillaries whose diameters are greater than or equal to r
N_t	total number of capillaries
D_f	pore fractal dimension
D_t	tortuosity fractal dimension
D_e	Euclidean dimension
ϕ	porosity
ϕ_c	critical porosity
L_0	straight-line length of a capillary
L	tortuous length of a capillary
τ	fractal tortuosity
$\tau(\phi - \phi_c)$	effective tortuosity of the electrical connected pores
V	total cumulative volume of pores
$V(< r_c)$	pore volume with pore diameter less than r_c
σ	electrical conductivity of water-saturated media
σ_w	electrical conductivity of water
F	formation factor
q	flow rate of single capillary
μ	fluid viscosity
ΔP	pressure gradient
A_0	cross section of rock
Q	total flow rate of porous media
K	permeability of porous media

References

1. Tahmasebi, P. Nanoscale and multiresolution models for shale samples. *Fuel* **2018**, *217*, 218–225. [CrossRef]
2. Liu, T.Y.; Tang, T.Z.; Du, H.H.; Zhang, H.N.; Wang, H.T. Study of rock conductive mechanism based on pore structure. *Chin. J. Geophys.* **2013**, *56*, 674–684.
3. Meng, H. Study on the rock-electric and the relative permeability characteristics in porous rocks based on the curved cylinder-sphere model. *J. Pet. Sci. Eng.* **2018**, *166*, 891–899. [CrossRef]
4. Mandelbrot, B.B. *The Fractal Geometry of Nature*; W.H. Freeman: New York, NY, USA, 1983.
5. Thompson, A.H. Fractals in rock physics. *Annu. Rev. Earth Planet. Sci.* **1991**, *19*, 237–262. [CrossRef]
6. Sahimi, M. Flow phenomena in rocks: From continuum models to fractals, percolation, cellular automata, and simulated annealing. *Rev. Mod. Phys.* **1993**, *65*, 1393–1534. [CrossRef]
7. Perfect, E.; Kay, B.D. Applications of fractals in soil and tillage research: A review. *Soil Till. Res.* **1995**, *36*, 1–20. [CrossRef]
8. Perrier, E.; Bird, N.; Rieu, M. Generalizing the fractal model of soil structure: The pore-solid fractal approach. *Dev. Soil Sci.* **2000**, *27*, 47–74.
9. Yu, B.; Liu, W. Fractal analysis of permeabilities for porous media. *AIChE J.* **2004**, *50*, 46–57. [CrossRef]
10. Perfect, E.; Pachepsky, Y.; Martin, M.A. Fractal and Multifractal Models Applied to Porous Media. *Vadose Zone J.* **2011**, *10*, 110–124. [CrossRef]
11. Coleman, S.W.; Vassilicos, J.C. Transport Properties of Saturated and Unsaturated Porous Fractal Materials. *Phys. Rev. Lett.* **2008**, *100*, 035504. [CrossRef]
12. Coleman, S.W.; Vassilicos, J.C. Tortuosity of unsaturated porous fractal materials. *Phys. Rev. E* **2008**, *78*, 016308. [CrossRef] [PubMed]
13. Ghanbarian, B.; Daigle, H. Fractal dimension of soil fragment mass-size distribution: A critical analysis. *Geoderma* **2015**, *245–246*, 98–103. [CrossRef]
14. Wu, Y.Q.; Tahmasebi, P.; Lin, C.Y.; Zahid, M.A.; Dong, C.M.; Golab, A.N.; Ren, L.H. A comprehensive study on geometric, topological and fractal characterizations of pore systems in low-permeability reservoirs based on SEM, MICP, NMR, and X-ray CT experiments. *Mar. Pet. Geol.* **2019**, *103*, 12–28. [CrossRef]
15. Dimri, V.P. Fractal Dimensional analysis of soil for flow studies. In *Application of Fractals in Earth Sciences*; Oxford and IBH publishing Co. Pvt. LTD: New Delhi, India, 2000; pp. 189–193.
16. Dimri, V.P. *Application of Fractals in Earth Sciences*; Oxford and IBH publishing Co. Pvt. LTD: New Delhi, India, 2000.
17. Feranie, S.; Latief, F.D.E. Tortuosity-porosity relationship in two-dimensional fractal model of porous media. *Fractals* **2013**, *21*, 1350013. [CrossRef]
18. Katz, A.J.; Thompson, A.H. Fractal sandstone pores: Implications for conductivity and pore formation. *Phys. Rev. Lett.* **1985**, *54*, 1325–1328. [CrossRef]
19. Krohn, C.E.; Thompson, A.H. Fractal sandstone pores: Automated measurements using scanning-electron-microscope images. *Phys. Rev.* **1986**, *33*, 6366–6374. [CrossRef]
20. Young, I.M.; Crawford, J.W. The fractal structure of soil aggregations: Its measurement and interpretation. *J. Soil Sci.* **1991**, *42*, 187–192. [CrossRef]
21. Yu, B.M.; Li, J. Some fractal characters of porous media. *Fractals* **2001**, *9*, 365–372. [CrossRef]
22. Roy, S.; Tarafdar, S. Archie's law from a fractal model for porous rocks. *Phys. Rev.* **1997**, *55*, 8038–8041. [CrossRef]
23. Nigmatullin, R.R.; Dissado, L.A.; Soutougin, N.N. A fractal pore model for Archie's law in sedimentary rocks. *J. Phys.* **1992**, *25*, 32. [CrossRef]
24. Wei, W.; Cai, J.; Hu, X.; Han, Q. An electrical conductivity model for fractal porous media. *Geophys. Res. Lett.* **2015**, *42*, 4833–4840. [CrossRef]
25. Pitchumani, R.; Ramakrishnan, B. A fractal geometry model for evaluating permeabilities of porous preforms used in liquid composite molding. *Int. J. Heat Mass Tran.* **1999**, *42*, 2219–2232. [CrossRef]
26. Xu, P.; Yu, B.M. Developing a new form of permeability and Kozeny-Carman constant for homogeneous porous media by means of fractal geometry. *Adv. Water Resour.* **2008**, *31*, 74–81. [CrossRef]
27. Mavko, G.; Nur, A. Effect of a percolation threshold in the Kozeny–Carman relation. *Geophysics* **1997**, *62*, 1480–1482. [CrossRef]
28. Bentz, D.P. Modelling cement microstructure: Pixels, particles, and property prediction. *Mater. Struct.* **1999**, *32*, 187–195. [CrossRef]

29. Mavko, G.; Mukerji, T.; Dvorkin, J. *The Rock Physics Handbook*, 2nd ed.; Cambridge University Press: Cambridge, UK, 2009.
30. Kirkpatrick, S. Percolation and Conduction. *Rev. Mod. Phys.* **1973**, *45*, 574–588. [CrossRef]
31. Wong, P.Z.; Koplik, J.; Tomanic, J.P. Conductivity and permeability of rocks. *Phys. Rev.* **1984**, *30*, 6606–6614. [CrossRef]
32. Thompson, A.H.; Katz, A.J.; Krohn, C.E. The microgeometry and transport properties of sedimentary rock. *Adv. Phys.* **1987**, *36*, 625–694. [CrossRef]
33. Sahimi, M. *Applications of Percolation Theory*; Taylor and Francis: Abingdon, UK, 1994.
34. Sahimi, M. *Flow and Transport in Porous Media and Fractured Rock: From Classical Methods to Modern Approaches*; Wiley-VCH: Weinheim, Germany, 2011.
35. Stauffer, D.; Aharony, A. *Introduction to Percolation Theory*, 2nd ed.; Taylor and Francis: Abingdon, UK, 1994.
36. Hunt, A.G. Percolative transport and fractal porous media. *Chaos Solitons Fractals* **2004**, *19*, 309–325. [CrossRef]
37. Hunt, A.G.; Ewing, R.P.; Ghanbarian, B. *Percolation Theory for Flow in Porous Media*; Lecture Notes Phys.; Springer: Berlin, Germany, 2014.
38. Ghanbarian, B.; Hunt, A.G.; Ewing, R.P.; Skinner, T.E. Universal scaling of the formation factor in porous media derived by combining percolation and effective medium theories. *Geophys. Res. Lett.* **2014**, *41*, 3884–3890. [CrossRef]
39. Wang, F.Y.; Lian, P.Q.; Jiao, L.; Liu, Z.C.; Zhao, J.Y.; Gao, J. Fractal analysis of microscale and nanoscale pore structures in carbonates using high-pressure mercury intrusion. *Geofluids* **2018**, *2*, 1–15. [CrossRef]
40. Perfect, E.; Kay, B.D.; Rasiah, V. Multifractal model for soil aggregate fragmentation. *Soil Sci. Soc. Am. J.* **1993**, *57*, 896–900. [CrossRef]
41. Smidt, J.M.; Monro, D.M. Fractal modeling applied to reservoir characterization and flow simulation. *Fractals* **1998**, *6*, 401–408. [CrossRef]
42. Yu, B.; Lee, L.J.; Cao, H. Fractal characters of pore microstructures of textile fabrics. *Fractals* **2001**, *9*, 155–163. [CrossRef]
43. Ojha, S.P.; Misra, S.; Sinha, A.; Dang, S.; Tinni, A.; Sondergeld, C.; Rai, C.A. Relative permeability estimates for Wolfcamp and Eagle Ford shale samples from oil, gas and condensate windows using adsorption-desorption measurements. *Fuel* **2017**, *208*, 52–64. [CrossRef]
44. Yun, M.; Yu, B.; Cai, J. Analysis of seepage characters in fractal porous media. *Int. J. Heat Mass Tran.* **2009**, *52*, 3272–3278. [CrossRef]
45. Wheatcraft, S.W.; Tyler, S.W. An explanation to scale dependent dispersivity in heterogeneous aquifers using concepts of fractal geometry. *Water Res. Res.* **1988**, *24*, 566–578. [CrossRef]
46. Ghanbarian, B.; Hunt, A.G.; Ewing, R.P. Tortuosity in Porous Media: A Critical Review. *Soil Sci. Soc. Am. J.* **2013**, *77*, 1461–1477. [CrossRef]
47. Wyllie, M.R.J.; Rose, W.D. Some theoretical considerations related to the quantitative evaluation of the physical characteristics of reservoir rock from electrical log data. *J. Petrol. Technol.* **1950**, *2*, 105–118. [CrossRef]
48. Archie, G.E. The electrical resistivity log as an aid in determining some reservoir characteristics. *Petrol. Trans. AIME* **1942**, *146*, 54–62. [CrossRef]
49. Bernabe, Y.; Brace, W.F.; Evans, B. Permeability, porosity, and pore geometry of hot-pressed calcite. *Mech. Mater.* **1982**, *1*, 173–183. [CrossRef]
50. Emelyanov, N.N. Analysis of the accuracy in the determination of collector properties of the ninth stratum at Ozek-Suat. In *Estimate of the Accuracy of Determining the Parameters of Oil and Gas Deposits*; Nedra: Moscow, Russia, 1965; pp. 135–152.
51. Daigle, H.; Johnson, A.; Thomas, B. Determining fractal dimension from nuclear magnetic resonance data in rocks with internal magnetic field gradients. *Geophysics* **2014**, *79*, D425–D431. [CrossRef]
52. Feng, Y.J.; Yu, B.M.; Zou, M.Q.; Zhang, D.M. A generalized model for the effective thermal conductivity of porous media based on self-similarity. *J. Phys.* **2004**, *37*, 3030–3040. [CrossRef]

© 2019 by the authors. Licensee MDPI, Basel, Switzerland. This article is an open access article distributed under the terms and conditions of the Creative Commons Attribution (CC BY) license (http://creativecommons.org/licenses/by/4.0/).

Article

Numerical Investigation of Fracture Compressibility and Uncertainty on Water-Loss and Production Performance in Tight Oil Reservoirs

Kai Liao *, Shicheng Zhang, Xinfang Ma and Yushi Zou

MOE Key Laboratory of Petroleum Engineering, China University of Petroleum (Beijing), Beijing 102249, China; zhangsc@cup.edu.cn (S.Z.); maxinfang@cup.edu.cn (X.M.); zouyushi@126.com (Y.Z.)
* Correspondence: 2015312060@student.cup.edu.cn; Tel.: +86-150-0125-7832

Received: 30 January 2019; Accepted: 25 March 2019; Published: 27 March 2019

Abstract: Multi-stage hydraulic fracturing along with horizontal wells are widely used to create complex fracture networks in tight oil reservoirs. Analysis of field flowback data shows that most of the fracturing fluids are contained in a complex fracture network, and fracture-closure is the main driving mechanism during early clean up. At present, the related fracture parameters cannot be accurately obtained, so it is necessary to study the impacts of fracture compressibility and uncertainty on water-loss and the subsequent production performance. A series of mechanistic models are established by considering stress-dependent porosity and permeability. The impacts of fracture uncertainties, such as natural fracture density, proppant distribution, and natural fracture heterogeneity on flowback and productivity are quantitatively assessed. Results indicate that considering fracture closure during flowback can promote water imbibition into the matrix and delay the oil breakthrough time compared with ignoring fracture closure. With the increase of natural fracture density, oil breakthrough time is advanced, and more water is retained underground. When natural fractures connected with hydraulic fractures are propped, well productivity will be enhanced, but proppant embedment can cause a loss of oil production. Additionally, the fracture network with more heterogeneity will lead to the lower flowback rate, which presents an insight in the role of fractures in water-loss.

Keywords: tight oil reservoirs; fracture compressibility; numerical simulation; flowback; fracture uncertainty

1. Introduction

The development of conventional petroleum reservoirs has been unable to keep pace with the increasing global energy demand, which has shifted industrial attention to low permeability or tight reservoirs [1]. Rapid development of horizontal wells and multi-stage hydraulic fracturing technology enables unconventional tight reservoirs to be exploited [2,3]. In the process of stimulation, hydraulic fractures may connect and reactivate the already existing natural fractures or generate new induced fractures near the wellbore. Well productivity could be substantially increased if complex fracture networks are created by hydraulic fracturing [4,5].

Different from conventional fracturing, slickwater is the most commonly used fracturing fluid in unconventional reservoirs. Slickwater has a relatively low cost and low viscosity, which may help to create complex fracture networks [5–7]. Horizontal wells that undergo multi-stage fracturing operations often require a large amount of water injected into the formation to create a large stimulated reservoir volume. After hydraulic fracturing is completed, the flowback treatment of fracturing fluid should be carried out, followed by long-term production [8–10]. However, field data show that the recovery efficiency of tight oil or gas wells is generally low, and a remarkable fraction of fracturing fluid remains in the reservoir even after long-term production [11–14]. Scholars have carried out some work to clarify the problems arising after hydraulic fracturing and to help improve the clean-up operation in low permeability formations [15,16].

If hydraulic fracturing fluid is trapped in the formation, then it must be in the rock matrix or fracture systems [17]. An analytical model was presented for spontaneous imbibition of capillary force, which explained the fracturing fluid was imbibed into the shale matrix by strong capillary forces [18]. However, Dutta et al. [19] showed that water-loss rate of low-permeability sands was far lower than that of high-permeability sands. Although the tight sands had strong capillary force, the low permeability of the formation also limited water imbibition. Similarly, experimental data showed that the water imbibition capacity of tight shale was very low, even for hydrophilic rock samples [20]. Furthermore, it was difficult to explain the reason for low water recovery efficiency when the spontaneous capillary forces were absent in oil-wet tight rocks [21,22] or the wells were immediately cleaned up without shut-in. By analyzing the pressure transient data, natural fractures that re-opened during fracturing were apparently closed in subsequent production processes [23]. In the process of fracture reopening and closing, water entering natural fractures continues to be imbibed into matrix, and only part of the water can flow back to the wellbore in the end. McClure [24] focused on the impact of fracture network complexity on water recovery efficiency. The results of his simulations indicated that the closure of unpropped fractures within fracture networks resulted in low water recovery. Fu et al. [25] collated and processed the early flowback data of seven multi-fractured horizontal wells in tight reservoirs. Data analysis suggested that most of the facture systems were un-propped after stimulation and held most of the fracturing water.

Whether the fracturing fluid remains in the matrix or fractures, it will affect the productivity of tight oil wells. Spontaneous imbibition of fracturing water into the matrix was considered as a possible mechanism to improve oil production [26,27]. Extensive experimental and mathematical studies were devoted to explaining the oil displacement efficiency by water imbibition with various rock wettabilities, physical parameters, pore size distribution, and fracture characteristics [28–30]. The results showed that water imbibition could drive oil out of the matrix pore, and the displacement effect would be better when the tight rocks contained fractures [29]. However, Wang and Leung [31,32] established a series of numerical models to investigate the water-loss mechanisms during flowback operations. They reported that although prolonging shut-in time could enhance water imbibition into the matrix and increase initial oil rate, there was no benefit for long-term production. Low recovery efficiency means that a large amount of water remains in secondary fractures, which reduced oil mobility and had a negative impact on long-term production. Similar negative impacts were also found in studies of tight shale [33–35], where water retention might cause water blocking or damage to hydrocarbon phase relative permeability. Moreover, for tight formations, the invasion of fracturing fluid into the matrix did not require much depth to cause enough damage [15,16].

Although the mechanism of fracturing fluid retention underground and its impact on well productivity have been widely studied, there are still many problems to be further explored. Firstly, existing models failed to accurately characterize the distribution of slickwater in matrix and fracture systems during fracturing. These works either directly assign high water saturation value to the fractures or simulate the injection process of fracturing fluid through stress-dependent permeability models [36,37]. Although the latter is slightly more reasonable than the former, its mechanism is unrealistic and overestimates the water absorption capacity of the matrix near fractures. The decrease of matrix permeability might be due to compaction, but it was hardly to reasonably increase matrix permeability [23]. Secondly, after stimulation, secondary fractures and re-opened natural fractures held most of the fracturing water [38], and fracture depletion was observed during early flowback [39]. The driving mechanism of fracture-closure is rarely considered in simulation and its impact on water-loss is unclear. Thirdly, the parameters of complex fracture networks are the key factors affecting the distribution of fracturing fluids in formations and subsequent productivity. Numerical studies mainly focus on the density and distribution of natural fractures [37,40], while the proppant distribution in complex fracture systems is seldom considered. The stress-dependent porosity and permeability correlations vary because of proppant concentrations, and the effect of proppant embedment on long-term production cannot be neglected [41,42]. Finally, the impact of natural fracture heterogeneity on water retention has not been reported.

This paper reports on simulations to better understand water loss and production from complex fracture networks. A triple-porosity model is established by using a numerical reservoir simulator, where matrix and fracture systems are explicitly discretized, so that we can examine imbibition hysteresis and stress-dependent porosity and permeability. The adaptability of a mechanistic model is validated, and the influence of fracture-closure on fracturing water retention is investigated. Next, we quantitatively analyze the effect of uncertain fracture parameters, such as natural fracture density, proppant distribution, and natural fracture heterogeneity. The results of this work can provide a better understanding of the impact of fracture compressibility and uncertainty on water-loss and production performance in tight oil reservoirs.

2. Methodology

A series of numerical models are constructed by fully exploiting the functions of commercial reservoir simulation software (CMG IMEX). The whole reservoir consists of triple media: matrix (M), hydraulic fracture (HF), and natural fracture (NF). The model is essentially a single-porosity medium, while a fracture network is set up by adopting logarithmic local grid refinement. This approach solves the problem of huge differences between fracture width and matrix grid size, and ensures the stability of numerical calculation [43]. Hydraulic fracture stages along horizontal wellbores are assumed to be uniformly spaced and symmetrical. Based on this symmetrical structure, only one hydraulic fracture segment is selected for simulation in this work, and its results can be mapped back to the whole horizontal well scale [44]. The grid structure of a segment is shown in Figure 1, where the horizontal wellbore is along the x-direction and the hydraulic fracture is along the y-direction.

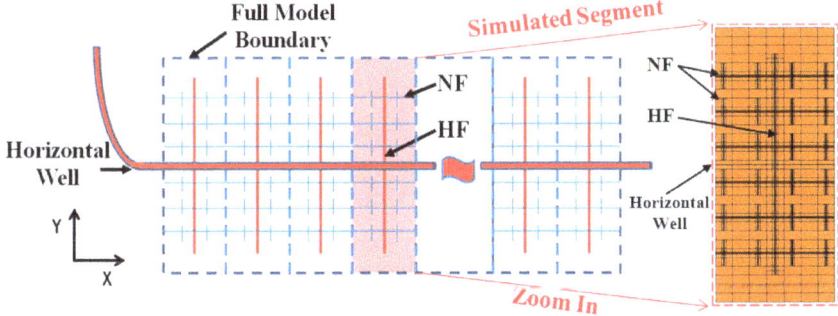

Figure 1. Schematic diagram of grid structure of numerical model (zoom scale: X = 2 times).

2.1. Model Description

For base case, a model with a size of 70 m × 300 m is established as shown in Figure 1, in which 70 m is the hydraulic fracture spacing and 300 m is the horizontal well spacing. Relevant known model parameters of Lucaogou Formation in Junggar Basin, obtained from field and literature reports [45,46], are summarized in Table 1. The parameters of complex fracture network formed after multi-stage fracturing along horizontal wells are unknown, so the typical values of other tight reservoirs [31,40] are assigned here as shown in Table 2. Hydraulic fractures are considered to be propped, while reactivated natural fractures with orthogonal distribution are considered to be un-propped [21]. Since the wellbore flow pressure is always higher than the bubble-point pressure during production, the oil-water two-phase flow can be considered in the simulation process. The two-phase relative permeability curves of matrix, or hydraulic fracture and natural fracture are assigned as presented in Figure 2a,b, respectively [37]. Additionally, the clay content in the Lucaogou Formation is low [47], therefore, the clay swelling has not been considered in this work.

Table 1. Known field data for the base case.

Parameters	Values	Parameters	Values
Initial reservoir pressure, MPa	37	Fracture stages	17
Reservoir Temperature, °C;	90	HF spacing, m	70
Reservoir thickness, m	50	Fracture-closure pressure, MPa	58
Reservoir depth, m	3030	Bubble-point pressure, MPa	3.95
Total compressibility, kPa^{-1}	2×10^{-6}	Wellbore flowing pressure, MPa	5
Matrix porosity	0.1	Volume of water injected per frac, m^3	720
Matrix permeability, mD	0.01	Injection rate per frac, m^3/min	6
Matrix initial water saturation	0.2	Maximum injection pressure, MPa	80

Table 2. Assumed fracture network parameters for the base case.

Parameters	Values
HF permeability, mD	5000
HF half-length, m	120
HF width, m	0.04
NF permeability, mD	50
NF width, m	0.01
NF (connection with HF) length, m	70
NF (connection with HF) spacing, m	40
NF (no connection with HF) length, m	30
NF (no connection with HF) spacing, m	20
HF/NF porosity	0.6
HF/NF initial water saturation	0

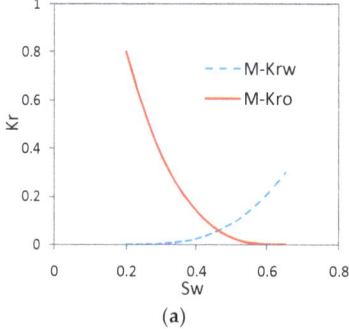
(a)

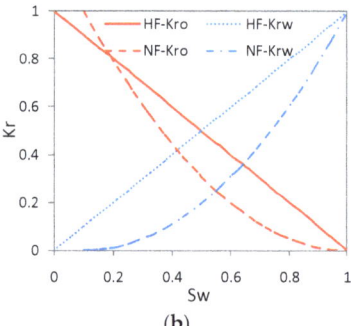
(b)

Figure 2. (a) Relative permeability for matrix; and (b) relative permeability for HF and NF.

2.1.1. Capillary Pressure Curve

The impact of imbibition hysteresis is considered in this study, so the capillary pressure curves are defined as functions of normalized water saturation (S_{wn}) according to Equations (1)–(3) [36]:

$$P_{cD} = a_1 + a_2(1 - S_{wn})^{a_3} \quad (1)$$

$$P_{cI} = b_1 + b_2(1 - S_{wn})^{b_3} - b_4 S_{wn}^{b_5} \quad (2)$$

The normalized water saturation is defined as:

$$S_{wn} = \frac{(S_w - S_{wc})}{(1 - S_{wc})} \qquad (3)$$

where P_{cD} is drainage capillary pressure; MPa; P_{cI} is imbibition capillary pressure; MPa; S_w is water saturation; S_{wc} is connate water saturation; and a_1, a_2, and a_3 are assumed to be equal to b_1, b_2, and b_3, respectively. In shale gas reservoirs, the rocks are usually more strongly water-wet. Therefore, only the positive part of the capillary force curve was considered in previous studies, by assuming both b_4 and b_5 as equal to zero. However, in tight oil reservoirs, the rocks are not all hydrophilic, but also show as intermediately wet or oil-wet. Thus, in this work, the negative capillary pressures are not neglected and hysteresis effect in the matrix is considered. The capillary pressure curve of the matrix is shown in Figure 3a, in which b_1, b_2, b_3, b_4, and b_5 are assigned to 0, 4.883, 20, 86, and 5, respectively [37].

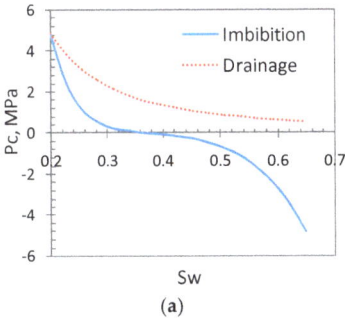 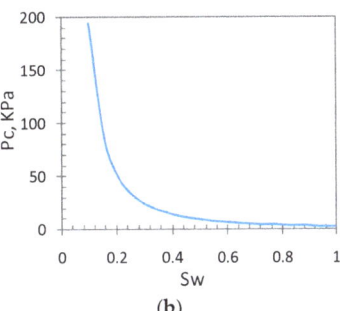

(a) (b)

Figure 3. (a) Capillary pressure for matrix; and (b) capillary pressure for NF.

For natural fractures, capillary pressure should not be zero, depending on wall roughness, fracture width and pore structure [48]. Here the J-function correlation in Equation (4) [49] is used to calculate capillary pressure for natural fractures, as shown in Figure 3b. Hydraulic fractures assume that capillary pressure is neglected because of high conductivity:

$$P_c = \frac{\sigma}{a_2'(S_w)^{a_1'}} \left(\frac{\varphi}{k}\right)^{a_3'} \times 6.895 \qquad (4)$$

In Equation (4), P_c is capillary pressure in NF, KPa; σ is interfacial tension and is equal to be 72 dynes/cm; φ is porosity; k is absolutely permeability, md; and a_1', a_2', and a_3' are equal to 1.86, 6.42, and 0.5, respectively.

2.1.2. Stress-Dependent Porosity and Permeability

During a hydraulic fracturing operation, a large amount of slickwater is pumped into a formation in a short time under high wellhead pressure, thereafter hydraulic fractures are created. Due to the limited fluid absorption capacity of the tight matrix, most of the high-pressure fluid leaks into natural fractures. With the increase of fluid pressure in natural fractures, the net closure stress within fractures decreases and the natural fracture width gradually increases until they are fully opened [50]. During the well shut-in, clean-up, and production periods, with the fracturing fluid imbibing into the matrix or being recovered to the surface, the net closure stress within fracture networks increases and the fracture gradually closes. Additionally, the reduction of hydraulic fracture width is effectively restrained because of proppant support, while un-propped natural fractures will be closed or even completely locked [23]. Previously, the matrix stress-dependent permeability was usually considered to enable rapid injection of so much fracturing fluid into the formation. However, it might not be realistic that the matrix permeability increases substantially with the increase of net pressure [23]. Therefore, in this work, stress dependence of porosity and permeability in fractures is considered by fully excavating the software [51,52], as shown in Figure 4. Four paths in the figure correspond to two processes: (1) Elastic path, and (2) plastic/dilation path is used to simulate the increase of porosity and permeability in fracture system during the injection process, (3) unloading path, and (4) recompaction path are used to simulate the decrease of porosity and permeability in fracture system during the flowback or production process.

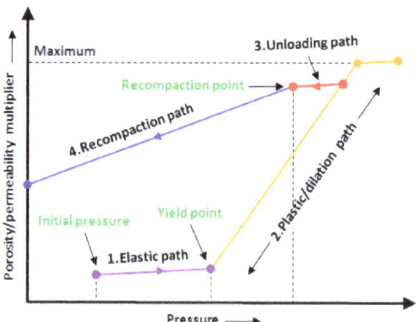

Figure 4. Diagram of model for HF and NF.

- During the injection process

The following empirical exponential formulas Equations (5) and (6) are used to describe the variation of porosity and permeability with the pressure in fracture system to realize the actual pumping rate and time [36,53]:

$$\frac{\varphi_f}{\varphi_{f0}} = e^{C_{fi} P_{net}} \tag{5}$$

$$\frac{k_f}{k_{f0}} = 10^{m_i P_{net}} \tag{6}$$

In Equations (5) and (6), φ_f is current fracture porosity; φ_{f0} is original fracture porosity; C_{fi} is the fracture compressibility during injection, KPa^{-1}; P_{net} is net pressure within fractures (difference between original pressure and current pressure), KPa; k_f is current fracture permeability, md; k_{f0} is original fracture permeability, md; m_i is the permeability changing factor, KPa^{-1} (exponent empirically determined). Tiab et al. [54] reported that the fracture compressibility was as much as two orders of magnitude higher than the matrix compressibility. Thus, C_{fi} is assigned a value of 5.97×10^{-5}, which is within a reasonable range. The values of m_i in HF and NF systems are assigned 6.67×10^{-5} and 1.00×10^{-4} [36,37]. The calculated results of stress-dependence in fractures are presented in Figure 5.

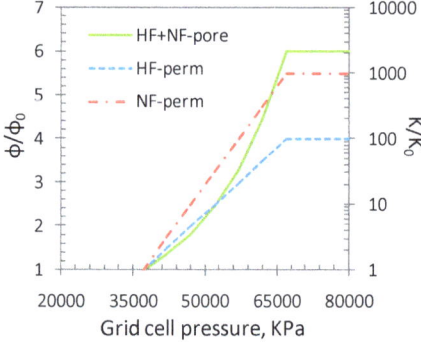

Figure 5. Stress-dependent porosity and permeability during injection.

- During the production process

In the process from well shut-in to production, fractures would gradually close, which is different from fracture dilation in the previous section. Many studies have been carried out to clarify the fracture-closure mechanism, among which the main difficulty lies in the determination of fracture compressibility [55,56]. For un-propped natural fractures, Equation (7) proposed by Jones [57] can be used:

$$C_{fp} = \frac{-1}{P_{net}\ln(P_{net}/P_h)} \qquad (7)$$

In Equation (7), C_{fp} is the fracture compressibility during production, KPa^{-1}; P_h is an apparent healing pressure and is equal to be 1.38×10^5 KPa. However, for propped fractures, the value of C_{fp} can refer to the chart of fracture compressibility inferred by Aguilera [41] as shown in Figure 6. Aguilera [41] emphasizes that the graph is only an approximation, which is not to replace good laboratory work. The correlation is useful only if laboratory data are not available. Figure 6 shows the results of filling mineral ratio from 0% to 50%, which represents the percentage of minerals in fractures. Fu [25] and Williams-Kovacs [56] had proposed that proppants in fractures could be assumed as minerals in Aguilera model. Therefore, the mineral ratio in this work can be the percentage of proppants in fractures, so that the compressibility of propped fractures can be obtained. When the fracture compressibility is determined, the stress-dependent porosity and permeability in the production process can be calculated by Equations (8) and (9) [55,58]. The calculation results of stress-dependence in fractures are presented in Figure 7a,b.

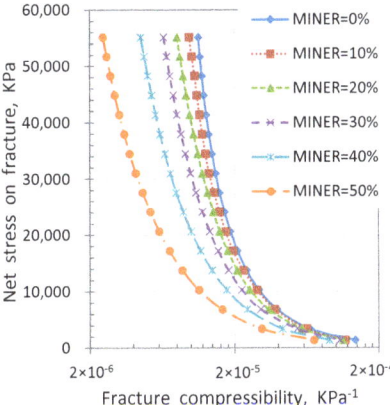

Figure 6. Chart for estimating fracture compressibility (Aguilera [41]).

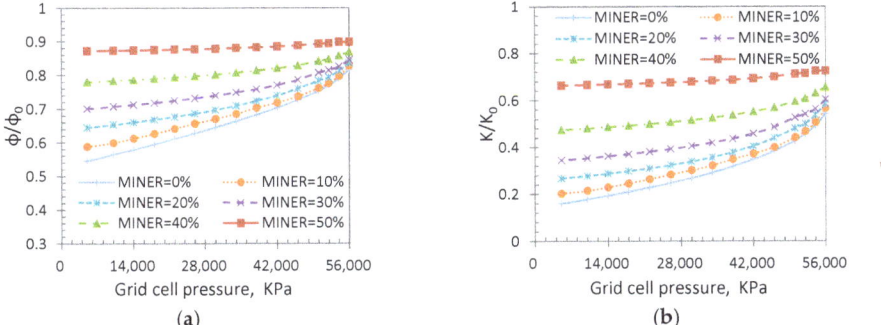

Figure 7. (a) Stress-dependent porosity during production; and (b) stress-dependent permeability during production.

$$\frac{\varphi_f}{\varphi_{f0}} = e^{C_{fp}P_{net}} \tag{8}$$

$$\frac{k_f}{k_{f0}} = \left(\frac{\varphi_f}{\varphi_{f0}}\right)^3 = e^{3C_{fp}P_{net}} \tag{9}$$

2.2. Model Validation

Based on the established basic model, the processes of fracturing fluid injection and flowback are sequentially simulated, in which the imbibition hysteresis and stress-dependent porosity and permeability are considered. The model is verified by checking the bottom-hole pressure curve at the injection stage and the rate transient analysis at the flowback stage.

2.2.1. Bottom-Hole Pressure Curve

The injection of slickwater during hydraulic fracturing is simulated through three different mechanisms. The cumulative injection volume is 720 m^3 and the pumping rate is 6 m^3/min, which are consistent with the field data reported by Chen et al. [59]. The simulation results under different mechanisms are shown in Figure 8. Figure 8a presents the variation of bottom-hole pressure with pumping time. When stress-dependent porosity is considered, the injection can smoothly proceed at an actual field scale. Pressure rises sharply to the threshold at the early stage, then drops to the vicinity of fracture-closure pressure and slowly increases with the extension of pumping time. Although proppant is absent from the simulation, the pressure variation characteristics conform to the fracturing curve in the field [59], considering stress-dependent porosity. By contrast, the other two mechanisms reflect their weaknesses. When stress-dependent permeability and porosity are neglected, it is very difficult to inject water into the formation, for example, it takes nearly a week to inject 720 m^3 water, which may unrealistically overestimate the sweep range of pressure in the reservoir. When stress-dependent permeability is the sole mechanism, then pumping rate can barely reach 1 m^3/min even though injection capacity is improved.

When stress-dependence is neglected, as shown in Figure 8b, more than half of the injected water leaks into the matrix. However, when stress-dependent porosity and permeability is considered, nearly 80% of water remains in the fracture systems. Fu et al. [24] analyzed flowback data from several wells in the field and predicted that an average of 77.7% of the injected water was in the fracture system. The distribution of fracturing water in our base case is very close to Fu's report, so the water-loss in the matrix should be overestimated when stress-dependent porosity is absent.

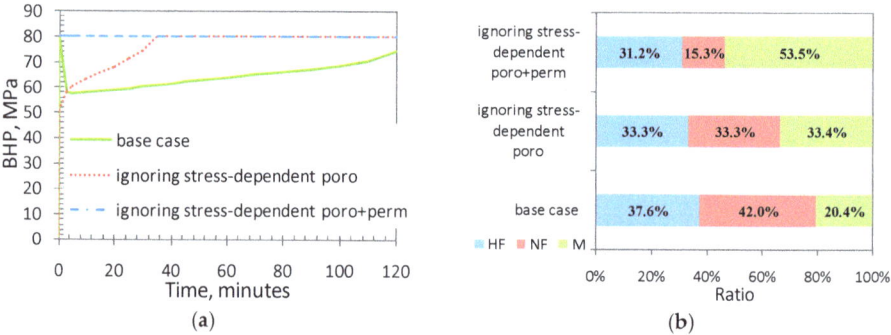

Figure 8. (a) Bottom-hole pressure curves during injection; and (b) the distribution ratio of injected water at the end of injection.

2.2.2. Rate Transient Analysis

Next, the flow-regime identification of fracturing fluid during early flowback is carried out for the base case. The water recovery is seldom at a constant rate, nor is it at a constant flowing pressure. The rate-normalized pressure (RNP) and its derivatives (RNP') are used to represent the flowback behavior that would be observed if the well were produced at a constant reference rate. The formulas are as follows [60]:

$$\text{RNP} = \frac{P_i - P_{wf}}{q_w} \quad (10)$$

$$\text{RNP}' = \frac{d\text{RNP}}{d\ln t_{MB}} \quad (11)$$

$$t_{MB} = Q_w / q_w \quad (12)$$

In Equations (10)–(12), P_i is initial pressure in fractures, KPa; P_{wf} is bottom-hole pressure, KPa; q_w is water rate, m³/day; t_{MB} represents the equivalent time, day; and Q_w is the cumulative water production volume, m³. Figure 9 uses a log-log plot of RNP and RNP' vs. tMB during early flowback to show the sequence of flow regimes observed with: (1) Transient linear flow (half slope) within fractures occurs when the flow regime presents water flow from the hydraulic fractures to the perforations; (2) fracture depletion (unit slope): occurs when the pressure response reaches the boundary of the fractures; (3) transient linear flow (half slope) in the matrix occurs when the flow regime begins with oil or gas breakthrough from matrix to fractures. This transition might be very short depending on the fracture conductivity and fracture complexity; and, finally, (4) matrix boundary dominated flow (unit slope) occurs when the density of natural fractures is large and the spacing between hydraulic fractures is small, the pressure response quickly propagates to the matrix boundary.

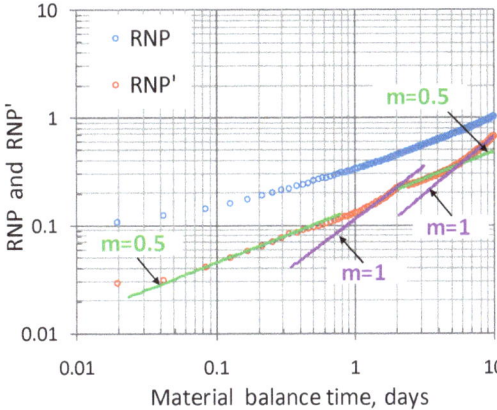

Figure 9. Water RNP and RNP' plot during early flowback for the base case.

The flow regimes observed in Figure 9 are consistent with field observations and analytical models for flowback analysis [61–63]. The first two flow regimes are well captured and last for more than two days.

3. Results and Discussion

Despite the good consistency of the base case results, an ideal historical match with actual flowback data is still difficult to achieve. There are two main reasons: (1) most of the flowback data obtained from the field are daily output, occasionally hourly data, which can hardly reflect the characteristics of early flowback [31]; and (2) properties of fracture systems are highly uncertain including fracture compressibility, density, and conductivity [37]. These uncertainties not only make the quantitative validation of simulation results extremely challenging, but also make it difficult to understand the retention mechanism of water and its impact on production. Therefore, a series of models were established to quantify the impact of various factors on water-loss and production performance.

3.1. Impact of Fracture Closure

Extensive analytical models have been used to reveal the fracture depletion process during early flowback. However, these models have some limitations, such as sequential flow and single-phase flow that make it difficult to explain the effect of fracture-closure on two-phase flow in the triple media after oil breakthrough. The flow region of fracture depletion was validated in the previous section, and the impact of fracture-closure on fracturing water retention is modeled here. Three models are built considering different mechanisms: (1) with stress-dependent porosity and permeability (base case), (2) ignoring stress-dependent porosity, and (3) ignoring stress dependence. The simulation result at the end of injection process in basic model is taken as the initial condition to better compare the impacts of three mechanisms on water-loss during flowback. The simulation process is as follows: shut-in for 22 h, then flowback for 10 days, followed by production for half a year.

Figure 10 shows the variation of bottom-hole pressure with time under different mechanisms during flowback. It can be seen from the figure that when considering stress-dependent porosity, the bottom-hole pressure is almost maintained near the fracture-closure pressure for the first half-day, then the pressure drop is gradually accelerated, even lower than the case without considering stress-dependent during the late flowback stage (from the insert figure in Figure 10). The drive mechanism of fracture-closure can be captured, and this mechanism dominates for nearly five days. The simulation results of production are shown in Table 3. The oil breakthrough time of the base case is the latest, and the single-phase flow duration is one day longer than that of the other two models, which further reflects the fractured storage effect. Moreover, the water recovery efficiency of the basic case is 55.3%, while ignoring stress-dependent porosity overestimates by 20.8% and ignoring stress dependence overestimates by 26.6%.

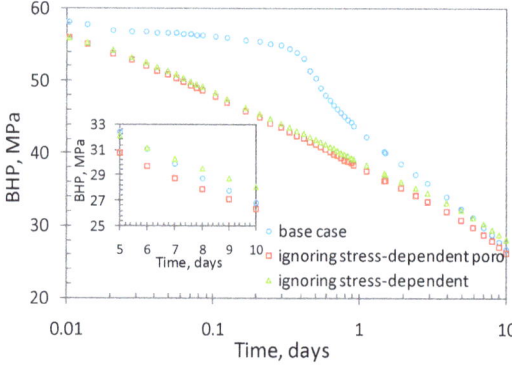

Figure 10. Bottom-hole pressure profiles with various mechanisms during flowback.

Table 3. Impact of various mechanisms on the half-year production process.

Simulation Cases	Base Case	Ignoring Stress Dependent Pore	Ignoring Stress Dependence
Cumulative oil, m^3	1971.3	2130.6	2267.7
Recovery efficiency, %	55.3	66.8	70.0
Oil breakthrough time, day	2.3	1.3	1.4

The reasons for these differences are illustrated in Figure 11: (1) the S_w near HF (0.04 m to the fracture surface) in base case remains above 50% during early flowback, and Sw in the deeper matrix also substantially increased. While in the other two models, the S_w near HF has been reduced to

43% and almost no water leaks far away; and (2) the equilibrium point S_w = 0.4 shown in Figure 3a. The S_w near HF is much higher than 0.4 at the end of fracturing, so the hysteresis results in a capillary force that is resistant to water imbibition. With the decrease of water in HF during flowback, if the fracture-closure is ignored, the water near a fracture will flow back into the fracture, which leads to an increase in water recovery efficiency. On the contrary, when fracture compressibility is considered, the reduced water volume in fractures will be equal to the volume of fracture-closure, which provides an opportunity for water near fractures to infiltrate into the deeper matrix.

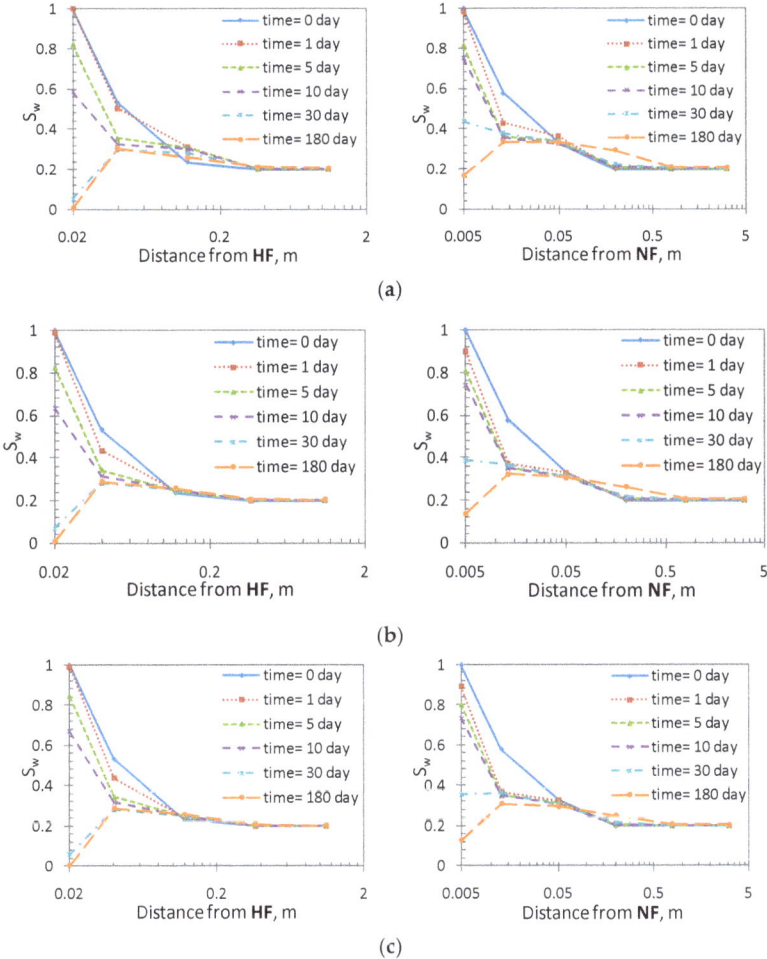

Figure 11. Water saturation distribution from HF or NF surface to deeper matrix under flowback and production periods. From top row to bottom row are: (**a**) basic case, (**b**) ignoring stress-dependent porosity, and (**c**) ignoring stress dependence.

From Figure 11, it can be found that the oil phase appears in NF after one day of flowback, when stress-dependent porosity is absent. This is also a good explanation for why the oil breakthrough time in the base case is delayed. By contrast, the cumulative oil production is the lowest for the basic case. Relatively high water saturation reduces the oil permeability because more water leaks into the matrix. The S_w within NF is relatively high (17%) for the base case, which limits the flow of oil in NF. However, the oil production gap between three models is not large because tight reservoirs do not require very high fracture conductivity, and proppant that is evenly distributed within a fracture should be more important for well productivity [64].

3.2. Impact of Natural Fracture Density

Extensive studies of natural fracture density have been carried out [21,24,37], but the relationship between fracture complexity and fracture width has not been considered. Zou et al. [65] reported that the average width of the fracture network decreased as the complexity of a fracture system increased. If this relationship is neglected, the total volume of a complex fracture network should be overestimated with the increase of natural fracture density, which will inevitably interfere with the simulation of water injection and the flowback process. Here, a fracture complexity index (FCI) is introduced to quantitatively evaluate natural fracture density, as shown in Equation (13) [66]:

$$\text{FCI} = \frac{V_{nf}}{V_{hf} + V_{nf}} \tag{13}$$

where FCI is fracture complexity index; V_{nf} is volume of natural fractures; and V_{hf} is volume of hydraulic fractures. The value of FCI increases with the augment of fracture complexity.

A series of models are set up as follows: (1) considering that the half-length of HF is constant for all cases, increasing the number of natural fractures increases the value of FCI; and (2) by adjusting the width of HF and NF, the initial volume of fracture system in each case is equal, as shown in Table 4, and the width of un-propped NF is assigned less than 2.5 mm [17]. The simulation process is as follows: (1) water injection for two hours and well shut-in for 22 h [32], then (2) clean-up at a constant rate for 10 days (in the field the choke size is gradually increasing reported by Clarkson [62], which is simplified here for simulation and analysis), followed by (3) production at constant pressure for one year.

Table 4. The setting of fracture complexity for cases.

Case Number	1	2	3 (Base Case)	4	5
HF initial effective width, mm	10	6	4	2.8	2
NF initial effective width, mm	-	1.5	1	0.7	0.5
Fracture-reservoir contact area, ×10^4 m^2	2.4	6.6	13.8	21.6	31
FCI	0	0.37	0.54	0.67	0.75
Fluid efficiency, %	89	86.3	79.6	75.8	70.4
Grid diagram					

As can be seen from Table 4, with the increase of natural fracture density, the contact area between fracture and reservoir increases, which reduces the fluid efficiency during injection and more fracturing fluid leaks into the matrix. Figure 12a shows that oil breakthrough occurs in all cases during flowback, and the oil rate increases with the increase of NF density. On the first day of converting to constant pressure production, the oil rate of case 5 reached 47.5 m^3/d, which was 4.5 times higher than that of case 1. However, Figure 12a illustrates that oil production rate decreases with the increase of NF density after half a year of production. The reasons for this phenomenon may be: (1) as the fractures become more complex, more water is imbibed into the deeper matrix, which reduces the oil permeability and interferes the long-term production; (2) due to stress-sensitivity in fractures, the longer production time is related to greater failure of un-propped NF conductivity; and (3) the simulated NFs here are completely connected, so the pressure response quickly reaches the matrix boundary, which limits the oil drainage area. As can be seen from Table 5, the cumulative oil of case 5 is 64% higher than that of case 1 in half a year of production, and the gap narrows to 40% after one year.

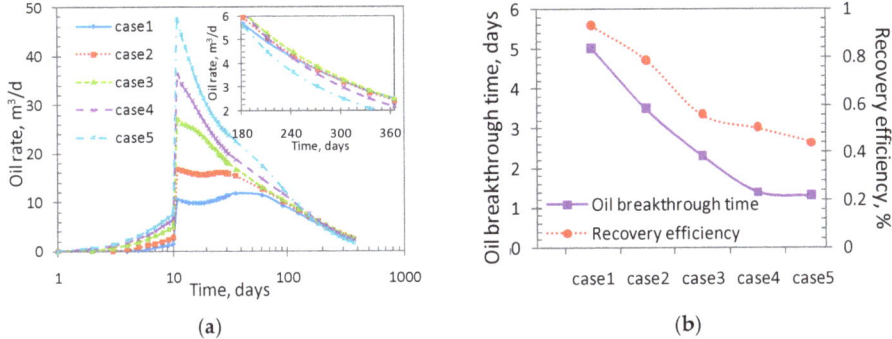

Figure 12. (a) Oil production profiles at various NF densities; and (b) oil breakthrough time and flowback rate profiles at various NF densities.

Table 5. Impact of various NF densities on Flowback and production performances.

Case Number	1	2	3 (Base Case)	4	5
Cumulative oil (half-year), m^3	1498.9	1764.0	1971.3	2208.2	2460.2
Cumulative oil (one-year), m^3	2149.6	2417.7	2646.6	2831.0	2992.4
Flowback rate, %	90.1	74.8	55.3	50.1	43.9

Figure 12b shows that with the increase of NF density, the oil breakthrough time is gradually shortened, and the water recovery efficiency decreases accordingly. For case 1, a single-phase flow lasts for five days under the drive mechanism of fracture closure, and more than 90% of water is recovered after long-term production. This is because of limited contact area between HF and the matrix, the effect of imbibition hysteresis, and high HF conductivity. With the increase of FCI, the drive of HF closure is limited, and transient linear flow in the matrix appears in advance, so the oil breakthrough time is advanced. In addition, even in the production process, spontaneous imbibition is continuing, especially for un-propped NFs with low fracture conductivity. For Case 5, although there is no long-term shut-in, the flowback rate is still less than 50%.

3.3. Impact of Proppant Distribution

Although high fracture conductivity is not required for tight reservoirs, open fractures are better than collapsed ones. Proppants with various grain diameters (ranging from 0.104 mm to 0.838 mm) are commonly used to pack fracture systems during fracturing [67]. As shown in Figure 6, the fracture-closure trend is related to proppant concentration, so it is necessary to assess the effect of proppant distribution on water and oil production. Moreover, proppant embedment depending on formation mechanical properties has been extensively studied [42,68–70]. The fracture width loss during production is associated with embedment, especially in soft or weakly consolidated formations, which may lead to a reduction of 60% in propped fracture width [68]. Therefore, the effect of compaction and embedment on fracture conductivity are considered comprehensively in this work.

Due to the lithology of Lucaogou Formation in this paper is similar to the core samples used in Chen's model [42], we introduce his formulas, Equations (14)–(16), for calculating proppant embedment that allows for modifying the stress-dependent porosity and permeability [42] with the results shown in Figure 13.

$$\frac{\varphi}{\varphi_0} = e^{C_{fp}P_{net}} \cdot \left(\frac{w_{f0} - h}{w_{f0}}\right) \tag{14}$$

$$\frac{k}{k_0} = \left(\frac{\varphi}{\varphi_0}\right)^3 = e^{3C_{fp}P_{net}} \cdot \left(\frac{w_{f0} - h}{w_{f0}}\right)^3 \tag{15}$$

$$h = \eta(P_{net})^\lambda \tag{16}$$

In Equations (14)–(16), w_{f0} is fracture width before embedment, m; h is the proppant embedment, m; $\eta = 2.1 \times 10^{-5}$, $\lambda = 2.8$, when the proppant is 20/40-mesh; and $\eta = 1.6 \times 10^{-5}$, $\lambda = 3.1$, when the proppant is 30/60-mesh.

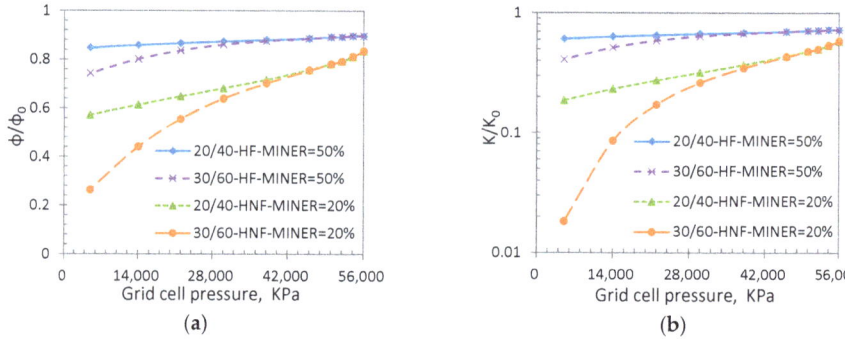

Figure 13. (a) Stress-dependent porosity considering both compaction and embedment; and (b) stress-dependent permeability considering both compaction and embedment.

A series of models are established and various proppant distributions are shown in Table 6. Figure 14a shows that the flowback rate is substantially increased when HNF is propped, which is 24.8% and 32.8% higher than un-propped. When NNF is propped, there is a slight increase of flowback rate because of the restraint of low conductivity and stress-dependence in NF. Water within HF and HNF is preferred to recover during early flowback, and a considerable portion of water in NNF is imbibed into the matrix. In addition, it can also be seen from the figure that the variation trends of oil production are similar to the results of water recovery, but there is an obvious difference at various proppant concentrations in HF. The oil production decreases by nearly 25% when proppant concentration in HF decreases, while water production decreases by less than 8% because water production is mainly at the flowback stage, which is mainly driven by fracture closure at the beginning. While oil production is mainly at the production stage, and stress dependence plays a significant role with the extension of time.

Figure 14b shows that when the proppant embedding is considered, the impact on flowback is negligible because water recovery mainly occurs in the early stage of production (flowback) when the main mechanism of fracture closure is compaction. Cumulative oil production is lower when the embedment occurs, and smaller proppant size results in greater loss of production. Furthermore, compared with the embedment in HF, there is greater impact of embedment in NF on oil production. Therefore, it is beneficial to use small size proppant to promote the propping of micro-fractures, but it is also necessary to examine its embedment in micro-fractures.

Table 6. Proppant distributions for construction models.

Fracture Types	HF	HNF	NNF	Embedment
Case 1	Miner = 50%	Miner = 20%	Miner = 20%	-
Case 2	Miner = 50%	Miner = 20%	Miner = 0%	-
Case 3 (base)	Miner = 50%	Miner = 0%	Miner = 0%	-
Case 4	Miner = 20%	Miner = 20%	Miner = 20%	-
Case 5	Miner = 20%	Miner = 20%	Miner = 0%	-
Case 6	Miner = 20%	Miner = 0%	Miner = 0%	-
Case 7	Miner = 50%	Miner = 20%	Miner = 0%	20/40-mesh
Case 8	Miner = 50%	Miner = 0%	Miner = 0%	20/40-mesh
Case 9	Miner = 50%	Miner = 20%	Miner = 0%	30/60-mesh
Case 10	Miner = 50%	Miner = 0%	Miner = 0%	30/60-mesh

HNF means an HF-connected NF, and NNF means a non-HF-connected NF.

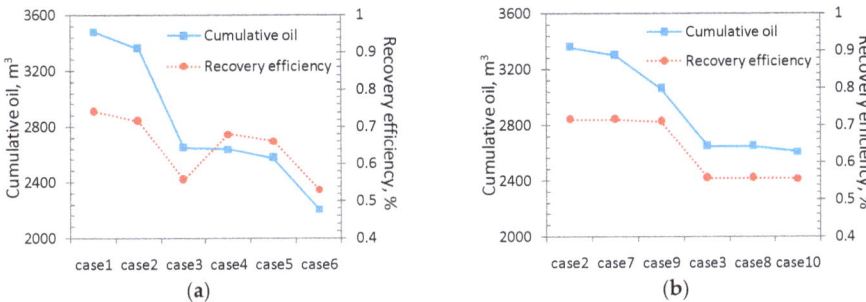

Figure 14. (a) Cumulative oil and flowback rate profiles at various propping patterns without embedment; and (b) cumulative oil and flowback rate profiles considering embedment.

3.4. Impact of Natural Fracture Heterogeneity

Natural fractures opened during fracturing cannot be uniform, and in the subsequent flowback or production process, excessive closure of some positions may be a key mechanism of water retention. As shown in Figure 15, NF width is generally non-uniform [7,50].

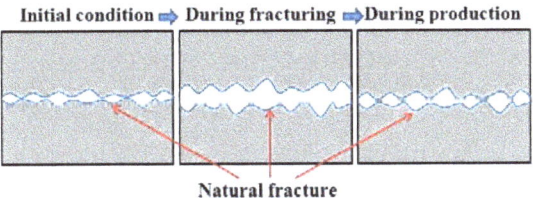

Figure 15. Illustration of stress sensitivity for natural fracture (modified from Liu et al. [50]).

In this work, the heterogeneity of NF conductivity is studied using the Yu et al. [71] approach to describe heterogeneous distribution of matrix permeability. Three variation coefficients of permeability are used to represent weak heterogeneity, medium homogeneity, and strong heterogeneity, as shown in Figure 16. Other model parameters and simulation processes are the same as the basic case. Figure 17 illustrates that, after one year of production, the amount of fracturing fluid remaining in reservoirs increases with the increase of heterogeneity of NF permeability. Figure 18a shows that the flowback rate decreases with the increase of heterogeneity, in which the rate of strong heterogeneity is 11.2% lower than that of homogeneity. For weak homogeneity, the water recovery efficiency is close to that of homogeneous natural fracture. Figure 18b shows that the closed natural fracture area still maintains high water saturation during production when there is strong heterogeneity. In addition, the heterogeneity of NF conductivity is 0.1%, 1.5%, and 4.1% less effective for oil production, compared with homogeneity. The reason is that high conductivity is not required for natural fractures in tight reservoirs, and apart from over-closed areas, connected fracture networks are still contributing to productivity.

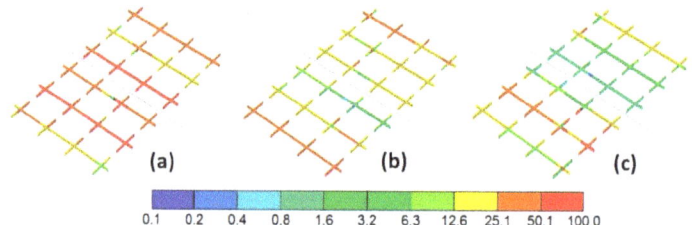

Figure 16. Three cases of NF permeability heterogeneity: (**a**) weak heterogeneity, (**b**) medium heterogeneity, and (**c**) strong heterogeneity (zoom scale: X = 2 times).

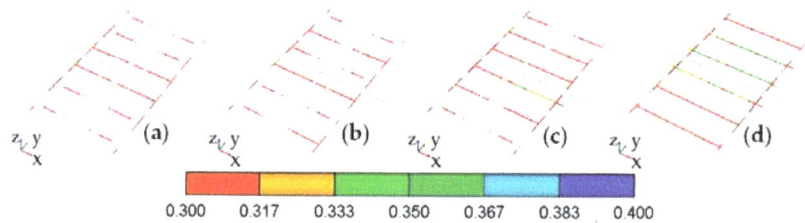

Figure 17. Water saturation after one-year production: (**a**) homogeneity (base case), (**b**) weak heterogeneity, (**c**) medium heterogeneity, and (**d**) strong heterogeneity (zoom scale: X = 2 times).

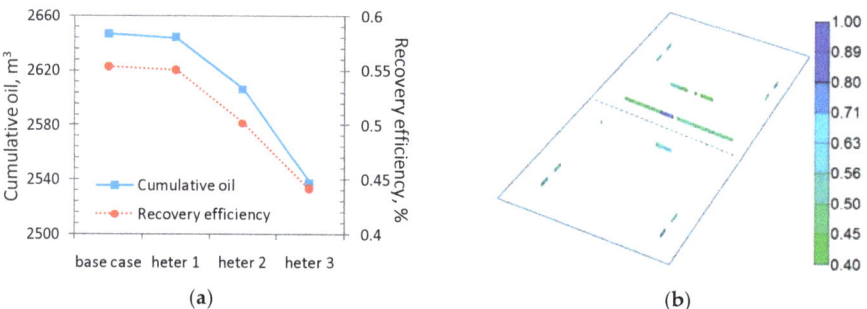

Figure 18. (**a**) Cumulative oil and flowback rate profiles for various NF heterogeneity (heter1: weak heterogeneity; heter2: medium heterogeneity; heter3: strong heterogeneity); and (**b**) water saturation within NFs after one-year production for strong heterogeneity (zoom scale: X = 2 times).

4. Conclusions

A stress-dependent porosity and permeability model is used to numerically simulate the opening and closing of fracture systems during fracturing and flowback operations. The simulation results indicate that fracture-closure is the main driving mechanism of fracturing water recovery in the early stage of clean-up, which promotes water imbibition into deeper matrix, resulting in increased water loss.

Uncertain parameters for the fracture network are assessed with a series of mechanistic models. With the increase of fracture complexity, more fracturing water remains underground, and the oil breakthrough time advances. Although tight reservoirs do not require high fracture conductivity, proppant packing in secondary fractures (opened natural fractures) can improve well productivity. The occurrence of proppant embedment will accelerate the decline of long-term production. In addition, the heterogeneity of natural fractures may be an important factor that results in low water recovery efficiency. Excessive closure of part positions results in water locking, while its impact on production is much less compared with other factors. This work provides a new insight into the mechanism of water retention in fracture systems.

Author Contributions: Methodology: K.L.; validation: K.L. and S.Z.; investigation: K.L. and X.M.; writing—original draft preparation: K.L., S.Z., and Y.Z.

Acknowledgments: This paper was jointly support by the National Natural Science Foundation Projects of China (51574255), the Major National Science and Technology Project of China (2015CB250903) and the Major National Science and Technology Projects of China (no. 2016ZX05046-004). The authors would also like to thank Computer Modeling Group (CMG) for their simulation software.

Conflicts of Interest: The authors declare no conflict of interest.

References

1. Greene, D.L.; Hopson, J.L.; Li, J. *Running out of and into Oil: Analyzing Global Depletion and Transition through 2050*; ORNL/TM-2003/259; Oak Ridge National Laboratory: Oak Ridge, TN, USA, 2003. Available online: http://www.osti.gov/scitech/biblio/885837 (accessed on 14 November 2003).
2. Ezulike, O.D.; Dehghanpour, H. A complementary approach for uncertainty reduction in post-flowback production data analysis. *J. Nat. Gas Sci. Eng.* **2015**, *27*, 1074–1091. [CrossRef]
3. Wei, Y.; Ran, Q.; Li, R.; Yuan, J.; Dong, J. Determination of dynamic reserves of fractured horizontal wells in tight oil reservoirs by multi-region material balance method. *Pet. Explor. Dev.* **2016**, *43*, 490–498. [CrossRef]
4. Fisher, M.K.; Heinze, J.R.; Harris, C.D.; Davidson, B.M.; Wright, C.A.; Dunn, K.P. Optimizing horizontal completion techniques in the barnett shale using microseismic fracture mapping. In Proceedings of the SPE Annual Technical Conference and Exhibition, Society of Petroleum Engineers, Houston, TX, USA, 26–29 September 2004. [CrossRef]
5. Jaripatke, O.A.; Chong, K.K.; Grieser, W.V.; Passman, A. A completions roadmap to shale-play development: A review of successful approaches toward shale-play stimulation in the last two decades. In Proceedings of the International Oil and Gas Conference and Exhibition in China, Beijing, China, 8–10 June 2010. [CrossRef]
6. Holditch, S.A.; Tschirhart, N. Optimal stimulation treatments in tight gas sands. In Proceedings of the SPE Annual Technical Conference and Exhibition, Dallas, TX, USA, 9–12 October 2005. [CrossRef]
7. Palisch, T.T.; Vincent, M.C.; Handren, P.J. Slickwater fracturing: Food for thought. In Proceedings of the SPE Annual Technical Conference and Exhibition, Denver, CO, USA, 21–24 September 2008. [CrossRef]
8. Balhoff, M.; Miller, M.J. Modeling fracture fluid cleanup in hydraulic fractures. In Proceedings of the SPE Annual Technical Conference and Exhibition, San Antonio, TX, USA, 29 September–2 October 2002. [CrossRef]
9. Crafton, J.W.; Gunderson, D.W. Stimulation flowback management—Keeping a good completion good. In Proceedings of the SPE Annual Technical Conference and Exhibition, Anaheim, CA, USA, 11–14 November 2007. [CrossRef]
10. Ezulike, O.D.; Dehghanpour, H. Modelling flowback as a transient two-phase depletion process. *J. Nat. Gas Sci. Eng.* **2014**, *19*, 258–278. [CrossRef]
11. King, G.E. Hydraulic fracturing 101: What every representative, environmentalist, regulator, reporter, investor, university researcher, neighbor and engineer should know about estimating frac risk and improving frac performance in unconventional gas and oil wells. In Proceedings of the SPE Hydraulic Fracturing Technology Conference, The Woodlands, TX, USA, 6–8 February 2012. [CrossRef]
12. Makhanov, K.; Dehghanpour, H.; Kuru, E. Measuring liquid uptake of organic shales: A workflow to estimate water loss during shut-in periods. In Proceedings of the SPE Unconventional Resources Conference Canada, Calgary, AB, Canada, 5–7 November 2013. [CrossRef]
13. Song, Z.; Hou, J.; Zhang, L.; Chen, Z.; Li, M. Experimental study on disproportionate permeability reduction caused by non-recovered fracturing fluids in tight oil reservoirs. *Fuel* **2018**, *226*, 627–634. [CrossRef]

14. Wang, G.; Zhao, Z.; Li, K.; Li, A.; He, B. Spontaneous imbibition laws and the optimal formulation of fracturing fluid during hydraulic fracturing in Ordos basin. *Procedia Eng.* **2015**, *126*, 549–553. [CrossRef]
15. Barati, R.; Hutchins, R.D.; Friedel, T.; Ayoub, J.A.; Dessinges, M.-N.; England, K.W. Fracture impact of yield stress and fracture-face damage on production with a three-phase 2D model. *SPE Prod. Oper.* **2009**, *24*, 336–345. [CrossRef]
16. Regina, T.A.; Mustafa, M.A.; Belladonna, M.; Reza, B. Simulating yield stress variation along hydraulic fracture face enhances polymer cleanup modeling in tight gas reservoirs. *J. Nat. Gas Sci. Eng.* **2019**, *65*, 32–44. [CrossRef]
17. Sharma, M.M.; Manchanda, R. The role of induced un-propped (IU) fractures in unconventional oil and gas wells. In Proceedings of the SPE Annual Technical Conference and Exhibition, Houston, TX, USA, 28–30 September 2015. [CrossRef]
18. Ren, W.; Li, G.; Tian, S.; Sheng, M.; Geng, L. Analytical modelling of hysteretic constitutive relations governing spontaneous imbibition of fracturing fluid in shale. *J. Nat. Gas Sci. Eng.* **2016**, *34*, 925–933. [CrossRef]
19. Dutta, R.; Lee, C.-H.; Odumabo, S.; Ye, P.; Walker, S.C.; Karpyn, Z.T.; Ayala, H.L.F. Experimental investigation of fracturing-fluid migration caused by spontaneous imbibition in fractured low-permeability sands. *SPE Reserv. Eval. Eng.* **2014**, *17*, 74–81. [CrossRef]
20. Yuan, W.; Li, X.; Pan, Z.; Connell, L.D.; Li, S.; He, J. Experimental investigation of interactions between water and a lower silurian Chinese shale. *Energy Fuels* **2014**, *28*, 4925–4933. [CrossRef]
21. Almulhim, A.; Alharthy, N.; Tutuncu, A.N.; Kazemi, H. Impact of imbibition mechanism on flowback behavior: A numerical study. In Proceedings of the Abu Dhabi International Petroleum Exhibition and Conference, Abu Dhabi, UAE, 10–13 November 2014. [CrossRef]
22. Jing, W.; Huiqing, L.; Genbao, Q.; Yongcan, P.; Yang, G. Investigations on spontaneous imbibition and the influencing factors in tight oil reservoirs. *Fuel* **2019**, *236*, 755–768. [CrossRef]
23. Ehlig-Economides, C.A.; Ahmed, I.A.; Apiwathanasorn, S.; Lightner, J.H.; Song, B.; Vera Rosales, F.E.; Xue, H.; Zhang, Y. Stimulated shale volume characterization: Multiwell case Study from the Horn River shale: II. Flow perspective. In Proceedings of the SPE Annual Technical Conference and Exhibition, San Antonio, TX, USA, 8–10 October 2012. [CrossRef]
24. McClure, M. The potential effect of network complexity on recovery of injected fluid following hydraulic fracturing. In Proceedings of the SPE Unconventional Resources Conference, The Woodlands, TX, USA, 1–3 April 2014. [CrossRef]
25. Fu, Y.; Dehghanpour, H.; Ezulike, D.O.; Jones, R.S., Jr. Estimating effective fracture pore volume from flowback data and evaluating its relationship to design parameters of multistage-fracture completion. *SPE Prod. Oper.* **2017**, *32*, 423–439. [CrossRef]
26. Balogun, A.S.; Kazemi, H.; Ozkan, E.; Al-kobaisi, M.; Ramirez, B. Verification and proper use of water-oil transfer function for dual-porosity and dual-permeability reservoirs. *SPE Reserv. Eval. Eng.* **2009**, *12*, 189–199. [CrossRef]
27. Barzegar Alamdari, B.; Kiani, M.; Kazemi, H. Experimental and numerical simulation of surfactant-assisted oil recovery in tight fractured carbonate reservoir cores. In Proceedings of the SPE Improved Oil Recovery Symposium, Tulsa, OK, USA, 14–18 April 2012. [CrossRef]
28. Ghanbari, E. Water Imbibition and Salt Diffusion in Gas Shales: A Field and Laboratory Study. Master's Thesis, University of Alberta, Edmonton, AB, Canada, 2015.
29. Javaheri, A.; Habibi, A.; Dehghanpour, H.; Wood, J.M. Imbibition oil recovery from tight rocks with dual-wettability behavior. *J. Pet. Sci. Eng.* **2018**, *167*, 180–191. [CrossRef]
30. Wang, C.; Cui, W.; Zhang, H.; Qiu, X.; Liu, Y. High efficient imbibition fracturing for tight oil reservoir. In Proceedings of the SPE Trinidad and Tobago Section Energy Resources Conference, Port of Spain, Trinidad and Tobago, 25–26 June 2018. [CrossRef]
31. Wang, M.; Leung, J.Y. Numerical investigation of fluid-loss mechanisms during hydraulic fracturing flow-back operations in tight reservoirs. *J. Pet. Sci. Eng.* **2015**, *133*, 85–102. [CrossRef]
32. Wang, M.; Leung, J.Y. Numerical investigation of coupling multiphase flow and geomechanical effects on water loss during hydraulic-fracturing flowback operation. *SPE Reserv. Eval. Eng.* **2016**, *19*, 520–537. [CrossRef]

33. Bennion, D.B.; Thomas, F.B.; Bietz, R.F.; Bennion, D.W. Water and hydrocarbon phase trapping in porous media-diagnosis, prevention and treatment. *J. Can. Pet. Technol.* **1996**, *35*, 8. [CrossRef]
34. Blasingame, T.A. The characteristic flow behavior of low-permeability reservoir systems. In Proceedings of the SPE Unconventional Reservoirs Conference, Keystone, CO, USA, 10–12 February 2008. [CrossRef]
35. Shaoul, J.R.; Van Zelm, L.F.; De Pater, H.J. Damage mechanisms in unconventional gas well stimulation—A new look at an old problem. In Proceedings of the SPE Middle East Unconventional Gas Conference and Exhibition, Muscat, Oman, 31 January–2 February 2011. [CrossRef]
36. Jurus, W.J.; Whitson, C.H.; Golan, M. Modeling water flow in hydraulically-fractured shale wells. In Proceedings of the SPE Annual Technical Conference and Exhibition, New Orleans, LA, USA, 30 September–2 October 2013. [CrossRef]
37. Zhang, T.; Li, X.; Li, J.; Feng, D.; Li, P.; Zhang, Z.; Chen, Y.; Wang, S. Numerical investigation of the well shut-in and fracture uncertainty on fluid-loss and production performance in gas-shale reservoirs. *J. Nat. Gas Sci. Eng.* **2017**, *46*, 421–435. [CrossRef]
38. Kanfar, M.S.; Clarkson, C.R. Reconciling flowback and production data: A novel history matching approach for liquid rich shale wells. *J. Nat. Gas Sci. Eng.* **2016**, *33*, 1134–1148. [CrossRef]
39. Clarkson, C.R.; Williams-Kovacs, J. Modeling two-phase flowback of multifractured horizontal wells completed in shale. *SPE J.* **2013**, *18*, 795–812. [CrossRef]
40. Yue, M.; Leung, J.Y.; Dehghanpour, H. Numerical investigation of limitations and assumptions of analytical transient flow models in tight oil reservoirs. *J. Nat. Gas Sci. Eng.* **2016**, *30*, 471–486. [CrossRef]
41. Aguilera, R. Recovery factors and reserves in naturally fractured reservoirs. *J. Can. Pet. Technol.* **1999**, *38*, 4. [CrossRef]
42. Chen, D.; Ye, Z.; Pan, Z.; Zhou, Y.; Zhang, J. A permeability model for the hydraulic fracture filled with proppant packs under combined effect of compaction and embedment. *J. Pet. Sci. Eng.* **2017**, *149*, 428–435. [CrossRef]
43. Rubin, B. Accurate simulation of non darcy flow in stimulated fractured shale reservoirs. In Proceedings of the SPE Western Regional Meeting, Anaheim, CA, USA, 27–29 May 2010. [CrossRef]
44. Cheng, Y. Impact of water dynamics in fractures on the performance of hydraulically fractured wells in gas-shale reservoirs. *J. Can. Pet. Technol.* **2012**, *51*, 143–151. [CrossRef]
45. Zheng, M.; Li, J.; Wu, X.; Li, P.; Wang, W.; Wang, S.; Xie, H. Physical modeling of oil charging in tight reservoirs: A case study of Permian Lucaogou Formation in Jimsar Sag, Junggar Basin, NW China. *Pet. Explor. Dev.* **2016**, *43*, 241–250. [CrossRef]
46. Sun, B.; Liu, L.; Ding, J. Main geologic factors controlling the productivity of horizontal wells in tight oil reservoirs. *Spec. Oil Gas Reserv.* **2017**, *24*, 190–201. [CrossRef]
47. Su, S. Imbibition Characteristics and Water Lock Damage Evaluation of Tight Oil Reservoir in Jimusar. Master's Thesis, China University of Petroleum, Beijing, China, 2017.
48. Pruess, K.; Tsang, Y.W. On two-phase relative permeability and capillary pressure of rough-walled rock fractures. *Water Resour. Res.* **1990**, *26*, 1915–1926. [CrossRef]
49. Gdanski, R.D.; Fulton, D.D.; Shen, C. Fracture face skin evolution during cleanup. In Proceedings of the SPE Annual Technical Conference and Exhibition, San Antonio, TX, USA, 24–27 September 2006. [CrossRef]
50. Liu, Y.; Guo, J.; Chen, Z. Leakoff characteristics and an equivalent leakoff coefficient in fractured tight gas reservoirs. *J. Nat. Gas Sci. Eng.* **2016**, *31*, 603–611. [CrossRef]
51. Tran, D.; Nghiem, L.; Buchanan, L. An overview of iterative coupling between geomechanical deformation and reservoir flow. In Proceedings of the SPE International Thermal Operations and Heavy Oil Symposium, Calgary, AB, Canada, 1–3 November 2005. [CrossRef]
52. *IMEX User's Guide, Version 2014.10*; Computer Modelling Group Ltd.: Calgary, AB, Canada, 2014.
53. Huang, X.; Wang, J.; Chen, S.; Gates, I.D. A simple dilation-recompaction model for hydraulic fracturing. *J. Unconv. Oil Gas Resour.* **2016**, *16*, 62–75. [CrossRef]
54. Tiab, D.; Restrepo, D.P.; Igbokoyi, A.O. Fracture porosity of naturally fractured reservoirs. In Proceedings of the International Oil Conference and Exhibition in Mexico, Cancun, Mexico, 31 August–2 September 2006. [CrossRef]
55. Aguilera, R. Effect of fracture compressibility on gas-in-place calculations of stress-sensitive naturally fractured reservoirs. *SPE Reserv. Eval. Eng.* **2008**, *11*, 307–310. [CrossRef]

56. Williams-Kovacs, J.D.; Clarkson, C.R. Stochastic modeling of multi-phase flowback from multi-fractured horizontal tight oil wells. In Proceedings of the SPE Unconventional Resources Conference Canada, Calgary, AB, Canada, 5–7 November 2013. [CrossRef]
57. Jones, F.O., Jr. A laboratory study of the effects of confining pressure on fracture flow and storage capacity in carbonate rocks. *J. Pet. Technol.* **1975**, *27*, 21–27. [CrossRef]
58. Lopez, B.; Piedrahita, J.; Aguilera, R. Estimates of stress dependent properties in tight reservoirs: Their use with drill cuttings data. In Proceedings of the SPE Latin American and Caribbean Petroleum Engineering Conference, Quito, Ecuador, 18–20 November 2015. [CrossRef]
59. Chen, X.; Li, S.; Liu, Y. Study on volume fracturing sliding water system of dense reservoir and its application in Changji oilfield. *Pet. Geol. Eng.* **2015**, *29*, 119–121. [CrossRef]
60. Song, B.; Ehlig-Economides, C.A. Rate-normalized pressure analysis for determination of shale gas well performance. In Proceedings of the North American Unconventional Gas Conference and Exhibition, Society of Petroleum Engineers, The Woodlands, TX, USA, 14–16 June 2011. [CrossRef]
61. Alkouh, A.; McKetta, S.; Wattenbarger, R.A. Estimation of effective-fracture volume using water-flowback and production data for shale-gas wells. *J. Can. Pet. Technol.* **2014**, *53*, 290–303. [CrossRef]
62. Clarkson, C.R.; Williams-Kovacs, J.D. A new method for modeling multi-phase flowback of multi-fractured horizontal tight oil wells to determine hydraulic fracture properties. In Proceedings of the SPE Annual Technical Conference and Exhibition, New Orleans, LA, USA, 30 September–2 October 2013. [CrossRef]
63. Zanganeh, B.; Soroush, M.; Williams-Kovacs, J.D.; Clarkson, C.R. Parameters affecting load recovery and oil breakthrough time after hydraulic fracturing in tight oil wells. In Proceedings of the SPE/CSUR Unconventional Resources Conference, Calgary, AB, Canada, 20–22 October 2015. [CrossRef]
64. Wilson, K. Analysis of drawdown sensitivity in shale reservoirs using coupled-geomechanics models. In Proceedings of the SPE Annual Technical Conference and Exhibition, Houston, TX, USA, 28–30 September 2015. [CrossRef]
65. Zou, Y.; Zhang, S.; Ma, X.; Zhou, T.; Zeng, B. Numerical investigation of hydraulic fracture network propagation in naturally fractured shale formations. *J. Struct. Geol.* **2016**, *84*, 1–13. [CrossRef]
66. Ghanbari, E.; Dehghanpour, H. The fate of fracturing water: A field and simulation study. *Fuel* **2016**, *163*, 282–294. [CrossRef]
67. Bose, C.C.; Fairchild, B.; Jones, T.; Gul, A.; Ghahfarokhi, R.B. Application of nanoproppants for fracture conductivity improvement by reducing fluid loss and packing of micro-fractures. *J. Nat. Gas Sci. Eng.* **2015**, *27*, 424–431. [CrossRef]
68. Lacy, L.L.; Rickards, A.R.; Bilden, D.M. Fracture width and embedment testing in soft reservoir sandstone. *SPE Drill. Complet.* **1998**, *13*, 25–29. [CrossRef]
69. French, S.W.; Economides, M.J.; Yang, M. Hydraulic fracture design flaws-proppant selection. In Proceedings of the SPE Western Regional & AAPG Pacific Section Meeting 2013 Joint Technical Conference, Monterey, CA, USA, 30 September–2 October 2013. [CrossRef]
70. Duenckel, R.; Barree, R.D.; Drylie, S.; O'Connell, L.G.; Abney, K.L.; Conway, M.W.; Moore, N.; Chen, F. Proppants—What 30 years of study have taught us. *SPE Prod. Oper.* **2018**, in press. [CrossRef]
71. Yu, W.; Lashgari, H.R.; Wu, K.; Sepehrnoori, K. CO_2 injection for enhanced oil recovery in Bakken tight oil reservoirs. *Fuel* **2015**, *159*, 354–363. [CrossRef]

© 2019 by the authors. Licensee MDPI, Basel, Switzerland. This article is an open access article distributed under the terms and conditions of the Creative Commons Attribution (CC BY) license (http://creativecommons.org/licenses/by/4.0/).

Article

A Semi-Analytical Method for Three-Dimensional Heat Transfer in Multi-Fracture Enhanced Geothermal Systems

Dongdong Liu and Yanyong Xiang *

School of Civil Engineering, Beijing Jiaotong University, Beijing 100044, China; dond_liu@126.com
* Correspondence: yyxiang@bjtu.edu.cn

Received: 14 February 2019; Accepted: 27 March 2019; Published: 28 March 2019

Abstract: Multiple fractures have been proposed for improving the heat extracted from an enhanced geothermal system (EGS). For calculating the production temperature of a multi-fracture EGS, previous analytical or semi-analytical methods have all been based on an infinite scale of fractures and one-dimensional conduction in the rock matrix. Here, a temporal semi-analytical method is presented in which finite-scale fractures and three-dimensional conduction in the rock matrix are both considered. Firstly, the developed model was validated by comparing it with the analytical solution, which only considers one-dimensional conduction in the rock matrix. Then, the temporal semi-analytical method was used to predict the production temperature in order to investigate the effects of fracture spacing and fracture number on the response of an EGS with a constant total injection rate. The results demonstrate that enlarging the spacing between fractures and increasing the number of fractures can both improve the heat extraction; however, the latter approach is much more effective than the former. In addition, the temporal semi-analytical method is applicable for optimizing the design of an EGS with multiple fractures located equidistantly or non-equidistantly.

Keywords: enhanced geothermal systems; multiple parallel fractures; semi-analytical solution

1. Introduction

Geothermal energy is a promising and clean renewable energy resource in the world. In addition to the wide use for heating done by ground source heat pumps (GSHP) [1], it can also be utilized for power generation from enhanced geothermal systems (EGS) [2]. Due to the low permeability of fractured rock reservoirs, it is a common challenge to extract significant amounts of energy. To improve the heat extraction from these reservoirs at a relatively lower cost, multiple fractures have been proposed to provide multiple water-conducting paths between the injection and production wells [3]. The thermal evolution in an EGS is associated with the spacing between the fractures and the number of fractures.

Owing to the nonlinear hydro-thermal coupling, a complex fracture network, irregular geometries, and heterogeneous materials, a wide range of commercial software is generally adopted for geothermal simulation. Those simulation tools are based on different numerical discretization schemes, i.e., finite element code [4–7], finite difference code [8], and finite volume code [9], etc. Although numerical methods have a great advantage in solving complicated problems, the increased storage memory to store the result of each time interval and the computation time to discretize the entire computational model compromise their application in practice to some extent, especially for geothermal systems with sparsely distributed fractures.

Regardless of the effects of temperature on water flow, analytical or semi-analytical methods can also be used to calculate the production temperature. For water flow and heat transfer in a reservoir with a single fracture, authors have proposed different conceptual models and computational methods. Lauwerier [10] established a mathematical model with an infinite line-shaped fracture and presented an analytical solution for temperature distribution considering the one-dimensional conduction of rocks and convective of fracture water. The two-dimensional heat transfer model for a reservoir with a finite line-shaped fracture was introduced by Cheng et al. [11] who formulated a Laplace transform semi-analytical method. Later, the three-dimensional heat transfer model with a finite disc-shaped fracture and its Laplace transform semi-analytical solution was proposed by Ghassemi et al. [12]. To accurately handle the singularities in the above integral, Xiang et al. [13] and Zhang et al. [14] applied analytical integration in the neighborhood of singularities. Unlike the Laplace transform semi-analytical method, the temporal semi-analytical method developed by Zhou et al. [15] was based on the Green function in the temporal domain instead of the one in the Laplace transform domain, and was simpler and more accurate because of avoiding the numerical Laplace inverse transformation process in which the result would be biased to some degree.

For water flow and heat transfer in a reservoir with multiple fractures, Abbasi et al. [16] considered parallel infinite line-shaped fractures at equal spacing and presented its analytical solution. Wu et al. [17] took into account equidistant, parallel, infinite disc-shaped fractures and derived its Laplace transform semi-analytical solution. However, both the analytical and semi-analytical solutions are based on the one-dimensional conduction in rocks and the infinite-scale fractures. To the best of the authors' knowledge, no analytical or semi-analytical methods are available for cases considering three-dimensional conduction in rocks and finite-scale fractures, and much less work has been conducted to investigate semi-analytically the effects of fracture spacing and fracture number on the response of a geothermal reservoir. Therefore, we attempt to develop a temporal semi-analytical model to predict the heat extracted from an EGS with multiple fractures.

2. Materials and Methods

In this study, a multi-fracture EGS model was built based on the geometry suggested by Ref. [17]. In Figure 1, multiple horizontal disc-shaped fractures located in the center of a reservoir connect the injection and production wells. Due to the low permeability of the rock matrix, the leak-off is assumed to be negligible; thus, the heat propagates by conduction in the rock matrix and by convection within fractures. When time $t > 0$, cold water with a constant temperature T_{in} and a constant injection rate Q_{in} is injected into the system with the rock's initial temperature of T_0. The hot water with a constant production rate Q_{ex} is pumped out at the same time. Some geometric parameters of the model are defined as follow: the radii of the wells are r_w; the separation between two wells is L; the number of fractures is F.

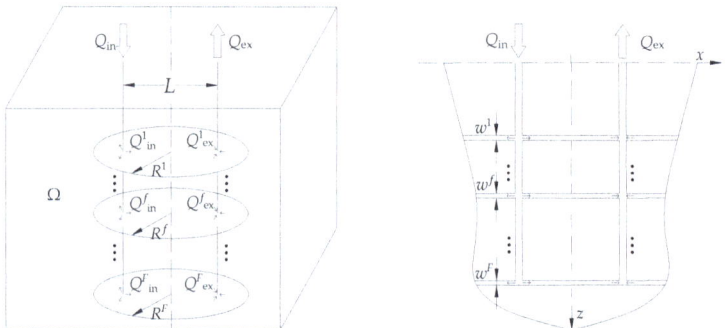

Figure 1. Geometry of the model with multi-parallel disc-shaped fractures and dipole wells.

We also mention some assumptions for the physical model.

- The fracture water is incompressible, and the rock matrix is impermeable and homogenous.
- The thermal properties of the rock matrix and fracture water are independent of temperature variation.

- The aperture of the fractures remains uniform and invariant without considering the deformation induced by hydro-mechanical coupling or thermo-mechanical interaction. Due to the fact that the fracture aperture is relatively small compared to the fracture surface, each fracture is simplified as a plane that coincides with its axisymmetric surface.
- The heat transfer coefficient between the rock matrix and fracture water is infinite, which means that the temperature of the fracture water is equal to that of the rock matrix at the fracture surfaces.
- There is steady-state water flow in the fractures because the heat exchange in a geothermal reservoir is a long-term process.

2.1. Water Flow in Fractures

The water flow in the fth fracture is assumed to obey Darcy's law [15]:

$$\nabla p^f\left(x^f\right) = -\frac{\pi^2 \mu^f}{w^{f3}} q^f\left(x^f\right), \quad x^f \in A^f \tag{1}$$

where the superscript f denotes the fracture plane f, ∇ is the gradient operator, x^f is the point (x^f, y^f, z^f), q^f is the water discharge, p^f is the hydraulic pressure in the fracture, μ^f is the water viscosity, w^f is the fracture aperture, and A^f is the fracture plane area.

The water continuity equation within the fth fracture is expressed as:

$$\nabla \cdot q^f\left(x^f\right) = Q_{ex}^f \delta\left(x^f - x_{ex}^f\right) - Q_{in}^f \delta\left(x^f - x_{in}^f\right) \tag{2}$$

where $\nabla \cdot$ is the divergence operator, $\delta(\cdot)$ is the Dirac delta function, Q_{in}^f is the flow rate at the injection well located at $x_{in}^f = (x_{in}^f, y_{in}^f, y_{in}^f)$, and Q_{ex}^f is the flow rate at the production well located at $x_{ex}^f = (x_{ex}^f, y_{ex}^f, z_{ex}^f)$.

Substituting Equation (1) to Equation (2) yields

$$\nabla^2 p^f\left(x^f\right) = \frac{\pi^2 \mu^f}{w^{f3}} \left[Q_{ex}^f \delta\left(x^f - x_{ex}^f\right) - Q_{in}^f \delta\left(x^f - x_{in}^f\right)\right]. \tag{3}$$

where ∇^2 is the Laplace operator.

For the water flow in the fth fracture, the following boundary condition is used:

$$\left. \begin{array}{l} \frac{\partial p^f(x^f)}{\partial n^f(x^f)} = 0, x^f \in \partial A^f \\ Q_{in}^f = Q_{ex}^f = \frac{Q_{in}}{F} = \frac{Q_{ex}}{F} \end{array} \right\} \tag{4}$$

where ∂A^f is the boundary of the fracture, and n^f the outward normal of ∂A^f

Based on the hypothesis that the fracture geometry including the radius and aperture is uniform, we can obtain the exact solution of the water discharge in the fth fracture by using the superposition procedure [18]:

$$q_x^f\left(x^f\right) = \sum_{i=1}^{2} \frac{Q_i^f}{2\pi} \left[\frac{x - x_i^f}{\left(x - x_i^f\right)^2 + \left(y - y_i^f\right)^2} + \frac{r_w^2\left(xr_w^2 - x_i^f R^{f2}\right)}{\left(xr_w^2 - x_i^f R^{f2}\right)^2 + \left(yr_w^2 - y_i^f R^{f2}\right)^2} \right], \tag{5}$$

$$q_y^f\left(x^f\right) = \sum_{i=1}^{2} \frac{Q_i^f}{2\pi} \left[\frac{y - y_i^f}{\left(x - x_i^f\right)^2 + \left(y - y_i^f\right)^2} + \frac{r_w^2\left(yr_w^2 - y_i^f R^{f2}\right)}{\left(xr_w^2 - x_i^f R^{f2}\right)^2 + \left(yr_w^2 - y_i^f R^{f2}\right)^2} \right]. \tag{6}$$

where Q_i^f is the flow rate at the ith well located at (x_i^f, y_i^f, z_i^f). The sign is negative for injection and positive for production. R^f is the radius of the fracture.

2.2. Heat Transfer in Fractures

According to Cheng et al. [11], the heat transfer within fractures contains four components, i.e., heat storage, dispersion, convection, and conduction between the fracture water and the rock matrix. Due to the relatively large convection velocity, the heat storage and dispersion can be ignored. Thus, the heat transport in the fth fracture can be expressed as:

$$\rho_w c_w q^f\left(x^f\right) \cdot \nabla T^f\left(x^f, t\right) + q_h^f\left(x^f, t\right) = 0 \tag{7}$$

where ρ_w is the water density, c_w is the water-specific heat, and q_h^f is the heat transfer rate between the fracture water and the rock matrix.

Assuming that temperature is continuous across fracture surfaces, the heat transfer rate in Equation (7) can be expressed as [19]:

$$q_h^f\left(x^f, t\right) = -2\lambda_r \left.\frac{\partial T(x,t)}{\partial z}\right|_{z=z^f} \tag{8}$$

where λ_r is the thermal conductivity of the rock matrix.

2.3. Heat Transfer in the Rock Matrix

In a low-permeability rock matrix, the heat transfer dominated by conduction can be expressed as:

$$\lambda_r \nabla^2 T(x, t) = \rho_r c_r \frac{\partial T(x,t)}{\partial t}, \quad x \in \Omega \tag{9}$$

where ρ_r is the rock matrix density and c_r is the rock matrix specific heat.

2.4. Initial and Boundary Condition

The initial and boundary conditions for the system are:

$$\left.\begin{array}{ll} T(x,0) = T_0, & x \subseteq \Omega \\ T(x,t) = T_{in}, & x = x_{in}^f \\ T(x,t) = T_0, & x \to \pm\infty \end{array}\right\}. \tag{10}$$

2.5. Numerical Formation

2.5.1. Integral Equation Method

The temperature increment at location x and time t induced by an instantaneous unit point source at location x' and time $t' < t$ is [20]:

$$\theta(x - x', t - t') = \sqrt{\frac{\rho_r c_r}{[4\pi\lambda_r(t-t')]^3}} \exp\left(\frac{-\rho_r c_r r^2}{4\lambda_r(t-t')}\right). \tag{11}$$

where $r = \sqrt{(x-x')^2 + (y-y')^2 + (z-z')^2}$.

Equation (11) is the temporal Green function of Equation (9). The heat transfer rate in Equation (8) is regarded as a heat sink when the fracture water is colder than the rock matrix. Taking into account heat sinks at all fracture planes, the resultant temperature in the reservoir at time t is:

$$T(x,t) = \sum_{f=1}^{F} \int_0^t \int_{A^f} q_h^f\left(x^f, t'\right) \theta\left(x - x^f, t - t'\right) dA^f dt' + T_0. \tag{12}$$

Applying Equation (12) at the fth fracture plane yields

$$T^f\left(x^f,t\right) = \sum_{i=1}^{F}\int_0^t\int_{A^i} q_h^i\left(x^i,t'\right)\theta\left(x^f - x^i, t - t'\right)\mathrm{d}A^i\mathrm{d}t' + T_0. \tag{13}$$

Using the convolution algorithm [21] for Equation (13) yields

$$T^f\left(x^f,t\right) = \sum_{i=1}^{F}\sum_{n=1}^{N} q_h^f\left(x^f,t_n\right)\int_{A^i}\theta_{N-n+1}^{ch}\left(x^f - x^i\right)\mathrm{d}A^i + T_0 \tag{14}$$

where the time from 0 to t is divided into N equal intervals of Δt, θ_n^{ch} is the integration of θ from time t_{N-n} to time t_{N-n+1}, and is expressed as:

$$\theta_n^{ch}\left(x - x'\right) = \int_{t_{N-n}}^{t_{N-n+1}}\theta(x - x', t - t')\mathrm{d}t' = \frac{1}{4\pi\lambda_r r}\left[\mathrm{erfc}\left(\frac{r}{\sqrt{4\alpha_r t_n}}\right) - \mathrm{erfc}\left(\frac{r}{\sqrt{4\alpha_r t_{n-1}}}\right)\right]. \tag{15}$$

where $\alpha_r = \lambda_r/\rho_r c_r$ is the heat diffusivity of the rock matrix and erfc (·) is the complementary error function, which approximates to zero when the expression in the brackets approaches infinity.

Equation (14) with the initial and boundary conditions in Equation (10) can be used to obtain the temperature and heat transfer rate at each fracture plane.

2.5.2. Numerical Calculation

Fracture planes are discretized into a number of four–noded quadrilateral elements. For the fth fracture plane, there are M^f elements and K^f nodes as shown in Figure 2.

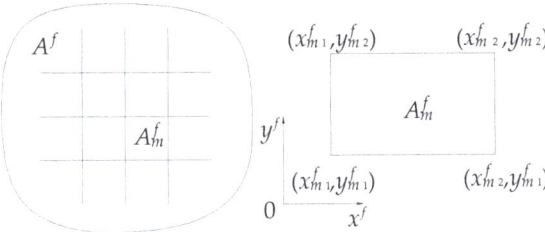

Figure 2. Discretization of the fth fracture plane.

The following interpolations for any element m are used:

$$T_m^f = \mathbf{N}\mathbf{T}_m^f, \quad q_{hm}^f = \mathbf{N}\mathbf{q}_{hm}^f \tag{16}$$

where the subscript m denotes the element m, $\mathbf{T}_m^f$ is the vector of nodal temperature, $\mathbf{q}_{hm}^f$ is the vector of the nodal heat transfer rate, and $\mathbf{N}$ is the vector of interpolation functions.

Substituting Equation (16) into Equation (14) yields

$$T^f\left(x^f,t\right) \approx \sum_{i=1}^{F}\sum_{n=1}^{N}\sum_{m=0}^{M^i}\int_{A_m^i}\mathbf{N}\theta_{N-n+1}^{ch}\left(x^f - x^i\right)\mathrm{d}A_m^i \mathbf{q}_{hm}^i(t_n) + T_0. \tag{17}$$

Applying Equation (17) to all element nodes at the fth fracture plane yields

$$\mathbf{T}^f(t) = \sum_{i=1}^{F}\left[\mathbf{B}^{f,i}\mathbf{q}_h^i(t) + \mathbf{C}^{f,i}\right] + \mathbf{D}^f \tag{18}$$

where, specifically,

$$\boldsymbol{B}^{f,i} = \sum_{m=1}^{M^i} \begin{bmatrix} \int_{A_m^i} \boldsymbol{N} \theta_1^{ch}\left(\boldsymbol{x}_1^f - \boldsymbol{x}^i\right) dA_m^i \\ \int_{A_m^i} \boldsymbol{N} \theta_1^{ch}\left(\boldsymbol{x}_2^f - \boldsymbol{x}^i\right) dA_m^i \\ \vdots \\ \int_{A_m^i} \boldsymbol{N} \theta_1^{ch}\left(\boldsymbol{x}_{K^f}^f - \boldsymbol{x}^i\right) dA_m^i \end{bmatrix}, \quad (19)$$

$$\boldsymbol{C}^{f,i} = \sum_{n=1}^{N-1} \sum_{m=1}^{M^i} \begin{bmatrix} \int_{A_m^i} \boldsymbol{N} \theta_{N-n+1}^{ch}\left(\boldsymbol{x}_1^f - \boldsymbol{x}^i\right) dA_m^i \\ \int_{A_m^i} \boldsymbol{N} \theta_{N-n+1}^{ch}\left(\boldsymbol{x}_2^f - \boldsymbol{x}^i\right) dA_m^i \\ \vdots \\ \int_{A_m^i} \boldsymbol{N} \theta_{N-n+1}^{ch}\left(\boldsymbol{x}_{K^f}^f - \boldsymbol{x}^i\right) dA_m^i \end{bmatrix} \boldsymbol{q}_h^i(t_n), \quad (20)$$

$$\boldsymbol{D}^f = \left[T_0\left(\boldsymbol{x}_1^f\right), T_0\left(\boldsymbol{x}_2^f\right), \cdots, T_0\left(\boldsymbol{x}_{K^f}^f\right)\right]^T. \quad (21)$$

where the superscript T denotes the transpose of a vector.

Usually, the conventional Galerkin finite element method for the convection-dominated problems is corrupted by spurious node-to-node oscillations. In order to reduce the numerical oscillations, the Streamline upwind/Petrov–Galerkin method [22] is used for Equation (7), which can be written as follows:

$$\boldsymbol{E}^f \boldsymbol{T}^f(t) + \boldsymbol{F}^f \boldsymbol{q}_h^f(t) = \boldsymbol{0}, \quad (22)$$

where

$$\boldsymbol{E}^f = \rho_w c_w \sum_{m=1}^{M^f} \int_{A_m^f} \sum_{i=1}^{2} \left(\boldsymbol{N}^T + \bar{k}_{x_i m}^f \frac{\partial \boldsymbol{N}^T}{\partial x_i^f}\right) \boldsymbol{N} q_{x_i}^f \frac{\partial \boldsymbol{N}}{\partial x_i^f} dA_m^f, \quad x_i = x, y, \quad (23)$$

$$\bar{k}_{x_i m}^f = \frac{q_{x_i m}^{cf}}{2} \sum_{k=1}^{2} \frac{q_{x_k m}^{cf} \Delta x_{km}^f}{q_{xm}^{cf2} + q_{ym}^{cf2}} \left(\coth\left(\frac{q_{x_k m}^{cf} \Delta x_{km}^f}{2w^f k_w}\right) - \frac{2w^f k_w}{q_{x_k m}^{cf} \Delta x_{km}^f}\right), \quad x_k = x, y, \quad (24)$$

$$\boldsymbol{F}^f = \sum_{m=1}^{M^f} \int_{A_m^f} \boldsymbol{N}^T \boldsymbol{N} dA_m^f. \quad (25)$$

in which Δx_{km}^f is the element length in x_k direction, $q_{x_k m}^{cf}$ is the water discharge at the center of the element, and k_w is the water diffusivity.

Substituting Equation (18) to Equation (22) yields:

$$\left(\boldsymbol{E}^f \boldsymbol{B}^{f,f} + \boldsymbol{F}^f\right) \boldsymbol{q}_h^f(t) + \sum_{i=1, i \neq f}^{F} \boldsymbol{E}^f \boldsymbol{B}^{f,i} \boldsymbol{q}_h^i(t) = -\boldsymbol{E}^f \left(\sum_{i=1}^{F} \boldsymbol{C}^{f,i} + \boldsymbol{D}^f\right). \quad (26)$$

The heat transfer rate $q_h^f(t)$ can be obtained by solving Equation (26). Thereafter, Equation (12) can be discretized as presented in this section to calculate the temperature of the fracture water and rock matrix.

3. Results and Discussion

3.1. Verification of the Temporal Semi-Analytical Method

3.1.1. Injection into an Infinite Rectangular Fracture

To verify the model proposed, we considered the problem of one-dimensional water injection into an infinite rectangular-shaped fracture as shown in Figure 3a. The length, width, and aperture of the fracture were 800 m, 80 m, and 0.003 m, respectively. The unit-width discharge was 1.5×10^{-5} m^2 s^{-1}. The parameters in Table 1 were used here. The temperature of line l located centrally in the fracture plane was calculated. Based on one-dimensional conduction in the rock matrix, the analytical solution for the normalized fracture water temperature is as follows [10]:

$$\frac{T - T_0}{T_{in} - T_0} = \text{erfc}\left[\left(\frac{\lambda_r x}{\rho_w c_w q} + \frac{w}{4}\right)\sqrt{\frac{\rho_r c_r}{\lambda_r t}}\right]. \quad (27)$$

where x is the perpendicular distance from the interest point to the injection well, q is the unit-width discharge, and w is the fracture aperture. The other variables are as described above.

Figure 3b shows the comparison of the results for injection times of 1, 4, and 10 years. We observe that the difference between the two solutions is small. However, with the elapse of time, the difference becomes noticeable due to the different assumptions about conduction in the rock matrix.

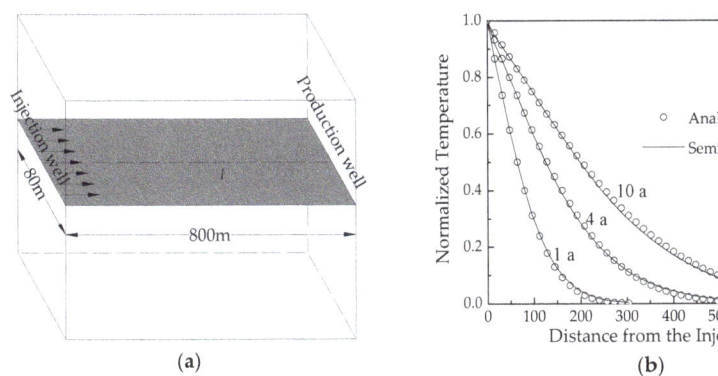

Figure 3. (a) Water flows through an infinite rectangular fracture in which the left and right sides are the positions of the injection and production wells, respectively [10]; (b) Normalized temperature increment distribution along line l.

Table 1. Parameters used.

Parameters	Values	Parameters	Values
Water density ρ_w (Kg m^{-3})	1000	Water diffusivity k_w (m^2 s^{-1})	0.01
Rock density ρ_r (Kg m^{-3})	2650	Rock conductivity λ_r (W m^{-1} °C^{-1})	2.59
Water-specific heat c_w (J kg^{-1} °C^{-1})	4180	Initial rock temperature T_0 (°C^{-1})	90
Rock-specific heat c_r (J kg^{-1} °C^{-1})	1000	Injected water temperature T_{in} (°C^{-1})	20

3.1.2. Injection into an Infinite Radial Fracture

To validate the model presented, the heat transfer of radial flow through an infinite fracture was modeled. In Figure 4a, cold water was injected into the center of the fracture. Based on one-dimensional conduction in the rock matrix, the analytical solution for the normalized fracture water temperature is given as [23]:

$$\frac{T - T_0}{T_{in} - T_0} = \text{erfc}\left[\left(\frac{\pi \lambda_r x^2}{\rho_w c_w Q_{in}} + \frac{w}{4}\right)\sqrt{\frac{\rho_r c_r}{\lambda_r t}}\right]. \quad (28)$$

where x is the radial distance from the interest point to the injection well and Q_{in} is the injection rate; the other variables are as described above.

The case that is shown in Figure 4a with an injection rate of 0.015 m^3 s^{-1}, wellbore radius of 0.1 m, and fracture aperture of 0.003 m was analyzed. In order to reduce the impact of truncated boundaries, the radius of the fracture was set at 600 m. Other parameters are listed in Table 1. Figure 4b illustrates the comparison of results for injection times of 1, 4, and 10 years. A good match is observed between the solutions from the present model and the analytical one.

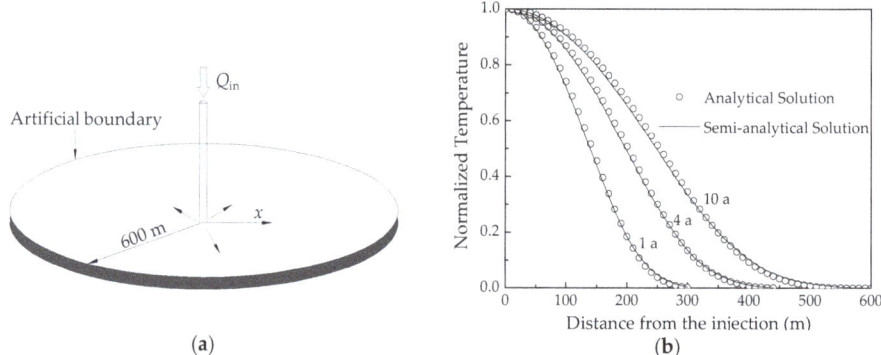

Figure 4. (a) Injection into an infinite radial fracture [23]; (b) Normalized temperature increment distribution in the radial fracture.

3.2. Multi-Fracture EGS

In this section, the temporal semi-analytical method was applied to investigate the effects of fracture spacing and fracture number on the response of an EGS. Without loss of generality, the fractures were equidistant. We assumed that the injection rate was 0.05 m^3 s^{-1}, wellbore radius 0.1 m, wellbore spacing 60 m, fracture radius 50 m, and fracture aperture 0.001 m. Each fracture plane was discretized spatially by 1992 quadratic elements. Other parameters are shown in Table 1.

3.2.1. Effects of the Fracture Spacing

Assuming that there were two parallel fractures in Figure 1, we aimed at probing the influence of fracture spacing on fracture water temperature. The spacing between the two fractures ranged from 20 m to 50 m. For three cases with fracture spacings of 20 m, 35 m, and 50 m, the fracture temperature distribution after 5 years of extraction and the production temperature variation within 10 years are shown in Figure 5. With increasing the fracture spacing, the production temperature rose because the thermal interplay between the two fractures decreased correspondingly, while its relative increment declined. When the fracture spacing was increased from 30 m to 50 m, the relative increasing rate of the production temperature was lower than 3%. We can indicate that a spacing larger than 50 m is too large for thermal interaction, which is compatible with the results obtained by Vik et al. [3].

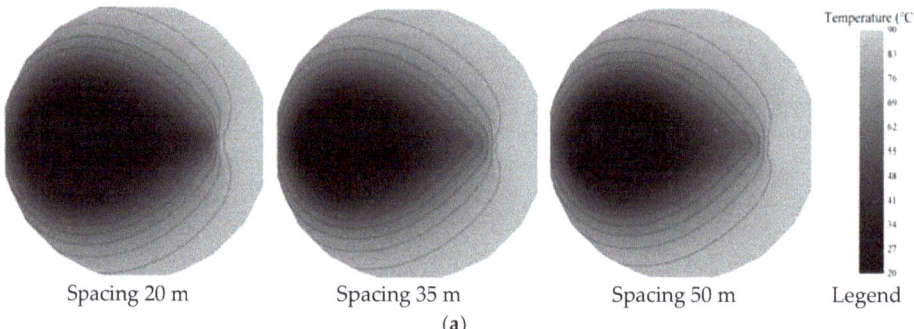

Figure 5. *Cont.*

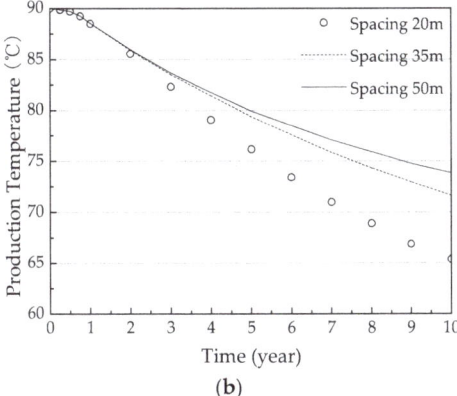

(**b**)

Figure 5. Effect of various fracture spacings on fracture temperature: (**a**) Fracture temperature distribution after 5 years of extraction; (**b**) Temperature breakthrough curves.

3.2.2. Effects of the Fracture Number

Assuming that the distance between the top fracture and the bottom one was 50 m in Figure 1, more horizontal fractures were added in the rock between them. Three cases were studied where the number of fractures was set to 2, 3, and 4, and correspondingly the uniform spacings were 50 m, 25 m, and 50/3 m. Figure 6 illustrates the average temperature of the water pumped from all fractures in each project, respectively. As the number of fractures increased, the production temperature rose because the contacting area between the fracture water and rock matrix expanded correspondingly, while its relative increment decreased. When the fracture number was increased from 3 to 4, the relative increasing rate of the production temperature was lower than 2%. We can observe that the optimal fracture number is 4, which approaches the smallest value of 6 proposed by Wu et al. [17]. In addition, the comparison between Figures 5b and 6 reveals that increasing the fracture number improves the heat extracted more efficiently than does enlarging the fracture spacing.

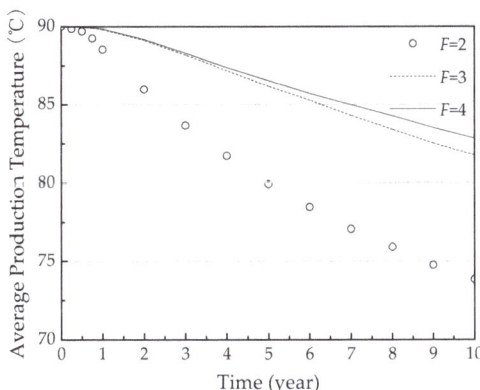

Figure 6. Average production temperature for a fracture number of 2, 3, and 4.

4. Conclusions

In this study, a temporal semi-analytical model was developed to predict the heat exploited from an EGS with multiple parallel fractures. This model considers the three-dimensional conduction in the rock matrix and the finite scale of fractures. The solution was applied to investigate the effects of fracture spacing and fracture number on the production temperature of an EGS. The results show the following:

- The temporal semi-analytical method provides an accurate solution and, thus, can serve as a benchmark for numerical methods.
- The fracture spacing and fracture number are two key factors controlling the heat extraction from an EGS. Increasing the fracture spacing maintains the production temperature by decreasing the thermal interaction between fractures. A multi-fracture also extends the life of a geothermal reservoir by increasing the contacting area between the fracture water and rock matrix. In terms of improving heat exploitation, increasing the fracture number is more efficient than increasing fracture spacing.
- The proposed method is efficient in calculations and is applicable to the design and optimization of an EGS.

Author Contributions: Conceptualization, L.D.; methodology, L.D.; software, L.D.; validation, L.D.; formal analysis, L.D.; investigation, L.D.; resources, L.D.; data curation, L.D.; writing—original draft preparation, L.D.; writing—review and editing, X.Y.; visualization, L.D.; supervision, X.Y.; project administration, X.Y.; funding acquisition, X.Y.

Funding: This research was funded by the Fundamental Research Funds for China Central Universities under grants, 2018YJS104. The APC was funded by Liu D.D.

Conflicts of Interest: The authors declare no conflict of interest.

References

1. Luo, J.; Luo, Z.Q.; Xie, J.H.; Xia, D.S.; Huang, W.; Shao, H.B.; Xiang, W.; Rohn, J. Investigation of shallow geothermal potentials for different types of ground source heat pump systems (GSHP) of Wuhan city in China. *Renew. Energy* **2018**, *118*, 230–244. [CrossRef]
2. Olasolo, P.; Juárez, M.C.; Morales, M.P.; D'Amico, S.; Liarte, I.A. Enhanced geothermal systems (EGS): A review. *Renew. Sustain. Energy Rev.* **2016**, *56*, 133–144. [CrossRef]
3. Vik, H.S.; Salimzadeh, S.; Nick, H.M. Heat Recovery from Multiple-Fracture Enhanced Geothermal Systems: The Effect of Thermoelastic Fracture Interactions. *Renew. Energy* **2018**, *121*, 606–622. [CrossRef]
4. Yao, J.; Zhang, X.; Sun, Z.X.; Huang, Z.Q.; Li, Y.; Xin, Y.; Yan, X.; Liu, W.Z. Numerical simulation of the heat extraction in 3D-EGS with thermal-hydraulic-mechanical coupling method based on discrete fractures model. *Geothemics* **2018**, *74*, 19–34. [CrossRef]
5. Xia, Y.D.; Plummer, M.; Mattson, E.; Podgorney, R.; Ghassemi, A. Design, modeling, and evaluation of a doublet heat extraction model in enhanced geothermal systems. *Renew. Energy* **2017**, *105*, 232–247. [CrossRef]
6. Galgaro, A.; Farina, Z.; Emmi, G.; Crali, M.D. Feasibility analysis of a Borehole Heat Exchanger (BHE) array to be installed in high geothermal flux area: The case of the Euganean Thermal Basin, Italy. *Renew. Energy* **2015**, *78*, 93–104. [CrossRef]
7. Kelkar, S.; Lewis, K.; Karra, S.; Zyvoloski, G.; Rapaka, S.; Viswanathan, H.; Mishra, P.K.; Chu, S.; Coblentz, D.; Pawar, R. A simulator for modeling coupled thermo-hydro-mechanical processes in subsurface geological media. *Int. J. Rock Mech. Min.* **2014**, *70*, 569–580. [CrossRef]
8. Zeng, Y.C.; Su, Z.; Wu, N.Y. Numerical simulation of heat production potential from hot dry rock by water circulating through two horizontal wells at Desert Peak geothermal field. *Energy* **2013**, *56*, 92–107. [CrossRef]
9. Huang, W.B.; Cao, W.J.; Jiang, F.M. Heat extraction performance of EGS with heterogeneous reservoir: A numerical evaluation. *Int. J. Heat Mass. Transf.* **2017**, *108*, 645–657. [CrossRef]
10. Lauwerier, H.A. The transport of heat in an oil layer caused by the injection of hot fluid. *Appl. Sci. Res. Sect. A* **1955**, *5*, 145–150. [CrossRef]
11. Cheng, A.H.D.; Ghassemi, A.; Detournay, E. Integral equation solution of heat extraction from a fracture in hot dry rock. Int. J. Numer. Anal. Methods Geomech. **2001**, *25*, 1327–1338. [CrossRef]
12. Ghassemi, A.; Tarasovs, S.; Cheng, A.H.D. An integral equation for three-dimensional heat extraction from planar fracture in hot dry rock. *Int. J. Numer. Anal. Methods Geomech.* **2003**, *27*, 989–1004. [CrossRef]
13. Xiang, Y.Y.; Guo, J.Q. A Laplace transform and Green function method for calculation of water flow and heat transfer in fractured rocks. *Rock Soil Mech.* **2011**, *32*, 333–340.

14. Zhang, Y.; Xiang, Y.Y. A semi-analytical method for calculation of three-dimensional water flow and heat transfer in single-fracture rock with distributed heat sources. *Rock Soil Mech.* **2013**, *34*, 685–695.
15. Zhou, X.X.; Ghassemi, A.; Cheng, A.H.D. A three-dimensional integral equation model for calculating poro- and thermoelastic stresses induced by cold water injection into a geothermal reservoir. *Int. J. Numer. Anal. Methods Geomech.* **2009**, *33*, 1613–1640. [CrossRef]
16. Abbasi, M.; Khazali, N.; Sharifi, M. Analytical model for convection-conduction heat transfer during water injection in fractured geothermal reservoirs with variable rock matrix block size. *Geothermics* **2017**, *69*, 1–14. [CrossRef]
17. Wu, B.S.; Zhang, X.; Jeffrey, R.G.; Bunger, A.P.; Jia, S.P. A simplified model for heat extraction by circulating fluid through a closed-loop multiple-fracture enhanced geothermal system. *Appl. Energy* **2016**, *183*, 1664–1681. [CrossRef]
18. Liggett, J.A.; Liu, P.L.F. *The Boundary Integral Equations Method for Porous Media Flow*; George Allen & Unwin: London, UK, 1983.
19. Ghassemi, A.; Zhou, X.X. A three-dimensional thermo-poroelastic model for fracture response to injection/extraction in enhanced geothermal systems. *Geothemics* **2011**, *40*, 39–49. [CrossRef]
20. Xiang, Y.Y. *Fundamental Theory of Groundwater Mechanics*; Science Press: Beijing, China, 2011.
21. Dargush, G.; Banerjee, P.K. A time domain boundary element method for poroelasticity. *Int. J. Numer. Meth. Eng.* **1989**, *28*, 2423–2449. [CrossRef]
22. Brooks, A.N.; Hughes, T.J.R. Streamline upwind/Petrov-Galerkin formulations for convection dominated flows with particular emphasis on the incompressible Navier-Stokes equations. *Comput. Method. Appl. M* **1982**, *32*, 19–259. [CrossRef]
23. Mossop, A.P. Seismicity, subsidence and strain at the Geysers geothermal field. Ph.D. Thesis, Stanford University, Location of University, Stanford, CA, USA, 2001.

 © 2019 by the authors. Licensee MDPI, Basel, Switzerland. This article is an open access article distributed under the terms and conditions of the Creative Commons Attribution (CC BY) license (http://creativecommons.org/licenses/by/4.0/).

Article

Mathematical Modeling of Unsteady Gas Transmission System Operating Conditions under Insufficient Loading

Vasyl Zapukhliak [1], Lyubomyr Poberezhny [1], Pavlo Maruschak [2], Volodymyr Grudz Jr. [1], Roman Stasiuk [1], Janette Brezinová [3,*] and Anna Guzanová [3]

1. Department of Construction and Renovation of Oil and Gas Pipelines and Oil and Gas Storage Facilities, Ivano-Frankivsk National Technical University of Oil and Gas, Ivano-Frankivsk, Karpatska str. 15, 76019 Ivano-Frankivsk, Ukraine; psmal@protonmail.com (V.Z.); rmax2019@bigmir.net (L.P.); ndl-pzism@meta.ua (V.G.J.); ndr-prism@meta.ua (R.S.)
2. Department of Industrial Automation, Ternopil National Ivan Pul'uj Technical University, Ruska str. 56, 46001 Ternopil, Ukraine; maruschak.tu.edu@gmail.com
3. Department of Engineering Technologies and Materials, Faculty of Mechanical Engineering, Technical University of Košice, Mäsiarska 74, 04001 Košice, Slovakia; anna.guzanova@tuke.sk
* Correspondence: janette.brezinova@tuke.sk; Tel.: +421-55-602-3542

Received: 12 March 2019; Accepted: 3 April 2019; Published: 6 April 2019

Abstract: Under insufficient loading of a main gas transmission system, high-amplitude fluctuations of pressure may occur in it. A mathematical model is proposed to estimate the amplitude of pressure fluctuations in a gas pipeline along its length. It has been revealed that the shutdown of compressor stations along the gas pipeline route has a significant impact on the parameters of the unsteady transient operating conditions. The possibility of minimizing oscillation processes by disconnecting compressor stations is substantiated for the "Soyuz" main gas pipeline.

Keywords: main gas pipeline; pressure fluctuations; unsteady process

1. Introduction

At its core, the design of any gas transmission system is based on the carrying capacity (throughput) of the gas pipeline defined as the maximum amount of gas that can be pumped per unit of time. This value is decisive when choosing the diameter and number of compressor stations, working pressures at their inlet/outlet limited by the pipe strength, and so on. The lower limit of these parameters is chosen from the normal operating conditions of gas pumping units [1–3]. Maximum and minimum pressures have an impact on the carrying capacity: the smaller the difference between them, the lower the performance, and vice versa.

Under conditions of insufficient loading of the gas transmission system, a significant decrease in performance compared with the throughput leads to a broader range of maximum and minimum pressures. Boundary variants of admissible regimes include the maximum one, which is characterized by the maximum pressure at the beginning of the linear section, and the minimum one, which is characterized by the minimum pressure at the end of the linear section. All other admissible regimes are in the range between the specified boundaries [3–5].

Papers which describe mathematical models of pressure fluctuations in the gas transmission system due to changes in performance characteristics under insufficient loading are known. The realization of such models under real operating conditions of gas pipelines allowed establishing the amplitude and frequency characteristics of the unsteady process [6,7]. It was found that in the low-frequency region, the pressure fluctuation amplitude can exceed 1.0 MPa. As a result, the absolute value of pressure goes beyond the admissible range. In addition, it should be noted that propagation rates of disturbances in the gas pipeline at high and low pressures differ significantly, which affects frequency characteristics of the unsteady process. Summarizing the foregoing, it should be emphasized that, despite the economy yielded by the transmission of gas at high pressures, it is advisable to leave a certain margin of possible amplitudes of pressure fluctuation in order to prevent the attainment of the limiting absolute values of pressure [8]. This paper focuses on determining permissible fluctuations of pressure in the insufficiently loaded gas transmission system based on mathematical models.

2. Research Technique

It should be noted that under conditions of insufficient loading of the gas transmission system, the only optimality criteria of operating conditions can be the minimum energy consumption in the process of gas transportation and the maximum reliability of the gas pipeline. Based on the first criterion, we define the principle of optimizing the above admissible regimes [9–11]. Consider a gas pipeline with an internal diameter d, which contains two compressor stations, with the pressure at the inlet and outlet P_{Hi}, P_{Bi}, respectively, and two linear sections with lengths L_i. The throughput of each section as a function of parameters of the operating conditions is the main indicator of performance, which characterizes the intended use of the pipe, and is determined from the basic equation of the gas pipeline [12–14]:

$$Q = 0.326 \cdot 10^{-6} \cdot d_i^{2.5} \cdot \sqrt{\frac{P_H^2 - P_K^2}{\lambda_i \Delta z T_{cp} L_i}}$$

$$\lambda_i = 0.067 \left(\frac{158}{Re_i} + \frac{2K_i}{d_i} \right)^{0.2}$$

$$Re_i = 1.81 \cdot 10^3 \frac{Q\Delta}{d_i \eta}$$

where λ_i is the coefficient of hydraulic resistance of the gas pipeline; T_{cp} is the average temperature of gas in the section; Re is the Reynolds number; k_e is the equivalent stiffness of pipes; η is the coefficient of dynamic viscosity; Δ is the ratio of gas density to air density; z is the gas compressibility coefficient.

For the given gas pipeline, we determine the total capacity of the compressor stations, based on the condition of isothermal gas compression:

$$N_\Sigma = N_1 + N_2 = Q_{B1} P_{B1} ln \frac{P_{H1}}{P_{B1}} + Q_{B2} P_{B2} ln \frac{P_{H2}}{P_{B2}} = Q_B P_B ln \frac{P_{H1}}{P_{B1}} \frac{P_{H2}}{P_{B2}} \quad (1)$$

where N_1 and N_2 refer to the capacities of the neighbouring compressor stations; N_Σ is the total capacity of the compressor stations.

In Equation (1), the product of all numbers in steady isothermal conditions is expressed as $Q_{B1} P_{B1} = Q_{B2} P_{B2} = Q_B P_B = const$. Expressing pressure P_{B2} by pressure P_{H1} from the main equation of gas pipelines, we obtain:

$$N_\Sigma = Q_B P_B ln \frac{P_{H2}}{P_{B1} \sqrt{1 - \frac{\lambda \Delta z T_{cp} L_1 Q^2}{(0.326 \cdot 10^{-6})^2 d^5 P_{H1}^2}}}. \quad (2)$$

From Equation (2) it is obvious that with an increase in P_{H1}, the total capacity of compressor stations decreases and reaches the minimum value at $P_{H1} = P_{max}$. Thus, from the viewpoint of minimizing energy costs on transporting gas under conditions of insufficient loading of the gas transmission system, it is necessary to choose operating conditions with the maximum allowable pressure at the outlet of the compressor stations. The appropriateness of transporting gas at high pressures, given the necessity to minimize energy costs on transport, has a physical explanation. High pressures enhance the density of gas (with other conditions being equal), while reducing the linear velocity due to the steady flow of gas, the magnitude of which influences the loss of hydraulic pressure on friction. A mathematical model of the unsteady, isothermal, one-way gas flow in a pipeline can be presented by the following equations:

$$\frac{\partial p}{\partial x} + \rho \alpha \frac{\partial}{\partial x}\left(\frac{\omega^2}{2\rho^2}\right) + \beta \rho g \frac{\partial h}{\partial x} + \frac{\lambda \omega^2}{2\rho D} + \gamma \frac{\partial \omega}{\partial t} = 0,$$
$$\frac{\partial \omega}{x} + \frac{1}{c^2}\frac{\partial p}{\partial t} = 0, \qquad (3)$$

where $p(x,t)$ is the pressure (p) as a function of the linear coordinate x and time t; ω is the linear velocity of gas; λ is the coefficient of hydraulic resistance; ρ is the gas density; D is the diameter; h is the geodetic mark of the profile; $c = \sqrt{kzRT}$ is the sound velocity in gas; α is the Coriolis coefficient ($\alpha = 2$ for the laminar flow, and $\alpha = 1.1$ for the turbulent flow). The first equation takes into account the friction forces, the difference between heights of the pipeline, and the inertial resistance. The second equation describes the quantitative balance of gas. In this case, changes in temperature depending on the pipeline length are taken into account by means of construction of the iterative algorithm. In system (3), coefficients β and γ are introduced in order to study the influence of the corresponding forces.

3. Obtained Results

3.1. Oscillatory Process of the Pressure Function in Space and Time

Neglecting the influence of gravitational and coriolis forces, we reduce system (3) to the following equation:

$$\frac{\partial^2 P}{\partial x^2} = \frac{2a}{c^2}\frac{\partial P}{\partial t} + \frac{1}{c^2}\frac{\partial^2 P}{\partial t^2}, \qquad (4)$$

where $2a$ is the linearization coefficient:

$$2a = \frac{\lambda \omega}{2D}$$

This equation describes the oscillatory process of the pressure function in space and time, and is known as telegraphic equation in mathematical physics. We note that pressure fluctuations in the gas flow may have different frequencies and amplitudes depending on their cause. Given the above, pressure fluctuations are conventionally divided into high-frequency, medium-frequency, and low-frequency ones. High-frequency fluctuations are characterized by frequencies in the range of 0.4–4.0 Hz, and usually result from a jump-like change in the parameter (pressure, flow) in a certain section of the gas pipeline. The amplitude of such fluctuations can reach 1 MPa. Fluctuations propagate along the gas pipeline with the speed of sound, while the amplitude and frequency decrease. The medium-frequency range is 0.5–1.0 Hz. Such fluctuations cause smooth changes in the flow parameters over time. They propagate along the pipeline with a significantly lower decrement of damping. Low-frequency fluctuations are caused by daily unevenness of gas consumption and are in the frequency range of 10^{-5}–0.5 Hz. The amplitude of the pressure variation depends on the nature of the disturbance factor and may be unrestricted (for example, under conditions of filling the gas pipeline with gas). In conditions of high frequency fluctuations, the inertial forces and hydraulic resistance forces in the flow of gas play a decisive role in the formation of the process. Forces of the hydraulic resistance of the pipeline are the main source of the medium- and low-frequency fluctuations. As regards the reliable operation of the gas transmission system, a crucial role is played by high-frequency fluctuations of pressure, given the unpredictable nature of the process. Since the frequency and amplitude of pressure fluctuations caused by disturbance of the gas flow parameters are characteristics of the unsteady process, there must be a link between the amplitude-frequency characteristics and the criterion of unsteadiness [12–14].

Thus, there is an optimization problem, which consists in determining the rational pressures of the steady process in the gas pipeline, under which minimum energy costs on transport would be achieved, on the one hand, and safe operation of pipeline systems would be ensured, on the other hand [15]. As noted, the maximum possible pressure in the gas pipeline allows minimizing the hydraulic losses during the transportation of gas, that is, achieving the minimum energy costs. However, pressure fluctuations in unsteady processes caused by a jump-like change in parameters (most often gas flow under insufficient loading) may go beyond the limits of permissible loads. Therefore, it is necessary to choose the maximum possible pressure of the steady process in the gas pipelines, at which the superposition of the pressure amplitude in the unsteady process would not force the load of pipe walls beyond the range of permissible values [16].

The above statement of the problem requires the solution of Equation (4) under the following initial and boundary conditions chosen for the following considerations. Prior to the beginning of an unsteady process caused by disturbance of gas flow, the gas pipeline worked in steady process conditions with the lengthwise distribution of pressure according to a parabolic law.

$$P(x,0) = \sqrt{P_H^2 - (P_H^2 - P_K^2)x/L}, \quad (5)$$

where $P(x,0)$ is the pressure at distance x from the beginning of the gas pipeline with length L; P_H, P_K is the pressure at the beginning and at the end of the gas pipeline, respectively.

At given pressures P_H, P_K, a certain mass productivity Q_0 of the gas pipeline is provided, which can be increased or decreased to a certain magnitude at any time under insufficient loading. We assume that starting from the moment $t > 0$, the supply of gas to the gas pipeline has not changed, whereas the offtake of gas at the end of the route has changed by ΔQ. Then the boundary conditions for the realization of Equation (2) will take the following form:

$$Q(0,t) = Q_0; \quad Q(L,t) = Q_1, \quad (6)$$

where $Q_1 = Q_0 + \Delta Q$. Align correctly

Using the first equation of system (3) and neglecting all types of power consumption except hydraulic resistance, we obtain:

$$-\frac{\partial P}{\partial x}\bigg|_{x=0} = \frac{2a}{F^2}Q_0; \quad -\frac{\partial P}{\partial x}\bigg|_{x=L} = \frac{2a}{F^2}Q_1; \quad (7)$$

where $F = \frac{\pi D^2}{4}$ is the cross-sectional area of the pipeline.

Equation (4) under the initial (5) and boundary (7) conditions is solved by the Fourier method:

$$P(x,t) = \frac{\lambda \rho w}{2dF^2}x(Q_0 - \frac{Q_0 - Q_L}{2L}x) +$$
$$+\frac{2}{L}\sum_{n=1}^{\infty}\left\{\int_0^L \sqrt{P_H^2 - (P_H^2 - P_K^2)x/L}\cos\frac{\pi n x}{L}dx - \frac{\lambda w}{\pi n F}[Q_0(1-(-1)^n)] - \right. \quad (8)$$
$$\left. -\frac{1}{2\pi n}(Q_0 - Q_L)(-1)^n]\right\}exp(-\frac{\lambda w}{4d}t)sin[\frac{\lambda w}{4d}t\sqrt{(\frac{4\pi mcd}{\lambda w})^2 - 1}]cos\frac{\pi n x}{L}.$$

3.2. Modeling of Unsteady Pressure Fluctuations in the Gas Pipeline with Enabled En-Route Offtake of Gas

The amplitudes of pressure fluctuations were estimated in time and along the route of the gas pipeline, when disturbances in the form of jump-like changes in productivity occurred at the beginning or at the end of the gas pipeline. We considered pressure fluctuations in the initial section $P(0,t)$, where the greatest values of the absolute pressure in superposition with amplitude fluctuations can exceed the permissible load. Pressure fluctuations depend on a jump-like change in the gas flow as a disturbance factor, as well as on the coordinates of disturbance, absolute values of pressure and temperature, physical properties of gas [12–14]. They were calculated from Equation (8) with different values of the above parameters as model parameters.

It is known that when temperature and other basic physical properties of gas change in the ranges that correspond to the real operating conditions of gas pipelines, their influence on the amplitude of pressure fluctuations is insignificant. The main parameters that determine the amplitude and frequency of pressure fluctuations in the unsteady process caused by a jump-like change in the gas flow are the working pressure, the value of the gas flow, and the linear coordinate of the gas offtake. Unsteady processes at different values of working pressure and linear coordinate of the gas offtake are simulated in Figure 1.

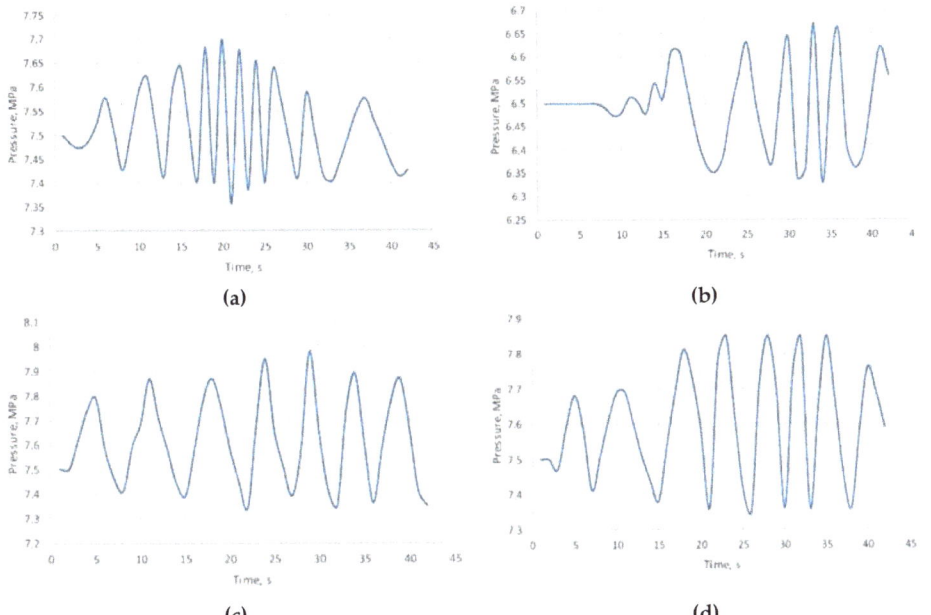

Figure 1. Results of unsteady process simulation with enabled en-route offtake of gas (pressure fluctuations at the beginning of the linear section): (**a**) at the initial pressure of 7.5 MPa; (**b**) at the initial pressure of 6.5 MPa; (**c**) at the initial pressure of 7.5 MPa, with offtake at the beginning of the section; (**d**) at the initial pressure of 7.5 MPa, with offtake in the middle of the section.

Let us denote the physical properties of gas by the gas constant R, and the thermal regime by the absolute temperature T. In this case, given the results of the model application, it can be stated that the product RT is a characteristic parameter of the dynamics of the unsteady process, which determines (although not to a large extent) the amplitude and frequency of pressure fluctuations. In other words, when changing the gas constant and the temperature of gas in such a way that their product remains unchanged, the nature of pressure fluctuations in the unsteady process will be the same. On the other hand, the RT product characterizes the propagation rate of small disturbances of gas, that is, the speed of sound $c = \sqrt{kRT}$. Hence, in order to characterize the unsteady process, it is necessary to choose the propagation rate of sound in the gas flow as the independent parameter.

The analysis of results presented in Figure 1 shows that during a jump-like change in the gas flow, the pressure wave propagates with the speed of sound to the initial cross section of the gas pipeline where the pressure is maximal, and causes the oscillatory process, the amplitude and frequency of which increase gradually and reach the maximum after (4–6) periods of oscillation. After that the amplitude and frequency of the process begin to decrease. The total duration of the oscillation process with high values of amplitude and frequency is 50–60 seconds, which is considered as a short-term overload of the pipeline. Subsequently, the amplitude and frequency of pressure fluctuations decrease significantly, and within 0.5–0.583 hours the oscillation process fades away completely, transferring the operation of the gas pipeline to a new steady regime.

A number of calculations made in accordance with the proposed mathematical model allowed establishing a number of regularities in the oscillation process caused by a jump-like change in the gas flow [16]. In particular, it was established that with a decrease in the working pressure in the gas pipeline, the total duration of the high-frequency oscillation process and the unsteady process in general increases, and the frequency and amplitude of oscillations decrease. Thus, when the working pressure is reduced from 7.5 MPa to 7.0 MPa (6.7%), the maximum amplitude of pressure fluctuations decreases from 0.199 MPa to 0.18 MPa, that is, by 9.5%, and when the working pressure is reduced to 6.5 MPa (by 13.3%), the amplitude of pressure fluctuations is 34.7%. The maximum frequency of the oscillation process at a working pressure of 7.5 MPa is 0.44 Hz, and with a decrease in pressure to 7.0 MPa it decreases by 21.8%, and with a further decrease in pressure to 6.5 MPa it decreases by 39.4%. From the physical point of view, this is due to a decrease in the elasticity of the medium, in which oscillating waves propagate, leading to a decrease in the propagation velocity of disturbances and, consequently, to an increase in the duration of the unsteady process and its high-frequency band [16,17].

The linear coordinate of the localized gas offtake has a significant impact on the nature of the unsteady process, in particular, on the amplitude and frequency of pressure fluctuations. When the localized offtake is close to the initial cross-section of the linear region, where the constant pressure is maximal, the amplitude of pressure fluctuations increases, and the frequency decreases. When the localized offtake is in the middle of the linear section of the gas pipeline, the maximum amplitude of the pressure fluctuation is 18.3% lower than that of the localized offtake located in the initial section of the gas pipeline. If the localized offtake is transferred to the final part of the linear section, the amplitude decreases by 48.5% under other identical conditions, and the frequency of oscillations varies in a smaller range. So, in the first case (when the localized offtake is in the middle of the linear section), the frequency increases by 7.2%, while in the second case (when the localized offtake is at the end of the linear section) – by 11.4%.

A change in the speed of sound has a lesser influence on the nature of pressure fluctuations in the unsteady process caused by a jump-like change in the gas flow. In addition, an increase in the speed of sound leads to an increase in the amplitude and frequency of oscillations. With an increase in the speed of sound from 400 m/s to 440 m/s (10%), the amplitude increases by 5.7%, and the frequency is 3.1%. With an increase in the speed of sound to 480 m/s (by 20%), the amplitude increases by 8.5%, and frequencies – by 4.9%.

The value of the localized gas offtake has the greatest influence on the amplitude and frequency of pressure fluctuations at the beginning of the linear section of the gas pipeline in the unsteady process caused by a jump-like change in gas flow. If a jump-like gas flow caused by the localized offtake is 10% of the total gas flow in the gas pipeline under steady conditions, the maximum amplitude of pressure fluctuations in the unsteady process caused by a sudden leakage will be 0.154 MPa. With an increase in the gas flow caused by the localized offtake to 20%, the amplitude of pressure fluctuations increases to 0.287 MPa, that is, by 2.45 times. If the gas flow due to the localized offtake is 30% of the total gas flow in the gas pipeline under steady conditions, the amplitude of pressure fluctuations will be 0.517 MPa, that is, it will increase by 3.55 times, while at a jump-like increase in the gas flow due to the localized offtake to 50% of the gas flow in the gas pipeline under steady conditions, the amplitude of pressure fluctuations will be 1.14 MPa, which may endanger the safe operation of the gas pipeline due to a short-term overload.

The maximum frequency of pressure fluctuations in an unsteady process decreases with an increase in the gas flow caused by the localized offtake. With an increase in the gas flow caused by the localized offtake from 10% to 20%, the maximum frequency of pressure fluctuations decreases by 5.8%, and with a further increase in the gas flow caused by the localized offtake to 30%, the maximum frequency of pressure fluctuations decreases by 12.3%.

3.3. Ability to Regulate Pressure by Disconnecting Individual Compressor Stations

The analysis of the results of simulation of an unsteady process in a gas pipeline caused by a jump-like change in the gas flow under conditions of insufficient loading shows that with the maximum allowable constant pressure at the beginning of the linear section of the gas pipeline, the amplitude of the pressure fluctuation can lead to a short-term overload of the pipe walls, that is, to an unsafe operation of the gas transmission system. Therefore, when controlling operating conditions, it is necessary to make decisions on providing the permissible pressure at the beginning of the linear section (at the outlet of the compressor station) in cases of a jump-like change in the gas flow. One of the options is disconnecting individual compressor stations. As shown in [18–20], depending on the quantity and serial numbers of the working stations, it is possible to achieve the required performance of the gas transmission system. At the same time, given the low efficiency of gas-pumping units with gas turbine drives, disconnecting individual CSs can be the most effective method of controlling performance from the energetic point of view. It is obvious that this method of control can be used for seasonal regulation of performance, and it should be borne in mind that shutdown and re-start of the CS will require additional energy costs.

However, from the technological point of view, shutdown and re-start of the compressor station may lead to an unsteady process, the duration of which should be predicted in order to provide consumers with gas. It is necessary to take into account the influence of unsteady processes in long-distance gas transmission systems with a large number of compressor stations [12–14]. Prediction and analysis of the said conditions in the gas transmission system, as well as the estimation of energy losses, are possible using the mathematical model of the unsteady gas flow in pipes taking into account an increase in pressure at compressor stations and discontinuity of flow:

$$-\frac{\partial P}{\partial x} + \sum_{i=1}^{m} \Delta P_{KCi}\delta(x - x_i) = \left(\frac{\partial(\rho w)}{\partial t} + \frac{\lambda \rho w^2}{2d}\right); \quad \frac{\partial P}{\partial t} = -c^2 \frac{\partial(\rho w)}{\partial x}, \quad (9)$$

where $P(x,t)$ is the pressure in the gas pipeline as a function of the linear coordinate x and time t; ΔP_{KCi} is an increase in pressure at the compressor station with coordinate x_i; $\delta(x - x_i)$ is the Dirac source function, which simulates an increase in pressure at the compressor station, ρ is the density of gas; w is the linear velocity of gas; d is the internal diameter of the gas pipeline; λ is the coefficient of hydraulic resistance.

We note that in order to simulate the fading vibration processes in the gas pipeline, the equation of gas flow should include inertial hydraulic losses and frictional losses. The given system of differential Equations (9) is reduced to the equation:

$$\frac{\partial^2 P}{\partial x^2} = \frac{1}{c^2}\frac{\partial^2 P}{\partial t^2} + \frac{2a}{c^2}\frac{\partial P}{\partial t} + \sum_{i=1}^{m} \Delta P_{KCi}\delta^*(x - x_i), \quad (10)$$

where $\delta^*(x - x_i)$ is the linear derivative of the Dirac function from the linear coordinate; c is the speed of sound in gas;

We assume that a gas transmission system with the length L has m intermediate compressor stations, which start to work simultaneously at the point of time $t = 0$, and that station k is disconnected at the point of time t_1. In this case, Equation (10) will take the following form:

$$\frac{\partial^2 P}{\partial x^2} = \frac{1}{c^2}\frac{\partial^2 P}{\partial t^2} + \frac{2a}{c^2}\frac{\partial P}{\partial t} + \sum_{i=1}^{m} \Delta P_{KCi}\delta^*(x - x_i) + \Delta P_{KCk}\delta^*(x - x_k)[\sigma(t) - \sigma(t - t_1)] \quad (11)$$

where $\sigma(t)$ is the single Heaviside function. We assume that at the initial moment of time, the gas pipeline was stopped, and constant pressure P_o was maintained throughout its length. Then the initial conditions will be:

$$t = 0, \quad P(x,0) = P_0, \quad \frac{\partial P}{\partial x} = 0.$$

Starting from the moment $t > 0$, constant pressure $P(0,t) = P_H$ is maintained at the beginning of the gas pipeline, and constant pressure $P(L,t) = P_K$ is maintained at its end.

For the solution of the mathematical model, integral transformations were used, in particular: Fourier sine-transformation and Laplace transformation [10,11]

The application of inverse transformations of Laplace and Fourier after simple transformations allows us to obtain the dependence of the pressure variation on the length and time of the unsteady process in the following form:

$$P(x,t) = P_0 + (P_H - P_K)\frac{x}{L} + \sum_{\substack{i=1 \\ i \neq k}}^{m} \Delta P_{KCi} \left\{ \begin{array}{l} (1-\frac{x}{L}) \; npu \; x > x_i \\ (-\frac{x}{L}) \; npu \; x < x_i \end{array} \right\} +$$

$$+ \Delta P_{KCk}[\sigma(t) - \sigma(t-t_1)] \left\{ \begin{array}{l} (1-\frac{x}{L}) \; npu \; x > x_i \\ (-\frac{x}{L}) \; npu \; x < x_i \end{array} \right\} + \sum_{n=1}^{\infty} C_n e^{-at} f(n,t) \sin\left(\frac{\pi n x}{L}\right) +$$

$$+ \frac{2}{\pi} \Delta P_{KCk} \sum_{n=1}^{\infty} \frac{1}{n} \cos\left(\frac{\pi n x_k}{L}\right) \sin\left(\frac{\pi n x}{L}\right) e^{-a(t-t_1)} f(n, t-t_1) \sigma(t-t_1).$$

The first four components of the solution of Equation (12) characterize steady operating conditions of the gas transmission system. The fifth component describes the unsteady process caused by the simultaneous connection of all the compressor stations at the moment $t = 0$. The last component modulates the unsteady process caused by the disconnection of the k-th compressor station from the moment t_1. If $t_1 >> 0$, that is, the process that occurs within the pipeline is considered after a significant period of time from the moment of connecting all CSs, the initial unsteadiness will not have a significant impact due to a higher order of smallness of the e^{-at} multiplier, and the solution to the problem of disconnecting the k-th compressor station can be represented as:

$$P(x,t) = P_0 + (P_H - P_K)\frac{x}{L} + \sum_{\substack{i=1 \\ i \neq k}}^{m} \Delta P_{KCi} \left\{ \begin{array}{l} (1-\frac{x}{L}) \; npu \; x > x_i \\ (-\frac{x}{L}) \; npu \; x < x_i \end{array} \right\} + \quad (12)$$

$$+ \frac{2}{\pi} \Delta P_{KCk} \sum_{n=1}^{\infty} \frac{1}{n} \cos\left(\frac{\pi n x_k}{L}\right) \sin\left(\frac{\pi n x}{L}\right) e^{-a(t-t_1)} f(n, t-t_1) \sigma(t-t_1).$$

The solution of Equation (13) describes an unsteady process caused by disconnecting the k-th compressor station and does not take into account the unsteadiness of the initial process of connecting all the CSs. Therefore, the countdown can start from the moment of disconnecting the k-th compressor station. In this case, we obtain:

$$P(x,t) = P_0 + (P_H - P_K)\frac{x}{L} + \sum_{\substack{i=1 \\ i \neq k}}^{m} \Delta P_{KCi} \left\{ \begin{array}{l} (1-\frac{x}{L}) \; npu \; x > x_i \\ (-\frac{x}{L}) \; npu \; x < x_i \end{array} \right\} + \quad (13)$$

$$+ \frac{2}{\pi} \Delta P_{KCk} \sum_{n=1}^{\infty} \frac{1}{n} \cos\left(\frac{\pi n x_k}{L}\right) \sin\left(\frac{\pi n x}{L}\right) e^{-at} \left(\cos\left(\sqrt{\left(\frac{\pi n c}{L}\right)^2 - a^2}\right)t + \frac{a}{\sqrt{\left(\frac{\pi n c}{L}\right)^2 - a^2}} \sin\left(\sqrt{\left(\frac{\pi n c}{L}\right)^2 - a^2}\right)t \right).$$

Equation (14) allows predicting the nature of the unsteady process in long-distance gas transmission systems with a large number of compressor stations caused by the shutdown and re-start of one of the stations.

To estimate the duration of the unsteady process, it is necessary to construct the dependence of fluctuations over the period of the mass flow of gas as the most inertial characteristic in the initial or final section of the gas pipeline [16,19,20].

To this end, we use the equation of the gas flow from system (9). Obviously, for the initial ($x = 0$) or finite cross section ($x = L$), the delta-function of Dirac is $\delta(x - x) = 0$, therefore:

$$-\frac{\partial P}{\partial x} = \frac{\partial (\rho w)}{\partial \iota} + \frac{\lambda \rho w^2}{2d}. \quad (14)$$

To simplify the computational process, we neglect inertial losses in the initial and final cross-sections, that is, we accept that $\frac{\partial(\rho w)}{\partial t} = 0$ [3]. This, of course, is associated with a certain error in the calculation of the mass flow of gas, however, in the forecast calculations, it is not important to determine the absolute value of gas consumption, but the dynamics of its change over time. In addition, by using the linearization of the equation of the gas flow, we obtain:

$$m(0,t) = -\frac{\pi d^3}{\lambda w} \frac{\partial P}{\partial x}|_{x=0}, \quad m(L,t) = -\frac{\pi d^3}{\lambda w} \frac{\partial P}{\partial x}|_{x=L}$$

Using Equation (14) after differentiation we obtain:

$$m(0,t) = -\frac{\pi d^3}{\lambda w}\left(\frac{P_H - P_K}{L} - \frac{\sum\limits_{\substack{i=1 \\ i \neq k}}^{m} \Delta P_{KCi}}{L} + \frac{2L}{\pi^2}\Delta P_{KCk}\sum_{n=1}^{\infty}\frac{1}{n^2}\cos\left(\frac{\pi n x_k}{L}\right)e^{-at}\left(\cos\left(\sqrt{\left(\frac{\pi n c}{L}\right)^2 - a^2}\,t\right)\right.\right.$$
$$\left.\left. + \frac{a}{\sqrt{\left(\frac{\pi n c}{L}\right)^2 - a^2}}\sin\left(\sqrt{\left(\frac{\pi n c}{L}\right)^2 - a^2}\,t\right)\right)\right), \quad (15)$$

$$m(L,t) = -\frac{\pi d^3}{\lambda w}\left(\frac{P_H - P_K}{L} - \frac{\sum\limits_{\substack{i=1 \\ i \neq k}}^{m} \Delta P_{KCi}}{L} + \frac{2L}{\pi^2}\Delta P_{KCk}\sum_{n=1}^{\infty}\frac{(-1)^n}{n^2}\cos\left(\frac{\pi n x_k}{L}\right)e^{-at}\left(\cos\left(\sqrt{\left(\frac{\pi n c}{L}\right)^2 - a^2}\,t\right)\right.\right.$$
$$\left.\left. + \frac{a}{\sqrt{\left(\frac{\pi n c}{L}\right)^2 - a^2}}\sin\left(\sqrt{\left(\frac{\pi n c}{L}\right)^2 - a^2}\,t\right)\right)\right).$$

The obtained dependencies, Figure 2, allow predicting the nature of the mass flow fluctuation over time at the beginning and end of a large-distance gas transmission system, which employs m intermediate compressor stations, caused by shutdown or re-start of the k-th compressor station ($k = 1, 2, \ldots, m$).

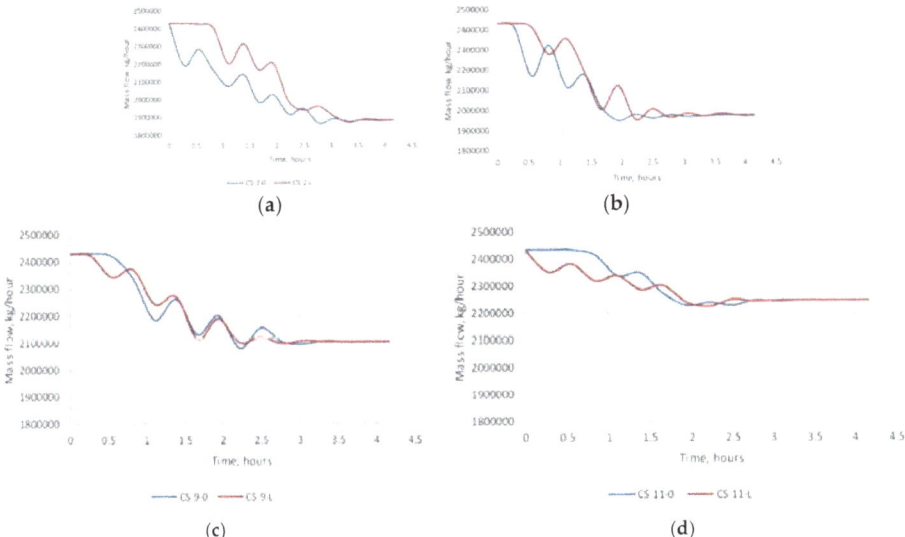

Figure 2. Regularities in the unsteady process upon shutdown of the CS (fluctuations of mass flow over time): (**a**) at the beginning of the gas pipeline; (**b**,**c**) at the middle of the gas pipeline; (**d**) at the end of the gas pipeline.

It is known that oscillatory processes in a gas pipeline create loading conditions different from normal operating condition, which may cause dynamic non-equilibrium processes in the pipe material [21], unpredictable accumulation of local micro-strains, and changes in the crack resistance of pipe steels. Therefore, minimizing such negative phenomena is not only a matter of optimizing the gas pumping technology, but also ensuring the integrity of the main gas pipeline.

4. Discussion of Results

The gas transportation system (GTS) of Ukraine is the second largest in Europe and one of the largest in the world. The GTS of Ukrtransgas PJSC consists of gas mains (37.6 thousand km long), distribution networks, gas storage facilities, compressor and gas measuring stations (71 compressor stations with a total capacity of 5405 MW). The throughput at the borders of Ukraine with Russia is 288 billion cubic meters per year, at the borders of Ukraine with Belarus, Poland, Slovakia, Hungary, Romania and Moldova - 178.5 billion cubic meters per year, and with the EU countries - 142.5 billion cubic meters per year [22].

At the same time, most of the gas mains have been in service for more than 30 years, Figure 3a, which requires optimizing their operating conditions and eliminating negative oscillation phenomena, which is particularly important for gas pipelines with a diameter of 1420 mm, the share of which is 16% of the total amount of gas pipelines, Figure 3b.

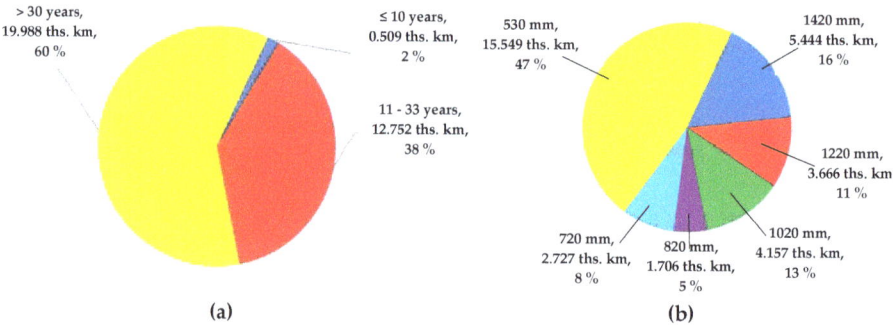

Figure 3. Structure of the system of gas mains and branches of Ukrtransgas PJSC (ths. km and %) in terms of the period of operation (**a**) and diameters (**b**).

Increasing the reliability of operation of gas mains, which have been in service for more than 30 years, is only possible taking into account the specific features of their laying, and optimization of performance parameters. The main parameters that determine the reliability of gas pipelines are their pressure and temperatures fluctuations.

Insufficient loading of a complex gas transportation system may cause pressure fluctuations in gas pipelines with a given constant productivity. The upper range of possible changes in pressure is limited by the depression line with the maximum initial pressure, and the lower range is limited by the depression line with the minimum final pressure. Violation of the specified range can lead to the pipeline failure due to excess pressure or to failure of centrifugal superchargers of the CS due to lowering of the pressure below the minimum permissible level. The formation of pressure depression at each moment of changing the magnitude of productivity is an unsteady process characterized by pressure fluctuations with a certain frequency and amplitude. The superposition of pressures at the upper limit of depression can lead to an excess of the initial pressure, and at the lower limit of depression – to lowering the pressure below the permissible level.

To study the impact caused by the serial number of the disconnected compressor station on the duration of the unsteady process, a computational experiment was conducted on the basis of the "Soyuz" gas main with a total length of 1567.3 km (on the territory of Ukraine), a diameter of 1420 mm, and a wall thickness of 20 mm, which has 13 compressor stations equipped with GTK-10I gas-pumping units along its route, Figure 4 [22]. Forecast calculations of the unsteady processes caused by shutdown of the CS were made under the conditions of the project regime, according to which the throughput of the gas pipeline is 26 billion m³ per year with the initial pressure (at the outlet of the CS) of 7.5 MPa, and the final pressure (at the inlet of the CS) of 5.0 MPa, which corresponds to the difference ΔP_{KC} = 2.5 MPa and is the same for all of the CSs.

Figure 4. The GTS of Ukraine with the designated "Soyuz" gas main with a total length of 1567.3 km (the territory of Ukraine), a diameter of 1420 mm, and a wall thickness of 20 mm, which has 13 compressor stations (CS) equipped with GTK-10I gas-pumping units along its route: 1. Novopskov; 2. Borova; 3. Pershotravneva; 4. Mashivka; 5. Kremenchuk; 6. Oleksandrivka; 7. Talne; 8. Haisyn; 9. Bar; 10. Husiatyn; 11. Bogorodchany; 12. Khust; 13. Uzhgorod.

The initial station CS-11 Novopskov was considered to be the main one, and the remaining 12 stations were intermediate. The task was to determine the nature of changes in the gas pipeline performance over time at the beginning and end of the gas pipeline using Equation (15) with the phased disconnection of each intermediate CS.

The analysis of the graphical dependences of the mass gas flow fluctuations at the beginning ($x = 0$) and at the end ($x = L$) of the gas transportation system has allowed to determine the duration of the unsteady process caused by the phased shutdown of each compressor station. Note that the unsteady processes in gas pipelines are unprofitable in terms of energy consumption for pipelines, as they cause the appearance of inertial forces in the flow of a continuous medium, the work of which leads to a decrease in the overall efficiency of the system. Therefore, the most favorable regime (under other identical conditions) is the one, under which the duration of the unsteady process is minimal.

Calculations using mathematical model (15) show that the longest duration of the unsteady process is possible when the CS-2 Borova is turned off, and is 3 h 24 min 12 s at the beginning of the gas transmission system (at the outlet of the Novopskov CS) and 3 hours 42 min at the end of the system. Thus, the duration of the unsteady process at the end of the system is 30.6% longer than at the beginning of the system, which is explained by a significant distance between the disconnected CS and the end of the route. When the CS Borova is disconnected, the performance of the new steady process is 22.5% lower than the throughput (for all working CSs it is 675 kg/s).

When disconnecting the CS Khust (the second one from the end of the route), the duration of the unsteady process is the shortest and is 2.55 h at the beginning of the gas transmission system and 2.26 hours at the end of the route. The duration of the unsteady process at the beginning of the system is 11.3% longer than at the end of the route, which is explained by the difference in the distance between the disconnected CS and the ends of the pipeline. A decrease in the gas pipeline's performance is 7.6% compared to the throughput.

Let us consider graphs showing the duration of the unsteady process at the beginning and at the end of the gas transportation system with the phased disconnection of each of the compressor stations, Figure 5. It is noticeable that the duration of the unsteady process caused by the disconnection of a CS decreases at the beginning and end of the gas pipeline with an increase in the serial number of the disconnected station. Moreover, at the beginning of the gas pipeline, a tendency towards decreasing the productivity is less pronounced than at its end. So, when disabling CS-2, the ratio of the duration of the unsteady process at the end of the gas pipeline to the corresponding duration at the beginning of the gas pipeline is 1.306, when disabling CS-5, this ratio is equal to 1.022, and when disabling CS-10, it is 0.935. This circumstance should be taken into account when planning control procedures for the gas transmission system by disconnecting certain CSs in order to ensure the uninterrupted supply of gas to consumers.

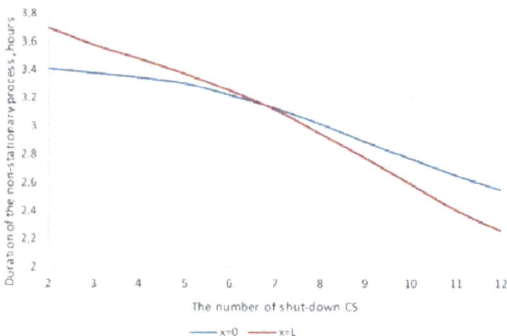

Figure 5. Duration of the unsteady process caused by disconnection of compressor stations for the initial ($x = 0$) and final ($x = L$) cross-section of the gas pipeline.

At present, the underutilization of the gas transportation system of Ukraine is 42.4% and it continues to grow [22,23]. Under such conditions, it is almost impossible to prevent unsteady vibrational changes in pressure within the gas pipeline. Therefore, it is necessary to study the amplitude-frequency characteristics of unsteady operating conditions and their duration in order to ensure reliable and efficient operation of gas mains. In [10,11] was found that the pressure difference in the cross-section of the gas pipeline under the obverse and reverse conditions reaches ~ 20%. This result should be taken into account when scheduling maintenance and in-pipe diagnostics. Planning should also be made taking into account the results of calculating more rigid operating conditions [24].

As of today, the Ukrainian GTS, and, in particular, the "Soyuz" gas main considered, is used for the transit of gas from Russia to the EU [25]. In the event the transit of the Russian gas through the territory of Ukraine is discontinued, this gas pipeline can be transferred to the reverse mode to supply gas to consumers in the east of the country. In this case, the gas pipeline will operate in conditions of insufficient loading, as discussed in this article.

5. Conclusions

The mathematical model is developed, which allows estimating the amplitude of pressure fluctuations in the gas pipeline along its length and in time in the presence of disturbances in the form of jump-like changes in productivity at the beginning or at the end of the gas pipeline.

It is proved that the location of the disconnected CS on the route of the gas pipeline has a significant effect on the duration of the unsteady transitional conditions. Moreover, with an increase in its serial number within the system, the duration of the unsteady process and the value of a decrease in productivity are reduced.

It is established that with the maximum allowable constant pressure at the beginning of the linear section of the gas pipeline, the amplitude of the pressure fluctuation can lead to a short-term overload of the pipe walls, that is, to an unsafe operation of the gas transmission system. One of the regulation options is disconnecting individual compressor stations.

The dependencies obtained in this research allow predicting the variations of mass flow in time at the beginning and at the end of a long-distance gas transmission system. The method of estimating the parameters of an unsteady process in long-distance gas transmission systems with a large number of compressor stations caused by the shutdown and re-start of one of the stations is proposed.

Author Contributions: Conceptualization, V.Z.; Formal analysis, J.B. and A.G.; Investigation, V.Z., L.P., P.M., V.G.J., R.S., J.B., A.G.; Methodology, V.G.J. and L.P.; Project administration, P.M.; Validation, P.M., R.S., V.Z., and J.B.; Writing—original draft, V.Z., P.M., J.B., L.P., and A.G.; Writing—review and editing, J.B.

Funding: This research was funded by Ministry of Education of the Slovak Republic VEGA No. 1/0424/17 and of the Slovak Research and Development Agency APVV-16-0359 and the APC was funded by Slovak Research and Development Agency.

Acknowledgments: This work was supported by scientific grant agency of the Ministry of Education of the Slovak Republic VEGA No. 1/0424/17 and of the Slovak Research and Development Agency APVV-16-0359.

Conflicts of Interest: The authors declare no conflict of interest.

References

1. Maruschak, P.; Danyliuk, I.; Prentkovskis, O.; Bishchak, R.; Pylypenko, A.; Sorochak, A. Degradation of the main gas pipeline material and mechanisms of its fracture. *J. Civil Eng. Manag.* **2014**, *20*, 864–872. [CrossRef]
2. Maruschak, P.; Poberezhny, L.; Pyrig, T. Fatigue and brittle fracture of carbon steel of gas and oil pipelines. *Transport* **2013**, *28*, 270–275. [CrossRef]
3. Doroshenko, Y.; Doroshenko, J.; Zapukhliak, V.; Poberezhny, L.; Maruschak, P. Modeling computational fluid dynamics of multiphase flows in elbow and T-junction of the main gas pipeline. *Transport* **2019**, *34*, 19–29. [CrossRef]
4. Vianello, C.; Maschio, G. Quantitative risk assessment of the Italian gas distribution network. *J. Loss Prev. Process Ind.* **2014**, *32*, 5–17. [CrossRef]
5. Al-Khalil, M.; Assaf, S.; Al-Anazi, F. Risk-based maintenance planning of cross country pipelines. *J. Perform. Construct. Facil.* **2005**, *19*, 124–131. [CrossRef]
6. Shcherbakov, S.G. *Problems of Pipeline Transportation of Oil and Gas*; Nauka: Moscow, Russia, 1982; 206p. (In Russian)
7. Grudz, V.Y.; Grudz, V.Y., Jr. Forecasting of non-stationary processes in gas transmission systems on condition of their incomplete loading. *Naftogazova Energetyka* **2017**, *28*, 62–68. (In Ukrainian)
8. Yakovlev, E.I.; Kazak, O.S.; Myhalkiv, V.B.; Timkov, D.F.; Grudz, V.Y. *Modes of Gas Transmission Systems*; Lviv: Svit, Ukraine, 1992; 170p. (In Ukrainian)
9. Filipchuk, O.; Grudz, V.; Marushchenko, V.; Myndiuk, V.; Savchuk, M. Development of cleaning methods complex of industrial gas pipelines based on the analysis of their hydraulic efficiency. *Eastern-Eur. J. Enterpr. Technol.* **2018**, *2*, 62–71. [CrossRef]
10. Grudz, Y. Formalization of the design model of gas-main pipelines infrastructure failure. *Metall. Mining Ind.* **2015**, *292*, 79–84.
11. Grudz, V.Y.; Grudz, V.Y., Jr.; Zapukhlyak, V.B.; Kyzymyshyn, Y.V. Non-stationary processes in the gas transmission systems at compressor stations shut-down. *J. Hydrocarbon Power Eng.* **2018**, *5*, 22–27.
12. Meshalkin, V.P.; Chionov, A.M.; Kazak, A.S.; Aristov, V.M. A computer model of the nonstationary gas flow in a long multilayer-insulated high-pressure subsea gas pipeline. *Doklady Chem.* **2016**, *469*, 241–244. [CrossRef]
13. Meshalkin, V.P.; Chionov, A.M.; Kazak, A.S.; Aristov, V.M. Applied computer model of the non-stationary gas flow in a long multilayer-insulated high-pressure subsea gas pipeline. *Doklady Chem.* **2016**, *470*, 279–282. [CrossRef]

14. Meshalkin, V.P.; Moshev, E.R. Modes of functioning of the automated system "pipeline" with integrated logistical support of pipelines and vessels of industrial enterprises. *J. Mach. Manuf. Reliab.* **2015**, *44*, 580–592. [CrossRef]
15. Rios-Mercado, R.Z.; Borraz-Sanchez, C. Optimization problems in natural gas transportation systems: A state-of-the-art review. *Appl. Energy* **2015**, *147*, 536–555. [CrossRef]
16. Prytula, N.M.; Gryniv, O.D.; Dmytruk, V.A. Simulation of nonstationary regimes of gas transmission system operation. *Math. Model. Comput.* **2014**, *1*, 224–233.
17. Szoplik, J. The steady-state simulations for gas flow in a pipeline network. *Chem. Eng. Trans.* **2010**, *21*, 1459–1464. [CrossRef]
18. Herrán-González, A.; De La Cruz, J.M.; De Andrés-Toro, B.; Risco-Martín, J.L. Modeling and simulation of a gas distribution pipeline network. *Appl. Math. Model.* **2009**, *33*, 1584–1600. [CrossRef]
19. Woldeyohannes, A.D.; Majid, M.A.A. Simulation model for natural gas transmission pipeline network system. *Simulat. Model. Pract. Theory* **2011**, *19*, 196–212. [CrossRef]
20. Chaczykowski, M.; Zarodkiewicz, P. Simulation of natural gas quality distribution for pipeline systems. *Energy* **2017**, *134*, 681–698. [CrossRef]
21. Chausov, M.G.; Maruschak, P.O.; Pylypenko, A.P.; Prentkovskis, O. Effect of impact-oscillatory loading on variation of mechanical properties and fracture toughness of pipe steel 17G1S-U. In Proceedings of the 20th International Conference 'Transport Means 2016', Vilnius, Lithuania, 5–7 October 2016; pp. 441–443.
22. Maruschak, P.; Prentkovskis, O.; Bishchak, R. Defectiveness of external and internal surfaces of the main oil and gas pipelines after long-term operation. *J. Civil Eng. Manag.* **2016**, *22*, 279–286. [CrossRef]
23. Eser, P.; Chokani, N.; Abhari, R. Impact of Nord Stream 2 and LNG on gas trade and security of supply in the European gas network of 2030. *Appl. Energy* **2019**, *238*, 816–830. [CrossRef]
24. Maruschak, P.O.; Bishchak, R.T.; Danyliuk, I.M. *Crack Resistance of Materials and Structures: Gas Mains after Long-Term Operation*; ZAZAPRINT: Ternopil, Ukraine, 2016; 182p.
25. Richter, P.M.; Holz, F. All quiet on the eastern front? Disruption scenarios of Russian natural gas supply to Europe. *Energy Policy* **2015**, *80*, 177–189. [CrossRef]

© 2019 by the authors. Licensee MDPI, Basel, Switzerland. This article is an open access article distributed under the terms and conditions of the Creative Commons Attribution (CC BY) license (http://creativecommons.org/licenses/by/4.0/).

Article

Fractal and Multifractal Analysis of Pore Size Distribution in Low Permeability Reservoirs Based on Mercury Intrusion Porosimetry

Penghui Su [1,*], Zhaohui Xia [1], Ping Wang [1], Wei Ding [1], Yunpeng Hu [1], Wenqi Zhang [1] and Yujie Peng [2]

1. PetroChina Research Institute of Petroleum Exploration and Development, Beijing 100083, China; xiazhui@petrochina.com.cn (Z.X.); wp2011@petrochina.com.cn (P.W.); dingwei-hw@petrochina.com.cn (W.D.); huyunpeng1129@petrochina.com.cn (Y.H.); zhang_wenqi@petrochina.com.cn (W.Z.)
2. Guizhou Polytechnic of Construction, Guizhou 551400, China; pyjcumt@126.com
* Correspondence: suphui317@petrochina.com.cn; Tel.: +86-150-105-669-36

Received: 3 March 2019; Accepted: 29 March 2019; Published: 8 April 2019

Abstract: To quantitatively evaluate the complexities and heterogeneities of pore structures in sandstone reservoirs, we apply single fractal theory and multifractal theory to explore the fractal characteristics of pore size distributions based on mercury intrusion porosimetry. The fractal parameters were calculated and the relationships between the petrophysical parameters (permeability and entry pressure) and the fractal parameters were investigated. The results show that the single fractal curves exhibit two-stage characteristics and the corresponding fractal dimensions D_1 and D_2 can characterize the complexity of pore structure in different sizes. Favorable linear relationships between $\log(\varepsilon)$ and $\log(\mu_i(\varepsilon))$ indicate that the samples satisfy multifractal characteristics and ε is the sub-intervals with size $\varepsilon = J \times 2^{-k}$. The multifractal singularity curves used in this study exhibit a right shape, indicating that the heterogeneity of the reservoir is mainly affected by pore size distributions in sparse regions. Multifractal parameters, $D(0)$, $D(1)$, and Δf, are positively correlated with permeability and entry pressure, while $D(0)$, $D(1)$, and Δf are negatively correlated with permeability and entry pressure. The ratio of larger pores volumes to total pore volumes acts as a control on the fractal dimension over a specific pore size range, while the range of the pore size distribution has a definite impact on the multifractal parameters. Results indicate that fractal analysis and multifractal analysis are feasible methods for characterizing the heterogeneity of pore structures in a reservoir. However, the single fractal models ignore the influence of microfractures, which could result in abnormal values for calculated fractal dimension. Compared to single fractal analysis, multifractal theory can better quantitatively characterize the heterogeneity of pore structure and establish favorable relationships with reservoir physical property parameters.

Keywords: multifractal theory; fractal theory; pore structure; mercury intrusion porosimetry; pore size distribution

1. Introduction

A reservoir consists of a complex porous medium composed of pores with different origins, irregular shapes, and self-similarities. The pore structure in a reservoir is mainly affected by three factors: sedimentation, diagenesis, and tectogenesis. Each of these factors result in the formation of pores with attributes which fall into different ranges. Thus, different types of pores are formed and exist in the reservoir in a certain distribution. This complex pore size distribution is dynamically nonlinear and is the result of numerous processes occurring at various scales [1].

Fractal theory, a promising tool for investigating complex structures, has been widely used to quantitatively characterize the complexities and heterogeneities of pore size distributions [2]. Fractal theory is considered to be an effective means of quantitatively depicting irregular shapes, and can accurately express the complexity and heterogeneity of geological bodies. Extensive research has proven that reservoirs exhibit fractal characteristics. Pfeifer found that the pore surface area of a reservoir exhibits fractal characteristics by the using molecular adsorption [3]. Katz investigated different types of sandstone using scanning electron microscopy; the results indicated that the pore spaces of sandstone possess fractal characteristics [4]. Friesen obtained the fractal dimensions of coal particles based on capillary injection data [5].

Mercury intrusion porosimetry is commonly used to determine the pore structure distributions of rocks [6]. Fractal studies of capillary pressure data mainly focus on obtaining fractal dimensions to establish the relationship between fractal dimension and reservoir physical properties [7,8]. There are several widely used fractal models to obtain the fractal dimension of mercury injection data based on single fractal theory [5,9–11]. Single fractal theory describes the integrity characteristics of pore structures [7]; it is suitable for application to a homogeneous reservoir, but cannot define the local pore structure or reflect other comprehensive and detailed information [12]. The segmented fractal phenomenon occurs when the fractal theory is applied to mercury injection data [13–15]. This fact demonstrates that describing the complexity of the whole pore system is difficult with a single fractal dimension.

Multifractal analysis was conducted to describe the complexity of pore size distribution because single fractal theory has limitations in describing the local characteristics of pore size distribution. Multifractal theory partitions objects at different scales to obtain distribution characteristics [16,17]. As a result, it likely characterizes the complex and heterogeneous behavior of a reservoir more effectively than the single fractal theory [7]. Multifractal theory has been widely applied to study effect of pore size changes on reservoir physical properties [18]. Multifractal analysis of soil pore structure has been widely studied to characterize soil structure stability and soil surface evolution stages [19]. Multifractal theory has also been confirmed as a useful tool for characterizing the internal complexity and amplifying the differences in pore size distributions between different coals [1]. Research on the quantitative characterization of the irregular microscopic pore structures of different rock types has been performed on the basis of 2-D images [20–23]. From the above literature review, we can conclude that multifractal analysis of pore size distribution measured by mercury intrusion porosimetry has not been extensively applied to sandstone reservoirs. Comparative studies of single fractal theory and multifractal theory based on mercury intrusion porosimetry have not been performed.

In the present study, single fractal theory and multifractal theory are applied to investigate variability and heterogeneity in pore structures based on mercury intrusion curves. Single fractal and multifractal parameters were analyzed for correlation with reservoir physical parameters. We compared single fractal theory and multifractal theory to test which method is more suitable for characterizing reservoir physical parameters. Multifractal analysis of pore structures can expand our understanding of the pore structures of reservoirs.

2. Materials and Methods

2.1. Single Fractal Theory

Several fractal models were established to obtain the fractal dimensions of mercury injection data. Su's fractal model, which considers both the fractal characteristics of pore space and pore length and was proposed based on the capillary bundle model, was applied in this paper [11]:

$$\log(S_{Hg}) \propto \left(D_f + D_T - 3\right) \log(P_c) \tag{1}$$

where S_{Hg} is the mercury saturation, D_f is the fractal dimension for pore space, with $2 < D_f < 3$ in three dimensions [24], D_T is the fractal dimension for tortuosity, with $1 < D_T < 3$ in three dimensions [25], and P_c is the capillary pressure. The sum of the fractal dimension of pore space and the fractal dimension of tortuosity can be obtained from the double logarithmic curve of mercury saturation and capillary pressure. Theoretically, the sum of the fractal dimensions of pore space and tortuosity should be between 3 and 6; the larger the sum of fractal dimension, the more complicated the pore structure [11].

2.2. Multifractal Theory

Multifractal theory describes the local conditions of a fractal structure through the singularity strength [26], and the overall characteristics are investigated from a local perspective. The prerequisite for applying the multifractal analysis method is that the measurement interval must be divided equally [27].

The mercury injection curves obtained in this study have fewer measurement points than is required for the multifractal method. To obtain enough data for multifractal analysis of mercury intrusion curves, we used cubic spline interpolation to interpolate mercury intrusion curves, allowing us to obtain more data points.

In this study, the whole measured pore size was defined as I, $I = [0.015, 35.6 \ \mu m]$. The difference of capillary pressure data in the aforementioned pore size range is measured over steps of $0.005 \ \mu m$. Thus, sub-intervals can be obtained. I_i represents the i-th sub-interval, and v_i represents the percentage of pore volume in the sub-interval I_i. A new measure can be obtained ($J = [\log 0.015, \log 35.6 \ \mu m]$) by plotting the aforementioned pore size range I on a logarithmic scale. J was divided into 2^k equal sub-intervals with size $\varepsilon = J \times 2^{-k}$. The whole interval is divided into reduced sub-intervals with increasing k; thus, the effects of pore space changes within a small interval can be investigated. J_i represents the i sub-interval in the interval J. $P_i(\varepsilon)$ represents the percentage of pore volume in the sub-interval J_i; it is equal to the sum of v_i that falls within the sub-interval J_i.

The partition function $\chi(q,\varepsilon)$ can be defined using $P_i(\varepsilon)$:

$$\chi(q,\varepsilon) = \sum_{i=1}^{N(\varepsilon)} P_i(\varepsilon)^q \tag{2}$$

where q is a real parameter that describes the moment order of the measure. For $q < 1$, $\chi(q,\varepsilon)$ emphasizes the regions determined by a small $P_i(\varepsilon)$ or minimally concentrated region of a measure. For $q > 1$, $\chi(q,\varepsilon)$ emphasizes the regions determined by a large $P_i(\varepsilon)$ or wide concentrated region of a measure. The q used in this study is between -20 and 20. $\chi(q,\varepsilon)$ and ε follow a power law relationship as follows:

$$\chi(q,\varepsilon) \propto \varepsilon^{\tau(q)} \tag{3}$$

where $\tau(q)$ is the mass exponent, which can also be expressed by the following formula:

$$\tau(q) = \lim_{\varepsilon \to 0} \frac{\lg(\sum_{i=1}^{N(\varepsilon)} P_i(\varepsilon)^q)}{\lg(\varepsilon)} \tag{4}$$

The mass exponent can also be expressed as follows, according to previous research results [19]:

$$\tau(q) = (q-1)D(q) \tag{5}$$

where $D(q)$ is the generalized dimension. Correspondingly, $D(q)$ can be expressed as follows:

$$D(q) = \lim_{\varepsilon \to 0} \frac{1}{q-1} \frac{\lg(\sum_{i=1}^{N(\varepsilon)} P_i(\varepsilon)^q)}{\lg(\varepsilon)} (q \neq 1), \tag{6}$$

For $q = 1$, $D(q)$ is defined as follows [26]:

$$D(1) = \lim_{\varepsilon \to 0} \left\{ \frac{\sum_{i=1}^{N(\varepsilon)} \mu_i(\varepsilon) \log(\mu_i(\varepsilon))}{\lg(\varepsilon)} \right\} \qquad (7)$$

If ε is sufficiently small, then $P_i(\varepsilon)$ is nearly evenly distributed within each subinterval, where $P_i(\varepsilon)$ and ε show the following relationship:

$$P_i(\varepsilon) \propto \varepsilon^\alpha \qquad (8)$$

where α is the singularity exponent. Different subintervals may have the same α. $N_\alpha(\alpha)$ represents the subinterval numbers of the singularity exponent between α and $\alpha + d\alpha$; it satisfies the following fractal power law relationship:

$$N_\alpha(\varepsilon) \propto \varepsilon^{-f(\alpha)} \qquad (9)$$

where $f(\alpha)$ is a multifractal spectrum with singularity exponent α. Different α values and corresponding $f(\alpha)$ constitute the multifractal spectrum that describes multifractal properties.

The singularity exponent can also be expressed as follows:

$$\alpha(q) = \lim_{\varepsilon \to 0} \frac{\sum_{i=1}^{N(\varepsilon)} \mu_i(q,\varepsilon) \lg(P_i(\varepsilon))}{\lg(\varepsilon)} \qquad (10)$$

The multifractal spectrum of pore distribution $f(\alpha)$ relative to α is defined as follows:

$$f[\alpha(q)] = \lim_{\varepsilon \to 0} \frac{\sum_{i=1}^{N(\varepsilon)} \mu_i(q,\varepsilon) \lg \mu_i(q,\varepsilon)}{\lg(\varepsilon)} \qquad (11)$$

The first step of the multifractal analysis of capillary pressure data is to interpolate mercury intrusion curves to obtain sufficient points. The equidistant division of a logarithmic pore size range is the basis for obtaining the probability density $P_i(\varepsilon)$ and the partition function $\chi(q,\varepsilon)$. $\tau(q)$, $D(q)$, $\alpha(q)$, and $f(q)$ can be obtained from Equations (2), (4), (6), (10) and (11), respectively. $\tau(q)$ and $D(q)$ describe the multifractal characteristics, whereas $\alpha(q)$, and $f(q)$ characterize the local characteristics of the multifractal structure.

3. Samples and Experiments

A total of 13 samples were obtained from a well located in Western Sichuan, China. The physical properties of the samples are relatively variable (Figure 1), which is convenient for comparison using multifractal analysis. All samples were tested for porosity, and permeability and subjected mercury injection experiments in accordance with Chinese Petroleum Industry Standards SY/T 6385-1999 and SY/T 5346-2005.

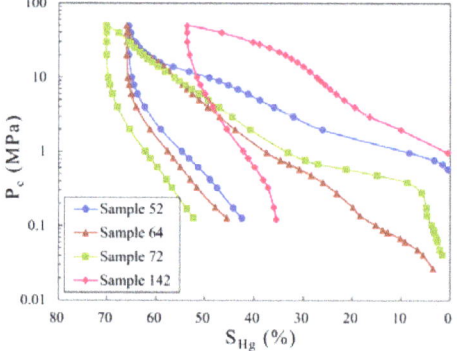

Figure 1. Four typical mercury intrusion curves.

Porosity and permeability were obtained using routine rock property measurement techniques. The average permeability is 6.11 mD; the range extends from 0.14 mD to 42.29 mD. The average porosity is 11.59%; it ranges from 6.31% to 16.65%. The experimental results are summarized in Table 1. The entry pressure (the point on the curve at which the mercury first enters the pores of the samples) varies from 0.037 MPa to 1.450 MPa, with an average of 0.726 MPa. r_{50} varies from 0.018 μm to 0.247 μm, with an average of 0.089 μm. An analysis of physical properties shows that the reservoir exhibits strong heterogeneity and complexity in its microscopic pore structure.

Table 1. Parameters of the pore throat structure obtained from the 13 samples.

Samples	Permeability	Porosity	Entry Pressure	Sorting Coefficient	r_{50}
	mD	%	MPa		um
9	0.3250	13.3547	1.0760	1.6867	0.0415
13	0.1420	7.2449	1.4070	1.5212	0.0186
22	5.0690	12.2603	0.2350	2.5778	0.1444
37	1.1000	14.0306	1.0030	1.8189	0.0956
46	0.5730	12.3959	1.2150	1.4568	0.2471
52	1.5060	8.1072	0.8450	1.9145	0.0706
64	42.2990	16.6477	0.0370	3.3595	0.1688
72	16.5960	11.2793	0.1900	2.6728	0.1367
81	0.8070	13.3353	0.6400	2.1086	0.0762
93	4.8510	14.9307	0.1970	2.6960	0.0285
105	5.3600	11.0967	0.1880	2.5586	0.0848
142	0.5420	6.3134	1.4500	1.4509	0.0165
146	0.2420	9.6940	0.9620	1.7438	0.0320

Note: r_{50} corresponds to pore throat diameter at 50% mercury saturation; Sorting coefficient is the dispersion degree of reservoir pores.

The pore size distributions of four samples are shown in Figure 2. Samples 64 and 72 have a wide range of pore sizes, with about half larger than 1 μm and half smaller than 1 μm, respectively. Samples 52 and 142 have a small range of pore size distributions, and the pores are mainly distributed below 1 μm.

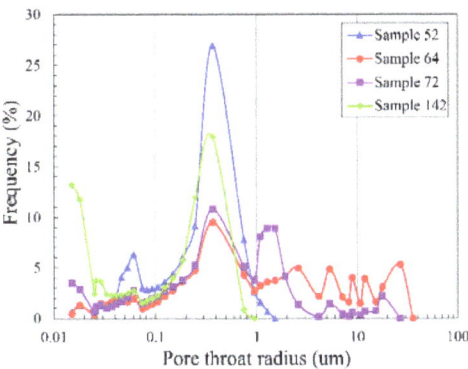

Figure 2. Pore size distributions of four samples obtained using mercury analysis.

4. Results and Discussion

4.1. Single Fractal Characteristics

Equation (1) was applied to obtain the fractal dimensions of the 13 mercury injection data. The results (Table 2) show that the high R-squared value demonstrates that fractal method is useful for mercury intrusion porosimetry (Figure 3). The fractal curves exhibit a two-stage characteristic and the corresponding fractal dimensions D_1 and D_2 can characterize the complexity of pore structure in different sizes. Despite the different pore size distributions, all 13 mercury intrusion curves exhibit a two-stage fractal characteristic. D_1 of 13 samples varies widely, while D_2 is mainly distributed around a value of 3.2.

Table 2. Parameters of the fractal dimension obtained from the 13 samples.

Samples	D_1	Correlation Coefficient R_2	D_2	Correlation Coefficient R_2	D_{stw}
9	5.1787	0.925	3.301	0.997	4.138
13	6.8716	0.946	3.302	0.985	4.592
22	5.9464	0.978	3.217	0.971	3.814
37	5.0279	0.920	3.319	0.957	4.092
46	7.1014	0.951	3.156	0.898	4.831
52	5.0715	0.948	3.276	0.939	3.989
64	4.1698	0.999	3.223	0.950	3.422
72	4.3472	0.927	3.195	0.968	3.638
81	4.8489	0.919	3.207	0.957	4.000
93	4.793	0.985	3.211	0.955	3.604
105	4.4449	0.998	3.216	0.989	3.636
142	6.3375	0.942	3.365	0.983	4.229
146	5.7515	0.992	3.281	0.999	4.367

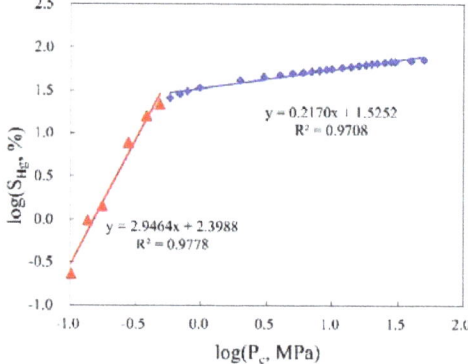

Figure 3. Plots of $\log(S_{Hg})$ vs. $\log(P_c)$ from the mercury intrusion curve of sample 22.

To clarify the factors controlling the fractal dimension, pore size distributions of four samples are shown in Figure 4. Samples 64 and 72 have a wide range of pore size distributions, with about half of the pore volume larger than 1 µm and half less than 1 µm. Samples 52 and 142 have a small range of pore size distributions and the pores are mainly distributed below 1 µm. The mercury pressure of macropores is lower, meaning that a smaller fractal dimension can be obtained under the same mercury saturation condition. Therefore, the D_1 values of samples 64 and 72—4.1698 and 4.3472, respectively—are significantly larger than the D_1 value of samples 52 and 142, which are 5.0715 and 6.3375, respectively. Under the same conditions, the larger the proportion of large pores, the easier it is to obtain smaller fractal dimensions; this can be confirmed by examining the correlation between D_1 and r_{50}. D_1 has a good negative correlation with r_{50} when two abnormal points affected by microfractures are neglected. The ratio of larger pores volumes to total pore volumes acts as a control on the fractal dimension over a specific pore size range.

The sums of fractal dimensions D_1 of samples 13, 46, and 142 are 6.87, 7.10 and 6.34, respectively; these values are all beyond the theoretical value. Despite simultaneously considering the fractal dimension for pore space and tortuosity, the sum of fractal dimension D_1 may be greater than the theoretical value. Friesen's model, Angulo's model, Shen's model, and Su's model assume that only porous media are present in the reservoir and ignore the influence of microfractures. However, there may be microfractures in the samples, and the existence of the microfractures causes the abnormalities in the sum of fractal dimension D_1 [11].

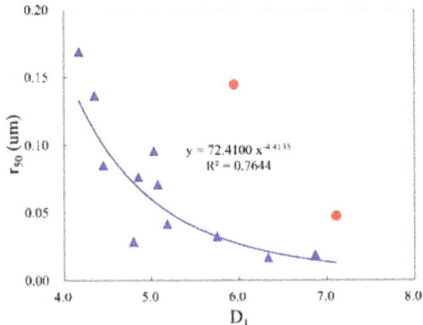

Figure 4. Plot of the sum of fractal dimension D_1 versus r_{50}.

4.2. Multifractal Characteristics

Multifractal analysis was conducted to describe the complexity of pore size distribution because single fractal theory has limitations in describing the local characteristics of pore size distribution. Figure 5 depicts the partition function for different q values in the double logarithmic coordinates of the four samples.

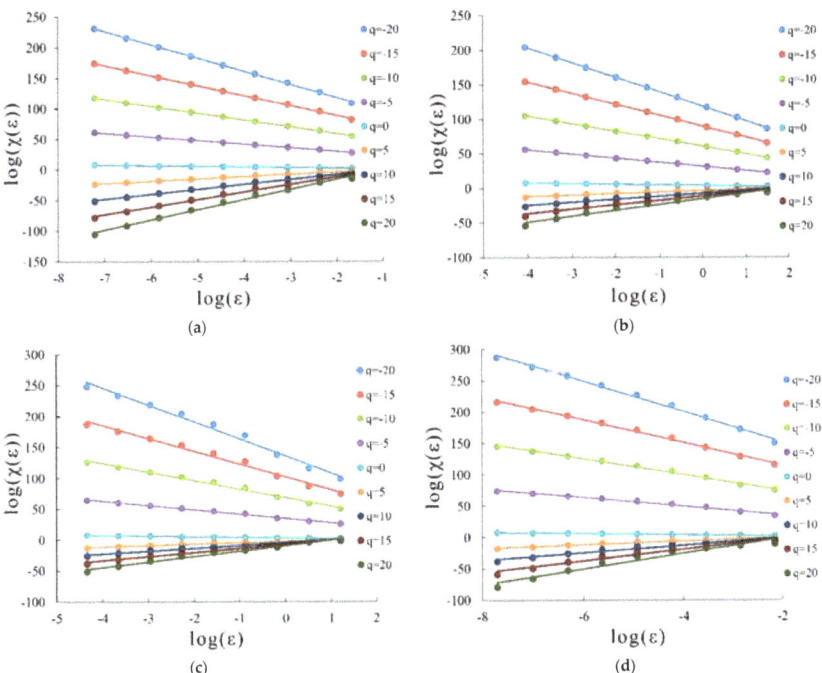

Figure 5. Plots of $\mu(\varepsilon)$ versus box size, ε, for the pore size distribution. (**a**) Sample 52; (**b**) Sample 64; (**c**) Sample 72; (**d**) Sample 142.

Linear relationships exist between $\log(\varepsilon)$ and $\log(\chi(\varepsilon))$ for the samples when $-20 \leq q \leq 20$ and the correlation coefficient is higher than 0.94 (Table 3). Favorable linear relationships between $\log(\varepsilon)$ and $\log(\chi(\varepsilon))$ indicate that the samples satisfy the multifractal characteristics. In accordance with Equation (4), the slopes of $\log(\varepsilon)$ and $\log(\chi(\varepsilon))$ are the mass exponent $\tau(q)$ The corresponding $\tau(q)$ increases when q increases from -20 to 20, indicating that the samples exhibit multifractal characteristics in a spatial distribution and that the multifractal method can be used to investigate the complexity and scale effects of pore size distribution.

Table 3. Coefficients for correlation determination R^2 of the fitting lines between $\log(\varepsilon)$ and $\log(\chi(\varepsilon))$.

q	Correlation Coefficient R^2			
	Sample 52	Sample 64	Sample 72	Sample 142
−20	0.9993	0.9999	0.9861	0.9947
−15	0.9994	0.9999	0.9870	0.9966
−10	0.9996	0.9999	0.9890	0.9978
−5	0.9998	1.0000	0.9945	0.9982
0	1.0000	1.0000	1.0000	1.0000
5	0.9957	0.9718	0.9829	0.9616
10	0.9937	0.9701	0.9820	0.9528
15	0.9928	0.9700	0.9820	0.9499
20	0.9924	0.9700	0.9820	0.9485

In accordance with Equation (6), the generalized dimension $D(q)$ was obtained in the range of $-20 \leq q \leq 20$. The relationship between the multifractal generalized dimension $D(q)$ and the moment order q is presented in Figure 6. The corresponding multifractal parameters are listed in Table 4. In the range of $-20 \leq q \leq 20$, the value of $D(q)$ when q is positive is less than the value of $D(q)$ when q is negative, thus indicating that regions with dense pore size distribution provides a better scale than the sparse regions. For a homogeneous fractal, the curves of $D(q)$ and q form a straight line, whereas those of non-uniform fractals have a certain width, and a large curvature indicates a poor homogeneity of samples. All four samples demonstrate a certain degree of curvature and exhibit a certain non-uniformity, but the curvatures were significantly greater in Samples 64 and 72 than in Samples 52 and 142, thereby demonstrating that Samples 64 and 72 are more heterogeneous in their pore size distribution.

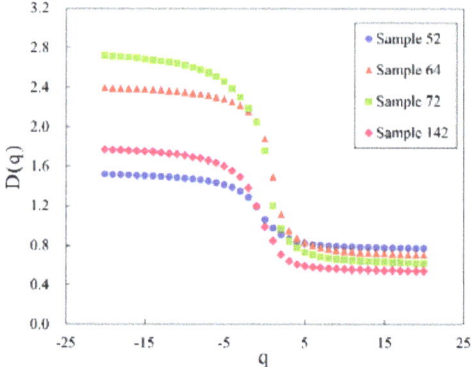

Figure 6. Plot of the multifractal generalized dimension $D(q)$ versus the moment order q.

Table 4. Multifractal dimension parameters obtained from the 13 samples.

Sample	D(0)	D(1)	D(2)	D(1)/D(2)	D_{min}	D_{max}	$\triangle D$
9	1.0082	0.9198	0.8314	0.9123	0.6934	1.3620	0.6686
13	0.9922	0.8695	0.7408	0.8764	0.5629	1.5441	0.9812
22	1.3816	1.1515	0.9762	0.8335	0.7323	2.1828	1.4506
37	1.0343	0.9346	0.8563	0.9036	0.7185	1.5000	0.7815
46	1.1469	1.0517	1.0171	0.9170	0.9096	1.6289	0.7193
52	1.0641	0.9773	0.9124	0.9185	0.7718	1.5227	0.7509
64	1.8827	1.4907	1.1157	0.7918	0.7073	2.3974	1.6901
72	1.7577	1.2061	0.9716	0.6862	0.6237	2.7235	2.0998
81	1.1421	0.9624	0.7910	0.8426	0.5907	1.6600	1.0692
93	1.3835	1.1169	0.8027	0.8073	0.5586	1.8413	1.2826
105	1.4816	1.1465	0.8809	0.7739	0.6233	2.1664	1.5431
142	0.9918	0.8520	0.7109	0.8590	0.5445	1.7734	1.2289
146	1.0658	0.9348	0.8195	0.8771	0.6215	1.4359	0.8144

Capacity dimension $D(0)$, information dimension $D(1)$, and correlation dimension $D(2)$ are listed in Table 4 [21]. A large capacity dimension $D(0)$ indicates a wide range of pore size distributions. The capacity dimensions of Samples 64 and 72 are 1.88 and 1.76, respectively, and are relatively larger than those for the 11 other samples. This indicates that the pore size distribution is large. A large pore size distribution suggests that large pores may be observed in the samples and may significantly improve the porosity and permeability of the reservoir; this condition can be confirmed by the large permeability and porosity of Samples 64 and 72.

The information dimension $D(1)$ reflects the degree of concentration of the pore size distribution, which represents the heterogeneity of pore structure. A large information dimension $D(1)$ indicates a highly heterogeneous pore size distribution. The information dimensions $D(1)$ of Samples 64 and 72 are relatively high, demonstrating that the unevenness of the pore size distribution is significant, and that pores are distributed over a wide range of pore sizes.

$D(1)/D(0)$ shows the dispersion of the pore size distribution. An added pore size is concentrated in the dense area when $D(1)/D(0)$ is close to 1, and the particle concentration in the sparse area is close to 0. The $D(1)/D(0)$ values of Samples 64 and 72 are relatively minimal. This result shows that the pore size distribution of Samples 64 and 72 is discrete and is biased toward the sparse areas of the pore size distribution. The sparsely-grained area mainly refers to the area with large pore sizes in this study. This area can improve the physical properties and increase the seepage and storage capacities of the fluid in a reservoir despite the relatively minimal volume.

The multifractal spectrum of the 13 samples was calculated in accordance with Equations (10) and (11). Figure 7 illustrates the multifractal spectrum curves of the four samples. The multifractal spectrum functions a-$f(a)$ denote a continuous distribution, indicating that multifractal theory is a common phenomenon of the pore size distribution. Curves a-$f(a)$ are asymmetrical upward convex curves, which demonstrate that the local superposition of the different degrees during the formation of pores leads to the occurrence of reservoir heterogeneity.

In calculating the multifractal spectrum, the calculation domain is divided into different scales, with considerable scale information in the reservoir pore size distribution. Δa describes the characteristics of different regions, levels, and local conditions in a fractal structure. A large Δa value indicates a highly uneven distribution. The parameters of the multifractal spectrum are listed in Table 5. The value of Δa ranges from 0.7167 to 2.2413, with an average value of 1.2361. The maximum Δa of Sample 72 suggests that its heterogeneity is robust. By contrast, the smallest Δa of Sample 9 suggest that its heterogeneity is relatively weak.

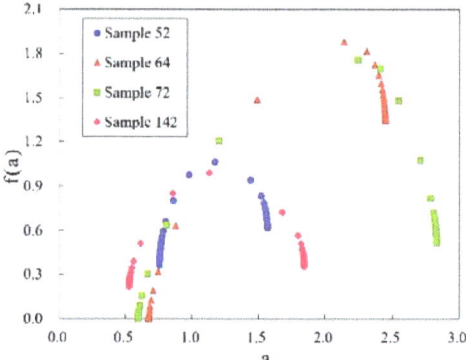

Figure 7. Plot of multifractal spectrum $f(a)$ versus the singularity strength a.

Table 5. Multifractal spectrum parameters obtained from the 13 samples.

Sample	a_{min}	a_{max}	$\triangle a$	$f(a_{min})$	$f(a_{max})$	$\triangle f(a)$	R	$f(a)_{max}$
9	0.6795	1.3962	0.7167	0.4158	0.6772	0.2614	0.1644	1.0082
13	0.5460	1.5946	1.0486	0.2234	0.5347	0.3113	0.1183	0.9922
22	0.7046	2.2587	1.5541	0.1782	0.6656	0.4874	0.2530	1.3816
37	0.7024	1.5436	0.8412	0.3967	0.6277	0.2311	0.0933	1.0343
46	0.8911	1.6651	0.7740	0.5395	0.9048	0.3653	0.0938	1.1469
52	0.7515	1.5680	0.8165	0.3654	0.6184	0.2530	0.0368	1.0641
64	0.6719	2.4498	1.7779	0.0003	1.3485	1.3482	0.6560	1.8827
72	0.5925	2.8338	2.2413	0.0000	0.5185	0.5185	0.4763	1.7577
81	0.5713	1.7032	1.1319	0.2018	0.7956	0.5938	0.3636	1.1421
93	0.5385	1.8867	1.3482	0.1557	0.9335	0.7778	0.5375	1.3835
105	0.5994	2.2303	1.6310	0.1442	0.8874	0.7432	0.4671	1.4816
142	0.5280	1.8444	1.3164	0.2142	0.3526	0.1384	-0.0823	0.9918
146	0.6001	1.4712	0.8711	0.1931	0.7297	0.5366	0.3952	1.0658

The equation Δf ($\Delta f = f(a_{min}) - f(a_{max})$) reflects the shape features of the multifractal spectrum. The shape of $f(a)$ depicts a right hook when the small probability subset dominates ($\Delta f < 0$). The shape of $f(a)$ illustrates a left hook when the large probability subset dominates ($\Delta f > 0$). The multifractal singularity curves in this study exhibit a right shape, indicating that the heterogeneity of the reservoir is mainly affected by the pore size distribution in the sparse region. This study emphasizes that large-scale pores contribute considerably to the spatial heterogeneity of a reservoir.

4.3. Relationship between Petrophysical and Single Fractal Parameters

The fractal dimensions D_1 and D_2 only characterize the complexity of pore structure in different sizes. The fractal dimension D_{sw} was introduced based on the weighted of the pore volume [28].

$$D_{sw} = D_1 \times S_{inf} + D_2 \times \left(S_{max} - S_{inf}\right) \quad (12)$$

where S_{inf} is the inflection point saturation and S_{max} is the maximum saturation. D_{sw} can characterize the complexity of the whole pore size, and has a better correlation with petrophysical parameters (Figures 8 and 9). Larger D_{sw} values indicate that macropores and microfractures have greater influence on reservoir physical properties. The D_{sw} values is between 3.42 and 4.83, with average value of 4.03.

To explore the meaning of saturation-weighted fractal dimension D_{sw}, the correlations between D_{sw} and petrophysical parameters were investigated. D_{sw} has a good negative correlation with permeability (Figure 8), while D_{sw} has a good positive correlation with entry pressure (Figure 9). Permeability is an important indicator of reservoir quality; larger permeabilities are typically associated with high-quality reservoirs. The entry pressure is mainly influenced by the pore size; the smaller the pore size, the greater the entry pressure. Larger D_{sw} values indicate that the macropores are more heterogeneous, but this does not guarantee a larger volume of macropores or better reservoir properties. The correlations between D_{sw} and petrophysical parameters show that the increase of D_{sw} is accompanied by the decrease of pore size and permeability, resulting in poorer reservoir properties. Therefore, D_{sw} is a good indicator of reservoir quality.

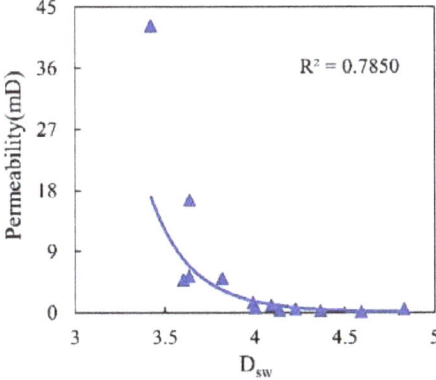

Figure 8. A plot of D_{sw} vs. permeability.

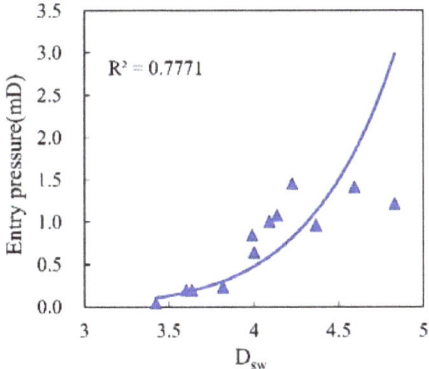

Figure 9. A plot of D_{sw} vs. entry pressure.

4.4. Relationship between Petrophysical and Multifractal Parameters

The relationships linking $D(0)$, $D(1)$, ΔD and Δf with permeability are determined to further explore the relationships between multifractal parameters and reservoir pore structure. Figure 10 shows that $D(0)$, $D(1)$, and ΔD are positively correlated with permeability, whereas Δf negatively correlates with permeability. $D(0)$ and $D(1)$ correlate well with permeability, with correlation coefficients of 0.8845 and 0.8665, respectively. $D(0)$ represents the range of the pore size distribution, and $D(1)$ characterizes the heterogeneity of the pore size distribution. The permeability improves with increasing $D(0)$ and $D(1)$. The more widely the reservoir particle size is distributed in the sparse region, the larger the reservoir size distribution range and the stronger the degree of heterogeneity. Therefore, the physical properties of the reservoir improve with increases in the range and heterogeneity of the pore size distribution.

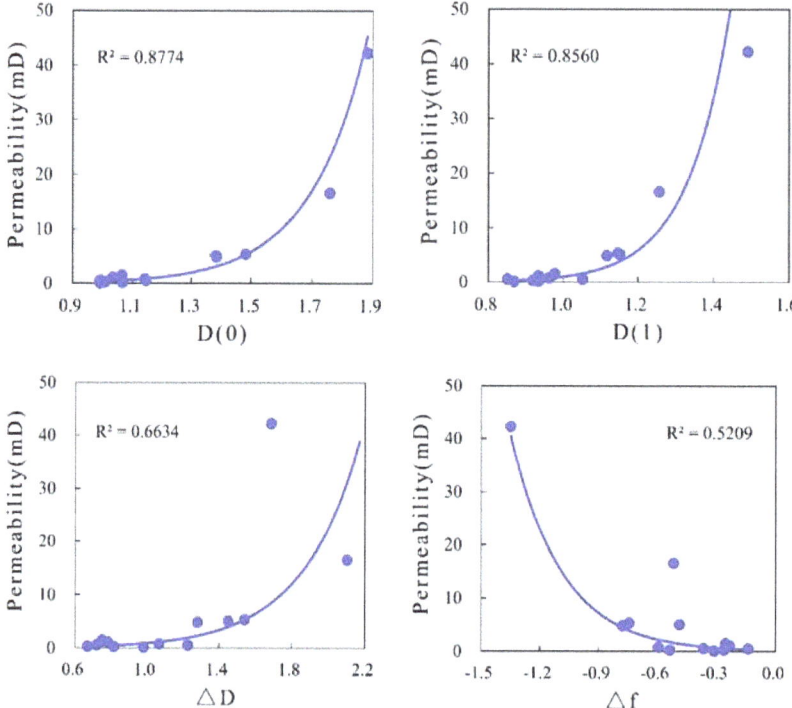

Figure 10. Plots of multifractal parameters and permeability.

Figure 11 presents the relationships linking $D(0)$, $D(1)$, ΔD, and Δf with the entry pressure. Favorable negative correlations exist between $D(0)$, $D(1)$, and ΔD and entry pressure, whereas Δf is positively correlated with entry pressure. Similarly, $D(0)$ and $D(1)$ correlate well with entry pressure, with correlation coefficients 0.9041 and 0.8971, respectively. $D(0)$ and $D(1)$ can be used as important parameters for characterizing and predicting physical properties of a reservoir.

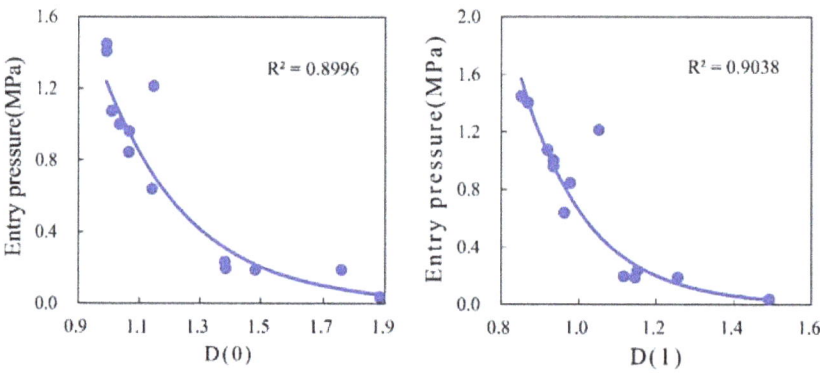

Figure 11. *Cont.*

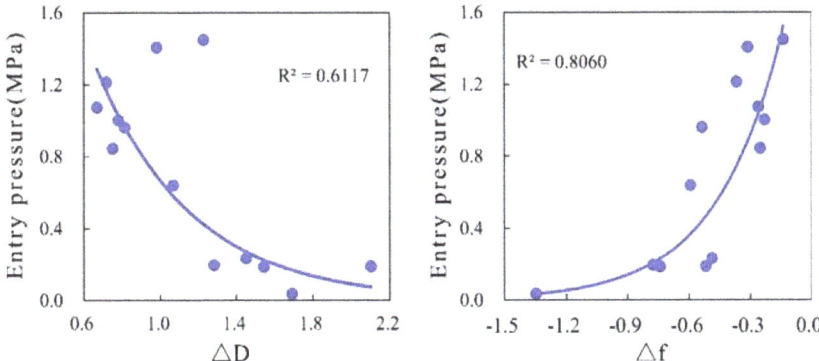

Figure 11. Plots of multifractal parameters and entry pressure.

4.5. Comparison between Single Fractal and Multifractal Analysis

Single fractal theory has been extensively used to characterize the heterogeneity of the pore structure, while multifractal analysis of mercury intrusion porosimetry in reservoir rocks is less commonly discussed. Comparative studies on single fractal and multifractal analysis are rare.

From the above analysis, we can see that there are some differences between single fractal and multifractal analysis. The D_{sw} values of samples 64 and 72 are lower than those of other samples, which generally implies that these two samples have low heterogeneity and good reservoir properties. However, samples 64 and 72 have bigger $D(0)$ and $D(1)$ values, as determined through multifractal analysis, demonstrating that these samples have a larger pore size distribution range and strong heterogeneity. Until now, this difference has not been observed or studied, since previous studies focus on the single or multifractal characteristics of mercury intrusion porosimetry. The difference is related to the fact that single fractal analysis and multifractal analysis characterize different aspects of the heterogeneity of a pore size distribution. The ratio of larger pores volumes to total pore volumes acts as a control on the fractal dimension over a specific pore size range, while the range of pore size distribution has the greatest effect on the multifractal analysis and multifractal parameters.

Single fractal theory characterizes the heterogeneity of pore structure from the entire range of pore sizes. The fractal curves show segmented fractal characteristics, and the fractal dimensions of each stage represents the complexity of the corresponding pore size range. The segmented fractal phenomenon demonstrates that multifractal theory is appropriate for describing fractal characteristics in a full-scale range. Multifractal theory is applied to explore pore size distribution in different scales. The pore structure exhibits multifractal characteristics in a local distribution, and the multifractal method can be used to investigate the complexity and scale effect of the pore size distribution.

It is worth noting that an intrinsic difference exists with respect to the single fractal models used to compute the fractal dimension. Friesen obtained a fractal model based on the Sierpinski carpet model [5]. The capillary bundle model was assumed in Shen's model [10]. Su set up a new fractal model considering of the fractal dimensions for pore space and tortuosity [11]. Different fractal dimensions may be obtained when these different fractal models are applied [29]. Moreover, the fractal dimension of the low pressure stage may exceed the theoretical value. This is because the current fractal models assume that the reservoir is composed only of pores and ignore the influence of microfractures. It has been confirmed that the presence of microfractures can significantly increase the fractal dimension [30]. Multifractal analysis explores pore size distribution characteristics without any a priori assumptions, considering the effect of microfractures.

We show that fractal analysis and multifractal analysis are feasible means for characterizing the heterogeneity of pore structures in a reservoir. By comparing the relationships between single fractal and multifractal parameters and reservoir physical parameters, we learn that multifractal parameters have better correlations with reservoir physical properties than single fractal parameters. The aforementioned factors combined mean that multifractal analysis is more suitable for characterizing the heterogeneity of a pore size distribution when the pore size distribution of the samples is addressed without any a priori assumptions.

5. Conclusions

In this study, single fractal and multifractal theory were applied to investigate pore size distribution characteristics of reservoir rocks as well as the influences of the single fractal and multifractal parameters on pore structure. The following conclusions can be drawn from our work:

(1) The single fractal curves exhibit segmented fractal characteristics and the fractal dimensions of the low-pressure section is greater than the fractal dimension of the high-pressure section. The saturation-weighted fractal dimension D_{sw} has a better correlation with permeability and entry pressure than D_1 or D_2.

(2) Linear relationships exist between $\log(\varepsilon)$ and $\log(\mu_i(\varepsilon))$ for the 13 samples when $-20 \leq q \leq 20$, suggesting that the pore structures of the 13 samples exhibit multifractal characteristics. Multifractal parameters $D(0)$ and $D(1)$ correlate well with permeability and entry pressure. The physical properties of the reservoir improve with increases in the range of pore size distribution.

(3) The ratio of larger pores volumes to total pore volumes acts as a control on the fractal dimension over a specific pore size range, while the range of pore size distribution has the greatest effect on the multifractal analysis and multifractal parameters.

(4) Fractal analysis and multifractal analysis are feasible methods for characterizing the complexity of pore size distribution in a reservoir. Multifractal analysis with parameters $D(0)$ or $D(1)$ produce better correlations with reservoir physical properties. In conclusion, multifractal analysis is more suitable for characterizing the heterogeneity of pore size distributions when the pore size distribution of samples is addressed without any a priori assumptions.

Author Contributions: P.S. analyzed the experiment data and wrote the paper; Z.X. provide core samples and designed the study; P.W. polished English; W.D. deduced formulas; Y.H. drawn figures; W.Z. searched literatures; Y.P. organize article format.

Funding: The research has been founded by the National Science and Technology Major Project of China (2016ZX05029005).

Conflicts of Interest: The authors declare no conflict of interest.

References

1. Li, W.; Liu, H.; Song, X. Multifractal analysis of Hg pore size distributions of tectonically deformed coals. *Int. J. Coal Geol.* **2015**, *144*, 138–152. [CrossRef]
2. Ghanbarian, B.; Hunt, A.G.; Daigle, H. Fluid flow in porous media with rough pore-solid interface. *Water Resour. Res.* **2016**, *52*, 2045–2058. [CrossRef]
3. Pfeifer, P.; Avnir, D. Chemistry in noninteger dimensions between two and three. I. Fractal theory of heterogeneous surfaces. *J. Chem. Phys.* **1983**, *79*, 3558–3565. [CrossRef]
4. Katz, A.J.; Thompson, A.H. Fractal sandstone pores: Implications for conductivity and pore formation. *Phys. Rev. Lett.* **1985**, *54*, 1325–1328. [CrossRef] [PubMed]
5. Friesen, W.I.; Mikula, R.J. Fractal dimensions of coal particles. *J. Colloid Interface Sci.* **1987**, *120*, 263–271.
6. Hu, J.; Tang, S.; Zhang, S. Investigation of pore structure and fractal characteristics of the Lower Silurian Longmaxi shales in western Hunan and Hubei Provinces in China. *J. Nat. Gas Sci. Eng.* **2016**, *28*, 522–535. [CrossRef]
7. Chen, X.; Yao, G.; Cai, J.; Huang, Y.; Yuan, X. Fractal and multifractal analysis of different hydraulic flow units based on micro-CT images. *J. Nat. Gas Sci. Eng.* **2017**, *48*, 145–156. [CrossRef]
8. Lei, G.; Cao, N.; Liu, D.; Wang, H. A Non-Linear Flow Model for Porous Media Based on Conformable Derivative Approach. *Energies* **2018**, *11*, 2986. [CrossRef]

9. Angulo, R.F.; Alvarado, V.; Gonzalez, H. Fractal dimensions from mercury intrusion capillary tests. In Proceedings of the SPE Latin America Petroleum Engineering Conference, Caracas, Venezuela, 8–11 March 1992.
10. Shen, P.; Li, K. A new method for determining the fractal dimension of pore structures and its application. In Proceedings of the 10th Offshore South East Asia Conference, Singapore, 6–9 December 1994.
11. Su, P.; Xia, Z.; Qu, L.; Yu, W.; Wang, P.; Li, D.; Kong, X. Fractal characteristics of low-permeability gas sandstones based on a new model for mercury intrusion porosimetry. *J. Nat. Gas Sci. Eng.* **2018**, *60*, 246–255. [CrossRef]
12. Stanley, H.E.; Meakin, P. Multifractal phenomena in physics and chemistry. *Nature* **1988**, *335*, 405. [CrossRef]
13. Lai, J.; Wang, G.; Fan, Z.; Chen, J.; Wang, S.; Zhou, Z.; Fan, X. Insight into the pore structure of tight sandstones using NMR and HPMI measurements. *Energy Fuels* **2016**, *30*, 10200–10214. [CrossRef]
14. Jiang, F.; Chen, D.; Chen, J.; Li, Q.; Liu, Y.; Shao, X.; Dai, J. Fractal analysis of shale pore structure of continental gas shale reservoir in the ordos basin, NW China. *Energy Fuels* **2016**, *30*, 4676–4689. [CrossRef]
15. Peng, C.; Zou, C.; Yang, Y.; Zhang, G.; Wang, W. Fractal analysis of high rank coal from southeast Qinshui basin by using gas adsorption and mercury porosimetry. *J. Pet. Sci. Eng.* **2017**, *156*, 235–249. [CrossRef]
16. Martínez, F.S.J.; Martín, M.A.; Caniego, F.J.; Tuller, M.; Guber, A.; Pachepsky, Y.; García-Gutiérrez, C. Multifractal analysis of discretized X-ray CT images for the characterization of soil macropore structures. *Geoderma* **2010**, *156*, 32–42. [CrossRef]
17. Jouini, M.S.; Vega, S.; Mokhtar, E.A. Multiscale characterization of pore spaces using multifractals analysis of scanning electronic microscopy images of carbonates. *Nonlinear Process. Geophys.* **2011**, *18*, 941–953. [CrossRef]
18. Subhakar, D.; Chandrasekhar, E. Reservoir characterization using multifractal detrended fluctuation analysis of geophysical well-log data. *Phys. A Stat. Mech. Its Appl.* **2016**, *445*, 57–65. [CrossRef]
19. Ferreiro, J.P.; Vázquez, E.V. Multifractal analysis of Hg pore size distributions in soils with contrasting structural stability. *Geoderma* **2010**, *160*, 64–73. [CrossRef]
20. Xie, S.; Cheng, Q.; Ling, Q.; Li, B.; Bao, Z.; Fan, P. Fractal and multifractal analysis of carbonate pore-scale digital images of petroleum reservoirs. *Mar. Pet. Geol.* **2010**, *27*, 476–485. [CrossRef]
21. Wang, P.; Jiang, Z.; Ji, W.; Zhang, C.; Yuan, Y.; Chen, L.; Yin, L. Heterogeneity of intergranular, intraparticle and organic pores in Longmaxi shale in Sichuan Basin, South China: Evidence from SEM digital images and fractal and multifractal geometries. *Mar. Pet. Geol.* **2016**, *72*, 122–138. [CrossRef]
22. Zhao, P.; Wang, Z.; Sun, Z.; Cai, J.; Wang, L. Investigation on the pore structure and multifractal characteristics of tight oil reservoirs using NMR measurements: Permian Lucaogou Formation in Jimusaer Sag, Junggar Basin. *Mar. Pet. Geol.* **2017**, *86*, 1067–1081. [CrossRef]
23. Liu, K.; Ostadhassan, M. Quantification of the microstructures of Bakken shale reservoirs using multi-fractal and lacunarity analysis. *J. Nat. Gas Sci. Eng.* **2017**, *39*, 62–71. [CrossRef]
24. Zheng, Q.; Yu, B. A fractal permeability model for gas flow through dual-porosity media. *J. Appl. Phys.* **2012**, *111*, 024316. [CrossRef]
25. Yu, B.; Li, J. Some fractal characters of porous media. *Fractals* **2001**, *9*, 365–372. [CrossRef]
26. Ge, X.; Fan, Y.; Li, J.; Zahid, M.A. Pore structure characterization and classification using multifractal theory—An application in Santanghu basin of western China. *J. Pet. Sci. Eng.* **2015**, *127*, 297–304. [CrossRef]
27. Wang, J.; Zhang, M.; Bai, Z.; Guo, L. Multi-fractal characteristics of the particle distribution of reconstructed soils and the relationship between soil properties and multi-fractal parameters in an opencast coal-mine dump in a loess area. *Environ. Earth Sci.* **2015**, *73*, 4749–4762. [CrossRef]
28. Liu, K.; Ostadhassan, M.; Kong, L. Fractal and multifractal characteristics of pore throats in the Bakken Shale. *Transp. Porous Media* **2019**, *126*, 579–598. [CrossRef]
29. Zhang, T.; Li, X.; Li, J.; Feng, D.; Wu, K.; Shi, J.; Han, S. A fractal model for gas–water relative permeability in inorganic shale with nanoscale pores. *Transp. Porous Media* **2018**, *122*, 305–331.
30. Lai, J.; Wang, G. Fractal analysis of tight gas sandstones using high-pressure mercury intrusion techniques. *J. Nat. Gas Sci. Eng.* **2015**, *24*, 185–196. [CrossRef]

© 2019 by the authors. Licensee MDPI, Basel, Switzerland. This article is an open access article distributed under the terms and conditions of the Creative Commons Attribution (CC BY) license (http://creativecommons.org/licenses/by/4.0/).

Article

An Efficiently Decoupled Implicit Method for Complex Natural Gas Pipeline Network Simulation

Peng Wang [1], Shangmin Ao [1], Bo Yu [1,*], Dongxu Han [1] and Yue Xiang [2]

[1] School of Mechanical Engineering, Beijing Key Laboratory of Pipeline Critical Technology and Equipment for Deepwater Oil & Gas Development, Beijing Institute of Petrochemical Technology, Beijing 102617, China; wangp@bipt.edu.cn (P.W.); aoshangmin@126.com (S.A.); handongxubox@bipt.edu.cn (D.H.)

[2] Research Center of Cloud Simulation and Intelligent Decision-making, Tsinghua Sichuan Energy Internet Research Institute, Chengdu 610042, China; cupxy@foxmail.com

* Correspondence: yubobox@vip.163.com; Tel.: +86-01-8129-2805

Received: 1 March 2019; Accepted: 18 April 2019; Published: 22 April 2019

Abstract: The simulation of a natural gas pipeline network allows us to predict the behavior of a gas network system under different conditions. Such predictions can be effectively used to guide decisions regarding the design and operation of the real system. The simulation is generally associated with a high computational cost since the pipeline network is becoming more and more complex, as well as large-scale. In our previous study, the Decoupled Implicit Method for Efficient Network Simulation (DIMENS) method was proposed based on the 'Divide-and-Conquer Approach' ideal, and its computational speed was obviously high. However, only continuity/momentum Equations of the simple pipeline network composed of pipelines were studied in our previous work. In this paper, the DIMENS method is extended to the continuity/momentum and energy Equations coupled with the complex pipeline network, which includes pipelines and non-pipeline components. The extended DIMENS method can be used to solve more complex engineering problems than before. To extend the DIMENS method, two key issues are addressed in this paper. One is that the non-pipeline components are appropriately solved as the multi-component interconnection nodes; the other is that the procedures of solving the energy Equation are designed based on the gas flow direction in the pipeline. To validate the accuracy and efficiency of the present method, an example of a complex pipeline network is provided. From the result, it can be concluded that the accuracy of the proposed method is equivalent to that of the Stoner Pipeline Simulator (SPS), which includes commercially available simulation core codes, while the efficiency of the present method is over two times higher than that of the SPS.

Keywords: natural gas; pipeline network; continuity/momentum and energy equations coupled; efficient simulation

---

1. Introduction

As a high quality and clean fossil fuel, natural gas plays an important role in global industry and economy [1]. Pipelines are the primary means by which the natural gas is transported. As more and more cities are using natural gas as the main energy source, the pipeline network is becoming extraordinarily complex. The characteristic of a complex topology has created many challenges for the design, monitoring, and operating of the pipeline network. The simulation of the natural gas pipeline network allows us to predict the behavior of a gas network system under different conditions. By matching the simulator's output with measured information from Supervisory Control and Data Administration (SCADA) systems, the gas pipelines can be estimated in real-time [2], the historical conditions can be reviewed, and the unexpected transient conditions can be analyzed [3]. Such predictions and analyses can be effectively used to guide decisions regarding the design and operation of the real pipeline network system. For example, Gssco improved his pipeline capacity with the help of a commercial tool known as the Pipeline Modeling System (PMS) [4].

Depending on the gas flow characteristics in the network system, there are two states in which the simulation can be: steady simulation (static simulation) and unsteady simulation (transient simulation). Steady simulation does not take into account the gas flow characteristics' variations over time. The goal of steady simulation is usually to compute the nodes' pressures, the nodes' loads, and the pipes' flow rates for the network design. The pressures at the nodes and the flow rates in the pipes must satisfy the general flow equations; the lodes' loads must fulfill the first Kirchhoff's law or Node formulation and the pressure drop around any closed loop must fulfill the second Kirchhoff's law or Loop formulation. The steady simulation was explained in detail by Osiadacz [5], and interested readers can refer to it. Unsteady simulation considers the facts that gas flow characteristics are mainly functions of time. In fact, the gas flow in a pipeline is actually a transient process, while a steady state is rare in practice. Thus, unsteady simulation is more practical for the operations of a real pipeline network. However, the mathematical model of unsteady simulation is a system of partial differential equations (PDEs) of mass conservation, momentum conservation, and energy conservation. The mathematical model is usually solved by numerical methods [6–15], and the frequently used numerical methods are method of characteristics (MOC), finite difference methods (FDM), and finite element methods (FEM). MOC firstly reduces PDEs to a family of ordinary differential equations (ODEs) along the characteristic curves, and the ODEs are then solved along the characteristic curves to obtain the flow and thermodynamic parameters. FDM firstly divides the long pipeline into many short sections, then converts PDEs into a system of difference Equations (DEs) on these small sections, and lastly solves DEs by matrix algebra techniques. FEM firstly divides the entire pipeline into many small sections that are so-called finite elements, then uses variational methods to derive the simple equations for these finite elements, and lastly solves the larger system of simple equations by minimizing an associated error function. Thorley and Tiley [16] have provided an excellent literature review of these methods, and interested readers can refer to it.

Among these methods, the implicit finite difference method is one of the most widely applied methods because its time step is not restricted by the stability criterion [6,17–19]. This means that the time step of the implicit finite difference method can be very large, which is very useful for simulating long-term transient flow in a natural gas pipeline. However, in the implicit finite difference method, one system of large-scale nonlinear discretized Equations needs to be solved [20]. When working with large-scale pipeline networks, its efficiency could be low and unsatisfactory, especially on a personal computer.

To improve the computational speed of the implicit finite difference method, a series of research projects have been carried out. Kiuchi [17,18] and Wylie et al. [19] directly ignored the convective inertia term in the governing Equations of a pipeline to avoid nonlinear equations. However, this process could reduce the accuracy of simulation [21]. Luskin [22] proposed a linearized method where the nonlinear governing equations were linearized about the previous time step based on the Taylor expansion. Zheng et al. [23] found that the linearized method could improve the computational speed by over five times. Barley [24] and Helgaker and Ytrehus [25] put forward a decoupled solution strategy, in which the continuity/momentum equations (flow equations) and energy equation (thermodynamics equation) were solved alternatively. This strategy could further increase the computational speed by about 20%. Wang et al. [26] found that the form of flow equations, which took the density and velocity as the solving variables, was the most efficient form. The computational speed was further improved by 50%. Many researchers, such as Wylie et al. [19] and Stoner [27], have recommended the sparse matrix technique to efficiently solve the large-scale discretized equations of network simulation. Madoliat et al. [13] proposed a novel approach based on intelligent algorithms, such as particle swarm optimization (PSO), for dynamic simulation of a gas pipeline network. These studies have improved the computational speed of natural gas pipeline network simulation to a considerable extent.

However, in the above studies, all pipelines in the network must be solved simultaneously, and one system of large-scale equations inevitably has to be solved. It is well-known that the computational speed of small-scale equations is faster than that of large-scale equations. Therefore, dividing the network into several pipelines and then solving them one by one is an effective way to further improve the computational speed of natural gas pipeline network simulation. This is the idea of the 'Divide-and-Conquer Approach'. Based on this idea, a fast method for the flow simulation of natural gas pipeline networks was proposed in our previous study [28], called the Decoupled Implicit Method for Efficient Network Simulation (DIMENS). In the DIMENS method, the flow equations of all the multi-pipeline interconnection nodes were firstly solved, and the flow parameters such as pressure and flow rate were known; the pipeline network was then divided into several independent pipelines, and the pipelines were efficiently solved one by one. Thus, this method is more efficient than the commercially available simulation core codes of Stoner Pipeline Simulator (SPS), whose computational speed can represent the highest level in the world.

In our previous work [28], the gas flow was considered isothermal, and the pipeline network was only composed of pipelines. Thus, the DIMENS method was only implemented in flow equations of the simple pipeline network. However, in practical engineering problems, the pipeline network is generally very complex, which includes not only pipelines, but also non-pipeline components, such as compressors, valves, supplies, and demands [12]. What is more, many researchers [24,25,29,30] have shown that the thermodynamic parameters have a major impact on the flow parameters of a pipeline network, and the effect of treating the gas in a non-isothermal manner was extremely necessary for pipeline flow calculation accuracies.

To overcome the above issue, the DIMENS method is extended to the flow and thermodynamic-coupled simulation of the complex pipeline network in this paper. The layout of this paper is as follows. First, the mathematical model of the pipeline network and its discretization is introduced. Second, the main idea of the DIMENS method is reviewed. Third, the implementation process of the extended DIMENS method is given, and the solution procedures of the thermodynamic equations are elaborated. Finally, a numerical case of the complex pipeline network is designed to test the performance of the present method.

2. Mathematical Model and Discretization

2.1. Mathematical Model

The pipeline network is mainly formed by interconnecting many kinds of components, and the basic components are pipelines, compressors, valves, supplies, demands, and so on [8]. The components of the pipeline network can be classified into four categories, namely, pipelines, non-pipelines, externals, and multi-component interconnection nodes. Therefore, the mathematical model of the natural gas pipeline network in this paper includes four parts: pipeline model, non-pipeline model, multi-component interconnection node model, and boundary conditions.

2.1.1. Pipeline Model

For natural gas pipeline network simulation, the flow in the pipeline is generally assumed to be homogeneous equilibrium flow with negligible fluid/structure interactions, and the influence of the non-uniformity of the velocity distribution is neglected. The pipeline model is a set of governing equations of homogeneous, geometrically one-dimensional flow in the pipeline, which consist of the continuity equation, momentum equation, and energy equation. The continuity and momentum equations are so-called flow equations, and the energy equation is a so-called thermodynamic equation. These equations can be written in a general form [24–26,31], as shown in Equation (1), and 50the parameters of the general form are given in Table 1. The detailed process of transformations and the parameter expressions of Equation (1) can be found in Appendix A:

$$\frac{\partial \mathbf{U}}{\partial t} + \mathbf{B} \cdot \frac{\partial \mathbf{U}}{\partial x} = \mathbf{F}, \tag{1}$$

Table 1. Parameters in the governing equations.

U	B	F
$\begin{bmatrix} p \\ m \end{bmatrix}$	$\begin{bmatrix} 0 & \frac{1}{A}\left(\frac{\partial p}{\partial \rho}\right)_T \\ [A - \frac{m^2}{A\rho^2}\left(\frac{\partial \rho}{\partial p}\right)_T] & \frac{2m}{A\rho} \end{bmatrix}$	$\begin{bmatrix} \left(\frac{\partial p}{\partial T}\right)_\rho \frac{\partial T}{\partial t} \\ -\frac{\lambda}{2}\frac{m|m|}{dA\rho} - A\rho g \sin\theta + \frac{m^2}{A\rho^2}\left(\frac{\partial \rho}{\partial T}\right)_p \frac{\partial T}{\partial x} \end{bmatrix}$
T	w	$\frac{1}{\rho c_v}[-T(\frac{\partial p}{\partial T})_\rho \frac{\partial w}{\partial x} + \frac{\lambda}{2}\frac{\rho|w|^3}{d} - \frac{4K(T-T_a)}{d}]$

It should be noted that natural gas is a compressible gas, and its thermodynamic properties (pressure, volume, temperature) are property relations. An equation of state which would express the density in terms of pressure and temperature needs to be close to Equation (1). Several different equations of state, including SRK, PR, BWRS, AGA-8, GERG 88, and GERG 2004, are applied in the industry. The sensitivity of the selection of the equation of state for pipeline gas flow models was investigated by Chaczykowski [32]. In this paper, the BWRS equation of state [20,33] is used, and the BWRS equation is formulated as

$$\begin{aligned} p = \rho RT + (B_0 RT - A_0 - \frac{C_0}{T^2} + \frac{D_0}{T^3} - \frac{E_0}{T^4})\rho^2 + (bRT - a - \frac{d}{T})\rho^3 \\ + \alpha(a + \frac{d}{T})\rho^6 + \frac{c\rho^3}{T^2}(1 + \gamma\rho^2)\exp(-\gamma\rho^2) \end{aligned} \tag{2}$$

In total, it contains 11 coefficients. Values and mixing rules can be found in the literature [20,33].

2.1.2. Non-Pipeline Model

The main non-pipeline components in the natural gas pipeline network are compressor and valve components. Their mathematical models can be found in much literature [12,23]. In this paper, the mathematical models of compressors and valves are presented as Equations (3) to (5) and Equations (6) to (7), respectively.

Compressors model:

$$m_{in} - m_{out} = 0, \tag{3}$$

$$p_{out} - \varepsilon p_{in} = 0, \tag{4}$$

$$T_{out} - T_{in}\varepsilon^{\frac{n-1}{n}} = 0, \tag{5}$$

where, $\varepsilon = a + bQ_{in} + cQ_{in}^2$.

Valves model:

$$m_{in} - m_{out} = 0, \tag{6}$$

$$m_{in} - C_v \rho_{in} \sqrt{\frac{p_{in}^2 - p_{out}^2}{Z\Delta T_{in}}} = 0, \tag{7}$$

$$T_{out} - T_{in} + (p_{out} - p_{in})\left\{\frac{1}{c_p}\left[\frac{T}{\rho^2}\frac{(\partial p/\partial T)_\rho}{(\partial p/\partial \rho)_T} - \frac{1}{\rho}\right]\right\}_{out} = 0, \tag{8}$$

2.1.3. Multi-Component Interconnection Node Model

In each multi-component interconnection node, the laws of mass conservation, the equality of pressure, and the equality of gas temperature must be observed [12,20,26,28,31]. The corresponding mathematical models are represented by Equations (9) to (10).

Mass conservation:
$$\sum m_{in} = \sum m_{out}, \tag{9}$$

Equality of pressure:
$$p_{in,1} = p_{in,2} \cdots = p_{out,1} = p_{out,2} \cdots, \tag{10}$$

Equality of gas temperature outflow from the node:
$$T_{out,1} = T_{out,2} \cdots = \sum |c_p m T|_{in} / \sum |c_p m|_{out}, \tag{11}$$

2.1.4. Boundary Conditions

The boundary conditions give the value of pressure, flow rate, and temperature of the external components [12,26,28,31]. The external components of the gas pipeline network include the supply and the demand. The supply is the component where gas is injected into the network, and the demand is the component where gas is extracted from the network. The boundary conditions are represented by Equations (12) to (14).

Pressure:
$$p = p(t), \tag{12}$$

Flow rate:
$$m = m(t), \tag{13}$$

Temperature:
$$T = T(t), \tag{14}$$

2.2. Discretization

The pipeline model, that is Equation (1), is a nonlinear partial differential equation. It can be linearized about the previous time step based on the Taylor expansion, as shown below [12,17,18,23,24,26]:

$$\frac{\partial \mathbf{U}}{\partial t} + \overline{\mathbf{B}} \cdot \frac{\partial \mathbf{U}}{\partial x} + \overline{\mathbf{G}} \cdot (\mathbf{U} - \overline{\mathbf{U}}) = \overline{\mathbf{F}} + \overline{\mathbf{S}} \cdot (\mathbf{U} - \overline{\mathbf{U}}), \tag{15}$$

where, $[\mathbf{G}]_{i,j} = \sum_{l=1}^{n} \left(\frac{\partial B}{\partial u_j}\right)_{i,l} \frac{\partial u_l}{\partial x}$, $[\mathbf{S}]_{i,j} = \frac{\partial F_i}{\partial u_j}$. The detailed expressions of $\mathbf{G}$ and $\mathbf{S}$ can be found in appendix A. It should be noted that the bar " " above the matrixes $\mathbf{B}$, $\mathbf{G}$, $\mathbf{F}$, $\mathbf{S}$, and $\mathbf{U}$ represents the previous time step, so $\overline{\mathbf{B}}$, $\overline{\mathbf{G}}$, $\overline{\mathbf{F}}$, $\overline{\mathbf{S}}$, and $\overline{\mathbf{U}}$ are calculated by the results of the previous time step.

The pipeline is divided into N sections, and thus, there are $N + 1$ grid points. The ith section is the section between the ith point and the $(i + 1)$th point. Figure 1 is the schematic diagram of the pipeline grid.

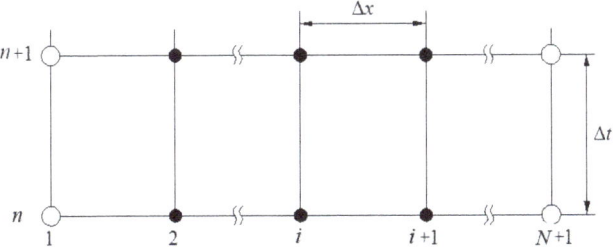

Figure 1. Grids of the pipeline

The flow equations of Equation (15) are discretized by the central difference scheme in the ith section in Figure 1, and the discretized flow equations are obtained as Equation (16) [26,28,31]:

$$\mathbf{CE}_i \cdot \mathbf{U}_i^{n+1} + \mathbf{DW}_i \cdot \mathbf{U}_{i+1}^{n+1} = \mathbf{H}_i, \qquad (16)$$

where, $\mathbf{CE}_i = \frac{1}{2\Delta t}\mathbf{I} - \frac{1}{\Delta x}\overline{\mathbf{B}} + \frac{1}{2}(\overline{\mathbf{G}} - \overline{\mathbf{S}}); \mathbf{DW}_i = \frac{1}{2\Delta t}\mathbf{I} + \frac{1}{\Delta x}\overline{\mathbf{B}} + \frac{1}{2}(\overline{\mathbf{G}} - \overline{\mathbf{S}}); \mathbf{H}_i = \overline{\mathbf{F}} + (\overline{\mathbf{G}} - \overline{\mathbf{S}} + \frac{1}{\Delta t}) \cdot \overline{\mathbf{U}}$.

The thermodynamic equation of Equation (15) is discretized by the upwind scheme at the ith point in Figure 1, and the discretized thermodynamic equation is obtained as Equation (17) [23,29,31]:

$$UP_i \cdot T_{i-1}^{n+1} + CE_i \cdot T_i^{n+1} + DW_i \cdot T_{i+1}^{n+1} = H_i, \qquad (17)$$

where, $UP_i = -\max(w_i^{n+1}, 0)\frac{1}{\Delta x}$; $CE_i = (\frac{1}{\Delta t} - \overline{S} + |w_i^{n+1}|\frac{1}{\Delta x})$; $DW_i = -\max(-w_i^{n+1}, 0)\frac{1}{\Delta x}$, $H_i = F(T_i^n) + (\frac{1}{\Delta t} - \overline{S})T_i^n$.

Equations (3) to (17) are the flow and thermodynamic-coupled discretized equations of the natural gas pipeline network. These equations are split into two parts: the system of flow equations and the system of thermodynamic equations.

The system of flow equations consists of Equations (3) to (4), Equations (6) to (7), Equations (9) to (10), Equations (12) to (13), and equation (16). The system of thermodynamic equations consists of Equation (5), Equation (6), Equation (11), Equation (14), and Equation (17).

The decoupled solution strategy [23–25] is adopted to efficiently solve the flow and thermodynamic-coupled discretized equations. In the decoupled solution strategy, the discretized flow equations are firstly solved using the interpolated temperature, and the discretized thermodynamic equations are then solved using the solved flow variables, such as the pressure and flow rate. The decoupled solution strategy is shown in Figure 2.

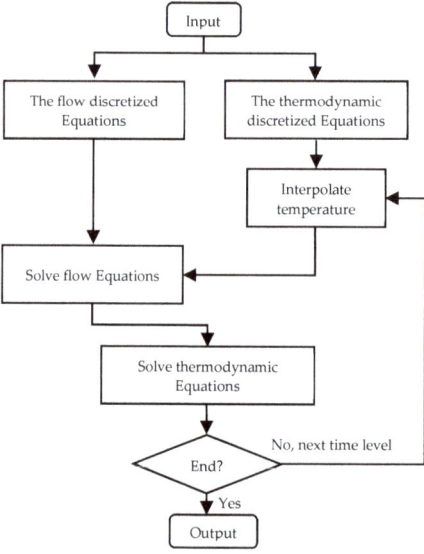

Figure 2. The decoupled strategy.

3. Main Idea of the DIMENS Method

In the DIMENS method, the network is firstly divided into several independent pipelines, and the pipelines are then efficiently solved one by one. Figure 3 shows an example, where a pipeline network is divided into four interconnection nodes and three pipelines.

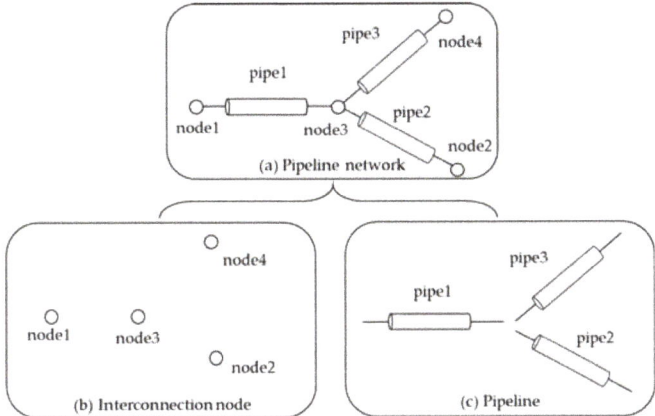

Figure 3. Divide-and-conquer approach for the natural gas pipeline network.

The key to decoupling the pipelines is solving the multi-pipeline interconnection nodes. However, the equation number of the multi-pipeline interconnection nodes is less than the number of the unknowns to be solved, so those equations of nodes cannot be solved directly.

Our previous work [28] is the first report on how to solve the above problem. The idea of solving the problem is that 'the supplemental equations for solving the multi-pipeline interconnection nodes are obtained from the discretized equations of the pipeline'. In other words, the discretized equations of pipelines are solved in advance to obtain supplemental equations, and the equations of multi-pipeline interconnection nodes are then solved with the supplemental equations. Figure 4 shows the procedure of the DIMENS method for the pipeline network.

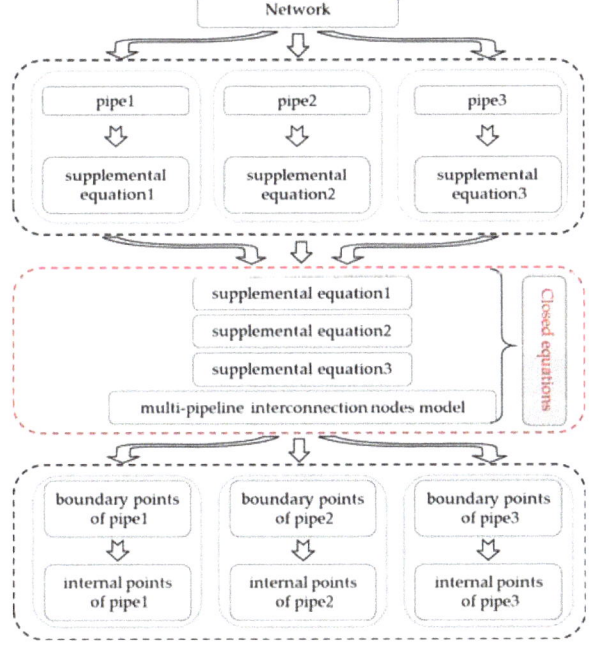

Figure 4. Procedure of the DIMENS method for the pipeline network.

In our previous work [23], the DIMENS method was only implemented in flow simulation of the simple pipeline network, which is only composed of pipelines. In the following section, the DIMENS method is extended to the flow and thermodynamic-coupled simulation of the complex pipeline network, which includes pipelines and non-pipeline components. Based on the flow and thermodynamic-decoupled solution strategy [23–25], the DIMENS method is implemented in the flow simulation and thermodynamic simulation, respectively.

4. Extended DIMENS for Complex Pipeline Network

4.1. Implementation of the Flow Simulation of the Complex Pipeline Network

4.1.1. Analysis of Flow Simulation

In our previous work [28], the process of solving flow Equations by the DIMNES method has been discussed in detail as step1, step2, and step3. However, the pipeline network does not involve non-pipeline components, such as compressor and valve components. In this paper, the non-pipeline components are appropriately solved as multi-component interconnection nodes in step 2 of the DIMENS method. The procedures of the DIMENS method in flow simulation are given below.

4.1.2. Procedures of the DIMENS Method in Flow Simulation

Step 1: Pre-solve pipeline

By taking the pressure of the starting point p_1 and the mass flow rate of the terminal point m_{N+1} as free variables, Equation (16) can be solved by the block three diagonal matrix algorithm (BTDMA). It's general solution can be written as Equation (18). The detailed solution process can be viewed in previous literature [28]:

$$[p_1, m_1, \cdots, p_{N+1}, m_{N+1}]^T = p_1 \alpha + m_{N+1} \beta + \gamma, \tag{18}$$

Equations (19) and (20) are the supplemental equations used to solve the multi-component interconnection nodes. These equations are the key to solving the multi-component interconnection nodes.

$$m_1 = \alpha_1^m p_1 + \beta_1^m m_{N+1} + \gamma_1^m, \tag{19}$$

$$p_{N+1} = \alpha_{N+1}^p p_1 + \beta_{N+1}^p m_{N+1} + \gamma_{N+1}^p, \tag{20}$$

Step 2: Solve the multi-component interconnection node

With the supplemental equations obtained from step 1, the equations of multi-component interconnection nodes are closed and can be solved. The closed equations are composed of four parts: (1) the flow equations of the non-pipeline model, Equations (3) to (4) and Equations (6) to (7); (2) the flow equations of the multi-component interconnection node model, Equations (9) to (10); (3) the flow equations of boundary conditions, Equations (12) to (13); and (4) the supplemental equations obtained from the first step, Equations (19) to (20). The unknown variables of the closed equations are the mass flow rates and pressures of the starting and terminal points of all components. The preconditioned conjugate gradient (CG) algorithm proposed in our previous work [28] is employed to solve the closed equations. For better understanding how the equations of multi-component interconnection nodes are solved, a simple pipeline network is given as the example in Appendix B.

Step 3: Solve the internal points of the pipeline

The values of the free variables (p_1 and m_{N+1}) solved in step 2 are substituted into Equations (19) to (20). The flow parameters of the internal grid points of the pipelines can be easily obtained to complete the flow simulation of the whole pipeline network. Through the above three steps, the flow simulation of the complex pipeline network can be solved by the DIMENS method.

4.2. Implementation of the Thermodynamic Simulation of the Complex Pipeline Network

4.2.1. Analysis of Thermodynamic Simulation

It is generally known that the downstream propagation of natural gas temperature is dominated by gas flow in the pipeline, and only upstream gas affects downstream gas. Therefore, only the temperature of the pipeline inlet can be chosen as the free variable for the DIMENS method to solve the thermodynamic equations. However, the direction of gas flow in the pipeline can unpredictably change during the transience. This means that gas could flow into the pipeline either from the starting or terminal point of the pipeline. Therefore, the flow states of gas in the pipeline should be analyzed to choose the free thermodynamic variables for the DIMENS method.

During the transience of the natural gas flow in the pipeline, there are four kinds of flow states: (1) natural gas flows out of the pipeline from both the starting and terminal points; (2) natural gas flows into the pipeline from the starting point and flows out of the pipeline from the terminal point; (3) natural gas flows into the pipeline from the terminal point and flows out of the pipeline from the starting point; and (4) natural gas flows into the pipeline from both the starting and terminal points. According to the corresponding gas flow state, pipelines can be classified as four types. The flow states and types of the pipeline are shown in Table 2.

Table 2. The flow states and the solution conditions of the pipeline.

Type	Flow State	Solution Requirement
First type pipeline	out ← ▯ → out	No need
Second type pipeline	in → ▯ → out	The temperature of starting point.
Third type pipeline	out ← ▯ ← in	The temperature of terminal point.
Fourth type pipeline	in → ▯ ← in	The temperature of both starting and terminal points.

For the first type of pipeline, the thermodynamic-discretized equation of the pipeline model, Equation (17), is a closed system. This means that there is no free variable, and the first type of pipeline can be solved directly. For the second type of pipeline, Equation (17) is not a closed system until the temperature of the starting point is known. That is to say, the temperature of the starting point should be chosen as the free variable. Similar to the second type of pipeline, the free variable of the third type of pipeline is the temperature of the terminal point. For the fourth type of pipeline, until the temperatures of the starting and terminal points are both known, the temperature of the internal grid point in the pipeline can be obtained by solving Equation (17). In other words, the fourth type of pipeline should be solved after the other three pipeline types. The solution requirements of these four types of pipelines are also shown in Table 2.

4.2.2. Procedures of the DIMENS Method in Thermodynamic Simulation

According to the above analysis of the pipeline flow state, the DIMENS method is extended to the thermodynamic simulation of the complex pipeline network, and the detailed procedures are presented as follows.

Step 1: Classify pipelines

Pipelines are classified as four types according to flow state, shown in Table 2.

Step 2: Solve the first type pipeline

The objective of this step is to obtain the thermodynamic parameters of the first type of pipeline. For the first type of pipeline, Equation (17) is closed, which can be rewritten as Equation (21). Then, Equation (21) is solved efficiently by the three diagonal matrix algorithm (TDMA) [34], and the temperature values of all points on the first type pipeline are obtained:

$$\begin{bmatrix} CE_1 & DW_1 & & & \\ UP_2 & CE_2 & DW_2 & & \\ & \ddots & \ddots & & \\ & & UP_N & CE_N & DW_N \\ & & & UP_{N+1} & CE_{N+1} \end{bmatrix} \begin{bmatrix} T_1 \\ T_2 \\ \vdots \\ T_N \\ T_{N+1} \end{bmatrix} = \begin{bmatrix} H_1 \\ H_2 \\ \vdots \\ H_N \\ H_{N+1} \end{bmatrix}, \quad (21)$$

Step 3: Pre-solve the second and third type pipelines

By taking the temperature of starting point T_1 of the second type of pipeline as the free variable, Equation (17) can be rewritten as Equation (22), and its general solution is Equation (23). Similar to the second type of pipeline, the temperature of terminal point T_{N+1} of the third type of pipeline is the free variable, and the corresponding discretized Equation and general solution are Equation (24) and Equation (25), respectively. In addition, the TDMA algorithm [34] is adopted too.

Thermodynamic-discretized equations of the second type of pipeline:

$$\begin{bmatrix} 1 & 0 & & & \\ UP_2 & CE_2 & DW_2 & & \\ & \ddots & \ddots & & \\ & & UP_N & CE_N & DW_N \\ & & & UP_{N+1} & CE_{N+1} \end{bmatrix} \begin{bmatrix} \alpha_1 & \gamma_1 \\ \alpha_2 & \gamma_2 \\ \vdots & \vdots \\ \alpha_N & \gamma_N \\ \alpha_{N+1} & \gamma_{N+1} \end{bmatrix} = \begin{bmatrix} 1 & 0 \\ 0 & H_2 \\ \vdots & \vdots \\ 0 & H_N \\ 0 & H_{N+1} \end{bmatrix}, \quad (22)$$

General solution of the second type of pipeline:

$$\begin{bmatrix} T_1 & T_2 & \cdots & T_N & T_{N+1} \end{bmatrix}^T = T_1 \boldsymbol{\alpha} + \boldsymbol{\gamma}, \quad (23)$$

Thermodynamic-discretized equations of the third type of pipeline:

$$\begin{bmatrix} CE_1 & DW_1 & & & \\ UP_2 & CE_2 & DW_2 & & \\ & \ddots & \ddots & & \\ & & UP_N & CE_N & DW_N \\ & & & 0 & 1 \end{bmatrix} \begin{bmatrix} \alpha_1 & \gamma_1 \\ \alpha_2 & \gamma_2 \\ \vdots & \vdots \\ \alpha_N & \gamma_N \\ \alpha_{N+1} & \gamma_{N+1} \end{bmatrix} = \begin{bmatrix} 0 & H_1 \\ 0 & H_2 \\ \vdots & \vdots \\ 0 & H_N \\ 1 & 0 \end{bmatrix}, \quad (24)$$

General solution of the third type of pipeline:

$$\begin{bmatrix} T_1 & T_2 & \cdots & T_N & T_{N+1} \end{bmatrix}^T = T_{N+1} \boldsymbol{\alpha} + \boldsymbol{\gamma}, \quad (25)$$

Equation (26) and Equation (27) are the supplemental equations used to solve the multi-component interconnection nodes. These equations are the key to solving the multi-component interconnection nodes.

$$T_{N+1} = T_1 \alpha_{N+1} + \gamma_{N+1}, \quad (26)$$

$$T_1 = T_{N+1} \alpha_1 + \gamma_1, \quad (27)$$

Step 4: Solve the multi-components interconnection node

In the DIMENS method, the fourth step of thermodynamic equations is similar to the second step of the flow equations. The thermodynamic parameters of multi-components interconnection nodes are solved via the closed equations of the starting and terminal points of all components. The complete system of closed equations is comprised of five parts: (1) thermodynamic equations of the non-pipeline model, Equation (5) and Equation (6); (2) thermodynamic equation of the multi-components interconnection node model, Equation (11); (3) thermodynamic equation of the boundary condition, Equation (14); (4) the temperature values of the starting and terminal points of the first type of pipeline that are solved in step 2; and (5) the supplemental equations obtained in step 3, Equation (26) and Equation (27). The unknown variables of these closed equations are the temperatures of the starting and terminal points of all components. The preconditioned CG algorithm that is proposed in the paper [34] is adopted to solve the closed equations. For better understanding how the equations of multi-component interconnection nodes are solved in thermodynamic equations, a simple pipeline network is given as the example in Appendix C.

Step 5: Solve the internal points of the second and third pipelines

The fifth step of thermodynamic equations is similar to the third step of the flow equations. The values of the free variables solved in step 4 are substituted into Equation (23) and Equation (25), and the temperature value of the internal grid points of the first and second types of pipelines can be easily obtained.

Step 6: Solve the fourth type pipeline

After step 5, the temperatures of the starting and terminal points of all pipelines are known. Then, Equation (17) of the fourth type of pipeline is rewritten as Equation (28), and is efficiently solved by the TDMA algorithm [34].

$$\begin{bmatrix} 1 & 0 & & & \\ UP_2 & CE_2 & DW_2 & & \\ & \ddots & \ddots & & \\ & & UP_N & CE_N & DW_N \\ & & & 0 & 1 \end{bmatrix} \begin{bmatrix} T_1 \\ T_2 \\ \vdots \\ T_N \\ T_{N+1} \end{bmatrix} = \begin{bmatrix} T_1^s \\ H(2) \\ \vdots \\ H(N) \\ T_{N+1}^s \end{bmatrix}, \quad (28)$$

Through the above six steps, thermodynamic simulation of the complex pipeline network can be solved by the DIMENS method.

5. Results and Analysis

An example of the complex pipeline is designed to validate the DIMENS method here. The calculation accuracy of the DIMENS method is investigated by comparing the numerical solution obtained by the DIMENS method with that of SPS. The computational speed of the DIMENS method is investigated by comparing the CPU time of the DIMENS method with that of SPS.

5.1. Description of the Numerical Test

The complex pipeline is comprised of 84 pipelines, four compressors, eight valves, and 51 externals, and the structure of the complex pipeline network is shown in Figure 5 and detailed data are given in Tables 3–6. All the pipelines are horizontal, with a roughness, thickness, and thermal conductivity of 0.02286 mm, 5 mm, and 0.0127 W/(m·K), respectively.

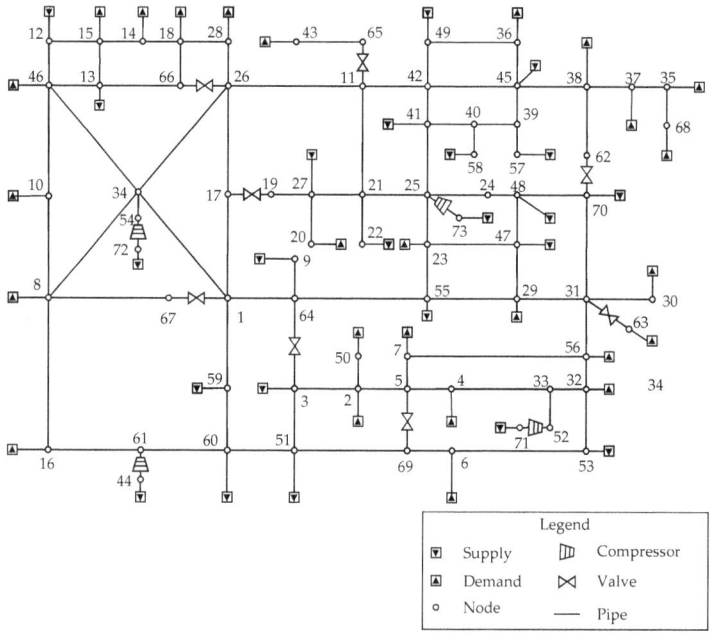

Figure 5. The schematic diagram of the topology structure of the complex pipeline network.

Table 3. Pipeline data.

Pipeline	1	2	3	4	5	6	7	8	9	10	11	12	13	14
Starting node	64	3	5	51	5	69	5	67	8	9	46	15	13	66
Ending node	1	2	2	3	4	6	7	8	10	64	10	12	46	13
Length (km)	52	78	66	40	23	22	39	58	50	60	35	59	57	56
Diameter (mm)	325	325	325	219	325	219	325	273	457	168	273	355	273	219
Pipeline	15	16	17	18	19	20	21	22	23	24	25	26	27	28
Starting node	13	14	1	17	66	18	19	27	27	22	25	25	24	26
Ending node	15	15	17	26	18	14	27	20	21	21	21	23	25	28
Length (km)	60	63	20	46	36	75	41	33	33	42	78	83	45	135
Diameter (mm)	273	168	457	457	273	325	219	219	168	168	355	355	355	457
Pipeline	29	30	31	32	33	34	35	36	37	38	39	40	41	42
Starting node	55	31	31	32	32	53	52	8	37	38	45	45	39	40
Ending node	64	29	30	56	33	6	33	61	35	37	38	39	40	41
Length (km)	92	46	74	50	40	77	25	85	63	85	69	44	76	67
Diameter (mm)	457	457	355	508	508	219	508	426	426	426	426	426	219	355
Pipeline	43	44	45	46	47	48	49	50	51	52	53	54	55	56
Starting node	11	65	56	47	11	26	2	41	45	12	53	54	29	31
Ending node	26	43	7	48	21	34	50	42	42	46	32	34	55	56
Length (km)	115	46	74	53	63	86	63	72	142	93	80	53	102	50
Diameter (mm)	426	273	273	219	219	355	219	426	426	325	325	508	426	426
Pipeline	57	58	59	60	61	62	63	64	65	66	67	68	69	70
Starting node	57	58	59	60	61	35	69	1	31	8	16	61	60	38
Ending node	39	40	1	59	1	68	51	34	70	16	61	60	51	62
Length (km)	30	68	43	56	52	36	45	52	52	80	84	101	95	33
Diameter (mm)	168	325	273	273	325	355	273	355	355	273	273	426	426	325
Pipeline	71	72	73	74	75	76	77	78	79	80	81	82	83	84
Starting node	41	23	33	18	49	45	49	8	46	11	24	48	47	23
Ending node	25	55	4	28	42	36	36	34	34	42	48	70	29	47
Length (km)	68	42	42	68	52	50	43	52	59	66	49	63	76	53
Diameter (mm)	355	273	457	273	273	273	219	355	325	355	355	325	325	219

Note: Diameter in Table 3 is outside diameter.

Table 4. Compressor data.

Compressor	1	2	3	4
Starting point	71	72	73	74
Ending point	52	54	25	61

Table 5. Valve data.

Valve	1	2	3	4	5	6	7	8
Starting point	70	31	17	64	11	1	26	5
Ending point	62	63	19	3	65	67	66	69

Table 6. External data.

Externals	1	2	3	4	5	6	7	8	9	10	11	12	13
Adjacent node	37	53	45	47	48	56	51	3	49	71	70	13	12
Externals	14	15	16	17	18	19	20	21	22	23	24	25	26
Adjacent node	72	55	22	58	9	57	73	41	59	60	27	44	6
Externals	27	28	29	30	31	32	33	34	35	36	37	38	39
Adjacent node	7	8	23	46	28	20	50	4	16	29	38	43	30
Externals	40	41	42	43	44	45	46	47	48	49			
Adjacent node	2	36	63	10	15	18	14	35	68	32			

The components of natural gas are listed in Table 7. The standard pressure and temperature are 101.325 kPa and 20 °C, respectively. The Colebrook Equation is used as the friction Equation [35]. During simulation, the ambient temperature is maintained at 15 °C.

Table 7. Main components of natural gas.

Component	CH_4	C_2H_4	C_3H_8	N_2	CO_2
Volume fraction (%)	97.07	0.17	0.02	0.71	2.03

The initial pressure is 2.8 MPa, the initial volume flow rate is zero, and the initial temperature is 15 °C. After $t = 0$ h, the pressure of some externals is maintained, the volume flow rate of some externals increases suddenly, the compressors are started, and the valves are opened. During the simulation, the volume flow rate of some externals is constant, while the volume flow rate of other externals and the valve opening (FR) of valves are changed, shown as Tables 8–10, respectively. It is assumed that: (1) all the compressors have the same performance curve as shown in Figure 6, the rated speed of the compressor is 6000 RPM, and the start time is 3 minutes; (2) all the valves are same, the flow coefficient of the valve in this paper is calculated by the Equation $C_v = 200 - 200FR$, and the travel time of the valve is one minute.

Table 8. Constant boundary conditions - pressure of the external.

Externals	10	14	20	25	46	48
Pressure (MPa)	2.8	2.8	2.8	2.8	2.5	2.5

Table 9. Constant boundary conditions flow rate of the external (flow in).

Externals	1	2	3	5	6	7	8	9	11	12
Volume flow rate (10^4 Nm^3/d)	−9	85	25	26	−100	44	15	36	55	68
Externals	15	16	17	18	19	21	22	23	24	27
Volume flow rate (10^4 Nm^3/d)	25	15	18	14	2	14	21	56	12	-34
Externals	28	29	30	31	32	33	35	36	37	38
Volume flow rate (10^4 Nm^3/d)	−142	−13	−150	−45	−13.7	−14	−35	−22	−13	−44
Externals	39	40	41	42	43	44	45	47	49	
Volume flow rate (10^4 Nm^3/d)	−21	−25	−50	−16	−56	−69.4	−63	−66	−85	

Table 10. Changed boundary conditions.

	External 4	External 13	External 26	External 34	Valves 1-8
	Volume flow rate (10^4 Nm^3/d)				FR
$0\,h \leq t < 5\,h$	22	32	−80	−68	1.0
$5\,h \leq t < 10\,h$	0	0	0	0	0.1
$t \geq 10\,h$	22	32	−80	−68	1.0

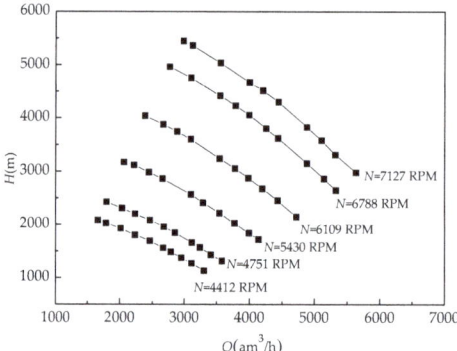

Figure 6. The performance curve of the compressor.

5.2. Comparison of Numerical Accuracy

The case of the complex pipeline network is simulated by the DIMENS and SPS, respectively. The time step and spatial step are 10 s and 0.5 km, respectively. The pressure, volume flow rate, and temperature of the starting point of some pipelines during 0–10 h are shown in Figures 7–9. The pressure, volume flow rate, and temperature of the compressor 1# are shown in Figure 10.

Figures 7–9 show that the numerical solutions of flow simulation, such as pressure and flow rate, obtained by the DIMENS method, are in good agreement with those obtained by SPS. The results of these comparisons are the same as in previous literature [28]. This can validate the DIMENS method for flow equations of the complex pipeline network. In addition, it is clearly seen from Figure 9 that the difference between the temperature obtained by the DIMENS method and that obtained by SPS is very small. The maximum deviation of temperature of the two methods is less than 1 °C, which can be found at the starting point of pipe 63 at t = 6 h, as shown in Figure 9b. This is acceptable for practical engineering problems of natural gas pipeline transportation. That is to say, the DIMENS method is also accurate for the thermodynamic simulation. Additionally, the results of pressure, flow rate, and temperature of the compressor obtained by the DIMENS method are also in good agreement with those of SPS, as shown in Figure 10. This means that the complex start-up process of the compressor can be accurately simulated by the DIMENS method too. The above analysis shows that the DIMENS method has a high accuracy comparable to the accuracy of the SPS and can meet the requirements of practical engineering problems.

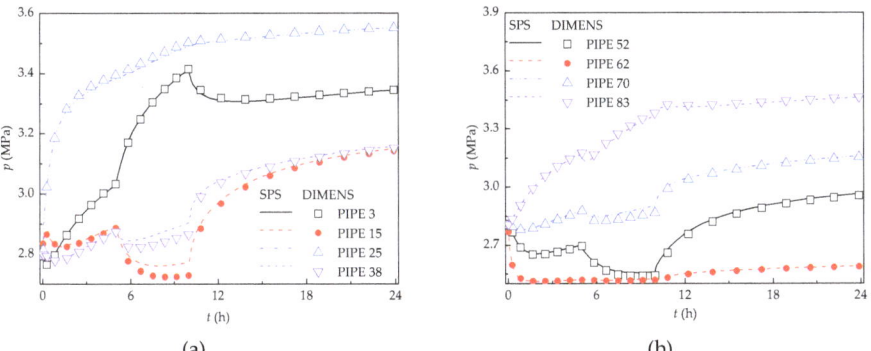

Figure 7. Pressure of the starting point of some pipelines: (**a**) Pipe 3, 15, 25, and 38; (**b**) Pipe 52, 62, 70, and 83.

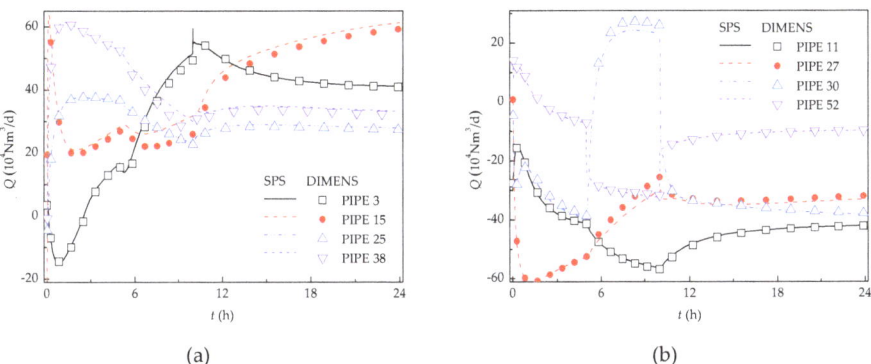

Figure 8. Volume flow rate of the starting point of some pipelines: (**a**) Pipe 1, 62, 72, and 81; (**b**) Pipe 11, 27, 30, and 52.

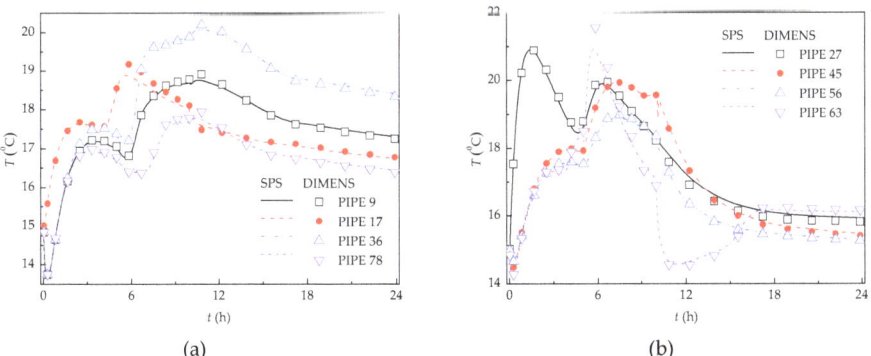

Figure 9. Temperature of the starting point of some pipelines: (**a**) Pipe 9, 17, 36, and 78; (**b**) Pipe 27, 45, 56, and 63.

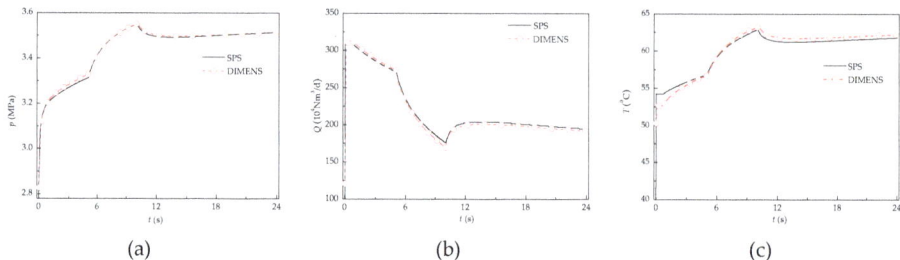

Figure 10. Pressure, flow rate, and temperature at the outlet of the 1#compressor outlet: (**a**) Pressure; (**b**) flow rate; (**c**) temperature.

To further test the simulation accuracy of the DIMENS method, the results of the pipeline network at $t = 24$ h obtained by the DIMENS method and SPS are compared, as shown in Figures 11–13.

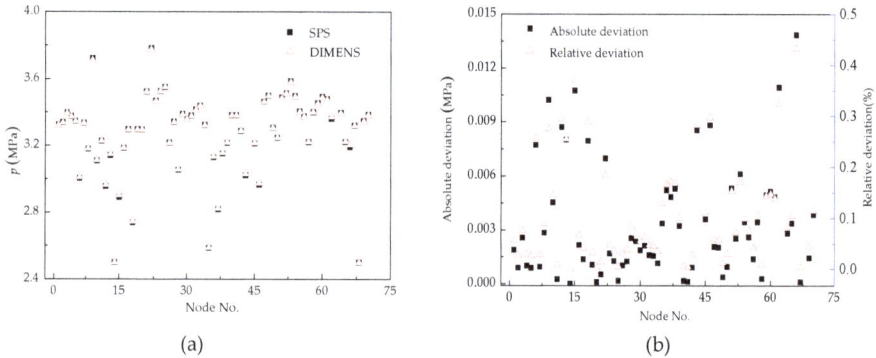

Figure 11. Comparison of pressure at the connection node: (**a**) Pressure value; (**b**) deviation.

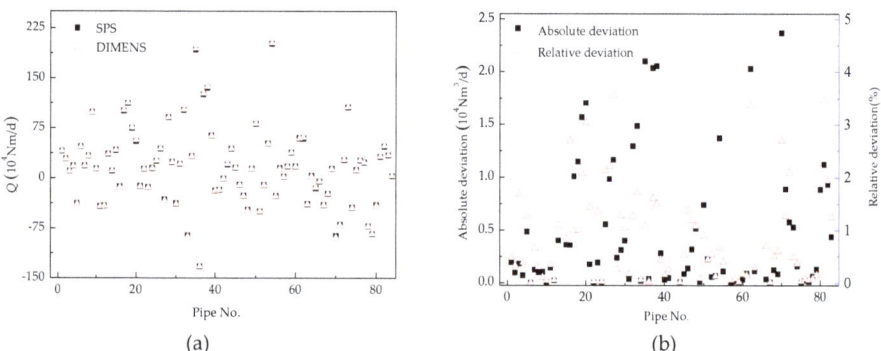

Figure 12. Comparison of flow rate of pipelines: (**a**) Flow rate value; (**b**) deviation.

From the comparison of the pressure of the connection nodes in Figure 11, it can be seen that the pressures obtained by the two methods are almost the same, and the maximum absolute and relative deviations of pressure are 0.015 MPa and 0.45%, respectively. From the comparison of the volume flow rate in Figure 12, the maximum absolute and relative deviations of the volume flow rate are 2.4×10^4 Nm^3/h and 3.5%, respectively. What is more, the maximum absolute deviation of temperature is less than 0.5 °C, as shown in Figure 13. It can be summarized that the deviation between the numerical solution calculated by the DIMENS method and SPS is sufficiently small. These analyses again imply that the calculation accuracy of the DIMENS method is comparable to that of SPS.

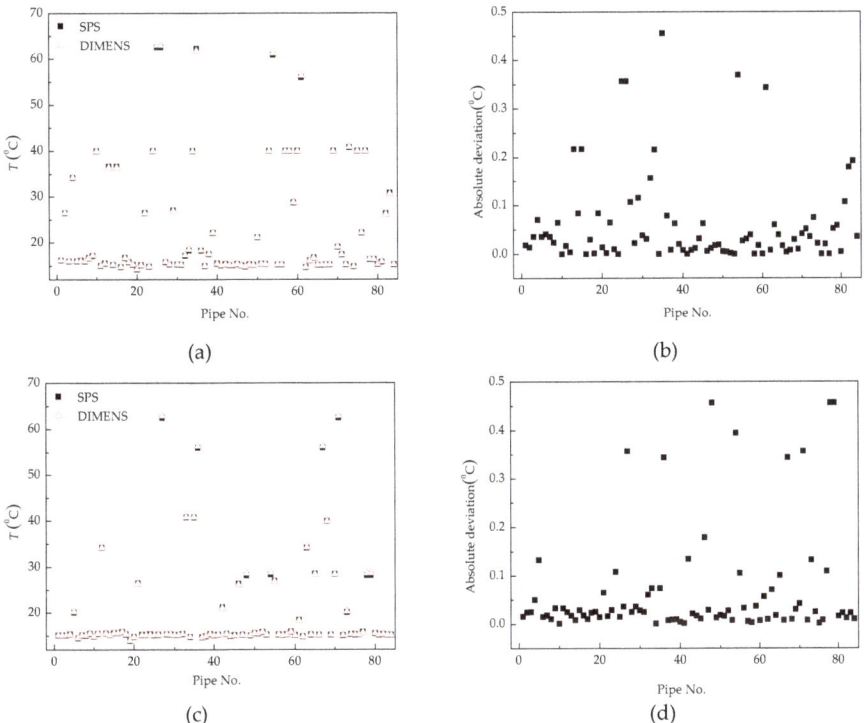

Figure 13. Comparison of temperature at the start and end nodes of pipelines: (**a**) Temperature value at the start node; (**b**) deviation at the start node; (**c**) temperature value at end node; (**d**) deviation at the end node.

5.3. Comparison of Computational Efficiency

The number of discretized grid points of the pipeline network and the number of pipeline sections of the pipeline network are two main factors that can affect the efficiency of pipeline network simulation. Therefore, the computational speed of the DIMENS method proposed in this paper is examined from three aspects: (1) the number of discretized grid points; (2) the number of components; and (3) the number of discretized grid points and the number of components. The restart process of the complex pipeline network is simulated by both the DIMENS method and SPS, and the CPU times are recorded and compared. The time step is 10 s, and the simulation time is 1 day. The computations are carried out on a computer equipped with a 2.6 GHz Intel Xeon E5-2640 CPU with 64 GB RAM.

First, the topology of the pipeline network in Figure 5 is unchanged, while the pipelines are discretized by different spatial steps. The adaptability of the DIMENS method for the number of discretized grid points is studied. The CPU time is shown in Figure 14.

It can be seen from Figure 14 that the CPU time of the DIMENS method is less than that of the SPS. This means that the DIMENS method is more efficient than SPS. What is more, the CPU times of these two methods are both almost linear with the number of discretized points. This indicates that the DIMNES method and the SPS both have a strong adaptability to the increase of discretized points of the pipeline network. However, with the same number of discretized points, the CPU time of SPS is always longer than that of DIMENS method. The slope of the line of the DIMENS method in Figure 14 is 0.01, while that of SPS is 0.036. This is means that the efficiency of the DIMENS method is about 0.0036/0.01 = 3.6 times higher than that of the SPS as the number of discretized points is increasing. In other word, the DIMENS method is more efficient than SPS.

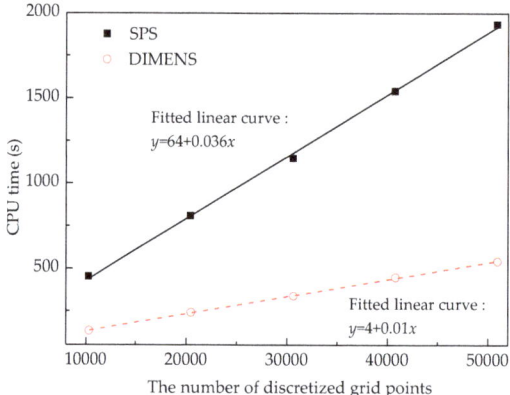

Figure 14. Change of CPU time with the number of discretized grid points.

Second, the total number of discretized grid points is unchanged, while the number of components of the pipeline network is increased. The adaptability of the DIMENS method for the number of components is studied. The CPU time is shown in Figure 15.

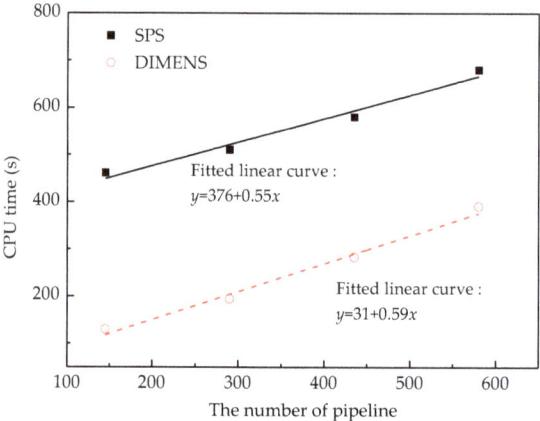

Figure 15. Change of computing time with the number of pipeline sections.

It is clearly shown in Figure 15 that the CPU time of the DIMENS method is also less than that of the SPS. This means that the computational efficiency of the DIMENS method is high. What is more, the CPU time of the DIMENS method and that of the SPS are both linear with the number of components. The fit linear curve of SPS is $y = 376 + 0.55x$, and that of DIMENS is $y = 31 + 0.59x$. In other words, the gradients of the two lines are similar. Thus, the DIMNES method and the SPS are both not only well adapted to the increase of discrete points, but also to the increase of components. This indicates that the DIMNES method and the SPS both have a strong adaptability to the number of pipeline sections.

Last, the pipe network, as shown in Figure 5, is copied multiple times and connected to each other to form a larger network. The adaptability of the DIMENS method for the couple of grid points and components is studied. The CPU time is shown in Figure 16.

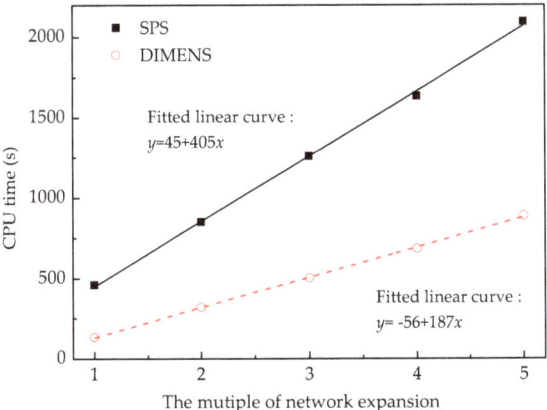

Figure 16. Change of computing time with multiple network expansions.

Figure 16 shows that the CPU time of the DIMENS method and that of the SPS are both linear with multiple network expansions. However, with the same multiple network expansions, the CPU time of SPS is always more than that of DIMENS method, and the slope of the line of the DIMENS method is smaller than that of SPS. That is to say, the CPU time of the DIMENS method is always less than that of SPS, and the DIMENS method would be more time-saving with the larger scale of the pipe network. This indicates that the DIMENS method is more efficient than SPS. It can be further seen that the fitted linear curve of SPS is $y = 45 + 405x$, while that of DIMENS is $y = -56 + 187x$. In other words, the slope of the line of the DIMENS method is about 187, while the slope of the line of SPS is about 405, which is over two times greater (405/187 = 2.16) than that of the DIMENS method. This means that the computing efficiency of the DIMENS method is over two times greater than that of SPS.

6. Conclusions and Future Work

In this paper, the DIMENS method has been extended to the flow and thermodynamic-coupled simulation of the complex pipeline network that includes pipelines and non-pipeline components. The simulation accuracy and efficiency of the present method have been investigated through an example of the complex pipeline network. The conclusions are as follows:

(1) DIMES shows a very good agreement with SPS (differences below 3.5%). This accuracy can meet the requirements of practical engineering;
(2) The CPU time of the DIMENS method is always less than that of SPS, and the computational speed of the DIMENS method is over two times higher than that of the SPS;
(3) The DIMENS method has a strong adaptability to the scale of the pipeline network. The CPU time of the DIMNES method is linear with the size of the pipeline network.

In recent years, as a rapidly developing role in High Performance Computing (HPC), Graphic Processing Unit (GPU) computing has been becoming a research hotspot in many research fields. In order to maximize the GPU computing speed, the parallel granularity of the computational task should be fine enough. Thus, the algorithm should be highly parallel. In the DIMENS method, the pipelines in the network are parallel and the solution process of one pipeline is parallel too. This means that the parallel characteristic of the DIMENS method greatly matches the GPU. Therefore, to improve the computational speed further, it would be meaningful and worthwhile to study the GPU-accelerated simulation for large-scale natural gas pipeline networks in the future.

Author Contributions: Conceptualization, B.Y.; Methodology, P.W.; Validation, S.A., D.H., and Y.X.; Writing—original draft, P.W.; Writing—review & editing, P.W. and B.Y.

Funding: The study is supported by the National Natural Science Foundation of China (No. 518060180), the Project of Construction of Innovative Teams and Teacher Career Development for Universities and Colleges under Beijing Municipality (No. IDHT20170507), and the Program of Great Wall Scholar (No. CIT&TCD20180313).

Conflicts of Interest: The authors declare no conflict of interest.

Nomenclature

Roman symbols

c_v	Specific heat capacity at constant volume, J/(kg·K)
c_p	Specific heat capacity at constant pressure, J/(kg·K)
e	Specific internal energy, J/(kg·K)
d	Internal diameter of pipe, m
g	Gravitational acceleration, m/s^2
h	Specific enthalpy, J/(kg·K)
m	Mass flow rate, kg/s
n	Polytropic index
p	Pressure, Pa
s	Elevation, m
t	Time, s
u	Corresponding component of general variable **U**
w	Flow velocity, m/s
x	Spatial coordinate, m
A	Cross-section area, m^2
C_v	Flow coefficient of valve
FR	Valve opening
K	Total heat transfer coefficient, W/(m^2·K)
N	Number of discretized sections of pipeline
Q	Volume flow, m^3/s
R	Specific gas constant, J/(kg·K)
T	Temperature, K
U	General variable
Z	Compressibility factor

Greek symbols

α, β	Fundamental set of solution
γ	Particular solution
ρ	Density, kg/m^3
θ	Inclination angle of a pipe, rad
λ	Friction factor
ε	Compression ratio
Δ	Specific density of natural gas relative to air
Δx	Spatial step, m
Δt	Time step, s

Superscripts

n	Time level
m	Related to mass flow
p	Related to pressure

Subscripts

npc	Non-pipeline components
in	Flow in
out	Flow out
p	Pipeline
c	Compressor
v	Valve
s	Supply
d	Demand

Appendix A

The pipeline model is a set of governing equations of homogeneous, geometrically one-dimensional flow in the pipeline, which consists of the continuity equation (A1), momentum equation (A2), and energy equation (A3):

$$\frac{\partial \rho}{\partial t} + \frac{\partial (\rho w)}{\partial x} = 0, \tag{A1}$$

$$\frac{\partial (\rho w)}{\partial t} + \frac{\partial (\rho w w)}{\partial x} + \frac{\partial p}{\partial x} = -\frac{\lambda}{2}\frac{\rho w|w|}{d} - \rho g \sin \theta, \tag{A2}$$

$$\frac{\partial}{\partial t}[(e + \frac{w^2}{2} + gs)\rho A] + \frac{\partial}{\partial x}[(h + \frac{w^2}{2} + gs)\rho w A] = -q\rho w A, \tag{A3}$$

With the following equations:
Cross-section area:

$$A = \frac{\pi}{4}d^2, \tag{A4}$$

Mass flux:

$$m = \rho w A, \tag{A5}$$

Specific heat increment:

$$q = \frac{\pi K d (T - T_g)}{m}, \tag{A6}$$

Relationship between elevation and inclination:

$$\theta = \frac{\partial s}{\partial x}, \tag{A7}$$

Derivation rule of composite functions:

$$\frac{\partial \rho}{\partial t} = \left(\frac{\partial \rho}{\partial p}\right)_T \frac{\partial p}{\partial t} + \left(\frac{\partial \rho}{\partial T}\right)_p \frac{\partial T}{\partial t}, \tag{A8}$$

$$\frac{\partial \rho}{\partial x} = \left(\frac{\partial \rho}{\partial p}\right)_T \frac{\partial p}{\partial x} + \left(\frac{\partial \rho}{\partial T}\right)_p \frac{\partial T}{\partial x}, \tag{A9}$$

$$\frac{d(\cdot)}{dt} = \frac{\partial(\cdot)}{\partial t} + w\frac{\partial(\cdot)}{\partial x}, \tag{A10}$$

Fundamental equations of thermodynamic functions:

$$\left(\frac{\partial \rho}{\partial T}\right)_p = -\left(\frac{\partial p}{\partial T}\right)_\rho \left(\frac{\partial \rho}{\partial p}\right)_T = -\frac{\left(\frac{\partial p}{\partial T}\right)_\rho}{\left(\frac{\partial p}{\partial \rho}\right)_T}, \tag{A11}$$

$$\left(\frac{\partial \rho}{\partial p}\right)_T = \frac{1}{\left(\frac{\partial p}{\partial \rho}\right)_T}, \tag{A12}$$

$$dh = de + d\left(\frac{p}{\rho}\right), \tag{A13}$$

$$de = c_v dT - \frac{1}{\rho^2}[T\left(\frac{\partial p}{\partial T}\right)_\rho - p]d\rho, \tag{A14}$$

Equations (A1) to (A3) can be transformed to mathematical forms with pressure, mass rate, and temperature as the dependent variables.

Continuity Equation

Substituting Equation (A5) and Equation (A8) into Equation (A1),

$$\left(\frac{\partial \rho}{\partial p}\right)_T \frac{\partial p}{\partial t} + \left(\frac{\partial \rho}{\partial T}\right)_p \frac{\partial T}{\partial t} + \frac{1}{A}\frac{\partial m}{\partial x} = 0, \tag{A15}$$

Substituting Equation (A11) and Equation (A12) into Equation (A13),

$$\frac{\partial p}{\partial t} - \left(\frac{\partial p}{\partial T}\right)_\rho \frac{\partial T}{\partial t} + \frac{1}{A}\left(\frac{\partial p}{\partial \rho}\right)_T \frac{\partial m}{\partial x} = 0, \tag{A16}$$

Momentum Equation

Substituting Equation (A5) into Equation (A2),

$$\frac{\partial m}{\partial t} + \frac{2m}{A\rho}\frac{\partial m}{\partial x} - \frac{m^2}{A\rho^2}\frac{\partial \rho}{\partial x} + A\frac{\partial p}{\partial x} = -\frac{\lambda}{2}\frac{m|m|}{dA\rho} - A\rho g \sin\theta, \tag{A17}$$

Substituting Equation (A9) into Equation (A17),

$$\frac{\partial m}{\partial t} + [A - \frac{m^2}{A\rho^2}\left(\frac{\partial \rho}{\partial p}\right)_T]\frac{\partial p}{\partial x} + \frac{2m}{A\rho}\frac{\partial m}{\partial x} = -\frac{\lambda}{2}\frac{m|m|}{dA\rho} - A\rho g \sin\theta + \frac{m^2}{A\rho^2}\left(\frac{\partial \rho}{\partial T}\right)_p \frac{\partial T}{\partial x}, \tag{A18}$$

Energy Equation

Substituting Equations (A4)–(A7) into Equation (A3),

$$\frac{\partial}{\partial t}[(h+\frac{w^2}{2})\rho A] - \frac{\partial p}{\partial t} + \frac{\partial}{\partial x}[(h+\frac{w^2}{2})\rho w A] = -\pi K d(T - T_g) - \rho w g A \sin\theta, \tag{A19}$$

Substituting Equation (A13) into Equation (A19),

$$\frac{\partial}{\partial t}[(h+\frac{w^2}{2})\rho] - \frac{\partial p}{\partial t} + w\frac{\partial}{\partial x}[(h+\frac{w^2}{2})\rho] + (h+\frac{w^2}{2})\rho\frac{\partial w}{\partial x} = -\frac{4K(T-T_g)}{d} - \rho w g \sin\theta, \tag{A20}$$

Equation (A20) can be rewritten in the form

$$\rho\frac{d}{dt}(h+\frac{w^2}{2}) + (h+\frac{w^2}{2})[\frac{\partial \rho}{\partial t} + w\frac{\partial \rho}{\partial x} + \rho\frac{\partial w}{\partial x}] - \frac{\partial p}{\partial t} = -\frac{4K(T-T_g)}{d} - \rho w g \sin\theta, \tag{A21a}$$

$$\rho\frac{dh}{dt} - \frac{\partial p}{\partial t} + \rho w\frac{dw}{dt} = -\frac{4K(T-T_g)}{d} - \rho w g \sin\theta, \tag{A21b}$$

Equation (A2) can be rewritten in the form

$$\rho\frac{\partial w}{\partial t} + \rho w\frac{\partial w}{\partial x} + \frac{\partial p}{\partial x} = -\frac{\lambda}{2}\frac{\rho w|w|}{d} - \rho g \sin\theta, \tag{A22a}$$

$$\rho w\frac{dw}{dt} = -w\frac{\partial p}{\partial x} - \frac{\lambda}{2}\frac{\rho|w|^3}{d} - \rho w g \sin\theta, \tag{A22b}$$

Substituting Equation (A22b) into Equation (A21b),

$$\rho\frac{dh}{dt} - \frac{\partial p}{\partial t} - w\frac{\partial p}{\partial x} = \frac{\lambda}{2}\frac{\rho|w|^3}{d} - \frac{4K(T-T_g)}{d}, \tag{A23a}$$

$$\rho \frac{dh}{dt} - \frac{dp}{dt} = \frac{\lambda}{2} \frac{\rho |w|^3}{d} - \frac{4K(T - T_g)}{d}, \quad \text{(A23b)}$$

Substituting Equation (A13) into Equation (A23b),

$$\rho \frac{de}{dt} - \frac{p}{\rho} \frac{d\rho}{dt} = \frac{\lambda}{2} \frac{\rho |w|^3}{d} - \frac{4K(T - T_g)}{d}, \quad \text{(A24)}$$

Substituting Equation (A14) into Equation (A24),

$$\rho c_v \frac{dT}{dt} - T \left(\frac{\partial p}{\partial T}\right)_\rho \frac{1}{\rho} \frac{d\rho}{dt} = \frac{\lambda}{2} \frac{\rho |w|^3}{d} - \frac{4K(T - T_g)}{d}, \quad \text{(A25)}$$

Equation (A1) can be rewritten in the form

$$\frac{\partial \rho}{\partial t} + w \frac{\partial \rho}{\partial x} + \rho \frac{\partial w}{\partial x} = 0, \quad \text{(A26a)}$$

$$\frac{d\rho}{dt} = -\rho \frac{\partial w}{\partial x}, \quad \text{(A26b)}$$

Substituting Equation (A26b) into Equation (A25),

$$\frac{\partial T}{\partial t} + w \frac{\partial T}{\partial x} = \frac{1}{\rho c_v} \left[-T \left(\frac{\partial p}{\partial T}\right)_\rho \frac{\partial w}{\partial x} + \frac{\lambda}{2} \frac{\rho |w|^3}{d} - \frac{4K(T - T_g)}{d} \right], \quad \text{(A27)}$$

Equation (A14), Equation (16) and Equation (A23) can be written in the general form

$$\frac{\partial \mathbf{U}}{\partial t} + \mathbf{B} \cdot \frac{\partial \mathbf{U}}{\partial x} = \mathbf{F}, \quad \text{(A28)}$$

The detailed expressions of **U**, **G**, and **S** in the pipeline model are as follows:
Flow Equation:

$$\mathbf{U} = \begin{bmatrix} u_1 \\ u_2 \end{bmatrix} = \begin{bmatrix} p \\ m \end{bmatrix}, \quad \text{(A29)}$$

$$\mathbf{G} = \begin{bmatrix} \frac{1}{A} \frac{\partial}{\partial p}\left(\frac{\partial p}{\partial \rho}\right)_T \frac{\partial m}{\partial x} & 0 \\ \left\{ \frac{m^2}{A} \left[\frac{2}{\rho^3}\left(\frac{\partial \rho}{\partial p}\right)_T^2 - \frac{1}{\rho^2} \frac{\partial}{\partial p}\left(\frac{\partial \rho}{\partial p}\right)_T \right] \frac{\partial p}{\partial x} - \frac{2m}{A\rho^2}\left(\frac{\partial \rho}{\partial p}\right)_T \frac{\partial m}{\partial x} \right\} & -\frac{2m}{A\rho^2}\left(\frac{\partial \rho}{\partial p}\right)_T \frac{\partial p}{\partial x} + \frac{2}{A\rho} \frac{\partial m}{\partial x} \end{bmatrix}, \quad \text{(A30)}$$

$$\mathbf{S} = \begin{bmatrix} \frac{\partial T}{\partial t} \frac{\partial}{\partial p}\left(\frac{\partial p}{\partial T}\right)_\rho & 0 \\ \left\{ \frac{\lambda}{2} \frac{m|m|}{dA\rho^2}\left(\frac{\partial \rho}{\partial p}\right)_T - Ag \sin\theta \left(\frac{\partial \rho}{\partial p}\right)_T + \frac{m^2}{A} \frac{\partial T}{\partial x} \frac{\partial}{\partial p}\left(\frac{1}{\rho^2}\left(\frac{\partial \rho}{\partial T}\right)_p\right) \right\} & -\lambda \frac{|m|}{dA\rho} + \frac{2m}{A\rho^2}\left(\frac{\partial \mu}{\partial T}\right)_p \frac{\partial l}{\partial x} \end{bmatrix}, \quad \text{(A31)}$$

Thermodynamic Equation:

$$U = T, \quad \text{(A32)}$$

$$G = 0, \quad \text{(A33)}$$

$$S = \left\{ \begin{array}{l} \frac{1}{\rho c_v}\left(-\left(\frac{\partial p}{\partial T}\right)_\rho \frac{\partial w}{\partial x} - T \frac{\partial w}{\partial x} \frac{\partial}{\partial T}\left(\frac{\partial p}{\partial T}\right)_\rho + \frac{\lambda}{2} \frac{|w|^3}{d} \frac{\partial \rho}{\partial T} - \frac{4K}{d}\right) \\ -\left[-T\left(\frac{\partial p}{\partial T}\right)_\rho \frac{\partial w}{\partial x} + \frac{\lambda}{2} \frac{\rho|w|^3}{d} - \frac{4K(T - T_a)}{d}\right]\left(\frac{1}{\rho c_v}\right)^2 \left(\rho \frac{\partial c_v}{\partial T} + c_v \frac{\partial \rho}{\partial T}\right) \end{array} \right\}, \quad \text{(A34)}$$

Appendix B

To describe in detail how the flow equations of multi-component interconnection nodes are solved, a pipeline network is given as the instance. The pipeline network is comprised of three pipes, one compressor, one valve, and two externals, as shown in Figure A1. The closed equations for the solution of multi-component interconnection nodes are Equations (A35)–(A58), shown as below.

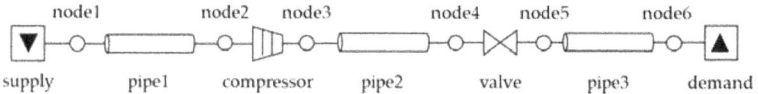

Figure A1. The schematic diagram of the structure of a pipeline network.

Flow equations of the compressor:
$$m_{in,c} - m_{out,c} = 0, \tag{A35}$$

$$\varepsilon p_{in,c} - p_{out,c} = 0, \tag{A36}$$

Flow equations of the valve:
$$m_{in,v} - m_{out,v} = 0, \tag{A37}$$

$$m_{in,v} - C_v \rho_{in,v} \sqrt{\frac{(p_{in,v}^2 - p_{out,v}^2)}{Z\Delta \overline{T_{in,v}}}} = 0, \tag{A38}$$

Flow equation of supply:
$$p_s = p(t), \tag{A39}$$

Flow equation of demand:
$$m_d = m(t), \tag{A40}$$

Flow equations of multi-component interconnection node 1:
$$m_s = m_{1,p1}, \tag{A41}$$

$$p_s = p_{1,p1}, \tag{A42}$$

Flow equations of multi-component interconnection node 2:
$$m_{N_1+1,p1} = m_{in,c}, \tag{A43}$$

$$p_{N_1+1,p1} = p_{in,c}, \tag{A44}$$

Flow equations of multi-component interconnection node 3:
$$m_{out,c} = m_{1,p2}, \tag{A45}$$

$$p_{out,c} = p_{1,p2}, \tag{A46}$$

Flow equations of multi-component interconnection node 4:
$$m_{N_2+1,p2} = m_{in,v}, \tag{A47}$$

$$p_{N_2+1,p2} = p_{in,v}, \tag{A48}$$

Flow equations of multi-component interconnection node 5:
$$m_{out,v} = m_{1,p3}, \tag{A49}$$

$$p_{in,v} = p_{1,p3}, \tag{A50}$$

Flow equations of multi-component interconnection node 6:
$$m_{N_3+1,p3} = m_d, \tag{A51}$$

$$p_{N_3+1,p3} = p_d, \tag{A52}$$

Supplemental equations of pipe 1:

$$m_{1,p1} = \alpha^m_{1,p1} p_{1,p1} + \beta^m_{1,p1} m_{N_1+1,p1} + \gamma^m_{1,p1}, \tag{A53}$$

$$p_{N_1+1,p1} = \alpha^p_{N_1+1,p1} p_{1,p1} + \beta^p_{N_1+1,p1} m_{N_1+1,p1} + \gamma^p_{N_1+1,p1}, \tag{A54}$$

Supplemental equations of pipe 2:

$$m_{1,p2} = \alpha^m_{1,p2} p_{1,p2} + \beta^m_{1,p2} m_{N_2+1,p2} + \gamma^m_{1,p2}, \tag{A55}$$

$$p_{N_2+1,p2} = \alpha^p_{N_2+1,p2} p_{1,p2} + \beta^p_{N_2+1,p2} m_{N_2+1,p2} + \gamma^p_{N_2+1,p2}, \tag{A56}$$

Supplemental equations of pipe 3:

$$m_{1,p3} = \alpha^m_{1,p3} p_{1,p3} + \beta^m_{1,p3} m_{N_3+1,p3} + \gamma^m_{1,p3}, \tag{A57}$$

$$p_{N_3+1,p3} = \alpha^p_{N_3+1,p3} p_{1,p3} + \beta^p_{N_3+1,p3} m_{N_3+1,p3} + \gamma^p_{N_3+1,p3} \tag{A58}$$

The number of the mass flow and pressure variables at the starting and terminal points of the pipeline is 4, and the numbers of the compressor and valve are both 4 too, while the number of the mass flow and pressure variables of the external component is 2. This means that there are a total of $4 \times 3 + 4 \times 1 + 4 \times 1 + 2 \times 2 = 24$ unknown variables for solving the multi-component interconnection nodes in the pipeline network in Figure 5. The number of equations in Equations (A35)–(A58) is 24. Therefore, these equations are closed and have a unique solution.

Appendix C

To describe how the thermodynamic equations of multi-component interconnection nodes are solved in detail, a pipeline network is given as the instance. For instance, the flow state of the pipeline network at a certain moment is shown in Figure A2. Pipe 1, pipe 2, and pipe 3 are respectively the second, fourth, and first type of pipeline. The complete system of linear equations of the multi-component interconnection nodes is presented as follows:

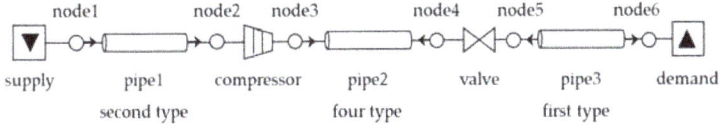

Figure A2. The flow status of the pipeline networks at a given time.

Thermodynamic-linearized equation of the compressor:

$$T_{out,c} - T_{in,c} \varepsilon^{\frac{m-1}{m}} = 0, \tag{A59}$$

Thermodynamic equation of the valve:

$$T_{out,v1} - T_{in,v1} + (p_{out,v1} - p_{in,v1}) \frac{1}{c_p} \left[\frac{T}{\rho^2} \frac{(\partial p/\partial T)_\rho}{(\partial p/\partial \rho)_T} - \frac{1}{\rho} \right]_{out,\, v1} = 0, \tag{A60}$$

Thermodynamic equation of the external component:

$$T_s = T_1(t), \tag{A61}$$

Thermodynamic equations of multi-component interconnection nodes 1:

$$T_{1,p1} = \frac{|c_p m|_{e1}}{|c_p m|_{1,p1}} T_s, \tag{A62}$$

Thermodynamic equations of multi-component interconnection nodes 2:

$$T_{in,c} = \frac{|c_p m|_{N_1+1,p1}}{|c_p m|_{in,c}} T_{N_1+1,p1}, \quad (A63)$$

Thermodynamic equations of multi-component interconnection nodes 3:

$$T_{1,p2} = \frac{|c_p m|_{out,c}}{|c_p m|_{1,p2}} T_{out,c}, \quad (A64)$$

Thermodynamic equations of multi-component interconnection nodes 4:

$$T_{N_2+1,p2} = \frac{|c_p m|_{in,v}}{|c_p m|_{N_2+1,p2}} T_{in,v}, \quad (A65)$$

Thermodynamic equations of multi-component interconnection nodes 5:

$$T_{1,p3} = \frac{|c_p m|_{out,v}}{|c_p m|_{1,p3}} T_{out,v}, \quad (A66)$$

Thermodynamic equations of multi-component interconnection nodes 6:

$$T_d = \frac{|c_p m|_{N_3+1,p3}}{|c_p m|_{2g}} T_{N_3+1,p3}, \quad (A67)$$

Temperature value of the starting and terminal points of pipe 3:

$$T_{1,p3} = T_{1,p3}^{solved}, \quad (A68)$$

$$T_{N+1,p3} = T_{N+1,p3}^{solved}, \quad (A69)$$

where, the superscript 'solved' is standing for having been solved in the second step.

Supplemental equation of pipe 1:

$$T_{N_1+1,p1} = T_{1,p1} \alpha_{N_1+1,p1} + \gamma_{N_1+1,p1}, \quad (A70)$$

The number of temperature variables at the starting and terminal points of the pipeline, compressor, and valve is 2, while the number of temperature variables of the external component is 1. This means that there are $5 \times 2 + 2 \times 1 = 12$ unknown temperature variables for solving the multi-component interconnection nodes solution in pipeline networks, as shown in Figure A2. There are also 12 Equations in Equations (A59) to (A70). Therefore, these equations are closed and have a unique solution.

References

1. Birol, F. Golden Rules for a Golden Age of Gas: World Energy Outlook Special Report on Unconventional Gas: Report. International Energy Agency. Available online: www.worldenergyoutlook.org/media/weowebsite/2012/goldenrules/WEO2012_GoldeGRulesReport.pdf (accessed on 22 October 2012).
2. Turner, W.J.; Kwon, P.S.J.; Maguire, P.A. Evaluation of a gas pipeline simulation program. *Math. Comput. Model.* **1991**, *15*, 1–14. [CrossRef]
3. Ouellette, L.; Hanks, H. Integrated simulation and SCADA to replay historical data for a local distribution company. In Proceedings of the PSIG Annual Meeting, Pipeline Simulation Interest Group, Tucson, AZ, USA, 15–17 October 1997.
4. Hendriks, P.H.G.M.; Postvoll, W.; Mathiesen, M.; Spiers, R.P.; Siddorn, J. Improved capacity utilization by integrating real-time sea bottom temperature data. In Proceedings of the PSIG annual Meeting, Pipeline Simulation Interest Group, Williamsburg, VA, USA, 23–25 October 1995.

5. Osiadacz, A. *Simulation and Analysis of Gas Networks*; E. & F.N. Spon Ltd.: London, UK, 1987; pp. 69–82.
6. Helgaker, J.F.; Müller, B.; Ytrehus, T. Transient flow in natural gas pipelines using implicit finite difference schemes. *J. Offshore Mech. Arct. Eng.* **2014**, *136*, 031701. [CrossRef]
7. Wang, H.; Liu, X.; Zhou, W. Transient flow simulation of municipal gas pipelines and networks using semi implicit finite volume method. *Procedia Eng.* **2011**, *12*, 217–223.
8. Ebrahimzadeh, E.; Shahrak, M.N.; Bazooyar, B. Simulation of transient gas flow using the orthogonal collocation method. *Chem. Eng. Res. Des.* **2012**, *90*, 1701–1710. [CrossRef]
9. Alamian, R.; Behbahani-Nejad, M.; Ghanbarzadeh, A. A state space model for transient flow simulation in natural gas pipelines. *J. Nat. Gas Sci. Eng.* **2012**, *9*, 51–59. [CrossRef]
10. Behbahani-Nejad, M.; Shekari, Y. The accuracy and efficiency of a reduced-order model for transient flow analysis in gas pipelines. *J. Pet. Sci. Eng.* **2010**, *73*, 13–19. [CrossRef]
11. Farzaneh-Gord, M.; Rahbari, H.R. Unsteady natural gas flow within pipeline network, an analytical approach. *J. Nat. Gas Sci. Eng.* **2016**, *28*, 397–409. [CrossRef]
12. Pambour, K.A.; Bolado-Lavin, R.; Dijkema, G.P.J. An integrated transient model for simulating the operation of natural gas transport systems. *J. Nat. Gas Sci. Eng.* **2016**, *28*, 672–690. [CrossRef]
13. Madoliat, R.; Khanmirza, E.; Moetamedzadeh, H.R. Transient simulation of gas pipeline networks using intelligent methods. *J. Nat. Gas Sci. Eng.* **2016**, *29*, 517–529. [CrossRef]
14. Wang, J.; Wang, T.; Wang, J. Application of π equivalent circuit in mathematic modeling and simulation of gas pipeline. *Appl. Mech. Mater.* **2014**, *496*, 943–946. [CrossRef]
15. Herrán-González, A.; Cruz, J.M.D.L.; De Andrés-Toro, B.; Risco-Martín, J.L. Modeling and Simulation of a Gas Distribution Pipeline Network. *Appl. Math. Model.* **2009**, *33*, 1584–1600. [CrossRef]
16. Thorley, A.R.D.; Tiley, C.H. Unsteady and transient flow of compressible fluids in pipelines—A review of theoretical and some experimental studies. *Int. J. Heat Fluid Flow* **1987**, *8*, 3–15. [CrossRef]
17. Kiuchi, T.T.; Izumi, H.H.; Huke, T.T. An operation support system of large city gas networks based on fluid transient model. *J. Energy Resour. Technol.* **1995**, *117*, 324–328. [CrossRef]
18. Kiuchi, T. An implicit method for transient gas flows in pipe networks. *Int. J. Heat Fluid Flow* **1994**, *15*, 378–383. [CrossRef]
19. Wylie, E.B.; Stoner, M.A.; Streeter, V.L. Network: System transient calculations by implicit method. *Soc. Pet. Eng. J.* **1971**, *11*, 356–362. [CrossRef]
20. Li, Y.; Yao, G. *Design and Operation of Gas Pipeline*; China University of Petroleum Press: Beijing, China, 2009; pp. 19–26. (In Chinese)
21. Abbaspour, M.; Chapman, K.S. Non-isothermal transient flow in natural gas pipeline. *J. Appl. Mech.* **2008**, *75*, 031018. [CrossRef]
22. Luskin, M. An approximation procedure for non-symmetric, nonlinear hyperbolic systems with integral boundary conditions. *Siam J. Numer. Anal.* **1979**, *16*, 145–164. [CrossRef]
23. Zheng, J.G.; Chen, G.Q.; Song, F.; Ai, M.Y.; Zhao, J.L. Research on simulation model and solving technology of large scale gas pipe network. *J. Syst. Simul.* **2012**, *14*, 133. (In Chinese)
24. Barley, J. Thermal decoupling: An investigation. In Proceedings of the PSIG annual Meeting, Pipeline Simulation Interest Group, Santa Fe, NM, USA, 16–19 April 2012.
25. Helgaker, J.F.; Ytrehus, T. Coupling between continuity/momentum and energy Equation in 1D gas flow. *Energy Procedia* **2012**, *26*, 82–89. [CrossRef]
26. Wang, P.; Yu, B.; Deng, Y.; Zhao, Y. Comparison study on the accuracy and efficiency of the four forms of hydraulic Equations of a natural gas pipeline based on linearized solution. *J. Nat. Gas Sci. Eng.* **2015**, *22*, 235–244. [CrossRef]
27. Stoner, M.A. Sensitivity analysis applied to a steady-state model of natural gas transportation systems. *Soc. Pet. Eng. J.* **1972**, *12*, 115–125. [CrossRef]
28. Wang, P.; Yu, B.; Han, D.; Sun, D. Fast method for the hydraulic simulation of natural gas pipeline networks based on the divide-and-conquer approach. *J. Nat. Gas Sci. Eng.* **2018**, *50*, 55–63. [CrossRef]
29. Keenan, P.T. Collation and upwinding for thermal flow in pipelines: The linearized case. *Int. J. Numer. Methods Fluids* **1996**, *22*, 835–849. [CrossRef]
30. Osiadacz, A.J.; Chaczykowski, M. Comparison of isothermal and non-isothermal pipeline gas flow models. *Chem. Eng. J.* **2001**, *81*, 41–51. [CrossRef]

31. Wang, P.; Yu, B.; Han, D.; Li, J.; Sun, D.; Xiang, Y.; Wang, L. Adaptive implicit finite difference method for natural gas pipeline transient flow. *Oil Gas Sci. Technol. Rev. D'ifp Energ. Nouv.* **2018**, *73*, 21. [CrossRef]
32. Chaczykowski, M. Sensitivity of Pipeline Gas Flow Model to the Selection of the Equation of State. *Chem. Eng. Res. Des.* **2009**, *87*, 1596–1603. [CrossRef]
33. Benedict, M.; Webb, G.B.; Rubin, L.C. An empirical Equation for thermodynamic properties of light hydrocarbons and their mixtures I. Methane, ethane, propane and n-butane. *J. Chem. Phys.* **1940**, *8*, 334–345. [CrossRef]
34. Scott, L.R. *Numerical Analysis*, 2nd ed.; Princeton University Press: Oxford, MS, USA, 2016; pp. 53–64, 145–149.
35. Colebrook, C.F. Turbulent flow in pipes with particular reference to the transition region between the smooth and rough pipe laws. *J. Inst. Civ. Eng.* **1939**, *11*, 133–156. [CrossRef]

© 2019 by the authors. Licensee MDPI, Basel, Switzerland. This article is an open access article distributed under the terms and conditions of the Creative Commons Attribution (CC BY) license (http://creativecommons.org/licenses/by/4.0/).

Article

Influence of Permeability and Injection Orientation Variations on Dispersion Coefficient during Enhanced Gas Recovery by CO_2 Injection

Muhammad Kabir Abba *, Athari Al-Otaibi, Abubakar Jibrin Abbas, Ghasem Ghavami Nasr and Martin Burby

Petroleum and Gas Research Group, The University of Salford, Manchester M4 5WT, UK; A.al-otaibi@edu.salford.ac.uk (A.A.-O.); a.j.abbas@salford.ac.uk (A.J.A.); g.g.nasr@salford.ac.uk (G.G.N.); m.burby@salford.ac.uk (M.B.)
* Correspondence: m.k.abba1@salford.ac.uk

Received: 5 May 2019; Accepted: 14 June 2019; Published: 18 June 2019

Abstract: This investigation was carried out to highlight the influence of the variation of permeability of the porous media with respect to the injection orientations during enhanced gas recovery (EGR) by CO_2 injection using different core samples of different petrophysical properties. The laboratory investigation was performed using core flooding technique at 1300 psig and 50 °C. The injection rates were expressed in terms of the interstitial velocities to give an indication of its magnitude and variation based on the petrophysical properties of each core sample tested. *Bandera Grey*, *Grey Berea*, and *Buff Berea* sandstone core samples were used with measured permeabilities of 16.08, 217.04, and 560.63 md, respectively. The dispersion coefficient was observed to increase with a decrease in permeability, with Bandera Grey having the highest dispersion coefficient and invariably higher mixing between the injected CO_2 and the nascent CH_4. Furthermore, this dispersion was more pronounced in the horizontal injection orientation compared to the vertical orientation with, again, the lowest permeability having a higher dispersion coefficient in the horizontal orientation by about 50%. This study highlights the importance of the permeability variation in the design of the injection strategy of EGR and provides a revision of the CO_2 plume propagation at reservoir conditions during injection.

Keywords: enhanced gas recovery; longitudinal dispersion coefficient; injection orientation; supercritical CO_2; CO_2 permeability

1. Introduction

The effects of greenhouse gas (GHG) emissions in the form of global warming with the resulting subsequent climate change cannot be overemphasized. The authors of [1] reported that the most significant of the GHG emissions are carbon dioxide (CO_2), methane (CH_4), dinitrogen oxide (N_2O), and other gases. Amongst all these gases, they iterated that CO_2 makes up about 76% of the total global emissions of GHGs. Incidentally, combustion of fossil fuels is the central source of the global CO_2 emissions and the oil and gas industry alone accounts for 65% of global CO_2 emissions [2]. Furthermore, CO_2 emissions from the oil and gas industry is ever increasing as a result of the high energy demand and at a rate of 1.7% per annum between the 1990s and early 2000s, and at an even higher rate of 3.1% per annum between 2000 and 2010 [3]. Therefore, the environmental consequences associated with CO_2 emissions have forced researchers in the oil and gas industry to come up with technologies to curb the proliferation of anthropogenic CO_2 due to the oil and gas activities. An avenue with a growing potential to address this issue is by the injection of the CO_2 emissions into geological formations like oil and gas reservoirs [4] using enhanced oil/gas recovery processes. Enhanced gas recovery (EGR) by CO_2 injection and sequestration is a simultaneous process whereby CO_2 is injected into a natural gas (CH_4) reservoir to displace CH_4 and store CO_2 in the reservoir.

EGR is still in its pilot/test phase and has not been widely accepted due to the nature of the gas-gas displacement whereby CO_2 disperses into CH_4 during the process and the recovered CH_4 will be heavily contaminated with the injected CO_2. This will affect the market value of the recovered CH_4 given that one of the reasons for the choice of CH_4 reservoirs as potential storage sites is that the recovered natural gas will offset part of the cost of the sequestration process [5]. Therefore, efforts are being made to stall the incessant mixing during EGR, and this is only achievable when the physics of the mixing are better understood. Pivotal to the adoption of the technique is understanding the mechanisms and factors which influence the interaction between the gases in situ, which leads to mixing at reservoir conditions. This will eventually provide an avenue to characterise gas systems for better injection scenarios and explore the potential and the viability of EGR as an adopted method of CO_2 sequestration.

In our previous works [6–9], we studied some of the factors that affect in situ mixing between the injected CO_2 and the displaced CH_4 which have not been previously considered. These factors include the connate water salinity, the injection orientation and, also, the flow behaviour of CO_2 in its supercritical state during EGR displacement. This research, however, focuses on the effect of injection orientation of the CO_2 on the dispersion of the gas in consolidated sandstone cores with an emphasis on permeability variation. The injection orientation determines the path of the CO_2 plume propagation through the pore spaces of the reservoir rock, which is also very important when it comes to the dip angle between the injector and producer wells. Consistent with our previous results, horizontal injections showed that the segregation of CO_2 to the bottom of the core sample—due to its higher density of supercritical CO_2 compared to that of CH_4 (showed in Figure 1)—led to higher residence time of the CO_2 traversing the length of the core sample, thereby increasing CO_2 dispersion. Invariably, dispersion in the vertical injection orientation of the CO_2 was significantly less than that in the horizontal orientation.

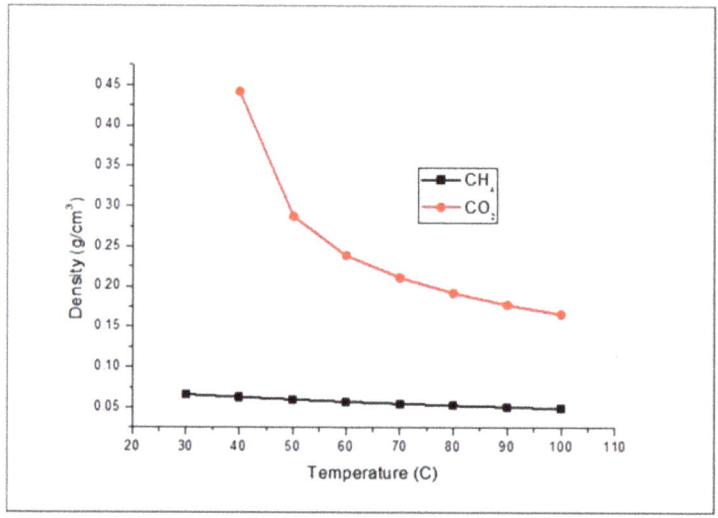

Figure 1. CH_4 and CO_2 density as a function of temperature at 1300 psi.

The authors of [10] carried out a horizontal dispersion of CO_2 in CH_4 in sand packs as the porous medium and also concluded that gravity plays a significant role in the dispersion of CO_2 during displacement applications. Their study included long and short sand packs with similar permeabilities, while here, different consolidated core samples with different petrophysical properties and equal length were employed to ascertain the effects of their variation on dispersion. The reasoning behind the use of equal length core samples was to minimise systemic disparities in our measurements. Given that permeability is a function of length, the pore structure variation which is responsible for the permeability difference between the core sample was explored, and the interplay between the core sample and the gases was evaluated. Therefore, the variation of the injection orientation and permeability was analysed to showcase their influence on dispersion coefficient.

2. Materials and Methods

The core samples were obtained from Kocurek Industries USA whose petrophysical properties are shown in Table 1. Gas permeability was obtained using a core flooding and the porosity was evaluated using Helium Porosimetry. Research grade CO_2 and CH_4 were obtained from BOC UK both with purity of >99.999%.

Table 1. Dimensions and petrophysical properties of core samples.

Core Samples	Length (mm)	Diameter (mm)	* Porosity (%)	* Permeability (md)
Grey Berea	76.27	25.22	19–20	200–315
Bandera Grey	76.00	25.47	21	30
Buff Berea	76.18	24.95	26	350–600

* Properties provided by suppliers.

Apparatus and Procedure

(1) Porosity and permeability measurements

The Helium Porosimeter PORG 200™ (Corelab, OK, USA) and Gas Permeater PERG 200 ™ (Corelab, OK, USA) were used to measure the porosity and permeability of the core samples respectively. Details of the equipment description, principles of operation, and procedure can be found in our previous works [8]. The results are shown in Table 2.

Table 2. Measured petrophysical properties.

Core Samples	Porosimetry Porosity (%)	Measured Permeability (md)
Bandera Grey	17	16.08
Grey Berea	20	217.04
Buff Berea	26.27	560.63

(2) Core flooding

Core flooding equipment, also from CoreLab Oklahoma, UFS 200 (details are presented in [7]), was used in this work. It simulated the displacement of CH_4 by supercritical CO_2 at reservoir conditions. The core holder was originally horizontally orientated; however, a stand was constructed to change this orientation to vertical in order to vary the injection orientation. The same procedure was adopted as the one employed in [9], the only difference being in the using different core samples of different permeabilities and porosities. The schematics of the setup are shown in Figure 2. After carrying out the same experiments at the same conditions in the horizontal orientation for each core sample, the core holder was installed vertically, and the same experiments were repeated using the same conditions as in the previous horizontal experiments.

Injection was made from the bottom of the core holder in the vertical orientation. The injection orientation adopted in this study was not based on the dip angle of injection during the flooding and recovery processes. The experiments were not field scale depictions of the injection strategy, rather, we were looking at core to pore scale investigation of the flow physics of CO_2 plume as it displaces nascent CH_4. Central to the experiments was the interaction and mixing between CO_2 in its supercritical state and the CH_4 in the reservoir given that the CO_2 plume occupies the lower echelon of the CH_4 zone due to its higher density at supercritical condition as we have iterated in our previous works. Vertical and horizontal flows of the CO_2 plume are the two extreme flow conditions in terms of propagation direction.

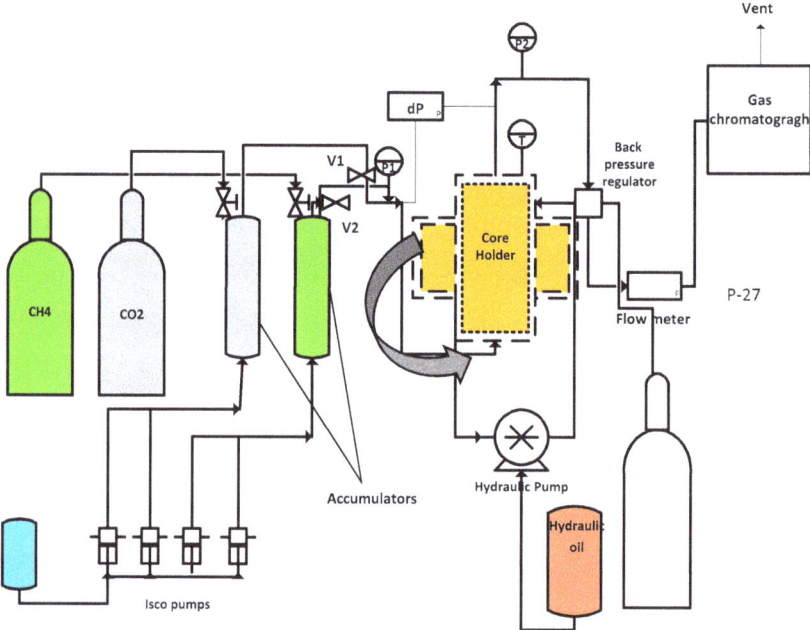

Figure 2. Core flooding set up schematics.

(3) Data analysis

A single parameter diffusion equation was used by [11] in the description of the longitudinal dispersion coefficient for gas flow in porous media which is shown in Equation (1):

$$K_l \frac{\partial^2 C}{\partial x^2} - u \frac{\partial C}{\partial x} = \frac{\partial C}{\partial t}, \tag{1}$$

where C-CO_2 concentration at time t, location x, K_l—the longitudinal dispersion coefficient, and u is the interstitial velocity. The dimensionless for of Equation (1) is written as

$$\frac{1}{P_e} \frac{\partial^2 C}{\partial x_D^2} - \frac{\partial C}{\partial x_D} = \frac{\partial C}{\partial t_D}, \tag{2}$$

where each parameter is defined in Table 3.

Table 3. Dimensionless parameters.

Parameter	Expression
P_e	$\frac{uL}{K_l}$
t_D	$\frac{tu}{L}$
x_D	$\frac{x}{L}$
u	$\frac{Q}{\pi r^2 \phi}$

This one-dimensional advection–dispersion (ADE) equation was used to measure/evaluate, analytically, the longitudinal dispersion of CO_2 into CH_4 using the solution of the equation and also assuming that the dispersion coefficient and interstitial velocity of the displacing species are not affected by the concentration. With the following boundary conditions, the initial condition—$C = 0$ at $t_D = 0$, boundary conditions—$C = 1$ at $x_D = 0$, $C \to 0$ as $x_D \to \infty$, the dimensionless solution to (Equation (1)) thus becomes [12–14]

$$C = \frac{1}{2}\left\{ erfc\left(\frac{x_D - t_D}{2\sqrt{t_D/P_e}}\right) + e^{P_e x_D} erfc\left(\frac{x_D + t_D}{2\sqrt{t_D/P_e}}\right) \right\}. \qquad (3)$$

Using the obtained concentration profiles from core flooding experiments, Equation (3) was used to fit the results from the experiments and the dispersion coefficient was obtained analytically using the least squares regression method with the dispersion coefficient as the fitting parameter.

Perkins and Johnston (1963) described the medium Péclet number denoted by P_{ex}, which defines the displacement mechanism that is dominant in gas dispersion in porous media as

$$P_{ex} = \frac{u_m d}{D}, \qquad (4)$$

where D—diffusion coefficient (m^2/s), P_{em}—medium Péclet number, u_m—mean interstitial velocity (m/s), and d is the characteristic length scale of mixing in the porous medium. Largely, at the values of $P_{em} < 0.1$, diffusion mixing dominates and, equally, at $P_{ex} > 10$, advective mixing dominates during the gas dispersion process. In this range of P_{ex}, [15] related diffusion to dispersion coefficients as presented in (Equation (6)):

$$\frac{K_l}{D} = \frac{1}{\tau} + \alpha \frac{u_m^n}{D}, \qquad (5)$$

where α represents the dispersivity of the porous medium in m, τ is the tortuosity of the porous medium, and n is an exponent.

Additionally, [16–18] reported a correlation between the molecular diffusion coefficient, pressure, and temperature to obtain accurate diffusivity at conditions relevant to EGR as follows:

$$D = \frac{\left(-4.3844 \times 10^{-13} p + 8.55440 \times 10^{-11}\right) T^{1.75}}{p}, \qquad (6)$$

where D (m^2/s) is the molecular diffusion coefficient of CO_2 in CH_4 pressure p (MPa) and at temperature T (K).

3. Results and Discussion

The core flooding experiments were carried out at 1300 psig, and 50 °C in both vertical and horizontal orientations for each core sample. The interstitial velocities were varied according to individual core samples given their different porosities. A range of interstitial velocities was based on the injection rates in our previous work [8], which were evaluated using Equation (7):

$$u = \frac{Q}{\phi A}, \qquad (7)$$

where Q—injection rate in m^3/s, A—cross-sectional area of core sample, and φ—porosity of core sample. The results are shown in Table 4.

Table 4. Interstitial velocities employed in each core sample.

SN	Core Sample	Porosity (%)	Interstitial Velocity (μm/s)
1	Grey Berea	20.10	33.5–83.8
2	Bandera Grey	17.20	39.3–98.3
3	Buff Berea	26.27	26.2–65.6

Accordingly, the results of each individual core sample will be presented, discussed, and analysed based on the injection orientation and the interstitial velocity. *Grey Berea* core sample will be presented first, followed by *Bandera Grey* and, finally, *Buff Berea*.

The mole fractions of the injected CO_2 were used to develop the concentration profile which assesses rate of mixing between the injected CO_2 and the CH_4, in situ, using Equation (3) to fit the equation to the experimental data and varying K_L, which is the longitudinal dispersion coefficient (keeping the interstitial velocity constant as assumed in the 1D ADE), until the analytical solution fits the experimental results. The L_{exp} was also adjusted in the regression to provide a good fit as carried out by [16] and adopted by [17]. Least square regression analysis was the method used in the curve fitting process.

Curve fitting was carried out using OriginPro 8 software and the curve-fitted concentration profiles for each u for *Grey Berea* core sample are shown in Figures 3 and 4 for horizontal and vertical orientation respectively. The dispersion coefficients were evaluated, and it was shown that the K_L increases with increase in the interstitial velocity. This was done for all core samples and injection orientation experiments in this investigation. There was early breakthrough at higher values of u run and a late breakthrough at lower values as expected in all the core samples. The fitting of the 1D ADE to the experimental results was meagre as a result of systemics like entry and exit effects as described in the works of [17]. This was noted for all the runs in the entire experiments. However, these do not affect the evaluation of the parameter i.e., dispersion coefficient. For all subsequent experiments, these systemic effects were noticed and are presented as such.

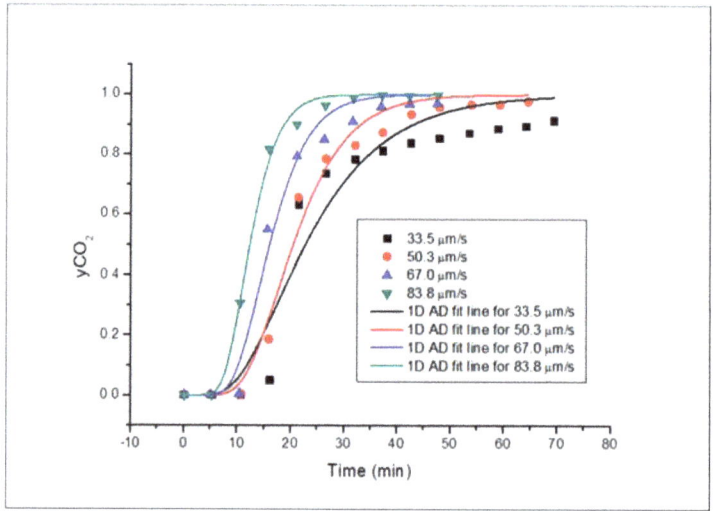

Figure 3. Concentration profile for Grey Berea in horizontal orientation.

Using Equation (6), the diffusion coefficients, D, were evaluated at the experimental conditions. This is essential when describing the dispersivity, α, and the P_{ex} of the core sample, and also when comparing the results to those in literature, which will further reaffirm the accuracy of the experiments.

The dispersivity can be analytically evaluated by fitting Equation (5) to the plot of u/D against k/D which is a straight line as shown in Figure 4.

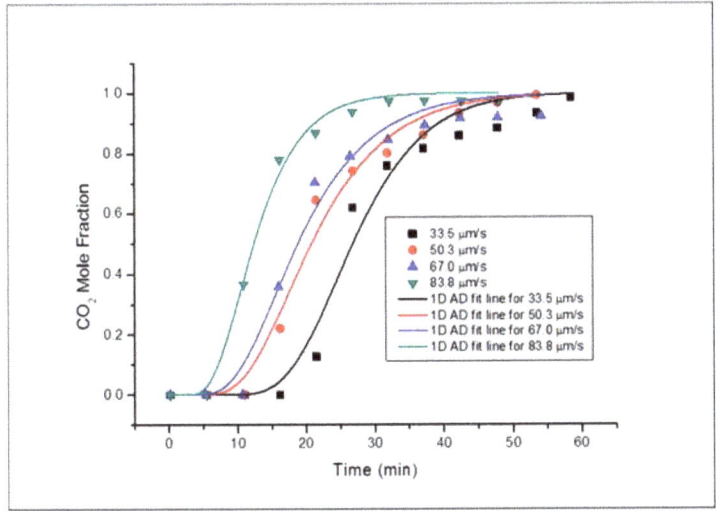

Figure 4. Concentration profile for all Grey Berea runs vertical orientation.

The fitted concentration profiles of the subsequent core sample (*Buff Bera*) is shown in Figures 5 and 6, for horizontal and vertical orientation respectively, which also showed meagre fitting as a result of systemic errors as seen in *Grey Berea*.

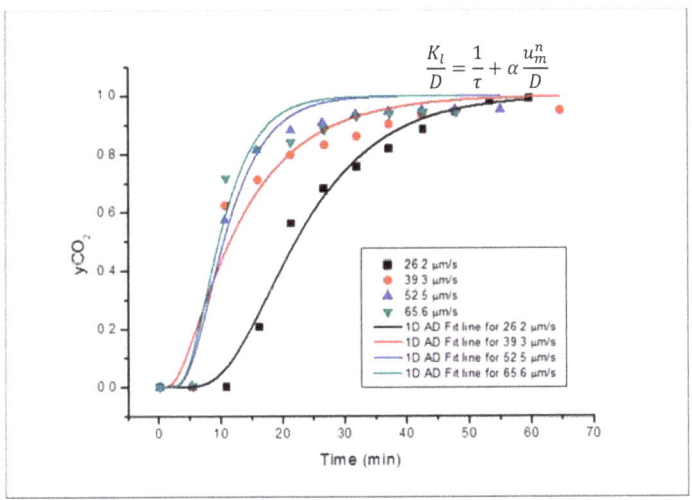

Figure 5. Concentration profile of all the runs for Buff Berea in horizontal orientation.

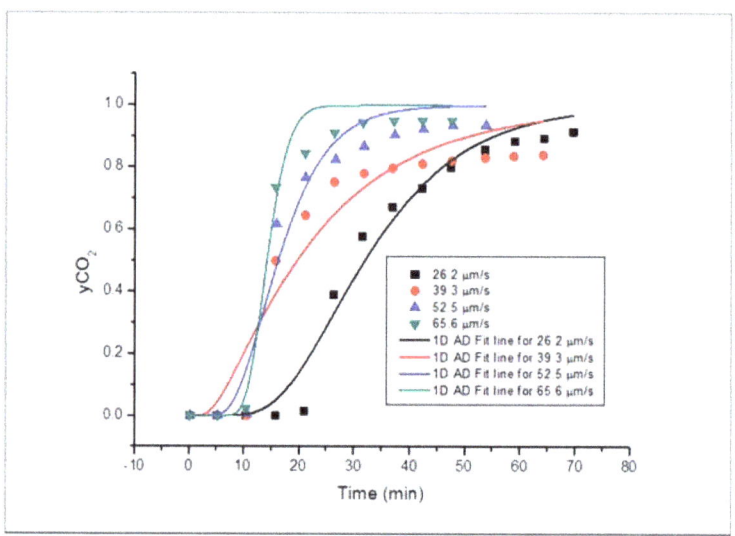

Figure 6. Concentration profile for all Buff Berea runs in vertical orientation.

Similarly, Bandera Grey core sample fitted horizontal and vertical orientation concentration profiles are shown in Figures 7 and 8. The profiles are steeper than the previous ones obtained for the core samples (Buff Berea and Grey Berea) because of the instant mixing during the displacement process due to higher interstitial velocities.

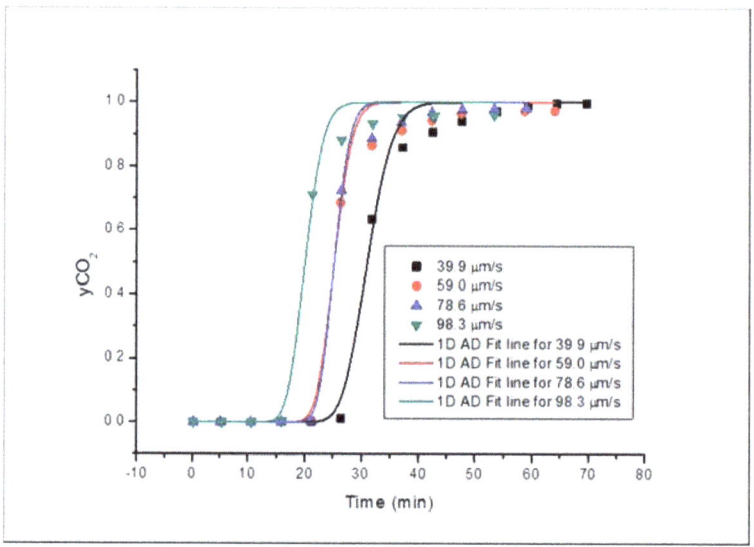

Figure 7. Concentration profile for Bandera Grey in horizontal orientation.

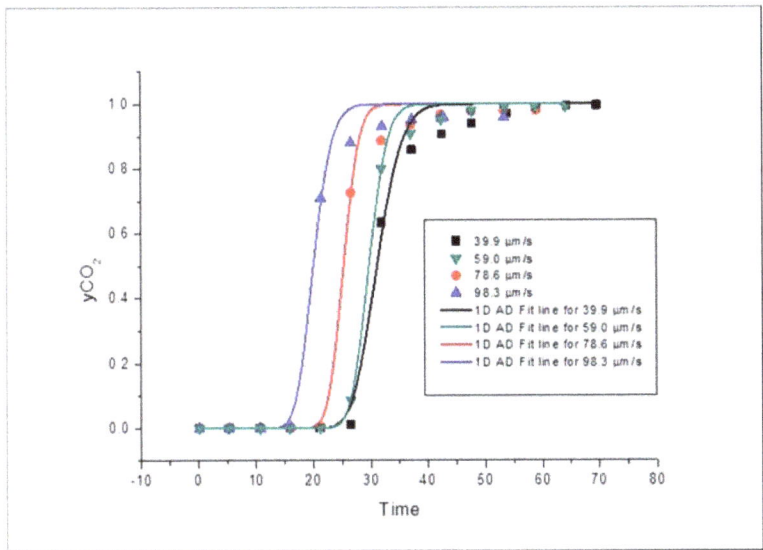

Figure 8. Concentration profile for Bandera Grey runs in vertical orientation.

The works of [16,18,19] presented values of the apparent dispersivity in consolidated porous media that were generally smaller than 0.01 ft (0.003 m). Hughes et al. (2012) obtained dispersivity in a range of 0.0001 to 0.0011 m using a core sample (Donny brook) with similar petrophysical properties as the ones used in this work. This provided a practical input variable for EGR simulations. As dispersivity is a very important porous media property, its accurate determination can provide a befitting technique to establish the optimum injection rate of the CO_2 for a better simulation of the fluid flow in the matrix of the reservoir during EGR. All the dispersivity obtained (from each experiment carried out in this study are well within those obtained from literature. This was evaluated from Equation (5) as earlier stated—a straight-line equation. This further demonstrates the reliability of the experiments.

The tortuosity (inverse of the intercept on the y-axis of the straight line) was similar in both cases (Figures 9–11), for *Grey Berea*, indicating that regardless of the injection orientation of the core sample, the tortuosity (a property of the core sample) remained unchanged. This shows that core orientation does not alter the pathways of the matrix of the porous media when fluids traverse through the porous medium, further attributing the fluid behaviour to mainly a function of the fluid properties and not the porous media. This was also true for *Buff Berea* depicted in Figures 12 and 13, and a comparison between the vertical and horizontal dispersivity is shown in Figure 14.

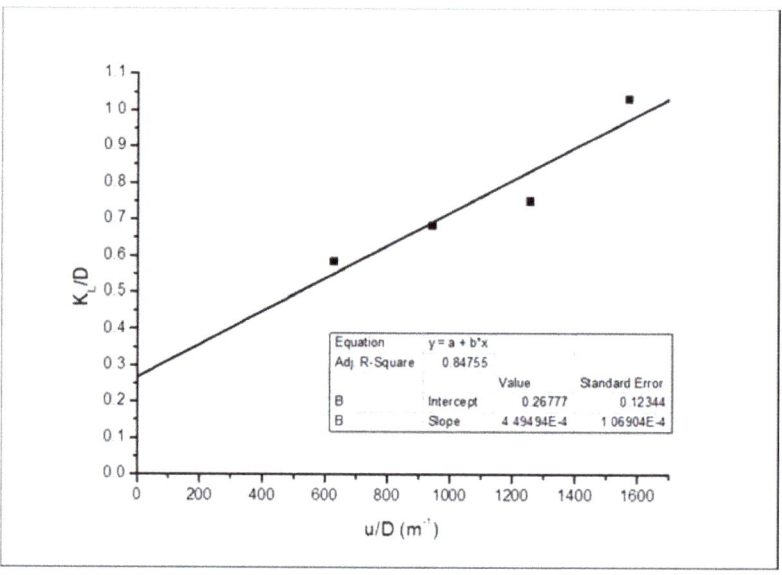

Figure 9. Dispersion to diffusion coefficient ratio against interstitial velocity for Grey Berea (vertical).

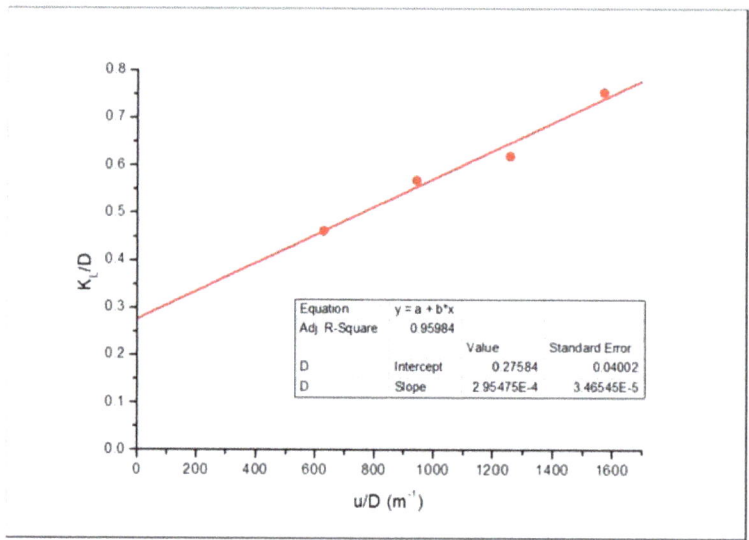

Figure 10. Dispersivity evaluation using the dispersion coefficient and interstitial velocity for Grey Berea (horizontal).

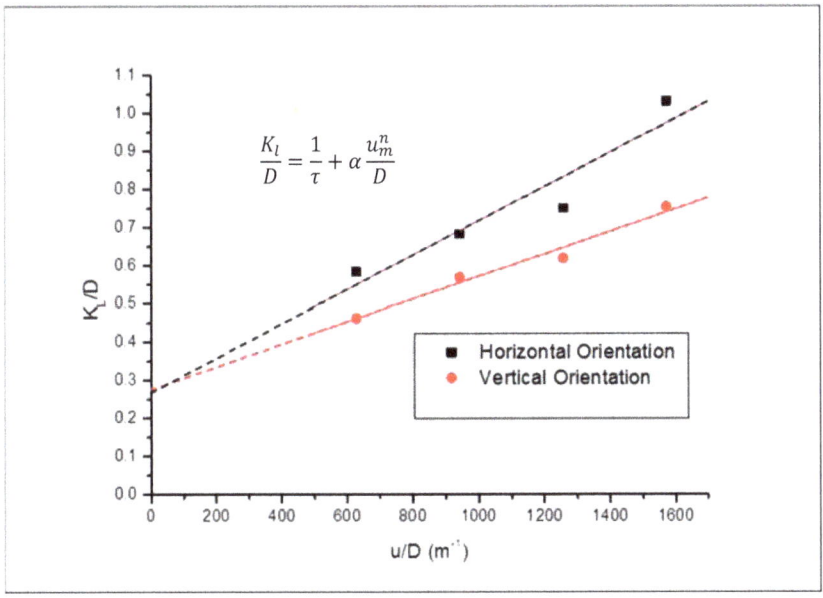

Figure 11. Comparison of the dispersivities in both orientations for Grey Berea.

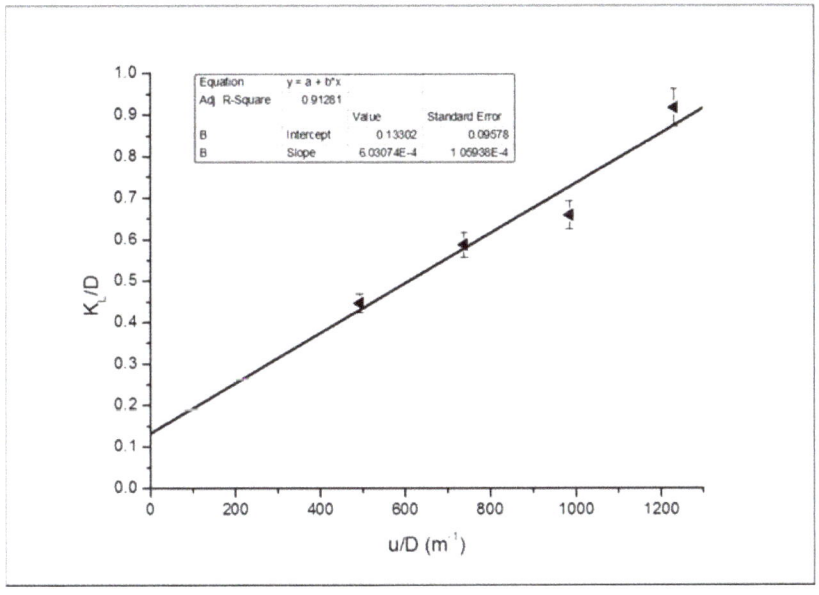

Figure 12. Dispersivity evaluation for Buff Berea in the horizontal orientation.

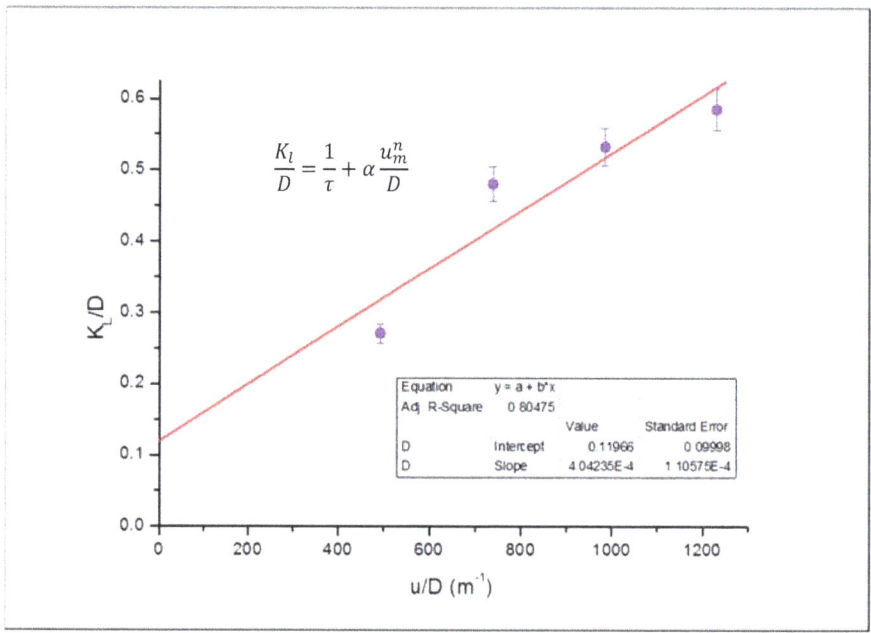

Figure 13. Dispersivity evaluation for Buff Berea in the vertical orientation.

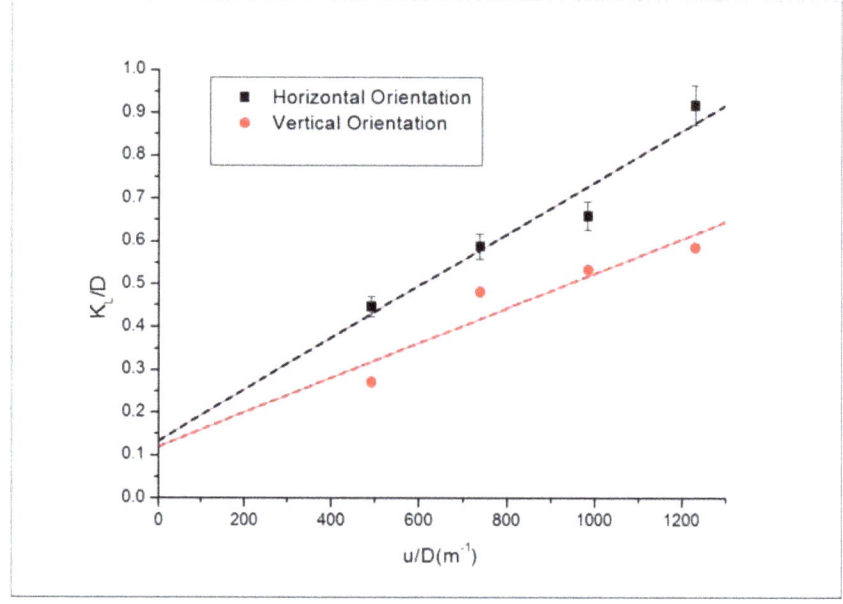

Figure 14. Comparison of the dispersivities in both orientations for Buff Berea.

Bandera Grey core sample exhibited similar trend as the previous core samples discussed in terms of dispersivity as shown in Figures 15–17. Dispersivity, however, is highest in *Bandera Grey* compared to the other core samples. This can be attributed to the grain arrangement of the core sample and the structure of the pore matrix which is tightly packed with its characteristic low permeability.

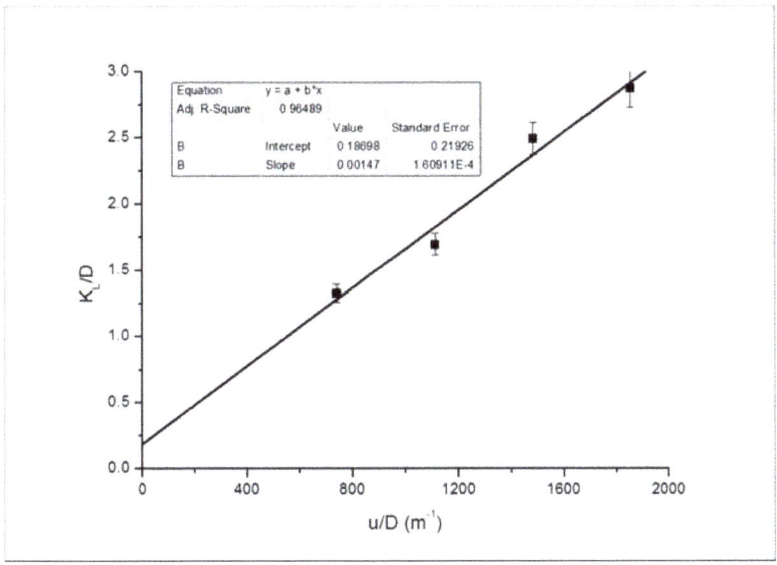

Figure 15. Dispersivity evaluation for Bandera Grey in the horizontal orientation.

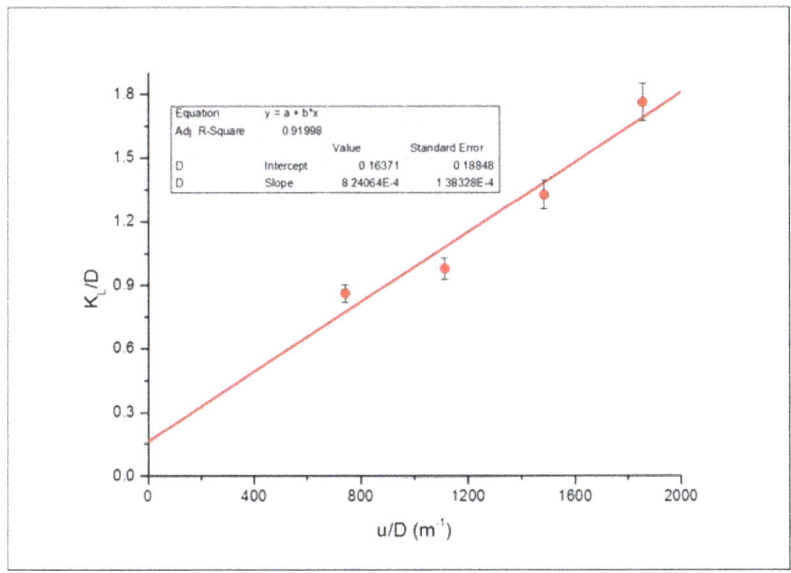

Figure 16. Dispersivity evaluation for Bandera Grey in the vertical orientation.

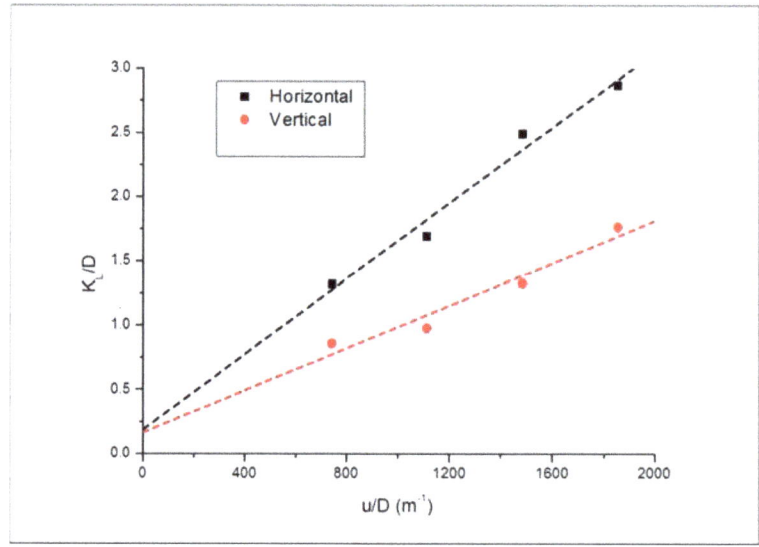

Figure 17. Comparison of the dispersivities in orientations for Bandera Grey.

There certainly are similarities between the vertical and horizontal orientations in all the core samples used in this study in terms of dispersion coefficient. This discussion is highlighted in the next section, which summarises the dispersion coefficient variation with the permeability.

Summary of the Dispersion Coefficient Investigation

A summary of the dispersion coefficients for all the core samples is shown in Table 5.

Table 5. Summary of all dispersion coefficients.

Core Sample	U (μm/s)	K_L (10^{-8} m^2/s)	
		Horizontal	Vertical
Bandera (k = 16.08 md)	39.3	7.01	4.56
	59.0	8.97	5.19
	78.6	13.20	7.03
	98.3	15.23	9.35
Grey Berea (k = 217.04 md)	33.5	3.11	2.46
	50.3	3.64	3.03
	67.0	4.01	3.70
	83.8	5.51	4.02
Buff Berea (k = 560.63 md)	26.2	2.38	1.44
	39.3	3.13	2.56
	52.5	3.51	2.84
	65.6	4.89	3.12

Dispersion coefficient generally decreases with increase in permeability as seen in Table 5. Hence, the core sample with the least permeability (*Bandera Grey*) showed a significantly higher dispersion coefficient. Another realisation from the Table is that in all the runs, those in the horizontal orientation appear to have the higher dispersion coefficient compared to the vertical counterparts. This can be attributed to the effect of gravity on the CO_2 as it traverses the core sample. Also, since the interstitial velocity is a function of porosity, the core sample with the most porosity had the lowest interstitial velocity and, hence, a lower dispersion coefficient at lower injection rates. The dispersion coefficient increases significantly at higher injection rates in all the runs regardless of the orientation. Furthermore, in the higher permeability core sample, the interaction between the injected CO_2 and the nascent CH_4 was limited as the CO_2 interstitial velocity was lower and, thus, the dispersion coefficient was lower compared to the subsequent core samples. The opposite was realised in the low permeability sample with the highest interstitial velocity was characterised by higher agitation of the gas molecules within the porous medium which eventually led to higher interaction and mixing and of course higher dispersion coefficient—the rate of mixing. Permeability is one of the vast factors that influence dispersion.

The dispersivity also increases with increase in permeability. This being a function of the core sample is evident with this trend, as the absolute permeability of the core sample is also a property of the core sample. Basically, the higher the permeability of the core sample, the higher the rate of mixing when CO_2 is injected to displace CH_4. However, dispersivity is scale-dependent [20,21] and, albeit being at laboratory scale, this finding is an indication of the effects of the petrophysical properties on the mixing taking place during EGR. Finding the right injection scenario is vital in achieving the best recovery efficiency whilst storing substantial volumes of CO_2.

4. Conclusions

The effects of permeability on the dispersion coefficient have been shown to be significant during the displacement of CH_4 by supercritical CO_2. This effect was more pronounced in the horizontal orientation compared to the vertical orientation which was attributed to the gravity effects on the supercritical CO_2 as it traverses the core sample. The magnitude difference in the dispersion coefficients was about 20%–30% higher in the horizontal orientation compared to the vertical orientation. Bandera Grey, with the lowest permeability and porosity, exhibited a wider contrast (40%–50%) between the dispersion coefficients in both orientations as a result of higher interstitial velocity and tortuosity of the core sample compared to the other core samples tested. Therefore, the effect of permeability and orientation on mixing between the displacing and displaced gas during EGR are vital. This study highlights the influence of the permeability variation and plume propagation orientation on the mixing between CO_2 and CH_4 during EGR. It provides reservoir engineers with an insight into characterising gas systems during the design of the EGR technology. The injection rates, injection pressures, and dip angle of injection from the injector to the producer can be evaluated accurately during simulation studies prior to field scale application and implementation. Inclusion of such hysteresis will provide a better representation of the injection process given that the main reason why EGR has not been widely adopted is the its economic drawback brought on by incessant mixing and contamination of the recovered CH_4.

Author Contributions: Conceptualization, M.K.A. and A.J.A.; Methodology, M.K.A.; Data software tools analysis, A.A.-O.; Investigation, M.K.A.; writing—original draft preparation, M.K.A.; writing—review and editing, M.B.; visualization, A.A.-O.; supervision, G.G.N.

Funding: This research received no external funding.

Acknowledgments: The authors would like to show gratitude to Petroleum Technology Development Fund Nigeria (PTDF) for the studentship.

Conflicts of Interest: The authors declare no conflict of interest.

References

1. Ziyarati, M.T.; Bahramifar, N.; Baghmisheh, G.; Younesi, H.; Younessi, H. Greenhouse gas emission estimation of flaring in a gas processing plant: Technique development. *Process. Saf. Environ. Prot.* **2019**, *123*, 289–298. [CrossRef]
2. Abas, N.; Kalair, A.; Khan, N. Review of fossil fuels and future energy technologies. *Futures* **2015**, *69*, 31–49. [CrossRef]
3. Wegener, M.; Amin, G.R. Minimizing Greenhouse Gas Emissions using Inverse DEA with an Application in Oil and Gas. *Expert Syst. Appl.* **2018**, *122*, 369–375. [CrossRef]

4. Vilcáez, J. Numerical modeling and simulation of microbial methanogenesis in geological CO_2 storage sites. *J. Pet. Sci. Eng.* **2015**, *135*, 583–595. [CrossRef]
5. Kalra, S.; Wu, X. CO_2 injection for Enhanced Gas Recovery. In Proceedings of the SPE Western North American and Rocky Mountain Joint Meeting, Denver, CO, USA, 17–18 April 2014; pp. 16–18.
6. Abba, M.K.; Abbas, A.J.; Nasr, G.G. Enhanced Gas Recovery by CO_2 Injection and Sequestration: Effect of Connate Water Salinity on Displacement Efficiency. In Proceedings of the SPE Abu Dhabi International Petroleum Exhibition & Conference, Abu Dhabi, UAE, 13–16 November 2017.
7. Abba, M.K.; Al-Othaibi, A.; Abbas, A.J.; Nasr, G.G. Effects of gravity on flow behaviour of supercritical CO2 during enhanced gas recovery and sequestration. In Proceedings of the Fifth CO2 Geological Storage Workshop, Utrecht, The Netherlands, 21–23 November 2018.
8. Abba, M.K.; Al-Othaibi, A.; Abbas, A.J.; Nasr, G.G.; Mukhtar, A. Experimental investigation on the impact of connate water salinity on dispersion coefficient in consolidated rocks cores during Enhanced Gas Recovery by CO_2 injection. *J. Nat. Gas Sci. Eng.* **2018**, *60*, 190–201. [CrossRef]
9. Abba, M.K.; Abbas, A.J.; Al-Othaibi, A.; Nasr, G.G. Enhanced Gas Recovery by CO_2 Injection and Sequestration: Effects of Temperature, Vertical and Horizontal Orientations on Dispersion Coefficient. In Proceedings of the Abu Dhabi International Petroleum Exhibition & Conference, Abu Dhabi, UAE, 12–15 November 2018.
10. Liu, S.; Song, Y.; Zhao, C.; Zhang, Y.; Lv, P.; Jiang, L.; Liu, Y.; Zhao, Y. The horizontal dispersion properties of CO_2-CH4in sand packs with CO2displacing the simulated natural gas. *J. Nat. Gas Sci. Eng.* **2018**, *50*, 293–300. [CrossRef]
11. Newberg, M.; Foh, S. Measurement of Longitudinal Dispersion Coefficients for Gas Flowing Through Porous Media. In Proceedings of the SPE Gas Technology Symposium, Dallas, TX, USA, 13–15 June 1988; pp. 5–9.
12. Mamora, D.D.; Seo, J.G. Enhanced Recovery by Carbon Dioxide Sequestration in Depleted Gas Reservoirs. In Proceedings of the SPE Annual Technical Conference and Exhibition, San Antonio, TX, USA, 29 September–2 October 2002; pp. 1–9.
13. Nogueira, M.; Mamora, D.D. Effect of Flue-Gas Impurities on the Process of Injection and Storage of CO_2 in Depleted Gas Reservoirs. *J. Energy Resour. Technol.* **2005**, *130*, 013301. [CrossRef]
14. Perkins, T.; Johnston, O. A Review of Diffusion and Dispersion in Porous Media. *Soc. Pet. Eng. J.* **1963**, *3*, 70–84. [CrossRef]
15. Coats, K.H.; Whitson, C.H.; Thomas, K. Modeling Conformance as Dispersion. *SPE Reserv. Eval. Eng.* **2009**, *12*, 33–47. [CrossRef]
16. Hughes, T.J.; Honari, A.; Graham, B.F.; Chauhan, A.S.; Johns, M.L.; May, E.F. CO_2 sequestration for enhanced gas recovery: New measurements of supercritical CO2–CH4 dispersion in porous media and a review of recent research. *Int. J. Greenh. Gas Control.* **2012**, *9*, 457–468. [CrossRef]
17. Liu, S.; Zhang, Y.; Xing, W.; Jian, W.; Liu, Z.; Li, T.; Song, Y. Laboratory experiment of CO2-CH4 displacement and dispersion in sandpacks in enhanced gas recovery. *J. Nat. Gas Sci. Eng.* **2015**, *26*, 1585–1594. [CrossRef]
18. Honari, A.; Hughes, T.J.; Fridjonsson, E.O.; Johns, M.L.; May, E.F. Dispersion of supercritical CO2 and CH4 in consolidated porous media for enhanced gas recovery simulations. *Int. J. Greenh. Gas Control* **2013**, *19*, 234–242. [CrossRef]
19. Coats, K.H.; Whitson, C.H. SPE 90390 Modeling Conformance as Dispersion. In Proceedings of the SPE Annual Technical Conference and Exhibition, Houston, TX, USA, 26–29 September 2004.
20. Bjerg, P. *Dispersion in Aquifers*; DTU Environment: Lyngby, Denmark, 2008.
21. Schulze-Makuch, D. Longitudinal dispersivity data and implications for scaling behaviour. *Groundwater* **2005**, *43*, 443–456. [CrossRef] [PubMed]

© 2019 by the authors. Licensee MDPI, Basel, Switzerland. This article is an open access article distributed under the terms and conditions of the Creative Commons Attribution (CC BY) license (http://creativecommons.org/licenses/by/4.0/).

Article

Correlational Analytical Characterization of Energy Dissipation-Liberation and Acoustic Emission during Coal and Rock Fracture Inducing by Underground Coal Excavation

Pengfei Shan *, Xingping Lai and Xiaoming Liu

School of Energy and Mining Engineering, Xi'an University of Science and Technology, Xi'an 710054, China; laixp@xust.edu.cn (X.L.); liuxiaoming@stu.xust.edu.cn (X.L.)
* Correspondence: shanpengfei@xust.edu.cn; Tel.: +86-1377-2190-561

Received: 4 April 2019; Accepted: 12 June 2019; Published: 20 June 2019

Abstract: This paper uses an acoustic emission (AE) test to examine the energy dissipation and liberation of coal and rock fracture due to underground coal excavation. Many dynamic failure events are frequently observed due to underground coal excavation. To establish the quantitative relationship between the dissipated energy and AE energy parameters, the coal and rock fracturing characteristics were clearly observed. A testing method to analyze the stage traits and energy release mechanism from damage to fracture of the unloading coal and rock under uniaxial compressive loading is presented. The research results showed that the relevant mechanical parameter discreteness was too large because the internal structures of the coal and rock were divided into multiple structural units (MSU) by a few main cracks. The AE test was categorized into four stages based on both the axial stress and AE event parameters: initial loading stage, elastic stage, micro-fracturing stage, and post-peak fracturing stage. The coal and rock samples exhibited minimum (maximum) U values of 60.44 J (106.41 J) and 321.19 J (820.87 J), respectively. A theoretical model of the dissipation energy during sample fracturing based on the AE event energy parameters was offered. The U decreased following an increase in $\Sigma E_{AE-II}/\Sigma E_{AE}$.

Keywords: Coal excavation; coal and rock fracture; multiple structural units (MSU); energy dissipation; AE energy

---

1. Introduction

An increasing number of coal mines exhibit deep-excavation status and require consideration attention due to the presence of mining problems that must be dealt with for increasing underground coal excavation depth increment [1]. More dynamic failure events, including rock burst in both the tunneling process and coal extraction, coal and gas outbursts, and mining-induced earthquakes, frequently occur and require the accurate prediction of all impending dynamic hazards [2]. A large number of engineering tests and experiments have indicated that coal fracture involves a sequence of micro- and macro-scale events, including initial crack deformation and propagation as well as macroscopic crack formation and propagation [3]. In addition, the dynamic failure of the coal is closely related to the mechanical mechanism of the coal under the mining unloading condition. Given the complexity of rock mechanics, research on the dynamic failure should emphasize energy concentration, storage, dissipation, and liberation. Many scholars have characterized the energy dissipation and liberation rules of void coal fracture using external loading. Zhao [4] proposed the minimum energy principle of the dynamic damage of the rock and indicated that authentic energy consumption during the dynamic failure was equal to the failure energy under the uniaxial stress status. Xie [5] indicated that rock deformation and damage combined the energy dissipation and liberation results. Xie also indicated the strength loss criterion based on the energy dissipation and the global failure criterion based on releasable strain energy. According to the dissipated energy by cyclic loading, Jin [6] offered a theoretical calculation equation for the damage variable and the determination of the damage threshold from the perspective of the material damage variable defined by the energy dissipation. Li [7] applied damage to the rock crushing stage and used a fatigue damage iterative relational expression under the action of stress waves to the rock post-destruction stage to generate a quantitative relationship between the impact energy, rock damage, and fragmentation distribution. Li [8] established the energy identification criterion of the mesoscopic rock failure by using the strain energy density theory. The criterion was also applied in the failure simulation of the Brazil split test and intermediate crack tensile test.

Previous research [9,10] has led to remarkable achievements in three aspects: (i) establishing the relationship between the rock unit damage and energy dissipation from a microscopic perspective; (ii) determining the rock constitutive relation based on energy dissipation; and (iii) understanding the failure and fracturing of the rock based on the releasable strain energy. However, few experimental studies on the quantitative rules of energy dissipation and the liberation of the coal and rock have accounted for the excavation conditions have been conducted.

As a specific monitoring method, the acoustic emission (AE) method [11–16] has obvious advantages in studying both the damage-fracturing characteristics and energy dissipation-liberation of the coal and rock under complex the excavation conditions. Tang [17] established a hypothesis and a frame of a numerical simulation based on the rock AE regularity, and offered the distribution of the temporal-spatial sequence. With the deep rock burst process simulation system, He [18] examined the AE waveform and frequency traits of limestone rock burst under the true triaxial unloading condition. Feng [19] investigated the chemical erosion characteristics of rock fracturing in various rock AE tests with different stress settings, the results of which revealed the mechanisms of the rock physical-mechanical properties and crack propagation at different stress settings and chemical circumstances. Ji [20] offered the AE signal frequency characteristics under various rock-fracturing stages based on granite uniaxial compressed testing. By using a AE monitoring system, Zhao [21] studied the granite failure with prefabricated cracks under uniaxial compressed loading, thereby quantitatively establishing the three-dimensional evolution model of the internal microcrack and the regularity of the AE events with respect to both the loading time and its values.

Previous related research focused on AE regularity during rock fracturing under diverse loading conditions, the Kaiser effect theory, and the AE application. From the perspective of macroscopic energy conservation, however, a correlation must be established between the energy dissipation of the coal and rock, and the accumulated values of the periodical acoustic wave energy during the coal and rock fracturing to generate a mechanical mechanism on the degree of coal and rock fracturing.

In this paper, an AE test was developed to examine energy dissipation and liberation of coal and rock fracture under complex excavation conditions to generate a quantitative regularity between the dissipated energy of the coal and rock and the AE energy parameters. This regularity would not only characterize the coal and rock fracturing characteristics, but also would uncover the catastrophic mechanism of coal and rock fracturing. The results can provide a theoretical basis for the design and construction of rock engineering excavation in deep well circumstances.

2. Test Scheme

2.1. Crustal Stress Setting and Physical Property Analysis

The Wudong colliery in the Urumchi coal field served as the study area. The Urumchi coal field is located in northern part of Tianshan Mountain, and there are 33 available coal-bearing strata there from the Xishanyao group of the Jurassic Period (Figure 1). The total thickness of the coal seams in the study area was 50 m, and the angle ranged from 65° to 87°, with an average angle of about 86°. Table 1 introduces the roof and floor values of the seams in B_{3-6}. Figure 2 shows the thicknesses and spans of the seams from B_3 to B_6. Thereafter, there were many interbeds in B_{4-5}, and their thicknesses were between 0.15 and 0.20 m. The petrographic compositions of B_{3-6} are also shown in Table 1. Over 95% of maceral composition of the coal are Vitrinite and Inertinite, so microlithotype of the coal is vitrinertite. On account of the vitrinertite and some pyrite being associated with the coal seams, the coal has a short spontaneous combustion cycle in the study area. However, both coal dust and methane pose low explosive hazards.

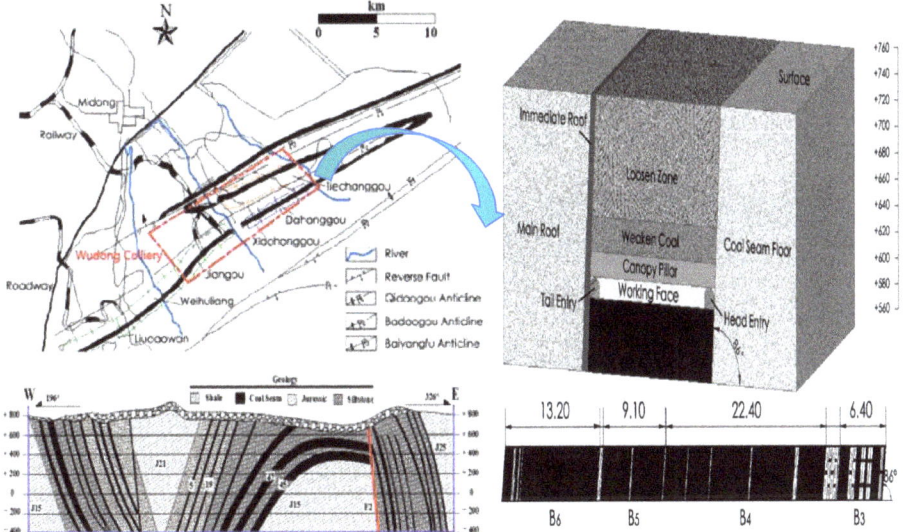

Figure 1. Profiles of the study area including the simplified geological conditions, sectional map with main lithostratigraphic units, and the coal excavation settings.

Table 1. Introductions to petrographic composition and roof and floor of B_{3-6} of study area.

Name	Lithology	Thickness (m)	Description
Main Roof	Shale	1.90	hard, dark grey, and lamellar
Immediate Roof	Siltstones	0.70	Lamellar and joint obviously
False Roof	Shale	0.10	Soft and joint obviously
Coal Seam Floor	Shale	0.55	Joint and fragile

Serial number	Maceral composition(%)					
	Vitrinite	Average	Inertinite	Average	Liptinite	Average
B_{3-6}	33.50–62.60	46.60	37.40–65.60	52.80	0.00–1.90	0.60
	Organic matter(%)		Inorganic matter(%)		Vitrinite reflectance(%)	Microlithotype
	86.35		13.65		0.67	vitrinertite

All the presented coal and rock samples in this paper were collected from the southern lane of the Wudong coal mine in the Urumchi coal field, specifically the top coal caving face at the No. 45 coal seam at the +500 m level. The KJ743 coal mine geostress monitoring system was used to collect real-time stress data. Figure 2a presents a plot of the stress data from November 11, 2014 to February 2, 2015. The vertical geostress remained fairly constant with small fluctuations within a narrow range due to the excavation disturbances in the area. The value of the vertical geostress varied from 6.8 MPa to 7.0 MPa. The distributions of the maximum and minimum principal horizontal stress are presented in Figure 2b,c. In particular, the minimum principal horizontal stress varied from 7.6 MPa to 8.4 MPa. From 340 m to 400 m, the minimum principal horizontal stress increased slightly due to a disturbance caused by the excavation activities. The maximum principal horizontal stress exhibited a uniform distribution. The maximum principal horizontal stress remained constant at 10.0 MPa, except locally at 480 m and 250 m, where it increased to 13.0 MPa and 28.8 MPa, respectively.

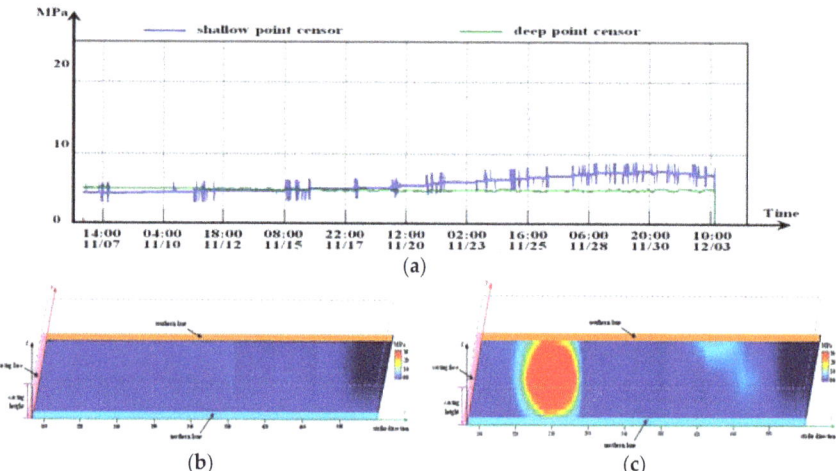

Figure 2. Results of the in-situ monitoring of the regional geostress: (**a**) vertical geostress; (**b**) minimum principal horizontal stress; (**c**) maximum principal horizontal stress.

Both the axial and radial stresses of the coal were zero following in-situ sampling. Following sampling, the samples were immediately returned to the Laboratory of Western Mines and Hazard Prevention, Ministry of Education of China for preparation with the plastic package. The rock was composed of siltstone with a small quantity of mudstone and argillaceous sandstone. The rock exhibited a relatively uniform particle size that varied from 0.30 to 0.90 mm with an average particle size of 0.70 mm. The fracture of the coal was like fine granulated sugar. The coal sample surface was fresh with a few apparent protogenetic cracks. The coal sample was prepared in strict accordance with the IRTM method recommended by the International Society for Rock Mechanics (ISRM) and the samples were cubes with side lengths equal to 60 mm following cutting in triplicate [22]. The surface was carefully ground without remarkable flaws. Moreover, both the nonparallelism and non-perpendicularity errors were less than 0.05 mm. Figure 3 presents the quartile statistical results of geometric dimensioning in the three orientations of the coal. Although a slightly discrepancy was observed in the finished sizes, the mean value of error did not exceed 1.0 mm.

In the AE test, seven coal and rock samples were employed, and the serial numbers of the samples were respectively from LXMT-01 to LXMT-07 for the coal samples and from LXYT-01 to LXYT-07 for the rock samples. The variable coefficients of all the parameters were within 5%, thereby indicating that the physical properties of the samples were well-consistent and validating the applicability of the samples in the follow-up experiments (refer to the Appendix A).

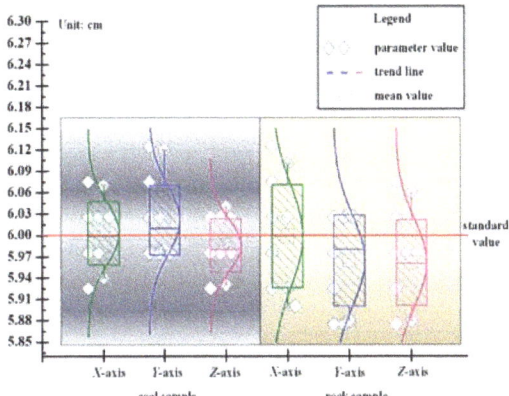

Figure 3. Stochastic statistics of the geometric dimension quartile of coal and rock samples.

2.2. Experimental Facility and Monitoring Technique

Figure 4 shows the AE testing scheme of coal and rock fracturing with uniaxial compression. The WE-10 rock mechanics testing machine was used in the experiment for both the AE and energy dissipation regularities during the coal fracturing process. The loading mode was the displacement control at a loading velocity of 0.1 mm/min, which complied with a series of ISRM experimental procedures. The prepared samples with the No. 40Cr alloy steel rigid cushions had a side length of 70 mm and a thickness of 10 mm. The top and bottom were placed in the testing machine. In the axial orientation, the sample was loaded until it exhibited failure due to axial unloading. The real-time loading values and the corresponding axial deformation were automatically acquired from the testing machine. This paper offers an SDAES7.5 AE instrument to record the acoustic wave energy in various stages during sample fracturing. All the energy data was applied to characterize the energy dissipation and liberation mechanism of the coal. The AE sensor was coupled with the sample using Vaseline. To minimize the interference of the external acoustic sources, the AE parameters were determined after adjusting the threshold value several times. In the experiment, real-time AE monitoring was employed to monitor synchronization with the external loading.

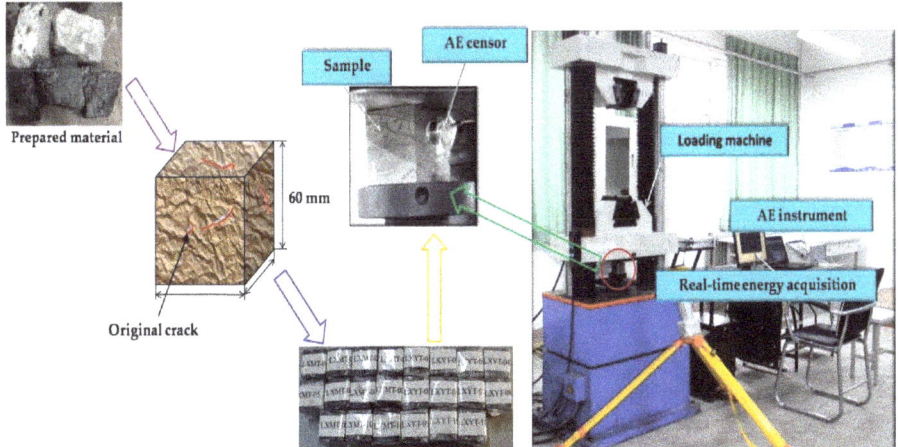

Figure 4. The acoustic emission testing scheme of coal and rock fracturing with uniaxial compression.

3. Coal Fracturing Mechanism

The characteristics of progressive failure were obvious with typical nonlinear deformation characteristics, especially the rock sample [23–27]. The basic mechanical parameters of all the samples are presented in Table 2. Here, σ_p and σ_r represent the peak and residual strengths, respectively. ε_p represents the strain value corresponding to the peak strength. E_t and E_b represent the elastic and deformation moduli, respectively. In particular, E_t is the average slope of the approximate linear part in the curves. In addition, E_b represents the stress-strain ratio at the point where the stress value was 50% that of the peak strength. All the presented strain values are the means of the experimental values. According to the Appendix A, the basic mechanical parameters exhibited large discreteness values due to a significant difference in the internal sample structures of the various sampling positions.

Table 2. Profile of the mechanical parameters and discreteness of the unloading coals.

Sample Serial Number	σ_p (MPa)	σ_r (MPa)	ε_p (10^{-3})	E_t (MPa)	E_b (MPa)	$d\varepsilon/dt$ (10^{-3}/s)
LXMT-01	18.57	11.53	98.30	595	110	
LXMT-02	14.37	7.55	85.75	956	91	
LXMT-03	12.45	6.44	69.60	410	122	
LXMT-04	11.74	6.11	52.62	660	142	0.56
LXMT-05	9.72	4.91	78.47	518	73	
LXMT-06	17.26	8.87	68.20	682	148	
LXMT-07	17.99	12.61	96.42	780	109	
LXYT-01	25.36	12.52	72.30	877	231	
LXYT-02	15.48	9.94	70.40	727	128	
LXYT-03	19.05	13.68	39.96	590	330	
LXYT-04	12.43	10.81	40.08	440	259	0.55
LXYT-05	37.26	18.48	34.41	764	668	
LXYT-06	43.13	24.97	54.83	716	525	
LXYT-07	39.03	19.87	52.50	689	622	
Lithology	Discrete parameter	σ_p	σ_r	ε_p	E_t	E_b	
Coal	$\overline{X}$	14.59	8.29	78.48	657	114	
	E	3.44	2.88	16.44	178	27	
	ζ	23.578	34.741	20.948	27.093	23.684	
Rock	$\overline{X}$	27.39	15.75	52.07	686	395	
	E	12.38	5.51	15.03	138	210	
	ζ	45.199	34.984	28.865	20.117	53.165	

As compared to normal isotropic materials, the internal structure of the coal and rock was divided into multiple structural units (MSU) by a few main cracks, which significantly affected the sample strength. The stress-strain curves exhibited a smooth monotonic increase at the onset of the AE test. The loading value of each MSU during the AE test ceaselessly exceeded its failure strength limit following an increase in the external loading. Therefore, essential differences were observed in the mechanical response characteristics and deformation characteristics of the coal and rock as compared to those of other materials. When the curves exhibited a smooth monotonic increase, the main cracks in both the coal and rock samples started to close due to the influence of external loading. Following a continuous increase in the external loading, the MSU with minimum of failure strength limit presented a yielding state, thereby generating a rapid decrease in the bearing capacity of the MSU until the bearing capacity was equal to zero, which was the threshold of the nonlinear constitutive relationship of the coal and rock.

A further increase in the axial loading resulted in the persistent destruction of the other MSU. The failure of samples was presented as progressive failure. When the external loading exceeded the maximum bearing capacity, the samples exhibited macroscopic fracturing and destabilization. We defined the MSU with a maximum bearing capacity as the primary MSU, of which the primary MSU exhibited the greatest influence on the characteristics of the coal and rock fracturing. Moreover, the bearing capacity of the primary MSU was equal to the peak strength in the stress-strain curve. Following the destruction of the primary MSU, the axial stress level of the samples kept decreasing and the MSU with minimum bearing capacity initially exhibited an integral yield. The other MSU then started to release without yielding phenomena. In conclusion, the post peak-point exhibited a weakened yield in the local area of the coal and rock, which were defined as the local distortion characteristics of the coal and rock. Therefore, the strain value after the peak point should only consider the plastic strain of the local yielding weakening region.

The yield, damage, and failure of the coal and rock essentially exhibited energy dissipation and liberation [28–30]. The energy being dissipated with the sample failure, U, may be defined as the work done to the coal by the mechanical testing machines under continuous axial loading condition, as defined by Equation (1) as follows:

$$U = \int F du = SL \int \sigma d\varepsilon = L^3 \Psi = V\Psi \tag{1}$$

where L, S, and V represent the side length, bottom area, and volume, respectively; and Ψ represents the dissipated energy per unit volume of the samples, where Ψ is equal to the corresponding area of the stress-strain curve and has a unit of MJ/m^3. Figure 5 presents the quantitative relation of the AE event parameters between the axial stress and time sequence. The whole AE test was categorized into four stages based on both the axial stress and AE event parameters:

(1) Initial loading stage (I). The stress-time curve exhibited a horizontal trend at the stage, and dU/dt, wherein the gaining rate of dissipated energy exhibited an obvious increase. The internal original cracks began to close under the loading condition and the mechanical characteristics tended to exhibit a quasi-isotropic status. The coal and rock samples exhibited stage durations of 25–30 s and 10–18 s, because the internal structure of the rock sample was simpler with fewer cracks and a structural plane. The AE counting number was smallest in all the stages and the Kaiser point of the partial coal samples was observed at this stage.

(2) Elastic stage (II). The samples were regarded as elastic medium at the beginning, and the σ-t curve exhibited an approximately linear trend at this stage. dU/dt was maintained steady on the whole. The coal and rock samples exhibited roughly similar stage durations of 25 s and 20 s, respectively. The samples started to transform from an elastomeric to elastic-plastic material following a gradual increase in the axial loading. Some partial mutations were observed in the σ-t curve, especially the curve of the rock sample. The possible sudden instability of the main crack in the rock may have generated overall structural distortion. In addition, the AE counting number increased in each subsequent stage and most of the samples exhibited the Kaiser point during the elastic stage.

(3) Micro-fracturing stage (III). The samples were regarded as plastic media and the σ-t curve exhibited a concave shape with a greater slope. However, dU/dt exhibited an initial decrease in the micro-fracturing stage following an increase in the time sequence. This stage exhibited the longest duration, and the coal and rock samples exhibited durations of about 70 s and 30 s, respectively. The loading resistance of the coal sample to the axial loading was poor as compared to the rock sample and the axial loading was proportional to the time sequence. In addition, dU/dt was maintained steady. Due to the specific structural characteristics of the rock, the primary MSU played a dominating role in resisting the axial loading. In addition, the progressive failure was observed in the rock σ-t curve. Before and after the arrival of the peak point, the curve exhibited dentate fluctuation because the secondary MSU opposed the axial loading. The AE counting number increased sharply near the peak point and presented the maximum AE counting number.

(4) Post-peak fracturing stage (IV). The structures of the samples were completely destroyed. The σ-t curve dropped suddenly and then exhibited persistent fluctuation near the peak point. On behalf of the more intact structure of the rock sample, the residual stress of the rock sample was so large that the rock sample easily opposed the external loading. An obvious reduction in dU/dt was observed at the peak point. A shorter duration of approximately 15–25 s was observed. Large-scale

mesoscopic deformation and macroscopic fracturing was observed within a short time, and the AE counting number was observed in the last increment at a lower increment rate.

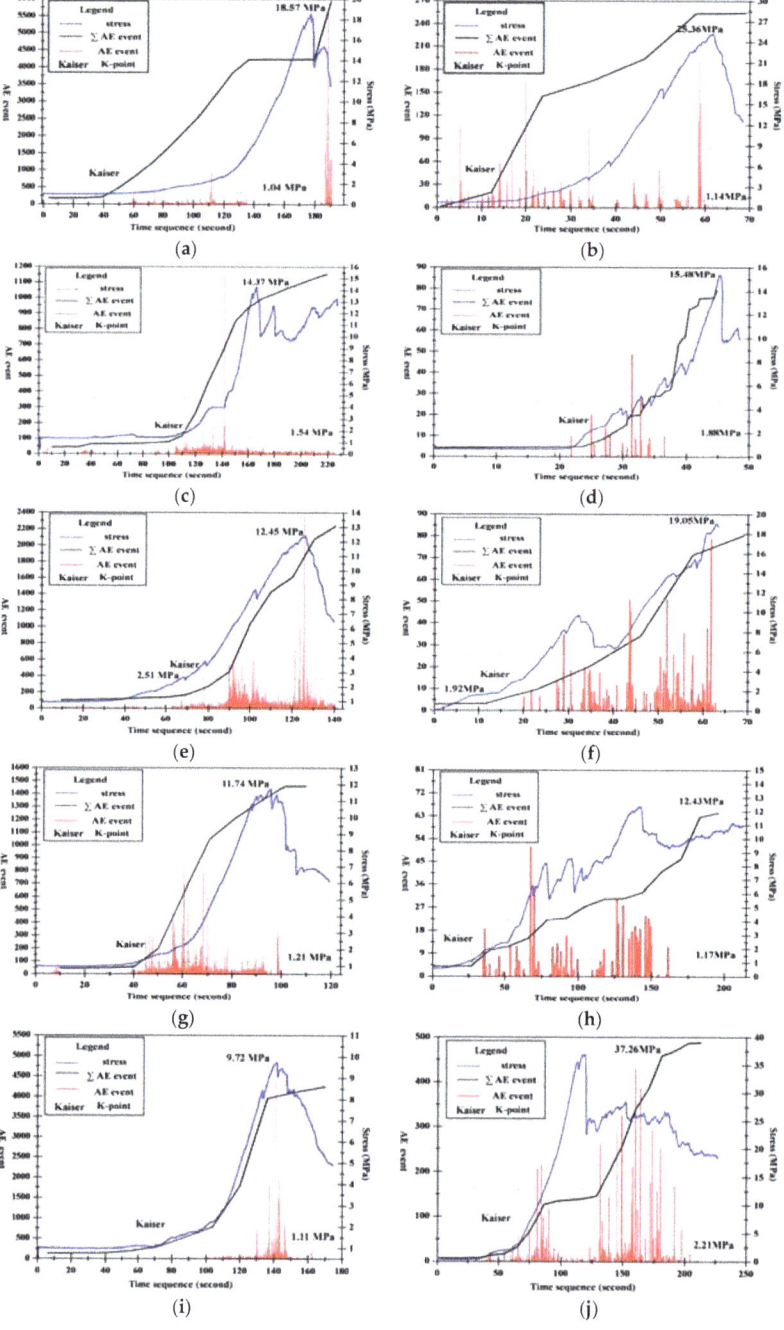

Figure 5. *Cont.*

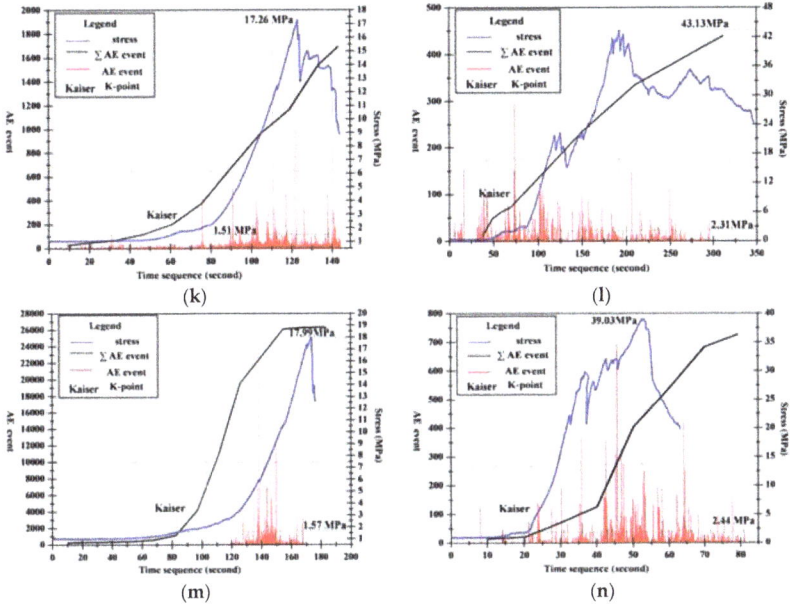

Figure 5. Plot for the time-based relation of the parameters, including the axial stress, AE counting number, and total counting number of coals: (**a**) LXMT-01; (**b**) LXYT-01; (**c**) LXMT-02; (**d**) LXYT-02; (**e**) LXMT-03; (**f**) LXYT-03; (**g**) LXMT-04; (**h**) LXYT-04; (**i**) LXMT-05; (**j**) LXYT-05; (**k**) LXMT-06; (**l**) LXYT-06; (**m**) LXMT-07; and (**n**) LXYT-07.

4. Analytical Discussion

Table 3 lists all the energy parameter data, including U, Ψ, K_t, and E_{AE}, wherein K_t represents the accumulated elastic strain energies when the axial stress was equal to the peak value, and K_t is equal to $\sigma_p^2/2E_t$. Here, E_t is defined as the elasticity modulus of the samples based on the description of the sample deformation trait. As an important mechanical index, K_t was usually provided to assess the dynamic failure events caused by underground coal excavation. Furthermore, E_{AE} represents the acoustic wave energy. According to analysis of the energy data in Table 2, the quantitative correlation between Ψ and $E_{AE-II}/\Sigma E_{AE}$ is defined as follows:

$$\Psi = K_t e^{a + \frac{E_{AE-II}}{\Sigma E_{AE}}/b} \tag{2}$$

where ΣE_{AE} defines the accumulated energy of the acoustic waves from the samples by AE (J); E_{AE-II} is the accumulated energy of the acoustic waves in the elastic stage (J); and $E_{AE-II}/\Sigma E_{AE}$ defines the percentage of the two above energy parameters ratios. In addition, a and b are both material parameters. Equation (2) was compiled as Equation (3). By means of the least square method, a linear relation between $\ln(\Psi/K_t)$ and $E_{AE-II}/\Sigma E_{AE}$ was established to define the material parameters.

$$\ln \frac{\Psi}{K_t} = a + \frac{E_{AE-II}/\Sigma E_{AE}}{b} \tag{3}$$

Table 3. Energy parameters of the coal-rock samples in continuous axial compressed testing.

Lithology	Sample Number	σ_{Kaiser} (MPa)	Ψ (MJ/m^3)	U (J)	K_t (MJ/m^3)	Ψ/K_t	ΣE_{AE} (J)	AE Stage	E_{AE} (J)	$E_{AE\text{-}II}/\Sigma E_{AE}$ (%)
Coal sample	LXMT-01	1.04	0.491	106.41	0.290	1.445	0.477	I	0.014	2.94
								II	**0.022**	**4.64**
								III	0.330	69.15
								IV	0.111	23.27
	LXMT-02	2.51	0.331	73.05	0.108	3.067	0.101	I	0.001	0.99
								II	**0.007**	**6.95**
								III	0.073	72.26
								IV	0.020	19.80
	LXMT-03	1.54	0.356	77.15	0.189	1.883	0.514	I	0.001	0.19
								II	**0.033**	**6.52**
								III	0.299	58.27
								IV	0.180	35.02
	LXMT-04	1.21	0.285	60.44	0.104	2.740	0.172	I	0.001	0.58
								II	**0.013**	**7.83**
								III	0.156	90.42
								IV	0.002	1.17
	LXMT-05	1.11	0.324	70.45	0.182	1.779	0.546	I	0.012	2.20
								II	**0.039**	**7.08**
								III	0.461	84.49
								IV	0.034	6.23
	LXMT-06	1.51	0.418	91.80	0.218	1.916	0.552	I	0.012	2.17
								II	**0.031**	**5.59**
								III	0.391	70.83
								IV	0.118	21.41
	LXMT-07	1.42	0.479	101.24	0.207	2.375	0.499	I	0.018	3.61
								II	**0.024**	**4.81**
								III	0.429	85.97
								IV	0.028	5.61
Rock sample	LXYT-01	1.14	2.066	446.31	0.367	1.98	0.293	I	0.007	2.39
								II	**0.078**	**26.55**
								III	0.206	70.37
								IV	0.002	0.69
	LXYT-02	1.88	1.917	419.41	0.165	4.158	0.215	I	0.002	0.93
								II	**0.059**	**27.57**
								III	0.142	66.38
								IV	0.011	5.12
	LXYT-03	1.92	1.848	397.22	0.308	2.165	0.240	I	0.001	0.42
								II	**0.067**	**28.07**
								III	0.157	66.09
								IV	0.013	5.42
	LXYT-04	1.17	1.551	321.19	0.176	3.237	0.114	I	0.001	0.88
								II	**0.035**	**30.48**
								III	0.070	61.63
								IV	0.008	7.01
	LXYT-05	2.21	3.904	820.87	0.455	2.364	0.242	I	0.001	0.41
								II	**0.043**	**17.83**
								III	0.109	44.98
								IV	0.089	36.78
	LXYT-06	2.31	3.537	783.21	0.650	1.570	0.384	I	0.006	1.57
								II	**0.074**	**19.18**
								III	0.238	62.07
								IV	0.066	17.18
	LXYT-07	2.44	3.662	801.47	0.557	1.866	0.194	I	0.001	0.52
								II	**0.036**	**18.71**
								III	0.076	39.02
								IV	0.081	41.75

Each point $(\ln(\Psi/K_t), E_{AE-II}/\Sigma E_{AE})$ was drawn in the coordinate system and observed using the relevant fitted curve. All the points exhibited a near-linear line, thereby allowing the adoption of the linear function. The square difference (χ) between the measured (Y_i) and calculated values (Y_j) fulfilled the optimization criterion following the application of the least square method. Supposing a_0 is equal to $1/b$, Equation (3) can be substituted into the optimization criterion as Equation (4).

$$\chi = \sum_{i=1}^{n}(Y_i - Y_j)^2 = \sum_{i=1}^{n}\left\{\left(\ln\frac{\Psi}{K_t}\right)_i - \left[a + a_0\left(\frac{E_{AE-II}}{\Sigma E_{AE}}\right)_i\right]\right\}^2 \tag{4}$$

Equation (4) exhibits the minimum, such that the partial derivatives of χ to both a and a_0 are equal to zero, as presented in Equations (5) and (6).

$$\frac{\partial \chi}{\partial a} = \sum_{i=1}^{n}\left[a + a_0\left(\frac{E_{AE-II}}{\Sigma E_{AE}}\right)_i - \left(\ln\frac{\Psi}{K_t}\right)_i\right] = 0 \tag{5}$$

$$\frac{\partial \chi}{\partial a_0} = \sum_{i=1}^{n}\left(\frac{E_{AE-II}}{\Sigma E_{AE}}\right)_i\left[a + a_0\left(\frac{E_{AE-II}}{\Sigma E_{AE}}\right)_i - \left(\ln\frac{\Psi}{K_t}\right)_i\right] = 0 \tag{6}$$

The following equations define an equation set for a and a_0,

$$na + \left[\sum_{i=1}^{n}\left(\frac{E_{AE-II}}{\Sigma E_{AE}}\right)_i\right]a_0 = \sum_{i=1}^{n}\left(\ln\frac{\Psi}{K_t}\right)_i \tag{7}$$

$$\left[\sum_{i=1}^{n}\left(\frac{E_{AE-II}}{\Sigma E_{AE}}\right)_i\right]a + \left[\sum_{i=1}^{n}\left(\frac{E_{AE-II}}{\Sigma E_{AE}}\right)_i^2\right]a_0 = \sum_{i=1}^{n}\left[\left(\frac{E_{AE-II}}{\Sigma E_{AE}}\right)_i\left(\ln\frac{\Psi}{K_t}\right)_i\right] \tag{8}$$

where n is equal to 7. The equation solutions are as follows:

$$a = \sum_{i=1}^{7}\left(\ln\frac{\Psi}{K_t}\right)_i/7 - a_0\left[\sum_{i=1}^{7}\left(\frac{E_{AE-II}}{\Sigma E_{AE}}\right)_i\right]/7 \tag{9}$$

$$a_0 = 7\sum_{i=1}^{7}\left[\left(\frac{E_{AE-II}}{\Sigma E_{AE}}\right)_i\left(\ln\frac{\Psi}{K_t}\right)_i\right] - \sum_{i=1}^{7}\left[\left(\frac{E_{AE-II}}{\Sigma E_{AE}}\right)_i\left(\ln\frac{\Psi}{K_t}\right)_i\right]/\left\{7\left[\sum_{i=1}^{7}\left(\frac{E_{AE-II}}{\Sigma E_{AE}}\right)_i^2\right] - \left[\sum_{i=1}^{7}\left(\frac{E_{AE-II}}{\Sigma E_{AE}}\right)_i\right]^2\right\} \tag{10}$$

The data from Table 3 were inserted into Equations (9) and (10) to obtain the material parameter solutions shown in Table 4. In addition, the relationship between $\ln(\Psi/K_t)$ and $E_{AE-II}/\Sigma E_{AE}$ for the coal and rock samples was defined as follows:

$$\begin{cases} \Psi_{coal-sample} = 0.162 e^{1.903 - 0.171\frac{E_{AE-II}}{\Sigma E_{AE}}} & (R \approx 1.051) \\ \Psi_{rock-sample} = 0.547 e^{3.267 - 0.073\frac{E_{AE-II}}{\Sigma E_{AE}}} & (R \approx 0.949) \end{cases} \tag{11}$$

Table 4. Values of the material parameters during coal and rock sample fracturing.

Parameter	a	a_0	$1/b$	K_t
Coal sample	1.903	−5.841	−0.171	0.162
Rock sample	3.267	−13.636	−0.073	0.547

Both sides of Equation (2) were multiplied with the volume (L^3). Equation (12) defines the theoretical model on the dissipation energy during sample fracturing based on AE event energy parameters.

$$U = L^3 K_t e^{a + \frac{E_{AE-II}}{\Sigma E_{AE}}} \quad (12)$$

The coal and rock samples exhibited minimum (maximum) values of U of 60.44 J (106.41 J) and 321.19 J (820.87 J), respectively. Supposing $U/L^3 K_t = Y$ and $\Sigma E_{AE-II}/\Sigma E_{AE} = X$, Equation (12) can be changed altered as Equation (13).

$$Y = e^{a + \frac{1}{b}X} \quad (13)$$

The derivative of both sides with respect to X was taken for Equation (13), and the result is defined in Equation (14).

$$\frac{dY}{dX} = \frac{e^{a + \frac{1}{b}X}}{b} \quad (14)$$

where dY/dX must be less than zero to allow Equation (13) to be defined as a monotone descending function, thereby indicating that a decrease in U generates an increase in $\Sigma E_{AE-II}/\Sigma E_{AE}$. In the elastic stage, the dissipation energy was much lower that the work done by the testing machine was used for crack initiation and propagation. When the anisotropy of the samples was further increased, the deformation and fracturing velocity of the samples accelerated due to external loading. In both the micro-fracturing and post-peak fracturing stages, much of the work done by the testing machine transformed into more acoustic waves that were received by the AE instrument. Moreover, the other work was dissipated by the macroscopic cracks and the structure friction.

5. Conclusions

(1) The relevant mechanical parameter discreteness was too large because the internal structure of the coal and rock was divided into multiple structural units (MSUs) due to the presence of a few main cracks. The MSU with the maximum bearing capacity was defined as the primary MSU, and the maximum bearing capacity was equal to the peak strength of the stress-strain curve. The post peak-point exhibited yielding weakening in the local area of the coal and rock, which defined the local distortion characteristics of the coal and rock. The strain value after the peak point should only consider the plastic strain of the local yielding weakening region.

(2) The entire acoustic emissions (AE) test was categorized into four stages based on both the axial stress and AE event parameters: initial loading, elastic, micro-fracturing, and post-peak fracturing stages. The coal and rock samples exhibited minimum (maximum) values of U of 60.44 J (106.41 J) and 321.19 J (820.87 J), respectively.

(3) A theoretical model on the dissipation energy during sample fracturing based on AE event energy parameters was offered. A decrease in U was observed following an increase in $\Sigma E_{AE-II}/\Sigma E_{AE}$.

Author Contributions: S.P. proposed the innovative points and conceived the study. S.P. and L.X. (Xiaoming Liu) wrote the paper. L.X. (Xingping Lai) revised the paper.

Acknowledgments: Financial support for this work was provided by the 973 Key National Basic Research Program of China (No. 2015CB251602), the Natural Science Foundation of Shaanxi Province (No. 2018JQ5194), the China Postdoctoral Science Foundation (No. 2017M623328XB), the Science and Technology Innovation Team Project of Shaanxi Province (No. 2018TD-038), the Hunan Province Engineering Research Center of Radioactive Control Technology in Uranium Mining and Metallurgy & Hunan Province Engineering Technology Research Center of Uranium Tailings Treatment Technology (No. 2018YKZX2001), and the Opening Project of Cooperative Innovation Center for Nuclear Fuel Cycle Technology and Equipment, University of South China (No. 2019KFY25). Support from these agencies is gratefully acknowledged.

Conflicts of Interest: The authors declare no conflict of interest.

Appendix A

In the AE testing, Table below presents the specific conditions of the physical propertiesAE tests, including the sample mass, sample dimension, and surface appearance. Equations (A1)–(A3) can be used to analyze the discreteness of all the physical parameter, of which the corresponding calculation results are listed in Table A1.

$$\overline{X} = \frac{1}{n}\sum_{i=1}^{n} X_i \tag{A1}$$

$$E = \sqrt{\frac{1}{n-1}\sum_{i=1}^{n}(X_i - \overline{X})^2} \tag{A2}$$

$$\zeta = \frac{E}{\overline{X}} \times 100\% \tag{A3}$$

where $\overline{X}$ is the mean value, E is the standard difference, and ζ is the variable coefficient.

Table A1. Profile of the basic physical parameters and discreteness of coal and rock.

Sample Number	Mass(g)	Geometric Dimension (mm)			Volume (cm³)	Density (g/cm⁻³)	Appearance Description
		X-axis	Y-axis	Z-axis			
LXMT-01	283	59.9	60.1	60.2	216.719	1.306	Apparent through cracks in Y-axis
LXMT-02	291	60.3	61.2	59.8	220.684	1.319	Apparent through cracks in all axes
LXMT-03	283	60.0	60.5	59.7	216.711	1.306	Apparent tiny cracks in all axes
LXMT-04	278	59.4	59.9	59.6	212.060	1.311	Apparent through cracks in Y-axis
LXMT-05	288	60.3	60.1	60.0	217.442	1.324	Apparent cracks in both X- and Z-axes
LXMT-06	289	60.7	59.9	60.4	219.610	1.316	Apparent tiny cracks in all axes
LXMT-07	275	59.6	59.8	59.3	211.350	1.301	Apparent tiny cracks in all axes
LXYT-01	580	59.9	60.3	59.8	215.996	2.685	Whole is more complete with small surface cracks
LXYT-02	540	59.2	59.8	61.8	218.782	2.638	Whole is more complete with small surface cracks
LXYT-03	573	60.0	60.2	59.5	214.914	2.666	Clear grain, through cracks in X,Y- and Y,Z-axis
LXYT-04	549	59.0	58.9	59.6	207.116	2.651	Clear grain, through cracks in X,Y- and Y,Z-axis
LXYT-05	561	60.1	59.3	59.0	210.272	2.668	Whole is more complete with small surface cracks
LXYT-06	594	60.7	60.2	60.6	221.441	2.682	Clear grain, through cracks in X,Y- and Y,Z-axis
LXYT-07	592	61.0	59.8	60.0	218.868	2.705	Apparent cracks in both X- and Z-axes

Lithology	Discrete parameter	L_X (mm)	L_Y (mm)	L_Z (mm)	Mass (g)	Density (g/cm⁻³)
Coal mass	$\overline{X}$	60.0	60.2	59.5	284	1.312
	E	0.447	0.491	0.537	5.902	0.008
	ζ	0.745	0.816	0.903	2.078	0.610
Rock	$\overline{X}$	60.0	59.6	59.6	570	2.671
	E	0.725	0.642	0.607	20.781	0.0223
	ζ	1.208	1.077	1.018	3.646	0.835

References

1. Lai, X.P.; Shan, P.F.; Cao, J.T.; Cui, F.; Sun, H. Simulation of asymmetric destabilization of mine-void rock masses using a large 3D physical model. *Rock Mech. Rock Eng.* **2016**, *49*, 487–502. [CrossRef]
2. Lai, X.P.; Sun, H.; Shan, P.F.; Cai, M.; Cao, J.T.; Cui, F. Structure instability forcasting and analysis of giant-rock-pillars in steeply dipping thick coal seams. *Int. J. Miner. Metall. Mater.* **2015**, *22*, 1–12. [CrossRef]
3. Sun, H.; Liu, X.L.; Zhu, J.B. Correlational fractal characterisation of stress and acoustic emission during coal and rock failure under multilevel dynamic loading. *Int. J. Rock Mech. Min. Sci.* **2019**, *117*, 1–10. [CrossRef]
4. Zhao, Y.S.; Feng, Z.C.; Wan, Z.J. Least energy principle of failure of rock. *Chin. J. Rock Mech. Eng.* **2003**, *22*, 1781–1783. (In Chinese)
5. Luo, Y.; Xie, H.P.; Ren, L.; Zhang, R.; Li, C.B.; Gao, C. Linear elastic fracture mechanics characterization of an anisotropic shale. *Sci. Rep.* **2018**, *8*, 8505. [CrossRef]
6. Jin, F.N.; Jiang, M.R.; Gao, X.L. Defining damage variable based on energy dissipation. *Chin. J. Rock Mech. Eng.* **2004**, *23*, 1976–1980. (In Chinese)

7. Hu, L.Q.; Li, X.B.; Zhao, F.J. Sudy on energy consumption in fracture and damage of rock induced by impact loadings. *Chin. J. Rock Mech. Eng.* **2003**, *21* (Suppl. 2), 2304–2308. (In Chinese)
8. Sun, Q.; Li, S.C.; Feng, X.D.; Li, W.T.; Yuan, C. Study of numerical simulation method of rock fracture based on strain energy density theory. *Rock Soil Mech.* **2011**, *32*, 1575–1582. (In Chinese)
9. Gong, W.L.; Peng, Y.Y.; Wang, H.; He, M.; e Sousa, L.R.; Wang, J. Fracture angle analysis of rock burst faulting planes based on true-triaxial experiment. *Rock Mech. Rock Eng.* **2015**, *48*, 1017–1039. [CrossRef]
10. Xu, Y.; Dai, F.; Xu, N.W.; Zhao, T. Numerical investigation of dynamic rock fracture toughness determination using a semi-circular bend specimen in split Hopkinson pressure bar testing. *Rock Mech. Rock Eng.* **2016**, *49*, 731–745. [CrossRef]
11. Stanchits, S.; Burghardt, J.; Surdi, A. Hydraulic fracturing of heterogeneous rock monitored by acoustic emission. *Rock Mech. Rock Eng.* **2015**, *48*, 2513–2517. [CrossRef]
12. Hamstad, M.A. Frequencies and amplitudes of AE signals in a plate as a function of source rise time. In Proceedings of the 29th European Conference on Acoustic Emission Testing, Vienna, Austria, 8–10 September 2010.
13. Zelenyak, A.M.; Hamstad, M.A.; Sause, M.G. Modeling of acoustic emission signal propagation in waveguides. *Sensors* **2015**, *15*, 11805–11822. [CrossRef] [PubMed]
14. Faisal Haider, M.; Giurgiutiu, V. Theoretical and numerical analysis of acoustic emission guided waves released during crack propagation. *J. Intell. Mater. Syst. Struct.* **2018**. [CrossRef]
15. Haider, M.F.; Giurgiutiu, V. Analysis of axis symmetric circular crested elastic wave generated during crack propagation in a plate: A Helmholtz potential technique. *Int. J. Solids Struct.* **2018**, *134*, 130–150. [CrossRef]
16. Wisner, B.; Kontsos, A. Investigation of particle fracture during fatigue of aluminum 2024. *Int. J. Fatigue* **2018**, *111*, 33–43. [CrossRef]
17. Zhuang, D.Y.; Tang, C.A.; Liang, Z.Z.; Ma, K.; Wang, S.Y.; Liang, J.Z. Effects of excavation unloading on the energy-release patterns and stability of underground water-sealed oil storage caverns. *Tunn. Undergr. Space Technol.* **2017**, *61*, 122–133. [CrossRef]
18. He, M.C.; Zhao, F.; Cai, M.; Du, S. A novel experimental technique to simulate pillar burst in laboratory. *Rock Mech. Rock Eng.* **2015**, *48*, 1833–1848. [CrossRef]
19. Xiao, Y.X.; Feng, X.T.; Li, S.J.; Li, S.J.; Feng, G.L.; Yu, Y. Rock mass failure mechanisms during the evolution process of rock bursts in tunnels. *Int. J. Rock Mech. Min. Sci.* **2016**, *83*, 174–181. [CrossRef]
20. Ji, H.G.; Wang, H.W.; Cao, S.; Hou, Z.; Jin, Y. Experimental research on frequency characteristics of acoustic emission signals under uniaxial compression of granite. *Chin. J. Rock Mech. Eng.* **2012**, *31* (Suppl. 1), 2900–2905. (In Chinese)
21. Zhao, X.D.; Tang, C.A.; Li, Y.H.; Yuan, R.F.; Zhang, J.Y. Study of AE activity characteristics under uniaxial compression loading. *Chin. J. Rock Mech. Eng.* **2006**, *25* (Suppl. 2), 3673–3678. (In Chinese)
22. Hudson, J.A.; Crouch, S.; Fairhurst, C.E. stiff and servo-controlled testing machines. *Int. J. Rock Mech. Min. Sci.* **1999**, *36*, 279–289.
23. Shan, P.F.; Lai, X.P. Numerical simulation of the samples in the fluid-solid coupling process during the failure of a fractured coal based on the regional geostress characteristics. *Transp. Porous Media* **2018**, *124*, 1061–1079. [CrossRef]
24. Wang, M.; Liu, L.; Zhang, X.Y.; Chen, L.; Wang, S.Q.; Jia, Y.H. Experimental and numerical investigations of heat transfer and phase change characteristics of cemented paste backfill with PCM. *Appl. Eng.* **2019**, *150*, 121–131. [CrossRef]
25. Liu, L.; Zhu, C.; Qi, C.C.; Zhang, B.; Song, K.-I. A microstructural hydration model for cemented paste backfill considering internal sulfate attacks. *Constr. Build. Mater.* **2019**, *211*, 99–108. [CrossRef]
26. Liu, L.; Yang, P.; Qi, C.C.; Zhang, B.; Guo, L.J.; Song, K.-I. An experimental study on the early-age hydration kinetics of cemented paste backfill. *Constr. Build. Mater.* **2019**, *212*, 283–294. [CrossRef]
27. Huan, C.; Wang, F.H.; Li, S.T.; Zhao, Y.J.; Liu, L.; Wang, Z.H.; Ji, C.F. A performance comparison of serial and parallel solar-assisted heat pump heating systems in Xi'an, China. *Energy Sci. Eng.* **2019**. [CrossRef]
28. Shan, P.F.; Lai, X.P. Influence of CT scanning parameters on rock and soil images. *J. Vis. Commun. Image R.* **2019**, *58*, 642–650. [CrossRef]

29. Shan, P.F.; Lai, X.P. Mesoscopic structure PFC similar to 2D model of soil rock mixture based on digital image. *J. Vis. Commun. Image R.* **2019**, *58*, 407–415. [CrossRef]
30. Shan, P.F. Image segmentation method based on K-mean algorithm. *EUR. J. Image Video Proc.* **2018**. [CrossRef]

© 2019 by the authors. Licensee MDPI, Basel, Switzerland. This article is an open access article distributed under the terms and conditions of the Creative Commons Attribution (CC BY) license (http://creativecommons.org/licenses/by/4.0/).

Article

Steady Flow of a Cement Slurry

Chengcheng Tao [1], Barbara G. Kutchko [1], Eilis Rosenbaum [1], Wei-Tao Wu [2] and Mehrdad Massoudi [1,*]

1. U. S. Department of Energy, National Energy Technology Laboratory (NETL), Pittsburgh, PA 15236, USA
2. School of Mechanical Engineering, Nanjing University of Science and Technology, Nanjing 210094, China
* Correspondence: Mehrdad.Massoudi@netl.doe.gov; Tel.: +1-412-386-4975

Received: 12 June 2019; Accepted: 5 July 2019; Published: 6 July 2019

Abstract: Understanding the rheological behavior of cement slurries is important in cement and petroleum industries. In this paper, we study the fully developed flow of a cement slurry inside a wellbore. The slurry is modeled as a non-linear fluid, where a constitutive relation for the viscous stress tensor based on a modified form of the second grade (Rivlin–Ericksen) fluid is used; we also propose a diffusion flux vector for the concentration of particles. The one-dimensional forms of the governing equations and the boundary conditions are made dimensionless and solved numerically. A parametric study is performed to present the effect of various dimensionless numbers on the velocity and the volume fraction profiles.

Keywords: cement; non-Newtonian fluids; rheology; variable viscosity; diffusion

1. Introduction

Portland cement is widely used as a construction material in civil engineering applications due to the widespread availability of its constituent materials [1]. It is produced from the grinding of clinker, which is produced by the calcination of limestone and other raw materials in a rotary kiln. The different phases in cement are: Alite (C_3S), belite (C_2S), aluminate phase (C_3A), ferrite phase (C_3AF), alkali sulfate, free lime, and gypsum [2,3]. Cement slurries are reactive systems, continuously changing their chemical and physical characteristics [4]. After the cement is mixed with water, a series of exothermal chemical reactions occur resulting in an increase in the strength and hardening [5]. The most reactive phases with water are C_3S and C_3A, the content of which affect the strength of cement developed at early stages. Water to cement ratio, which is defined as the ratio of the weight of water to the weight of cement, also plays an important role in the strength development and flow behavior of cement slurry.

According to a recent study by the United States Geological Survey (USGS), U.S. cement and clinker production is about 80 million metric tons per year. The production of cementitious materials consumes a significant amount of energy (20–40% total energy cost) [6,7]. Cement production also contributes to 4% of the global industrial carbon dioxide (CO_2) emissions [8]. Therefore, it is important for the cement industry to seek energy efficient technologies and improve the performance of cement productions for sustainability purpose. Physical testing is widely applied to study the cement behavior. This requires time, energy, and material resources. In the past few decades, the cement research community has sought advanced computational modeling for cement hydration processes, flow and mechanical properties in order to eliminate the substantial cost for physical testing of cement. Bentz [9] developed a three-dimensional computational model for the cement microstructure and cement hydration process. Haekck et al. [10] studied the physical and the chemical properties of cement such as the heat of hydration, the elastic modulus, and the pore concentrations using the software Virtual Cement and Concrete Testing Laboratory (VCCTL). Bullard et al. described the details of this software [11,12]. Watts et al. [13] and Tao et al. [14] validated the software optimization framework to characterize and evaluate different computational models for cement.

In petroleum-related applications, cement slurries are pumped down the wellbore and up the annular space between the casing and the geological formations surrounding the wellbore to provide zonal isolation in oil, gas, and water wells [15]. It is a great challenge for petroleum industry to prevent gas entry into the cement and achieve the annular cement seal for a long term [16]. Researchers have investigated the possible mechanisms for fluid migration during cementing by using experimental and computational models, although the exact failure mechanism of this problem is very complicated. Monitoring the conditions of cement slurry in realtime is a critical issue where wireless sensor network-based monitoring system can be used [17]. Cement must remain as a fluid long enough while it is being pumped to the anticipated location; it should also have sound compressive-strength within a specific time after placement. Gel strength is related to the resisting shear stress before the cement can flow, and is considered to be one of the major factors for hydrostatic pressure loss and gas migration [18]. Chenevert and Jin [19] suggested that the rheological properties of the cement affect the static gel strength. Stiles [20] indicated that the rheological properties of the cement slurry are related to the annular fluid displacement. Brandt et al. [21] investigated a deep-water operation for drilling fluid and well cementing, and suggested that the slurry properties at low temperatures and high pressures should be taken into consideration. During the cementing, when the slurry is pumped into the oilwell, it flows to the bottom of the wellbore through the casing and begins to develop more strength from sedimentation. To develop computational models for cement slurry, many researchers assume that the cement slurry is a suspension with non-Newtonian characteristics [22]. Foroushan et al. [23] modeled the instability of the interface and the mixing of cement slurry and drilling mud during cementing operation in oil and gas wells in three dimensions by using commercial Computational Fluid Dynamics (CFD) software, and compared the results with experiments. Skadsem et al. [24] studied the flow of a non-Newtonian fluid in an inclined wellbore with concentric and eccentric configurations numerically and experimentally, using a finite element approach in OpenFOAM. Liu et al. [25] modeled the multi-phase pipe flow and considered the hydration effect of cemented paste backfill slurry by applying a CFD model. Murphy et al. [26] simulated the shear flow of two Bingham-type plastic cement slurries containing Portland cement and fly ash particles by applying the fast lubrication dynamics and discrete element model with LAMMPS.

In this paper, the flow of a cement slurry between two flat plates at different tilt angles is studied. It is assumed that the viscosity of the cement depends on the shear rate and the volume fraction of the particles. A convection–diffusion equation is used to study the effect of the particle concentration. Section 2 presents the governing equations. In Section 3, we describe the constitutive relations for the viscous stress and the diffusive particle flux vector. Section 4 defines the geometry of the problem and provides the dimensionless forms of the equations and the boundary conditions. In Section 5, numerical results are presented, and a parametric study is performed for different dimensionless numbers. Section 6 provides some concluding remarks.

2. Governing Equations

In this paper, cement slurry is assumed to behave as a non-homogenous nonlinear suspension. If the electromagnetic and the thermochemical effects are ignored, then the governing equations of motion are the conservations of mass, linear momentum, angular momentum, and the equation for the flux of concentration [27].

2.1. Conservation of Mass

$$\frac{\partial \rho}{\partial t} + div(\rho \boldsymbol{v}) = 0, \tag{1}$$

where $\partial/\partial t$ is the partial derivative with respect to time, div is the divergence operator, $\boldsymbol{v}$ is the velocity vector, and ρ is the density of the slurry. For an isochoric motion $div(\boldsymbol{v}) = 0$.

2.2. Conservation of Linear Momentum

$$\rho \frac{d\mathbf{v}}{dt} = div\mathbf{T} + \rho \mathbf{b}, \tag{2}$$

where $\mathbf{b}$ is the body force vector, $\mathbf{T}$ is the Cauchy stress tensor, and d/dt is the total time derivative given by $d(.)/dt = \partial(.)/\partial t + [grad(.)]\mathbf{v}$.

2.3. Conservation of Angular Momentum

The conservation of angular momentum indicates that in the absence of couple stresses the stress tensor is symmetric, that is

$$\mathbf{T} = \mathbf{T}^T. \tag{3}$$

2.4. Convection–Diffusion Equation

In flows of suspensions, a convection–diffusion equation [28] is often used for the particle concentration ϕ,

$$\frac{\partial \phi}{\partial t} + div(\phi \mathbf{v}) = -div\mathbf{N}, \tag{4}$$

where $\frac{\partial \phi}{\partial t}$ is the rate of accumulation of particles, $div(\phi \mathbf{v})$ is the term representing particle migration and movement due to the flow, and $div\mathbf{N}$ is the diffusive particle flux. The function ϕ, called the volume fraction (related to concentration), has the property $0 \leq \phi(\mathbf{x},t) \leq \phi_{max} < 1$. In reality, ϕ is either one or zero at any position and time, depending upon whether one is pointing to a particle or to the void space (fluid) at that location. The density of the cement slurry, in general, can be related to the density of water and the cement particles, via the following relation: $\rho = (1-\phi)\rho_{f0} + \phi \rho_{s0}$, where ϕ is the volume fraction (concentration) of the cement particles, ρ_{f0} and ρ_{s0} are the pure density of water and the cement particles in the reference configuration (before mixing), respectively. The assumption that the particle and the fluid densities are the same, is a special case of the above equation. In this paper, we use ρ as the bulk density of the cement slurry.

Looking at Equations (1)–(4), we can see that we need constitutive relations for $\mathbf{T}$ and $\mathbf{N}$. We will discuss these in the next section.

3. Constitutive Relations

A cement slurry, in general, behaves as a (nonlinear) fluid. Once cement particles are mixed with water, after a series of hydration reactions, the slurry begins to develop solidlike behavior [29]. Cement-based materials could stand under their own weight without flowing and develop strength and stiffness during setting [30]. The flow behavior of slurry plays an important role on the cement quality [31]. Understanding the rheological behavior of a cement slurry is important in industry for easy pumping and filling the annulus without excessive separation of water and cement [15,30]. Rheological measurements for cement-based materials are well established [32]. The Bingham viscoplastic fluid model is widely used to describe the yield stress of cement slurries. The yield stress is often related to particle concentration, shear rate history, time, and temperature. The static yield stress (also known as the static gel strength) affects the pumping of the cement. Moon and Wang [33] suggested that during the gelation process, the cement shows non-Newtonian behavior. In general, nonlinear fluids exhibit characteristics such as yield stress, viscoelasticity, normal stress effects, shear-rate dependent viscosity, etc. Constitutive relations can be obtained or derived in different ways, for example, by using: (a) Techniques in continuum mechanics, (b) models based on physical and experimental observations, (c) numerical simulations, (d) statistical mechanics approaches, and (e) ad-hoc approaches. Next, we briefly discuss the constitutive modeling of the stress tensor and the diffusive flux, using a continuum mechanics approach.

3.1. Stress Tensor

Many cement-based materials exhibit a yield stress. The Bingham viscoplastic model is widely used to describe the behavior of cement [30]. It has also been noticed that in many situations, the Herschel–Bulkley model predicts the sedimentation tendencies more accurately than the Bingham model [31]. In general, a constitutive relation for a cement slurry should have a yield stress component as well as a viscous stress part. Thus, the stress tensor can be written as:

$$T = T_y + T_v, \tag{5}$$

where T_y is the yield stress (which in theory can be measured and can depend on parameters, such as the solid volume fraction [34]) and T_v is the viscous stress, where the viscosity is assumed to depend on the shear rate, the particle concentration, and possibly on temperature and chemical composition. In addition, if viscosity and other rheological parameters depend on time, we can consider the thixotropic nature of cement by introducing a structural parameter (see [35,36]). For the remainder of this paper, we will only focus on the viscous stress tensor T_v. We plan to study the effect of the yield stress (see [37]) in future. One of the simplest models which can show the shear-rate dependency of the viscosity, is the power-law model or the generalized Newtonian fluid (GNF) model [38]

$$T_v = -pI + \mu_0 (tr A_1^2)^m A_1, \tag{6}$$

where p is the pressure, I is the identity tensor, μ_0 is the coefficient of viscosity, m is the power-law exponent, a measure of non-linearity of the fluid, related to the shear-thinning ($m < 0$) or shear-thickening ($m > 0$) effects of the fluid, tr is the trace operator, and A_1 is related to the velocity gradient. For additional information about other power-law models such as Carreau-type fluid, we refer the reader to [38,39].

It has been observed that the viscosity of a cement slurry increases with increasing concentration of the solid particles [40]. Researchers have studied the relation between the shear viscosity and the concentration extensively and have proposed various empirical models. The Einstein expression was first applied for the relation between viscosity and particle concentration for dilute suspensions [41]: $\mu = \mu_0(1 + 2.5\phi)$, where μ is the viscosity of suspension, μ_0 is the viscosity of pure liquid (the base fluid), and ϕ is the volume fraction of the particles. Other studies [42,43] have shown that the relationship between viscosity and volume fraction can be expressed more accurately as $\mu/\mu_0 = (1 - 1.35\phi)^{-2.5}$. These relationships are unable to predict the behavior of a suspension at high particle concentration, and more complicated models are required [24]. Krieger and Dougherty's model is widely applied for nonflocculated suspensions such as cement slurries [40,44–49]:

$$\mu/\mu_0 = \left(1 - \frac{\phi}{\phi_m}\right)^{-\beta}, \tag{7}$$

where β is a fitting experimental parameter and is usually assumed to be between 1.5–2 for cement [48,49]. ϕ_m is the maximum solid concentration packing, which is about 0.65 for suspension with spherical particles [40].

Although most experimental studies related to cement slurry focus on the measurement of viscosity and yield stress, it is not known whether cement slurry, similar to dense granular materials and some suspensions/polymers, would exhibit normal stress effects related to phenomena such as 'die-swell' and 'rod-climbing' (see [50,51]). One of the simplest models that can capture the normal stress effects is the second grade fluid, or the Rivlin–Ericksen fluid of grade two [52,53]. Based on the brief discussion above, in this paper, we will focus our attention on the modeling of the viscous stress, by assuming that the slurry behaves as a modified second grade (Rivlin–Ericksen) fluid model, where,

$$T_v = -pI + \mu_{eff}(\phi, A_1) A_1 + \alpha_1 A_2 + \alpha_2 A_1^2. \tag{8}$$

The kinematical tensors A_1 and A_2 are defined through

$$A_1 = \text{grad}v + (\text{grad}v)^T, \tag{9}$$

$$A_2 = \frac{dA_1}{dt} + A_1(\text{grad}v) + (\text{grad}v)^T A_1, \tag{10}$$

where α_1 and α_2 are the normal stress coefficients and μ_{eff} is the effective viscosity, which is dependent on the volume fraction and the shear rate. Equation (8) is a possible generalization of the second grade fluid model (for a detailed discussion of this see [54,55]). Using the Clausius–Duhem inequality, Dunn and Fosdick [56] showed:

$$\begin{aligned} \mu &\geq 0, \\ \alpha_1 &\geq 0, \\ \alpha_1 + \alpha_2 &= 0. \end{aligned} \tag{11}$$

For further details on this and other relevant issues in fluids of differential type, we refer the reader to the review article by Massoudi and Vaidya [57]. Furthermore, the effective viscosity is given by the equation:

$$\mu_{eff}(\phi, A_1) = \mu^*(\phi)\left[1 + \alpha \text{tr}A_1^2\right]^m. \tag{12}$$

where Krieger and Dougherty's correlation for $\mu^*(\phi)$ is used in this paper:

$$\mu^*(\phi) = \mu_0\left(1 - \frac{\phi}{\phi_m}\right)^{-\beta}, \tag{13}$$

where β is the experimental parameter, assumed to be 1.82 in our paper [58], and ϕ_m is the maximum volume fraction of particles. This will be an input to the problem, i.e., we will solve the equations for different values of ϕ_m. By substituting Equations (12) and (13) into (8), we have the equation for the viscous stress tensor:

$$T_v = -pI + \mu_0\left(1 - \frac{\phi}{\phi_m}\right)^{-\beta}\left[1 + \alpha \text{tr}A_1^2\right]^m A_1 + \alpha_1 A_2 + \alpha_2 A_1^2. \tag{14}$$

Equation (14) is used in our analysis. It is noticed that this model has 6 material parameters, namely: μ_0, β, α, m, α_1, and α_2.

3.2. Particle Fluxes

The particle transport fluxes could be affected by various mechanisms such as particles collision, body force, Brownian motion, etc. [58]. For a description of additional flux terms, we refer the reader to the recent paper by Li et al. [59]. In this paper, the particle transport flux is assumed as.

$$N = -a^2\phi K_c \nabla(\dot{\gamma}\phi) - a^2\phi^2\dot{\gamma}K_\mu \nabla(\ln\mu_{eff}) - D\nabla\phi, \tag{15}$$

where the terms on the right hand side are the transport flux contributions due to particles collisions, spatially varying viscosity, and the Brownian diffusive flux, respectively. In the above equation, a is the characteristic particle length—for example, the particle radius—and $\dot{\gamma}$ is the local shear rate,

$$\dot{\gamma} = \left(2A_{ij}A_{ij}\right)^{1/2}. \tag{16}$$

where K_c and K_μ are empirical coefficients, and D is the diffusion coefficient (diffusivity). It is assumed that D has two contributions, one related to the shear rate and the other to concentration. Thus,

$$D = D(\dot{\gamma}, \phi) = D_1(\dot{\gamma})D_2(\phi). \tag{17}$$

For $D_1(\dot{\gamma})$, Bridges and Rajagopal's assumption is used [60]:

$$D_1(\dot{\gamma}) = \eta \|A_1^2\|, \qquad (18)$$

where η is a constant. For $D_2(\phi)$, the ideas of Garboczi and Benz [61] who studied the dependence of cement diffusivity on pore structure are used. During the hydration process, the capillary pore space is gradually filled by water. The main product of the cement hydration is calcium silicate hydrate (C-S-H). Figure 1 shows the schematic of the capillary pores φ. The hydration products have larger volumes than the cement reactants, which explains why the cement particles in the viscous suspension develop strength from the hydration process. Two different phases contribute to the diffusivity, namely, the capillary pore space and the C-S-H gel phase. After the capillary pores close off, the smaller C-S-H gel micropores start to dominate the transport. Capillary pore space is filled with water, thus, the volume fraction of the cement particles is $\phi = 1 - \varphi$. The minimum capillary porosity is about 18% (known as the percolation threshold); therefore, the maximum volume fraction of cement particles is $\phi_m = 1 - 0.18 = 0.72$.

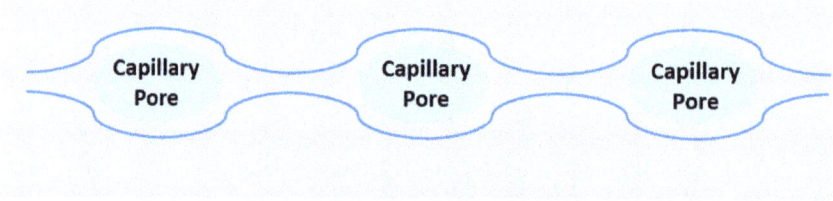

Figure 1. Schematic of capillary pores φ, lined with C-S-H [62].

By considering the effects of both the capillary porosity and the C-S-H contribution on cement diffusivity, the authors proposed a relation between diffusivity and capillary pore space. After adjusting the volume fraction ϕ of cement particles with the capillary porosity φ, their equation becomes:

$$D_2(\phi) = D_0 \left[K_1 + K_2(1-\phi)^2 + K_3(\phi_m - \phi)^2 H(\phi_m - \phi) \right], \qquad (19)$$

where D_0 is the diffusivity parameter; K_1, K_2, and K_3 are fitting coefficients with suggested values 0.001, 0.07, and 1.8; and H is the Heaviside function $H(x) = 1$ for $x > 0$, $H(x) = 0$ for $x \leq 0$.

If we consider a steady-state condition and substitute Equation (15) into (4) with $div(\phi v) = 0$, we have [59]:

$$div\, N = 0. \qquad (20)$$

At solid boundaries, no particles can penetrate the walls, which indicates that the particle flux should be zero at the walls [63]

$$0 = N|_{wall}. \qquad (21)$$

By integrating Equation (20), considering the boundary condition (21), we notice that for this flow field, the total flux equals zero everywhere in the flow, that is:

$$0 = N. \qquad (22)$$

In summary, the following equation is used for the diffusive flux N:

$$\begin{aligned}N = &-a^2\phi K_c \nabla(\dot{\gamma}\phi) - a^2\phi^2\dot{\gamma}K_\mu \nabla\left[\ln\left(\mu_0\left(1 - \frac{\phi}{\phi_m}\right)^{-\beta}\left[1 + \alpha tr A_1^2\right]^m\right)\right] \\ &- \eta A_1^2 D_0\left[K_1 + K_2(1-\phi)^2 + K_3(\phi_m - \phi)^2 H(\phi_m - \phi)\right]^2 \nabla\phi.\end{aligned} \qquad (23)$$

Equations (14) and (23) form the basic constitutive relations used in this paper.

4. Flow between Two Plates

To test the models proposed in the previous section, we will solve a simple problem with relevant industrial application. As most drilling operations are done either in a vertical or a horizontal arrangement (shown in Figure 2a), we consider a tilted channel, as shown in Figure 2b where θ is measured from the horizontal direction. The motion is assumed to be steady and fully developed.

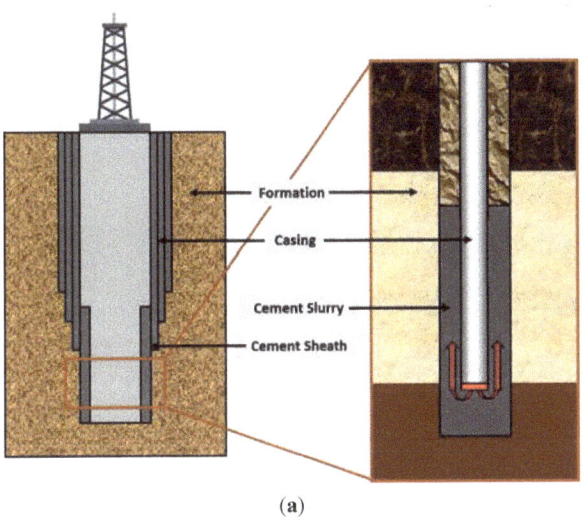

(a)

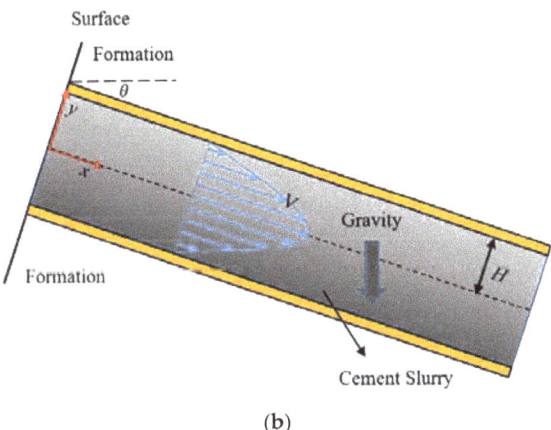

(b)

Figure 2. Schematic diagram of (**a**) oil well cementing and (**b**) cement slurry flow in a tilted channel.

The velocity and the volume fraction fields are assumed to be of the form:

$$\begin{cases} \phi = \phi(y) \\ v = v(y)e_x \end{cases} . \tag{24}$$

Using Equation (24), the conservation of mass (Equation (1)) is automatically satisfied. By substituting Equation (8) into Equation (2), the equations of linear momentum in component form become:

$$\frac{\partial}{\partial x}\left[-p + \alpha_2\left(\frac{dv}{dy}\right)^2\right] + \frac{\partial}{\partial y}\left[\mu_0\left(1 - \frac{\phi}{\phi_m}\right)^{-\beta}\left[1 + 2\alpha\left(\frac{dv}{dy}\right)^2\right]^m \frac{dv}{dy}\right] + \rho g \sin\theta = 0 \qquad (25a)$$

$$\frac{\partial}{\partial y}\left[-p + (2\alpha_1 + \alpha_2)\left(\frac{dv}{dy}\right)^2\right] - \rho g \cos\theta = 0 \qquad (25b)$$

$$\frac{\partial p}{\partial z} = 0 \qquad (25c)$$

Let us define a modified pressure $\hat{p}$ (see [64])

$$\hat{p} = p - (2\alpha_1 + \alpha_2)\left(\frac{dv}{dy}\right)^2. \qquad (26)$$

Then, the Equations (25) are simplified to

$$-\frac{\partial \hat{p}}{\partial x} + \frac{\partial}{\partial y}\left[\mu_0\left(1 - \frac{\phi}{\phi_m}\right)^{-\beta}\left[1 + 2\alpha\left(\frac{dv}{dy}\right)^2\right]^m \frac{dv}{dy}\right] + \rho g \sin\theta = 0 \qquad (27a)$$

$$-\frac{\partial \hat{p}}{\partial y} - \rho g \cos\theta = 0 \qquad (27b)$$

$$\frac{\partial \hat{p}}{\partial z} = 0 \qquad (27c)$$

Equations in (27) provide the basic equations for the solution of volume fraction, velocity field, and pressure distribution in the x, y, and z-directions. In this paper, Equation (27a) is used to solve for the velocity, which is coupled to the volume fraction equation. We specify the pressure gradient in the x-direction. In a sense, we do not use Equation (27b) to solve for pressure, which will be affected by the normal stress coefficients. In more complicated flow situations and geometries, all three components of the momentum equation, along with the convection–diffusion equation, should be solved simultaneously. To obtain the expanded form of the convection–diffusion equation, we substitute Equations (12) and (15) into (22):

$$a^2 K_c\left(\phi^2 \frac{d}{dy}\left|\frac{dv}{dy}\right| + \phi\left|\frac{dv}{dy}\right|\frac{d\phi}{dy}\right) + a^2\phi^2 K_\mu \frac{\frac{\partial}{\partial y}\left\{\mu_0\left(1 - \frac{\phi}{\phi_m}\right)^{-\beta}\left[1 + 2\alpha\left(\frac{dv}{dy}\right)^2\right]^m\right\}}{\mu_0\left(1 - \frac{\phi}{\phi_m}\right)^{-\beta}\left[1 + 2\alpha\left(\frac{dv}{dy}\right)^2\right]^m}\left|\frac{dv}{dy}\right| + 2\eta$$

$$\cdot D_0\left[K_1 + K_2(1 - \phi)^2 + K_3(\phi_m - \phi)^2 H(\phi_m - \phi)\right]\frac{d\phi}{dy}\left(\frac{dv}{dy}\right)^2 = 0. \qquad (28)$$

Before solving Equations (27a) and (28), we nondimensionalize the equations by introducing the dimensionless length $\bar{y}$ and velocity $\bar{v}$ as:

$$\bar{y} = \frac{y}{H}; \; \bar{v} = \frac{v}{V}, \qquad (29)$$

where H is the distance between the two plates and V is a reference velocity. The dimensionless forms for Equations (27a) and (28) become

$$\frac{\partial}{\partial \bar{y}}\left\{\left(1 - \frac{\phi}{\phi_m}\right)^{-\beta}\left[1 + R_0\left(\frac{d\bar{v}}{d\bar{y}}\right)^2\right]^m \frac{d\bar{v}}{d\bar{y}}\right\} = R_1 - R_2 \sin\theta, \qquad (30)$$

$$\frac{K_c}{K_\mu}\left(\phi^2 \frac{d}{dy}\left|\frac{d\bar{v}}{dy}\right| + \phi\left|\frac{d\bar{v}}{dy}\right|\frac{d\phi}{dy}\right) + m\phi^2\left|\frac{d\bar{v}}{dy}\right|\left[1 + R_0\left(\frac{d\bar{v}}{dy}\right)^2\right]^{-1} \cdot 2R_0 \frac{d\bar{v}}{dy}\frac{d^2\bar{v}}{dy^2}$$
$$+ \frac{\beta}{\phi_m}\phi^2\left(1 - \frac{\phi}{\phi_m}\right)^{-1}\frac{d\phi}{dy}\left|\frac{d\bar{v}}{dy}\right| \qquad (31)$$
$$+ \left[R_3 + R_4(1-\phi)^2 + R_5(\phi_m - \phi)^2 H(\phi_m - \phi)\right]\frac{d\phi}{dy}\left(\frac{d\bar{v}}{dy}\right)^2 = 0,$$

where the following dimensionless numbers are obtained:

$$R_0 = 2\alpha \frac{V^2}{H^2}; R_1 = \frac{\partial p}{\partial x}\frac{H^2}{\mu_0 V}; R_2 = \frac{\rho_f g H^2}{\mu_0 V},$$
$$R_3 = \frac{2\eta D_0 V K_1}{a^2 K_\mu H}; R_4 = \frac{2\eta D_0 V K_2}{a^2 K_\mu H}; R_5 = \frac{2\eta D_0 V K_3}{a^2 K_\mu H}. \qquad (32)$$

The physical meanings of these dimensionless numbers are: R_0 is related to the coefficient of the shear rate, related to the magnitude of shear-thinning/shear-thickening effect, R_1 is a measure of the importance of the force due to the pressure gradient and the viscous effects, R_2 is related to the gravity, and R_3, R_4, and R_5 are parameters related to the coefficients in the diffusion equation related to the volume concentration $D_2(\phi)$. In addition to these six dimensionless numbers, we also have parameters such as ϕ_m, which can be varied independently (see Table 1). Equations (30) and (31) are to be solved numerically, subject to appropriate physical boundary conditions. We will perform a limited parametric study to look at the effects of these dimensionless parameters on the flow and concentration profiles.

Table 1. Values of the dimensionless numbers and other parameters.

Parameters	Range of Values
ϕ_m	0.45, 0.5, 0.55, 0.6, 0.65
K_c/K_μ	0, 0.02, 0.04, 0.06, 0.08
θ	0°, 30°, 45°, 60°, 90°
m	−0.3, −0.1, 0, 0.1, 0.3, 0.7
R_0	0.01, 0.1, 1, 10
R_1	0, −1.5, −2.5, −3.5
R_2	0, 0.5, 1, 1.5
R_3	0.01, 0.1, 1
R_4	0.01, 0.1, 1
R_5	0.01, 0.1, 1

We also notice that we need two boundary conditions for $\bar{v}$ and one condition for ϕ. A no-slip boundary condition for velocity at the two plates is assumed. These are:

$$\bar{v}(\bar{y} = -1) = 0; \bar{v}(\bar{y} = 1) = 0. \qquad (33)$$

For the volume fraction, we specify an average quantity ϕ_{avg} given in terms of an integral taken across the cross section of the flow (see [65]) namely:

$$\int_{-1}^{1} \phi d\bar{y} = \phi_{avg}. \qquad (34)$$

Note that whenever a second grade fluid or any higher grade fluid models is used, the order of the differential equations (the linear momentum equation) is raised, and additional boundary conditions are needed [66,67]; although in this problem, we are not concerned with this issue since we are not solving Equation (27b).

5. Numerical Results and Discussion

The dimensionless nonlinear ordinary differential Equations (30) and (31) with the boundary conditions (33) and (34) are solved using the MATLAB solver *bvp4c* for boundary value problem. The step size is set as default value in the solver. The tolerance for the maximum residue is set as 0.001. The shooting method is applied to implement ϕ_{avg} in the integral form. Table 1 shows the designated values of dimensionless numbers and parameters used in this case study.

In this section, we do a basic parametric study by varying these dimensionless numbers and paramters to seetheir effects on the velocity and the concentration profiles.

5.1. Effect of ϕ_m

Recall that ϕ_m is the maximum volume fraction of the cement particles, which is usually about 0.5–0.6 [40]. Figure 3 shows the effect of ϕ_m on the velocity distribution and the volume fraction. Five values between 0.45 and 0.65 are selected for the parametric study. The velocity shows a parabolic distribution, and the particles tend to concentrate at the center mainly due to the effects of the first term in the particle flux Equation (15). As ϕ_m increases, the velocity tends to increase, and more particles tend to concentrate at the center. From Figure 4, we can see that both the velocity and the volume fraction distributions become more nonuniform for larger values of ϕ_m, indicating a higher packing. In a qualitative way, Phillips et al. and Wu et al. obtained similar results for the velocity profiles (see Figure 10 in [68]) and the concentration profiles (see Figure 6 in [68]).

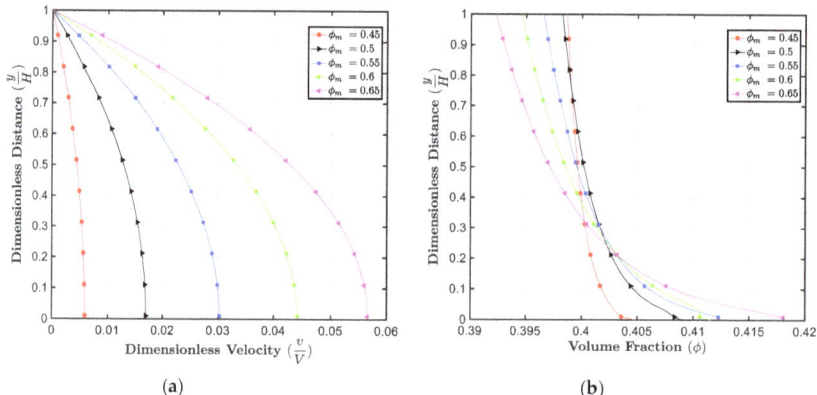

Figure 3. Effect of ϕ_m on (**a**) the velocity; and (**b**) the volume fraction profiles, with $\beta = 1.82$, $\phi_{avg} = 0.4$, $R_0 = 0.1$, $R_1 = -2.5$, $R_2 = 0.1$, $R_3 = 0.01$, $R_4 = 0.07$, $R_5 = 1.8$, $\frac{K_c}{K_\mu} = 0.05$, $m = 1$, $\theta = 45°$.

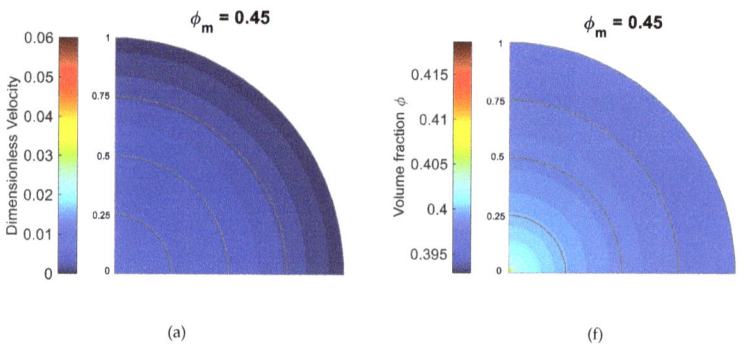

Figure 4. *Cont.*

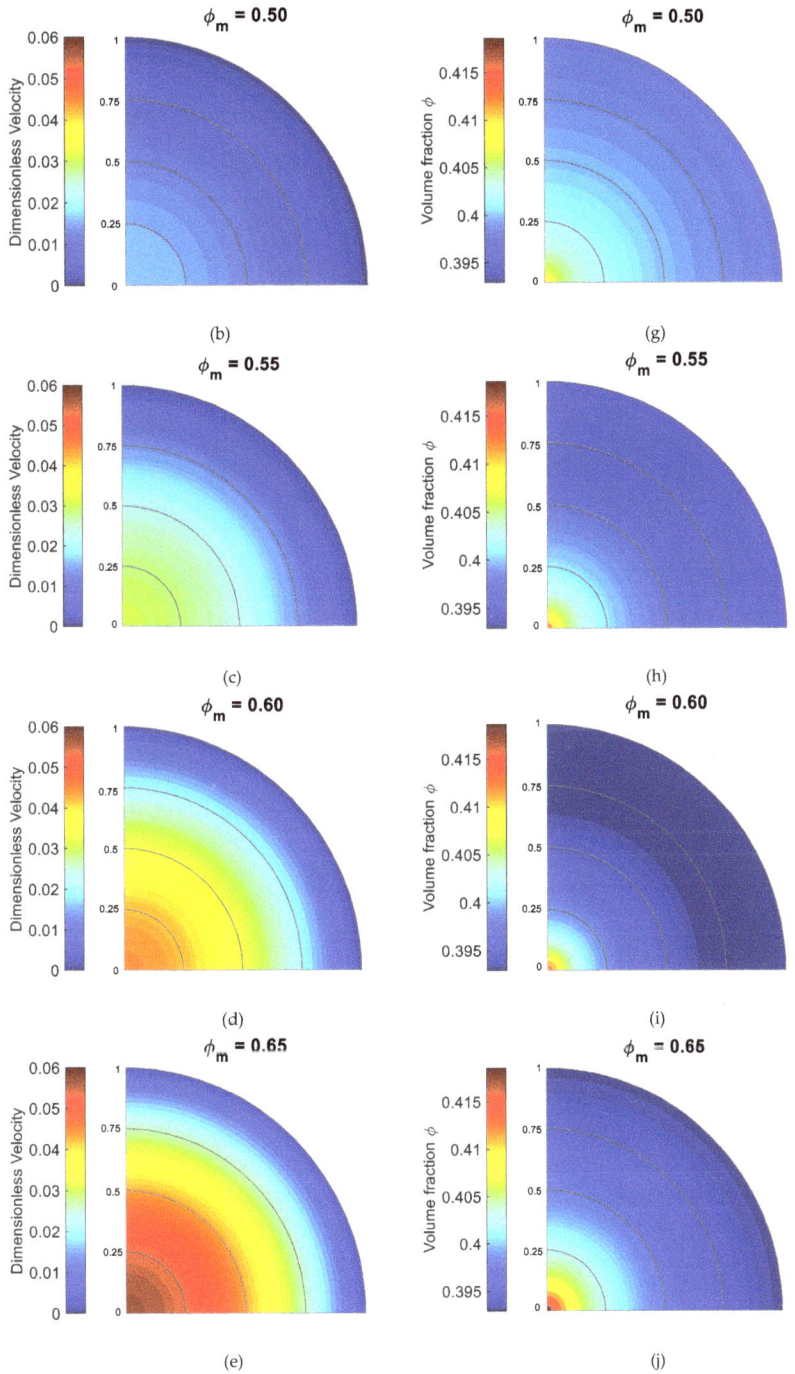

Figure 4. Distribution of the velocity for (**a**) $\phi_m = 0.45$; (**b**) $\phi_m = 0.50$; (**c**) $\phi_m = 0.55$; (**d**) $\phi_m = 0.60$; (**e**) $\phi_m = 0.65$; and the cement volume concentration for (**f**) $\phi_m = 0.45$; (**g**) $\phi_m = 0.50$; (**h**) $\phi_m = 0.55$; (**i**) $\phi_m = 0.60$; (**j**) $\phi_m = 0.65$.

5.2. Effect of $\frac{K_c}{K_\mu}$

Figure 5 shows the effect of $\frac{K_c}{K_\mu}$. This ratio is an indication of the effect of particles collision to the spatial variation of the viscosity. As $\frac{K_c}{K_\mu}$ increases, the velocity at the centerline increases a bit and the volume fraction distribution becomes more nonuniform, as shown in Figure 6. When $\frac{K_c}{K_\mu}$ is zero, the distribution of the volume fraction is nearly uniform. Larger values of $\frac{K_c}{K_\mu}$ indicate that the particles migrate towards the centerline. In a qualitative way, Wu et al. [68] obtained similar results for the effect of $\frac{K_c}{K_\mu}$ (see Figures 17 and 18 in [68]).

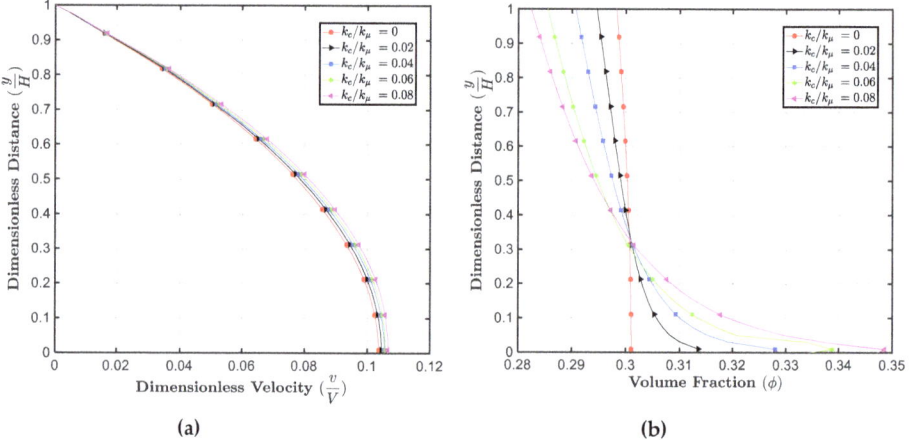

Figure 5. Effect of $\frac{K_c}{K_\mu}$ on (**a**) the velocity; and (**b**) the volume fraction profiles, with $\beta = 1.82$, $\phi_{avg} = 0.3$, $R_0 = 0.1$, $R_1 = -2.5$, $R_2 = 0.1$, $R_3 = 0.01$, $R_4 = 0.07$, $R_5 = 1.8$, $\phi_m = 0.65$, $m = 1$, $\theta = 45°$.

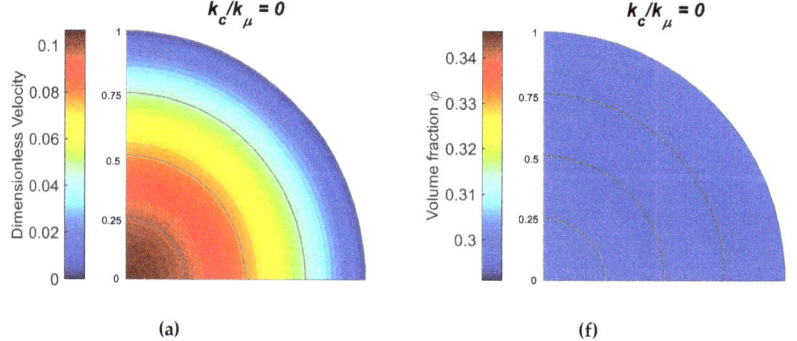

Figure 6. *Cont.*

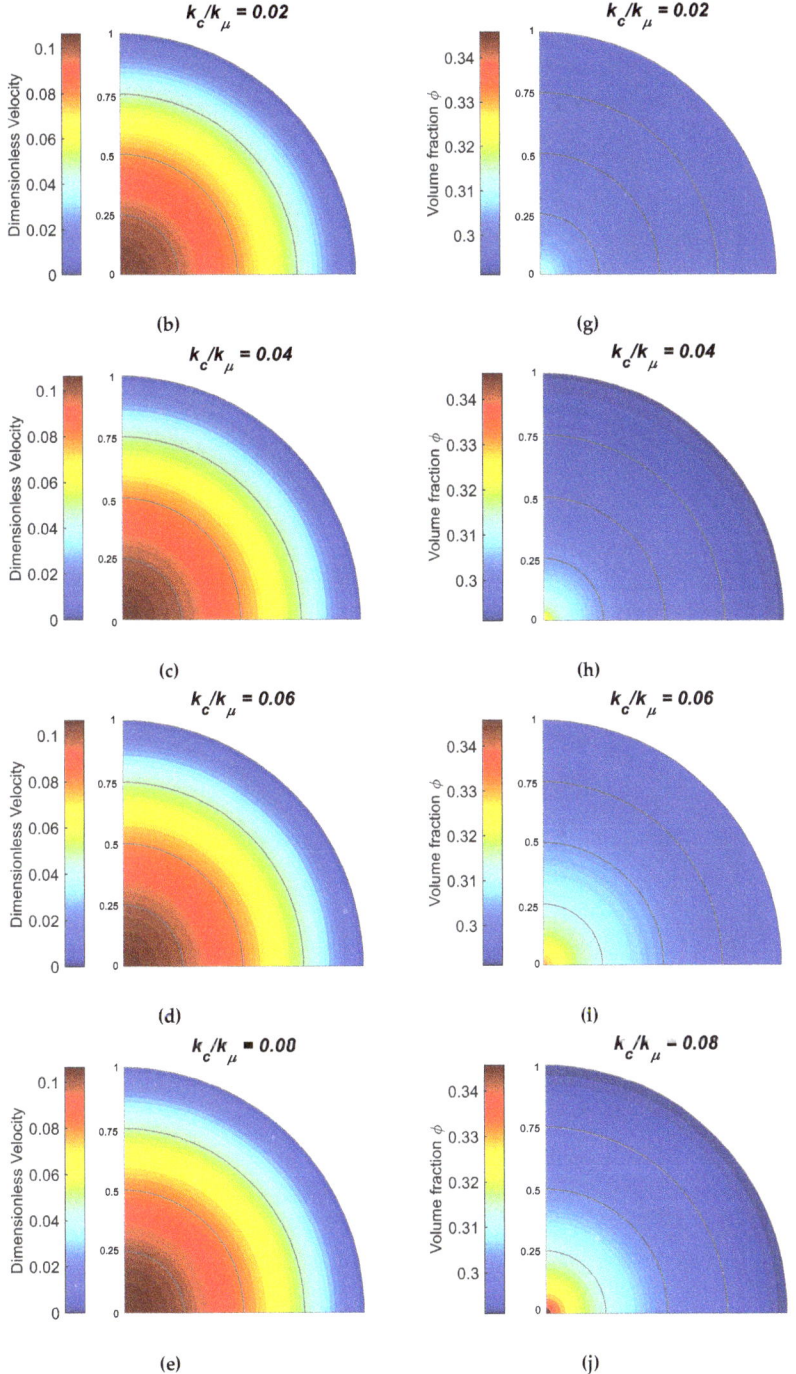

Figure 6. Distribution of the velocity for (**a**) $\frac{K_c}{K_\mu} = 0$; (**b**) $\frac{K_c}{K_\mu} = 0.02$; (**c**) $\frac{K_c}{K_\mu} = 0.04$; (**d**) $\frac{K_c}{K_\mu} = 0.06$; (**e**) $\frac{K_c}{K_\mu} = 0.08$; and the cement volume concentration for (**f**) $\frac{K_c}{K_\mu} = 0$; (**g**) $\frac{K_c}{K_\mu} = 0.02$; (**h**) $\frac{K_c}{K_\mu} = 0.04$; (**i**) $\frac{K_c}{K_\mu} = 0.06$; (**j**) $\frac{K_c}{K_\mu} = 0.08$.

5.3. Effect of θ

Figure 7 shows the effect of the inclination angle θ. For horizontal flow, $\theta = 0°$, and when the two plates are in vertical arrangment $\theta = 90°$. As θ increases, the value of the gravitational force in the x-direction increases. Larger values for θ result in faster flows and nonuniform velocity and volume fraction distributions, as shown in Figure 8. In other words, when the plates are inclined at a sharper angle from the horizontal direction, the particles tend to move faster and have lower concentration at the plates.

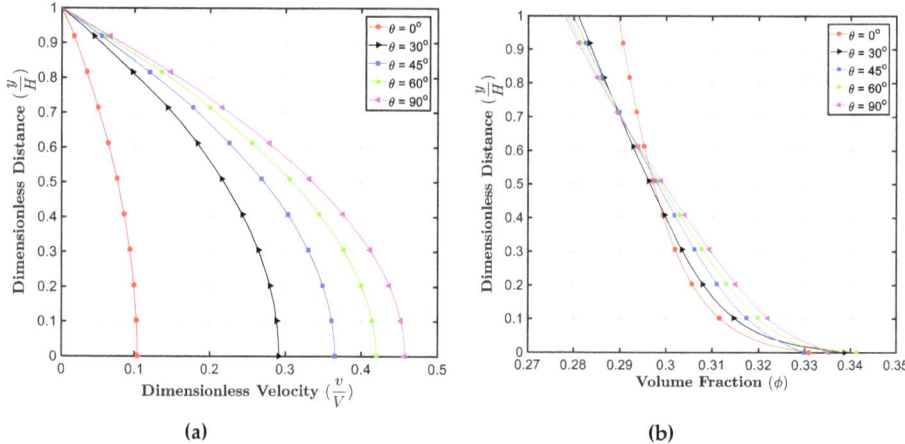

Figure 7. Effect of θ on (**a**) the velocity; and (**b**) the volume fraction profiles, with $\beta = 1.82$, $\phi_{avg} = 0.3$, $R_0 = 0.1$, $R_1 = -2.5$, $R_2 = 10$, $R_3 = 0.01$, $R_4 = 0.07$, $R_5 = 1.8$, $\frac{K_c}{K_\mu} = 0.05$, $\phi_m = 0.65$, $m = 1$.

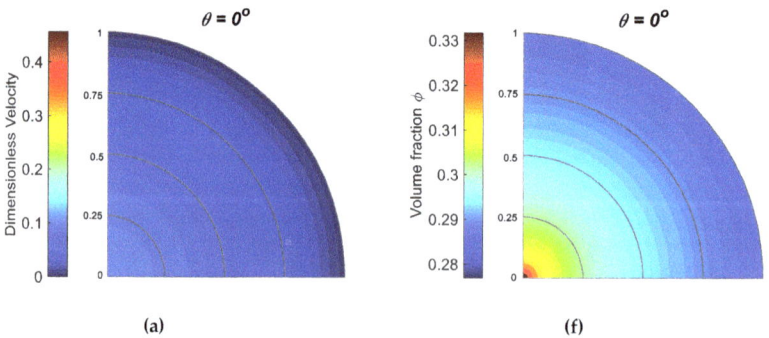

Figure 8. Cont.

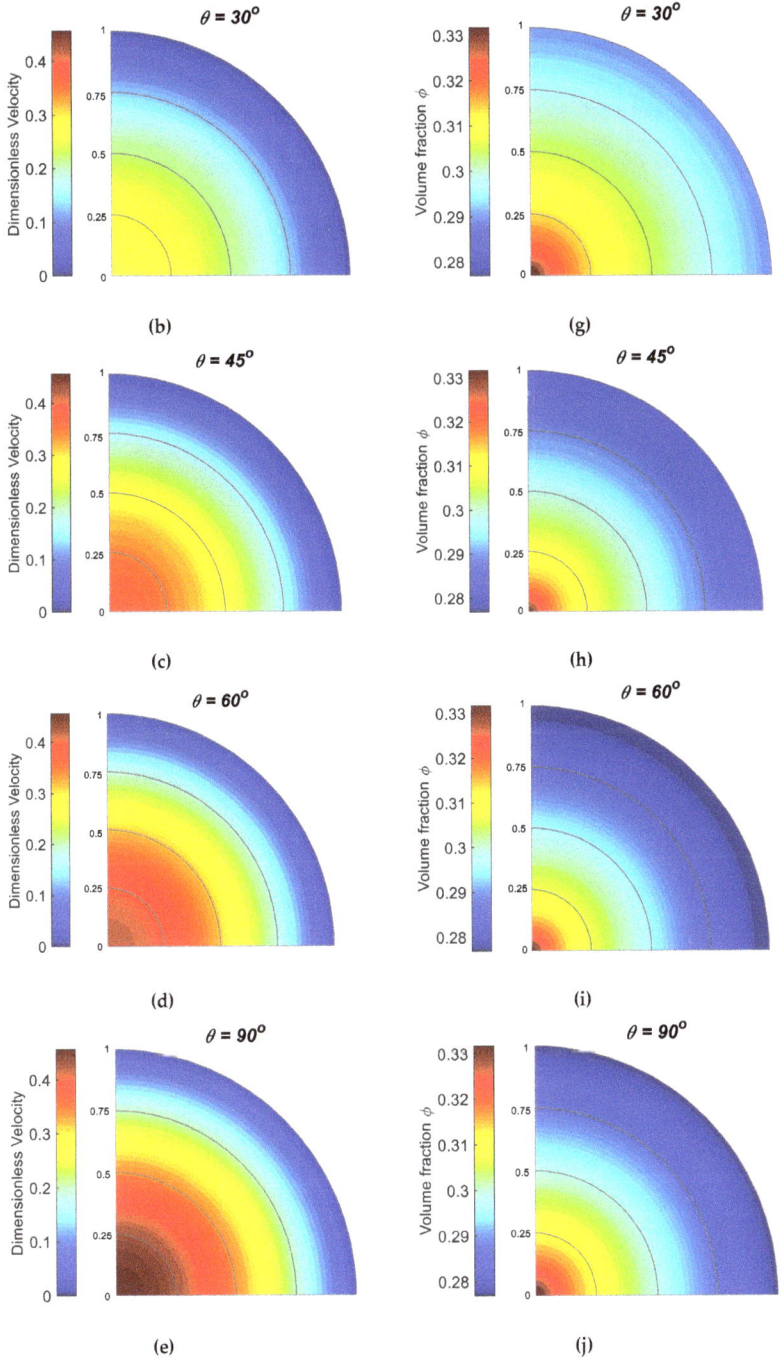

Figure 8. Distribution of the velocity for (**a**) $\theta = 0°$; (**b**) $\theta = 30°$; (**c**) $\theta = 45°$; (**d**) $\theta = 60°$; (**e**) $\theta = 90°$; and the cement volume concentration for (**f**) $\theta = 0°$; (**g**) $\theta = 30°$; (**h**) $\theta = 45°$; (**i**) $\theta = 60°$; (**j**) $\theta = 90°$.

5.3.1. Effect of m

When $m = 0$, the slurry behaves as Newtonian fluid (when the viscosity does not depend on the volume fraction); when $m < 0$, the slurry is shear-thinning; and when $m > 0$, the slurry is shear-thickening. Cement slurry shows shear-thinning behavior if there are no dispersing agents while it exhibits shear-thickening behavior with the addition of dispersing acrylic polyelectrolyte [69]. Thus, we select both positive and negative values of m to study the rheological behavior of cement slurry. Figure 9 shows the effect of m on the velocity and volume fraction profiles. From the above figure, we can see that if the fluid changes from a shear-thinning fluid to a shear-thickening one (m changing from negative to positive), the velocity at the centerline decreases and the distribution of velocity becomes more linear. Similar trends could be found in Figure 2 in [68]. Larger values of m indicate more nonuniform distribution of the particles. In a qualitative way, a similar trend is observed by other researchers (see Figure 3 in [68] and Figure 2a,b in [70]).

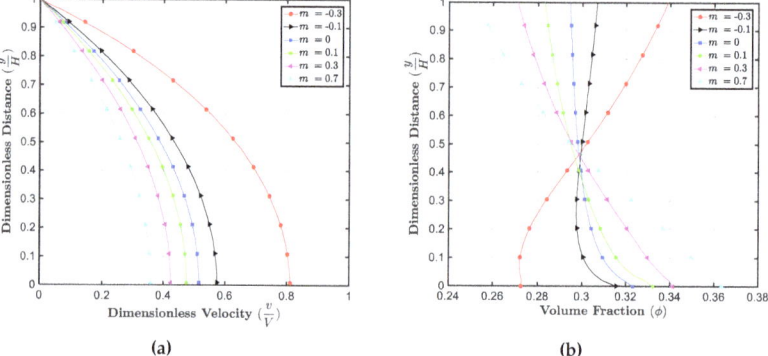

Figure 9. Effect of m on (**a**) the velocity; and (**b**) the volume fraction profiles, with $\beta = 1.82$, $\phi_{avg} = 0.3$, $R_0 = 0.1$, $R_1 = -2.5$, $R_2 = 10$, $R_3 = 0.01$, $R_4 = 0.07$, $R_5 = 1.8$, $\frac{K_c}{K_\mu} = 0.05$, $\phi_m = 0.65$, $\theta = 90°$.

5.4. Effect of R_0

Figure 10 shows that as R_0 increases, the shear-thickening behavior becomes weaker ($m = 1$), while the magnitude of the velocity decreases and more particles tend to concentrate near the centerline and the distribution of volume fraction becomes more nonuniform, as shown in Figure 11.

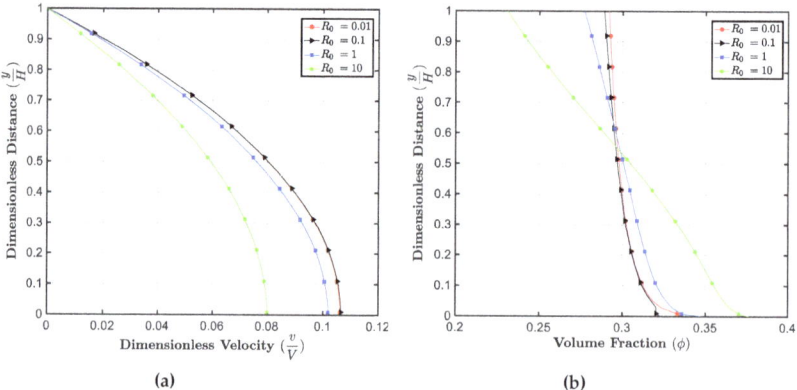

Figure 10. Effect of R_0 on (**a**) the velocity; and (**b**) the volume fraction profiles, with $\beta = 1.82$, $\phi_{avg} = 0.3$, $R_1 = -2.5$, $R_2 = 0.1$, $R_3 = 0.01$, $R_4 = 0.07$, $R_5 = 1.8$, $\frac{K_c}{K_\mu} = 0.05$, $\phi_m = 0.65$, $m = 1$, $\theta = 45°$.

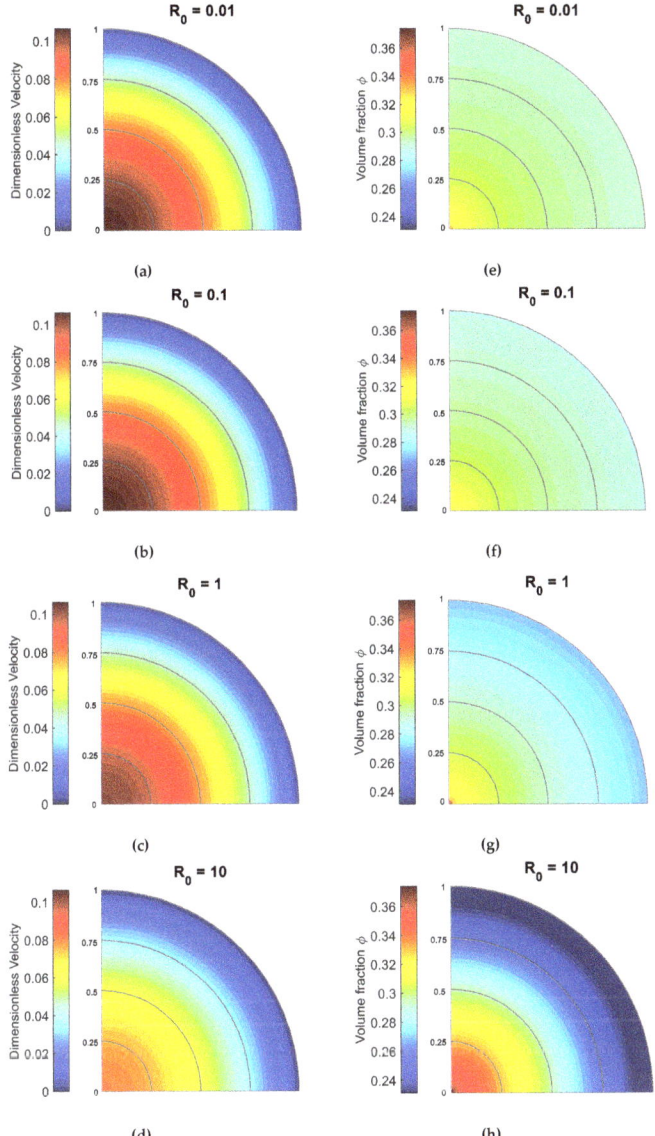

Figure 11. Distribution of the velocity for (**a**) $R_0 = 0.01$; (**b**) $R_0 = 0.1$; (**c**) $R_0 = 1$; (**d**) $R_0 = 10$; and the cement volume concentration for (**e**) $R_0 = 0.01$; (**f**) $R_0 = 0.1$; (**g**) $R_0 = 1$; (**h**) $R_0 = 10$.

5.5. Effect of R_1

Recall that R_1 is related to the effect of pressure gradient in the x-direction. Figure 12 shows that when R_1 becomes more negative, the flow becomes faster and more particles are concentrated at the center. When $R_1 = 0$, indicating no pressure gradient, a constant volume fraction profile is noticed. For larger (absolute) values of R_1 more nonuniform velocity as well as nonuniform volume fraction profiles are noticed, shown in Figure 13. Larger values for the pressure gradient indicate smaller values for the volume fraction at the plate and larger values for the volume fraction at the center. Similar trends for the effect of pressure gradient can also be found in [70] (Figure 4a,b) and [71] (Figures 2 and 3).

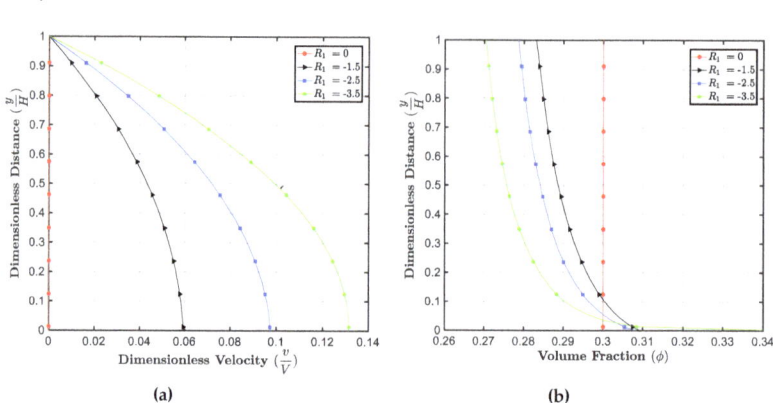

Figure 12. Effect of R_1 on (**a**) the velocity; and (**b**) the volume fraction profiles, with $\beta = 1.82$, $\phi_{avg} = 0.3$, $R_0 = 0.1$, $R_2 = 0.1$, $R_3 = 0.01$, $R_4 = 0.07$, $R_5 = 1.8$, $\frac{K_c}{K_\mu} = 0.05$, $\phi_m = 0.65$, $m = 1$, $\theta = 45°$.

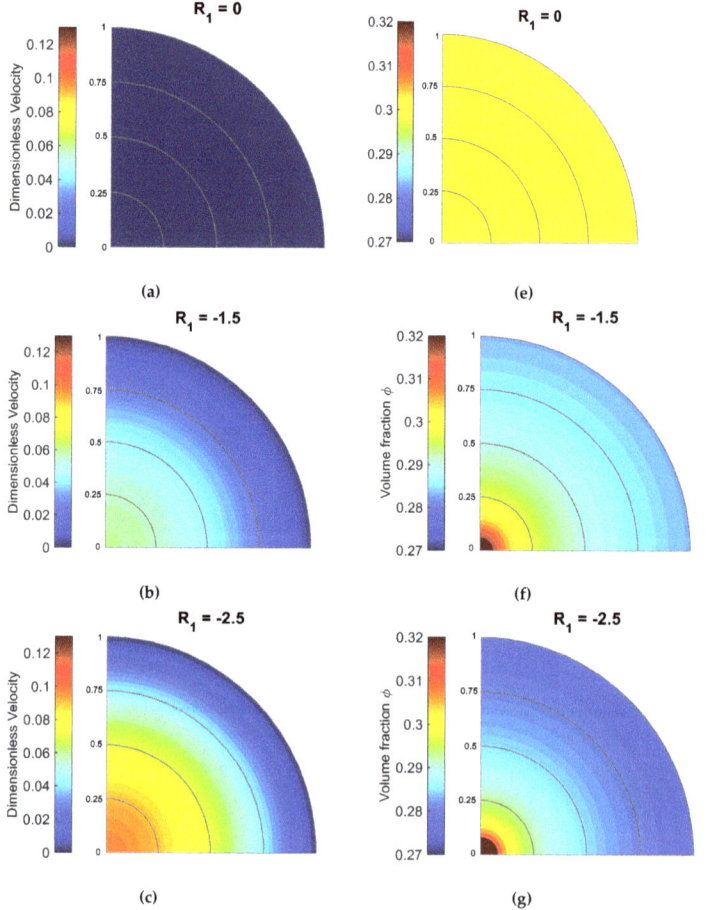

Figure 13. *Cont.*

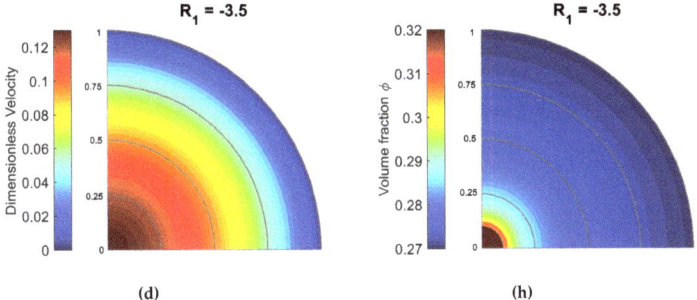

Figure 13. Distribution of the velocity for (**a**) $R_1 = 0$; (**b**) $R_1 = -1.5$; (**c**) $R_1 = -2.5$; (**d**) $R_1 = -3.5$; and the cement volume concentration for (**e**) $R_1 = 0$; (**f**) $R_1 = -1.5$; (**g**) $R_1 = -2.5$; (**h**) $R_1 = -3.5$.

5.6. Effect of R_2

Recall that R_2 is related to the effect of the gravity term (related to the weight of the particles). As shown in Figure 14, with an increase in R_2, the slurry has a higher centerline velocity and more particles tend to concentrate at the centerline. Both distributions of the velocity and the volume fraction become more non-uniform when R_2 increases. In a qualitative way, Miao and Massoudi [70] show similar trends for the effect of R_2 (see Figure 6a,b in their paper).

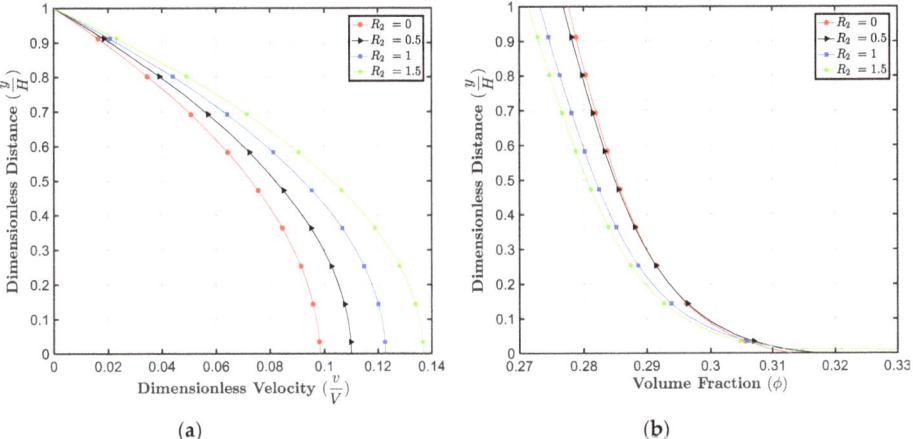

Figure 14. Effect of R_1 on (**a**) the velocity; and (**b**) the volume fraction profiles, with $\beta = 1.82$, $\phi_{avg} = 0.3$, $R_0 = 0.1$, $R_1 = -2.5$, $R_3 = 0.01$, $R_4 = 0.07$, $R_5 = 1.8$, $\frac{K_\kappa}{K_\mu} = 0.05$, $\phi_m = 0.65$, $m = 1$, $\theta = 45°$.

5.7. Effects of R_3, R_4, and R_5

The three dimensionless numbers, R_3, R_4, and R_5 reflect the dependence of cement diffusivity on the volume fraction directly, as can be seen from Equation (31). They also influence the velocity profiles indirectly, as seen from Equation (30). We selected three values 0.01, 0.1, and 1 for R_3, R_4, and R_5, respectively. From Figure 15, we can see that R_3, R_4, and R_5 have similar effects on the velocity and the volume fraction profiles. As the values of these parameters are increased, the velocity at the centerline does not change much and fewer particles tend to concentrate at the center. The volume fraction distribution becomes more uniform for larger values of R_3, R_4, and R_5.

Finally, we should mention that due to the kinematical assumptions made (see Equation (24)), many of the coupling effects and the nonlinear effects in the momentum equations have disappeared.

In three dimensional unsteady flows, we anticipate more interesting results due to these effects, such as the contributions from the normal stress effects, or additional flux terms, etc.

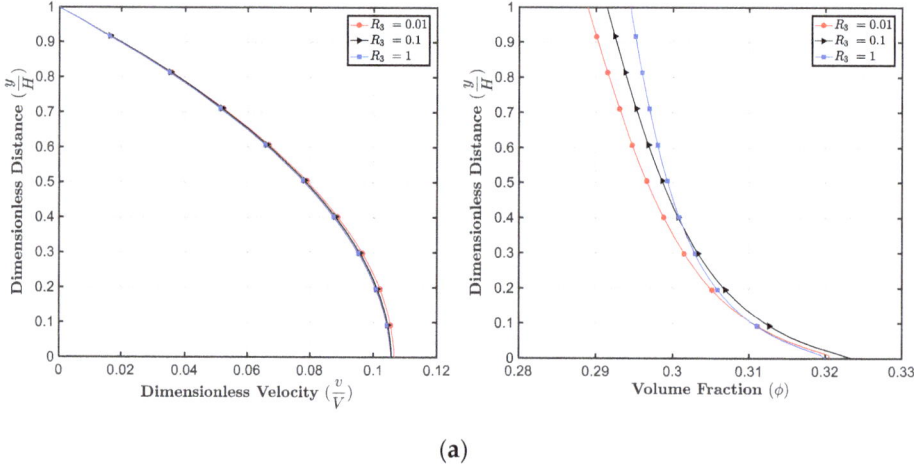

(**a**)

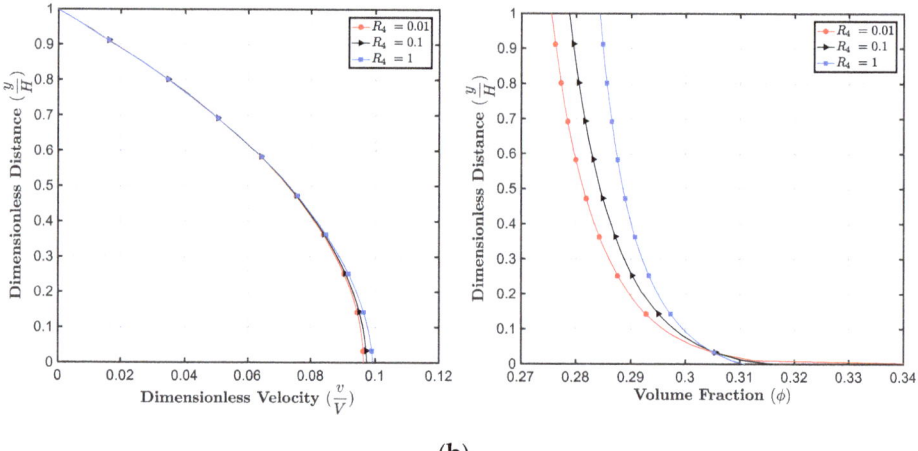

(**b**)

Figure 15. *Cont.*

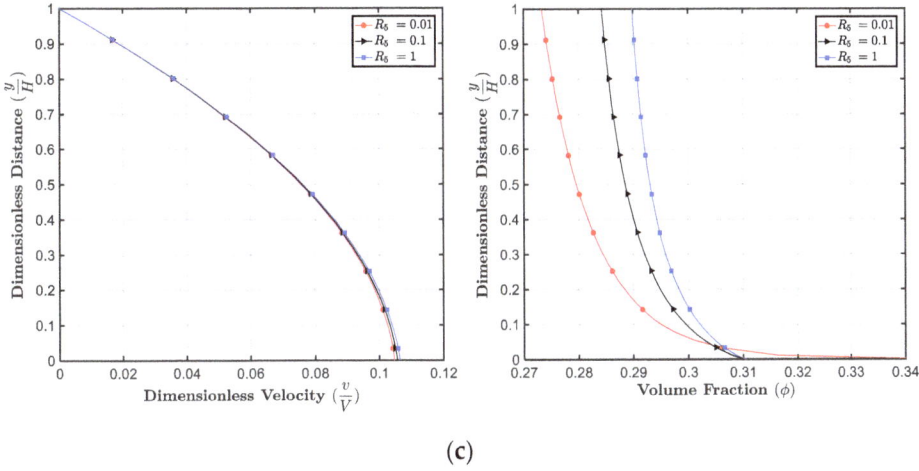

(c)

Figure 15. (a) Effect of R_3 on the velocity and the volume fraction profiles, with $\beta = 1.82$, $\phi_{avg} = 0.3$, $R_0 = 0.1$, $R_1 = -2.5$, $R_2 = 0.1$, $R_4 = 0.07$, $R_5 = 1.8$, $\frac{K_c}{K_\mu} = 0.05$, $\phi_m = 0.65$, $m = 1$, $\theta = 45°$; (b) effect of R_4 on the velocity and volume fraction profile, with $\beta = 1.82$, $\phi_{avg} = 0.3$, $R_0 = 0.1$, $R_1 = -2.5$, $R_2 = 0.1$, $R_3 = 0.01$, $R_5 = 1.8$, $\frac{K_c}{K_\mu} = 0.05$, $\phi_m = 0.65$, $m = 1$, $\theta = 45°$; and (c) effect of R_5 on the velocity and volume fraction profile, with $\beta = 1.82$, $\phi_{avg} = 0.3$, $R_0 = 0.1$, $R_1 = -2.5$, $R_2 = 0.1$, $R_3 = 0.01$, $R_4 = 0.07$, $\frac{K_c}{K_\mu} = 0.05$, $\phi_m = 0.65$, $m = 1$, $\theta = 45°$.

6. Conclusions

The space between the well casing and the geological formation surrounding the wellbore must be filled with cement slurry before the cement hardens. Poor understanding of rheological properties will cause the failure of the zonal isolation and can cause problems related to gas and fluid migration. Disasters, financial loss, and other serious consequences may occur from unsuccessful cementing design. Rheological behavior of cement slurry is important in the petroleum industry. In this paper, we have modeled the cement slurry as a non-Newtonian fluid (a generalized second grade fluid), where the viscosity depends on the shear rate and the particle concentration. To consider the particle transport, we use a concentration flux equation, where the coefficients of diffusivity and other fluxes depend on the shear rate and viscosity. The governing equations and the boundary conditions for flow between two plates are nondimensionalized and solved numerically. We performed a parametric study for different dimensionless numbers and the results indicate that the velocity and the volume fraction profiles are affected by the shear rate dependent viscosity and the parameters in the concentration flux equation. We also notice that the maximum packing ϕ_m, concentration flux parameters $\frac{K_c}{K_\mu}$, the angle of inclination θ, pressure and gravity terms affect the velocity and particle distributions significantly. For example, we can see that the velocity and the volume fraction distributions become more nonuniform for larger values of ϕ_m, which indicates a higher packing. In a qualitative way, Phillips et al. and Wu et al. obtained similar results for the velocity profiles (see Figure 10 in [68]) and concentration profiles (see Figure 6 in [68]). We also notice that larger values for the pressure gradient indicate smaller values for the volume fraction at the plate but bigger values for the volume fraction at the center. Similar trends for the effect of pressure gradient can also be found in [70] (Figure 4a,b) and [71] (Figures 2 and 3). For future studies, we will look at unsteady flows. Two-dimensional and three-dimensional geometries of the oil well annulus will also be studied with more advanced numerical software. More effects such as yield stress, heat transfer, and cement hydration parameters will also be considered. We should mention that in general, cement is a multi-component material, and methods of multiphase flows and mixture theories, which are more complicated can also be used to study cement slurry (for details see for example the recent papers by [72,73]).

Author Contributions: M.M. developed the framework of the paper and C.T. did most of the derivations (with help from MM) and all of the numerical simulations with some help from WTW. All authors contributed to the writing of the paper.

Funding: This research was supported by the U. S. Department of Energy, National Energy Technology Laboratory.

Acknowledgments: This paper is supported by an appointment to the U.S. Department of Energy Postgraduate Research Program for C.T. at the National Energy Technology Laboratory, administrated by the Oak Ridge Institute for Science and Education.

Disclaimer: This report was prepared as an account of work sponsored by an agency of the United States Government. Neither the United States Government nor any agency thereof, nor any of their employees, makes any warranty, express or implied, or assumes any legal liability or responsibility for the accuracy, completeness, or usefulness of any information, apparatus, product, or process disclosed, or represents that its use would not infringe privately owned rights. Reference herein to any specific commercial product, process, or service by trade name, trademark, manufacturer, or otherwise does not necessarily constitute or imply its endorsement, recommendation, or favoring by the United States Government or any agency thereof. The views and opinions of authors expressed herein do not necessarily state or reflect those of the United States Government or any agency thereof.

Conflicts of Interest: The authors declare no conflict of interest.

Nomenclature

Symbol	Explanation
ρ	Density
g	Acceleration due to gravity
H	Characteristic length
V	Reference velocity
a	Particle radius
θ	Inclination angle
t	Time
D	Diffusion coefficient
D_0	Cement diffusivity parameter
D_1	Shear-rate-dependent diffusion coefficient
D_2	Volume fraction-dependent diffusion coefficient
η	Constant in volume fraction-dependent diffusion coefficient
$K_1, K_2 K_3$	Cement diffusivity fitting coefficients
$H(.)$	Heaviside function
K_c, K_μ	Empirical coefficients for transport flux
$\dot{\gamma}$	Local shear rate
ϕ	Volume fraction of particles
φ	Capillary pore in cement
μ	Viscosity
μ_0	Coefficient of viscosity
μ_{eff}	Effective viscosity
$\mu^*(\phi)$	Concentration-dependent viscosity
β	Experimental (fitting) parameter for viscosity
ϕ_m	Maximum solid concentration
ϕ_{avg}	Average value for the volume fraction
p	Pressure
$\hat{p}$	Modified pressure
α_1, α_2	Normal stress coefficients
α, m	Material parameters
$R_0, R_1, R_2, R_3, R_4, R_5$	Dimensionless numbers
v	Velocity vector
b	Body force vector
N	Particle transport flux

N_c	Flux contribution due to particles collision
N_μ	Flux contribution due to variation in viscosity
N_b	Brownian diffusive flux
T	Cauchy stress tensor
T_y	Yield stress tensor
T_v	Viscous stress tensor
I	Identity tensor
L	Gradient of the velocity vector
A_n	n-th order Rivlin–Ericksen tensor
∇	Gradient symbol
div	Divergence operator

References

1. Mindess, S.; Young, J.F. *Concrete*; Prentice Hall: Englewood, NJ, USA, 1981.
2. Taylor, H.F. *Cement Chemistry*; Thomas Telford: London, UK, 1997.
3. Hanehara, S.; Yamada, K. Interaction between cement and chemical admixture from the point of cement hydration, absorption behaviour of admixture, and paste rheology. *Cem. Concr. Res.* **1999**, *29*, 1159–1165. [CrossRef]
4. Vlachou, P.-V.; Piau, J.-M. Physicochemical study of the hydration process of an oil well cement slurry before setting. *Cem. Concr. Res.* **1999**, *29*, 27–36. [CrossRef]
5. Barbic, L.; Tinta, V.; Lozar, B.; Marinkovic, V. Effect of Storage Time on the Rheological Behavior of Oil Well Cement Slurries. *J. Am. Ceram. Soc.* **1991**, *74*, 945–949. [CrossRef]
6. Worrell, E.; Kermeli, K.; Galitsky, C. *Energy Efficiency Improvement and Cost Saving Opportunities for Cement Making an ENERGY STAR®Guide for Energy and Plant Managers*; EPA: Washington, DC, USA, 2013.
7. Chatziaras, N.; Psomopoulos, C.S.; Themelis, N.J. Use of waste derived fuels in cement industry: A review. *Manag. Environ. Qual. Int. J.* **2016**, *27*, 178–193. [CrossRef]
8. Benhelal, E.; Zahedi, G.; Shamsaei, E.; Bahadori, A. Global strategies and potentials to curb CO_2 emissions in cement industry. *J. Clean. Prod.* **2013**, *51*, 142–161. [CrossRef]
9. Bentz, D.P. Three-Dimensional Computer Simulation of Portland Cement Hydration and Microstructure Development. *J. Am. Ceram. Soc.* **1997**, *80*, 3–21. [CrossRef]
10. Haecker, C.; Bentz, D.; Feng, X.; Stutzman, P. Prediction of cement physical properties by virtual testing. *Cem. Int.* **2003**, *1*, 86–92.
11. Bullard, J.W.; Ferraris, C.; Garboczi, E.J.; Martys, N.; Stutzman, P. Virtual cement. *Chapt* **2004**, *10*, 1311–1331.
12. Thomas, J.J.; Biernacki, J.J.; Bullard, J.W.; Bishnoi, S.; Dolado, J.S.; Scherer, G.W.; Luttge, A. Modeling and simulation of cement hydration kinetics and microstructure development. *Cem. Concr. Res.* **2011**, *41*, 1257–1278. [CrossRef]
13. Watts, B.; Tao, C.; Ferraro, C.; Masters, F. Proficiency analysis of VCCTL results for heat of hydration and mortar cube strength. *Constr. Build. Mater.* **2018**, *161*, 606–617. [CrossRef]
14. Tao, C.; Watts, B.; Ferraro, C.C.; Masters, F.J. A Multivariate Computational Framework to Characterize and Rate Virtual Portland Cements. *Comput.-Aided Civ. Infrastruct. Eng.* **2019**, *34*, 266–278. [CrossRef]
15. Banfill, P.F.G.; Kitching, D.R. 14 Use of a Controlled Stress Rheometer to Study the Yield Stress of Oilwell Cement Slurries. In *Rheology of Fresh Cement and Concrete: Proceedings of an International Conference, Liverpool, 1990*; CRC Press: Boca Raton, FL, USA, 1990; p. 125.
16. Bonett, A.; Pafitis, D. Getting to the root of gas migration. *Oilfield Rev.* **1996**, *8*, 36–49.
17. Guan, S.; Rice, J.A.; Li, C.; Li, Y.; Wang, G. Structural displacement measurements using DC coupled radar with active transponder. *Struct. Control Health Monit.* **2017**, *24*, e1909. [CrossRef]
18. Prohaska, M.; Ogbe, D.O.; Economides, M.J. Determining wellbore pressures in cement slurry columns. In Proceedings of the SPE Western Regional Meeting, Anchorage, AK, USA, 26–28 May 1993.
19. Chenevert, M.E.; Jin, L. Model for predicting wellbore pressures in cement columns. In Proceedings of the SPE Annual Technical Conference and Exhibition, San Antonio, TX, USA, 8–11 October 1989.
20. Stiles, D.A. Successful Cementing in Areas Prone to Shallow Saltwater Flows in Deep-Water Gulf of Mexico. In Proceedings of the Offshore Technology Conference, Houston, TX, USA, 5–8 May 1997.

21. Brandt, W.; Dang, A.S.; Magne, E.; Crowley, D.; Houston, K.; Rennie, A.; Hodder, M.; Stringer, R.; Juiniti, R.; Ohara, S.; et al. Deepening the search for offshore hydrocarbons. *Oilfield Rev.* **1998**, *10*, 2–21.
22. Wallevik, J.E. Thixotropic investigation on cement paste: Experimental and numerical approach. *J. Non-Newton. Fluid Mech.* **2005**, *132*, 86–99. [CrossRef]
23. Foroushan, H.K.; Ozbayoglu, E.M.; Miska, S.Z.; Yu, M.; Gomes, P.J. On the Instability of the Cement/Fluid Interface and Fluid Mixing (includes associated erratum). *SPE Drill. Complet.* **2018**, *33*, 63–76. [CrossRef]
24. Skadsem, H.J.; Kragset, S.; Lund, B.; Ytrehus, J.D.; Taghipour, A. Annular displacement in a highly inclined irregular wellbore: Experimental and three-dimensional numerical simulations. *J. Pet. Sci. Eng.* **2019**, *172*, 998–1013. [CrossRef]
25. Liu, L.; Fang, Z.; Qi, C.; Zhang, B.; Guo, L.; Song, K.I.-I.L. Numerical study on the pipe flow characteristics of the cemented paste backfill slurry considering hydration effects. *Powder Technol.* **2019**, *343*, 454–464. [CrossRef]
26. Murphy, E.; Lomboy, G.; Wang, K.; Sundararajan, S.; Subramaniam, S. The rheology of slurries of athermal cohesive micro-particles immersed in fluid: A computational and experimental comparison. *Chem. Eng. Sci.* **2019**, *193*, 411–420. [CrossRef]
27. Slattery, J.C. *Advanced Transport Phenomena*; Cambridge University Press: Cambridge, UK, 1999.
28. Probstein, R.F. *Physicochemical Hydrodynamics: An Introduction*; John Wiley & Sons: Hoboken, NJ, USA, 2005.
29. Struble, L.; Sun, G.-K. Viscosity of Portland cement paste as a function of concentration. *Adv. Cem. Based Mater.* **1995**, *2*, 62–69. [CrossRef]
30. Banfill, P.F. The rheology of fresh cement and concrete-a review. In Proceedings of the 11th international cement chemistry congress, Durban, South Africa, 11–16 May 2003; Volume 1, pp. 50–62.
31. Gandelman, R.; Miranda, C.; Teixeira, K.; Martins, A.L.; Waldmann, A. On the rheological parameters governing oilwell cement slurry stability. *Annu. Trans. Nord. Rheol. Soc.* **2004**, *12*, 85–91.
32. Tattersall, G.H.; Banfill, P.F. *The Rheology of Fresh Concrete*; Pitman Advanced Publishing Program: London, UK, 1983.
33. Moon, J.; Wang, S. Acoustic method for determining the static gel strength of slurries. In Proceedings of the SPE Rocky Mountain Regional Meeting, Gillette, WY, USA, 15–18 May 1999.
34. Bellotto, M. Cement paste prior to setting: A rheological approach. *Cem. Concr. Res.* **2013**, *52*, 161–168. [CrossRef]
35. Barnes, H.A. Thixotropy—A review. *J. Non-Newton. Fluid Mech.* **1997**, *70*, 1–33. [CrossRef]
36. Mewis, J.; Wagner, N.J. *Colloidal Suspension Rheology*; Cambridge University Press: Cambridge, UK, 2012.
37. Barnes, H.A. The yield stress—A review or '$\pi\alpha\nu\tau\alpha\ \rho\epsilon\iota$'—Everything flows? *J. Non-Newton. Fluid Mech.* **1999**, *81*, 133–178. [CrossRef]
38. Carreau, P.J.; de Kee, D.C.R.; Chhabra, R.P. *Rheology of Polymeric Systems: Principles and Applications*; Hanser: New York, NY, USA, 1997.
39. Carreau, P.J.; de Kee, D. Review of some useful rheological equations. *Can. J. Chem. Eng.* **1979**, *57*, 3–15. [CrossRef]
40. Justnes, H.; Vikan, H. Viscosity of cement slurries as a function of solids content. *Ann. Trans. Nord. Rheol. Soc.* **2005**, *13*, 75–82.
41. Asaga, K.; Roy, D.M. Rheological properties of cement mixes: IV. Effects of superplasticizers on viscosity and yield stress. *Cem. Concr. Res.* **1980**, *10*, 287–295. [CrossRef]
42. Vand, V. Viscosity of solutions and suspensions. I. Theory. *J. Phys. Chem.* **1948**, *52*, 277–299. [CrossRef]
43. Roscoe, R. The viscosity of suspensions of rigid spheres. *Br. J. Appl. Phys.* **1952**, *3*, 267. [CrossRef]
44. Bonen, D.; Shah, S.P. Fresh and hardened properties of self-consolidating concrete. *Prog. Struct. Eng. Mater.* **2005**, *7*, 14–26. [CrossRef]
45. Chougnet, A.; Palermo, T.; Audibert, A.; Moan, M. Rheological behaviour of cement and silica suspensions: Particle aggregation modelling. *Cem. Concr. Res.* **2008**, *38*, 1297–1301. [CrossRef]
46. Tregger, N.A.; Pakula, M.E.; Shah, S.P. Influence of clays on the rheology of cement pastes. *Cem. Concr. Res.* **2010**, *40*, 384–391. [CrossRef]
47. Bentz, D.P.; Ferraris, C.F.; Galler, M.A.; Hansen, A.S.; Guynn, J.M. Influence of particle size distributions on yield stress and viscosity of cement–fly ash pastes. *Cem. Concr. Res.* **2012**, *42*, 404–409. [CrossRef]
48. Ouyang, J.; Tan, Y. Rheology of fresh cement asphalt emulsion pastes. *Constr. Build. Mater.* **2015**, *80*, 236–243. [CrossRef]
49. O'Neill, R.; McCarthy, H.O.; Montufar, E.B.; Ginebra, M.P.; Wilson, D.I.; Lennon, A.; Dunne, N. Critical review: Injectability of calcium phosphate pastes and cements. *Acta Biomater.* **2017**, *50*, 1–19. [CrossRef] [PubMed]

50. Massoudi, M. Constitutive modelling of flowing granular materials: A continuum approach. In *Granular Materials: Fundamentals and Applications*; The Royal Society of Chemistry: Cambridge, UK, 2004; Volume 63.
51. Massoudi, M.; Mehrabadi, M.M. A continuum model for granular materials: Considering dilatancy and the Mohr-Coulomb criterion. *Acta Mech.* **2001**, *152*, 121–138. [CrossRef]
52. Rivlin, R.S. Further remarks on the stress-deformation relations for isotropic materials. *J. Ration. Mech. Anal.* **1955**, *4*, 681–702. [CrossRef]
53. Truesdell, C.; Noll, W. *The Non-linear Field Theories of Mechanics*; Springer: Berlin, Germany, 1992.
54. Man, C.-S.; Sun, Q.-X. On the significance of normal stress effects in the flow of glaciers. *J. Glaciol.* **1987**, *33*, 268–273. [CrossRef]
55. Man, C.-S. Nonsteady channel flow of ice as a modified second-order fluid with power-law viscosity. *Arch. Ration. Mech. Anal.* **1992**, *119*, 35–57. [CrossRef]
56. Dunn, J.E.; Fosdick, R.L. Thermodynamics, stability, and boundedness of fluids of complexity 2 and fluids of second grade. *Arch. Ration. Mech. Anal.* **1974**, *56*, 191–252. [CrossRef]
57. Massoudi, M.; Vaidya, A. On some generalizations of the second grade fluid model. *Nonlinear Anal. Real World Appl.* **2008**, *9*, 1169–1183. [CrossRef]
58. Phillips, R.J.; Armstrong, R.C.; Brown, R.A.; Graham, A.L.; Abbott, J.R. A constitutive equation for concentrated suspensions that accounts for shear-induced particle migration. *Phys. Fluids A Fluid Dyn.* **1992**, *4*, 30–40. [CrossRef]
59. Li, Y.; Wu, W.-T.; Liu, X.; Massoudi, M. The effects of particle concentration and various fluxes on the flow of a fluid-solid suspension. *Appl. Math. Comput.* **2019**, *358*, 151–160. [CrossRef]
60. Bridges, C.; Rajagopal, K.R. Pulsatile flow of a chemically-reacting nonlinear fluid. *Comput. Math. Appl.* **2006**, *52*, 1131–1144. [CrossRef]
61. Garboczi, E.J.; Bentz, D.P. Computer simulation of the diffusivity of cement-based materials. *J. Mater. Sci.* **1992**, *27*, 2083–2092. [CrossRef]
62. Snyder, K.A.; Bentz, D.P. Suspended hydration and loss of freezable water in cement pastes exposed to 90% relative humidity. *Cem. Concr. Res.* **2004**, *34*, 2045–2056. [CrossRef]
63. Wu, W.-T.; Aubry, N.; Antaki, J.; McKoy, M.; Massoudi, M. Heat transfer in a drilling fluid with geothermal applications. *Energies* **2017**, *10*, 1349. [CrossRef]
64. Gupta, G.; Massoudi, M. Flow of a generalized second grade fluid between heated plates. *Acta Mech.* **1993**, *99*, 21–33. [CrossRef]
65. Massoudi, M. Boundary conditions in mixture theory and in CFD applications of higher order models. *Comput. Math. Appl.* **2007**, *53*, 156–167. [CrossRef]
66. Dunn, J.E.; Rajagopal, K.R. Fluids of differential type: Critical review and thermodynamic analysis. *Int. J. Eng. Sci.* **1995**, *33*, 689–729. [CrossRef]
67. Rajagopal, K.R.; Kaloni, P.N. Some remarks on boundary conditions for flows of fluids of the differential type. *Contin. Mech. Appl.* **1989**, *48*, 935–942.
68. Wu, W.-T.; Massoudi, M. Heat transfer and dissipation effects in the flow of a drilling fluid. *Fluids* **2016**, *1*, 4. [CrossRef]
69. Lootens, D.; Hébraud, P.; Lécolier, E.; van Damme, H. Gelation, Shear-Thinning and Shear-Thickening in Cement Slurries. *Oil Gas Sci. Technol.* **2004**, *59*, 31–40. [CrossRef]
70. Miao, L.; Massoudi, M. Heat transfer analysis and flow of a slag-type fluid: Effects of variable thermal conductivity and viscosity. *Int. J. Non-Linear Mech.* **2015**, *76*, 8–19. [CrossRef]
71. Gudhe, R.; Yalamanchili, R.C.; Massoudi, M. The flow of granular materials in a pipe: Numerical solutions. *ASME Appl. Mech. Div.-Publ.-AMD* **1993**, *160*, 41.
72. Massoudi, M. A note on the meaning of mixture viscosity using the classical continuum theories of mixtures. *Int. J. Eng. Sci.* **2008**, *46*, 677–689. [CrossRef]
73. Massoudi, M. A Mixture Theory formulation for hydraulic or pneumatic transport of solid particles. *Int. J. Eng. Sci.* **2010**, *48*, 1440–1461. [CrossRef]

 © 2019 by the authors. Licensee MDPI, Basel, Switzerland. This article is an open access article distributed under the terms and conditions of the Creative Commons Attribution (CC BY) license (http://creativecommons.org/licenses/by/4.0/).

Article

Techno-Economic Comparison of Onshore and Offshore Underground Coal Gasification End-Product Competitiveness

Natalie Nakaten [1,*] and Thomas Kempka [1,2]

1. GFZ German Research Centre for Geosciences, Fluid Systems Modelling, Telegrafenberg, 14473 Potsdam, Germany
2. University of Potsdam, Institute of Geosciences, Karl-Liebknecht-Str. 24-25, 14476 Potsdam, Germany
* Correspondence: natalie.christine.nakaten@gfz-potsdam.de; Tel.: +49-331-288-28722

Received: 29 May 2019; Accepted: 28 June 2019; Published: 23 August 2019

Abstract: Underground coal gasification (UCG) enables utilization of coal reserves, currently not economically exploitable due to complex geological boundary conditions. Hereby, UCG produces a high-calorific synthesis gas that can be used for generation of electricity, fuels, and chemical feedstock. The present study aims to identify economically-competitive, site-specific end-use options for onshore- and offshore-produced UCG synthesis gas, taking into account the capture and storage (CCS) and/or utilization (CCU) of produced CO_2. Modeling results show that boundary conditions favoring electricity, methanol, and ammonia production expose low costs for air separation, low compression power requirements, and appropriate shares of H_2/N_2. Hereby, a gasification agent ratio of more than 30% oxygen by volume is not favorable from the economic and CO_2 mitigation viewpoints. Compared to the costs of an offshore platform with its technical equipment, offshore drilling costs are marginal. Thus, uncertainties related to parameters influenced by drilling costs are negligible. In summary, techno-economic process modeling results reveal that air-blown gasification scenarios are the most cost-effective ones, while offshore UCG-CCS/CCU scenarios are up to 1.7 times more expensive than the related onshore processes. Hereby, all investigated onshore scenarios except from ammonia production under the assumed worst-case conditions are competitive on the European market.

Keywords: underground coal gasification (UCG); economics; cost of electricity (COE); techno-economic model; methanol; ammonia; carbon capture and storage (CCS); carbon capture and utilization (CCU); electricity generation; process simulation

1. Introduction

Underground coal gasification (UCG) can provide an economic approach to increase worldwide coal reserves by utilization of coal deposits that are currently not mineable by conventional methods. Hereby, the target coal seam is converted into a synthesis gas within a controlled, sub-stoichiometric gasification process [1–5]. After processing, UCG synthesis gas is applicable for different end-uses as, e.g., provision of chemical raw materials, liquid fuels, hydrogen, fertilizers, or electricity (cf. Figure 1). The early idea of UCG and its evolution has a long history that was picked up especially at times of hydrocarbon scarcity. Since the 1930s, more than 50 pilot-scale UCG operations have been carried out worldwide, e.g., in the former Union of Soviet Socialist Republics (USSR), Europe, the U.S., South Africa, Australia, and China. Predominantly, these tests have been undertaken at shallow depths, e.g., at Angren (110 m) in Uzbekistan, Chinchilla (140 m) in Australia, and Hanna (80 m) and Hoe Creek (30–40 m) in the U.S [1–4,6]. One recent pilot-scale UCG installation in Poland was constructed in 2010 at about a 400 m depth in order to gasify 1,300 metric tons of coal at an average rate of 600 kg/h [7–10].

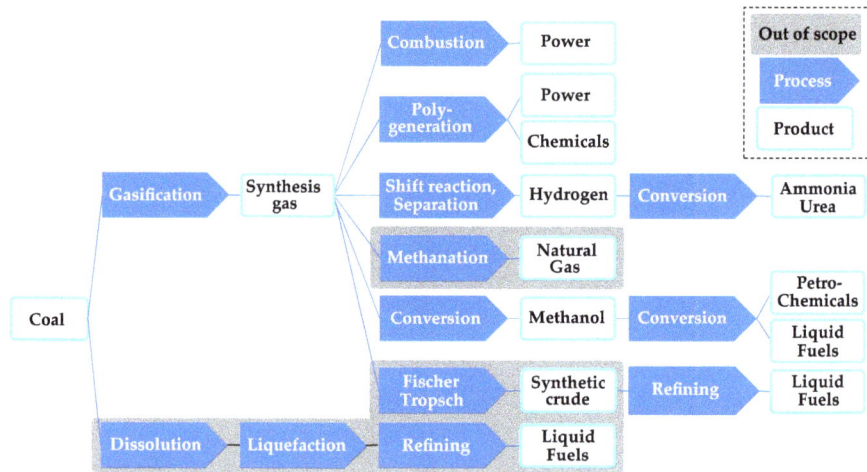

Figure 1. Supply chains for different underground coal gasification end-use options analyzed in the present study (electricity generation, hydrogen production, and its utilization for methanol and ammonia synthesis), modified from Kinaev et al. [11].

The in-situ UCG trial at the Wieczorek mine (cf. Figure 2) was part of the Polish research project "Elaboration of coal gasification technologies for a high efficiency production of fuels and electricity" [12]. The results of the pilot-scale research at the Wieczorek mine serve as the basis for the development of prospective commercial-scale UCG operations in the Upper Silesian Basin.

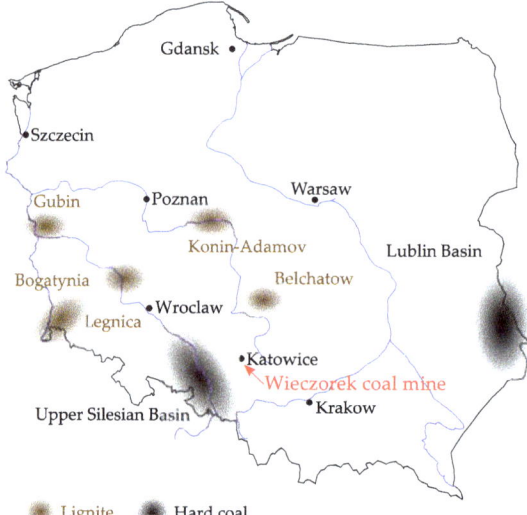

Figure 2. Location of the selected study area (Wieczorek coal mine) in Poland. Modified from the European Association for Coal and Lignite (EURACOAL) [13].

In view of the underlying techno-economic study, the wide data basis available from the Polish UCG pilot test, as well as representative coal samples offer a solid basis to parametrize the techno-economic process and life cycle assessment models [14,15]. Hence, the present study aims at a transparent documentation of the techno-economic analysis to identify economically-competitive site-specific UCG synthesis gas based end-utilization options. In order to handle carbon emissions resulting from the combustion of coal, the UCG process is linked to the capture and subsequent storage of CO_2 (CCS) in saline aquifers and/or its utilization (CCU) as a raw material for fuel production. Taking into account the comprehensive domestic coal resources, the EU environmental guidelines claiming a reduction of import dependency, a reduction of CO_2 emissions, and an increase in energy efficiency [16], integrated UCG-CCS/CCU processes may offer approaches to meet all of these criteria for Poland.

The tool applied for economic assessment is the techno-economic model developed by Nakaten et al. [5]. This model is applicable to calculate the costs of using UCG synthesis gas for electricity generation in a combined cycle gas turbine (CCGT) power plant coupled with CCS, taking into account site-specific geological, chemical, technical, and market-dependent boundary conditions. It consists of six sub-models (cf. Figure 3) and is controlled by more than 130 model input parameters, adaptable to site-specific boundary conditions for any study area. For the present study, the techno-economic model was extended to include UCG-based methanol and ammonia production (cf. Figure 3), as well as a corresponding offshore UCG setup (cf. Section 2.2.2).

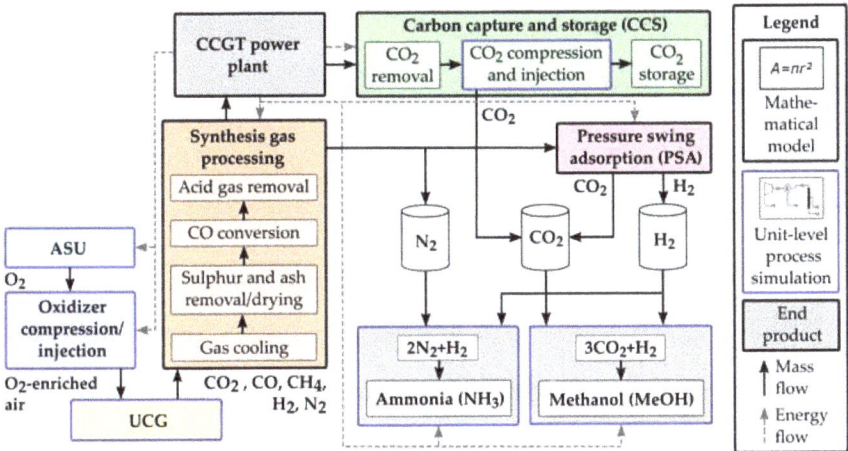

Figure 3. Techno-economic model for cost determination of integrated underground coal gasification (UCG), carbon capture and storage (CCS)/carbon capture, and utilization (CCU) scenarios based on Nakaten et al. [5] and extended to fulfill the requirements of the present study.

Innovations provided within the present study are:

- The implementation of a mass and energy balance assessment related to air separation (ASU), CO_2 compression, CO_2 and gasification agent injection, UCG synthesis gas production, and methanol and ammonia synthesis (cf. Figure 3, blue bordered boxes) to identify process steps requiring energy-saving measures, and
- the assessment of individual synthesis gas composition components and the related applied gasification agent, impacting overall levelized onshore and offshore costs.

Results obtained from quantification of cost bandwidths by single model parameter variation and comparing offshore to onshore levelized costs for selected UCG-CCS/CCU scenarios show that offshore UCG-based methanol and ammonia production are up to 1.7 times more expensive than the related onshore processes. Nevertheless, considering the average market prices in 2018, offshore methanol production is yet competitive on the European market, whereby the offshore ammonia production scenario costs exceed the current ammonia market price by 34%. With regard to the onshore scenarios, all investigated UCG-CCS/CCU processes except for ammonia production at the assumed worst-case conditions are economically competitive. Furthermore, techno-economic modeling results revealed that compared to the offshore platform with its technical equipment, offshore drilling costs have a minor impact of up to 3% on total levelized costs, only. Hence, uncertainties related to parameters influenced by drilling costs are negligible. Besides, results point out approaches for a cost-effective commercial-scale UCG-CCS/CCU implementation. The most important aspect required to maintain low production costs is a precise management of the gasification agent composition, whereby an oxygen ratio of a maximum of 30% by volume is suggested to maintain low ASU costs. Besides, methanol and ammonia production also significantly benefit from high H_2/N_2 and low CO_2 amounts. In the present study, the synthesis gas composition with the lowest CO_2 share sums up to 6.43% (reference scenario). Here, 26% CO_2 are sufficient to supply methanol synthesis at a yield of about 490 kt/year, determined constantly as output for all investigated methanol production scenarios.

2. Materials and Methods

Within the underlying section, basic model assumptions, the calculation methods applied, and intermediate calculation results, fundamental for the involved onshore and offshore UCG-CCS/CCU scenarios, are discussed.

2.1. Dimensioning of Assessed UCG-Based Production Chains

Selected onshore UCG synthesis gas end-use options in line with the present study are electricity generation, as well as methanol and ammonia production. Hereby, the dimensions of involved facilities were adapted to representative commercial-scale plant capacities. For electricity generation, an installed plant capacity of 100 MW net generation was chosen. Gross generation amounts were up to 286 MW, ensuring energy supply for air separation, gas processing, and compression. Following the model setup presented by Perez et al. [17], methanol production amounts were up to about 490 kt per year. The amount of produced ammonia was determined as 314 kt per year, representing an average Polish ammonia plant capacity [18]. Hereby, the product outputs listed above were maintained constant for all investigated scenarios to allow for their comparability.

2.2. Applied UCG Technologies and Well Design

2.2.1. Onshore UCG Technology Implementation

For the target coal seam utilization, we considered the Controlled Retraction Injection Point (CRIP) technology [1–3]. Key data related to geological boundary conditions (cf. Table 1) and coal seam extraction (cf. Table 2) were adapted from the coupled thermo-mechanical 3D UCG model setup introduced by Otto et al. [19], as well as data elaborated by Otto et al. [20,21].

Table 1. Geological data of the target coal seam, adapted from Otto et al. [19].

Model Input Parameters	Value
Average coal seam thickness (m)	11.0
Average coal calorific value (MJ/kg)	29.0
Average dip angle of the coal seam (°)	8.6
Average coal seam depth (m)	475.0
Average coal density (t/m^3)	1.281

The suggested optimized vertical well layout for injection and production, as well as the network of lined deviated injection boreholes, drilled horizontally into the coal seam, are presented in Figure 4.

To achieve an individual control of the UCG process by managing gasification agent injection rates and liner retraction, each gasification channel is operated by a separate liner, introduced by a separate injection well.

Aiming at an energy-efficient injection process, we iteratively determined inner liner well diameters, using thermodynamic process simulations integrated into the techno-economic model [22] to maintain effective flow velocities and related pressure losses. For thermodynamic modeling, we applied the DWSIM software package [22] with the ChemSepdatabase [23], whereby interfaces between techno-economic and thermodynamic models were established via a DWSIM control module (cf. Figure 5).

Allowing for pressure losses of a max. of 2 bar, inner-well diameters for injection and production wells were aligned according to a gasification agent mass flow of up to 13.7 kg/s per injection well, a synthesis gas mass flow of up to 18.5 kg/s per production well, and a CO_2 mass flow of up to 28.3 kg/s per CO_2 injection well. Adhering to the chosen inner-well diameters of 7 inches for gasification agent injection, 9 inches for all production, and CO_2 injection wells, as well as significantly differing mass flow rates in the three model setups (cf. Table 2), the number of required simultaneously-operated injection and production wells varied significantly (cf. Table 2). However, we neglected to develop separate well layouts that would hardly differ from an economical point of view. Hereby, smaller inner-liner diameters allowed for higher achievable build-up rates, and thus less drilling meters. Further simulation constraints that influence the density and viscosity of the gas mixtures, and thus pressure losses in wells, were determined by the hydraulic and thermal well profiles, as well as temperature and pressure conditions at the well heads or at the production well bottomhole. The hydraulic well profile comprises information on the well casing material, length and diameter.

Based on the inner liner diameters of the injection wells, the achievable build-up rates of the deviated drillings were determined as 6° per 30.48 m (100 ft) according to Godbolt [24]. Referring to thermomechanical modeling results on channel stability in the absence of subseismic faults, a pillar width of 60 m (cf. Figure 4) is sufficient to avoid inter-channel hydraulic short circuits [19]. The achievable gasification channel width in the Wieczorek target coal seam amounts to 20 m (cf. Table 2) according to Otto et al. [19]. In order to ensure the required synthesis gas feed in the reference scenario for an overall operational lifetime of 20 years, the operation of 1.2 UCG panels (15 gasification channels) is required for electricity generation. For methanol and ammonia production, up to four UCG panels are required (cf. Table 2) to ensure energy supply for a 20-year operational lifetime.

Table 2. UCG design for a 2 km × 1 km-sized target area adapted from Otto et al. [19] and adjusted to the electricity, methanol, and ammonia production scenarios.

Key Data	Electricity	Methanol	Ammonia	Reference
UCG panel width (km)	1.9	1.9	1.9	[19–21]
UCG panel length (km)	1.0	1.0	1.0	[19–21]
UCG panel coal resources (Mt)	7.3	7.3	7.3	Calculated
Channel width-to-height ratio (-)	1.8	1.8	1.8	[19–21]
Number of simultaneously-operated production wells (-)	4.0	13.0	8.0	Calculated
Number of injection wells per UCG panel (-)	13.0	13.0	13.0	Calculated
Number of simultaneously-operated injection wells (-)	4.0	13.0	8.0	Calculated
Gasification channel width (m)	20.0	20.0	20.0	[19–21]
Distance between gasification channels (m)	60.0	60.0	60.0	[19–21]
Total coal consumption (Mt)	8.5	26.5	16.4	Calculated
Total required gasification agent mass (Mt)	34.3	110.0	68.8	Calculated
UCG panels required (-)	1.2	3.6	2.2	Calculated

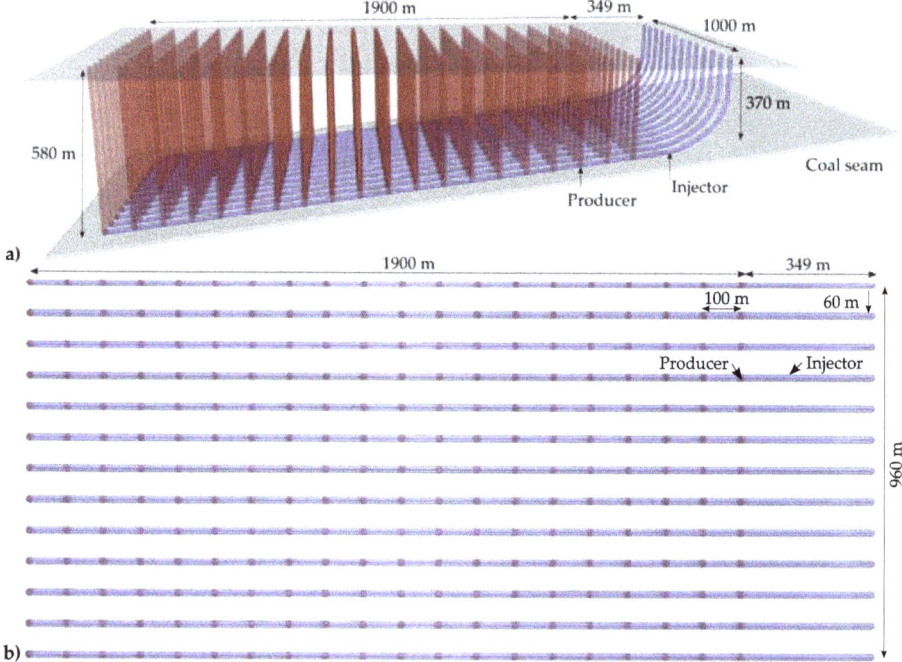

Figure 4. Computer-aided design model visualizing (**a**) 3D perspective and (**b**) plan views of the conceptual technical onshore well design for one underground coal gasification panel with 13 reactor channels and 260 production wells.

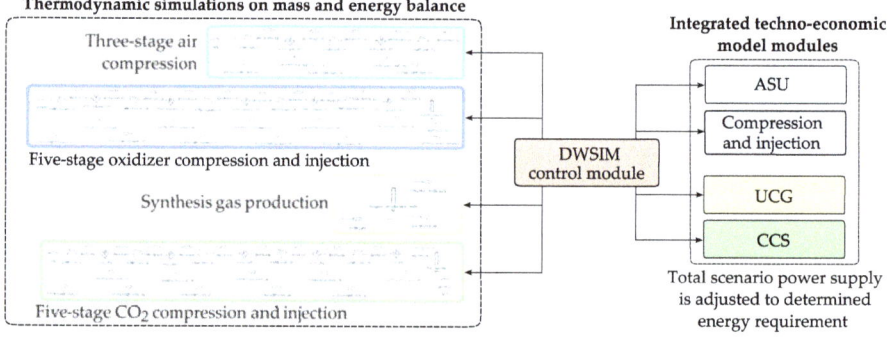

Figure 5. Integration of the techno-economic model with thermodynamic process simulations by means of the DWSIM process simulator [22], considering air compression, oxidizer compression and injection, synthesis gas production, as well as CO_2 compression and injection.

Production wells are drilled vertically from the surface and spaced at a distance of 100 m along the horizontal extension of each gasification channel (cf. Figure 4). This distance is considered appropriate to avoid consumption of high-value product gas components (synthesis gas cannibalism) along the gasification channels [1,4,25]. In line with the extraction of a complete UCG panel at a width of 2 km, 260 production wells are required. Taking into account the number of simultaneously-operated gasification channels (cf. Table 2), the total coal consumption, as well as geometrical gasification channel data (cf. Tables 1 and 2), an average daily horizontal gasification front progress of up to one meter can be achieved [19,26]. The overall coal yield using the chosen UCG well layout was about 25%. Drilling costs (cf. Table 3) were calculated based on the total number of required injection and production wells, as well as the target coal seam dimensions (cf. Tables 1 and 2). Within the UCG sub-model (cf. Figure 3), all cost positions associated with gasification agent production, its compression and injection (Section 2.3), synthesis gas processing (Section 2.4), drilling, land acquisition, piping, measuring, control equipment costs, as well as land concession fees and permissions were combined. Table 4 lists all synthesis gas production costs for electricity, methanol, and ammonia production.

Table 3. Calculated drilling meters and resulting costs for injection and production wells for the onshore electricity, methanol, and ammonia production scenarios.

Cost Position	Electricity	Methanol	Ammonia	Reference
Vertical drilling length (km)	188.50	590.80	364.50	CAD model
Deviated drilling length (km)	0.60	1.90	1.20	CAD model
Horizontal drilling length (km)	30.30	95.10	58.70	CAD model
Costs vertical drilling meter (€/m)	80.00	80.00	80.00	[12,14,27,28]
Costs deviated drilling meter (€/m)	480.00	480.00	480.00	[12,14,27,28]
Costs horizontal drilling meter (€/m)	230.00	230.00	230.00	[12,14,27,28]
Cumulative costs vertical drilling (M€)	15.08	47.26	29.16	Calculated
Cumulative costs deviated drilling (M€)	0.30	0.93	0.57	Calculated
Cumulative costs horizontal drilling (M€)	6.98	21.87	13.49	Calculated
Total drilling costs (M€)	**22.36**	**70.06**	**43.22**	**Calculated**

Table 4. Total onshore UCG costs for a 20-year operation of onshore electricity, methanol, and ammonia production scenarios.

Cost Position	Electricity	Methanol	Ammonia	Reference
Fees for area, permission, exploration (M€)	1.96	2.48	2.22	[12,14,27,28]
Total drilling costs (M€)	22.36	70.05	43.22	Calculated
Land acquisition costs (M€)	59.29	118.58	88.93	[12,14,27,28]
Piping, measuring, control equipment costs (M€)	14.20	60.71	28.52	[29]
Gasification agent production/injection costs (M€)	350.55	725.55	514.94	[30]
Synthesis gas processing costs (M€)	279.96	261.33	248.30	[29]
Pressure swing adsorption (PSA) costs (M€)	-	374.20	234.38	[31]
Staff salaries (M€)	104.37	129.44	81.08	[29]
Total costs (M€)	**832.69**	**1742.34**	**1241.59**	**Calculated**

2.2.2. Offshore UCG Technology Implementation

We applied a modified version of the model developed by Nakaten et al. [5] to assess the costs of two offshore UCG-CCS/CCU production chains aiming at methanol and ammonia generation. For comparability, the generic offshore model parametrization was equal to that of the previously-discussed onshore model. Revised model assumptions were primarily varying due to the adapted UCG well design and increased costs accounted for the offshore infrastructure. Furthermore, we did not consider a power generation scenario for the offshore UCG-CCS/CCU economic assessment in favor of methanol and ammonia production. Electricity production exceeding the offshore UCG platform's own demand was of limited applicability, since installation of costly offshore transport networks would be required. This is not considered to be as economic as, e.g., the transport of methanol or ammonia by ships. While varying methanol and/or ammonia production capacities do not significantly impact the overall production costs, electricity generation economics strongly depend on a continuous power supply to an electricity network [32,33].

Figure 6 shows the suggested radial development design for offshore coal seam extraction, taking into account the parallel controlled retracting injection point (P-CRIP) UCG approach [34–37].

To maximize the coal extraction in one offshore panel using 25 gasification channels, injection and production wells were drilled in-seam next to each other [37]. Applying this well layout, a maximum coal yield of 46% can be achieved. Hereto, we implemented alternating "long" (about 1000 m) and "short" (650 m) gasification channel systems to optimize the coal yield (cf. Figure 6), considering a safety distance (pillar width) of at least 100 m between the single gasification channels. We validated the chosen safety distance in view of potential hydraulic short circuit formation between single UCG reactors and seabed level subsidence by numerical geomechanical simulations [37].

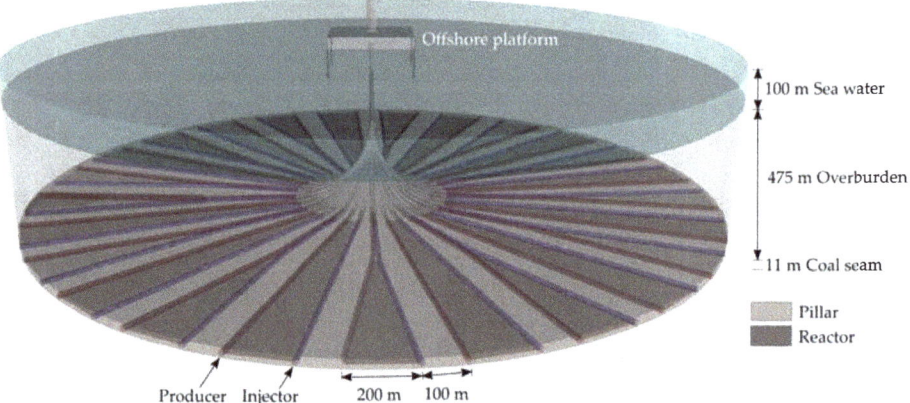

Figure 6. Computer-aided design model visualizing the proposed parallel controlled retracting injection point underground coal gasification offshore well design with twelve "long" and 13 "short" channel systems. Figure copyright [36] (modified), licensed under CC BY-NC-ND 4.0.

For the reference scenario, we assumed a maximum gasification channel width of 200 m, decreasing to 45 m ("long" channel system) and about 20 m ("short" channel systems) at the narrow parts close to the circle center. We further implemented different scenarios with maximum gasification channel widths from 50–200 m and quantified the resulting cost bandwidth (cf. Section 3.4.3) to consider economic uncertainties resulting from the assumed maximum achievable gasification channel width. The proposed well design allowed for the extraction of about 34 Mt coal per panel. Thus, taking into account an average daily coal consumption rate of 3.6 kt for methanol and 2.2 kt for ammonia production in the reference scenario, one coal panel was sufficient to ensure the operation for 20 years. Data on the geological boundary conditions of the offshore UCG design are listed in Table 5.

Table 5. Boundary conditions for the radial offshore UCG well design in the reference scenario [37].

Model Input Parameters	Value
Average coal seam thickness (m)	11.0
UCG panel radius (km)	1.3
Overall UCG panel extension (km^2)	5.3
Extractable coal resources per UCG panel (Mt)	33.8
Pillar width (m)	100.0

Due to the lack of data, offshore drilling costs were determined based on the previously-introduced onshore drilling costs (cf. Table 3). Since offshore drilling (cf. Table 6) is significantly more expensive than onshore drilling, we conservatively assumed five times higher drilling costs in the offshore reference scenario (cf. Table 7). Uncertainties related to offshore drilling costs were further assessed and quantified in the scope of the sensitivity analysis. Total UCG costs for offshore methanol production summed up to 2.9 G€ and to 2.4 G€ for offshore ammonia production.

Table 6. Drilling meters in the offshore methanol and ammonia production scenarios.

Drilling Meters (m)	Methanol	Ammonia
Deviated	15,521	9,130
Horizontal	34,006	19,989
Vertical	6,180	3,635
Total drilling meters (m)	**55,707**	**32,754**

Table 7. Drilling costs in the offshore methanol and ammonia production scenarios.

Cost Position	Methanol	Ammonia
Cumulated costs for deviated drilling meters (M€)	37.3	21.9
Cumulated costs for horizontal drilling meters (M€)	39.1	23.0
Cumulated costs for vertical drilling meters (M€)	2.5	1.5
Total drilling costs (M€)	**78.9**	**46.4**

2.3. Costs for UCG Offshore Platform

According to Boeing [38], we considered CAPEX of about 620 M€ for the construction of an offshore platform (corrected for inflation) and further 620 M€ for technical services and material costs for 20 years of operation. As for the onshore scenarios, transportation was not considered in the UCG offshore cost determination.

2.4. Gasification Agent Compression and Injection

For comparability, process dimensions of the offshore and onshore models were equalized, thus parametrization related to gasification agent production, compression, and injection, as well as the related intermediate results were comparable. To provide a gasification agent mixture consisting of 30% O_2 and 70% N_2 by volume, a cryogenic air separation unit (ASU) was considered. Air compression in the ASU process was implemented within three stages, whereby modeling results on the total power requirement, obtained from thermodynamic process simulations using the DWSIM software package [22], summed up to about 16 MW for onshore electricity generation. Air compression power demand of methanol production amounted to 51 MW and 32 MW for ammonia production. Gasification agent compression before its injection was achieved within five stages, resulting in a total compression power requirement of 28 MW (electricity), 88 MW (methanol), and 55 MW (ammonia), respectively. Total CAPEX and OPEX for the ASU process, as well cost positions related to gasification agent compression and injection in the onshore and offshore reference scenarios are listed in Table 8.

Electricity costs were neglected, since the power requirement for all process stages is autonomously provided by the integrated CCGT power plant (onshore) or a gas turbine at the offshore platform.

Table 8. Air separation costs in the onshore/offshore reference scenarios with an air feed demand of 70 kg/s for onshore electricity production, as well as 222 and 139 kg/s for onshore/offshore methanol and ammonia production for 20 years of operation, modeled with IECM [30].

CAPEX	Electricity	MeOH	NH$_3$
Process facilities capital (M€)	64.2	101.5	77.8
General facilities capital (M€)	9.6	15.2	11.7
Staff costs (M€)	6.4	10.2	7.8
Project and process contingency costs (M€)	12.8	20.3	15.6
Interest charges (M€)	5.0	7.9	6.1
Royalty fees (M€)	0.3	0.5	0.4
Pre-production (start-up) costs (M€)	2.3	3.6	2.8
Inventory (working) capital (M€)	0.5	0.7	0.6
Total ASU CAPEX (M€)	**101.1**	**159.9**	**122.8**
OPEX (variable and fixed costs)	**Electricity**	**MeOH**	**NH$_3$**
Variable costs (M€)	52.6	168.6	105.6
Operating labor (M€)	31.0	49.1	37.6
Maintenance labor (M€)	14.9	23.5	18.1
Maintenance material (M€)	22.3	35.3	27.1
Admin and support labor (M€)	15.9	25.2	19.3
Total ASU OPEX (M€)	**136.7**	**301.7**	**207.7**
Total ASU CAPEX and OPEX (M€)	**237.8**	**461.6**	**330.5**

2.5. Synthesis Gas Processing

For the techno-economic modeling of the selected UCG end-product utilization options, we considered a synthesis gas composition of 15.90% H$_2$, 1.20% CH$_4$, 6.43% CO$_2$, 17.54% CO, and 58.93% N$_2$ including 0.30% of minor constituents such as C$_2$H$_2$ and H$_2$S (all data given in % by volume). This composition was achieved during the first stage of a five-stage in-situ UCG gasification test at the Wieczorek mine [8], in which different gasification agent compositions at each gasification stage were analyzed. The oxygen fraction in the gasification agent applied in the present study amounted to 25% by volume. The average synthesis gas calorific value achieved during this stage was 4.4 MJ/sm^3. After its transportation to the surface via production wells, the synthesis gas was cooled, scrubbed to remove trace elements, and excess water was separated. Subsequently, the synthesis gas was processed in a gas cleaning section, whereby CO was converted into CO$_2$ in a CO-shift reactor. Thereafter, sulfur components and CO$_2$ were removed during a physical absorption process. In line with methanol and ammonia production, CO$_2$ and H$_2$ were separated from the synthesis gas by means of physical adsorption (PSA) [39–41], considering a subsequent tail gas disposition block for separation of sulfur-containing compounds.

CAPEX and OPEX for synthesis gas processing (cf. Table 9) using the Selexol process were determined by the Integrated Environmental Control Model (IECM)

modeling tool [30] and scaled linearly to the dimensions of the underlying operational UCG setup by three scaling factors (net output, operating hours, gas flow rate). For electricity generation, synthesis gas processing CAPEX summed up to 132 M€ and variable and fixed costs to 7 M€/year. Synthesis gas processing costs for methanol and ammonia production with tail gas cleaning (Beavon–Stretford) summed up to CAPEX of 115.5 M€ and 102.5 M€, respectively; as well as OPEX of up to 6.9 M€/year for each of both. Costs for the PSA process (cf. Table 4) were calculated according to Barranon [31], taking into account an interest rate of 10%, as well as an operational lifetime of 20 years. For improved cost comparability, generic model assumptions of the offshore and onshore models were equalized.

Table 9. Synthesis gas processing costs for a synthesis gas mass flow of 2.3 Mt/year in the electricity production scenario with 20 years of operation, modeled with IECM [30].

CAPEX	Value
Selexol sulfur removal unit (M€)	46.2
Process facilities capital (M€)	49.4
General facilities capital (M€)	7.4
Staff fees (M€)	4.9
Project and process contingency cost (M€)	10.8
Interest charges (M€)	8.1
Royalty fees (M€)	0.3
Pre-production (start-up) costs (M€)	4.2
Inventory (working) capital (M€)	0.7
Total CAPEX (M€)	**132.0**
OPEX (variable and fixed costs)	**Value**
Selexol solvent (M€)	2.3
Sulfur by-product (M€)	7.5
Operating labor (M€)	81.3
Maintenance labor (M€)	11.6
Maintenance material (M€)	17.4
Administrative and support labor (M€)	27.9
Total OPEX (M€)	**148.0**
Total CAPEX and OPEX (M€)	**280.0**

2.6. CO₂ Capture and Storage and Utilization

Three existing technologies can be employed to capture CO_2 from fossil-fueled power plants: post-combustion, pre-combustion, and oxy-fuel combustion capture [40,42,43]. In the present study, we chose a typical physical absorption pre-combustion process by means of a Selexol solvent, suitable for high CO_2 capture ratios (95% considered in our calculations) and common for IGCC power plants [44]. Hereby, the Selexol procedure requires less equipment for dehydration, solvent recovery, and CO_2 compression compared to the Rectisol process [41]. Nevertheless, purification of hydrogen and carbon dioxide, crucial for methanol and ammonia syntheses, requires the application of the pressure swing adsorption (PSA) method with a subsequent tail gas disposition block. The PSA procedure is based on a physical binding of gas molecules to adsorbent material. Most of the modern plants used multi-bed PSA to remove water, ethane, and CO for the recovery of high-purity hydrogen (99.99%). Carbon dioxide is also separated during this process with a capture ratio of 90% [45,46].

In view of CO_2 storage, one approach discussed by Burton et al. [47] and Kempka et al. [25] was to reinject captured CO_2 into the abandoned UCG reactors. However, we assumed the availability of a geological storage reservoir with sufficient capacity to completely trap the captured CO_2 of up to 42 Mt during a 20-year operational lifetime. In the present study, this storage reservoir was assumed to be close to the UCG site at a depth of about 800 m, considered as minimum depth to ensure economic storage [47].

2.6.1. Onshore CCS/CCU Costs

Selexol capture costs referring to UCG-based electricity production were determined according to data on energy consumption (0.108 kWh/kg CO_2 [41]), chemical process simulation results on power demand for CO_2 compression, as well as Selexol CAPEX (0.67 €/kW) and OPEX (0.10 €/kW) adapted from Mohammed et al. [41]. Scaled to the dimensions of the underlying operational setup (net power generation 100 MW$_{el}$) with emissions of about 0.6 t CO_2/MWh, the total energy demand for CCS summed up to 22 MW. Due to the assumed short distance between the UCG site and CO_2 storage reservoir, intermediate CO_2 compression was not considered. As levelized costs for transportation (0.002 €/t CO$_2$) have no relevant impact on total levelized CCS costs, they were neglected in this study. Taking into account one well for CO$_2$ injection, CAPEX for injection amounted to 1.57 M€ and OPEX to 0.31 M€/year, summing up to total levelized CO$_2$ injection and storage costs of 0.48 €/t CO$_2$. The estimated discounted monitoring costs for a 40-year period (20 years operational lifetime and additional 20 years for post-closure monitoring) were taken from the "Characterization of European CO$_2$ storage" (SiteChar) project report [48] and amounted to 28 M€ (0.79 €/t CO$_2$). Cumulative levelized CCS costs (cf. Table 10), considering the underlying boundary conditions applied for the electricity generation setup, amounted to about 15 €/t CO$_2$.

Table 10. Levelized costs for the reference scenario (electricity generation setup) in terms of onshore Selexol capture, injection, storage, and monitoring considering an emission rate of 2.2 t CO$_2$/t coal.

Selexol Capture Cost	Value
Energy costs (€/MWh)	3.97
Selexol CAPEX (€//MWh)	6.71
Selexol OPEX (€//MWh)	1.00
CO$_2$ Storage Costs	**Value**
Injection and storage costs (€//MWh)	1.06
Monitoring costs (€//MWh)	1.91
Total levelized CCS costs (€//MWh)	14.65

Methanol output of about 490 kt methanol per year was accompanied by a total CO$_2$ production of almost 88 kg CO$_2$/s resulting from UCG, power generation, in addition to a CO$_2$ mass flow of 0.58 kg/s purged during methanol synthesis. Deducting the amount of 22.4 kg/s CO$_2$ required for methanol synthesis resulted in an excess CO$_2$ stream of 65 kg/s, whereby 26% of CO$_2$ produced throughout the entire UCG-MeOH production chain was utilized. Consequently, carbon capture and its subsequent utilization offer a new economy for CO$_2$, since captured CO$_2$ does not have to be considered as a waste product, only, but is required as a raw material for other processes [17]. In the present study, costs for pre-combustion CO$_2$ capture and utilization (CCU) that arise in line with methanol synthesis were not charged separately. Thus, CO$_2$ capture costs considering the PSA process were associated with hydrogen production CAPEX and OPEX (cf. Section 3.2.1, Table 15), since carbon dioxide was purified from the synthesis gas as by-product of hydrogen. Costs for capturing CO$_2$ that result from internal power supply, its compression before entering the methanol recycle loop, as well as costs for excess CO$_2$ pre-injection compression, injection, and storage were part of the methanol synthesis process and thus charged directly with levelized costs (9% on overall levelized costs). The power requirement to compress the excess CO$_2$ (65 kg/s) before storage amounted to about 23 MW.

As for methanol synthesis, PSA CO$_2$ capture costs in line with ammonia production were not charged separately, but associated with hydrogen costs (cf. Section 3.3). The power required to compress the CO$_2$ mass flow (55 kg/s) before storage was determined by chemical process simulations and amounts to about 19 MW. As for the electricity generation model setup, CO$_2$ transportation costs were neglected in this study. Taking into account the boundary conditions for the ammonia model setup, costs for CO$_2$ capture resulting from internal power supply, CO$_2$ pre-injection compression, as well as injection, storage, and monitoring summed up to about 18 €/t ammonia.

2.6.2. Offshore CCS/CCU Costs

Since offshore electricity generation was not considered in the present study, offshore CCS costs arise from offshore methanol and ammonia production, only. Hereby, similar assumptions referring to monitoring were taken into account to maintain comparability with the onshore models. Thus, related cost positions were identical in the onshore and offshore scenarios. Furthermore, we used identical assumptions for the CO$_2$ storage reservoir in all models. However, assuming five times higher injection well drilling costs than in the onshore scenario, total annualized capital, and operational costs for injection and monitoring were about 100 k€ higher for all offshore model setups.

2.7. Methanol Production Energy Balance

In line with the assessment related to the economic competitiveness of UCG-based methanol on the European market, mass and energy flows, as well as the amount of CO_2 that is not emitted, we considered a commercial-scale methanol production scenario based on UCG synthesis gas at an average scale of already operating MeOH synthesis plants in Europe. For the present study, we chose to investigate methanol production due to its importance as fuel that is blendable with gasoline, transformable, and blendable with diesel, and applicable in fuel cells. Hereby, existing gasoline distribution and storage infrastructure like pipelines, road tankers, and filling stations would require little modification to operate with methanol. Besides, methanol can function as storage for hydrogen or as a feedstock to synthesize olefins [17,49–51].

Usually, methanol is produced by means of the Fischer–Tropsch process, where CO and $2H_2$ react to form CH_3OH [17]. In the present study, we followed methanol production by direct hydrogenation of CO_2 with H_2, since CO contained in the UCG synthesis gas was considered to support the internal power supply. Hereby, direct CO_2 hydrogenation with H_2 is governed by two reactions taking place in the reactor. While Reaction (1) was the one that produced MeOH, CO formed in Reaction (2) was recycled together with unreacted H_2 to increase the output, once MeOH and H_2O were separated.

$$CO_2 + 3H_2 \rightleftharpoons CH_3OH + H_2O \tag{1}$$

$$CO_2 + H_2 \rightleftharpoons CO + H_2O \tag{2}$$

Thermodynamic process simulations to analyze mass and energy flows were implemented using the DWSIM software tool [22], whereby applied boundary conditions and subsequent model validation followed the reference process according to [17,52]. In summary, methanol synthesis consisted of the following process steps. After CO_2 pressure was increased from 1 bar up to 23.2 bar, H_2 (30 bar, 25 °C) and CO_2 (23.2 bar, 28 °C) feeds were mixed, compressed to 76.4 bar and heated by a heat exchanger to 210 °C before the reactor inlet. Hereby, the reactor was modeled as an adiabatic Gibbs reactor. The high temperature was crucial, even though the methanol yield had an inverse dependence on temperature and a positive dependence on pressure. However, the catalyst efficiency was more intense at higher temperatures [49]. At the reactor outlet, before methanol and water were separated in a component separator, stream temperature was cooled down from 294 °C to 35 °C in a heat exchanger. About 99% of unreacted H_2, CO_2, and CO was recycled, whereby 1% is purged to avoid accumulation of inert gases [17]. In the present study, total electricity consumption for the synthesis of 489 kt MeOH/year resulted from the power demand for CO_2/H_2 compression (34.4 MW), as well as methanol and water separation (1.4 MW). Additional heat exchanged among the streams amounted to 71.2 MW, compared to 81 MW according to Perez et al. [17]. This heat can be further used to generate electricity; however, in order not to exceed the focus of this study, we did not consider conversion of surplus heat into electricity in our simulations. Compared to identical model boundary conditions applied by Perez et al. [17], compression power demand in the present study was about twice as high. This significant difference was related to the energy efficient process setup applied by Perez et al. [17], using electricity generated by ad hoc steam turbine systems, as well as four expanders that take advantage of the calorific value of the purge gases. In this way, electricity demand can be reduced by 46% [17]. Consequently, considering the utilization of the heat produced, power demand in the present study offered opportunities to be significantly optimized. Discrepancy in the methanol production rate of our model implementation compared to Perez et al. [17] amounted to 1.6%. This was likely attributable to the application of different reactor types (Gibbs reactor versus plug flow reactor) in both studies. Besides, we did not apply any catalysts, whereby Perez et al. [17] used a productive copper catalyst ($Cu/ZnO/Al_2O_3$).

Combining thermodynamic process simulations to quantify the energy balance of methanol synthesis with the techno-economic process model, gross power generation for the entire UCG-MeOH production chain including air and gasification agent compression, CO_2 capture after electricity generation, as well as methanol synthesis summed up to 609 MW. Hereby, the gross generation and thus the overall model setup was adjusted iteratively to provide the required H_2 amount of at least 3.05 kg/s in the synthesis gas mass flow, so that the envisaged methanol output of 15.5 kg MeOH/s was achieved. The electric power output not required for methanol production summed up to about 128 MW_{el} and can be fed into the local power grid. As a result of H_2 consumption for methanol production, the excess synthesis gas CV was decreased by about 28%.

2.8. Ammonia Production Energy Balance

Ammonia is a key ingredient for fertilizer production, whereby the widespread use of ammonia in agriculture was initiated by the "green revolution", which also involved development of high-yield crops and advances in pesticides. Thus, ammonia synthesis optimization is a topic of high interest in the industry, as the market continues to expand and demand increases [53]. Ammonia is produced from hydrogen and nitrogen at operating pressures above 100 bar (Haber–Bosch process), necessary to achieve a high reaction rate [53]. Reduction of expenses for the required compression power can be achieved by the implementation of compressors driven by steam turbines that take advantage of steam produced elsewhere in the process, as applied in the methanol model. According to the approach of Villesca et al. [53], compression power reduction during ammonia synthesis was achieved by increasing reactor efficiency through the use of an innovative high activity ruthenium-based synthesis catalyst being twenty times more active than conventional iron catalysts, so that lower synthesis pressures and temperatures can be applied.

For the assessment of costs, mass, and energy flows in view of ammonia production, we chose an average scale of existing ammonia plants in Poland with a production rate of about 314 kt/year [53]. The DWSIM ammonia synthesis model was parametrized according to the optimized process setup by Villesca et al. [53] and scaled to the envisaged output. Hereby, hydrogen and nitrogen derived from the UCG synthesis gas along with small fractions of methane and argon were compressed from 20 bar up to 100 bar and fed to the ammonia process at a molar ratio of 3:1. Mixed with yet not reacted components entrained in a recycling process, nitrogen and hydrogen entered the synthesis reactor at a pressure of 100 bar and at the temperature of 11 °C. The reactor was modeled as an adiabatic Gibbs reactor, governed by the exothermic equilibrium Reaction (3). The stream (vapor phase) left the reactor at a temperature of 375 °C and was cooled down to −30 °C before the end-product ammonia was separated. Unreacted components (H_2, N_2), making up less than 1% of the purge stream, were recycled to increase the ammonia yield.

$$N_2 + 3H_2 \Leftrightarrow 2NH_3 \tag{3}$$

Before being scaled to the applied dimensions, our simulation results exhibited just insignificant differences of about 1% compared to the ammonia yield achieved by Villesca et al. [53]. Waste heat resulting from the heat exchanged between the product streams amounted to 29 MW. The total power demand for ammonia synthesis summed up to about 26 MW, where about 99% was required for fluid compression. Villesca et al. [53] did not present the total power requirement for ammonia synthesis. However, compared to the findings on different ammonia synthesis loop efficiencies [54], our energy balance simulation results appeared to be representative. To determine finally the economic competitiveness of UCG-based NH_3 production on the European market, thermodynamic simulation results on the power required to synthesize ammonia were integrated into the techno-economic model by Nakaten et al. [55]. Thus, taking into account all process steps related to the entire UCG-NH_3 production chain, the required gross generation added up to about 382 MW. Hereby, the gross generation was iteratively adjusted in order to provide at least 1.91 kg/s H_2 within the synthesis gas mass flow, necessary to achieve the envisaged ammonia output. The generated electric output not required for ammonia production summed up to about 66 MW_{el} and can be fed into the local power grid. Excess synthesis gas CV was up to 28% lower than the initial one, as 100% of H_2 from the synthesis gas was consumed to synthesize ammonia.

3. Results

In the present section, we discuss the techno-economic calculation results for UCG-based electricity, methanol, and ammonia production.

3.1. Levelized UCG-CCGT-CCS Costs

Costs of electricity (COE) are the total costs required for conversion of a fuel into electricity. For the current study, we used the COE calculation according to Nakaten et al. [5], based on UCG synthesis gas electrification in a combined cycle gas turbine power plant (CCGT) with an efficiency of 58% and a power output of 100 MW_{el}. The calculated gross power generation amounted to 286 MW in the reference scenario. Hereby, gross power generation for all electricity production model setups was automatically adjusted to the energy demand that changes with the underlying synthesis gas composition, while power output was maintained constant at 100 MW_{el} to ensure comparability between the different scenarios assessed within the sensitivity analysis. Further power plant-related boundary conditions that were considered within the economic assessment were a calculated interest rate on the planning horizon of 7.5% and a real operating cost increase of 1.5%. Electricity production costs were calculated as the average costs on a full-cost basis, and all costs were adapted to the reference year 2018. Equations to determine total investment costs, annual capital costs, operating costs, the capital value of the overall costs, and the levelized total annual costs with and without demolition were taken from Nakaten et al. [5], Hillebrand [32], and Schneider [33]. Synthesis gas costs contained all UCG-related costs, such as gasification agent production and injection, synthesis gas processing, drilling, land acquisition, piping-, measuring-, control equipment costs, as well as concession fees. The levelized synthesis gas costs (4.6 €/GJ) were calculated by dividing UCG costs by the amount of synthesis gas produced (9 PJ/year). All above-mentioned cost positions, relevant for the determination of COE (32.2 €/MWh), were shown in Table 11. Hereby, COE was the quotient of the levelized total annual costs and the amount of electricity produced (1.5 TWh/year). In summary, COE and CCS costs summed up to 46.9 €/MWh. Compared to COE of other Polish power plants amounting to 45–75 €/MWh without CO_2 emission charges [56–60] and to 91 €/MWh for gas-fired power plants with CO_2 emission charges [57], the application of UCG synthesis gas for electricity production was competitive at the Polish energy market. Offshore electricity production was not considered in the present study. However, considering the same amount of synthesis gas produced as in the related onshore model (286 MW gross power generation), offshore synthesis gas production costs would be about twice as high.

Table 11. Costs of all integrated UCG and CCGT process stages for 20 years of operation, combined as levelized costs of electricity (COE).

Cost Position	Value
CCGT investment costs (M€)	30.3
CCGT interest payments (M€)	20.2
CCGT fixed operating costs (M€)	24.5
CCGT variable operating costs (M€)	8.2
UCG synthesis gas production costs (M€)	832.7
CCGT levelized total annual costs with demolition (M€)	0.9
Total costs (M€)	916.8
Levelized UCG-CCGT costs of electricity (€/MWh)	32.2

3.2. Levelized UCG-MeOH-CCU Costs

3.2.1. Onshore UCG-MeOH-CCU Costs

Techno-economic process modeling results related to total methanol production costs are listed in Table 12, whereby levelized methanol production costs summed up to 206 €/t MeOH (cf. Table 13).

Table 12. Cost positions for methanol production process stages, combined as UCG-based methanol production costs for a 20-year operational lifetime (plant size of about 490 kt/year).

CAPEX	Value	Reference
Plant equipment, civil work, site preparation (M€)	72.2	[17]
Staff (engineering) costs, infrastructure modification (M€)	38.9	[17]
Further costs for plant designing, constructing, building (M€)	66.7	[17]
Working capital (M€)	22.2	[17]
CO_2/H_2 provision by PSA (M€)	34.0	[31]
Total CAPEX (M€)	**234.0**	**Calculated**
OPEX	Value	Reference
Operating labor (M€)	34.2	[17]
CO_2/H_2 provision by PSA (M€)	340.2	[31]
Process water, other materials (k€)	22.8	[17]
Total OPEX (M€)	**397.2**	**Calculated**
Total UCG costs excluding PSA (M€)	1368.1	Calculated
Total costs for CO_2 injection, storage, monitoring (M€)	35.1	[48,61]
Total MeOH synthesis costs including UCG (M€)	**2034.4**	**Calculated**

Compared to the average European market price of 419 € per tonne MeOH in 2018 [62], UCG-based methanol production was competitive. Costs for the production of synthesis gas in line with onshore methanol production amounted to 4.5 €/GJ. Costs for CO_2 compression, injection, storage, and monitoring summed up to about 14 €/t MeOH. The share of individual cost positions on total levelized onshore methanol production costs is shown in Table 13.

Table 13. Share of individual cost positions on levelized onshore UCG-based methanol production costs for 20 years of operation.

Cost Position	Costs (€/t MeOH)	Percentage Share (%)
Total drilling costs	8.3	4.0
Fees, land acquisition, piping, measuring, control equipment	67.6	32.9
Synthesis gas processing	44.2	21.5
Gasification agent production (ASU) and injection	85.7	41.6
Total levelized costs/total percentage	**205.8**	**100.0**

3.2.2. Offshore UCG-MeOH-CCU Costs

Offshore UCG-based methanol production costs were composed of all UCG-related process costs taken into account for the onshore model, as well as the aforementioned CAPEX and OPEX for the offshore platform. Hereby, total levelized offshore methanol production costs exceeded the onshore costs by 64% (cf. Table 14). Nevertheless, compared to the current methanol market price, offshore methanol production was yet economically competitive. Synthesis gas production costs under the given boundary conditions summed up to 7.4 €/GJ. The share of the main cost positions on total levelized UCG-based offshore methanol production costs is shown in Table 14.

Table 14. Share of individual cost positions on levelized offshore UCG-based methanol production costs for 20 years of operation.

Cost Position	Costs (€/t MeOH)	Percentage Share (%)
Total drilling costs	8.9	2.8
Offshore platform, piping, measuring, control equipment	189.2	58.7
Synthesis gas processing	42.2	13.1
Gasification agent production (ASU) and injection	81.9	25.4
Total levelized costs/total percentage	322.2	100.0

3.3. Levelized UCG-NH$_3$-CCS Costs

3.3.1. Onshore UCG-NH$_3$-CCS Costs

Ammonia synthesis costs (cf. Table 15) were adapted from Bartels [63] and scaled to the dimensions of the present study.

Table 15. Ammonia production CAPEX/OPEX for 20 years of operation (plant size of 314 kt/year).

Cost Position	Value	Reference
Capital charge without ASU/gas turbine (M€)	461.5	[63]
Haber–Bosch synthesis loop (M€)	131.7	[63]
Costs for injection, storage, and monitoring (M€)	34.6	[48,61]
Total UCG costs with PSA (M€)	1241.6	Calculated
Total NH$_3$ synthesis costs including UCG (M€)	1869.4	Calculated

Adding total UCG costs related to the underlying model setup, total levelized ammonia production costs cumulated to about 298 €/t NH$_3$ (cf. Table 16). Costs for CO$_2$ injection, storage, and monitoring summed up to about 18 €/t NH$_3$. Synthesis gas production costs accompanying onshore ammonia synthesis amounted to 5.2 €/GJ. Compared to the current European ammonia market price for 2018 with 320 €/t NH$_3$ [64], UCG-based ammonia production was economically competitive. The share of individual cost positions on total levelized onshore ammonia production costs is shown in Table 16.

Table 16. Share of individual cost positions on levelized onshore ammonia production costs.

Cost Position	Costs (€/t NH$_3$)	Percentage Share (%)
Total drilling costs	10.4	3.5
Fees, land acquisition, piping, measuring, control equipment	107.7	36.2
Synthesis gas processing	56.2	18.8
Gasification agent production (ASU) and injection	123.5	41.5
Total levelized costs/total percentage	297.8	100.0

3.3.2. Offshore UCG-NH$_3$-CCS Costs

The setup for offshore ammonia production was equal to that used in the onshore scenarios. Thus, costs for the onshore and offshore ammonia synthesis processes were the same. However, due to higher CAPEX and OPEX for the technical offshore equipment, five times higher drilling costs compared to the onshore scenarios, as well as the application of a different UCG exploitation scheme in the offshore scenarios, offshore ammonia production was 60% more expensive than its onshore variant (cf. Table 17). Synthesis gas production accompanying offshore ammonia production amounted to 10 €/GJ. Compared to the average European ammonia market price, offshore UCG-based ammonia production was not competitive. Table 17 lists the share of main cost positions on total levelized UCG-based offshore ammonia production costs.

Table 17. Share of individual cost positions on levelized offshore ammonia production costs.

Cost Position	Costs (€/t NH_3)	Percentage Share (%)
Total drilling costs	9.3	1.9
Offshore platform, piping, measuring, control equipment	323.6	66.9
Synthesis gas processing	47.1	9.7
Gasification agent production (ASU) and injection	103.5	21.5
Total levelized costs/total percentage	483.5	100.0

3.4. Sensitivity Analysis

We applied a one-at-a-time (OAT) sensitivity analysis to assess uncertainties related to the UCG synthesis gas end-utilization option costs. This kind of sensitivity analysis was used to assess the range of possible outcomes imposed by the variation of one model input parameter across a plausible range of uncertainty. Hereby, the focus was on quantifying the influence of single model input parameters on overall costs to, e.g., assess which process steps benefit most from optimization. In the underlying study, we further investigated the impact of different synthesis gas compositions and calorific values, as well as gasification agent compositions for the onshore realization. For offshore realization, uncertain model input data such as offshore drilling costs and the maximum achievable gasification channel width were considered in view of the well design.

3.4.1. Impact of Synthesis Gas Composition, CV, and Gasification Agent Compositions on Total Costs

We further investigated the impact of different synthesis gas compositions and related CVs, as the CV impacts UCG synthesis gas end-utilization options in view of its economical suitability for commercialization. Besides, CV is one of the parameters that can be controlled with little effort as even minor modifications of the model assumptions may induce its increase or decrease by, e.g., different applied gasification agent compositions and gasification stages [65,66]. To analyze potential cost variations, we focused on four synthesis gas compositions (cf. Table 18). The chosen compositions were derived from in-situ and ex-situ gasification tests at the Wieczorek mine and ex-situ tests on coals sampled at the Bielszowice mine, whereby different gasification agent compositions at varying p/T conditions were investigated [8]. In order to maintain a site-specific techno-economic assessment in the sensitivity analysis, we implemented two scenarios with data directly related to the selected target area (Scenarios I and II). Furthermore, two scenarios representing the results of gasification tests on coals from the Bielszowice mine (Scenarios III and IV) were included into the analysis due to the availability of extensive analysis data. The considered synthesis gas compositions exhibited significant differences in their CO_2 content, CV, and UCG-to-synthesis gas conversion efficiency. Thus, the chosen synthesis gas compositions allowed quantifying the impact of the parameters expected to have a major impact on the total levelized costs.

Table 18. Synthesis gas compositions and calorific values investigated in one-at-a-time sensitivity analyses; values given in % by volume; data derived from Stańczyk et al. [8,10]; CVs calculated.

Scenario	CO_2 (%)	H_2 (%)	N_2 (%)	CH_4 (%)	CO (%)	CV (MJ/sm³)	Oxidizer Composition (%)
Wieczorek (I)	6.4	15.9	58.9	1.2	17.5	4.4	O_2: 35, N_2: 65
Wieczorek (II)	9.2	10.7	63.7	2.0	14.5	3.7	Air: 100
Bielszowice (III)	14.8	11.9	60.1	2.8	10.4	3.6	Air: 100
Bielszowice (IV)	23.2	18.9	36.2	4.2	17.5	5.8	O_2: 51.3, N_2: 48.7

Onshore UCG-CCGT-CCS Scenario

OAT sensitivity analysis results showed that cost effectiveness in the electricity generation scenarios was rather dominated by technical constraints (power plant net capacity, gasification agent production) than by geological or chemical boundary conditions, such as the synthesis gas CV and UCG-to-synthesis gas conversion efficiency. UCG-to-synthesis gas conversion efficiency was determined by synthesis gas and coal CVs, as well as the theoretical synthesis gas-to-coal ratio, whereby the theoretical synthesis gas-to-coal ratio was the ratio of the carbon content in the coal and that in the synthesis gas. Data on the coal carbon content and that in the tars for the Wieczorek and Bielszowice coal types taken into account were provided by Stańczyk et al. [8,10] and Mocek et al. [67]. Thus, despite the high CV (cf. Table 18) and the highest UCG-to-synthesis gas conversion efficiency achieved among the CCGT scenarios (cf. Table 19), costs for electricity generation and CCS in Scenario I exceeded those calculated for the other three scenarios by 11–37%. The notable cost increase in Scenario I compared to the other scenarios was attributable to the high oxygen share in the gasification agent resulting in high ASU costs and to the lowest installed net capacity in that scenario. Thus, although ASU costs in Scenario IV exceeded those in Scenario I, the higher installed net capacity induced an overall cost decrease by up to 2.7%. Hereby, the installed net capacities for the different scenarios were iteratively adjusted to the energy requirement of all process steps involved, while a previously-determined fixed electrical power output of 100 MW$_{el}$ was maintained for comparability. With regard to the four investigated scenarios, the synthesis gas composition most suitable for electricity generation was represented by Scenario III. The benefit of gasification with pure air (cf. Table 18) and the relatively high installed net power capacity overcame the disadvantages associated with the low synthesis gas CV and UCG-to-synthesis gas conversion efficiency. Synthesis gas production costs for electricity generation are listed in Table 19. Cost effectiveness of the energy-intensive CO_2 capture correlated with the available CO_2 amount in the given synthesis gas composition. Since the synthesis gas composition in Scenario I showed the lowest CO_2 and CO ratios among the CCGT scenarios, CCS costs for Scenario I were about 26 to 34% above those for the other scenarios.

In summary, the OAT sensitivity analysis revealed that under the given model assumptions and even in the case of the assumed worst-case conditions in Scenario I, the application of UCG synthesis gas for electricity production was competitive for the Polish energy market.

Table 19. One-at-a-time sensitivity analysis results for the onshore UCG-CCGT-CCS scenario considering different synthesis gas compositions and calorific values.

Scenario	COE (€/MWh)	CCS Costs (€/MWh)	Synthesis Gas Costs (€/GJ)	UCG Synthesis Gas Efficiency (MJ/MJ)	P_{gross} (MW)	P_{units} (MW)	P_{el} (MW)	P_{net} (MW)
I	32.2	14.7	4.7	0.84	285	66	100	166
II	20.8	10.9	2.8	0.70	374	117	100	217
III	19.7	9.7	2.6	0.53	500	191	100	291
IV	31.5	10.4	4.5	0.53	411	138	100	238

P_{gross} = total gross power generation, P_{units} = power required to operate all integrated process units, P_{el} = electric power output, P_{net} = net power generation consisting of P_{units} and P_{el}.

Onshore UCG-MeOH-CCU Scenario

While high oxygen contents in the gasification agent were favorable by increasing the overall synthesis gas CV, the UCG-to-synthesis gas conversion efficiency was reduced, since more carbon, and thus more coal, was required to maintain a consistent synthesis gas mass flow. From an economic point of view, air separation and compression increased the required power demand, and thus gross power generation and related costs significantly, whereby the resulting methanol yield remained unchanged. Gross power generation in the methanol scenarios was iteratively adjusted to cover the required H_2 amount (3.05 kg/s) in the underlying synthesis gas composition, as well as the resulting energy requirement of all process steps involved. Despite the high H_2 content in the synthesis gas (cf. Table 18), Scenario IV represented the worst case with costs of 268 €/t methanol induced by a share of about 36% ASU costs on overall costs and a low UCG-to-synthesis gas conversion efficiency. Due to H_2 utilization in the methanol synthesis, the synthesis gas CV in Scenario IV was reduced from initially 5.8 to 4.6 MJ/sm^3. Among the four investigated synthesis gas compositions, the one most suitable for methanol production was that in Scenario III. With regard to Scenario II, where air separation was not required and even more favorable geological and chemical boundary conditions were given, the gross power generation was 127 MW above that in the best case. This resulted in higher CAPEX and OPEX, while the methanol production rate was constant. Nevertheless, one option to increase the competitiveness of Scenario II compared to the best case is to make use of excess electric power output, which was not required for methanol production in Scenario II (116 MW_{el}). However, the impact of cost revenues for providing surplus energy to the public was not considered in the underlying study. Synthesis gas production costs that occur in the context of methanol production are listed in Table 20. The lowest levelized synthesis gas production costs were achieved in Scenario II because of the high gross power generation, the increased synthesis gas mass flow rate, and a higher installed net power capacity. Taking into account the amount of CO_2 that is required for hydrogenation in comparison to the amount of excess CO_2, Scenario I represented the best CO_2 utilization rate for the investigated scenarios (cf. Table 20).

Table 20. One-at-a-time sensitivity analysis results for the onshore UCG-MeOH-CCU scenario considering different synthesis gas compositions and calorific values.

Scenario	Levelized Costs (€/t)	Synthesis Gas Costs (€/GJ)	Excess CV (MJ/sm^3)	UCG synthesis Gas Efficiency (MJ/MJ)	Utilized CO_2 (%)	P_{gross} (MW)	P_{units} (MW)	P_{el} (MW)	P_{net} (MW)
I	205.8	4.5	3.2	0.61	25.5	609	226	128	354
II	189.8	2.9	2.9	0.54	16.4	869	388	116	504
III	177.8	3.1	2.7	0.39	15.8	742	429	1	430
IV	268.2	5.1	4.6	0.42	16.0	721	358	60	418

P_{gross} = total gross power generation, P_{units} = power required to operate all integrated process units, P_{el} = electric power output, P_{net} = net power generation consisting of P_{units} and P_{el}.

Due to the additionally required coal amount resulting from the lower carbon content of the Bielszowice coals, and thus a lower UCG-to-synthesis gas conversion efficiency, the least appropriate synthesis gas composition in view of CO_2 emission mitigation was represented by Scenario III. In summary, OAT sensitivity analysis results showed that for all investigated methanol production scenarios, UCG-based methanol production was profitable and could compete on the Polish market.

Onshore UCG-NH$_3$-CCS Scenario

As for methanol production, gross power generation in the ammonia scenarios was iteratively adjusted to maintain a H_2 mass flow of 1.91 kg/s in the synthesis gas composition and the energy requirement. Comparing OAT sensitivity analysis results related to ammonia production economics, the levelized costs of the four selected synthesis gas compositions differed by up to 28%. Hereby, the lowest costs were achieved in Scenarios II and III, whereby Scenario IV represented the worst case. Unlike competitiveness in the context of methanol production, sensitivity analysis results revealed that the most economic ammonia production was achieved in Scenario II (cf. Table 21). For methanol production, geological benefits of Scenario II were overlapped by a significantly higher gross generation, and thus synthesis gas mass flow to cover the required H_2 amount. However, in the context of ammonia production, gross power generation in Scenario II exceeded that achieved in Scenario III by 4 MW. However, due to the more favorable coal type considered in Scenario II, synthesis gas mass flow in Scenario II was about 4% lower than in Scenario III, resulting in slightly lower CAPEX/OPEX. Besides, ammonia production in the best case benefited from the higher installed net capacity for autonomous power supply. Providing surplus energy would further advantage the best case, since excess electric power not required for ammonia production in Scenario II significantly exceeded that of Scenario III. After H_2 separation, the synthesis gas CV in Scenario III was reduced by 23% compared to the initial CV. With regard to synthesis gas costs, Scenario II showed the most and Scenario IV the least favorable synthesis gas composition. Increased synthesis gas costs in the worst case resulted from high air separation costs. Synthesis gas production competitiveness in the best case was rather impacted by the higher installed net capacity than by economies of scale, since the differences in the synthesis gas mass flows in Scenarios II and III were negligible.

In summary, OAT sensitivity analysis results on ammonia production costs showed that except from the underlying assumed worst-case constraints, all onshore UCG-based ammonia production scenarios were competitive with the average European ammonia market price of 320 €/t NH_3, [64].

Table 21. One-at-a-time sensitivity analysis results for the onshore UCG-NH_3-CCS scenario considering different synthesis gas compositions and calorific values.

Scenario	Levelized Costs (€/t)	Synthesis Gas Costs (€/GJ)	Excess CV (MJ/sm^3)	UCG Synthesis Gas Efficiency (MJ/MJ)	P_{gross} (MW)	P_{units} (MW)	P_{el} (MW)	P_{net} (MW)
I	297.77	5.17	3.20	0.61	382	156	66	222
II	271.89	3.15	2.90	0.54	544	257	58	315
III	274.18	3.20	2.70	0.41	541	311	3	314
IV	353.60	5.60	4.60	0.42	451	238	23	261

P_{gross} = total gross power generation, P_{units} = power required to operate all integrated process units, P_{el} = electric power output, P_{net} = net power generation consisting of P_{units} and P_{el}.

Sensitivity analysis results for all assessed UCG-based end products revealed that applying gasification agent oxygen ratios above 30% by volume, and thus increasing the CO_2 share in the resulting synthesis gas, was not favorable from economic and CO_2 emission mitigation perspectives. Furthermore, an oxygen ratio above 30% by volume in the gasification agent was likely to significantly increase abrasion effects in pipelines and well tubings due to the resulting higher flow velocities.

3.4.2. Variation of Offshore Drilling Costs

For the assessment of offshore drilling costs, we assumed a 5-(reference scenarios), 10-, and 15-fold (worst case) cost increase, compared to the drilling costs applied in the onshore calculations. Our calculation results showed that increasing costs for offshore drilling up to the 15-fold induced an almost linear methanol production cost increment by 8.3 €/t MeOH compared to the offshore reference scenario. The increase in ammonia production costs was also linear and amounted to 7.8 €/t NH_3. The cost bandwidths for methanol and ammonia production resulting from varying drilling costs were low, since the share of drilling costs on total costs in the offshore reference scenarios was about 3% for methanol and 2% for ammonia production. Hereby, the share of offshore drilling costs related to total levelized offshore costs was about 1% lower than the respective share of onshore drilling costs to total levelized onshore costs. This resulted from the radial shape of the offshore well layout, offering a more optimized design than the applied onshore well layout. However, this design optimization in the offshore scenarios was associated with higher operational risks related to the technical implementation of the P-CRIP approach, which has not yet been as widely applied as the CRIP-based UCG scheme.

3.4.3. Impact of Technically-Achievable Gasification Channel Width

Based on the gasification channel width in the reference scenario (200 m, best case), we further assessed costs taking into account channel widths of 50 m (worst case), 100 m, and 150 m (at the outer boundary, cf. Figure 7).

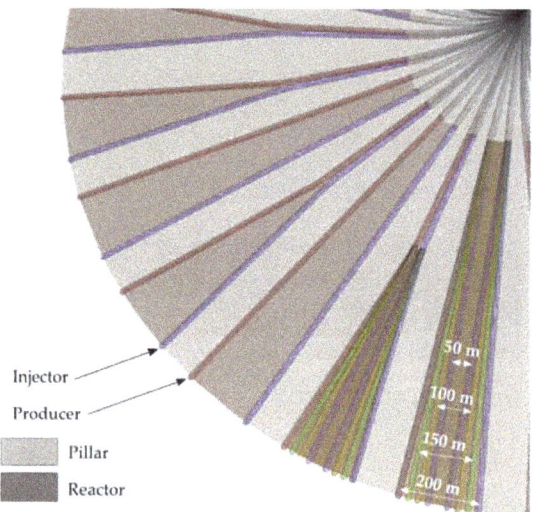

Figure 7. Gasification channel design for different channel widths (50 m–200 m).

In the present study, the model parameter mainly affected by a variation of gasification channel width was the coal yield per gasification channel, also determining the drilling meters. Consequently, drilling costs depended on the number of additional wells to be drilled to maintain the required daily coal supply when channel width was decreasing. OAT sensitivity analysis results listed in Table 22 showed that varying the gasification channel width in the offshore methanol and ammonia production scenarios by 150 m caused a difference in drilling meters of up to 68%. However, this obviously significant difference in drilling meters induced only a minor difference in the total levelized methanol (7.5%) and ammonia (4.3%) production costs due to the low impact of drilling costs on total offshore UCG-based end-product costs. Figure 8 presents the percentage variation in the worst and best cases for the investigated onshore and offshore UCG-CCS/CCU end-uses, determined within the sensitivity analysis and compared to current market prices.

Table 22. Different UCG channel widths and resulting levelized methanol and ammonia costs.

Channel Width (m)	Required Channels (-)		Drilling Length (km)		Drilling Costs (M€)		MeOH Costs (€/t MeOH)	NH$_3$ Costs (€/t NH$_3$)
	MeOH	NH$_3$	MeOH	NH$_3$	MeOH	NH$_3$		
50	53	31	173.7	101.6	245.8	143.8	348.5	505.3
100	32	20	104.9	65.7	148.4	92.8	333.9	490.9
150	25	12	82.0	39.3	116.0	55.6	326.0	485.0
200	17	10	55.7	32.8	78.8	46.4	322.2	483.5

In summary, sensitivity analysis results for the onshore scenarios showed that UCG synthesis gas-based electricity and methanol production can compete on the market even under the worst-case assumptions. With regard to ammonia production, only the onshore worst-case scenario was not economically competitive. Sensitivity analysis results on drilling costs and gasification channel widths for the offshore scenarios reveal that UCG-based methanol production was competitive, whereby ammonia production was not.

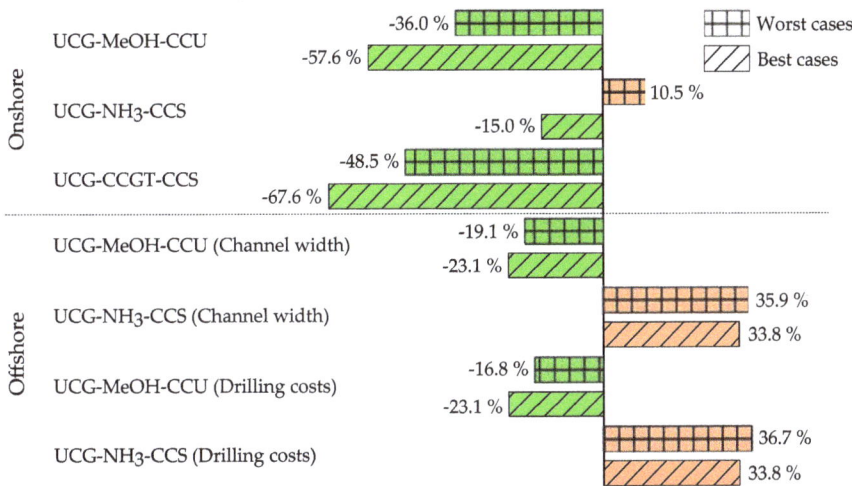

Figure 8. Percentage variation of the best- and worst-case costs in the underground coal gasification-based onshore (electricity, methanol, ammonia production) and offshore scenarios (methanol and ammonia production) to the current market prices.

4. Discussion and Conclusions

In the present study, we applied an enhanced techno-economic model based on Nakaten et al. [5] to determine levelized costs for integrated onshore UCG-CCS/CCU end-use options. Electricity generation (47 €/MWh with CCS), as well as methanol (206 €/t MeOH) and ammonia (298 €/t ammonia) production were considered in the reference scenario. Modeling results showed that with the exception of the worst-case ammonia production scenario, the investigated onshore UCG synthesis gas end-use options can compete on the energy market, even with CCS costs considered. An offshore UCG-CCS/CCU model was applied to determine the levelized costs for methanol (322 €/t MeOH) and ammonia production (484 €/t ammonia). Modeling results exhibited that offshore methanol and ammonia production costs were up to 1.6 times higher than in the onshore scenarios. Hereby, only UCG-based methanol production was economically competitive.

OAT sensitivity analyses were applied to investigate the techno-economics in view of model input data variation and availability, model parametrization and boundary conditions. Hereby, we assessed four synthesis gas compositions, and consequently different synthesis gas CVs (3.6–5.8 MJ/sm^3) within the scope of four onshore scenarios. This model input parameter was selected for further investigation within the sensitivity analysis, since even small changes in process conditions were likely to induce significant changes in the resulting synthesis gas composition and its suitability as a UCG end-product. For the proposed offshore UCG well layout, we investigated the impact of varying gasification channel widths and drilling costs in seven offshore scenarios. For that purpose, we chose offshore drilling costs and gasification channel widths as objective parameters in the OAT sensitivity analysis, especially due to the lack of data on offshore drilling costs and practical experience related to the assumed maximum achievable gasification channel width.

Supplying different coal types, synthesis gas compositions, calorific values, and gasification agent compositions, OAT sensitivity analysis results revealed five factors that mainly impacted the competitiveness of the investigated UCG-CCS/CCU process chains. These comprised the share of oxygen in the gasification agent, availability of required synthesis gas components (H_2, N_2), installed net capacity, the economy of scale effect, as well as the required power. Hereby, ASU costs had the strongest effect on total levelized costs, since scenarios with an O_2 share above 30% were the most expensive ones, while the highest synthesis gas CVs were achieved and all synthesis gas components required for the end-uses provided. Other factors exhibited only minor impacts in the underlying study. The economy of scale takes effect only for products without a fixed output (e.g., synthesis gas production) in contrast to methanol and ammonia production, where the produced output was maintained constant for all scenarios for comparability. Hence, taking into account all considered synthesis gas compositions with a maximum deviation of 43% for H_2, 72% for CO_2, and 43% for the N_2 content, as well as 37% in view of the CV, cost variations of 23% (ammonia production) up to 39% (electricity generation) were determined. Hereby, the most favorable synthesis gas composition for power generation (30 €/MWh with CCS) and methanol production (178 €/t methanol with CCU/CCS) was represented by Scenario III. Optimum ammonia production (272 €/t ammonia with CCS) was achieved within Scenario II. Requirements for production at low-costs were pure air gasification, high synthesis gas CV, high installed net capacity, and the availability of the required synthesis gas components (H_2, N_2). The least favorable synthesis gas composition in view of power generation (47 €/MWh with CCS) was that in Scenario I and for methanol (268 €/t methanol with CCU/CCS) and ammonia production (354 €/t ammonia with CCS) that in Scenario IV.

Sensitivity analysis results on offshore drilling costs showed that due to the small share of drilling costs on the total costs, total levelized methanol and ammonia production costs were only marginally affected. Thus, increasing drilling costs by the 10- and 15-fold, compared to the reference scenario, caused a linear methanol and ammonia production cost increment by 8.3 €/t methanol and by 7.8 €/t ammonia. Different gasification channel widths impacted the extractable coal amount per channel and, hence, the required number of channels to maintain the daily coal demand. Varying the gasification channel width by 150 m resulted in a difference in drilling meters of up to 68%. However, due to the low impact of drilling costs on total costs, levelized methanol and ammonia production costs differed by up to 7.5%, only.

Based on the findings elaborated in the present study, we drew the following conclusions:

- Except from ammonia production under the assumed worst-case conditions, the costs of the investigated onshore UCG-CCS/CCU scenarios were economically competitive on the European market.
- Boundary conditions supporting cost-effective electricity generation as well as methanol and ammonia production were characterized by air-blown gasification, and thus by lower power requirements for air separation and compression in the first place. In order not to exceed the synthesis gas CO_2 share, an oxygen-based gasification agent ratio of more than 30% by volume was not favorable; neither from an economic point of view, nor for CO_2 emission mitigation. Besides, synthesis gas compositions that favored methanol and ammonia production exhibited adequate shares of H_2 and N_2.
- Offshore UCG-based methanol and ammonia production costs were about 1.6 times higher than the respective onshore costs, whereby only UCG-based methanol production was economically competitive on the EU market.
- Compared to the offshore platform with its technical equipment, drilling costs had a minor impact on total levelized costs. Thus, uncertainties in relation to parameters influenced by drilling costs were negligible. A parameter of high uncertainty was the maximum achievable channel width in P-CRIP UCG operations, which has to be further investigated in UCG field tests.
- The impact of boundary conditions and synthesis gas compositions that favored or hampered UCG-based end-product cost-effectiveness in the present study may change, if the methanol and ammonia outputs are not constant for all scenarios and economies of scale take effect. In the underlying study, economies of scale only occurred in the context of synthesis gas production, which was not fixed, but iteratively adjusted to the overall required gross generation.

Aiming at an improved comparability between the different UCG-CCS/CCU end-product costs in the scope of the present study, total levelized costs were determined separately from each other. However, future investigations will aim at the integration of different production chains to quantify cost savings resulting from the synergies of similar processes (e.g., H_2/CO_2 separation), shared infrastructure (e.g., compression, piping systems), as well as the utilization of excess heat in heat exchangers or additional electricity supply. Besides, future research activities will focus on the implementation of costs taking into account the potential environmental impacts of UCG-CCS/CCU.

Author Contributions: N.N. and T.K. conceived of and designed research; N.N. performed the research; N.N. and T.K. analyzed the data; N.N. and T.K. wrote the paper.

Funding: The authors gratefully acknowledge the funding received for the EU-FP7 TOPS project (Grant 608517) by the European Commission (EC).

Acknowledgments: We thank all TOPS project partners for many fruitful discussions. Special thanks go to our colleagues Krzysztof Kapusta, who contributed with knowledge on coal-related data for the study area, as well as Dorota Burchart-Korol (Główny Instytut Górnictwa, Katowice, Poland), who contributed Polish onshore drilling data. We further would like to thank Robert Schmidt, who was involved in the ammonia synthesis model development.

Conflicts of Interest: The authors declare no conflict of interest.

Abbreviations

The following abbreviations are used in this manuscript:

ASU	Air separation unit
CAPEX	Capital expenditure
CCGT	Combined cycle gas turbine
CCS	Carbon capture and storage
CCU	Carbon capture and utilization
COE	Costs of electricity
CRIP	Controlled retraction injection point
EOS	Equation of state
IECM	Integrated environmental control model
OAT	One-at-a-time
OFS	Offshore
ONS	Onshore
OPEX	Operational expenditure
P-CRIP	Parallel controlled retracting injection point
UCG	Underground coal gasification

References

1. Hewing, G.; Hewel-Bundermann, H.; Krabiell, K.; Witte P. *Post-1987 Research and Development Studies of Underground Coal Gasification*; Research Association for Second-Generation Coal Extraction: Essen, Germany, 1988.
2. Klimenko, A. Early Ideas in Underground Coal Gasification and Their Evolution. *Energies* **2009**, *2*, 456–476. [CrossRef]
3. Prabu, V.; Jayanti, S. Simulation of cavity formation in underground coal gasification using bore hole combustion experiments. *Energy* **2011**, *36*, 5854–5864. [CrossRef]
4. Prabu, V.; Jayanti, S. Integration of underground coal gasification with a solid oxide fuel cell system for clean coal utilization. *Hydrog. Energy* **2012**, *37*, 1677–1688. [CrossRef]
5. Nakaten, N.C.; Schlüter, R.; Azzam, R.; Kempka, T. Development of a techno-economic model for dynamic calculation of COE, energy demand and CO_2 emissions of an integrated UCG-CCS process. *Energy* **2014**, *66*, 779–790. [CrossRef]
6. Otto, C.; Kempka, T. Prediction of Steam Jacket Dynamics and Water Balances in Underground Coal Gasification. *Energies* **2017**, *10*, 739. [CrossRef]
7. Kapusta, K.; Stańczyk, K. Pollution of water during underground coal gasification of hard coal and lignite. *Fuel* **2011**, *90*, 1927–1934. [CrossRef]
8. Stańczyk, K, Howaniec, N.; Smolinski, A.; Swiadrowski, J.; Kapusta, K.; Wiatowski, M.; Grabowski, J.; Rogut, J. Gasification of lignite and hard coal with air and oxygen enriched air in a pilot scale ex-situ reactor for underground gasification. *Fuel* **2011**, *90*, 1953–1962. [CrossRef]

9. Kapusta, K.; Stańczyk, K.; Wiatowski, M.; Ćhećko, J. Environmental aspects of a field-scale underground coal gasification trial in a shallow coal seam at the Experimental Mine Barbara in Poland. *Fuel* 2013, *113*, 196–208. 2013.05.015. [CrossRef]
10. Stańczyk, K. Experience of Central Mining Institute in Underground Coal Gasification–Research and Pilot Test in "KWK Wieczorek" Mine. 2015. Available online: http://www.fossilfuel.co.za (accessed on 26 September 2017).
11. Kinaev, N.; Belov, A.; Bongers, G.; Grebenyuk, I.; Vinichenko, I. Integrated assessment of feasibility of coal-to-chemical projects. In Proceedings of the 8th International Freiberg Conference on IGCC & XtL Technologies: Innovative Coal Value Chains, Cologne, Germany, 12–26 June 2016.
12. NCBiR Project. *Development of Coal Gasification Technology for Highly Efficient Production of Fuels and Electricity*; Project Report; 2015; unpublished.
13. EURACOAL. Mineable Coal and Lignite Reserves Poland. 2016. Available online: http://euracoal2.org (accessed on 9 December 2016).
14. Czaplicka-Kolarz, K.; Krawczyk, P.; Ludwik-Pardala, M.; Burchart-Korol, D. Cost-effectiveness of underground coal gasification by the shaft method. *Przem. Chem.* 2015, *94*, 1708–1713. [CrossRef]
15. Burchart-Korol, D.; Korol, J.; Czaplicka-Kolarz, K. Life cycle assessment of heat production from underground coal gasification. *Int. J. Life Cycle Assess.* 2016, *21*, 1391–1403. [CrossRef]
16. EC. *Energy Roadmap 2050*; European Commission: Brussels, Belgium, 2011. Available online: http://eur-lex.europa.eu (accessed on 15 February 2017).
17. Pérez-Fortes, M.; Schöneberger, J.C.; Boulamanti, A.; Tzimas, E. Methanol synthesis using captured CO_2 as raw material: Techno-economic and environmental assessment. *Appl. Energy* 2016, *161*, 718–732. [CrossRef]
18. EC. *Large Volume Inorganic Chemicals-Solids and Others*; European Commission: Brussels, Belgium, August 2007. Available online: https://www.mpo.cz (accessed on 22 September 2017)
19. Otto, C.; Kempka, T.; Kapusta, K.; Stańczyk, K. Fault Reactivation Can Generate Hydraulic Short Circuits in Underground Coal Gasification-New Insights from Regional-Scale Thermo-Mechanical 3D Modeling. *Minerals* 2016, *6*, 101. [CrossRef]
20. Otto, C.; Kempka, T. Thermo-mechanical Simulations Confirm: Temperature-dependent Mudrock Properties are Nice to have in Far-field Environmental Assessments of Underground Coal Gasification. *Energy Procedia* 2015, *76*, 582–591. [CrossRef]
21. Otto, C.; Kempka, T. Thermo-Mechanical Simulations of Rock Behavior in Underground Coal Gasification Show Negligible Impact of Temperature-Dependent Parameters on Permeability Changes. *Energies* 2015, *8*, 5800–5827. [CrossRef]
22. DWSIM. Open-Source CAPE-OPEN Compliant Chemical Process Simulator. Available online: http://dwsim.inforside.com.br (accessed on 12 December 2016).
23. ChemSep. Modeling Separation Processes, Databases. Available online: http://www.chemsep.org (accessed on 12 December 2016).
24. Godbolt, B. Scientific Drilling UCG Training School, Directional Drilling in Coal. 2011. Available online: http://repository.icse.utah.edu (accessed on 12 December 2016).
25. Kempka, T.; Fernandez-Steeger, T.; Li, D.; Schulten, M.; Schlüer, R.; Krooss, B. Carbon dioxide sorption capacities of coal gasification residues. *Environ. Sci. Technol.* 2011, *45*, 1719–1723. [CrossRef] [PubMed]
26. Najafi, M.; Jalali, S.M.E.; KhaloKakaie, R. Thermal-Mechanical Numerical Analysis of Stress Distribution in the vicinity of Underground Coal Gasification (UCG) Panels. *Int. J. Coal Geol.* 2014, *134–135*, 1–16. [CrossRef]
27. Świadrowski, J.; Mocek, P.; Jedrysik, E.; Stańczyk, K.; Krzemień, J.; Krawczyk, P.; Ćhećko, J. Demonstration facility for underground coal gasification. *CHEMIK* 2015, *69*, 815–826.
28. KOPEX. *The Draft Technical Installations PZW on KWK Wieczorek*; Project Report; KOPEX Construction Company Szybów, SA: Ibadan, Nigeria, 2012.
29. Acheick, A.M.; Batto, S.F.; Changmoon, Y.; Chien, S.C.; Choe, J.I.; Cole, K.R.; Engel, K.; Gardner, W.; Gilbert, S.N.; Hui, Y.; et al. *Viability of Underground Coal Gasification with Carbon Capture and Storage in Indiana*; Indiana University-Bloomington School of Public and Environmental Affairs: Bloomington, IN, USA, 2011. Available online: http://www.indiana.edu (accessed on 20 January 2017).
30. Integrated Environmental Control Model (IECM). A Tool for Calculating the Performance, Emissions, and Cost of a Fossil-Fueled Power Plant (Version 8.0.1 Beta). Available online: http://www.iecm-online.com (accessed on 12 December 2016).

31. Cardenas Barranon, D.C. Methanol and Hydrogen Production. Master's Thesis, 2006. Available online: http://ltu.diva-portal.org (accessed on 18 September 2017).
32. Hillebrand, B. *Stromerzeugungskosten Neu zu Errichtender Konventioneller Kraftwerke*; RWI-Papiere Nr. 47; Rheinisch-Westfälisches Institut für Wirtschaftsforschung, Essen, Germany, 1997.
33. Schneider, L. *Stromgestehungskosten von Großkraftwerken*; Öko-Institut e.V.: Berlin, Germany, 1998. Available online: http://www.oeko.de (accessed on 12 December 2016).
34. Chen, K.D.; Yu, L. Experimental study on long-tunnel large-section two-stage underground coal gasification. In *Mining Science and Technology*; T.S. Golosinski, Guo Yuguan; Balkema, Rotterdam 1996; pp. 313–316.
35. Creedy, D.P.; Garner, K. *Clean Energy from Underground Coal Gasification in China*; COAL R250 DTI/Pub URN 03/1611; Department of Trade and Industry: London, UK, 2004.
36. Couch, G.R. *Underground Coal Gasification*; CCC/151; IEA Clean Coal Centre: London, UK, 2009.
37. Nakaten, N.; Kempka, T. Radial-symmetric well design to optimize coal yield and maintain required safety pillar width in oshore underground coal gasification. *Energy Procedia* **2017**, *125*, 27–33. [CrossRef]
38. Boeing, N. Bohrende Fragen. Tech. Review. Magazin für Innovation. 2010. Available online: https://www.heise.de (accessed on 12 December 2016).
39. Gräbner, M.; Morstein, O.; Rappold, D.; Gunster, W.; Beysel, G.; Meyer, B. Constructability study on a German reference IGCC power plant with and without CO_2 capture for hard coal and lignite. *Energy Convers. Manag.* **2010**, *51*, 2179–2187. [CrossRef]
40. Zero emissions platform (ZEP). The Costs of CO_2 Capture. 2011. Available online: http://www.zeroemissionsplatform.eu (accessed on 12 December 2016).
41. Mohammed, I.Y.; Samah, M.; Mohamed, A.; Sabina, G. Comparison of SelexolTM and Rectisol Technologies in an Integrated Gasification Combined Cycle (IGCC) Plant for Clean Energy Production. *IJER* **2014**, *3*, 742–744. [CrossRef]
42. Hammond, G.P.; Ondo Akwe, S.S.; Williams, S. Techno-economic appraisal of fossil-fuelled power generation systems with carbon dioxide capture and storage. *Energy* **2011**, *36*, 975–984. [CrossRef]
43. Li, H.; Ditaranto, M.; Berstad, D. Technologies for increasing CO_2 concentration in exhaust gas from natural gas-fired power production with post-combustion, amine-based CO_2 capture. *Energy* **2011**, *36*, 1124–1133. [CrossRef]
44. Ausfelder, F.; Bazzanella, A. *Diskussionspapier Verwertung und Speicherung von CO_2*; DECHEMA e.V.: Frankfurt am Main, Germany, 2008. Available online: https://dechema.de (accessed on 12 December 2016).
45. Katofsky, R.E.*The Production of Fluid Fuels From Biomass*; Princeton University: Princeton, NJ, USA, 1993. Available online: http://acee.princeton.edu (accessed on 18 September 2017).
46. Komiyama, H.; Mitsumori, T.; Yamaji, K.; Yamada, K. Assessment of energy systems by using biomass plantation. *Fuel* **2001**, *80*, 707–715. [CrossRef]
47. Burton, E.; Friedmann, J.; Upadhye, R. *Best Practices in Underground Coal Gasification*; Contract No. W-7405-Eng-48; Lawrence Livermore National Laboratory: Livermore, CA, USA, 2006. Available online: http://www.purdue.edu (accessed on 6 January 2014).
48. SiteChar Project. *Characterisation of European CO_2 Storage*; Project Report, Deliverable D2.2 Economic Assessment. Unpublished work, 2013.
49. Kunkes, E.; Behrens, M. Methanol Chemistry. In *Chemical Energy Storage*; Walter de Gruyter: Berlin, Germany, 2013. Available online: http://pubman.mpdl.mpg.de (accessed on 20 September 2017).
50. Moffat, A.S. Methanol-powered. *Science* **1991**, *251*, 515–515.
51. Olah, G.A.; Goeppert, A.; Prakash, G.K.S. *Beyond Oil and Gas: The Methanol Economy*; Wiley-VCH. XIV: Weinheim an der Bergstrasse, Germany, 2006.
52. Van-Dal, E.; Bouallou, C. CO_2 abatement through a methanol production process. *Chem. Eng. Trans.* **2012**, *29*, 463–468. [CrossRef]
53. Villesca, J.; Bala, V.; Garcia, A. Reactor Project: Ammonia Synthesis. Available online: http://www.owlnet.rice.edu/~ceng403/nh3syn97.html (accessed on 20 August 2019)
54. Penkuhn, M.; Tsatsaronis, G. Comparison of different ammonia synthesis loop configurations with the aid of advanced exergy analysis. In Proceedings of the ECOS, 29th International Conference on Efficiency, Cost, Optimization, Simulation and Environmental Impact of Energy Systems, Portoroz, Slovenia, 19–23 June 2016. Available online: http://www.owlnet.rice.edu (accessed on 20 September 2017).

55. Nakaten, N.C.; Islam, R.; Kempka, T. Underground Coal Gasification with Extended CO_2 Utilization—An Economic and Carbon Neutral Approach to Tackle Energy and Fertilizer Supply Shortages in Bangladesh. *Energy Procedia* **2014**, *63*, 8036–8043. [CrossRef]
56. Ernst&Young. Wplyw Energetyki Wiatrowej Na Wzrost Gospodarczy w Polsce. 2012. Available online: http://domrel.pl (accessed on 12 December 2016).
57. Zaporowski, B. Koszty Wytwarzania Energii Elektrycznej Dla Perspektywicznych Technologii WytwóRczych Polskiej Elektroenergetyki. *Polityka Energetyczna* **2012**, *15*, 43–55. Available online: https://www.min-pan.krakow.pl (accessed on 12 December 2016).
58. Central Statistical Office of Poland (CSOP). *The Economy of Fuel and Energy in 2013 and 2014*; Central Statistical Office of Poland: Warszawa, Poland, 2015. Available online: http://stat.gov.pl/ (accessed on 12 December 2016).
59. ERO. Information of the President of the Energy Regulatory Office No. 3/2015. 2015. Available online: https://www.kpmg.com (accessed on 29 February 2014).
60. ERO. Information of President of Energy Regulatory Office No. 46/2015. 2015. Available online: https://www.ure.gov.pl (accessed on 4 November 2016).
61. McCollum, D. and Ogden, J. *Techno-Economic Models for Carbon Dioxide Compression, Transport, and Storage. Correlations for Estimating Carbon Dioxide Density and Viscosity*; Institute of Transportation Studies (ITS), University of California: Berkeley, CA, USA, 2006. Available online: http://www.its.ucdavis.edu (accessed on 30 January 2019).
62. Methanex. The Power of Agility. 2018. Available online: https://www.methanex.com (accessed on 28 September 2018)
63. Bartels, J.R. A feasibility study of implementing an Ammonia Economy. Master's Thesis, Iowa State University, Ames, Iowa, USA, 2008. Available online: http://lib.dr.iastate.edu (accessed on 20 September 2017).
64. AMIS. Fertilizer Outlook. Market Monitor. 2018. Available online: http://www.fao.org (accessed on 1 October 2018).
65. Klebingat, S.; Kempka, T.; Schulten, M.; Azzam, R.; Fernández-Steeger, T.M. Innovative thermodynamic underground coal gasification model for coupled synthesis gas quality and tar production analyses. *Fuel* **2016**, *183*, 680–686. [CrossRef]
66. Klebingat, S.; Kempka, T.; Schulten, M.; Azzam, R.; Fernández-Steeger, T. M. Optimization of synthesis gas heating values and tar by-product yield in underground coal gasification. *Fuel* **2018**, *229*, 248–261. [CrossRef]
67. Mocek, P.; Pieszczek, M.; Swiadrowski, J.; Kapusta, K.; Wiatowski, M.; Stańczyk, K. Pilot-scale underground coal gasification (UCG) experiment in an operating Mine "Wieczorek" in Poland. *Energy* **2016**, *111*, 313–321. [CrossRef]

© 2019 by the authors. Licensee MDPI, Basel, Switzerland. This article is an open access article distributed under the terms and conditions of the Creative Commons Attribution (CC BY) license (http://creativecommons.org/licenses/by/4.0/).

Article

Multiscale Apparent Permeability Model of Shale Nanopores Based on Fractal Theory

Qiang Wang, Yongquan Hu, Jinzhou Zhao *, Lan Ren, Chaoneng Zhao and Jin Zhao

State Key Laboratory of Oil-Gas Reservoir Geology & Exploitation, Southwest Petroleum University, Chengdu 610500, China
* Correspondence: 201811000142@stu.swpu.edu.cn

Received: 4 August 2019; Accepted: 27 August 2019; Published: 2 September 2019

Abstract: Based on fractal geometry theory, the Hagen–Poiseuille law, and the Langmuir adsorption law, this paper established a mathematical model of gas flow in nano-pores of shale, and deduced a new shale apparent permeability model. This model considers such flow mechanisms as pore size distribution, tortuosity, slippage effect, Knudsen diffusion, and surface extension of shale matrix. This model is closely related to the pore structure and size parameters of shale, and can better reflect the distribution characteristics of nano-pores in shale. The correctness of the model is verified by comparison with the classical experimental data. Finally, the influences of pressure, temperature, integral shape dimension of pore surface and tortuous fractal dimension on apparent permeability, slip flow, Knudsen diffusion and surface diffusion of shale gas transport mechanism on shale gas transport capacity are analyzed, and gas transport behaviors and rules in multi-scale shale pores are revealed. The proposed model is conducive to a more profound and clear understanding of the flow mechanism of shale gas nanopores.

Keywords: fractal; slippage effect; Knudsen diffusion; surface diffusion; apparent permeability

1. Introduction

Shale gas reservoirs have different reservoir formation modes and reservoir physical properties from conventional gas reservoirs, which leads to the complexity and multi-scale characteristics of shale gas percolation. Therefore, it is necessary to systematically analyze the gas migration mechanism in pore structures of different scales. Due to the nanoscale pore size of shale and the coexistence of multiple gas migration mechanisms, it is very challenging to simulate gas flow in the nanoscale pores of shale [1,2]. Most scholars have established the flow mathematical model and deduced the shale gas apparent permeability model mainly by dividing the flow pattern and considering the multiple migration mechanism of shale gas. Kuuskraa systematically analyzed the transport mechanism of shale gas and concluded that shale gas reservoirs are triple porous media. Shale gas transport mechanisms include gas desorption diffusion and Darcy flow. However, he did not propose a three-hole model considering multiple seepage mechanisms [3]. Roy and Raju defined the Knudsen number expression by using the average free path of gas molecules and pore radius of shale and other parameters, divided the flow of gas in shale pores into slip-off flow and transition flow, and believed that Darcy's flow law was not applicable to describe the gas migration of shale nanoscale pores [4]. Javadpour further established a mathematical model for gas flow considering diffusion and adsorption. By comparing this model with the conventional Darcy flow equation, the expression of apparent permeability is obtained [5]. This permeability can be applied directly to reservoir numerical simulation. It is shown that the ratio of apparent permeability to Darcy permeability increases with smaller pore size. Therefore, the influence of various migration mechanisms on the study of gas migration law of nano-pores in shale cannot be ignored. However, this theoretical model is based on the single-pipe model, without considering the actual complex pore structure of shale and the high-temperature and high-pressure characteristics of shale reservoir. Freeman established a multi-component fluid flow model based on micron-nanometer scale pore media by considering the interaction mechanism of convection, Knudsen diffusion and molecular diffusion [6]. In the same year, Curtis studied and described the structural characteristics of micron and nanometer shale by focusing ion beam and scanning electron microscope. The research results show that the physical characteristics of shale gas flow change with the change of pore radius, and the classic Darcy's law is no longer adaptive [7]. Under the pore size of shale observed by scanning electron microscope, it is necessary to consider the introduction of new flow mechanism when shale gas is transferred in reservoir. Ebrahim considered shale as a nano-pore medium, and proposed to study the flow of gas in nano-scale pore materials by using the Lattice Boltzmann method. It was believed that particle transport included the slippage of free gas molecules and the interfacial transport of adsorbed gas molecules. Studies show that there is a critical numeral, beyond which molecular slippage and interfacial transport may lead to the generation of molecular flow, which will improve the gas migration ability in the nanopores of organic matter [8]. Shabro considered slippage effect, Knudsen diffusion and adsorption and desorption, established pore scale seepage model, and quantitatively analyzed the contribution of Knudsen diffusion, adsorption and desorption to the whole flow. Studies show that slip flow and Knudsen diffusion have important effects on the apparent permeability of shale reservoirs. Gas desorption reduces shale reservoir pressure slowly, and slippage and Knudsen diffusion increase apparent permeability, which explains why actual production is higher than expected [9]. Michel modified Beskok and Karniadakis' equations to describe flow problems in nanoscale pores of shale reservoirs and extended the influence of arbitrary pore size distribution on apparent permeability under real gas conditions [10]. Freeman proposed a numerical model that can describe the porous size function of the diffusion and desorption process in shale. Combining macroscopic (reservoir fracture) and microscopic (diffusion of nanopores) physical phenomena, the model shows how gas composition varies over time and space during production [11]. Based on the research results of Javadpour and Swami and Settari established a single-tube flow model of shale gas in nanoscale pores by considering multiple seepage mechanisms such as Darcy flow, Knudsen diffusion and slippage effect. The finite difference method is used to solve the model, and the influence of Knudsen diffusion and slippage effect on gas production is analyzed and studied. The results show that the influence of the above factors on the production of shale gas reservoir must be taken into account [12,13]. Guo established the mathematical model of shale nano-pore seepage based on the convection-diffusion model, and deduced a new calculation formula of relative permeability. The reliability and accuracy of the model are verified by comparing with experimental data [14]. Singh ignored the slippage effect and derived the analytical formula of apparent permeability. The formula is only related to pore radius and pore morphology. It is considered that this model is applicable to all shale reservoir conditions where Knudsen number is less than 1 [15]. Mohammad et al. established the shale gas apparent permeability model based on gas dynamics principle. The results show that the gas molecular weight and temperature have significant effects on the apparent permeability, while the Tangential Momentum Accommodation Coefficient (TMAC) coefficient has the least effect. Through experimental study, it is found that there is no linear relationship between the apparent permeability and the adsorption concentration [16]. With the increase of the adsorption concentration, the influence of the adsorption concentration on the apparent permeability increases. Zhang established a new shale gas apparent permeability model by comprehensively considering slippage effect, Knudsen diffusion and surface diffusion, and verified it through experimental data. The research shows that when the local layer pressure decreases, the influence of surface diffusion on gas transportation in shale gradually increases, which cannot be ignored. When the local layer pressure is high, the effect of surface diffusion can be ignored [17]. Wang established a shale gas apparent permeability calculation model in real gas state by considering adsorption and desorption, stress sensitivity, non-Darcy effect and surface diffusion. Single-layer adsorption and multi-layer adsorption were described by uniform equation. The contributions of different flow mechanisms to apparent permeability and the sensitive parameters of apparent permeability are analyzed [18]. Wu established a model for calculating shale apparent permeability by considering the transport mechanism of free gas and air suction. It is considered that the total mass flux cannot be represented by the sum of Darcy flow mass flux and Knudsen diffusion mass flux, and the weight factors of Darcy flow mass flux and Knudsen diffusion mass flux are proposed based on the probability of molecular collision. The Langmuir isothermal adsorption equation and the material balance equation were used to calculate the apparent permeability [19]. Based on the fractal theory, Wang et al. deduced an analytical model for the apparent permeability of tight gas/shale gas under the conditions of gas convection and diffusion, and compared the experimental data to verify the accuracy of the model. But the flow characteristics of shale gas are not fully considered [20]. Wang et al. used logarithmic normal distribution function and cylindrical capillary to characterize pore size distribution in porous media and nanopores in matrix respectively, and proposed a real shale matrix pore gas transport model [21]. Zhang et al. established a random apparent liquid permeability model that considers the wettability and pore size related liquid slip effect, total organic carbon content, and the structural parameters. The results reveal the transport mechanism of dual wettability nano-porous shale [22]. Li et al. used the 3D intermingled-fractal model to derive a new permeability model for the organic-rich shale nanopore matrix and Shen et al. also developed an apparent permeability model of shale nanopores, which describes the adsorptive gas flow behavior in shale by considering the effects of gas adsorption, stress dependence, and non-Darcy flow [23,24]. In addition, there has been recent work showing that the constitutive equations of classical density functional theory, molecular dynamic simulations, Lattice Boltzmann and multi-continuum are able to reproduce more complex molecular dynamics simulations fluids in nanopores [25–28].

In previous studies on apparent permeability model in nano-pores of shale gas, the influences of pore structure, slippage effect, Knudsen diffusion and surface expansion on pore flow of shale matrix have not been comprehensively considered [29–31]. However, as shale reservoirs are different from conventional reservoirs in physical characteristics and gas occurrence characteristics, it is necessary to comprehensively consider each influencing factor when establishing flow mechanism. To achieve this goal, a new shale gas apparent permeability model is established based on fractal geometry theory, Hagen–Poiseuille law, and Langmuir adsorption law, which can be well matched with experimental data. This model integrates all characteristics of shale gas and is an important innovation in studying shale gas flow mechanism. This model can be a good reference for scientific research and engineering application.

2. Establishment of Multi-Scale Flow Mechanism of Shale Gas

2.1. Fractal Porous Media Model

Previous mathematical models of gas flow in shale matrix are based on the tortuous bundle model with uniform radius. However, the pore structure of shale matrix is extremely complex and random. The pore radius of shale is distributed at 0.3–300 nm [32], while the complex structure of matrix pores can be described by fractal geometry theory within a certain scale. Therefore, based on the fractal geometry theory, the tortuous bundle model with uneven pore radius distribution is adopted to characterize shale [19,33,34]. The physical model is shown in Figure 1.

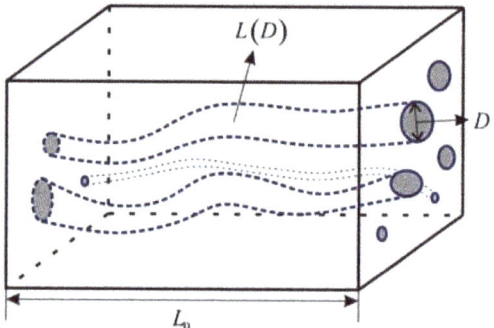

Figure 1. Shale tortuous bundle model.

The fractal porous media is composed of capillary bundles or pores with different sectional radii, and the tube radii of capillary bundles satisfy the fractal scale law, so the cumulative distribution of pore sizes in the unit section of the matrix is [35–38].

$$N(L \geq D) = \left(\frac{D_{max}}{D}\right)^{D_p} \tag{1}$$

where N is the number of channels; L is the length scale; D_{max} is the pipe diameter of the largest capillary tube bundle, m; D is the fractal dimension of hole area.

Assuming that the distribution of pore size is continuous, the differential equation of pore diameter D in Equation (1) can be obtained

$$-dN = D_p D_{max}^{D_p} D^{-(D_p+1)} dD \tag{2}$$

where $-dN > 0$ represents the cumulative number of pores increases with the decrease of pore diameter, so the total pore area on the flow section A_p is

$$A_p = -\int_{D_{min}}^{D_{max}} \frac{1}{4}\pi D^2 dN = \int_{D_{min}}^{D_{max}} \frac{1}{4}\pi D^2 D_p D_{max}^{D_p} D^{-(D_p+1)} dD = \frac{\pi D_p D_{max}^2}{4(2-D_p)}\left[1 - \left(\frac{D_{min}}{D_{max}}\right)^{2-D_p}\right] \tag{3}$$

where D_{min} is the smallest capillary tube bundle pipe diameter, m.

Assuming that the porosity of the fractal porous media surface is equal to the volume porosity, the flow cross-sectional area (A) is

$$A = \frac{A_p}{\phi} = \frac{\pi D_p D_{max}^2}{4\phi(2-D_p)}\left[1 - \left(\frac{D_{min}}{D_{max}}\right)^{2-D_p}\right] \tag{4}$$

where ϕ is the porosity.

The relationship between length of tortuous capillary tube and radius of capillary tube can be expressed as fractal power law [39]:

$$L(D) = D^{1-D_t} L_0^{D_t} \tag{5}$$

where D_t is the fractal dimension of tortuosity; L_0 is the characteristic length of the capillary along the flow direction, m.

2.2. Shale Gas Multi-Scale Flow Model

Due to the random distribution of pore size in the shale matrix and different storage mechanisms (including viscous flow, free gas slip flow, transition flow and surface diffusion generated by adsorption and desorption) [25,40,41], there are multiple gas migration mechanisms in the shale reservoir, as shown in Figure 2.

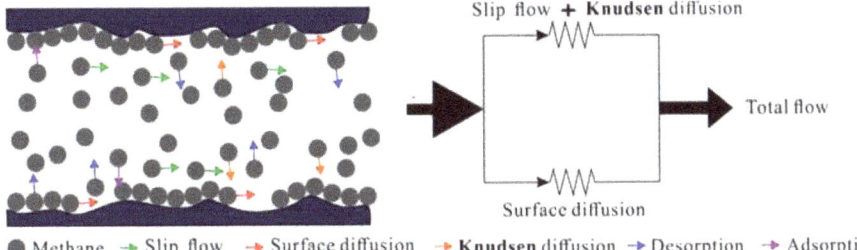

Figure 2. Multi-scale flow mechanism in shale nanopores.

2.2.1. Slip Flow

According to Hagen-Poiseuille law, the mass flux of mass flow through a single circular pipe cross section is [19,41,42]

$$J_H = \frac{D^2 p M_g}{32 \mu Z R T} \frac{\Delta p}{L(D)} \tag{6}$$

where μ is gas viscosity, Pa·s; p is pore pressure, MPa; Z is the gas compression factor; R is the gas constant, 8.314 J/(mol·K); T is temperature, K; M_g is the molecular molar mass of the gas, kg/mol.

When K_n (Knudsen number) is between 0.001 and 0.1, the collision between gas molecules controls gas migration, and there is slippage effect. At this time, the flow state changes from viscous flow to slip flow, so Equation (6) needs to be corrected [43,44]. Considering the effects of slippage effect and rarefied gas effect, the expression of the mass flux (J_H) of the real gas flowing through the section of a single circular pipe after coefficient modification is

$$J_H = (1 + \alpha K_n)\left(1 + \frac{4K_n}{1 - bK_n}\right)\frac{D^2 p M_g}{32 \mu Z R T} \frac{\Delta p}{L(D)} \tag{7}$$

where

$$\alpha = \alpha_0 \frac{2}{\pi} \tan^{-1}\left(\alpha_1 K_n^\beta\right) \tag{8}$$

where α is the effect coefficient of rarefied gas, dimensionless; b is the slip factor; is the effect coefficient of rarefied gas as the number approaches infinity, dimensionless; α_0 is the effect coefficient of rarefied gas as K_n approaches infinity, dimensionless; α_1, β is the fitting coefficient, dimensionless.

2.2.2. Knudsen Diffusion

When $10 \leq K_n$, the gas flow pattern is molecular free flow, and the gas mass flux (J_N) of a single circular tube under Knudsen diffusion is

$$J_N = \frac{D}{3}\left(\frac{8M_g}{\pi R T}\right)^{1/2} \frac{\Delta p}{L(D)} \tag{9}$$

By correcting Equation (9), the mass flux of real gas (J_N) through single circular pipe Knudsen diffusion is

$$J_N = \frac{D}{3} C_g \left(\frac{8ZM_g}{\pi R T}\right)^{1/2} \frac{p}{Z} \frac{\Delta p}{L(D)} \tag{10}$$

where C_g is compression coefficient, MPa^{-1}.

A weighting factor was introduced to characterize the contribution rate of slip flux and Knudsen diffusion flux. The slip flow and Knudsen diffusion weighting factors were calculated by the ratio of the collision frequency between the gas molecules and the total collision frequency, and the ratio of collision frequency between gas molecules and nanopore wall to total collision frequency [19]. The relationship between the number of gas molecule collisions per unit time in nanopores is as follows

$$\frac{1}{t_T} = \frac{1}{t_M} + \frac{1}{t_S} \tag{11}$$

where t_T is the average time of gas molecules colliding once, s; t_M is the average time of one collision between gas molecules, s; t_S is the average time of a collision between gas molecules and nanopore wall, s.

The number of collisions can be expressed as

$$\frac{1}{t_T} = \frac{\bar{v}}{\lambda_T}\frac{1}{t_M} = \frac{\bar{v}}{\lambda}\frac{1}{t_S} = \frac{\bar{v}}{D} \tag{12}$$

where $\bar{v}$ is the average velocity of gas molecules, m/s; λ_T is the average free path of gas molecules, m; λ is the real gas mean free path, m.

Substitute Equation (12) into Equation (11) to get

$$\frac{1}{\lambda_T} = \frac{1}{\lambda} + \frac{1}{D} \tag{13}$$

The contribution of slip flow and Knudsen diffusion to free gas transport can be obtained by the ratio of the collision frequency between gas molecules and the total collision frequency, and the ratio of the collision frequency between gas molecules and the nanopore wall surface to the total collision frequency, respectively, then

$$\varepsilon_H = \frac{1}{\lambda} / \frac{1}{\lambda_T} = \frac{1}{1+K_n} \tag{14}$$

$$\varepsilon_N = \frac{1}{D} / \frac{1}{\lambda_T} = \frac{1}{1+1/K_n} \tag{15}$$

Then, the total mass flux of free gas through a single circular pipe cross section can be expressed as

$$J_F = \varepsilon_H J_H + \varepsilon_N J_N \tag{16}$$

and the total mass flow rate of free gas through a single circular pipe section is

$$q_F(D) = J_F \frac{\pi D^2}{4}. \tag{17}$$

Substitute Equations (7), (10) and (16) into Equation (17) to get

$$q_F(D) = \left[\varepsilon_H(1+\alpha K_n)\left(1+\frac{4K_n}{1-bK_n}\right)\frac{D^2 p M_g}{32\mu ZRT}\frac{\Delta p}{L(D)} + \varepsilon_N \frac{D}{3}C_g\left(\frac{8ZM_g}{\pi RT}\right)^{1/2}\frac{p}{Z}\frac{\Delta p}{L(D)}\right]\frac{\pi D^2}{4}. \tag{18}$$

By integrating the total mass flow rate of free gas through single circular pipe section in the interval of minimum pore diameter and maximum pore diameter $[D_{min}, D_{max}]$, the total mass flow rate of free gas in shale gas porous media can be obtained

$$\begin{aligned} Q_F(D) &= -\int_{D_{min}}^{D_{max}} q_F(D)dN \\ &= \int_{D_{min}}^{D_{max}}\left(\varepsilon_H(1+\alpha K_n)\left(1+\frac{4K_n}{1-bK_n}\right)\frac{D^2 pM_g}{32\mu ZRT}\frac{\Delta p}{L(D)} + \varepsilon_N\frac{D}{3}C_g\left(\frac{8ZM_g}{\pi RT}\right)^{1/2}\frac{p}{Z}\frac{\Delta p}{L(D)}\right)\frac{\pi D^2}{4}D_p D_{max}^{D_p}D^{-(D_p+1)}dD \\ &= \frac{\pi}{4}\frac{p}{Z}\frac{D_p D_{max}^{D_p}\Delta p}{L_0^{D_t}}\int_{D_{min}}^{D_{max}}\left(\varepsilon_H(1+\alpha K_n)\left(1+\frac{4K_n}{1-bK_n}\right)\frac{M_g}{32\mu ZRT}D^{2-D_p+D_t} + \varepsilon_N\frac{1}{3}C_g\left(\frac{8ZM_g}{\pi RT}\right)^{1/2}D^{1-D_p+D_t}\right)dD \end{aligned} \tag{19}$$

2.2.3. Adsorption and Desorption

In the pressure attenuation process of shale gas reservoir exploitation, shale gas adsorption and desorption is a very fast physical process compared with surface diffusion, so the adsorption amount can be calculated by Langmuir's adsorption law

$$q_{ai} = \frac{q_L p}{p_L + p} \tag{20}$$

where q_L is Langmuir volume, m³/kg; p_L is Langmuir pressure, MPa.

Considering the influence of real gas, the adsorption capacity is

$$q_a = \frac{q_L p/Z}{p_L + p/Z} \tag{21}$$

Gas coverage was defined as the ratio of adsorbed gas volume to Langmuir volume, and the gas coverage of ideal gas and real gas was respectively

$$\theta_i = \frac{p}{p_L + p} \tag{22}$$

$$\theta = \frac{p/Z}{p_L + p/Z} \tag{23}$$

where θ_i is the gas coverage of ideal gas, dimensionless; θ is the gas coverage of real gas, dimensionless.

Due to the adhesion of adsorbed gas molecules to the pore surface, the nano-pore space of shale decreases, and the effective diameter of ideal gas and real gas flow is

$$D_{ei} = D - 2d_m \theta_i \tag{24}$$

$$D_e = D - 2d_m \theta \tag{25}$$

where d_m is the molecular diameter of methane, m.

2.2.4. Surface Diffusion

In shale gas, the concentration of adsorbed gas is much higher than that of free gas. Due to the adsorption and desorption of shale gas, the influence of surface diffusion on gas migration cannot be ignored. The mass flux (J_B) under the surface diffusion of a single circular pipe can be expressed as

$$J_B = D_B^0 \frac{C_{sc}}{p} \frac{\Delta p}{L(D)} \tag{26}$$

where C_{sc} is the adsorption gas concentration, kg/m³; D_B^0 is the surface diffusion coefficient when gas coverage is 0.

According to Chen and Yang, the surface diffusion coefficient considering gas coverage is [45]

$$D_B = D_B^0 \frac{(1-\theta) + \frac{\kappa}{2}\theta(2-\theta) + [H(1-\kappa)](1-\kappa)\frac{\kappa}{2}\theta^2}{\left(1 - \theta + \frac{\kappa}{2}\theta\right)^2} \tag{27}$$

where

$$H(1-\kappa) = \begin{cases} 0 & \kappa \geq 1 \\ 10 & \leq \kappa < 1 \end{cases} \tag{28}$$

Combined with Equations (22) and (23), the ideal adsorbed gas concentration and the real adsorbed gas concentration in Langmuir monolayer adsorbed shale nano-pores are respectively expressed as

$$C_{sci} = \frac{4\theta_i M_g}{\pi d_m^3 N_A} \tag{29}$$

$$C_{sc} = \frac{4\theta M_g}{\pi d_m^3 N_A} \tag{30}$$

where κ is the diffusion capacity coefficient of gas molecules; θ is the true gas hole wall coverage; N_A is Avogadro constant, 6.022×10^{23} mol^{-1}.

By introducing Equations (29) and (30) into Equation (26), the mass flux under surface diffusion of desired (J_{Bi}) and real (J_B) adsorbed gas can be expressed as follows

$$J_{Bi} = D_B^0 \frac{4\theta_i M_g}{\pi d_m^3 N_A p} \frac{\Delta p}{L(D)} \tag{31}$$

$$J_B = D_B \frac{4\theta M_g}{\pi d_m^3 N_A p} \frac{\Delta p}{L(D)} \tag{32}$$

In real state, the surface diffusion mass flow rate of adsorbed gas through single circular pipe section is

$$q_B(D) = D_B \frac{4\theta M_g}{\pi d_m^3 N_A p} \frac{\Delta p}{L(D)} \frac{\pi D^2}{4} \tag{33}$$

By integrating the mass flow rate of adsorbed gas diffusing on the surface of single circular pipe section in the interval of minimum pore diameter and maximum pore diameter $[D_{min}, D_{max}]$, the mass flow rate of adsorbed gas diffusing on the surface of shale fractal porous media is

$$\begin{aligned} Q_B(D) &= \int_{D_{min}}^{D_{max}} D_B \frac{4\theta M_g}{\pi d_m^3 N_A p} \frac{\Delta p}{D^{1-D_t} L_0^{D_t}} \frac{\pi D^2}{4} D_p D_{max}^{D_p} D^{-(D_p+1)} dD \\ &= D_B \frac{\theta M_g}{d_m^3 N_A p} \frac{\Delta p}{L_0^{D_t}} \frac{D_p D_{max}^{D_p}}{1-D_p+D_t} \left(D_{max}^{1-D_p+D_t} - D_{min}^{1-D_p+D_t} \right) \end{aligned} \tag{34}$$

2.3. Apparent Shale Gas Permeability

Gas migration mechanism in nanoscale pores of shale matrix includes free gas slip flow, Knudsen diffusion and surface diffusion of adsorbed gas, so the total mass flow rate of gas flowing through nanoscale pores of shale matrix is

$$Q = Q_F + Q_B \tag{35}$$

According to Darcy's law, the mass flow rate of porous media is expressed as

$$Q = \frac{K_{app} A}{\mu} \frac{p M_g}{ZRT} \frac{\Delta p}{L_0} \tag{36}$$

By substituting Equations (19), (34), and (35) into Equation (36), the apparent shale permeability under various flow mechanisms can be written as

$$\begin{aligned} K_{app} &= \frac{\mu RT}{M_g} \frac{\phi(2-D_p)}{L_0^{D_t-1}\left(D_{max}^{2-D_p}-D_{min}^{2-D_p}\right)} \int_{D_{min}}^{D_{max}} \left(\varepsilon_H (1+\alpha K_n)\left(1+\frac{4K_n}{1-bK_n}\right) \frac{M_g}{32\mu RT} D^{2-D_p+D_t} + \varepsilon_N \frac{C_g}{3}\left(\frac{8ZM_g}{\pi RT}\right)^{1/2} D^{1-D_p+D_t} \right) dD \\ &+ \mu D_B \frac{\theta ZRT}{d_m^3 N_A p^2} \frac{D_{max}^{1-D_p+D_t} - D_{min}^{1-D_p+D_t}}{L_0^{D_t-1}(1-D_p+D_t)} \frac{4\phi(2-D_p)}{\pi \left(D_{max}^{2-D_p}-D_{min}^{2-D_p}\right)} \end{aligned} \tag{37}$$

By integrating and differentiating Equation (37), the analytical formula of apparent permeability can be obtained as follows:

$$K_{app} = \frac{\mu RT}{M_g} \frac{\phi(2-D_p)}{L_0^{D_t-1}\left(D_{max}^{2-D_p}-D_{min}^{2-D_p}\right)} \left(\frac{\varepsilon_H}{3-D_p+D_t} \frac{M_g}{32\mu RT}(1+\alpha K_n)\left(1+\frac{4K_n}{1-bK_n}\right)\left(D_{max}^{3-D_p+D_t}-D_{min}^{3-D_p+D_t}\right) + \frac{\varepsilon_N}{2-D_p+D_t} \frac{C_g}{3}\left(\frac{8ZM_g}{\pi RT}\right)^{1/2}\left(D_{max}^{2-D_p+D_t}-D_{min}^{2-D_p+D_t}\right) \right) + \mu D_B \frac{\theta ZRT}{d_m^3 N_A p^2} \frac{D_{max}^{1-D_p+D_t}-D_{min}^{1-D_p+D_t}}{L_0^{D_t-1}(1-D_p+D_t)} \frac{4\phi(2-D_p)}{\pi\left(D_{max}^{2-D_p}-D_{min}^{2-D_p}\right)} \quad (38)$$

It can be seen from Equation (38) that the apparent permeability of shale considering multiple gas migration mechanism is composed of three parts, which can be respectively regarded as the apparent permeability of slip flow (K_H), the apparent permeability of Knudsen diffusion (K_N) and the apparent permeability of surface diffusion (K_B). The form can be expressed as

$$K_H = \frac{1}{32}(1+\alpha K_n)\left(1+\frac{4K_n}{1-bK_n}\right)\frac{\phi \varepsilon_H}{L_0^{D_t-1}}\frac{2-D_p}{3-D_p+D_t}\frac{D_{max}^{3-D_p+D_t}-D_{min}^{3-D_p+D_t}}{D_{max}^{2-D_p}-D_{min}^{2-D_p}} \quad (39)$$

$$K_N = \frac{\mu C_g}{3}\left(\frac{8ZM_g}{\pi RT}\right)^{1/2}\frac{\phi \varepsilon_N}{L_0^{D_t-1}}\frac{2-D_p}{2-D_p+D_t}\frac{D_{max}^{2-D_p+D_t}-D_{min}^{2-D_p+D_t}}{D_{max}^{2-D_p}-D_{min}^{2-D_p}} \quad (40)$$

$$K_B = 4\mu D_B \frac{\theta ZRT}{\pi d_m^3 N_A p^2}\frac{\phi}{L_0^{D_t-1}}\frac{2-D_p}{1-D_p+D_t}\frac{D_{max}^{1-D_p+D_t}-D_{min}^{1-D_p+D_t}}{D_{max}^{2-D_p}-D_{min}^{2-D_p}} \quad (41)$$

Through Equations (39)–(41), the relationship between slip apparent permeability, Knudsen diffusion apparent permeability, surface diffusion apparent permeability and geometric fractal parameters is established, and the relationship between matrix apparent permeability of shale gas reservoir and three flow mechanisms and geometric fractal theory is also established through Equation (38).

According to fractal geometry theory and Darcy flow equation, Darcy permeability (K_D) of shale can also be obtained as

$$K_D = \frac{\pi D_p D_{max}^{D_p}\left(D_{max}^{3-D_p+D_t}-D_{min}^{3-D_p+D_t}\right)}{128 L_0^{D_t-1} A(3-D_p+D_t)} = \frac{1}{32}\frac{\phi}{L_0^{D_t-1}}\frac{2-D_p}{D_{max}^{2-D_p}-D_{min}^{2-D_p}}\frac{D_{max}^{3+D_t-D_p}-D_{max}^{3+D_t-D_p}}{3+D_t-D_p} \quad (42)$$

Equations (38)–(42) contain many parameters. Among them, porosity, physical parameters of shale gas, temperature, and diameter of maximum and minimum capillary bundle are all obtained by experimental methods. Fractal dimension (D_p and D_t) can be referred to Yu's study [35]. The value of Knudsen's number is obtained by Equation (49).

3. Model Validation

Firstly, fractal characterization of shale pore characteristics is needed to obtain integral shape dimension of shale pore surface and fractal dimension of pore tortuosity. According to the 2D model proposed by Yu et al. [35], the relationship between the integral shape dimension of the pore surface and the distribution of pore size is

$$D_p = 2 - \frac{\ln \phi}{\ln(D_{min}/D_{max})} \quad (43)$$

It is noteworthy that Equation (46) needs to satisfy the precondition of $D_{min}/D_{max} < 0.01$, which is obviously satisfied in shale reservoirs. Because the fractal dimension of pore tortuosity of shale has not yet been measured, the fractal dimension of pore tortuosity is determined by the model proposed by Yu et al. [35].

$$D_t = 1 + \frac{\ln \tau_{av}}{\ln(L_0/D_{av})} \quad (44)$$

where τ_{av} is the average tortuosity, dimensionless; D_{av} is the average pore diameter, m.

The shale permeability test data of Letham et al. were used for model verification [46]. In its experimental test, helium and methane were used to measure shale permeability, and the basic input parameters of model verification were shown in Table 1.

Table 1. Basic input parameters of the shale apparent permeability model.

Parameters	Value	Unit
Minimum pore diameter	0.4	nm
Maximum pore diameter	100	nm
Temperature	303	K
Langmuir pressure	5	MPa
Gas constant	8.314	J/(mol·K)
Avogadro constant	6.022×10^{23}	Mol^{-1}

Since gas viscosity is a function of pressure, the widely used formula is adopted to calculate, namely [47–50]

$$\mu = 10^{-4} H_0 \exp\left[X\left(10^{-3} \frac{28.96\gamma_g p}{ZRT}\right)^{0.2(12-X)}\right] \quad (45)$$

$$X = 0.01\left(350 + \frac{54777.78}{T} + 28.96\gamma_g\right) \quad (46)$$

$$H_0 = \frac{(9.379 + 0.01607\gamma_g)T^{1.5}}{209.2 + 19.26\gamma_g} \quad (47)$$

where γ_g is gas molar weight, kg/mol.

Since shale has no adsorption effect on helium gas, the effect of surface diffusion on apparent permeability is not considered when calculating the apparent permeability of helium gas. At this time, the expression of apparent permeability is simplified as

$$K_{app} = \frac{\mu RT}{M_g} \frac{\phi(2-D_p)}{L_0^{D_t-1}\left(D_{max}^{2-D_p}-D_{min}^{2-D_p}\right)} \left(\frac{\varepsilon_H}{3-D_p+D_t} \frac{M_g}{32\mu RT}(1+\alpha K_n)\left(1+\frac{4K_n}{1-bK_n}\right)\left(D_{max}^{3-D_p+D_t}-D_{min}^{3-D_p+D_t}\right) + \frac{\varepsilon_N}{2-D_p+D_t} \frac{C_g}{3}\left(\frac{8ZM_g}{\pi RT}\right)^{1/2}\left(D_{max}^{2-D_p+D_t}-D_{min}^{2-D_p+D_t}\right) \right) \quad (48)$$

As can be seen from Figure 3, the calculated results of the model in this paper are very close to the experimental data. The apparent permeability of shale ranges from 400 nD to 1300 nD, and decreases with the increase of pressure. This is because with the decrease of pressure, the molecular mean free path increases and the K_n number increases, the influence of microscopic seepage on permeability increases, and the deviation from Darcy permeability increases. In addition, under the same reservoir conditions, even if the apparent permeability of methane is contributed by surface diffusion, the apparent permeability of helium is always greater than the apparent permeability of methane. It can be explained that the collision radius of helium molecule (0.26 nm) is smaller than that of methane molecule (0.38 nm), so the helium K_n number is larger than methane K_n number, and the apparent permeability is higher. The following formula can be used for calculating K_n number [51]:

$$K_n = \frac{\mu}{pD} \sqrt{\frac{\pi ZRT}{2M_g}} \quad (49)$$

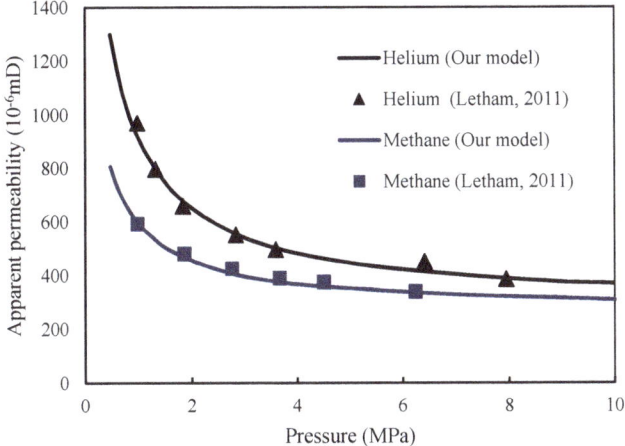

Figure 3. Comparison of apparent permeability calculated by the model in this paper with experimental data.

4. Discussion of Results

Based on fractal theory, this paper establishes a shale nano-pore permeability model considering the integral shape dimension of pore surface, fractal dimension of tortuosity, surface diffusion, Knudsen diffusion and slip flow. Based on the parameters in Table 2, we analyzed the apparent permeability and the contribution rate of different components to the apparent permeability. These results are of great significance to the study of microscopic seepage mechanism of nano-pore shale gas.

Table 2. Basic input parameters of the shale apparent permeability model.

Parameters	Value	Unit
Minimum pore diameter	1	nm
Maximum pore diameter	100	nm
Temperature	350	K
Langmuir pressure	5	MPa
Gas constant	8.314	J/(mol·K)
Avogadro constant	6.022×10^{23}	Mol^{-1}
Tortuosity	5	
Porosity	3	%
Surface diffusion coefficient	1×10^{-7}	m/s^2
Gas viscosity	0.015	mPa·s

The variation trend of various dimensionless apparent permeability with pressure is shown in Figure 4. As can be seen from the Figure 4, slip flow permeability, Knudsen diffusion permeability, surface diffusion permeability and apparent permeability all decrease with the increase of pressure. When the pressure increases, the average free path of gas molecules decreases, and the gas flow under the action of slip flow, Knudsen diffusion and surface diffusion (from Equation (40)) decreases. Therefore, the permeability decreases. When the pressure is less than 10 MPa, the slip flow permeability, surface diffusion permeability and apparent permeability rapidly decrease with the increase of pressure, and then tend to flatten. Therefore, under low pressure, pressure has a great influence on the apparent permeability of shale, while under high pressure, pressure has a relatively small influence on the apparent permeability. In essence, these performances can explain the phenomenon that in shale gas production, with the decrease of reservoir pressure, the apparent permeability gradually increases, and the rate of production decline gradually decreases. The effect of temperature on dimensionless apparent permeability is shown in Figure 5. The dimensionless apparent permeability increases nonlinearly with temperature.

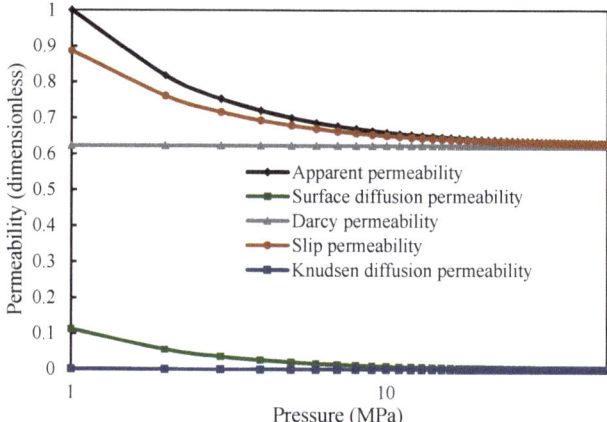

Figure 4. Relationship between dimensionless permeability and pressure.

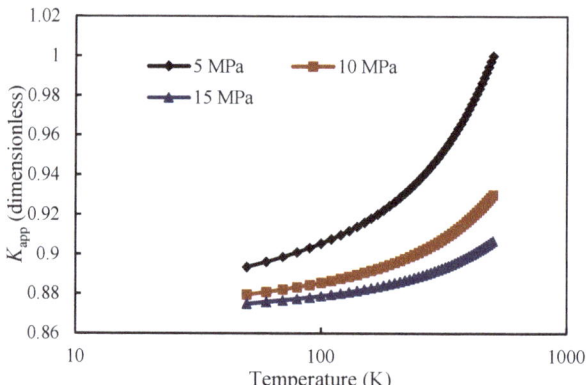

Figure 5. Relationship between temperature and dimensionless apparent permeability.

The contribution rates of different gas migration mechanisms to apparent permeability (K_{app}) are shown in Figure 6. It can be seen from the figure that the slip flow contributes the most to the apparent permeability, followed by the surface diffusion ($k_B = K_B/K_{app}$), and finally the Knudsen diffusion ($k_N = K_N/K_{app}$). The contribution rate of slip flow ($k_H = K_H/K_{app}$) to apparent permeability increases rapidly and then tends to flatten with the increase of pressure, while the contribution rate of surface diffusion to apparent permeability decreases rapidly and then tends to flatten with the increase of pressure. Under low pressure, the contribution of surface diffusion to apparent permeability is more significant, but even at pressure of 1 MPa, it is less than 10%. The contribution of slip flow to the apparent permeability is dominant, while the contribution of Knudsen diffusion to the apparent permeability is small, which can be ignored.

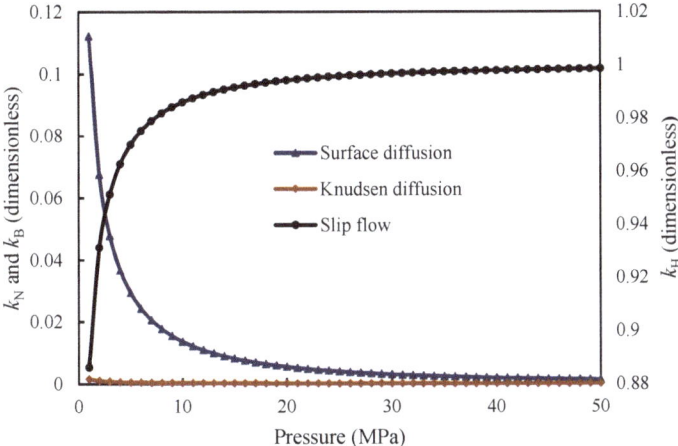

Figure 6. Relationship between contribution rate and pressure.

The overall effect of fractal dimension on dimensionless apparent permeability is shown in Figure 7. The apparent permeability increases with the increase of the integral shape dimension of the pore surface, and the change relation is strongly nonlinear. The relationship between tortuosity fractal dimension and apparent permeability is linearly negative. The influences of the integral shape dimension of pore surface and the fractal dimension of tortuosity on the contribution rate of permeability of different mechanisms are shown in Figures 8 and 9. Figure 8 shows that the surface diffusion contribution rate (k_B) and Knudsen diffusion contribution rate (k_N) are positively correlated with the integral shape dimension of the pore surface, while the slip flow is negatively correlated with the integral shape dimension of the pore surface. The integral shape dimension of the pore surface has the least effect on Knudsen diffusion. However, the influence of fractal dimension of tortuosity on contribution rate is opposite to that of integral shape dimension of pore surface on contribution rate, as shown in Figure 9. The fractal dimension of tortuosity is negatively correlated with surface diffusion contribution rate (k_B) and Knudsen diffusion contribution rate (k_N), and positively correlated with slip flow contribution rate (k_H). It comes down to the nature of fractal dimension. The larger the integral shape dimension of pore surface, the smaller the resistance of Knudsen diffusion and surface diffusion of gas molecules in shale porous media, and the greater the contribution of Knudsen diffusion and surface diffusion to apparent permeability. Although the contribution rate of slip flow decreases, the decrease is less than the total increase of Knudsen diffusion and surface diffusion, so the apparent permeability is higher. However, the larger the fractal dimension of tortuosity is, the more complex the pore structure is and the greater the diffusion resistance of gas molecules is. The contribution of Knudsen diffusion and surface diffusion to the apparent permeability decreases and the apparent permeability decreases.

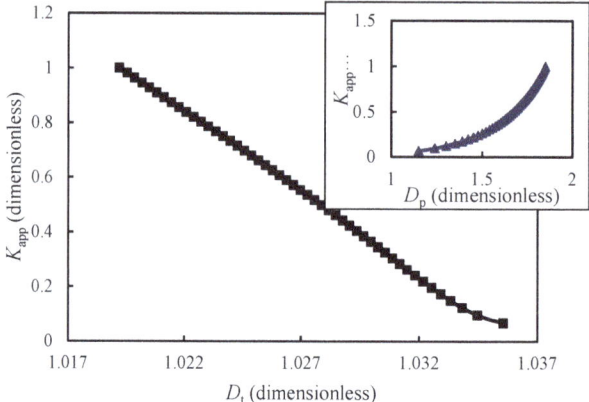

Figure 7. Effect of fractal dimension on dimensionless apparent permeability.

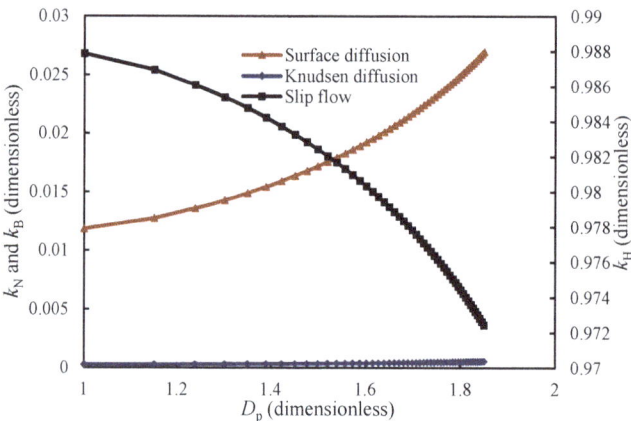

Figure 8. Relationship between contribution rate and fractal dimension of pore surface.

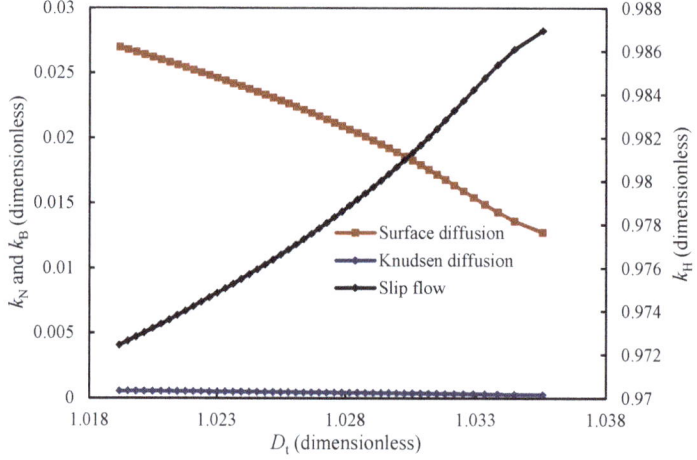

Figure 9. Relationship between contribution rate and fractal dimension of tortuosity.

5. Conclusions

In this paper, based on fractal geometry theory, the Hagen Poiseuille law, the Langmuir adsorption law, and Darcy law, a mathematical model of gas flow in nano-pores of shale was established. A new shale apparent permeability model is derived. The model comprehensively considers the influence of slip flow, Knudsen diffusion and surface diffusion, and is closely related to the size parameters of shale nano-pore structure, which can better reveal the behavior and law of multi-scale gas nonlinear flow in shale reservoir than traditional models. The nonlinear flow law of shale gas and the influencing factors of shale apparent permeability are analyzed. The following conclusions and understandings are obtained:

(1) The smaller the pressure, the higher the temperature, the larger the integral dimension of pore surface, the smaller the fractal dimension of tortuosity, the larger the apparent permeability, and the stronger the nonlinear flow of shale gas.

(2) The apparent permeability changes rapidly when the pressure is less than 10 MPa, and slowly when the pressure exceeds 10 MPa. Under any conditions, the contribution of slip flow to apparent permeability is the largest, followed by surface diffusion and Knudsen diffusion. The contribution rate of slip flow to apparent permeability increases rapidly and then tends to flatten with the increase of pressure, while the contribution rate of surface diffusion to apparent permeability decreases rapidly and then tends to flatten with the increase of pressure. Surface diffusion contributes significantly to apparent permeability at low pressure. The contribution of slip flow to the apparent permeability is dominant, while the contribution of Knudsen diffusion to the apparent permeability is small, which can be ignored.

(3) Fractal dimension of pore surface is positively correlated with apparent permeability, surface diffusion and Knudsen diffusion. The fractal dimension of tortuosity is negatively correlated with apparent permeability, surface diffusion and Knudsen diffusion. The fractal dimension of pore surface and the fractal dimension of tortuosity have the opposite effect on gas flow of shale nano-pores.

Author Contributions: Q.W. and J.Z. (Jinzhou Zhao) deduced the formula, verified the results and analyzed the influencing factors. Y.H. and L.R. provided validation data and examined the format and grammar of the paper. C.Z. and J.Z. (Jin Zhao) revised the manuscript for submission.

Funding: The authors would like to acknowledge the financial support of the National Natural Science Foundation of China (51404204); the National Science and Technology Major Project of the Ministry of Science and Technology of China (2016ZX05060).

Conflicts of Interest: The authors declare no conflict of interest.

References

1. Nguyen, V.H.; Rohan, E.; Naili, S. Multiscale simulation of acoustic waves in homogenized heterogeneous porous media with low and high permeability contrasts. *Int. J. Eng. Sci.* **2016**, *101*, 92–109. [CrossRef]
2. Pia, G.; Sanna, U. An intermingled fractal units model and method to predict permeability in porous rock. *Int. J. Eng. Sci.* **2014**, *75*, 31–39. [CrossRef]
3. Kuuskraa, V.A.; Wicks, D.E.; Thruber, J.L. Geologic and reservoir mechanisms controlling gas recovery from the Antrim shale. In Proceedings of the SPE Annual Technical Conference and Exhibition, Washington, DC, USA, 4–7 October 1992. Paper SPE 24883.
4. Roy, S.; Raju, R. Modeling gas flow through microchannels and nanopores. *J. Appl. Phys.* **2003**, *93*, 4870–4878. [CrossRef]
5. Javadpour, F. Nanopores and apparent permeability of gas flow in mudrocks (shales and siltstone). *J. Gas Can. Pet. Technol.* **2009**, *48*, 16–21. [CrossRef]
6. Freeman, C.M. A numerical study of microscale flow behavior in tight gas and shale gas. In Proceedings of the SPE Annual Technical Conference, Florence, Italy, 19–22 September 2010. Paper SPE 141125.
7. Curtis, M.E.; Ambrose, R.J.; Sondergeld, C.H. Structural characterization of gas shales on the micro- and nano-scales. In Proceedings of the Canadian Unconventional Resources and International Petroleum Conference, Calgary, AB, Canada, 19–21 October 2010; Paper SPE 137693.

8. Ebrahim, F.I.; Yucel, A. Lattice Boltzmann method for simulation of shale gas transport in kerogen. In Proceedings of the SPE Annual Technical Conference, Denver, CO, USA, 30 October–2 November 2011; Paper SPE 146821.
9. Shabro, V.; Torres-Verdin, C.; Javadpour, F. Numerical simulation of shale-gas production: From pore-scale modeling of slip-flow, Knudsen diffusion, and Langrnuir desorption to reservoir modeling of compressible fluid. In Proceedings of the North American Unconventional Gas Conference, Woodlands, TX, USA, 14–16 June 2011; Paper SPE 144355.
10. Michel, G.G.; Sigal, R.F.; Civan, F.; Devegowda, D. Parametric investigation of shale gas production considering nano-scale pore size distribution, formation factor, and non-darcy flow mechanisms. In Proceedings of the SPE Annual Technical Conference, Denver, CO, USA, 30 October–2 November 2011; Paper SPE 147438.
11. Freeman, C.M.; Moridis, G.J.; Michael, G.E. Measurement modeling, and diagnostics of flowing gas composition changes in shale gas wells. In Proceedings of the SPE Latin America and Caribbean Petroleum Engineering Conference, Mexico City, Mexico, 16–18 April 2012; Paper SPE 153391.
12. Javadpour, F.; Fisher, D.; Unsworth, M. Nanoscale gas flow in shale gas sediments. *J. Gas Can. Pet. Technol.* **2007**, *46*, 55–61. [CrossRef]
13. Swami, V.; Settari, A. A pore scale gas flow model for shale gas reservoir. In Proceedings of the Americas Unconventional Resources Conference, Pittsburgh, PA, USA, 5–7 June 2012; Paper SPE 155756.
14. Guo, C.H.; Bai, B.J.; Wei, M.Z.; He, X.; Wu, Y.S. Study on gas permeability in nano-pores of shale gas reservoirs. In Proceedings of the SPE Unconventional Resources Conference Canada, Calgary, AB, Canada, 5–7 November 2013; Paper SPE 167179.
15. Singh, H.; Javadpour, F.; Ettehadtavakkol, A. Nonempirical apparent permeability of shale. *SPE Reserv. Eval. Eng.* **2013**, *7*, 414–424. [CrossRef]
16. Mohammad, K.; Ali, T.B. An analytical model for shale gas permeability. *Int. J. Coal Geol.* **2015**, *146*, 188–197.
17. Zhang, L.; Li, D.; Lu, D.; Zhang, T. A new formulation of apparent permeability for gas transport in shale. *J. Nat. Gas Sci. Eng.* **2015**, *23*, 221–226. [CrossRef]
18. Wang, J.; Liu, H.Q.; Wang, L.; Zhang, H.; Luo, H.; Gao, Y. Apparent permeability for gas transport in nanopores of organic shale reservoirs including multiple effects. *Int. J. Coal Geol.* **2015**, *152*, 50–62. [CrossRef]
19. Wu, K.; Chen, Z.X.; Li, X.F.; Guo, C.; Wei, M. A model for multiple transport mechanisms through nanopores of shale gas reservoirs with real gas effect-adsorption-mechanic coupling. *Int. J. Heat Mass Transf.* **2016**, *93*, 408–426. [CrossRef]
20. Wang, J.; Kang, Q.; Wang, Y.; Pawar, R.; Rahman, S.S. Simulation of gas flow in micro-porous media with the regularized lattice Boltzmann method. *Fuel* **2017**, *205*, 232–246. [CrossRef]
21. Wang, S.; Shi, J.; Wang, K.; Sun, Z.; Miao, Y.; Hou, C. Apparent permeability model for gas transport in shale reservoirs with nanoscale porous media. *J. Nat. Gas Sci. Eng.* **2018**, *55*, 508–519. [CrossRef]
22. Zhang, T.; Li, X.; Shi, J.; Sun, Z.; Yin, Y.; Wu, K.; Li, J.; Feng, D. An apparent liquid permeability model of dual-wettability nanoporous media: A case study of shale. *Chem. Eng. Sci.* **2018**, *187*, 280–291. [CrossRef]
23. Li, C.; Lin, M.; Ji, L.; Jiang, W.; Cao, G. Rapid evaluation of the permeability of organic-rich shale using the 3D intermingled-fractal model. *SPE J.* **2018**, *23*, 2175–2187. [CrossRef]
24. Shen, Y.; Li, C.; Ge, H.; Guo, X.; Wang, S. Spontaneous imbibition process in micro-nano fractal capillaries considering slip flow. *Fractals* **2017**, *26*, 184002. [CrossRef]
25. Lee, J.W.; Nilson, R.H.; Templeton, J.A.; Griffiths, S.K.; Kung, A.; Wong, B.M. Comparison of Molecular Dynamics with Classical Density Functional and Poisson–Boltzmann Theories of the Electric Double Layer in Nanochannels. *J. Chem. Theory Comput.* **2012**, *8*, 2012–2022. [CrossRef] [PubMed]
26. Shen, Y.; Pang, Y.; Shen, Z.; Tian, Y.; Ge, H. Multiparameter analysis of gas transport phenomena in shale gas reservoirs: Apparent permeability characterization. *Sci. Rep.* **2018**, *8*, 2601. [CrossRef]
27. Miao, T.; Chen, A.; Zhang, L.; Yu, B. A novel fractal model for permeability of damaged tree-like branching networks. *Int. J. Heat Mass Transf.* **2018**, *127*, 278–285. [CrossRef]
28. Takbiri-Borujeni, A.; Fathi, E.; Kazemi, M.; Belyadi, F. An integrated multiscale model for gas storage and transport in shale reservoirs. *Fuel* **2019**, *237*, 1228–1243. [CrossRef]
29. Tan, X.H.; Liu, J.Y.; Li, X.P.; Zhang, L.H.; Cai, J.C. A simulation method for permeability of porous media based on multiple fractal model. *Int. J. Eng. Sci.* **2015**, *95*, 76–84. [CrossRef]

30. Hosa, A.; Curtis, A.; Wood, R. Calibrating Lattice Boltzmann flow simulations and estimating uncertainty in the permeability of complex porous media. *Adv. Water Resour.* **2016**, *94*, 60–74. [CrossRef]
31. Zhao, J.; Wang, Q.; Hu, Y.; Ren, L.; Zhao, C. Numerical investigation of shut-in time on stress evolution and tight oil production. *J. Pet. Sci. Eng.* **2019**, *179*, 716–733. [CrossRef]
32. Chalmers, G.R.; Bustin, R.M.; Power, I.M. Characterization of gas shale pore systems by porosimetry, pycnometry, surface area, and field emission scanning electron microscopy/transmission electron microscopy image analyses: Examples from the Barnett, Woodford, Haynesville, Marcellus, and Doig units. *AAPG Bull.* **2012**, *96*, 1099–1119.
33. Wang, F.; Zhao, J. A mathematical model for co-current spontaneous water imbibition into oil-saturated tight sandstone: Upscaling from pore-scale to core-scale with fractal approach. *J. Pet. Sci. Eng.* **2019**, *178*, 376–388. [CrossRef]
34. Su, P.; Xia, Z.; Qu, L.; Yu, W.; Wang, P.; Li, D.; Kong, X. Fractal characteristics of low-permeability gas sandstones based on a new model for mercury intrusion porosimetry. *J. Nat. Gas Sci. Eng.* **2018**, *60*, 246–255. [CrossRef]
35. Yu, B.; Lee, L.; Cao, H. Fractal characters of pore microstructures of textile fabrics. *Fractals* **2001**, *9*, 155–163. [CrossRef]
36. Wu, J.S. *Analysis of Resistance for Flow through Porous Media*; Zhejiang University: Hangzhou, China, 2006.
37. Chen, Z.L.; Wang, N.T.; Sun, L.; Tan, X.H.; Deng, S. Prediction method for permeability of porous media with tortuosity effect based on an intermingled fractal units model. *Int. J. Eng. Sci.* **2017**, *121*, 83–90. [CrossRef]
38. Zhao, J.; Zhao, J.Z.; Hu, Y.Q.; Zhao, C.N.; Wang, Q.; Zhao, X.; Zhang, Y. Coiled tubing friction reduction of plug milling in long horizontal well with vibratory tool. *J. Pet. Sci. Eng.* **2019**, *177*, 452–465.
39. Yu, B.; Li, J. Some characters of porous media. *Fractals* **2011**, *8*, 365–372.
40. Fan, W.; Sun, H.; Yao, J.; Fan, D.; Zhang, K. A new fractal transport model of shale gas reservoirs considering multiple gas transport mechanisms, multi-scale and heterogeneity. *Fractals* **2018**, *26*, 1850096. [CrossRef]
41. Sun, Z.; Shi, J.; Wu, K.; Zhang, T.; Feng, D.; Huang, L.; Shi, Y.; Ramachandran, H.; Li, X. An analytical model for gas transport through elliptical nanopores. *Chem. Eng. Sci.* **2019**, *199*, 199–209. [CrossRef]
42. Wu, K.; Li, X.F.; Wang, C.C.; Chen, Z.; Yu, W. Apparent permeability for gas flow in shale reservoirs coupling effects of gas diffusion and desorption. In Proceedings of the Unconventional Resources Technology Conference, Denver, CO, USA, 25–27 August 2014. Paper SPE 1921039.
43. Wu, K.; Li, X.; Wang, C.; Chen, Z.; Yu, W. A model for gas transport in microfractures of shale and tight gas reservoirs. *AIChE J.* **2015**, *61*, 2079–2088. [CrossRef]
44. Wu, K.; Chen, Z.X.; Li, X.F. Real gas transport through nanopores of varying cross-section type and shape in shale gas reservoirs. *Chem. Eng. J.* **2015**, *281*, 813–825. [CrossRef]
45. Chen, Y.D.; Yang, R.T. Concentration dependence of surface diffusion and zeolitic diffusion. *AIChE J.* **1991**, *37*, 1579–1582. [CrossRef]
46. Letham, E.A. *Matrix Permeability Measurements of Gas Shales: Gas Slippage and Adsorption as Sources of Systematic Error*; University of British Columbia: Vancouver, BC, Canada, 2011.
47. Song, W.; Yao, J.; Li, Y.; Sun, H.; Zhang, L.; Yang, Y.; Zhao, J.; Sui, H. Apparent gas permeability in an organic-rich shale reservoir. *Fuel* **2016**, *181*, 973–984. [CrossRef]
48. Wang, J.; Luo, H.; Liu, H.; Cao, F.; Li, Z.; Sepehrnoori, K. An integrative model to simulate gas transport and production coupled with gas adsorption, non-Darcy flow, surface diffusion, and stress dependence in organic-shale reservoirs. *SPE J.* **2017**, *22*, 244–264. [CrossRef]
49. Wang, S.; Tao, W.U.; Cao, X.; Zheng, Q.; Ai, M. A fractal model for gas apparent permeability in microfracture of tight/shale reservoirs. *Fractals* **2017**, *25*, 1750036. [CrossRef]
50. Sun, Z.; Li, X.; Shi, J.; Zhang, T.; Sun, F. Apparent permeability model for real gas transport through shale gas reservoirs considering water distribution characteristic. *Int. J. Heat Mass Tran.* **2017**, *115*, 1008–1019. [CrossRef]
51. Civan, F. Effective correlation of apparent gas permeability in low-permeability porous media. *Transp. Porous Media* **2010**, *82*, 375–384. [CrossRef]

© 2019 by the authors. Licensee MDPI, Basel, Switzerland. This article is an open access article distributed under the terms and conditions of the Creative Commons Attribution (CC BY) license (http://creativecommons.org/licenses/by/4.0/).

Article

Analysis of Coupled Wellbore Temperature and Pressure Calculation Model and Influence Factors under Multi-Pressure System in Deep-Water Drilling

Ruiyao Zhang [1,*], Jun Li [1,*], Gonghui Liu [1,2], Hongwei Yang [1] and Hailong Jiang [1]

1. College of petroleum engineering, China University of Petroleum-Beijing, Beijing 102249, China; lgh1029@163.com (G.L.); yhw520991@163.com (H.Y.); 18701077624@163.com (H.J.)
2. College of petroleum engineering, Beijing University of Technology, Beijing 100192, China
* Correspondence: 2018312029@student.cup.edu.cn (R.Z.); lijun446@vip.163.com (J.L.)

Received: 22 July 2019; Accepted: 12 September 2019; Published: 14 September 2019

Abstract: The purpose of this paper is to discuss the variation of wellbore temperature and bottom-hole pressure with key factors in the case of coupled temperature and pressure under multi-pressure system during deep-water drilling circulation. According to the law of energy conservation and momentum equation, the coupled temperature and pressure calculation model under multi-pressure system is developed by using the comprehensive convective heat transfer coefficient. The model is discretized and solved by finite difference method and Gauss Seidel iteration respectively. Then the calculation results of this paper are compared and verified with previous research models and field measured data. The results show that when the multi-pressure system is located in the middle formation, the temperature of the annulus corresponding to location of the system is the most affected, and the temperature of the other areas in annulus is hardly affected. However, when the multi-pressure system is located at the bottom hole, the annulus temperature is greatly affected from bottom hole to mudline. In addition, the thermo-physical parameters of the drilling fluid can be changed by overflow and leakage. When only overflow occurs, the annulus temperature increases the most, but the viscosity decreases the most. When only leakage occurs, the annulus temperature decreases the most and the viscosity increases the most. However, when the overflow rate is greater than the leakage rate, the mud density and bottom-hole pressure increase the most, and both increase the least when only leakage occurs. Meanwhile, bottom-hole pressure increases with the increase of pump rate but decreases with the increase of inlet temperature. The research results can provide theoretical guidance for safe drilling in complex formations such as multi-pressure systems.

Keywords: wellbore temperature; bottom-hole pressure; multi-pressure system; comprehensive heat transfer model; leakage and overflow

1. Introduction

Deep water drilling faces problems such as narrow formation pressure window and difficult pressure control. The pressure in wellbore is affected not only by properties of drilling fluid but also by formation and annulus temperature in deep water drilling. At the same time, the pressure variation makes the drilling fluid density change, further affecting the annulus and formation temperature distribution. Therefore, to ensure efficient and safe drilling, the study of coupled wellbore temperature and pressure distribution is also very important in deep water drilling. In addition, the tectonic steep, formation fracture, fault, fracture, and hole development are often encountered in deep water drilling. Under the condition of multi-pressure system, it is easy for overflow and leakage to occur simultaneously in the formation. Because the overflow and leakage exist at the same time, part of the fluid from the annulus goes into the formation, and the fluid from the formation enters the annulus.

Then, the corresponding thermo-physical property parameters of drilling fluid in annulus also change, which affects the distribution of temperature and pressure in the wellbore. Therefore, in order to ensure safe and efficient drilling operations, the research on coupled wellbore temperature and distribution under multi-pressure system has highly practical significance.

In recent years, the research methods related to wellbore and formation temperature have been mainly analytical and numerical methods [1]. Ekaterina Wiktorski [2] derived temperature-dependent thermo-physical correlations, which were applied in a wellbore heat transfer model for oil production scenario by considering a complex well architecture. Javad Abdollahi and Stevan Dubljevic [3] simplified the heat transfer model of wellbore fluid into a set of hyperbolic partial differential equations. Then the observability of temperature distribution was discussed by using the methods of characteristic functions and Riemann invariants. R. Hasan and C. S. Kabir [4] discussed a unified approach for modeling heat transfer in various situations that result in physically sound solutions. This modeling approach depends on many common elements, such as temperature profiles surrounding the wellbore and any series of resistances for the various elements in the wellbore. Based on the energy conservation principle and the Modified Raymond, Yang M et al. [5] developed simplified and full-scale models, and the results indicated that wellbore and formation temperatures were significantly influenced at the connection points between the drill collar and drill pipe, as well as the casing shoe. Jia HJ, Meng YF, Li G, Su G, et al. [6] established the thermal conductivity model of liquid-filled annulus, and the influence of casing annulus fluid on temperature distribution was simulated numerically. Ramey [7] established temperature calculation model under the condition that the formation and wellbore heat transfer be considered as unsteady and steady respectively. The significant difference between Holmes Swift's [8] temperature prediction model and other models is that the accuracy of the model can only be reflected under the condition of a long-circulation time. Hasan–Kabir [9,10] established a model where variation of properties of drilling fluid with temperature and pressure was not considered, nor was the heat generated by drilling fluid flow taken into account, so there will be obvious errors in the calculation results. Raymond's [11] model does not take into account the heat source generated by friction during drilling fluid flow, so the final temperature calculation result is smaller than the actual value. Marshall Bentsen [12] established the wellbore and formation temperature calculation model by using the comprehensive convective heat transfer coefficient without considering the wellbore structure. Li Mengbo et al. [13] established the coupled temperature and pressure calculation model of wellbore in multiphase flow during normal circulation, and the effect of lost circulation or kick on annulus temperature and pressure was not considered. García et al. [14] established the wellbore temperature calculation model in geothermal wells and that during shut-in under the condition of lost circulation. Espinosa et al. [15] obtained the wellbore temperature distribution prediction model according to the law of energy conservation under the conditions of circulation and stopping circulation. Yang Mou et al. [16] considered the axial and radial heat conduction of wellbore and formation simultaneously; the results showed that the axial heat conduction exert almost no effect on distribution of wellbore temperature compared with that of the radial temperature. Zhang Zheng et al. established the wellbore and formation temperature calculation model when leakage was located in different position of stratum, but without considering the comprehensive influence of leakage and overflow and the coupled temperature and pressure [17].

Although there are many models for calculating wellbore and formation temperature under different conditions, there are few models that the effects of multi-pressure systems and coupled temperature and pressure are taken into account on wellbore temperature and pressure during circulation in deep water drilling.

According to the law of energy conservation, when the multi-pressure system is located at different positions of the formation, such as the middle open-hole formation or at the bottom hole, the mathematically coupled temperature and pressure model of the wellbore and the formation is established. Then, the first-order windward scheme is used for the first-order space, three-point central difference is used for the second-order space, and two-point backward difference is used for the first-order time; finally, Gauss Seidel iteration is used for numerical calculation for the discrete model. The distribution law of coupled wellbore temperature and pressure under multi-pressure system is obtained, and the key factors affecting wellbore temperature and bottom-hole pressure are analyzed, which provides theoretical support for adjusting drilling parameters and ensuring safe and efficient drilling in deep water.

2. Mathematical Model of Each Heat Transfer Region

2.1. Physical Model of All Heat Transfer Regions

During the normal circulation in deep water drilling, drilling fluid first enters the pipe, then passes through the drill bit into the annulus and finally returns to the surface ground. In this process, the drilling fluid generates convection heat transfer with the interior wall and exterior wall of the drill pipe as well as the borehole wall. Meanwhile, the interior wall and exterior wall of the drill pipe, casing, cement sheath, seawater and inside of the formation generate radial heat conduction [18]. Hence all heat transfer regions can be divided into three major regions such as inside drill string, in annulus (including formation segment annulus and seawater segment annulus), and inside of the formation [19]. The physical model of all heat transfer regions is shown in Figure 1.

All areas were meshed and divided into layers at intervals of 50 m along the axis of the wellbore, denoted by i, and each zone in the radial direction denoted by j. According to the layer of overflow and leakage, the multi-pressure system is divided into several sub-layers in the axial direction, which are represented by l and k respectively.

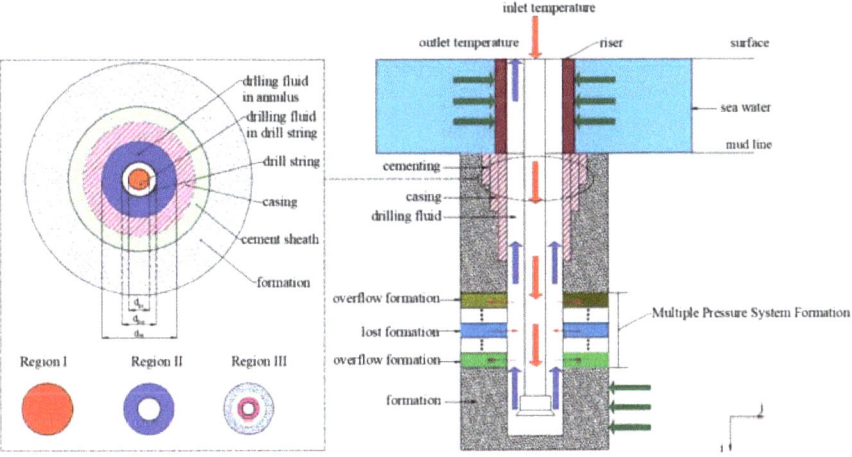

Figure 1. Heat transfer model of wellbore and formation under multi-pressure system.

2.2. Mathematical Model

2.2.1. Model Hypothesis

(1) All fluids in the model are incompressible, other thermo-physical parameters do not change with temperature and pressure variation except viscosity and density.
(2) Radial gradient of the temperature, pressure, and flow rate inside the wellbore are neglected and the radial heat conduction inside seawater and formation is considered.
(3) The layer of overflow and the leakage alternate with each other; the flow rate of the same type of sublayer is the same, and the fluid entering the annulus is evenly mixed with the drilling fluid.

2.2.2. Heat Transfer Mathematic Models for Each Region

According to the model hypothesis and the first law of thermodynamics, the mathematical model of heat transfer in each region is established because the seawater flow is affected by multiple factors, which cannot be listed simply by hypothesis conditions. In addition, the seawater flow has little influence on wellbore temperature [20], so the flow of seawater is ignored in this model.

(1) Drill string, the casing, riser, and the cement sheath.

During the drilling circulation, drill string, casing, cement ring, and riser have the same type of heat transfer [21]. The heat transfer physical model in control element is shown in Figure 2, therefore, according to the convection heat transfer between the drilling fluid and the interior and exterior wall of the drill string and the heat conduction between the two walls, the heat transfer Equation (1) of drill string can be written as follows: According to the law of energy conservation, the first term of the equation is the convection heat transfer between the annulus drilling fluid and the outer wall of the drill string, the second term is the convection heat transfer between the drilling fluid and the inner wall of the drill string, and the third term is the heat conduction between the inner wall of the drill pipe and the outer wall.

$$h_{po}\pi d_{po}\Delta y \Delta t[T_a(y,t) - T_{po}(y,t)]\Delta t = h_{pi}\pi d_{pi}\Delta y \Delta t(T_p(y,t) - T_{pi}((y,t)) = 2\pi \lambda_p \Delta y \Delta t \frac{[T_{po}(y,t) - T_{pi}(y,t)]}{\ln(\frac{d_{po}}{d_{pi}})} \quad (1)$$

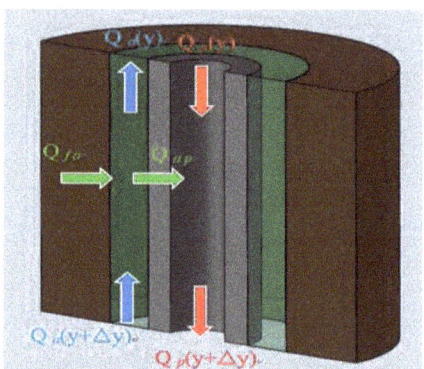

Figure 2. Heat transfer physical model in control element.

According to the above equation established by the Energy conservation relation, the total heat transferred between the region inside the drill string and in annulus can be obtained, and the relationship is as follows:

$$Q_{ap} = \frac{\pi d_{pi}[T_a(y,t) - T_p(y,t)]}{\frac{1}{h_{pi}} + \frac{d_{pi}}{h_{po}d_{po}} + \frac{d_{pi}}{2\lambda_p}\ln(\frac{d_{po}}{d_{pi}})}\Delta y \Delta t, \frac{1}{U_{ap}} = \frac{1}{h_{pi}} + \frac{d_{pi}}{h_{po}d_{po}} + \frac{d_{pi}}{2\lambda_p}\ln(\frac{d_{po}}{d_{pi}}) \quad (2)$$

According to the heat transfer mechanism of drill string, the heat transfer relationship among casing (or wellbore wall), cement sheath, and formation can be obtained similarly:

$$\frac{2\pi \lambda_{cemi}\Delta y \Delta t[T_f(y,t) - T_{cemi}(y,t)]}{\ln(\frac{d_{cemo}}{d_{cemi}})} = \frac{2\pi \lambda_w \Delta y \Delta t[T_{cemi}(y,t) - T_w(y,t)]}{\ln(\frac{d_{cemi}}{d_w})} = h_w \pi d_w \Delta y \Delta t (T_w - T_a) \quad (3)$$

According to the Equation (3), the comprehensive heat transfer from annulus to outside wall of cement sheath (or formation) can be obtained:

$$Q_{af} = \frac{\pi d_w [T_f(y,t) - T_a(y,t)]}{\frac{1}{h_w} + \frac{d_w}{2\lambda_w} \ln(\frac{d_{cemi}}{d_w}) + \frac{d_w}{2\lambda_{cem}} \ln(\frac{d_{cemo}}{d_{cemi}})} \Delta y \Delta t, \frac{1}{U_{af}} = \frac{1}{h_w} + \frac{d_w}{2\lambda_w} \ln(\frac{d_{cemi}}{d_w}) + \frac{d_w}{2\lambda_{cem}} \ln(\frac{d_{cemo}}{d_{cemi}}) \quad (4)$$

(2) Inside the drill string.

Heat transfer in these regions include heat convection between drilling fluid and interior wall of drill string, heat gone into the control element during Δt time and heat generated by drilling fluid flow friction. According to the law of the energy conservation, the following Equation (5) can be obtained. The first item of the equation is the change of internal energy in element during Δt time, the second item is the convection heat transfer Δt, the third item is the heat entering the element during Δt, and the fourth item is the heat generated by friction.

$$\frac{\pi}{4} d_{pi}^2 \Delta y \Delta t \frac{\partial [\rho_p(y,t) c_p(y,t) T_p(y,t)]}{\partial t} = h_{pi} \pi d_{pi} (T_p - T_{pi}) \Delta y \Delta t + Q_m \frac{\partial [\rho_p(y,t) c_p(y,t) T_p(y,t)]}{\partial y} \Delta y \Delta t + Q_{cp} \Delta y \Delta t \quad (5)$$

According to the comprehensive convective heat transfer coefficient of drilling fluid from drill string to annulus, the Equation (5) can be modified as:

$$\frac{\pi}{4} d_{pi}^2 \frac{\partial [\rho_p(y,t) c_p(y,t) T_p(y,t)]}{\partial t} = U_{ap} \pi d_{pi} (T_a(y,t) - T_p(y,t)) + Q_m \frac{\partial [\rho_p(y,t) c_p(y,t) T_p(y,t)]}{\partial y} + Q_{cp} \quad (6)$$

(3) In annulus.

Heat transfer in annulus mainly includes convection heat transfer between drilling fluid and exterior wall of drill string as well as wellbore wall, heat gone into the control element during Δt time, and heat generated by drilling fluid flow friction.

a. When the multi-pressure system is located in the middle open hole formation:

The lower annulus

When the multi-pressure system is located in the middle open-hole formation, the annulus is divided into two parts by taking the multi-pressure system section as a marker. The first part is the lower annulus below the multi-pressure system. The second part is the annulus where the multi-pressure system and its upper segment formation as well as seawater segment. The physical model is shown in Figure 3.

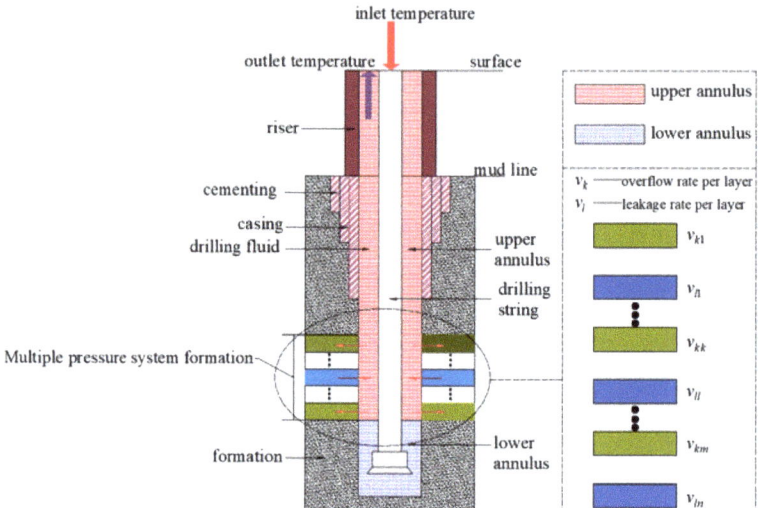

Figure 3. Physical model of multi-pressure system in the middle open hole formation.

The thermo-physical parameter of the drilling fluid in the lower annulus, which is located upstream of the multi-pressure system, is not affected by overflow and leakage. The heat transfer in this region is the same as that of the annulus during normal circulation. According to the law of energy conservation, the following Equation (7) can be obtained in the control element: The first item of the equation is the change of internal energy in element during Δt time, the second and third items are the convection heat transfer Δt, the fourth item is the heat entering the element during Δt, and the fifth item is the heat generated by friction.

$$\begin{aligned}\frac{\pi}{4}(d_w - d_{po}^2)\Delta y \Delta t \frac{\partial [\rho_a(y,t)c_a(y,t)T_a(y,t)]}{\partial t} &= h_{po}\pi d_{po}[T_a(y,t) - T_{po}(y,t)]\Delta y \Delta t \\ &+ h_w \pi d_w [T_a(y,t) - T_w(y,t)]\Delta y \Delta t \\ &+ Q_m \frac{\partial [\rho_a(y,t)c_a(y,t)T_a(y,t)]}{\partial y}\Delta y \Delta t + Q_{ca}\Delta y \Delta t\end{aligned} \quad (7)$$

According to the comprehensive convective heat transfer coefficient from annulus to formation and its counterpart from the drill string to the annulus, Equation (7) can be changed as:

$$\begin{aligned}\frac{\pi}{4}(d_w - d_{po}^2)\frac{\partial [\rho_a(y,t)c_a(y,t)T_a(y,t)]}{\partial t} &= U_{ap}\pi d_{po}[T_a(y,t) - T_p(y,t)] + U_{af}\pi d_w[T_a(y,t) - T_w(y,t)] \\ &+ Q_m \frac{\partial [\rho_a(y,t)c_a(y,t)T_a(y,t)]}{\partial y} + Q_{ca}\end{aligned} \quad (8)$$

The upper annulus

When the drilling fluid returns to the annulus segment where the multi-pressure system is located, some drilling fluid enters the multi-pressure system formation, and the fluid from the multi-pressure system formation also goes into the annulus because of simultaneous existence of leakage and overflow in this multi-pressure system formation. When drilling fluid from the lower annulus is mixed with fluid from the multi-pressure system formation, then the thermo-physical property of drilling fluid entering the upper annulus is changed. It can be known from the assumed conditions that the formation of leakage and overflow formation are staggered, and the fluid in annulus is evenly mixed. Now, each layer of formation of overflow and leakage is divided into several sub-layers, so the heat transfer equation of drilling fluid in the upper annulus can be described as follows. The physical meaning of terms in the equation is the same as that in Equation (7), except that the value of thermo-physical parameters changes due to the influence of overflow and leakage.

$$\frac{\pi}{4}(d_w - d_{po}^2)\frac{\partial[\rho_a'(y,t)c_a'(y,t)T_a(y,t)]}{\partial t}\Delta y\Delta t = h_{po}'\pi d_{po}[T_a(y,t) - T_{po}(y,t)]\Delta y\Delta t$$
$$+ h_w'\pi d_w[T_a(y,t) - T_w(y,t)]\Delta y\Delta t + Q_{ca}\Delta y\Delta t$$
$$+ (v_a + \sum_{k=1}^{n} v_{kk} - \sum_{l=2}^{n-1} v_{ll})\frac{\partial[\rho_a'(y,t)c_a(y,t)'T_a(y,t)]}{\partial y}\Delta y\Delta t \quad (9)$$

Similarly, according to the comprehensive heat transfer coefficient, the Equation (9) can be changed as Equation (10), but the comprehensive convective heat transfer coefficient of the mixed fluid is different from the former in Equation (8):

$$\frac{\pi}{4}(d_w - d_{po}^2)\frac{\partial[\rho_a'(y,t)c_a'(y,t)T_a(y,t)]}{\partial t} = U_{ap}'\pi d_{po}[T_a(y,t) - T_{po}(y,t)] + U_{fa}'\pi d_w[T_f(y,t) - T_a(y,t)]$$
$$+ (v_a + \sum_{k=1}^{n} v_{kk} - \sum_{l=2}^{n-1} v_{ll})\frac{\partial[\rho_a'(y,t)c_a(y,t)'T_a(y,t)]}{\partial y} + Q_{ca}' \quad (10)$$

where, the $k = 1, 3, 5, \ldots, n, l = 2, 4, 6, \ldots, n-1$.

b. The multi-pressure system is located at the bottom hole

When the multi-pressure system formation is located at bottom hole, as shown in Figure 4, the drilling fluid in the whole annulus is mixed with the fluid from the multi-pressure system formation because of the co-existence of overflow and leakage. According to the law of energy conservation, the heat transfer equation of the annulus can be obtained as follows:

$$\frac{\pi}{4}(d_w - d_{po}^2)\frac{\partial[\rho_a^*(y,t)c_a^*(y,t)T_a(y,t)]}{\partial t} = U_{ap}^*\pi d_{po}[T_a(y,t) - T_{po}(y,t)] + U_{fa}^*\pi d_w[T_f(y,t) - T_a(y,t)]$$
$$+ (v_a + \sum_{k=1}^{n} v_{kk}^* - \sum_{l=2}^{n-1} v_{ll}^*)\frac{\partial[\rho_a^*(y,t)c_a(y,t)^*T_a(y,t)]}{\partial y} + Q_{ca}^* \quad (11)$$

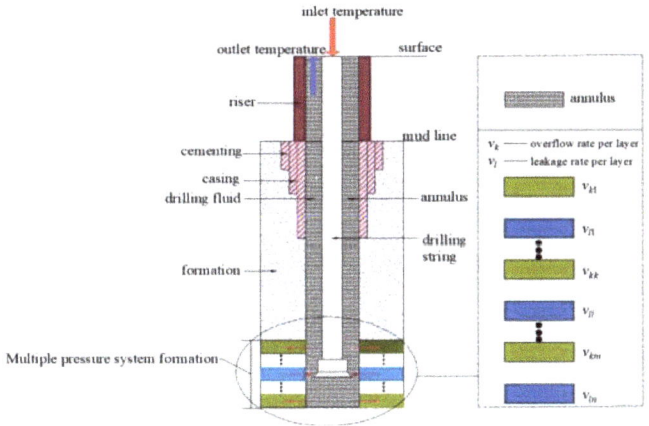

Figure 4. Physic model of multi-pressure system segment located at bottom hole.

Due to the combined effects of overflow and leakage, the composition of drilling fluid in the upper annulus was finally changed [22,23]. According to the literature, the formula for calculating the density of mixed liquid is as follows:

$$\rho_m^{(-2/3)} = a + bw_1 + cw_1^2 \quad (12)$$

The thermal conductivity of the mixed liquid is a function of density, so according to the mathematical function expansion theorem:

$$\lambda_2' = k_0 + k\rho_m, \text{ making } \rho_m \to 0, \lambda_2' \to 0, \text{ so } \lambda_2' = k\rho_m \tag{13}$$

where, k_0, k is constant, in combination with Equations (12) and (13),

$$\lambda_2'^{(-2/3)} = B_1 + B_2 w_1 + B_3 w_1^2, (w_1 + w_2 = 1) \tag{14}$$

According to the binary liquid mixture proportion relationship, binary mixed fluid thermal conductivity is obtained and extended to multiple mixed fluid as follows:

$$\lambda_2'^{(-2/3)} = (B_1 + B_2)w_1 + (B_1 + B_3)w_2 - B_3 w_1 w_2 \tag{15}$$

where B_1, B_2, B_3 is the constant, $A_i = \lambda_1^{(-2/3)}$, $B_{ij} = 0.0055 - 0.011|\ln(\lambda_i/\lambda_j)|$.

(4) Heat transfer model of formation

Whether the presence of fluid is in the formation or not leads to great difference in the heat transfer of the formation. This paper only considers the condition of the presence of fluid, and the heat transfer equation is as follows:

$$(\rho c)_{eff} \frac{\partial T_i(x,y,t)}{\partial t} = \lambda_{eff} \frac{\partial^2 T_f(x,y,t)}{\partial^2 x} + \frac{\lambda_{eff}}{x} \frac{\partial T_f(x,y,t)}{2x} \tag{16}$$

According to the equilibrium volume method and energy balance equation, the thermal and physical parameters and the fluid in the formation can be calculated [24]. Where, $\lambda' = \lambda_l^\phi + \lambda_f^{(1-\phi)}$, $(\rho c)' = \phi(\rho c)_l + (1-\phi)(\rho c)_f$.

The fluid flow in the formation can be regarded as unidirectional incompressible plane radial steady seepage [25]. According to its corresponding differential equation and Darcy's law, the seepage velocity can be obtained as follows [26]:

$$v_r = -\frac{K}{\mu}\frac{\partial P}{\partial r}, \frac{K_1}{\mu}(\frac{\partial^2 P}{\partial r^2} + \frac{1}{r}\frac{\partial P}{\partial r}) + \frac{\overline{q}}{\rho} = 0 \tag{17}$$

(5) Heat transfer model of seawater

$$(\rho_s c_s) \frac{\partial T_i(x,y,t)}{\partial t} = \lambda_s \frac{\partial^2 T_f(x,y,t)}{\partial^2 x} + \frac{\lambda_s}{x} \frac{\partial T_f(x,y,t)}{2x} \tag{18}$$

2.2.3. Momentum Equation and the Relationship between Pressure and Density

$$\frac{\partial(\rho_i v_i A_i)}{\partial t} + \frac{\partial(\rho_i v_i^2 A_i)}{\partial z} = \frac{\partial(\rho_i A_i)}{\partial z} + A_i \frac{\partial P_{fi}}{\partial z} + \rho_i g A_i \tag{19}$$

$$\frac{\partial P}{\partial y} = \rho_0 g e^{\chi_P(P-P_0) + \chi_{PP}(P-P_0)^2 + \chi_T(T-T_0) + \chi_{TT}(T-T_0)^2 + \chi_{PT}(P-P_0)(T-T_0)} \tag{20}$$

2.3. Initial and Boundary Conditions

2.3.1. Initial Conditions

Since the temperature of seawater is affected by multiple factors such as season, current, depth, and so on, Locarnini described the distribution of seawater temperature in the vertical direction as follows, in consideration of coupling factors [27]:

$$T_3 = (y, t = 0) = T_{surf}(200 - y) + 13.68y/200, \ y < 200, \quad (21)$$

$$T_3 = (y, t = 0) = m_2 + (m_1 - m_2)/(1 + e^{(y-m_0)/m_3}), \ 200 < y < y_{ml} \quad (22)$$

The temperature of the formation is mainly related to the temperature gradient and depth, so the temperature distribution model along the vertical direction can be expressed as [28]:

$$T_f(y, t = 0) = T_3(y = y_{ml}, t = 0) + Gh, \ y_{ml} < y < h \quad (23)$$

2.3.2. Boundary Conditions

Wellhead temperature can be regarded as the initial temperature inside drill string, so its boundary condition can be expressed as:

$$T_0(y = 0, t = 0) = T_{in} \quad (24)$$

According to the continuity of drilling fluid flow at bottom hole, the bottom hole temperatures of the drilling fluid inside the drill string, the wall of drill string, and drilling fluid in the annulus are equal [29]. The boundary condition can be expressed as:

$$T_0(y = H, t) = T_1(y = H, t) = T_2(y = H, t) \quad (25)$$

The boundary condition of ambient temperature distribution far from the wellbore is as follows:

$$T(x \to \infty, y, t) = T_{surf} + Gh \quad (26)$$

For the meshing of the physical model, the discrete process of the mathematical model and the overall calculation flow chart, all of them are shown in Appendix A.

3. Model Validation

3.1. Comparison and Verification with Theoretical Calculation Model

According to drilling data from Tables 1–3, which comes from a well in the South China Sea [30], the calculated results of this paper are obtained and then compared with other scholar's models; the obtained comparison result is shown in Figure 5a. The model of Zhang Zheng is the closest to the model in this paper, while Yang Mou's model has a large gap. Marshall's model also used the comprehensive convective heat transfer coefficient and considered the heat generated by drilling fluid flow friction during circulation. However, the influence of wellbore structure, pressure, and other factors on the temperature distribution was not considered [12]. Because the well structure has a great influence on wellbore heat transfer [12], the temperature in deep formation is generally higher than that in the annulus, while the temperature in shallow formation (including seawater) is lower than that in the annulus.

Table 1. The Basic parameters of the well.

Parameters	Value	Parameters	Value
Well depth, m	5094	Drilling fluid flow rate, m³/s	0.02
Water depth, m	1521	ROP, m/h	3.05
Open hole diameter, mm	215.9	Surface temperature, °C	20
Inlet temperature, °C	15	Geothermal gradient, °C	0.024
Drilling fluid viscosity, mPa·s	10	Fluid density in formation, kg/m³	1100

Table 2. Thermo-physical parameters of different materials or fluid.

Medium	Density, kg/m^3	Specific Heat, J/(kg·K)	Coefficient of Heat Conduction W/(m·K)
Drilling fluid	1180	3935	1.75
Drill string	7850	400	43.7
Drill collars	8910	400	43.7
Fluid in formation	1150	4200	0.8
Seawater	1050	4128	0.6
Casing	8300	400	43.7
Cement	2140	900	0.85
Rock	2640	853	2.3

Table 3. Inner diameter and outer diameter of drill string and casing.

Parameters	Inside Diameters (mm)	Outside Diameters (mm)
Drill pipe	108	127
Drill collar	80	171
First casing	486	508
Second casing	317	339
Third casing	221	245

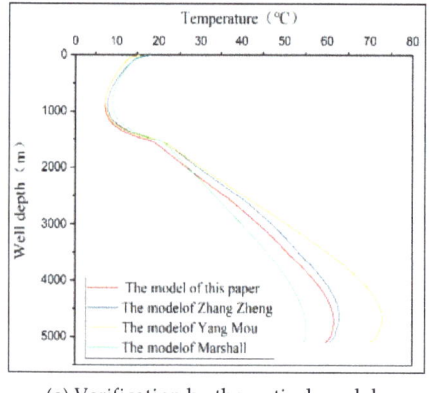

(a) Verification by theoretical models

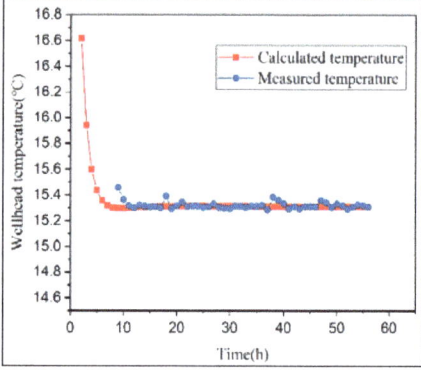

(b) Verification by measurement data

Figure 5. Model verification by using previous models and measurement data.

Therefore, compared with Marshall model, the well structure and pressure was taken into account, and the comprehensive convective heat transfer coefficient was used to solve the temperature distribution in this paper. On the one hand, the deep formation transfers less heat to the lower annulus during same circulation time, so the formation temperature cools more slowly and the temperature of the lower annulus is higher. On the other hand, the drilling fluid from the lower annulus carries less heat into the upper annulus, so the temperature in upper annulus is cooler. Yang Mou's model is mainly aimed at traditional land drilling [31]. Compared with Yang's model, the model in this paper is about deep-water drilling and has different assumptions, so there are some errors in the solution. Zhang Zheng's temperature calculation model is the closest to the solution results of the model in this paper. In the entire annulus temperature distribution, the maximum calculated temperature difference between two models is not more than 3.31 °C, which has little impact on the whole wellbore temperature distribution, thus basically proving the reliability of the temperature calculation model. In addition, compared with Zhang Zheng's model, the temperature prediction model in this paper not only considers overflow but also considers overflow and leakage occurring simultaneously under multi-pressure system. This temperature prediction model in this paper is more consistent with the actual temperature distribution in deep water drilling and provides a better reference for managed pressure drilling in deep water.

3.2. Verification by Comparison with Measured Data

The wellhead temperature calculated by using the data from Tables 1–3 is compared with its counterpart measured in a well in the South China Sea over a period of circulation time. As Figure 5b shows, wellhead temperature drops in calculated results, firstly, because the wellhead temperature is higher than that of the seawater. Therefore, there is heat transfer between drilling fluid near wellhead and seawater, which can lead to wellhead temperature decreases. After circulation for a period of time, the heat absorption of seawater increases and the temperature of seawater increases gradually, while the wellhead temperature decreases gradually and finally tends to stabilize. During the period when the wellhead temperature is stable, although the measured data have some deviation from the theoretical calculation at some time, the overall results are basically consistent, and the calculation and measurement error is less than 5%, so the validity of the model is verified.

4. Analysis of Key Factors Influencing Wellbore Temperature Distribution

The calculation data of this part comes from Tables 1–3. If the multi-pressure system is located in the middle open-hole formation from 2500 m to 3500 m, or located at the bottom hole, it is divided into multiple sub-layers, one per 50 m. Then the total number of sub-layers is 20, and the number of the sub-layers of overflow or leakage is 10, respectively, and distributed alternately. According to the assumptions of the model, the overflow rate or leakage rate in each sub-layer is the same. Therefore, on the one hand, if the flow rate of overflow is greater than that of the leakage, take the flow rate of each sub-layer of overflow as 0.6 L/s and the flow rate of sub-layer of each leakage as 0.2 L/s for calculation, then overflow and leakage rates are 6 L/s and 2 L/s, respectively. On the other hand, the flow rate of sub-layer of each leakage is 0.6 L/s, and the flow rate of each sub-layer of overflow is 0.2 L/s, so the total leakage and overflow rates are 6 L/s and 2 L/s, respectively, when the overflow rate is greater than the leakage rate.

4.1. Annulus Temperature Distribution When Overflow and Leakage Rates Are Different under the Multi-Pressure System

In deep water drilling, the deep formation temperature is higher than that in lower annulus during normal circulation, while the upper formation temperature (including seawater) is lower than that in upper annulus. Before the temperature reaches equilibrium, heat from the deep formation is transferred to lower annulus, the temperature of which increases gradually. Then some heat from the drilling fluid in the lower annulus is transferred to the drilling fluid in the upper annulus. The heat from the upper annulus is transferred to the formation or seawater, and the drilling fluid temperature starts to decrease. The temperature gradually increases as it approaches the wellhead; the temperature reaches an equilibrium state after a period of circulation. Under the multi-pressure system, the fluid carrying formation heat goes into the annulus when overflow occurs, and the temperature of the annulus increases. Some drilling fluid enters the formation when leakage occurs, and temperature of the annulus drilling fluid decreases. When the mixed fluid is uniformly mixed in the annulus where the multi-pressure system is located, the thermo-physical properties and the temperature of drilling fluid in the annulus change.

As shown in Figure 6a, the multi-pressure system is located from 2500 m to 3500 m. If the overflow rate is less than that of leakage, the annulus temperature is less than that during the normal circulation. On the contrary, the annulus temperature is higher. Regardless of whether the overflow rate or leakage rate is greater, the area indicating the largest temperature variation is between 2500 m and 3500 m in the annulus. However, the temperature in the annulus deeper than 3500 m changes slightly and is almost unaffected. Moreover, the temperature in the upper part of the annulus, shallower than 2500 m, is affected and gradually decreases as well-depth decreases. Temperature distribution is almost no longer affected in the upper part of the annulus shallower than 1500 m. Therefore, when the multi-pressure system is located in the formation from 2500 m to 3500 m, the annulus temperature is mainly affected from 1500 m to 3500 m.

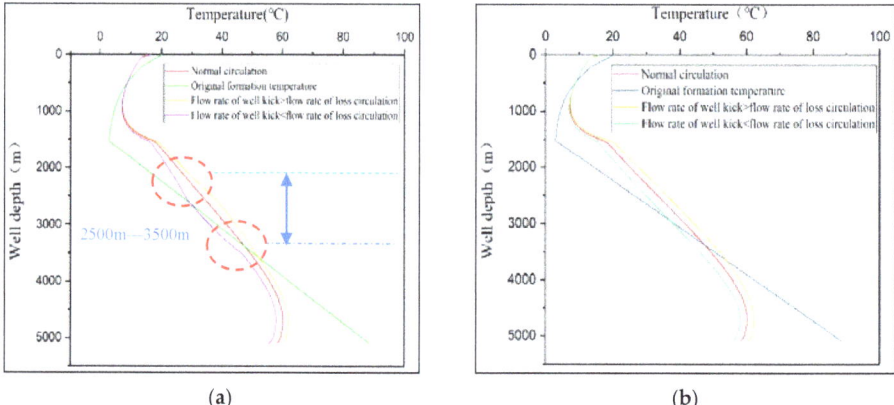

Figure 6. Annulus temperature distribution when overflow and leakage rates are different under the multi-pressure system; (**a**,**b**) show conditions when the multi-pressure system is located in middle formation and at bottom hole, respectively.

As shown in Figure 6b, the multi-pressure system is located at bottom hole; because the leakage and overflow occur near the bottom hole, the thermo-physical property of drilling fluid returning from the bottom hole is affected. As drilling fluid returns from the bottom hole to the surface, then the thermo-physical property of drilling fluid in the entire annulus is also changed. However, the influence on the temperature in annulus shallower than 1500 m gradually decreases. Similarly, when the multi-pressure system is located in the middle open-hole formation, the annulus drilling fluid temperature is higher compared with that during the normal circulation when the overflow rate is greater than leakage rate. On the contrary, the annulus temperature is lower than that during normal circulation when the overflow rate is less than leakage rate.

4.2. Annulus Temperature Distribution during Different Circulation Time under Multi-Pressure System When Overflow Rate Is Greater Than Leakage Rate

When the multi-pressure system is located in the middle open formation from 2500 m to 3500 m, the annulus temperature distribution during the circulation for 0 h, 1 h, 2 h, 4 h, and 8 h is shown in Figure 7a. Since the deeper annulus temperature is lower than that of the corresponding deeper formation, the longer the circulation time of drilling fluid is, the more heat in the deeper formation is taken away by the drilling fluid. Then the formation temperature is gradually cooled down, so the annulus temperature gradually decreases at the same depth. The heat in the deeper annulus then is transferred to the shallower annulus, and the longer time the drilling fluid circulates, the higher the upper annulus temperature is.

As shown in Figure 7b, when the multi-pressure system is located at the bottom hole, the annulus temperature distribution during circulation for 0 h, 1 h, 2 h, and 4 h. The trend is basically consistent with the temperature distribution law shown in Figure 8a.

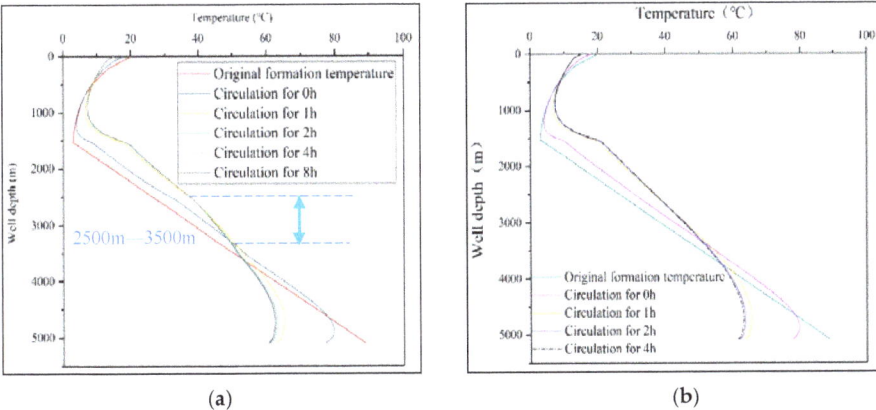

Figure 7. Annulus temperature during different circulation time under multi-pressure system. (**a**) The multi-pressure system is located in the middle open formation; (**b**) The multi-pressure system is located at bottom hole.

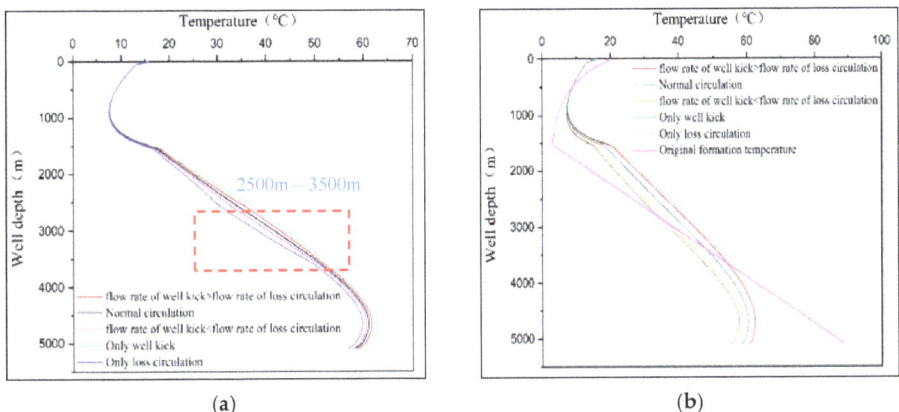

Figure 8. Annulus temperature distribution under multi-pressure system, only overflow or only leakage; (**a**,**b**) show conditions when the multi-pressure system is located in middle formation and at bottom hole, respectively.

4.3. Annulus Temperature Distribution under Multi-Pressure System, Only Overflow or Only Leakage

As shown in Figure 8a, the multi-pressure system is located in open-hole formation from 2500 m to 3500 m. As the Figure 9a shows, the annulus temperature increases when the overflow rate is greater than leakage rate under the multi-pressure system compared with that during normal circulation. On the contrary, it is smaller than that during normal circulation. The temperature variation is most significant in the annulus where the multi-pressure system is located; however, the temperature in the deeper part of the annulus is almost unaffected, and the temperature in the shallower part of the annulus decreases as the well depth decreases. Figure 8b shows annulus temperature when the multi-pressure system is located at the bottom hole. Since drilling fluid from the lower annulus circulates into the upper annulus gradually, so the thermo-physical properties of the whole annulus drilling fluid is changed. Therefore, compared with the annulus temperature during the normal circulation, the counterpart of the entire annulus is affected. Moreover, the annulus drilling fluid temperature changes more in this case than when the multi-pressure system is located in the middle open-hole formation. In both cases (a) and (b), only leakage or only overflow creates greater influence on the annulus temperature than that when leakage and overflow occur synchronously.

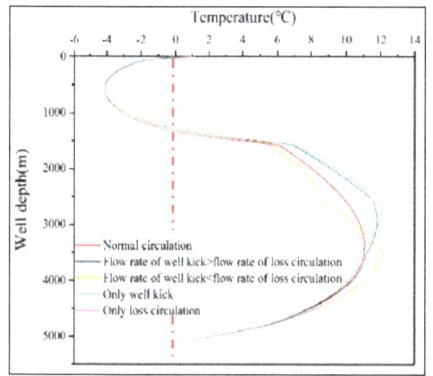
(a) Verification by theoretical models

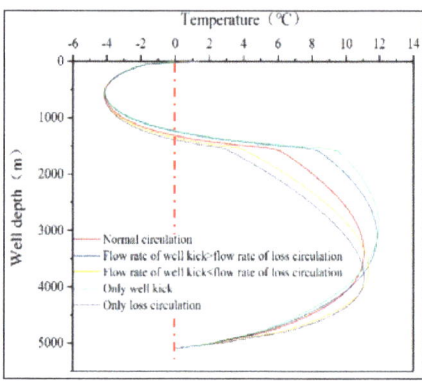
(b) Verification by measurement data

Figure 9. The temperature difference between in annulus and inside drill string when multi-pressure system, only leakage or only overflow occurs; (**a**,**b**) show conditions when the multi-pressure system is located in middle formation and at bottom hole, respectively.

4.4. The Temperature Difference between inside the Drill String and in Annulus When Multi-Pressure System, Only Leakage or Only Overflow Occurs

As shown in Figure 9a,b, the drilling fluid temperature difference distribution was obtained after 2-hour circulation when the multi-pressure system was located in the formation from 2500 m to 3500 m and at bottom hole, respectively. Because the annulus temperature is the same as that inside the drill string at the bottom hole, the temperature difference is 0. When the well depth is from bottom hole to 1500 m, the temperature difference is positive, indicating that the annulus temperature is higher than that inside the drill string. When the well depth is less than 1500 m, the temperature difference is less than 0, indicating that the annulus drilling fluid temperature is lower than that inside the drill string. If overflow rate is greater than leakage rate under multi-pressure system, the temperature difference in this case is higher than that during the normal circulation at the same well depth. On the contrary, the temperature difference is smaller. Obviously, the temperature difference is the largest when only overflow or only leakage occurs. As shown in Figure 9a,b, the obvious temperature difference variation, compared with that during the normal circulation, is mainly in annulus from 1500 to 3500 meters. However, the temperature difference in other parts of the annulus is basically the same when the multi-pressure system is located in the middle formation. As shown in Figure 10b, temperature difference when multi-pressure system is located at bottom hole changes significantly from the bottom hole to the vicinity of the wellhead compared with that during normal circulation.

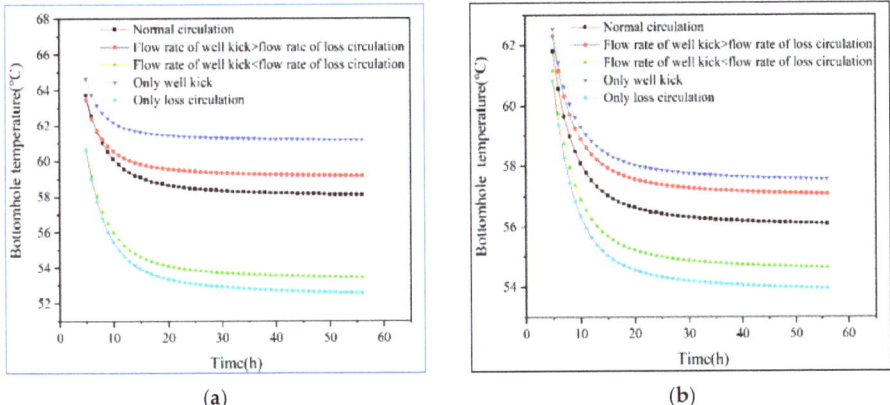

Figure 10. Bottom-hole temperature when multi-pressure system, only overflow or only leakage occurs; (**a**,**b**) show conditions when multi-pressure system is located in middle open formation and at bottom hole, respectively.

4.5. Bottom Hole Temperature Distribution When Multi-Pressure System, Only Overflow or Only Loss Circulation Occurs

As shown in Figure 10a,b, the bottom-hole temperature decreases firstly and then gradually tends to stabilize as the circulation time increases. Because the deep formation temperature is higher than that at the bottom hole, the drilling fluid absorbs heat from the deep formation. As the circulation time increases, the formation temperature is cooled gradually and the bottom hole temperature decreases. When the drilling fluid temperature at the bottom hole reaches an equilibrium state with formation temperature, the former basically remains unchanged. For the same circulation time, only overflow increases the bottom hole temperature the most, and only leakage decreases the bottom hole temperature the most the influence on bottom-hole temperature when multi-pressure system occurs is between them.

4.6. Annulus Temperature Distribution at Different Overflow and Leakage Rates When Multi-Pressure System Exists

Figure 11a–d shows the annulus temperature distribution at different overflow and leakage rates after 2 hours of circulation. As shown in Figure 11a,b, it is annulus temperature distribution when overflow rate is 6 L/s and leakage rate is 1 L/s, 2 L/s and 4 L/s, respectively. Some drilling fluid is lost into the formation because of leakage; then, the heat from the drilling fluid is carried away during the process. As the leakage rate increases, so does the heat loss, then the annulus temperature decreases lower than that in normal condition at the same depth. When the leakage rate continues to increase, the abrupt change point occurs at 2500 m and 3500 m, and the annulus temperature in this region is more affected than that in other annulus regions. Figure 11b also shows the annulus temperature distribution when the overflow rate is 6 L/s, the leakage rates are 1 L/s, 2 L/s, and 4 L/s respectively. As the overflow rate increases, the flow carries heat from the formation into the annulus, which leads to temperature increase in the annulus at the same depth. As shown in Figure 11c, when flow rate of overflow is 6 L/s and the flow rate of leakage gradually increases from 1 L/s, 2 L/s, and 4 L/s, the annulus temperature gradually decreases at the same well depth. As the leakage rate increases, the more heat the drilling fluid carries into the formation, the lower the annulus temperature is at the same depth. Figure 11d shows the annulus temperature distribution when the leakage rate is 6 L/s and overflow rate gradually increases from 1 L/s, 2 L/s, and 4 L/s. The annulus temperature increases gradually at the same depth because the fluid from the formation carries heat into the annulus and circulates with the drilling fluid into the shallower part of the annulus from the bottom hole, then the temperature throughout the annulus increases.

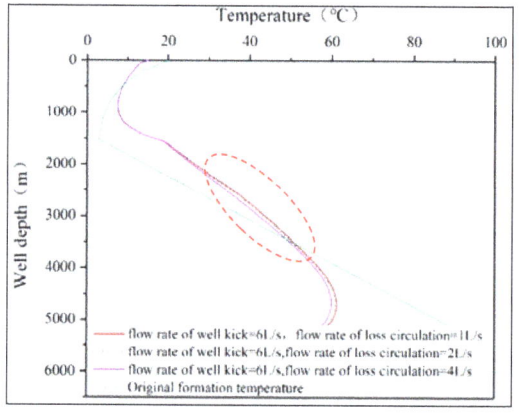

(a)

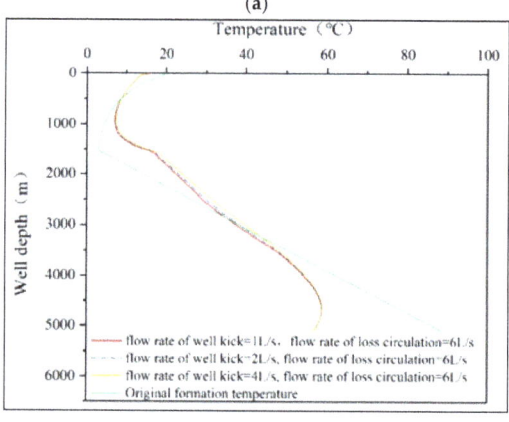

(b)

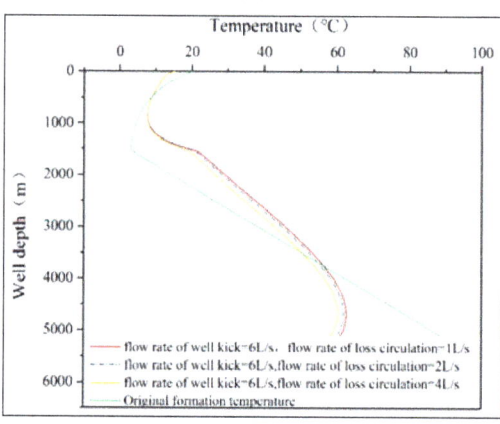

(c)

Figure 11. *Cont.*

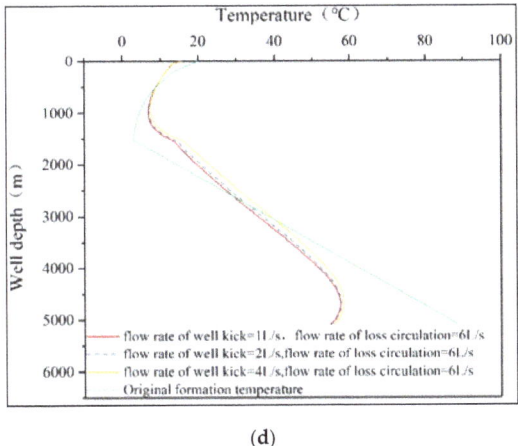

(d)

Figure 11. Annulus temperature distribution at different overflow and leakage rates when multi-system occurs. (**a**) The flow rate of overflow is greater than that of leakage when the multi-pressure system is located in the middle open hole formation; (**b**) The flow rate of overflow is less than that of leakage when the multi pressure system is located in middle open hole formation; (**c**) The overflow rate is greater than leakage rate when the multi pressure system is located at bottom hole; (**d**) The flow rate of overflow is less than that of leakage when the multi pressure system is located at bottom hole.

5. Analysis of the Influence of Coupled Temperature and Pressure on Viscosity and Density of Drilling Fluid

5.1. Normal Circulation

Density and viscosity of drilling fluid are positively correlated with temperature and negatively correlated with pressure. As shown in Figure 12a,b, During normal circulation, with the increase of circulation time, the pressure variation is not obvious for the same well depth, but the temperature of annulus near borehole gradually decreases, so the density and viscosity increase gradually, especially the annulus drilling fluid near the bottom hole. But from the bottom hole up, the density and viscosity vary less and less. Because, the temperature of the shallower annulus has a tendency to increase gradually with the increase of circulation time, so the density and viscosity gradually decrease.

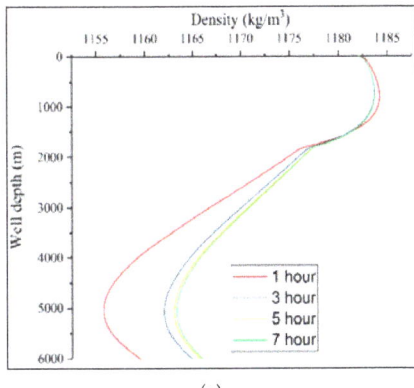

(a)

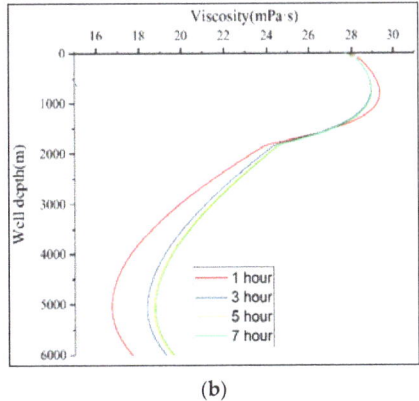

(b)

Figure 12. Variation of density and viscosity with time during normal circulation. (**a**) Drilling fluid density changes over time; (**b**) Drilling fluid viscosity changes over time.

5.2. Under Multi-Pressure System

After circulating for 3 h, when the multi-pressure system is located in the middle formation or at the bottom hole, the variation of density and viscosity at different overflow and leakage rates is shown in Figure 13a–d, respectively. No matter where the multi-pressure system is located, the density increases with the different overflow and leakage rates. When only leakage occurs, the density variation is the smallest. In other cases, the density variation is more obvious than that in the former case. However, when the overflow rate is greater than the leakage rate, the density variation is the largest. When the density of formation fluid is higher than that of drilling fluid, the density of mixed fluid in this part of the annulus is reduced because of leakage and is increased because of overflow. So the density variation is greatest when the overflow rate is greater than the leakage rate; the condition that only leakage occurs hardly affects the density of the drilling fluid, and the influencing degree of other cases is between them.

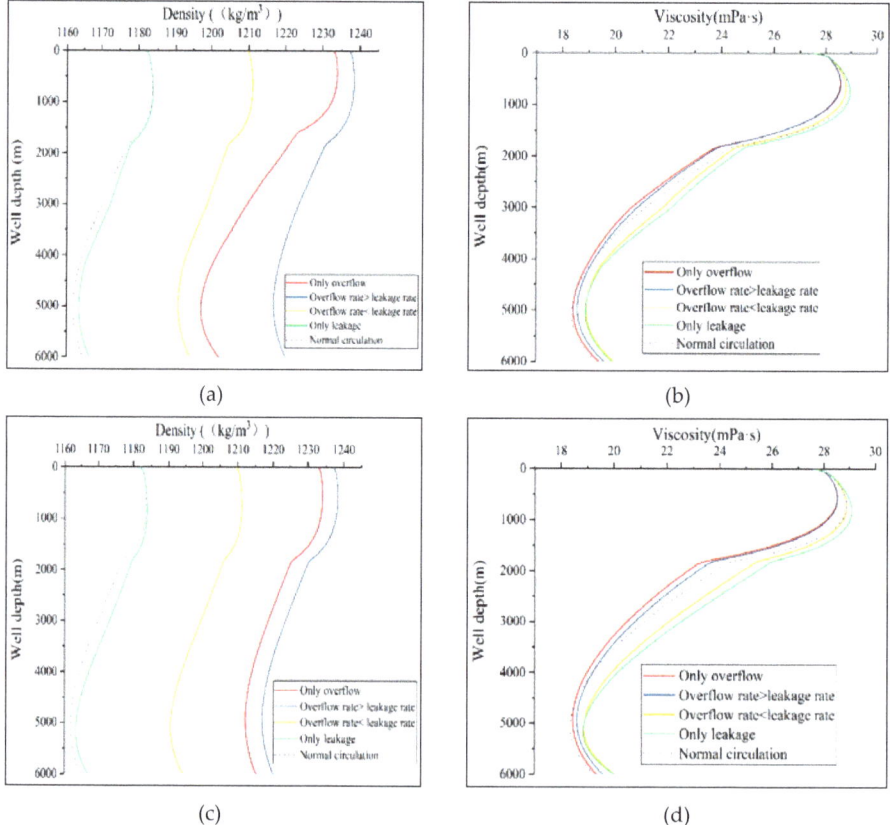

Figure 13. Variation of density and viscosity at different overflow and leakage rates. (**a**) Variation of density when the multi-pressure system is located in the middle formation; (**b**) Variation of viscosity when the multi-pressure system is located in the middle formation; (**c**) Variation of density when the multi-pressure system is located at bottom hole; (**d**) Variation of viscosity when the multi-pressure system is located at bottom hole.

Regardless of where the multi-pressure system is located in the middle formation or at the bottom hole or how the leakage rate and overflow rate change according to the four cases, the viscosity of the drilling fluid decreases. In addition, at the same depth, the viscosity of the drilling fluid is the maximum when only leakage occurs, and the minimum when only overflow occurs. Otherwise, the viscosity is between them. Since the viscosity of formation fluid is smaller than that of drilling fluid, the only leakage increases the solid phase ratio and significantly increases the viscosity of the drilling fluid, while the only overflow dilutes the drilling fluid directly in this part of the annulus resulting in a significant decrease in the viscosity.

6. Analysis of Key Factors Influencing Bottom-Hole Pressure under Multi-Pressure System

6.1. Normal Circulation

As shown in Figure 14a, when the pump rate remains unchanged, the bottom-hole pressure increases rapidly with the increase of circulation time; then, the bottom-hole pressure tends to be constant after circulation for 3 h. When the circulation time is the same, if the pump rate continues to increase, the bottom-hole pressure starts to increase significantly and then increases slowly when the pump rate increases from 25 L/s to 30 L/s. As the circulation time increases, the temperature and density of the deep annulus gradually decrease, so the bottom-hole pressure gradually increases. When the annulus temperature reaches equilibrium, the bottom-hole pressure basically stays constant. For the same circulation time, if the pump rate is higher, the more heat is taken away from the annulus near the bottom hole, the greater the temperature drop is, so the bottom-hole pressure is higher. As shown in Figure 14b, the variation of bottom-hole pressure over time is the same as that in Figure 14a. When the circulation time is the same, but the inlet temperature increases, the bottom-hole pressure decreases gradually. This is because when the inlet temperature remains the same, the longer the circulation time is, the annulus temperature near the bottom hole gradually decreases, then the density gradually increases, so the bottom-hole pressure is higher. For the same circulation time, the higher the inlet temperature is, the lower the density of the annulus near the bottom hole is, so the bottom-hole pressure at the bottom of the well is higher.

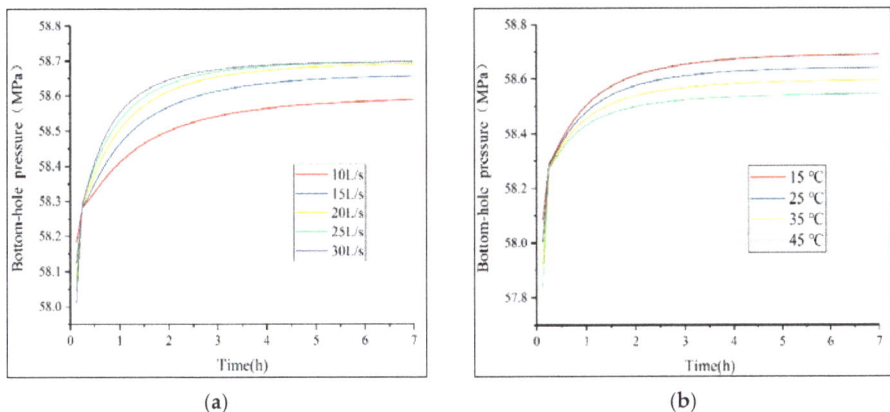

Figure 14. Variation of bottom-hole pressure at different pump rate and inlet temperature. (**a**) Variation of bottom-hole pressure at different pump rate; (**b**) Variation of bottom-hole pressure with different inlet temperature.

6.2. Under Multi-Pressure System

As shown in Figure 15a,b, when the multi-pressure system is located in the middle formation or at the bottom hole, the bottom-hole pressure increases with the circulation time under different overflow and leakage rates. The impact on bottom-hole pressure is greatest when the overflow rate is greater than the leakage rate, because the overflow significantly increases the density of the mixed fluid, resulting in increased bottom-hole pressure. Several other conditions affect bottom-hole pressure to a degree between the only leakage and the overflow rate being greater than the leakage rate.

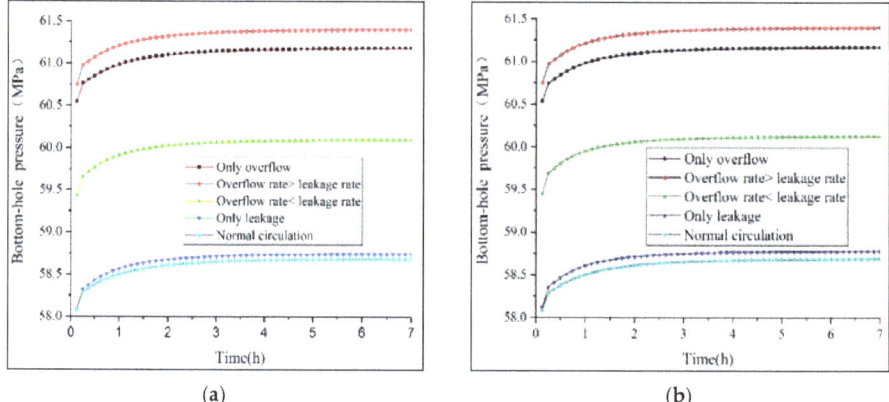

Figure 15. Variation of bottom-hole pressure at different overflow and leakage rates. (a) Variation of bottom-hole pressure when multi-pressure system is located in the middle formation; (b) Variation of bottom-hole pressure when multi-pressure system is located at bottom hole.

7. Conclusions

According to the law of energy conservation, the coupled temperature distribution prediction model of wellbore and formation is established. Meanwhile, the key factors affecting annulus temperature distribution and bottom-hole pressure are analyzed, and the following conclusions are obtained:

(1) As the circulation time increases, annulus temperature decreases gradually in the deeper part and increases gradually in the shallower part, no matter where the multi-pressure system is located.

(2) When the multi-pressure system is located in the middle formation from 2500 m to 3500 m, annulus temperature in this area is most affected; however, the annulus temperature above 2000 m and below 4000 m is almost unaffected. If the multi-pressure system is located at the bottom hole, the annulus temperature from the bottom hole to the mud line is affected, and the temperature above the mud line is basically unchanged.

(3) Whenever overflow occurs, the annulus temperature increases, or leakage occurs, the annulus temperature decreases. However, the annulus temperature increases the most when only overflow occurs and decreases the most when only leakage occurs.

(4) Compared with the normal circulation, the temperature difference between inside drill string and in annulus in the middle well depth is more affected than in other areas along the borehole no matter where the multi-pressure system is located. What is more, only overflow or only leakage has the largest influence on temperature difference.

(5) If the overflow rate is constant and leakage rate keeps increasing, then the annulus temperature gradually decreases; otherwise, it keeps rising at the same depth.

(6) During normal circulation, bottom-hole pressure increases with increase of pump rate and decreases with the increase of inlet temperature. When the overflow rate is greater than leakage rate, the density of drilling fluid and bottom-hole pressure increases the most, and the only leakage has the least increase; the effect of the other cases listed on bottom-hole pressure is between them.

Author Contributions: Conceptualization, J.L. and G.L.; methodology, R.Z.; software, R.Z. and H.Y; validation, R.Z., H.Y. and H.J; formal analysis, R.Z.; investigation, R.Z. and H.J; resources, H.Y.; data curation, R.Z.; writing—original draft preparation, R.Z.; writing—review and editing, R.Z.; visualization, R.Z.; supervision, R.Z.; project administration, J.L.; funding acquisition, J.L.

Funding: This research was supported by the National natural science foundation of China, "basic research on wellbore pressure control in deep water oil and gas drilling and production" (51734010).

Conflicts of Interest: The authors declare no conflict of interest.

Nomenclature

h_{pi}	convective heat transfer coefficient of drill string inside wall, W/(m²·°C)
h_{po}	convective heat transfer coefficient of drill string outside wall, W/(m²·°C)
h_w	convective heat transfer coefficient of the borehole, W/(m²·°C)
U_{ap}	the comprehensive convective heat transfer coefficient between drilling fluid in the drill string and drilling fluid in the annulus, W/(m²·°C)
U_{af}	the comprehensive convective heat transfer coefficient between drilling fluid in the annulus and the formation, W/(m²·°C)
λ_w	heat conductivity coefficient of borehole, W/(m·°C)
λ_p	heat conductivity coefficient of drill string, W/(m·°C)
λ_f	heat conductivity coefficient of formation, W/(m·°C)
λ_{cem}	heat conductivity coefficient of cement sheath, W/(m·°C)
λ_{eff}	effective heat conductivity coefficient of formation, W/(m·°C)
λ_1	heat conductivity coefficient of fluid in the formation, W/(m·°C)
λ_2	heat conductivity coefficient of mixed fluid in annulus, W/(m·°C)
d_{pi}	inside diameter of drill string, mm
d_{cemi}	inside diameter of cement sheath, mm
d_{cemo}	outside diameter of cement sheath, mm
d_{po}	outside diameter of drill string, mm
T_p	drilling fluid temperature in drill string, °C
T_{pi}	inside wall temperature of drill string, °C
T_{po}	outside wall temperature of drill string, °C
T_a	drilling fluid temperature in annulus, °C
T_w	wall temperature of the well, °C
T_f	temperature of the formation, °C
T_{surf}	surface temperature, °C
T_0	bottom-hole temperature inside the drill string, °C
T_1	bottom-hole temperature of the drill string, °C
T_2	bottom-hole temperature in annulus, °C
T_3	seawater temperature, °C
T_{in}	inlet temperature of the drill string, °C
ρ_s	density of seawater, kg/m³
ρ_p	density of the drilling fluid inside the drill string, kg/m³
ρ_a	density of the drilling fluid in annulus, kg/m³
ρ_{eff}	effective fluid density with in the formation, kg/m³
ρ_m	mixed fluid density within the formation, kg/m³
ρ_0	original density of the drilling fluid inside the drill string, kg/m³
ρ_i	density of each fluid phase, kg/m³
c_s	specific heat capacity of seawater, J/(kg·°C)
c_p	specific heat capacity of the drilling fluid inside the drilling string, J/(kg·°C)
c_a	specific heat capacity of the drilling fluid in the annulus, J/(kg·°C)
c_{eff}	specific heat capacity of fluid information, J/(kg·°C)

Q_{cp} friction heat source of the drilling fluid inside the drill string, W/m³
Q_{ca} friction heat source of the drilling fluid in the annulus, W/m³
Q'_{ca} friction heat source of the drilling fluid in the annulus of the multi-system pressure, W/m³
Q_m pump rate of the drilling fluid inside the drill string, m³/s
Q_a pump rate of the drilling fluid in the annulus, m³/s
Q_k flow rate of overflow of fluid in the formation, m³/s
Q_l flow rate of leakage of fluid in the formation, m³/s
φ porosity of formation rock
v_a flow rate of drilling fluid in annulus, m/s
v_{kk} the flow rate of overflow of each sublayer in a multi-pressure system, m/s
v_{ll} the flow rate of leakage of each sublayer in a multi-pressure system, m/s
v_i the flow rate of each fluid phase, m/s
A_i the area of the flow cross section, m²
K absolute permeability of isotropic porous medium
K_1 relative permeability
P intrinsic average pressure of formation, Pa
P_{fi} Annulus friction pressure loss
H well depth, m
h the depth of a well at a given location, m
G geothermal gradient, °C/m
y_{ml} depth of mudline
μ velocity in the x direction, m/s
v_r seepage velocity of fluid information, m/s
χ experimental measurement coefficient

Appendix A. Model Solution

The model established in this paper is discretized by finite difference method. The first-order space uses the first-order windward scheme, the first-order time uses two points for backward difference, and the second-order space uses three points for central difference. The meshing method of wellbore and formation is shown in Figure A1.

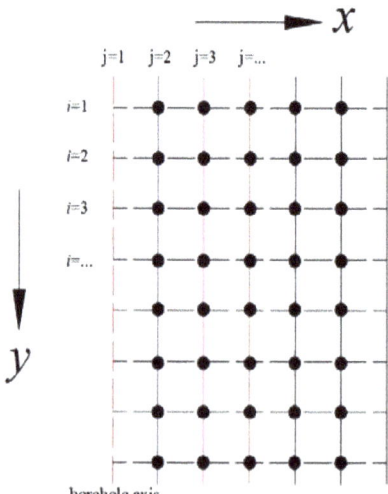

Figure A1. Schematic diagram of mesh grids of wellbore and formation.

Appendix A.1. Heat Transfer Model Discretization of Drilling Fluid in Drill String

$$\frac{\pi}{4}d_{pi}^2\frac{\partial[\rho_p c_p T_p]}{\partial t} = U_{ap}\pi d_{pi}(T_a - T_p) + Q_m\frac{\partial[\rho_p c_p T_p]}{\partial y} + Q_{cp} \quad (A1)$$

Making $A_1 = \frac{\pi d_{pi}^2 \rho_p c_p}{4\Delta t}$, $B = \pi U_{ap}d_{pi}$, $C = \frac{\rho_p c_p Q_m}{\Delta y}$, Then,

$$A_1(T_{i,j}^n - T_{i,j}^{n-1}) = B(T_{i,j+1}^n - T_{i,j}^n) - C_1(T_{i,j}^n - T_{i-1,j}^n) + (Q_{cp})_i^n \quad (A2)$$

Equation (A2) can be rewritten as:

$$aT_{i,j}^n + bT_{j,j+1}^n + cT_{i-1,j}^n = U_{i,j}^n \quad (A3)$$

where, $a = A_1 + B + C_1$, $b = -B$, $c = -C_1$, $U_{i,j}^n = A_1 T_{i,j}^n + (Q_{cp})_i^n$.

Appendix A.2. Heat Transfer Model Discretization of Drilling Fluid in Annulus

Appendix A.2.1. The Multi-Pressure System Is Located in the Middle Section of the Formation

(1) Heat transfer equation of the lower annulus can be written as follows:

$$\frac{\pi}{4}(d_w - d_{po}^2)\frac{\partial[\rho_a c_a T_a]}{\partial t} = U_{ap}\pi d_{po}[T_a - T_p] + h_w \pi d_w[T_a - T_w] + Q_m\frac{\partial[\rho_a c_a T_a]}{\partial y} + Q_{ca} \quad (A4)$$

Making $D_1 = \frac{\pi(d_w^2 - d_{po}^2)\rho_a c_a}{4\Delta t}$, $E = \pi d_{po}U_{ap}$, $F = \pi d_w U_{af}$, $G_1 = \frac{Q_a \rho_a c_a}{\Delta y}$, Then

$$D_1(T_{i,j}^n - T_{i,j}^{n-1}) = E(T_{i,j-1}^n - T_{i,j}^n) + F(T_{i,j+1}^n - T_{i,j}^n) + G_1(T_{i+1,j}^n - T_{i,j}^n) + (Q_{ca})_i^n \quad (A5)$$

Equation (A5) can be rewritten as follows:

$$a_1 T_{i,j}^n + b_1 T_{i+1,j}^n + c_1 T_{i,j+1}^n + d T_{i,j-1}^n = U_{i,j}^n \quad (A6)$$

where, $a_1 = D_1 + E + F + G_1$, $b_1 = -G_1$, $c_1 = -F$, $d_1 = -E$, $U_{i,j}^n = D_1 T_{i,j}^{n-1} + (Q_{ca})_i^n$.

(2) Heat transfer equation of the upper annulus can be written as follows:

$$\frac{\pi}{4}(d_w - d_{po}^2)\frac{\partial[\rho_a'c_a'T_a]}{\partial t} = U_{ap}'\pi d_{po}[T_a - T_{po}] + U_{fa}'\pi d_w[T_f - T_a] + (v_a + \sum_{k=1}^{n}v_{kk} - \sum_{l=2}^{n-1}v_{ll})\frac{\partial[\rho_a'c_a'T_a]}{\partial y} + Q_{ca}' \quad (A7)$$

Making $D_1 = \frac{\pi(d_w^2 - d_{po}^2)\rho_a'c_a'}{4\Delta t}$, $E = \pi d_{po}U_{ap}'$, $F = \pi d_w U_{af}'$, $G_1 = \frac{\rho_a'c_a'(Q_a + Q_k - Q_l)}{\Delta y}$, then

$$D_1(T_{i,j}^n - T_{i,j}^{n-1}) = E(T_{i,j-1}^n - T_{i,j}^n) + F(T_{i,j+1}^n - T_{i,j}^n) + G_1(T_{i+1,j}^n - T_{i,j}^n) + (Q_{ca}')_i^n \quad (A8)$$

Equation (A5) can be rewritten as follows:

$$a_1 T_{i,j}^n + b_1 T_{i+1,j}^n + c_1 T_{i,j-1}^n + d_1 T_{i,j+1}^n = U_{i,j}^n \quad (A9)$$

where, $a_1 = D_1 + E + F + G_1$, $b_1 = -G_1$, $c_1 = -E$, $d_1 = -F$, $U_{i,j}^n = D_1 T_{i,j}^{n-1} + (Q_{ca}')_i^n$.

Appendix A.2.2. When the Multi-Pressure System Is Located at Bottom Hole, the Whole Annulus Is Affected by the Multi-Pressure System. Therefore, Its Heat Transfer Model and Discrete Method Are the Same as the Upper Annulus Model when the Multi-Pressure System Appears in the Middle Open Hole Formation

Appendix A.3. The Heat Transfer Mode in the Formation and Seawater

The heat transfer mode in the formation and seawater is radial heat conduction, so the heat transfer model and discrete method are the same. Therefore, only the heat transfer equation inside the formation is discretized to illustrate.

$$(\rho c)_{eff} \frac{\partial T_i(x,y,t)}{\partial t} = \lambda_{eff} \frac{\partial^2 T_f(x,y,t)}{\partial^2 x} + \frac{\lambda_{eff}}{x} \frac{\partial T_f(x,y,t)}{2x} \tag{A10}$$

Making $H_1 = \frac{(\rho c)_{eff} \Delta r}{\lambda_{eff} \Delta t}$, $K = \frac{1}{\Delta r}$, then

$$H_1(T_{i,j}^n - T_{i,j}^{n-1}) = K(T_{i,j+1}^n - 2T_{i,j}^n + T_{i,j-1}^n) + \frac{1}{r_j}(T_{i,j+1}^n - T_{i,j}^n) \tag{A11}$$

Equation (A11) can be rewritten as follows:

$$a_2 T_{i,j}^n + b_2 T_{i+2,j}^n + c_2 T_{i-1}^j = U_{i,j}^n \tag{A12}$$

where, $a_2 = H_1 + 2K + \frac{1}{r_j}$, $b_2 = -(K + \frac{1}{r_j})$, $c_2 = -K$, $U_{i,j}^n = H_1 T_{i,j}^{n-1}$.

Appendix A.4. Heat Transfer Equation of Boundary between Strata and Annulus Can Be Written as Follows

$$\pi d_w U_{af}(T_f - T_a) = \pi d_w \lambda_{eff}(\frac{\partial T_f}{\partial r}), \text{making } M = \frac{U_{af} \Delta r}{\lambda_{eff}} \tag{A13}$$

Equation (A13) can be rewritten as follows:

$$(M+1)T_{i,j}^n - MT_{i,j-1}^n - T_{i,j+1}^n = 0 \tag{A14}$$

Appendix A.5. The Dispersion of the Momentum Equation

$$\frac{P_i^n A_i - P_{i-1}^n A_{i-1}}{\Delta z} = \frac{(\rho v)_i^n A_i - (\rho v)_i^{n-1} A_i}{\Delta t} + \frac{(\rho v^2)_i^n A_i - (\rho v^2)_{i-1}^n A_{i-1}}{\Delta z} - \frac{(P_{fi})_i^n A_i - (P_{fi})_{i-1}^n A_{i-1}}{\Delta z} - gA_i \rho_i^n \tag{A15}$$

Making the $\psi = A_{i-1}/A_i$, Equation (A15) can be changed as:

$$P_i^n - \psi P_{i-1}^n = \frac{\Delta z}{\Delta t}[(\rho v)_i^n - (\rho v)_i^{n-1} + (\rho v^2)_i^n - \psi(\rho v^2)_i^{n-1} - ((P_{fi})_i^n - \psi(P_{fi})_{i-1}^n)] - g\rho_i^n \tag{A16}$$

Figure A2 shows the flow chart of solution process.

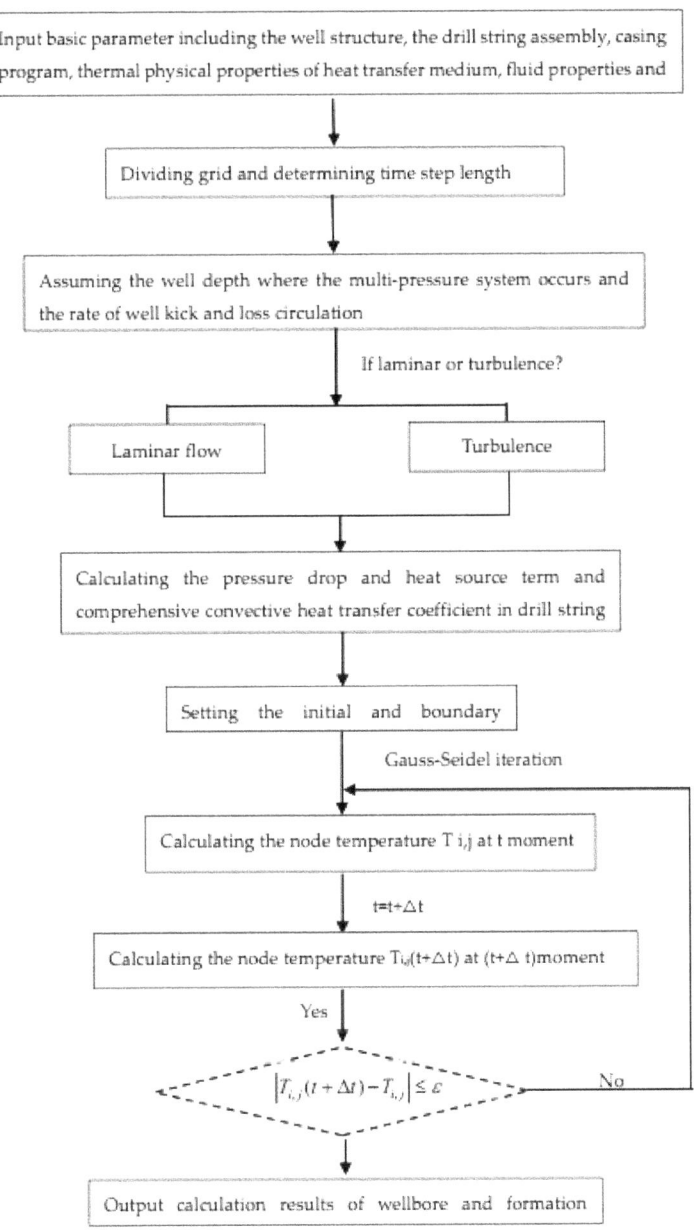

Figure A2. Flowchart of wellbore and formation temperature calculation.

References

1. Zhang, Z.; Xiong, Y.M.; Gao, Y.; Liu, L.M.; Wang, M.; Peng, G. Wellbore temperature distribution during circulation stage when well-kick occurs at bottom-hole from the bottom-hole. *Energy* **2018**, *164*, 964–977. [CrossRef]
2. Wiktorski, E.; Cobbah, C.; Sui, D.; Khalifeh, M. Experimental study of temperature effects on wellbore material properties to enhance temperature profile modeling for production wells. *J. Pet. Sci. Eng.* **2019**, *176*, 689–701. [CrossRef]
3. Abdollahi, J.; Dubljevic, S. Transient Fluid Temperature Estimation in Wellbores. *IFAC Proc. Vol.* **2013**, *46*, 108–113. [CrossRef]
4. Hasan, R.; Kabir, C.S. Wellbore heat-transfer modelling and applications. *J. Pet. Sci. Eng.* **2012**, *86–87*, 127–136. [CrossRef]
5. Yang, M.; Zhao, X.; Meng, Y.; Li, G.; Zhang, L.; Xu, H.; Tang, D. Determination of transient temperature distribution inside a wellbore considering drill string assembly and casing program. *Appl. Therm. Eng.* **2017**, *118*, 299–314. [CrossRef]
6. Jia, H.J.; Meng, Y.F.; Li, G.; Su, G.; Zhao, X.Y.; Wei, N.; Song, W. Research on overflow accompanied with lost circulation during drilling of gas formation with multi-pressure system. *Fault-Block Oil Gas.* **2012**, *19*, 360–363.
7. Ramey, H.J., Jr. Wellbore heat transmission. *J. Pet. Technol.* **1962**, *14*, 427–435. [CrossRef]
8. Holmes, C.S.; Swift, S.C. Calculation of circulating mud temperatures. *J. Pet. Technol.* **1970**, *22*, 670–674. [CrossRef]
9. Kabir, C.S.; Hasan, A.R.; Kouba, G.E.; Ameen, M. Determining circulating fluid temperature in drilling, workover, and well control operations. *SPE Drill. Complet.* **1996**, *11*, 74–79. [CrossRef]
10. You, J.; Rahnema, H.; McMillan, M.D. Numerical modeling of unsteady-state wellbore heat transmission. *J. Nat. Gas Sci. Eng.* **2016**, *34*, 1062–1076. [CrossRef]
11. Raymond, L.R. Temperature distribution in a circulating drilling fluid. *J. Pet. Technol.* **1969**, *21*, 333–341. [CrossRef]
12. Marshall, D.W.; Bentsen, R.G. A computer model to determine the temperature distributions in a wellbore. *J. Can. Pet. Technol.* **1982**, *21*, 63–75. [CrossRef]
13. Santoyo-Gutierrez, E.R. Transient Numerical Simulation of Heat Transfer Processes During Drilling of Geothermal Wells. Ph.D. Thesis, University of Salford, Salford, UK, 1 September 1997.
14. García, A.; Santoyo, E.; Espinosa, G.; Hernandez, I. Estimation of temperatures in geothermal wells during circulation and shut-in in the presence of lost circulation. *Transp. Porous Media* **1998**, *33*, 103–127. [CrossRef]
15. Espinosa-Paredesa, G.; Garcia, A.; Santoyo Hernandez, E. A computer program for estimation of fully transient temperatures in geothermal wells during circulation and shut-in. *Comput. Geosci.* **2001**, *27*, 327–344. [CrossRef]
16. Yang, M.; Tang, D.; Chen, Y.; Li, G.; Zhang, X.; Meng, Y. Determining initial formation temperature considering radial temperature gradient and axial thermal conduction of the wellbore fluid. *Appl. Therm. Eng.* **2019**, *147*, 876–885. [CrossRef]
17. Zheng, Z.; Xiong, Y.M.; Guo, F. Analysis of Wellbore Temperature Distribution and Influencing Factors During Drilling Horizontal Wells. *J. Energy Resour. Technol.* **2018**, *140*, 092901. [CrossRef]
18. Wang, X.R.; Sun, B.J.; Luo, B.Y. Transient temperature and pressure calculation model of a wellbore for dual gradient drilling. *Appl. Therm. Eng.* **2018**, *30*, 701–714. [CrossRef]
19. Farahani, H.S.; Yu, M.; Miska, S.; Takach, N.; Chen, G. Modeling Transient Thermo-Poroelastic Effects on 3D Wellbore Stability. In Proceedings of the SPE Technical Conference and Exhibition, San Antonio, TX, USA, 24–27 September 2006.
20. Song, X.C.; Guan, Z.C. Full Transient Analysis of Heat Transfer During Drilling Fluid Circulation in Deep-Water Wells. *Acta Pet. Sin.* **2011**, *32*, 704–708.
21. Zheng, Z.; Xiong, Y.M.; Mao, L.J.; Lu, J.S.; Wang, M.H.; Peng, G. Transient temperature prediction models of wellbore and formation in wellkick condition during circulation stage. *J. Pet. Sci. Eng.* **2019**, *175*, 266–279. [CrossRef]

22. Chen, Y.H.; Yu, M.J.; Miska, S.; Zhou, S.H.; Al-Khanferi, N.; Ozbayoglu, E. Fluid flow and heat transfer modeling in the event of lost circulation and its application in locating loss zones. *J. Pet. Sci. Eng.* **2017**, *148*, 1–9. [CrossRef]
23. Wang, L.Y.; Wang, K.Q. A new method for calculating thermal conductivity of mixed liquid. *Chem. Eng.* **1999**, *27*, 45–51.
24. Yang, H.W.; Li, J.; Liu, G.H.; Wang, C.; Li, M.; Jiang, H. Numerical analysis of transient wellbore thermal behavior in dynamic deepwater multi-gradient drilling. *Energy* **2019**, *179*, 138–153. [CrossRef]
25. Arnold, F.C. Temperature variation in a circulating wellbore fluid. *J. Energy Res. Technol.* **1990**, *112*, 79–83. [CrossRef]
26. Edwardson, M.J.; Girner, H.M.; Parkinson, H.R.; Williamson, C.D.; Matthews, C.S. Calculation of formation temperatures disturbances caused by mud circulation. *J. Pet. Technol.* **1962**, *14*, 416–426. [CrossRef]
27. Deng, S.; Fan, H.; Tian, D.; Liu, Y.; Zhou, Y.; Wen, Z.; Ren, W. Calculation and application of safe mud density window in deepwater shallow layers. In Proceedings of the Offshore Technology Conference, Houston, TX, USA, 2–5 May 2016.
28. Li, M.B.; Liu, G.H.; Li, J.; Zhang, T.; He, M. Thermal performance analysis of drilling horizontal wells in high temperature formations. *Appl. Therm. Eng.* **2015**, *78*, 217–227. [CrossRef]
29. Zahedi, G.; Karami, Z.; Yaghoobi, H. Prediction of hydrate formation temperature by both statistical models and artificial neural network approaches. *Energy Convers. Manag.* **2009**, *50*, 2052–2059. [CrossRef]
30. Yang, H.W.; Li, J.; Liu, G.H. Development of transient heat transfer model for controlled gradient drilling. *Appl. Therm. Eng.* **2019**, *148*, 331–339. [CrossRef]
31. Yang, M.; Li, X.X.; Deng, J.M.; Meng, Y.F.; Li, G. Prediction of wellbore and formation temperatures during circulation and shut-in stages under kick conditions. *Energy* **2015**, *91*, 1018–1029. [CrossRef]

© 2019 by the authors. Licensee MDPI, Basel, Switzerland. This article is an open access article distributed under the terms and conditions of the Creative Commons Attribution (CC BY) license (http://creativecommons.org/licenses/by/4.0/).

Article

Three-Dimensional Numerical Simulation of Geothermal Field of Buried Pipe Group Coupled with Heat and Permeable Groundwater

Xinbo Lei [1], Xiuhua Zheng [1,*], Chenyang Duan [1], Jianhong Ye [2] and Kang Liu [1]

1. School of Engineering and Technology, China University of Geosciences (Beijing), Beijing 100083, China; Lxb@cugb.edu.cn (X.L.); dcy1990@sina.com (C.D.); liukang2@cugb.edu.cn (K.L.)
2. Institute of Rock and Soil Mechanics, Chinese Academy of Sciences, Wuhan 430071, China; yejianhongcas@gmail.com
* Correspondence: xiuhuazh@cugb.edu.cn; Tel.: +86-10-8232-2624

Received: 18 July 2019; Accepted: 26 September 2019; Published: 27 September 2019

Abstract: The flow of groundwater and the interaction of buried pipe groups will affect the heat transfer efficiency and the distribution of the ground temperature field, thus affecting the design and operation of ground source heat pumps. Three-dimensional numerical simulation is an effective method to study the buried pipe heat exchanger and ground temperature distribution. According to the heat transfer control equation of non-isothermal pipe flow and porous media, combined with the influence of permeable groundwater and tube group, a heat-transfer coupled heat transfer model of the buried pipe group was established, and the accuracy of the model was verified by the sandbox test and on-site thermal response test. By processing the layout of the buried pipe in the borehole to reduce the number of meshes and improve the meshing quality, a three-dimensional numerical model of the buried pipe cluster at the site scale was established. Additionally, the ground temperature field under the thermal-osmotic coupling of the buried pipe group during groundwater flow was simulated and the influence of the head difference and hydraulic conductivity on the temperature field around the buried pipe group was calculated and analyzed. The results showed that the research on the influence of the tube group and permeable groundwater on the heat transfer and ground temperature field of a buried pipe simulated by COMSOL software is an advanced method.

Keywords: GSHP (ground source heat pump); heat transfer; coupled heat conduction and advection; nest of tubes; three-dimensional numerical simulation

1. Introduction

Ground heat exchanger (GHE), which is also called a geothermal heat exchanger, has emerged as a promising and globally accepted way of exploiting shallow geothermal energy. Typically, circulating heat-carrying fluids flow in a U-shaped channel inserted in a vertical borehole and absorb or discharge heat from or to the ground [1]. The performance of ground source heat pump (GSHP) systems, constructed with many GHEs, is determined by ground stratigraphy in which thermal conductivity, groundwater flow, and the initial temperature play an essential role [2,3]. Knowledge of the thermal processes in the ground is not only critical for the optimal design and operation of systems (GSHP) [4–6], but is also a prerequisite in evaluating their relevance and the hazards related to the local climatic conditions and geological context [7].

In order to estimate the heat transfer of borehole heat exchangers (BHEs), based on the pioneering work of Ingersoll et al. [4], diverse numerical and analytical methods have been proposed [8]. Plenty of research has focused on the system design, penalty temperature prediction [9], parameter estimation [10,11], quasi-steady heat transfer around a borehole [12], and progress in conventional models for borehole GHEs [13]. The buried depth of BHEs is about 60–300 m, where groundwater exists and its flow benefits the heat transfer and reduces or eliminates the accumulating imbalance effects of heat transfer [14–19].

The thermal response test (TRT), designed based on the infinite line source model (ILS), is the most common approach to describe the heat conduction process in homogeneous media with a constant temperature at infinity [20]. Several studies have been conducted to analyze the importance of groundwater on heat transfer in TRTs. Diao et al. [21,22] obtained an analytical solution of two-dimensional temperature response with uniform permeability through a line heat source in the infinite field by means of the Green function analysis. Molina-Giraldo et al. [23] deduced an analytical solution through the use of the moving finite line source model (MFLS) with the consideration of the groundwater flow and axial effects. Wagner et al. [24,25] proposed a parameter estimation method which was sensitive to conduction and advection to calculate the actual Darcy velocity based on MFLS. Aranzabal et al. [26] complemented the standard TRT analysis and estimated the thermal conductivity profile from the borehole temperature profile during the test.

With the wider utilization of GSHPs, the scale of projects is increasing with hundreds, even thousands, of groups of borehole heat exchangers [27]. The heat transfer in the underground heat exchanger groups is coupled and interacts with each other in the actual operation process, so some single hole research conclusions cannot be directly applied [28]. The long-term research models for buried pipe groups mainly include the two-dimensional heat transfer model and three-dimensional heat transfer model. Most researchers have often used two-dimensional models to study the long-term variation of the temperature field in a buried tube region [29–31]. Jia constructed 2D and 3D unsteady heat transfer models of vertical U-tube GHEs based on practical projects [32]. Yang et al. established a 3D heat transfer model of a ground heat exchanger that couples thermal conduction with groundwater flow, and pointed out that the heat transfer capacity of staggered arrangement was higher than that of an aligned arrangement, which is made more obvious by the groundwater flow [33].

In summary, the research on the heat transfer model of buried pipes has always been a difficult point in the technology of ground source heat pump systems. The influence of groundwater flow on the heat transfer of buried pipes has been of high concern, and relevant research has made great progress. However, many of the buried pipe models have been simplified, based on the mobile line heat source. The true line heat source can give an exact solution, and the splitting of the flow field and the temperature field cannot obtain a more accurate analytical solution model. It is more practical to establish a physical model that is exactly the same as the shape of the actual buried pipe [34].

There are few studies on the interaction between long-running buried pipe groups, and the impact of groundwater flow on buried pipe groups has rarely been considered. Numerical simulation is very time consuming in solving the formation temperature field of the three-dimensional buried tube group thermal-permeability coupling in the whole year or the whole life cycle, especially in calculating the temperature distribution of all-time nodes and spaces of the large buried tube group.

In the research work of this paper, based on the non-isothermal pipe flow module in the finite-element numerical simulation software COMSOL, the buried pipe was simplified into a three-dimensional curve, and the fluid heat transfer, velocity, and pressure in the buried pipe were solved. A new simplified model of the buried pipe was proposed to improve the efficiency and accuracy of the finite element numerical model calculation. The porous medium heat transfer module and the groundwater flow module were used to solve the heat-transfer coupling heat transfer process in the soil. The influence of the buried pipe group and the thermal transfer coupling on the formation temperature field was studied after one year of operation. The simplified model could accurately solve the temperature response of the formation in the long-term heat transfer process of the buried tube group with limited computational resources.

This paper is mainly composed of the following parts: First, the governing equation of the heat transfer in the formation coupling heat transfer of the buried pipe was given, then the simplified physical model of the buried pipe heat transfer was established. Second, based on the sandbox test data and on-site thermal response test data, the finite element simulation software COMSOL was used to simulate and verify the full-scale three-dimensional buried tube thermal-permeability coupling model, which confirmed the accuracy and practicability of the simplified model. Third, a numerical model of the thermal infiltration coupling of the full-scale buried pipe group was established. When the different hydraulic conductivities were calculated, the temperature field distribution in the formation after one year of system operation was obtained. The effects of different hydraulic conductivity on the formation temperature field were obtained.

2. Establishment of Heat Transfer Models for A Single Well Buried Pipe Coupled with Flow of Groundwater

The heat exchange process between the buried pipe and the soil can be divided into heat conduction between the circulating fluid in the U-shaped pipe and the pipe wall, heat conduction between the buried pipe wall and the backfill material, and heat conduction between the backfill material and the soil. The non-isothermal pipe flow module based on the finite element numerical simulation software COMSOL Multiphysics 5.3 simulates the heat transfer process of the circulating fluid in the buried pipe; the heat transfer coupling heat transfer process in the soil is simulated by the porous medium heat transfer module and the Darcy permeable module. A schematic diagram of the buried pipe is shown in Figure 1.

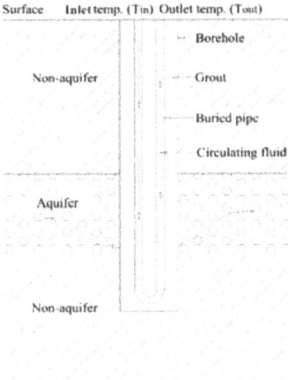

Figure 1. Schematic diagram of the buried pipe.

2.1. Establishment of Governing Equations for Coupled Model of Buried Pipe with Heat and Groundwater

The control equations of heat transfer between the pipeline and porous formation with Darcy's fluid are deduced according to the principle of the conservation of energy.

(1) Energy equation for pipeline flow

It is assumed that the fluid in each section of the pipe is in a fully developed state and is transient-state heat transfer. The velocity, pressure, and temperature at the same section are the same, and change only along the axial direction of the pipe. The circulating fluid in the buried pipe is considered as incompressible flow. The heat balance equation of the fluid in the pipe is as follows:

$$\rho A C_p \frac{\partial T}{\partial t} + \rho A C_p \mathbf{u}_{et} \cdot \nabla_t T = \nabla_t \cdot (A \lambda \nabla_t T) + \frac{1}{2} f_D \frac{\rho A}{2 d_h} |u| u^2 + Q_p + Q_{wall} \quad (1)$$

$$A = \frac{\pi}{4} d_i^2 \quad (2)$$

$$f_D = 8\left[\left(\frac{8}{R_e}\right)^{12} + (C_A + C_B)^{-1.5}\right]^{\frac{1}{12}} \tag{3}$$

$$R_e = \frac{\rho u d_i}{\mu} \tag{4}$$

$$C_A = \left[-2.457 \ln\left(\left(\frac{7}{R_e}\right)^{0.9} + 0.27\left(\frac{e}{d_i}\right)\right)\right]^{16} \tag{5}$$

$$C_B = \left(\frac{37530}{R_e}\right)^{16} \tag{6}$$

$$Q_{wall} = (hZ)_{eff}(T_{ext} - T) \tag{7}$$

$$Z = \pi d_i \tag{8}$$

where ρ is the fluid density in the pipe, kg/m³; A is the pipe cross-sectional area, m²; C_p is the isobaric heat capacity of fluid, J/(kg·K); u_{et} is the tangential velocity, m/s; T is the temperature of the fluid, K; λ is the thermal conductivity of fluid, W/(m·K); f_D is the friction coefficient; d_i is the pipe inner diameter, m; u is the mean velocity of fluid, m/s; Q_p is the heat source/sink item in the pipe, W/m (as the pipe in this study does not have an extra heat source, Q_p is zero); Q_{wall} is the heat exchange on the pipe wall, W/m; $(hZ)_{eff}$ is the effective total thermal resistance of the pipe wall, W/(m·K), which includes the thermal resistance of the pipe and the thermal resistances of the inner and outer pipe wall with the convection layer [35]; h is the heat transfer coefficient, W/(m²·K); Z is the wetted perimeter of the pipe, m; T_{ext} is the exterior temperature outside pipe, K; and T is the fluid temperature inside the pipe, K [36,37].

(2) Energy equation for pipeline flow for porous formation

Heat transfer in porous water-bearing formations can be expressed by the following equation:

$$Q_f + Q_{geo} = (\rho C_p)_{eq}\frac{\partial T_f}{\partial t} + \rho C_p v \cdot \nabla T_f - \nabla \cdot (\lambda_{eq} \nabla T_f) \tag{9}$$

where Q_f is the heat source/sink item in the formation, W/m; Q_{geo} is the special geothermal heating function (expressed as domain conditions) in the porous media heat transfer interface, W/m. Since there was no extra heat source or sink and special geothermal heating in the ground of this study, the terms of Q_f and Q_{geo} were set as zero; $(\rho C_p)_{eq}$ is the equivalent volumetric specific heat, J/(m³·K); v is the groundwater flow velocity, m/s; ∇T_f is the incremental temperature of formation, K; and λ_{eq} is the equivalent thermal conductivity, W/(m·K).

The average value of the thermal parameters of the soil is expressed as a weighting factor, expressed as matrix volume fraction θ. In saturated formation, the pore is occupied by water, while in unsaturated formation, water and air may exist in the pore. Using the volume averaging method, the volumetric specific heat of the heat transfer equation in formation is:

$$(\rho C_p)_{eq} = \sum_i (\theta_i \rho_i C_{pi}) + \left(1 - \sum_i \theta_i\right)\rho_s C_{ps} \tag{10}$$

where θ_i is the volume ratio of the ith non-solid in the formation; ρ_i is the density of the ith non-solid material, kg/m³; C_{pi} is specific heat capacity of the ith non-solid material, J/(kg·K); ρ_s is the density of solid, kg/m³; and C_{ps} is the specific heat capacity of the solid, J/(kg·K).

The equivalent thermal conductivity is described as:

$$\lambda_{eq} = \sum_i \theta_i \lambda_i + \left(1 - \sum_i \theta_i\right)\lambda_s \tag{11}$$

where λ_i is the thermal conductivity of the ith non-solid material, W/(m·K) and λ_s is the thermal conductivity of solid, W/(m·K).

(3) The permeable equation of fluid flowing in porous formation

According to the universal conservation principle proposed, the mass conservation equation describing the underground hot water system can be written as follows:

$$\frac{\partial}{\partial t}(\rho_w \varphi) + \nabla \cdot (\rho_w v) = \rho_q q_s \tag{12}$$

where t is the time, s; ρ_w is the groundwater density, kg/m³; φ is the effective porosity, dimensionless; v is the permeable velocity of groundwater, m/s; ρ_q is the density of source sink source, kg/m³; and q_s is the source sink term, 1/s.

Darcy's law expresses different permeable flow as follows:

$$v = -\frac{k\rho g}{\mu}\nabla H \tag{13}$$

where k is permeability, m²; g is the gravity acceleration, m/s²; μ is the dynamic viscosity of groundwater, Pa·s; and H is the hydraulic head, m.

2.2. Initial Condition and Boundary Conditions

2.2.1. Initial Condition

Before the fluid flows into the pipe, the temperature of the pipe and formation is distributed as the initial formation temperature.

$$T|_{t=0} = T_f|_{t=0} = T_{ei} \tag{14}$$

where T_{ei} is the initial formation temperature, °C.

2.2.2. Boundary Conditions

The temperature of fluid in the pipe at the surface is the inlet temperature:

$$T|_{z=0} = T_{in} \tag{15}$$

where T_{in} is inlet temperature of fluid in the pipe, °C.

The formation temperature does not change along the horizontal at infinity, i.e.:

$$\left.\frac{dT_f}{dx}\right|_{x\to\infty} = \left.\frac{dT_f}{dy}\right|_{y\to\infty} = 0 \tag{16}$$

The formation at the surface and the bottom is adiabatic, then,

$$-\lambda_{eq}\left.\frac{dT_f}{dz}\right|_{z=0} = -\lambda_{eq}\left.\frac{dT_f}{dz}\right|_{z=L} = 0 \tag{17}$$

where, L is the depth of the formation, m.

2.3. Simplification of the Buried Pipe Model

As the length of the buried pipe often exceeds 100 m, and the radial dimension of the buried pipe is within 36 mm, conducting the length of the buried pipe is much larger than the radial dimension, which is not conducive to the division of the buried pipe mesh and the network is in poor quality.

The non-isothermal pipe flow module in COMSOL can simplify the problem of pipe flow with a finite diameter to the problem of fluid flow velocity, temperature, and pressure in the calculation curve, which is beneficial to mesh segmentation and greatly improves the efficiency of the pipe fluid simulation calculation.

The drilling hole was simplified into a very long cuboid with a long length of borehole depth and a square end with sides equal to the borehole diameter. The buried pipe was simplified as curves arranged along the four long sides of the cuboid. The buried pipe was modeled and the model was simplified as shown in Figure 2.

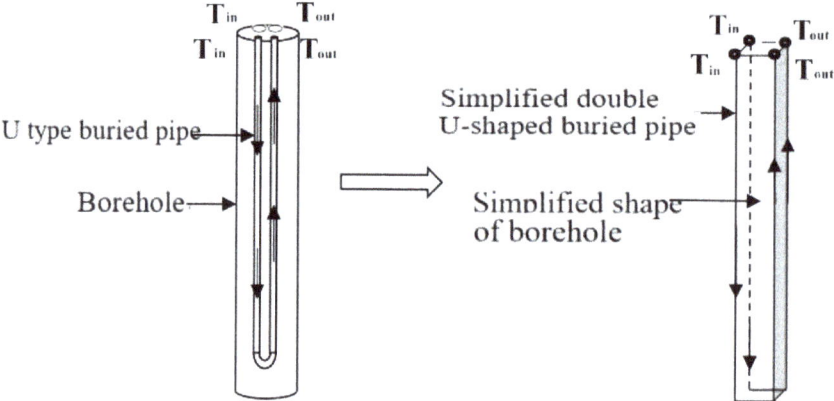

Figure 2. Borehole with U-pipe and its simplified model showing the relative position of the buried pipes.

2.4. Simplified Model Verification

Based on the heat-induced coupled heat transfer control equation in Sections 2.1 and 2.3, a three-dimensional numerical model of a full-scale buried tube thermal-permeability coupling was established by using the non-isothermal pipe flow module of COMSOL, the porous medium heat transfer module, and Darcy permeable module. The non-isothermal pipe flow module simulates the heat transfer of the circulating fluid in the buried pipe, and the porous medium and the groundwater permeable module simulate the heat-induced coupled heat transfer in the soil.

In order to verify the accuracy of the model, this paper first used the sandbox test data to verify the small size model. Then, a full-scale model verification was performed based on the on-site thermal test. The boundary conditions around the soil were set to adiabatic boundaries, the temperature values were the same as the initial values of the formation, and the top surface temperature of the model was set to 25 °C. The inlet of the buried pipe was set to a known flow velocity boundary, and the outlet was set to a constant pressure boundary. The following assumptions were made in the model: (1) The temperature around the tube is equal; (2) The velocity direction of the fluid in the buried tube is along the tube axis; and (3) the heat transfer in the tube wall is quasi-static.

2.4.1. Sandbox Test Verification

The sandbox test presented by Beier was used to verify the model above. The size of the sand box was 18.3 m × 1.8 m × 1.8 m, an aluminum pipe was regarded as the borehole wall where the inner diameter was 12.6 cm and the thickness was 0.2 cm. The length, inner radius, and outer radius of the U-tube pipe were 18.3 m, 2.733 cm, and 3.340 cm, respectively. The volumetric flow rate was 0.197 L/s. The thermal conductivities of the soil and grout were 2.82 W/(m·K) and 0.73 W/(m·K), respectively, and the sand porosity was 0.89, the specific heat capacity was 1900 J/(kg·K), and the density was 1200 kg/m^3 [38]. Based on the simplified buried pipe model in Section 2.2 and the sandbox size and test conditions, a numerical model was established. The numerical simulation of the sandbox test was carried out by using the non-isothermal pipe flow module, porous medium heat transfer module, and groundwater permeable module in COMSOL. The results are shown in Figure 3. The solid blue line is the simulated value of the outlet temperature, and the red hollow dot is the recorded value of the outlet temperature. It can be seen that the two agreed well, and the measured outlet temperature value differed from the simulated outlet temperature by a maximum of 0.2 °C, confirming the accuracy of the simplified model.

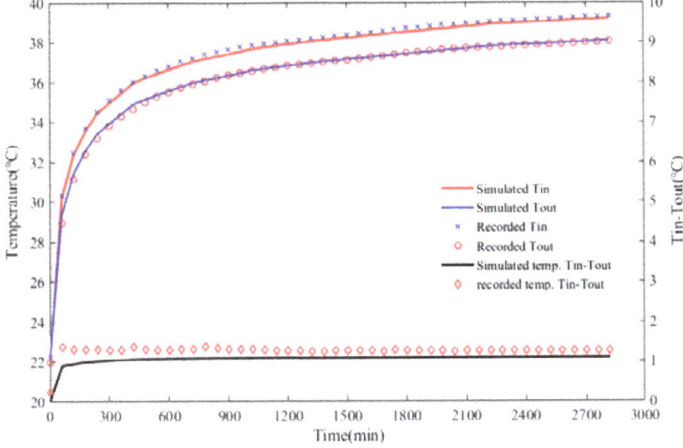

Figure 3. Sandbox test verification.

2.4.2. On-Site Thermal Response Test Verification

In order to further verify the simplified model to simulate the mesh length quality, calculation efficiency, and accuracy of the buried pipe with a large aspect ratio, the on-site thermal response test was simulated and verified. The outdoor test address was located in Baimiao Village, Tongzhou District, Beijing, with a total of nine drill holes. The initial temperature of the formation was first measured and then an on-site thermal response test was carried out. Twelve temperature measuring points were arranged in each of the #1 hole and the #2 hole, and the depth and temperature of the temperature measuring point in the hole were as shown in Table 1.

Table 1. Initial formation temperature.

Depth (m)	−5	−10	−15	−20	−30	−40
#1 temperature (°C)	17.11	14.25	14.81	14.87	14.50	14.80
#2 temperature (°C)	17.13	14.63	14.81	14.88	14.44	14.81
Depth (m)	−50	−60	−70	−80	−90	−100
#1 temperature (°C)	14.87	15.01	15.24	15.37	15.24	15.87
#2 temperature (°C)	14.88	15.06	15.25	15.38	15.19	15.88

Then, the on-site thermal response test was carried out on the #1 hole and the #2 hole, respectively. The #1 hole was used for the constant temperature condition. The inlet and outlet temperature data of the buried pipe was used to measure the thermal property parameters of the formation; the #2 hole was used for the constant power condition and buried. The tube inlet and outlet temperature data were used to validate the simplified buried tube model. To obtain a more accuracy thermal conductivity, two situations where the inlet temperature was 30 °C and 35 °C were adopted during the test. The on-site thermal response test is shown in Figure 4.

Figure 4. Thermal response tester.

Based on the on-site thermal response test data, a numerical model of the full-scale three-dimensional thermal response test was established based on the simplified buried pipe physical model. The parameters in the model were the same as the on-site thermal response test parameters, as shown in Table 2.

Table 2. Thermal response test and formation thermal property parameters.

Name	Unit	Number
The depth of the hole	m	102
The depth of ground pipe	m	100
The diameter of Drilling	mm	150
The outer diameter of buried tube	mm	32
The inner diameter of buried tube	mm	26
U-tube spacing	mm	100
The flow rate of circulating fluid	m^3/h	1.1
Effective thermal conductivity	W/(m·K)	1.86
Formation specific heat capacity	kJ/(kg·K)	2.80
U-tube thermal conductivity	W/(m·K)	0.44
The roughness of wall surface	mm	0.0015
The conductivity of backfill material thermal	W/(m·K)	2.3
The capacity of backfill material specific heat capacity	kJ/(kg·K)	0.84

The bottom plane of the borehole was divided according to the free-distributed triangle mesh, and then the borehole mesh was generated by sweeping the bottom surface of the source's surface along the drill depth direction. The total number of mesh was 321,157, the minimum unit mass was 0.28, the average unit mass was 0.83, the minimum unit size was 0.05 m, the maximum unit size was 0.44 m, and the time step was 5 min. The thermal response test simulation results are shown in Figure 5.

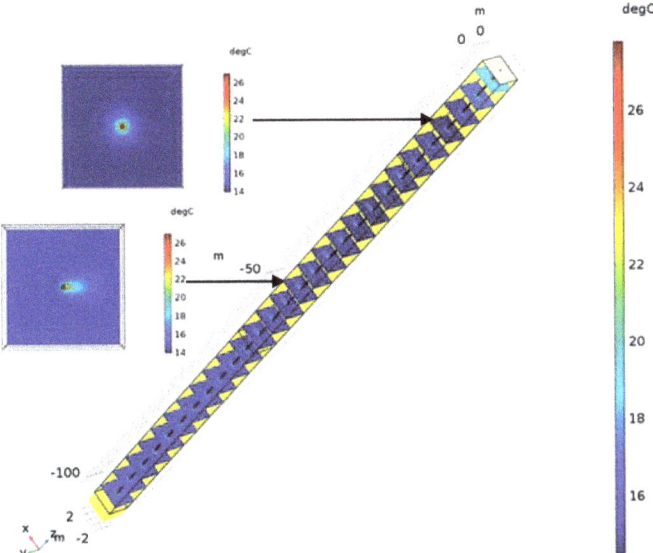

Figure 5. The results of the on-site thermal response test simulation.

We compared the measured outlet temperature value of the buried pipe with the outlet temperature calculated by the numerical model, as shown in Figure 6. The solid line in red in the figure is the simulated outlet temperature, and the purple solid line is the recorded outlet temperature in the thermal response experiment. The outlet temperature calculated by the numerical model was basically consistent with the outlet temperature recorded in the field, and the accuracy of the established three-dimensional numerical model of the buried tube thermal-osmotic coupling was verified. The simplified buried pipe physical model greatly reduced the total number of meshes on the basis of ensuring the quality of meshing. In order to establish a thermal-osmotic coupling model of the buried pipe group in accordance with the actual size under limited computing resources, the foundation was laid.

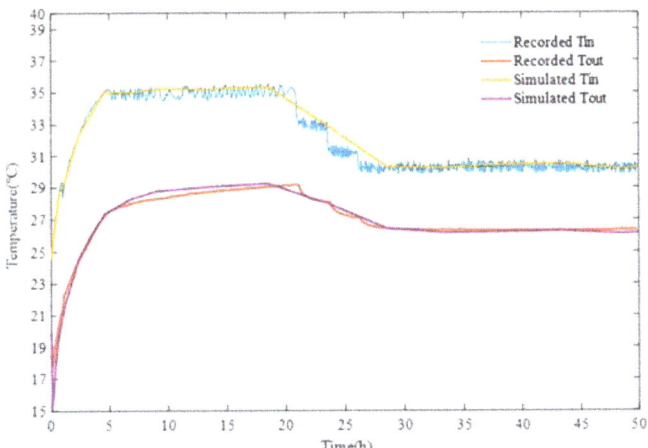

Figure 6. On-site thermal response test verification.

3. Three-Dimensional Numerical Simulation of Influence of Thermal Permeable Coupling of Ground Tube Group on Geothermal Field

Based on the control equation in Section 2.1 of the ground buried tube thermal permeable coupling model and the simplified model of the buried pipe in Section 2.3, the geothermal field coupled with the thermal permeable coupling of the buried pipe group was simulated by the misaligned tube group model.

According to the previous research results, in the case of groundwater permeability, the heat transfer capacity of the misaligned tube group is higher than that of the smoothing tube group, and the permeable effect of the groundwater makes it more obvious. In this paper, a thermal-permeability coupling model of the buried tube group with misaligned arrangement was established. The model length × width × depth = 21 m × 21 m × 102 m (see Figure 7a). The tube group consisted of nine buried pipes. The buried pipe was buried at a depth of 100 meters and the buried pipes were arranged at a distance of 4.5 m (see Figure 7b). The buried pipe passed through three layers, where the first layer was clay, and the second layer was a water-bearing homogeneous pore sandstone. The water flow direction flowed from the hole 1 side to the hole 7 side (Figure 7b), and the third layer was clay.

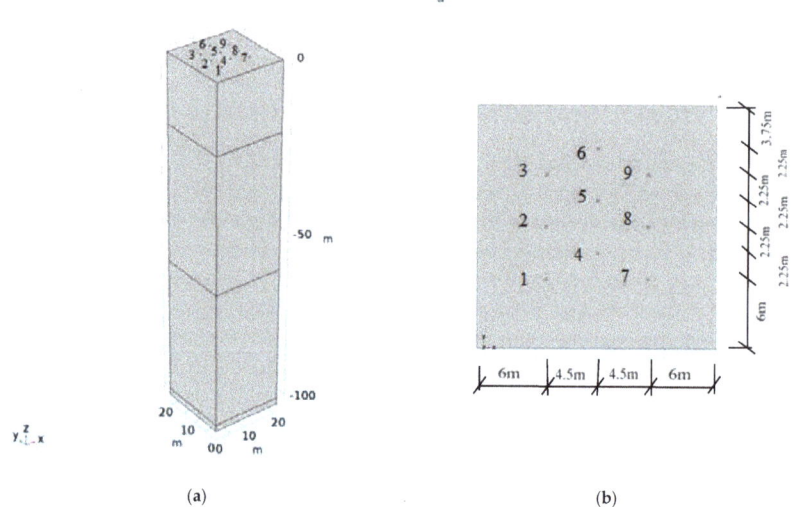

Figure 7. The model of buried pipe group physical model: (**a**) three dimensions and (**b**) cross section.

The initial values of the formation temperature of the buried pipe group thermal permeable coupling model are shown in Table 1. The drilling dimensions and thermal properties of the formation are shown in Table 2.

According to Equations (5) and (6), the groundwater permeable velocity can be calculated by the hydraulic conductivity, k, and the water head, ∇H. The simulation system ran for 360 days, where the 0–120 days circulating fluid flow was 0.16 L/s, the inlet temperature was set to 30 °C, and the ground source heat pump cooling condition was simulated; the 120–240 days circulating fluid flow rate was 0, simulating the shutdown condition; and for 240–360 days, the circulating fluid flow rate was 0.16 L/s and the inlet temperature was set to 6 °C to simulate the heating conditions. The size and thermal property parameters of the formation, grout, and U-tube were the same as the thermal response test in Section 2.4.2.

Discretize the model and the total mesh generation was 850,000. In the borehole domain, it had a minimum mesh size of 0.02 m and a maximum mesh size of 0.15 m. In the formation domain, the mesh sizes were larger than those in the borehole domain, where the minimum mesh size was 0.2 m and the maximum mesh size was 2.5 m. The time step was one day.

4. Results and Discussion

The head difference $\Delta H = 0.5$ m in the second aquifer, at a different hydraulic conductivity condition of $k = 5 \times \frac{10^{-6} m}{s}$, $k = 5 \times \frac{10^{-5} m}{s}$, $k = 5 \times 10^{-4} m/s$, after which the simulation calculated the heat removal to the formation for 120 days, stopped for 120 days, and took heat from the formation for 120 days, where the system continuously ran the temperature field in the 360 day formation. Additionally, the effect of the hydraulic conductivity on the temperature field around the buried pipe was studied.

When $k = 5 \times 10^{-6} m/s$, the 120 days, 240 days, and 360 days' formation temperature fields are as shown as Figure 8a–c.

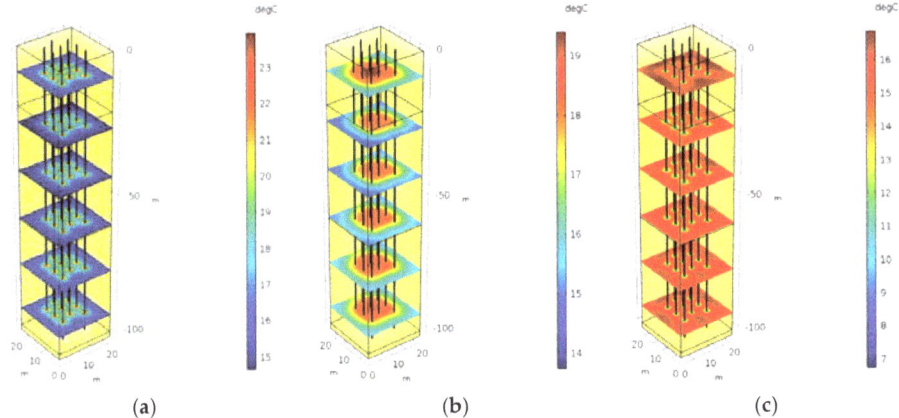

Figure 8. When $k = 5 \times 10^{-6} m/s$, the formation temperature field at different times: (**a**) 120 days; (**b**) 240 days; (**c**) 360 days.

When $k = 5 \times 10^{-5} m/s$, the 120 days, 240 days and 360 days' formation temperature fields are shown in Figure 9a–c.

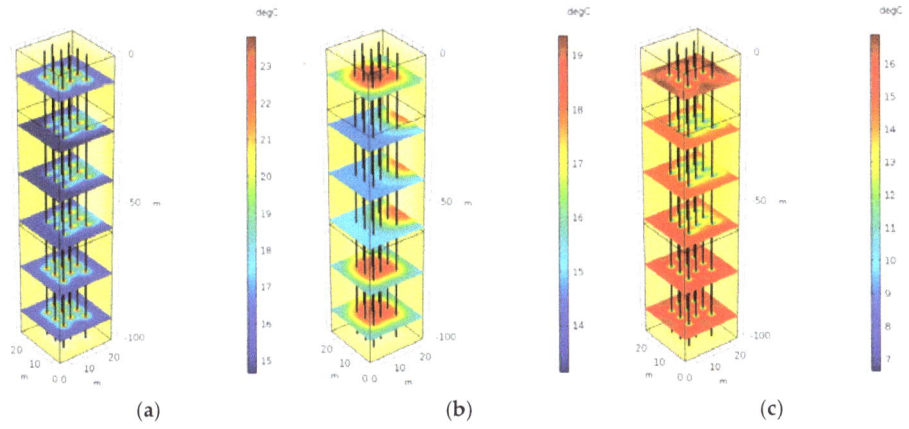

Figure 9. When $k = 5 \times 10^{-5} m/s$, the formation temperature field at different times: (**a**) 120 day; (**b**) 240 days; (**c**) 360 days.

When $k = 5 \times 10^{-4} m/s$, the 120 days, 240 days and 360 days' formation temperature fields are as shown in Figure 10a–c.

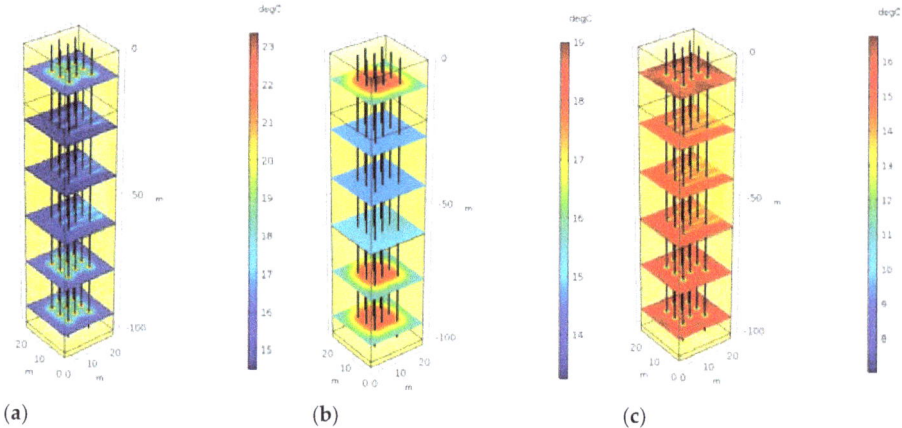

Figure 10. When $k = 5 \times 10^{-4} m/s$, formation temperature field at different times: (**a**) 120 days; (**b**) 240 days; (**c**) 360 days.

Comparing and analyzing Figures 8–10, we observed the following. (1) When the hydraulic conductivity was $k = 5 \times 10^{-6} m/s$, the temperature field in the permeable stratum was symmetrically distributed around the borehole. The heat discharged after shutdown was mainly concentrated around the buried pipe, which is beneficial to the formation heat storage. (2) When $k = 5 \times 10^{-5} m/s$, the temperature field around the buried pipe in the permeable stratum was offset, and it was distributed in an elliptical shape around the buried pipe. The influence between the buried pipes was small. The heat was transferred to the outside of 9 m with the permeable groundwater. (3) When $k = 5 \times 10^{-4} m/s$, the temperature around the buried pipe in the permeable stratum was mainly concentrated in the downstream of the buried pipe, and the heat was transferred to 21 m after the shutdown.

Under different hydraulic conductivity, the temperature value of the connection between the #2 hole and the #8 hole (see Figure 6b) of the 40-meter permeable stratum in the depth of the borehole pipe system after 120 days, 240 days, and 360 days operation was studied and Figure 11 was obtained.

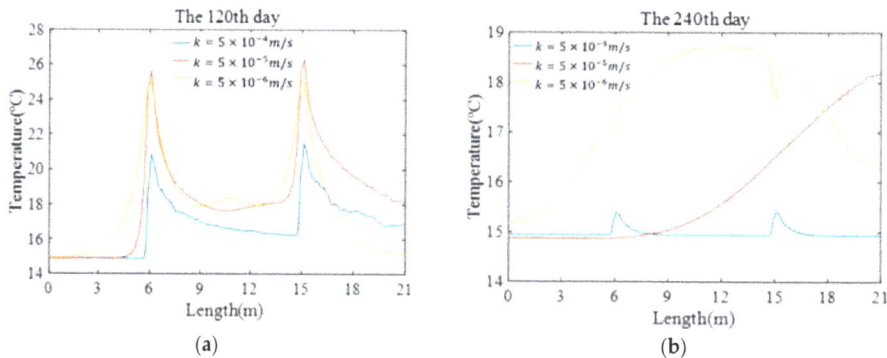

Figure 11. *Cont.*

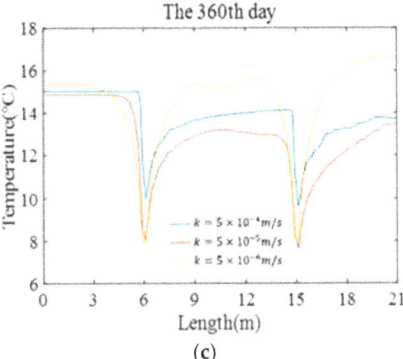

(c)

Figure 11. Temperature at the junction of #2 and #8 holes in the −40 m section at different times: (a) 120 days; (b) 240 days; (c) 360 days.

Comparing the analysis of Figure 11a–c, it can be concluded that the misalignment can reduce the mutual influence between the buried pipes. When the hydraulic conductivity is $k = 5 \times 10^{-4} m/s$, the temperature outside the buried pipe wall is significantly reduced under the influence of permeable groundwater, and the temperature in the formation returns to the original formation temperature after shutdown.

When the head difference VH is 0.5 m and the hydraulic conductivity k is $5 \times 10^{-6} m/s$, $5 \times 10^{-5} m/s$, and $5 \times 10^{-4} m/s$, respectively, the outlet temperature value of the buried pipe located at the middle of the buried pipe group and Figure 12 can be obtained. When the hydraulic conductivity was $k = 5 \times 10^{-4} m/s$, regardless of the heat exhaust condition or the heat extraction condition, the outlet temperature of the buried pipe was more than 1.8 °C lower than the hydraulic conductivity $k = 5 \times 10^{-5} m/s$, and the permeable groundwater significantly provided the heat exchange efficiency of the buried pipe.

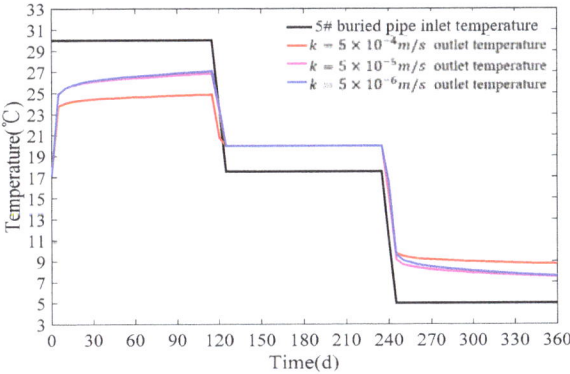

Figure 12. Effect of hydraulic conductivity on outlet temperature.

5. Conclusions

In this paper, a new simplified model of buried pipe was proposed. Based on the finite element numerical simulation software COMSOL and the simplified buried pipe model, the same numerical model as the sandbox test and the on-site thermal response test was established using the sandbox test data and the on-site thermal response test data. The results of simulation showed that the simplified ground tube model simulation results agreed well with the experimental data, and the simplified buried tube model improved the meshing efficiency and significantly improved the simulation efficiency.

Based on the simplified buried pipe model, the temperature field in the formation after one year of thermal-permeability coupled heat transfer operation of the buried pipe group under different hydraulic conductivity conditions was studied, and the temperature field distribution of the system under different operating conditions was obtained. By analyzing and comparing the formation temperature field, the influence of groundwater can be neglected only when the head difference is less than 0.5 m and the hydraulic conductivity is less than $k = 5 \times 10^{-6}$ m/s. When the head difference is more than 0.5 m and the hydraulic conductivity is more than $k = 5 \times 10^{-5}$ m/s, the influence of permeable groundwater can effectively avoid the heat accumulation in the formation. The misplaced arrangement of the buried pipes can reduce the mutual influence between the buried pipes. The distribution of temperature fields in the different strata was different, which is closely related to the thermal properties of the formation and the flow of groundwater. When the hydraulic conductivity is more than $k = 5 \times 10^{-4}$ m/s, the flow of groundwater can significantly improve the heat exchange efficiency of the buried pipe.

Author Contributions: Conceptualization, X.Z. and X.L.; Methodology, X.L. and C.D.; Software X.L. and J.Y.; Validation, X.Z; Formal Analysis, X.L.; Investigation, X.L. and C.D.; Resources, X.Z; Data Curation, J.Y.; Writing-Original Draft Preparation, X.L.; Writing-Review & Editing, X.Z., C.D. and K.L.; Visualization, X.L.; Supervision, X.Z.; Project Administration, X.Z.; Funding Acquisition, X.Z.

Funding: This research was funded by National Natural Science Foundation of China, Grant No. 41572361, Grant No. 41872184 and Science and technology department of Guizhou province, China, Grant No. QianKehe SY [2015] 3059.

Acknowledgments: The author warmly thanks to Tongren Kowloon Including Geological Mining Investment and Development Co. Ltd. and Beijing Taili new energy technology development co. LTD for supporting and providing all necessary resources to complete this research.

Conflicts of Interest: The authors declare no conflict of interest.

References

1. Lund, J.W.; Sanner, B.; Rybach, L.; Curtis, R.; Hellstrom, G. Geothermal (groundsource) heat pumps. A world overview. *GHC Bulletin* **2004**, *25*, 1–10.
2. Bernier, M. Closed-loop ground-coupled heat pump systems. *ASHRAE J.* **2006**, *48*, 13–24.
3. Nguyen, H.V.; Law, Y.L.E.; Alavy, M.; Walsh, P.R.; Leong, W.H.; Dworkin, S.B. An analysis of the factors affecting hybrid ground-source heat pump installation potential in North America. *Appl. Energy* **2014**, *125*, 28–38. [CrossRef]
4. Ingersoll, L.R.; Zobel, O.J.; Ingersoll, A.C. *Heat Conduction with Engineering, Geological, and Other Applications, Revised ed.*; The University of Wisconsin Press: Madison, WI, USA, 1954.
5. Li, M.; Lai, A.C.K. Review of analytical models for heat transfer by vertical ground heat exchangers (GHEs): A perspective of time and space scales. *Appl. Energy* **2015**, *151*, 178–191. [CrossRef]
6. Witte, H.J.L. Geothermal response test with heat extraction and heat injection: Examples of application in research and design of geothermal ground heat exchangers. In *Europaischer Workshop uber Geothermische Response Tests*; Ecole Poly Technique Federal de Lausanne: Lausanne, Switzerland, 25–26 October 2001.
7. Santa, G.D.; Galgaro, A.; Tateo, F.; Cola, S. Modified compressibility of cohesive sediments induced by thermal anomalies due to a borehole heat exchanger. *Eng. Geol.* **2016**, *202*, 143–152. [CrossRef]
8. Eskilson, P. Thermal Analysis of Heat Extraction Boreholes. Ph.D. Thesis, Lund University, Lund, Sweden, 1987.
9. Capozza, A.; De Carli, M.; Zarrella, A. Design of borehole heat exchangers for ground-source heat pumps: A literature review, methodology comparison and analysis on the penalty temperature. *Energy Build.* **2012**, *55*, 369–379. [CrossRef]
10. Raymond, J.; Therrien, R.; Gosselin, L.; Lefebvre, R. A review of thermal response test analysis using pumping test concepts. *Groundwater* **2011**, *49*, 932–945. [CrossRef] [PubMed]
11. Zhang, C.X.; Guo, Z.J.; Liu, Y.F.; Cong, X.C.; Peng, D.G. A review on thermal response test of ground-coupled heat pump systems. *Renew. Sustain. Energy Rev.* **2014**, *40*, 851–867. [CrossRef]
12. Lamarche, L.; Kajl, S.; Beauchamp, B. A review of methods to evaluate borehole thermal resistances in geothermal heat-pump systems. *Geothermics* **2010**, *39*, 187–200. [CrossRef]

13. Yang, H.X.; Cui, P.; Fang, Z.H. Vertical-borehole ground-coupled heat pumps: A review of models and systems. *Appl. Energy* **2010**, *87*, 16–27. [CrossRef]
14. Wang, F.H.; Yu, Bin.; Yan, L. Heat transfer analysis of groundwater flow for multi-pipe heat exchanger of ground source heat pump. *CIHESC J.* **2010**, *61*, 57–62.
15. Wang, H.; Qi, C.; Du, H.; Gu, J. Thermal performance of borehole heat exchanger under groundwater flow: A case study from Baoding. *Energy Build.* **2009**, *41*, 1368–1373. [CrossRef]
16. Chiasson, A.D.; Rees, S.J.; Spitler, J.D. A preliminary assessment of the effects of groundwater flow on closed-loop ground-source heat pump systems. *ASHRAE Trans.* **2000**, *106*, 380–393.
17. Fan, R.; Jiang, Y.; Yao, Y.; Deng, S.; Ma, Z. A study on the performance of a geothermal heat exchanger under coupled heat conduction and ground-water advection. *Energy* **2007**, *32*, 2199–2209. [CrossRef]
18. Lu, G. Analysis of effect of groundwater seepage on heat transfer in under ground heat exchanger. *Fujian Construct. Sci. Technol.* **2011**, *1*, 62–63, 74.
19. Luo, J.; Tuo, J.; Huang, W.; Zhu, Y.; Jiao, Y.; Xiang, W.; Rohn, J. Influence of groundwater levels on effective thermal conductivity of the ground and heat transfer rate of borehole heat exchangers. *Appl. Therm. Eng.* **2018**, *128*, 508–516. [CrossRef]
20. Spitler, J.D.; Gehlin, S.E.A. Thermal response testing for ground source heat pump systems – An historical review. *Renew. Sustain. Energy Rev.* **2015**, *50*, 1125–1137. [CrossRef]
21. Diao, N.R.; Li, Q.Y.; Fang, Z.H. An analytical solution of the temperature response in geothermal heat exchangers with groundwater advection. *J. Shandong Univ. Archit. Eng.* **2003**, *18*, 1–5.
22. Diao, N.; Li, Q.; Fang, Z. Heat transfer in ground heat exchangers with groundwater advection. *Int. J. Therm. Sci.* **2004**, *43*, 1203–1211. [CrossRef]
23. Molina-Giraldo, N.; Bayer, P.; Blum, P. Evaluating the influence of thermaldispersion on temperature plumes from geothermal systems using analytical solutions. *Int. J. Therm. Sci.* **2011**, *50*, 1223–1231. [CrossRef]
24. Wagner, V.; Blum, P.; Kubert, M.; Bayer, P. Analytical approach for groundwaterinfluencedthermal response tests of grouted borehole heat exchangers. *Geothermics* **2013**, *46*, 22–31. [CrossRef]
25. Wagner, V.; Bayer, P.; Bisch, G.; Kubert, M.; Blum, P. Hydraulic characterizationof aquifers by thermal response testing: Validation by large scale tank laboratoryand field experiments. *Water Resour. Res.* **2014**, *50*, 1–15. [CrossRef]
26. Aranzabal, N.; Martos, J.; Montero, Á.; Monreal, L.; Soret, J.; Torres, J.; García-Olcina, R. Extraction of thermal characteristics of surrounding geological layers of a geothermal heat exchanger by 3D numerical simulations. *Appl. Therm. Eng.* **2016**, *99*, 92–102. [CrossRef]
27. Guan, X. China's ground source heat pump development and utilization prospects. *China Sci. Technol. Inf.* **2010**, *21*, 21–23.
28. Yavuzturk, G. *Modeling of Vertical Ground Loop Heat Exchangers for Ground Source Heat Pump Systems*; Oklahoma State University: Stillwater, OK, USA, 1999.
29. Li, X.G.; Zhao, J.; Zhou, Q. Numerical simulation on the ground temperature field around U pipe underground heat exchanger. *J. Acta Energ. Sol. Sin.* **2004**, *25*, 703–707.
30. Gao, Q.; Li, M.; Yan, Y. Effect on ground source field heat transfer of original earth temperature and configuration of multi-borehole. *J. Therm. Sci. Technol.* **2005**, *4*, 34–40.
31. Zhao, J.; Wang, H.J. Numerical simulation on heat transfer characteristics of soil around compact pile-buried underground heat exchangers. *J. HV AC* **2006**, *36*, 11–14.
32. Jia, Y.B.; Wang, Y. Comparative analyses on 2D and 3D buried tube-group heat transfer models in GCHP system. *J. Grad. Univ. Chin. Acad. Sci.* **2013**, *30*, 311–316.
33. Yang, G.J.; Guo, M. Research on group arrangement method of ground heat exchanger under coupled thermal conduction and groundwater seepage conditions. *Renew. Energy Resour.* **2014**, *32*, 1182–1187.
34. Li, P.; Li, M.; Ma, W.; Xiao, L. Temperature responses of ground heat exchanger clusters based on full-scale line-source solution. *J. Central South Univ.* **2017**, *48*, 2818–2822.
35. Han, C.J.; Yu, X. Sensitivity analysis of a vertical geothermal heat pump system. *Appl. Energy* **2016**, *170*, 148–160. [CrossRef]
36. Oberdorfer, P.; Hu, R.; Rahman, M.; Holzbecher, E.; Sauter, M.; Mercker, O.; Pärisch, P. *Coupled Heat Transfer in Borehole Heat Exchangers and Long Time Predictions of Solar Rechargeable Geothermal Systems*; COMSOL: Milan, Italy, 2013.

37. Yu, Y.; Liu, G.; Jiang, B. Finite element analysis of ground source heat pump based on non-isothermal pipe flow. *J. Zhejiang Univ. Sci. Technol.* **2015**, *27*, 218–224.
38. Beier, R.A.; Smith, M.D.; Spitler, J.D. Reference data sets for vertical borehole ground heat exchanger models and thermal response test analysis. *Geothermics* **2011**, *40*, 79–85. [CrossRef]

© 2019 by the authors. Licensee MDPI, Basel, Switzerland. This article is an open access article distributed under the terms and conditions of the Creative Commons Attribution (CC BY) license (http://creativecommons.org/licenses/by/4.0/).

Article

Numerical Study of Highly Viscous Fluid Sloshing in the Real-Scale Membrane-Type Tank

Shuo Mi [1], Zongliu Huang [2], Xin Jin [3,*], Mahdi Tabatabaei Malazi [1] and Mingming Liu [3]

[1] State Key Laboratory of Hydraulics and Mountain River Engineering, Sichuan University, Chengdu 610065, China; 2017223060039@stu.scu.edu.cn (S.M.); m.tabatabaei.malazi@scu.edu.cn (M.T.M.)
[2] Key Laboratory of Fluid and Power Machinery, Ministry of Education, Xihua University, Chengdu 610039, China; hzl27@mail.xhu.edu.cn
[3] College of Energy, Chengdu University of Technology, Chengdu 610059, China; liumingming19@cdut.edu.cn
* Correspondence: jinx@cdut.edu.cn; Tel.: +86-158-2800-6890

Received: 20 September 2019; Accepted: 4 November 2019; Published: 7 November 2019

Abstract: The highly viscous liquid (glycerin) sloshing is investigated numerically in this study. The full-scale membrane-type tank is considered. The numerical investigation is performed by applying a two-phase numerical model based on the spatially averaged Navier-Stokes equations. Firstly, the numerical model is validated against the available numerical model and a self-conducted experiment then is applied to systematically investigate the full-scale sloshing. In this study, two filling levels (50% and 70% of the tank height) are considered. The fluid kinematic viscosity is fixed at a value being 6.0×10^{-5} m^2/s with comparative value to that of the crude oil. A wide range of forcing periods varying from 8.0 s to 12.0 s are used to identify the response process of pressures as well as free surface displacements. The pressures are analyzed along with breaking free surface snapshots and corresponding pressure distributions. The slamming effects are also demonstrated. Finally, the frequency response is further identified by the fast Fourier transformation technology.

Keywords: sloshing; real-scale; highly viscous fluids; Navier-Stokes equations; impact pressure

1. Introduction

Liquid sloshing in partially filled LNG (Liquefied Natural Gas)/crude-oil carriers may occur under different sea conditions. The highly nonlinear phenomenon can produce localized high impact loads on tank walls, potentially leading to structural damage. Hence, it is necessary to investigate the sloshing phenomena and associated structural behavior in the design of tanks.

For the analysis of the sloshing phenomenon, the model test can be undertaken under different sea conditions and filling levels. However, there exists uncertainty when the measured impact load from the model test is scaled up to the real size [1], and the liquid viscosity also has important effects on sloshing pressure [2–4]. The analytical works are mainly divided into two scenarios, one the linear potential flow theory including ideal and quasi-viscous liquid sloshing [5], and the other is nonlinear potential flow theory [6–8]. The real-viscous sloshing was also considered by Wu et al. [9] by neglecting the nonlinear advection terms, and the linear sloshing was obtained. However, the nonlinear viscous sloshing remains to be well solved.

Besides the model tests and analytical analysis, numerical methods are alternative tools in understanding the relevant sloshing problems. Rudman et al. used the smoothed particle hydrodynamics (SPH) to simulate sloshing in a 2-D water model that is a representation of a scaled LNG tank [10]. Luo et al. used the Consistent Particle Method (CPM) to study water sloshing in a scaled LNG tank under translational excitation [11]. Zhao et al. [12] investigated the motion induced by 3D sloshing in a partially-filled LNG tank using a new coupled Level-Set and Volume-of-Fluid (CLSVOF) method incorporated into Finite-Analytic Navier–Stokes (FANS) method. Kim et al. [13]

developed a three-dimensional finite-element method to calculate the impact pressure due to liquid sloshing in the LNG tank. The effects of liquid viscosity on sloshing were obtained by Zou et al. [3], and the results also revealed that the boundary layer had significant influences on the response pressures. Xin and Lin [14] adopted the spatially averaged Navier-Stokes turbulence model to study the viscous effects on horizontally and multi-degree freedom excited sloshing, and the threshold of liquid viscosity that the response regularity shifts were obtained numerically.

Although there are many studies on viscous effects, the full-scale sloshing is rarely reported. Considering this situation, the present work focuses on the full-scale sloshing of highly viscous fluids. The 3D numerical model called NEWTANK developed by Liu and Lin [15] will be used to perform numerical investigations, and the full-scale prismatic tank is chosen. The two-phase fluid flow model solves the spatially averaged Navier-Stokes equations. The second-order accurate Volume-of-Fluid (VOF) method is used to track the distorted and broken free surface. The large-eddy simulation (LES) is used for turbulence modeling. The numerical validations of the sloshing especially the highly viscous fluids will be carried out with a self-conducted experiment test and available numerical data, and then the sloshing of the real-scale membrane-type tank was studied. Numerical experiments will be conducted for two different filling levels (50% and 70% of the tank height). The kinematic viscosity of the liquid is selected from the crude oil and its value is 6.0×10^{-5} m^2/s, which is 60 times the water. A wide range of forcing periods varying from 8.0 s to 12.0 s are used to identify the response process of pressures as well as free surface displacements. The free surface displacements and pressures are analyzed with the nonlinearity being further discussed. The frequency responses are also identified by the fast Fourier transformation technology.

2. Numerical Methodology

In the present model, the motion of an incompressible fluid is described by Navier–Stokes equations in which turbulence is modeled LES (Large Eddy Simulation) method. The non-inertial reference frame that follows the tank motion is adopted to avoid the moving complicated boundary. The Navier-Stokes equations in this study are summarized below:

$$\frac{\partial \bar{u}_i}{\partial x_i} = 0 \tag{1}$$

$$\frac{\partial \bar{u}_i}{\partial t} + \frac{\partial \bar{u}_i \bar{u}_j}{\partial x_j} = -\frac{1}{\rho}\frac{\partial \bar{p}}{\partial x_i} + f_i + \frac{1}{\rho}\frac{\partial \bar{\tau}_{ij}}{\partial x_j} + \frac{1}{\rho}\frac{\partial \tau^r_{ij}}{\partial x_j} \tag{2}$$

where $\bar{u}_i$ is the spatially averaged flow velocity in i direction, $\bar{p}$ is the effective pressure and ρ is the liquid density, f_i is the i-th component of the external acceleration.

The LES method is employed to capture turbulence transport and dissipation in this model. After being filtered by the spatial filter top-hat function, the sub-grid stress terms appear in the momentum equations, which can be modeled by the Smagorinsky sub-grid scale model, where $\bar{\tau}_{ij} = 2\rho \nu \bar{\sigma}_{ij}$ is the molecular viscous stress tensor with ν being the kinematic viscosity and $\tau^r_{ij} = 2\rho \nu_t \bar{\sigma}_{ij}$ is SGS Reynolds stress tensor. In the above definition, $\bar{\sigma}_{ij} = \frac{1}{2}\left(\frac{\partial \bar{u}_i}{\partial x_j} + \frac{\partial \bar{u}_j}{\partial x_i}\right)$ is the rate of strain of the filtered flow. ν_t represents the eddy viscosity and is modeled as:

$$\nu_t = l_s^2 \sqrt{(2\bar{\sigma}_{ij}\bar{\sigma}_{ij})} \tag{3}$$

where l_s is the characteristic length scale which equals $C_s \Delta$ with $C_s = 0.5$ [16] and Δ is written as:

$$\Delta = \sqrt[3]{\Delta x \Delta y \Delta z} \tag{4}$$

where Δx, Δy and Δz are the grid lengths in the three directions.

The external force f_i includes the gravitational acceleration, translational and rotational inertia forces, whose expression can be found in Liu and Lin [15] and Liu [17], and the components are given as follows:

$$f_x = g_x - \frac{du}{dt} - \left[\frac{d\Omega_y}{dt}(z-z_0) - \frac{d\Omega_z}{dt}(y-y_0)\right] - \{\Omega_y[\Omega_x(y-y_0) - \Omega_y(x-x_0)] \\ - \Omega_z[\Omega_z(x-x_0) - \Omega_x(z-z_0)]\} - \left(2\Omega_y\frac{d(z-z_0)}{dt} - 2\Omega_z\frac{d(y-y_0)}{dt}\right) \quad (5)$$

$$f_y = g_y - \frac{dv}{dt} - \left[\frac{d\Omega_z}{dt}(x-x_0) - \frac{d\Omega_x}{dt}(z-z_0)\right] - \{\Omega_z[\Omega_y(z-z_0) - \Omega_z(y-y_0)] \\ - \Omega_x[\Omega_x(y-y_0) - \Omega_y(x-x_0)]\} - \left(2\Omega_z\frac{d(x-x_0)}{dt} - 2\Omega_x\frac{d(z-z_0)}{dt}\right) \quad (6)$$

$$f_z = g_z - \frac{dw}{dt} - \left[\frac{d\Omega_x}{dt}(y-y_0) - \frac{d\Omega_y}{dt}(x-x_0)\right] - \{\Omega_x[\Omega_z(x-x_0) - \Omega_x(z-z_0)] \\ - \Omega_y[\Omega_y(z-z_0) - \Omega_z(y-y_0)]\} - \left(2\Omega_x\frac{d(y-y_0)}{dt} - 2\Omega_y\frac{d(x-x_0)}{dt}\right) \quad (7)$$

where $\vec{g} = \vec{g}(g_x, g_y, g_z)$, $\vec{U} = \vec{U}(u, v, w)$ and $\vec{\Omega} = \vec{\Omega}(\Omega_x, \Omega_y, \Omega_z)$ are gravitational vector, translational velocity, and rotational velocity vector, respectively. $\vec{r}$ and $\vec{R}$ are the position vector of the considered point and the rotational motion origin, respectively.

The above governing equations are solved by the two-step projection method originally proposed by Chorin [18]. The free surface is tracked by the second-order accurate Volume-Of-Fluid (VOF) method [19]. A combination of the upwind scheme and the central difference scheme is adopted in the discretization of the convection terms. A second-order central difference scheme is used for the diffusion terms. Readers are referred to [15] for more details of numerical implementation.

3. Model Validations

3.1. Validation of Sloshing of Glycerin under Surge Excitation

The low viscous liquid has been extensively investigated. Higher viscosity has also attracted much attention. To verify the accuracy of the present numerical model in handling the sloshing of highly viscous liquid, the experimental model test is performed in the Laboratory of Vibration Test and Liquid Sloshing at Hohai University of China [20] and Glycerin at 35 °C in degrees Celsius unit is chosen as the experimental liquid. The corresponding physical properties-viscosity and density are 0.000408 m²/s and 1252.2 kg/m³, respectively. The experimental tank is rectangular with dimensions 0.6 m in length, 0.3 m in width and 0.6 m in height. In the numerical simulation, the computational domain 0.6 m × 0.3 m × 0.35m is discretized into 64 × 40 × 76 non-uniform grids with a minimal grid size of 0.0025m near the bottom and walls. The no-slip velocity boundary condition is adopted accordingly. The displacement of the tank follows the harmonic function $s = A\cos(\omega_h t)$, where $A = 0.03\ m$ and $\omega_h = \omega_{1,0} = 5.3483\ rad/s$ according to the dispersion relationship:

$$\omega_{mn}^2 = \sqrt{(\frac{m g \pi}{L})^2 + (\frac{n g \pi}{W})^2} \tanh\sqrt{(\frac{m \pi h}{L})^2 + (\frac{n \pi h}{W})^2} \quad (m, n = 0, 1, 2 \cdots) \quad (8)$$

where L and W are tank length and tank width, respectively.

The comparisons of dynamic pressures between the numerical result and experimental data at the left wall 0.03 m and 0.07 m away from the bottom are shown in Figure 1. The numerical results overestimate the experimental data a little in the crests. However, there exists no phase shift between the two results. The difference may due to the present numerical results neglect the unevenness of the tank walls. Overall, the difference between the two results is negligible.

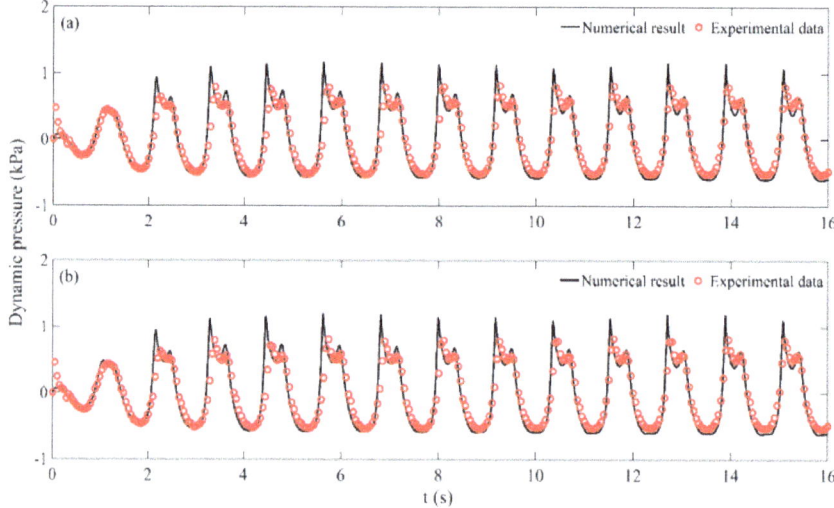

Figure 1. Comparisons of time histories of the dynamic pressures between the present numerical result (solid line) and experimental data (circle): (**a**) 3 cm away from the bottom, (**b**) 7 cm away from the bottom.

Besides the dynamic pressures, the free surface snapshots of the numerical results at 2.5 s, 5.0 s, 10.0 s, and 15.0 s are further demonstrated in Figure 2. The wavelength is twice the tank length. The free surfaces of glycerin break significantly under the excitation. Due to strong viscous effect, the liquid has adhered to side walls, obvious 3D features were guaranteed although the excitation was only one dimension.

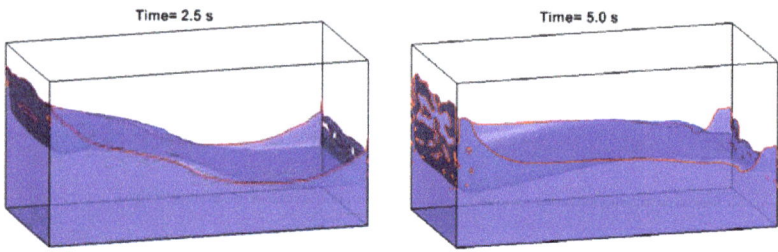

Figure 2. *Cont.*

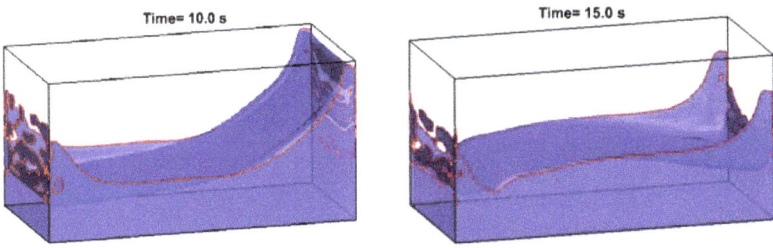

Figure 2. Numerical snapshots of the free surfaces at 2.5 s, 5.0 s, 10.0 s, and 15.0 s.

3.2. Validation of Membrane-Type Tank Sloshing under Roll Excitation

Except for the prismatic tank, the membrane-type tank is more preferred in engineers. Luo et al. [11] did experiments to study the sloshing in a membrane-type tank which was scaled down from the real scale [21]. The scale tank and its corresponding dimensions are shown in Figure 3, which P1 is the pressure measurement point. The filling level 50% of the tank height was considered by Luo et al. [11], and the forcing frequency of 6.618 rad/s was adopted. The tank displacement followed sinusoidal function: $s = A(t)\sin(\omega_h t + \pi)$, where $A(t)$ and ω_h were the amplitude and forcing frequency, respectively. In this case, $A(t)$ linearly increases in the first 10 s and finally reaches 0.005 m. In the numerical simulations, the 3D computation domain is discretized as 80 × 40 × 60 uniform grids, with $\Delta x = 0.007$ m, $\Delta y = 0.001$ m, and $\Delta z = 0.0066$ m. The time step is automatically adjusted through CFL (Courant-Friedrichs-lewy) conditions to ensure numerical stability.

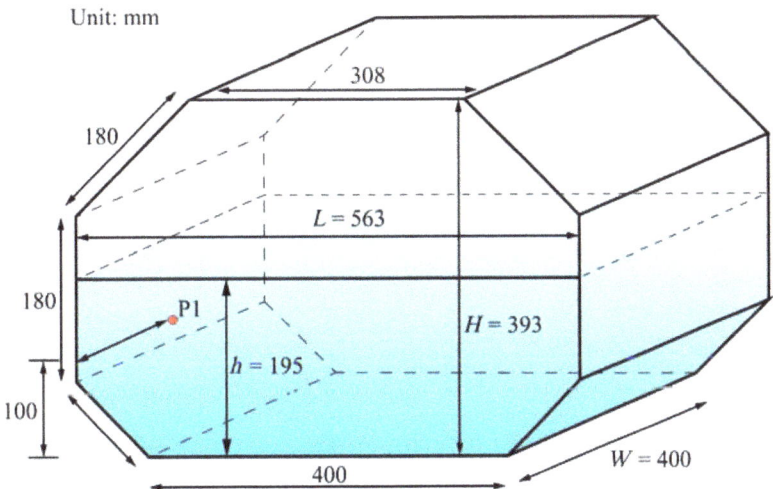

Figure 3. Diagram of the membrane-type tank and corresponding arrangements.

In the paper of Luo et al. [11], only the results from 8 s to 18 s were demonstrated. The full results of the first 18 s are presented in this study. The comparison of the dynamic pressures at P1 between experimental data of Luo et al. [11] and the present numerical result is shown in Figure 4. The present result generally matches the experimental data at the first 13 s, after that, the comparison among the crests agrees well enough, but there exists some discrepancy at the troughs with the maximal error less than 7%. On a whole, the results reveal that the present numerical model is of good accuracy in modeling sloshing in the membrane-type tank.

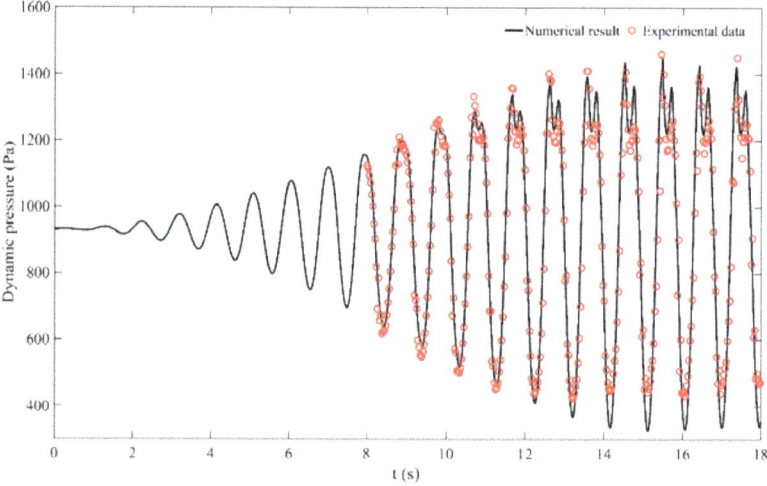

Figure 4. Comparison of time histories of the dynamic pressures between the present numerical result (solid line) and experimental data (circle) at P1.

From the time history of dynamic pressures, it follows that there exist two obvious peaks, in order to better understand the two-peak features (already been found by Zou et al. [3] and Jin and Lin [14]), the free surface profiles at moments 15.8 s, 16.0 s, 16.2 s, 16.3 s, 16.7 s, and 17.0 s are shown in Figure 5, of which 15.8 s and 16.7 s stand for the second peaks which are induced by the falling water hitting the underline water, the moments 16.0 s and 17.0 s are close to the troughs which are nearly close to the minimal excitation. The moments 16.2 s and 16.0 s represent the first peak and trough within the first and second peaks. Besides, considering excitation is near-resonant, the free surface is violently broken.

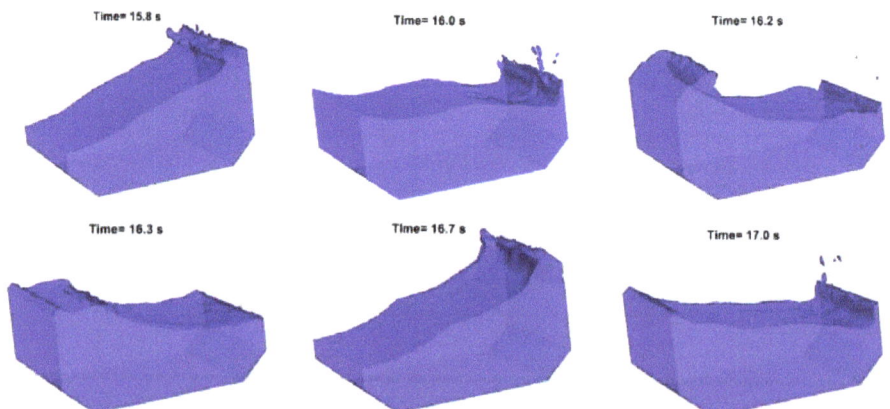

Figure 5. Numerical snapshots of the free surfaces at 15.8 s, 16.0 s, 16.2 s, 16.3 s, 16.7 s, and 17.0 s.

4. Results and Discussion

In real cases, the tank dimensions are of tens or hundreds of meters in length, width as well as height. Bass et al. [1] have concluded that small-scale physical model tests would overestimate slosh pressures as a result of improper liquid compressibility scaling, using Froude-scaled slosh and ullage pressures. To well predict the full-scale loads, the real-scale membrane-type tank [22] will be discussed. For fundamental research, here the 2D sloshing in the length direction with dimensions being length

$L = 46.605\ m$ and height $H = 26.651\ m$ is simulated as shown in Figure 6. Other parameters h_l and h_u are equal to 6.317 m and 8.523 m, respectively. Six pressure transducers are set in the left tank wall, and the detail coordinates and arrangement of the tank can be found in Figure 6. In the transport industry, the filling level is strictly restricted and suggested with the range between 10% of L and 80% of H. However, several filling levels are evitable in real applications. In this study, two filling levels 50% and 70% of H are considered, and the corresponding two filling depths h are 13.3255 m and 18.6557 m, respectively, which are also equal to 28.6% and 40% of L, respectively.

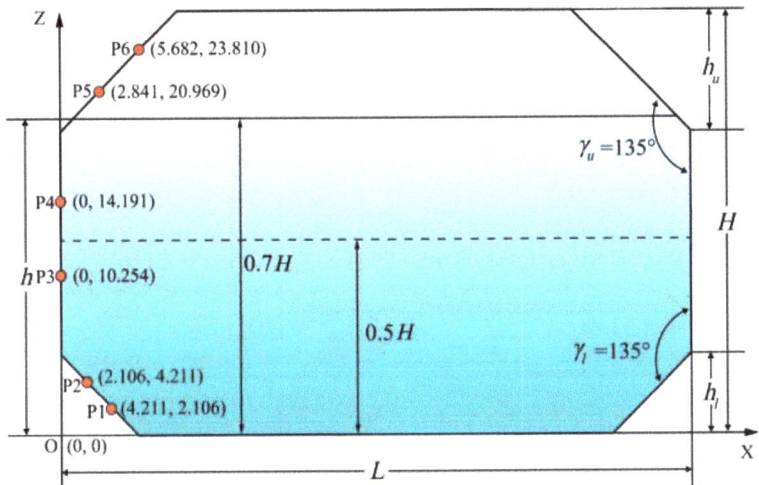

Figure 6. Diagram of the membrane-type tank and corresponding arrangements.

According to the available literature, the prevailing sea states generally have a wide range of periods between 4.0 s and 15.0 s, and the most serious periods are between 6.0 s to 12.0 s [23–26]. To reveal the realistic responses of sloshing in real-scale tanks in the sea conditions, the forcing periods 8.0 s, 9.0 s, 10.0 s, 11.0 s, 12.0 s are adopted in this study. The single freedom of roll motion is considered, with a fixed roll angle of 8.0° being used. For a preliminary study, the rolling center is fixed at (23.3025, 18.6557) throughout the present work.

In the numerical simulations, the 2D computation domain 46.6 m × 26.65 m is discretized as 400 × 160 uniform grids, with $\Delta x = 0.1165$ m and $\Delta z = 0.1666$ m, respectively. To the numerical stability, the time step is automatically adjusted through CFL conditions. The sloshing of two different levels will be investigated first, and the nonlinearity will be discussed in detailed. Finally, the frequency response will be followed.

4.1. Sloshing Responses of Lower Filling Level 50% of Tank Height

Firstly, the filling level 50% of h is discussed, with the corresponding ratio of filling depth to tank length being 28.6% which is also above the traditional shallow filling level (lower than 20% of tank length). The comparisons of dynamic pressures among various forcing periods at P2 and P4 are shown in Figure 7. The left column stands for the results of dynamic pressures at P2, and the right column represents those of P4. Both results reveal that the maximal dynamic pressures occur around the forcing periods 10.0 s, when the forcing period derivates from 10.0 s, the dynamic pressure reduces especially for the cases 8.0 s and 9.0 s. As for P4 which is initially located above the still filling depth, the results in the right column can also stand for the total pressures. When the forcing period increases, the dynamic pressure goes to a comparative value compared to the results of P2 especially for the case

T = 10.0 s. We can also find that dynamic pressures in peaks have only one peak which reveals that the slamming effect is weak.

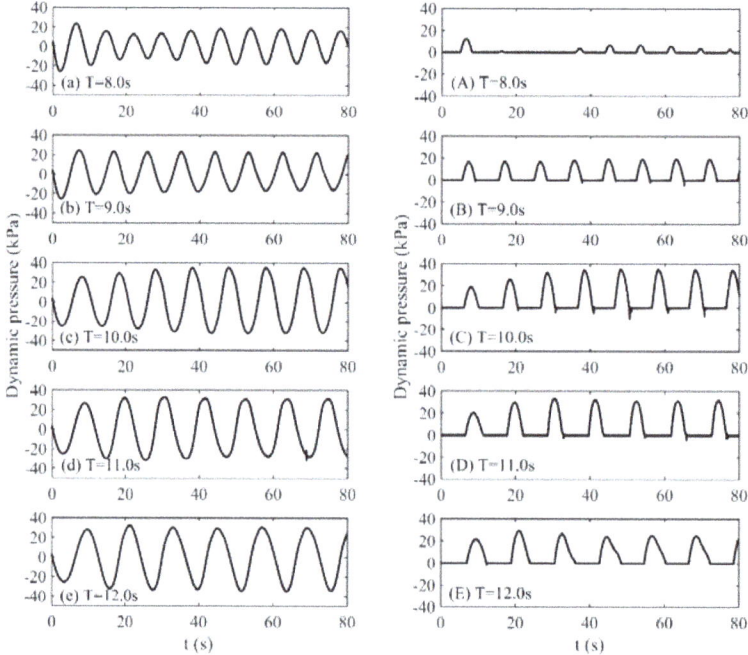

Figure 7. Comparisons of dynamic pressures among various forcing periods 8.0 s, 9.0 s, 10.0 s, 11.0 s and 12.0 s at P2 (**left column**) and P4 (**right column**).

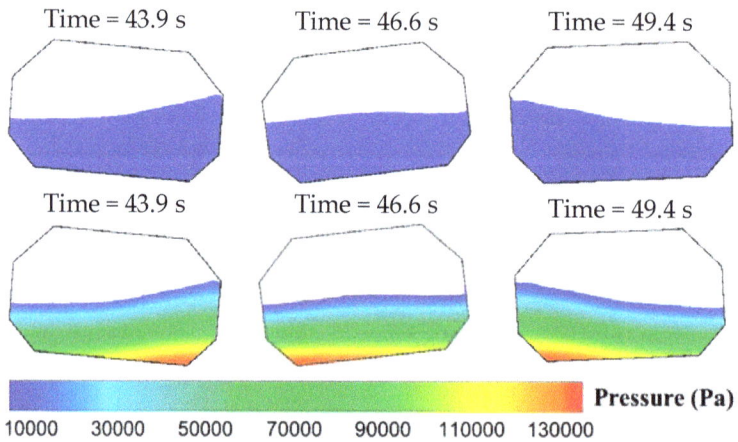

Figure 8. Snapshots of free surfaces (**first line**) and corresponding pressures (**second line**) in the case T = 10.0 s at various moments.

The free surface snapshots of forcing periods 10.0 s at various moments corresponding to three successive troughs and peak in Figure 7 are demonstrated in the first line of Figure 8. From the snapshots, we can find that the maximal free surface displacements at the walls can approach the upper inclined walls. The wave propagates back and forth and behaves like a traveling wave which is an obvious phenomenon in relatively shallow water sloshing. The corresponding snapshots of pressure distributions are displayed below the free snapshots. Due to slightly breaking and slamming, large dynamic loads impact on the upper corners. The maximal pressures are generally located near the tank bottom, especially in the lower corners.

The corresponding results of the forcing period 11.0 s are also demonstrated in Figure 9. The maximal pressures are much close to the case of forcing period 10.0 s than the other two cases with a relatively smaller difference. The free surface snapshots reveal that the liquid is close to the inclined walls. From the two cases, we can conclude that the most serious and dangerous sloshing can occur around 10.0 s and 11.0 s. If this kind of tank is adopted in sea conditions with the peak period around 10.0 s and 11.0 s, moderately sloshing can occur and the safety problem should be paid attention with proper suppression devices being further considered.

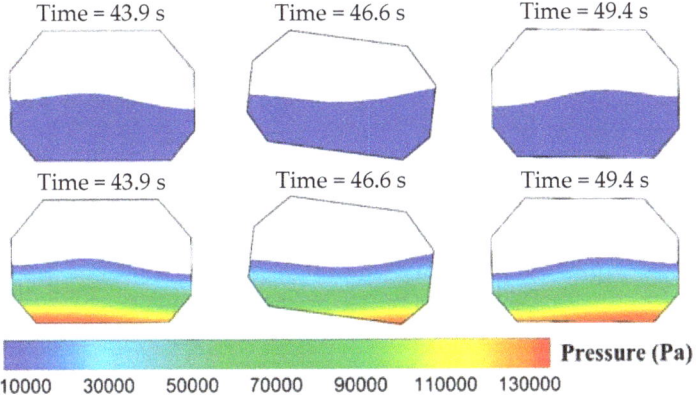

Figure 9. Snapshots of free surfaces (**first line**) and corresponding pressures (**second line**) in the case T = 11.0 s at various moments.

4.2. Sloshing Responses of Higher Filling Level 70% of Tank Height

Except for the 50% filling level, a higher filling level 70% of H is also applicable in engineering, and the corresponding ratio of filling depth to tank length is 40.0%, which is an intermediate filling level and also within the operational category. To explore the response characteristics of the sloshing, the dynamic pressures among various forcing periods at P4 and P5 are shown in Figure 10. The left column stands for the results of dynamic pressures at P4, and the right column represents those at P5.

Similar to the former 50% filling case, the maximal dynamic pressures also appear in terms of the case with the forcing period 10.0 s, implying the most violent sloshing can be triggered under the forcing period 10.0 s. The left column reveals the crest of the case with the forcing period 10.0 s derivates from the balance more significantly than the trough compared to other four cases, which is an obvious nonlinear phenomenon. Considering that P5 is also above the still filling depth, the results in the right column can also represent the total pressure as well as dynamic pressure acting on the tank. The sloshing of the case with forcing period 8.0 is mild, and the glycerin cannot reach P5. Also, unlike the filling level 50% of H, the pressures above the still filling depth as shown in the right column of Figure 10, has several peaks resulting in a quick rising and falling which threats the safety of tanks. Although the present pressures in P5 are lower than those of P4 in the case of 50% filling of H, the quickly rising and falling can result in a high-frequency percussion which can cause fatigue damage and threat the safety of the long-distance transport. Besides, the quickly rising and falling slamming effect can also cause the instability of the storage tank. Thus, the present higher filling cases may have large safety issues than the lower filling level to some extent.

For a better understanding of the violent sloshing, the case with forcing period 10.0 s is discussed first, and the free surface snapshots at various moments corresponding to three successive troughs and peak in Figure 10 are demonstrated in the first line of Figure 11, and the corresponding pressure distributions are displayed below. From the snapshots, we can find that the liquid can reach the upper inclined wall with moderately breaking at some moments, little droplets climbing through the walls, and the breaking wave results in quickly slamming load on the tank wall. The quickly rising and falling of the pressures in P5 are due to the maximal free surface rising and receding of glycerin that pushes down [3,14].

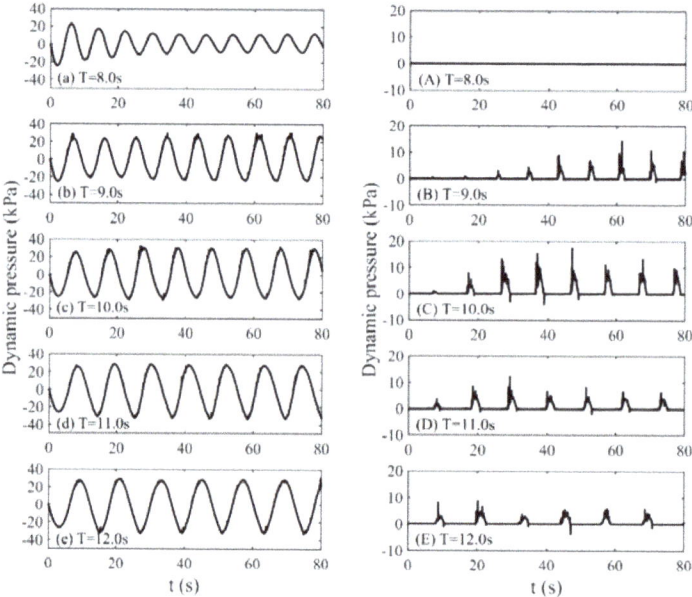

Figure 10. Comparisons of dynamic pressures among various forcing periods 8.0 s, 9.0 s, 10.0 s, 11.0 s and 12.0 s at P4 (**left column**) and P5 (**right column**).

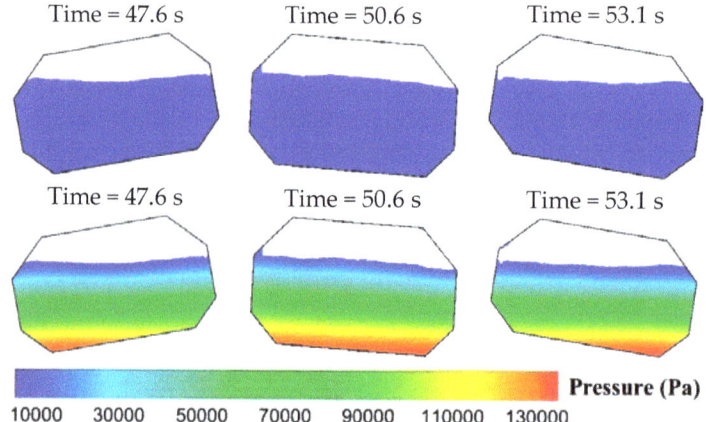

Figure 11. Snapshots of free surfaces (**first line**) and corresponding pressures (**second line**) in the case T = 10.0 s at various moments.

Like the case with forcing period 10.0 s, the corresponding results of the forcing period 11.0 s at the same moments are demonstrated in Figure 12. We can find that both cases are violent and breaking, and there also exist some slashing droplets resulting in large loads on the upper inclined walls. The comparisons between the two cases also reveal that the case with the forcing period being 10.0 s is relatively much violent. To guarantee safety, the violent slamming should be avoided, and the most popular devices-internal horizontal, vertical, ring and performed baffles [27,28] should be introduced.

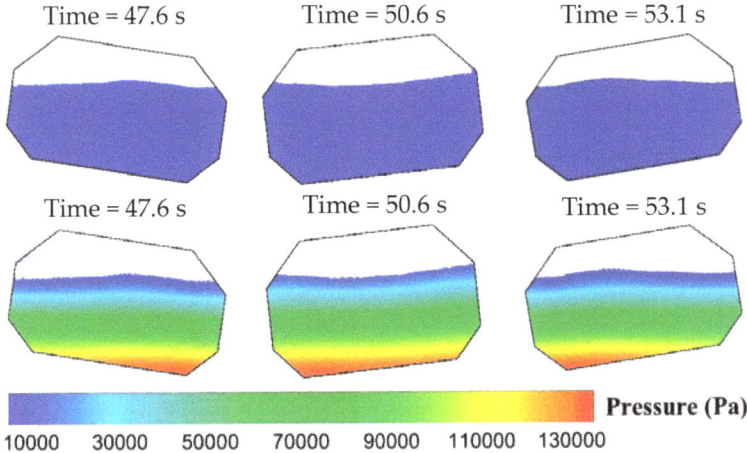

Figure 12. Snapshots of free surfaces (**first line**) and corresponding pressures (**second line**) in the case T = 11.0 s at various moments.

4.3. Frequency Responses of the Sloshing

Besides the maximal response pressures, the frequency response is also of great importance in applications, which can guide the designers and manipulators to avoid incidents. To reveal the relationships among the forcing period and the response period, the fast Fourier transform (FFT) technique will be adopted to perform further investigation. Here three results under forcing periods 8.0 s, 10.0 s, and 12.0 s are demonstrated and discussed in detail.

The dynamic pressures of the case of 50% filling of tank height at P2 and P4 and the corresponding FFT (Fast Fourier Transform) results are shown in Figure 13a,b and Figure 13c,d respectively. It can be seen that the main dominant frequencies are 8.0 s, 10.0 s, and 12.0 s for the individual three cases which are also in accordance with the forcing periods for both pressures at P2 and P4. As for the results at P4, some obvious peaks at lower periods also exist especially for the cases T = 10.0 s and T = 12.0 s, which are contributed to slamming and breaking effects with shorter impact periods as shown and discussed in the former section. The spectra densities focused at lower frequencies of the case T = 12.0 s is relatively smaller than that of the case T = 10.0 s, which reveals that the slamming effect is weaker and it identifies that the case T = 10.0 s is more violent. From Figure 13c,d, it is obvious that the dominant frequency is the forcing period in all cases, however, we can find that there exist additional frequencies in Figure 13d in comparing to Figure 13c, equal to half, quarter, one-fourth and one-fifth of the forcing period, respectively, which is apparently different from the traditional low viscous fluid-water, with the response frequencies being the forcing period and the lowest natural frequency.

Except for the case of 50% filling of tank height, the dynamic pressures of the case of 70% filling of tank height at P4 and P5 and the corresponding FFT results are shown in Figure 14a,b and Figure 14c,d, respectively. Similar to the results of the case of 50% filling of tank height, the dominant frequency of all three cases is the forcing period. From Figure 14b, it is obvious that there exist several peaks, to better understand the response process, the dash section is zoomed. From the section zoomed in, we can find that the time interval between two peaks is small which results in a fast impact process. The corresponding FFT results of the dynamic pressures at P5 shown in Figure 14d display several peaks at lower periods. Taking the case of forcing period 10.0 s for example, besides the dominant frequency 10.0 s-equal to the forcing period, more peaks with values being 5.0 s, 2.5 s, and 2.0 s exist, close to half, a quarter and one-fourth of the forcing period, respectively. Except the mentioned four peaks, there also exists a period 3.5 s equal to half of the summation of 5.0 s and 2.0 s, contributing to the strongly nonlinear effect. Recalling the results of the 50% filling case in Figure 13, there exist fewer peaks with higher frequencies than that of the 70% filling case, which also reveals that the nonlinearity of 70% filling case is stronger than the 50% filling case.

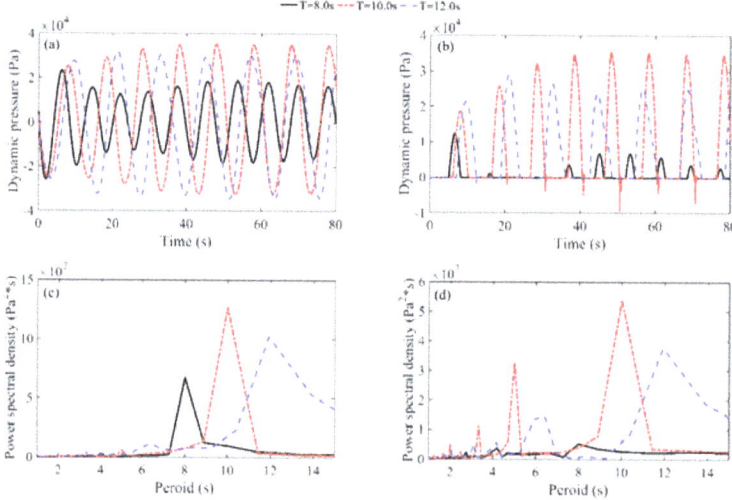

Figure 13. Dynamic pressures for the case of 50% filling of tank height under forcing period 8.0 s (black line), 10.0 s (dash-dot line), and 12.0 s (dash line) at (**a**): P2, (**b**): P4. The corresponding power spectra (**c**) and (**d**).

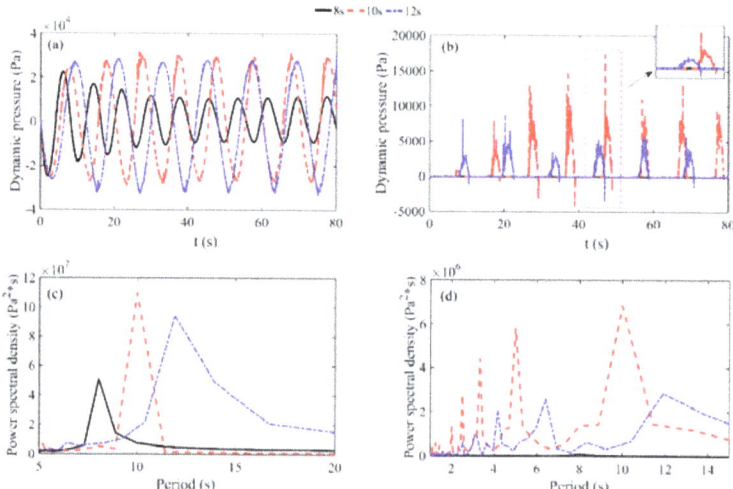

Figure 14. Dynamic pressures for the case of 70% filling of tank height under forcing period 8.0 s (black line), 10.0 s (dash-dot line), and 12.0 s (dash line) at (**a**): P4, (**b**): P5. The corresponding power spectra (**c**) and (**d**).

5. Conclusions

Based on the Navier-Stokes numerical model, the sloshing of a highly viscous fluid in the full-scale prismatic tank is numerically investigated. The numerical model was validated against a self-conducted experiment and available numerical data, and rather good agreements have been guaranteed. Then the proposed numerical model is adopted to systematically study the full-scale sloshing of highly viscous fluid. Two different filling levels (50% and 70% of the tank height) are considered. The liquid kinematic viscosity 6.0×10^{-5} m^2/s is chosen throughout this work. A wide range of forcing frequencies are used to identify the response process of pressures as well as free surface displacements. The frequency responses are further identified by the fast Fourier transformation technology. Through the discussions, some conclusions can be drawn:

1. The responses of dynamic pressures of the 50% filling height of the tank length behave larger values than that of the 70% filling height of the tank length for the same forcing period. The sloshing of the lower filling level behaves like a traveling wave and the sloshing of the higher filling level generally moves like a standing wave.
2. The free surface of the 70% filling case is more breaking than the lower filling case. There exist several peaks in the crests of the dynamic pressures for the 70% filling case, and the time intervals between peaks are very shot, which also reveals that the slamming effect is more obvious than the lower filling case.
3. Due to viscous effects, the nonlinearity is largely reduced compared to that of water but also exists by recalling the existence of the combinations of various frequencies. The dominant response frequencies of glycerin sloshing turn from the forcing period and lowest natural period to the forcing period and score times of the forcing period.

Author Contributions: Conceptualization, X.J. and S.M.; methodology, X.J.; software, X.J.; validation, S.M. and M.T.M.; investigation, S.M.; data curation, M.T.M.; writing—original draft preparation, M.L. and M.T.M.; writing—review and editing, X.J., S.M. and Z.H.; visualization, S.M. and M.L.; supervision, X.J.

Funding: This research received no external funding.

Acknowledgments: The experiment was conducted in Hohai University and supported by Mi-An Xue.

Conflicts of Interest: The authors declare no conflict of interest.

References

1. Bass, R.L.; Bowles, E.B.; Trudell, R.W.; Navickas, J.; Peck, J.C.; Yoshimura, N.; Endo, S.; Pots, B.F.M. Modeling Criteria for Scaled LNG Sloshing Experiments. *J. Fluid Eng.* **1985**, *107*. [CrossRef]
2. Kim, Y.; Shin, Y.-S.; Lee, K.H. Numerical study on slosh-induced impact pressures on three-dimensional prismatic tanks. *Appl. Ocean Res.* **2004**, *26*, 213–226. [CrossRef]
3. Zou, C.-F.; Wang, D.-Y.; Cai, Z.-H.; Li, Z. The effect of liquid viscosity on sloshing characteristics. *J. Mar. Sci. Technol.* **2015**, *20*, 765–775. [CrossRef]
4. Baudry, V.; Rousset, J.-M. Experimental Study of Viscous Cargo Behaviour and Investigation on Global Loads Exerted on Ship Tanks. In Proceedings of the ASME 2017 36th International Conference on Ocean, Offshore and Arctic Engineering, Trondheim, Norway, 25–30 June 2017; p. V07BT06A015.
5. Faitinsen, O.M. A numerical nonlinear method of sloshing in tanks with two-dimensional flow. *J. Ship Res.* **1978**, *22*, 193–202.
6. Faltinsen, O.M.; Rognebakke, O.F.; Lukovsky, I.A.; Timokha, A.N. Multidimensional modal analysis of nonlinear sloshing in a rectangular tank with finite water depth. *J. Fluid Mech.* **2000**, *407*, 201–234. [CrossRef]
7. Hill, D.F. Transient and steady-state amplitudes of forced waves in rectangular basins. *Phys. Fluids* **2003**, *15*, 1576–1587. [CrossRef]
8. Faltinsen, O.M.; Timokha, A.N. Resonant three-dimensional nonlinear sloshing in a square-base basin. Part 4. Oblique forcing and linear viscous damping. *J. Fluid Mech.* **2017**, *822*, 139–169. [CrossRef]
9. Wu, G.; Taylor, R.E.; Greaves, D.J.J.o.E.M. The effect of viscosity on the transient free-surface waves in a two-dimensional tank. *J. Eng. Math.* **2001**, *40*, 77–90. [CrossRef]

10. Rudman, M.; Cleary, P.W.; Prakash, M. Simulation of liquid sloshing in model LNG tank using smoothed particle hydrodynamics. *Int. J. Offshore Polar Eng.* **2009**, *19*, 286–294.
11. Luo, M.; Koh, C.; Bai, W.J.O.E. A three-dimensional particle method for violent sloshing under regular and irregular excitations. *Ocean Eng.* **2016**, *120*, 52–63. [CrossRef]
12. Zhao, Y.; Chen, H.-C.J.O.E. Numerical simulation of 3D sloshing flow in partially filled LNG tank using a coupled level-set and volume-of-fluid method. *Ocean Eng.* **2015**, *104*, 10–30. [CrossRef]
13. Kim, J.K.W.; Shin, Y.; Sim, I.; Kim, Y.; Bai, K. Three-Dimensional Finite-Element Computation for the Sloshing Impact Pressure In LNG Tank. In Proceedings of the Thirteenth International Offshore and Polar Engineering Conference, Honolulu, HI, USA, 25–30 May 2003.
14. Jin, X.; Lin, P. Viscous effects on liquid sloshing under external excitations. *Ocean Eng.* **2019**, *171*, 695–707. [CrossRef]
15. Liu, D.; Lin, P. A numerical study of three-dimensional liquid sloshing in tanks. *J. Comput. Phys.* **2008**, *227*, 3921–3939. [CrossRef]
16. Lin, P.; Li, C. Wave–current interaction with a vertical square cylinder. *Ocean Eng.* **2003**, *30*, 855–876. [CrossRef]
17. Dongming, L. *Numerical Modeling of Three-Dimensional Water Waves and Their Interaction with Structures*; National University of Singapore: Singapore, 2008.
18. Chorin, A.J. Numerical solution of the Navier-Stokes equations. *Math. Comput* **1968**, *22*, 745–762. [CrossRef]
19. Gueyffier, D.; Li, J.; Nadim, A.; Scardovelli, R.; Zaleski, S.J.J.o.C.p. Volume-of-fluid interface tracking with smoothed surface stress methods for three-dimensional flows. *J. Comput. Phys.* **1999**, *152*, 423–456. [CrossRef]
20. Yu, L.; Xue, M.-A.; Zheng, J. Experimental study of vertical slat screens effects on reducing shallow water sloshing in a tank under horizontal excitation with a wide frequency range. *Ocean Eng.* **2019**, *173*, 131–141. [CrossRef]
21. Register, L. *Guidance on the Operation of Membrane LNG Ships to Reduce the Risk of Damage Due to Sloshing*; Technical report; Lloyds Register: London, UK, 2008.
22. Cai, Z.-h.; Wang, D.-y.; Li, Z. Influence of excitation frequency on slosh-induced impact pressures of liquefied natural gas tanks. *J. Shanghai Jiao Tong Univ.* **2011**, *16*, 124–128. [CrossRef]
23. Toba, Y.; Iida, N.; Kawamura, H.; Ebuchi, N.; Jones, I.S. Wave dependence of sea-surface wind stress. *J. Phys. Oceanogr.* **1990**, *20*, 705–721. [CrossRef]
24. Earle, M.D. Extreme wave conditions during Hurricane Camille. *J. Geophys. Res.* **1975**, *80*, 377–379. [CrossRef]
25. Pickrill, R.; Mitchell, J. Ocean wave characteristics around New Zealand. *N. Zeal. J. Mar. Fresh* **1979**, *13*, 501–520. [CrossRef]
26. Araújo, C.E.; Franco, D.; MELO, E.; Pimenta, F. Wave regime characteristics of the southern Brazilian coast. In Proceedings of the Sixth International Conference on Coastal and Port Engineering in Developing Countries, COPEDEC VI, Colombo, Sri Lanka, 5–19 September 2003; p. 15.
27. Biswal, K.; Bhattacharyya, S.; Sinha, P. Free-vibration analysis of liquid-filled tank with baffles. *J. Sound Vib.* **2003**, *259*, 177–192. [CrossRef]
28. Xue, M.-A.; Lin, P. Numerical study of ring baffle effects on reducing violent liquid sloshing. *Comput. Fluids* **2011**, *52*, 116–129. [CrossRef]

© 2019 by the authors. Licensee MDPI, Basel, Switzerland. This article is an open access article distributed under the terms and conditions of the Creative Commons Attribution (CC BY) license (http://creativecommons.org/licenses/by/4.0/).

Article

Semi-Analytical Model for Two-Phase Flowback in Complex Fracture Networks in Shale Oil Reservoirs

Yuzhe Cai and Arash Dahi Taleghani *

Department of Energy and Mineral Engineering, Pennsylvania State University, State College, PA 16801, USA; ybc5081@psu.edu
* Correspondence: arash.dahi@psu.edu; Tel.: +1-814-865-5421

Received: 28 October 2019; Accepted: 5 December 2019; Published: 12 December 2019

Abstract: Flowback data is the earliest available data for estimating fracture geometries and the assessment of different fracturing techniques. Considerable attention has been paid recently to analyze flowback data quantitively in order to obtain fracture properties such as effective half-length and effective conductivity by simply assuming fractures having bi-wing planar geometries and constant fracture compressibility. However, this simplifying assumption ignores the complexity of fracture networks. To overcome this limitation, we proposed a semi-analytical method, which can be used as a direct model for fast inverse analysis to characterize complex fracture networks generated during hydraulic fracturing. A two-phase oil–water flowback model with a matrix oil influx for wells with bi-wing planar fractures is also presented to identify limitations of the former solution. Since most available flowback studies use constant fracture properties and the assumption of planar fractures, considering variable fracture properties and complex fracture geometries gives this model more robustness for modeling fracture flow during flowback, more realistically. The proposed models have been validated by numerical simulations. The presented procedure provides a simple way for modeling early flowback in complex fracture networks and it can be used for inverse analysis.

Keywords: flowback; complex fracture network; shale oil

---

1. Introduction

Reservoir simulation is a powerful tool to predict reservoir performance under different operating conditions, however, when it comes to fractured wells, it requires detailed knowledge of induced fractures. Routinely, well testing and long-term rate-transient analysis (RTA) are used for post-fracture analysis to determine fracture properties [1,2]. Recent advances in field data acquisition and development of powerful data analytic techniques promise new venues for unlocking fracture properties such as fracture conductivity and complexity of induced fracture networks by analysis of flowback data if a good physical model for flowback is available. Flowback data is the earliest part of production data that operators may acquire to determine fracture properties for building reservoir models and assess the stimulation design.

The first attempt to quantitively analyze flowback data uses a single-phase reciprocal productivity Index (RPI) procedure [3]. This method provides an estimation of the effective permeability-thickness, apparent fracture half-length or effective wellbore radius for vertical wells. The model is derived based on the bi-wing planar fracture geometries and assumes that fracture volume does not change during flowback. Flowback production time in shale wells is divided into three periods: (1) the pre-breakthrough water dominated region, (2) the transition region, and (3) the post-breakthrough hydrocarbon dominated region [4]. The authors developed a single-phase flowback model for the first period only and assumed a linear correlation between rate-normalized pressure and material balance flowrate to calculate total storage coefficients and effective fracture permeability. A semi-analytical model for flowback in tight oil reservoirs uses the concept of a dynamic drainage area (DDA) to capture the transient behaviors [5]. A bi-wing planar fracture is assumed in the model. However, due to the presence of natural fractures in the formation and their interaction with advancing hydraulic fractures, the geometry of induced fractures can be very complicated [6]. The outcrop of cross-cutting joints found in the Marcellus shale exposed in Oatka Creek also shows well-developed natural fracture systems [7]. To accommodate the reactivation of pre-existing natural fractures, they defined an enhanced fracture region (EFR) along the primary fracture, inside which the permeability is enhanced. However, this approach indicates that overall permeability is enhanced in the direction perpendicular to the primary fracture, which misleads the true directionality of fracture permeability in complex fracture networks [8]. A complex fracture network is desirable from the design perspective as it increases the effective surface area of the wellbore and enable more-efficient linear flow to predominate over the radial flow. Production from reservoirs with complex fracture networks is often modeled by a numerical simulator, which can be computationally expensive for use in inverse analyses, unless some meshless methods are utilized for flow through fractures [9]. Most available flowback models incorporate fracture closure by using constant fracture compressibility for wells with bi-wing planar fractures [4,5,8,10]. However, the outcome can be misleading for two main reasons: (1) compressibility of propped fracture is expected to evolve with effective pressure based on the analytical solution for compliance of proppant beds. Especially during early flowback, rapid pressure drawdown may lead to significant decrease in fracture compressibility. Flowback analysis based on constant compressibility will overestimate the fracture storage capacity at the end of flowback. (2) Fracture width is decreasing from the wellbore to the tip, knowing that fracture compressibility is a function of proppant density, we do not expect that fracture compressibility remains constant. This assumption can introduce misleading impacts on determining fracture properties. To incorporate compressibility changes over time, we initially proposed analytical solutions for single and two-phase water–oil flowback using eigenfunction expansion methods. Then, we develop a semi-analytical procedure to incorporate variations of fracture width and fracture compressibility along the fracture, which can be utilized for complex fracture networks and long-planar fractures. We also proposed a two-phase oil-water flowback solution with matrix oil influx for wells with bi-wing planar fractures, which can be applied to formations without significant natural fractures. The proposed model is verified against numerical simulations implemented in COMSOL Multiphysics (COMSOL, Burlington, Massachusetts, USA) Multiphysics and CMG (Computer Modelling Group LTD., Calgary, Alberta, Canada). The goals of this paper are set to (1) provide a quick and accurate direct flowback model for fluid flow through the fracture network, (2) investigate the influence of non-uniform fracture closure and variable fracture compressibility on flowback characteristics.

2. Governing Equations for Flowback Models

Flowback from shale oil wells often initially show only a typical single-phase region. Assuming there is no matrix flow, the dimensionless form of the governing equation for 1D single-phase flowback is given by

$$\frac{\partial P_D}{\partial t_{aD}} = \frac{\partial^2 P_D}{\partial x_D^2}, \quad (1)$$

where dimensionless variables are defined as

$$P_D = \frac{P}{P_i}, \quad (2)$$

$$x_D = \frac{x}{L}, \quad (3)$$

$$t_{aD} = \frac{1}{\mu_w L^2} \int_0^t \frac{k_f(\bar{p})}{w_{fD}(\bar{p})\phi_f(\bar{p})c_f(\bar{p})} dt, \quad (4)$$

where P is fluid pressure along the fracture, P_i is the initial fracture pressure, $\bar{p}$ is the average pressure in the fracture, L is the length of the fracture segment, ϕ is porosity, μ_w is water viscosity, k_f is fracture permeability, w_{f_D} is dimensionless fracture width which is defined as $\frac{w_f}{w_{f_i}}$, x is the coordinate along the fracture direction. c_t is the total compressibility which is the summation of fracture compressibility c_f and water compressibility c_w, which are defined as follows

$$c_f = \frac{1}{V_f} \frac{dV_f}{dP}, \quad (5)$$

$$c_w = -\frac{1}{V_w} \frac{dV_w}{dP}. \quad (6)$$

In the case of wells with an extended shut-in period, a small amount of oil may enter the fracture network. The following assumptions should be considered to develop the two-phase oil–water model: (1) capillary pressure is neglected inside the fractures, (2) oil and water are both incompressible, and (3) no matrix flow is considered. The average saturation of the fracture domain does not change due to the assumption of incompressible fluids as the coefficients of $\frac{dS_w}{dt}$ terms collapse to zero in the two-phase governing equation. The mass balance equation for the fracture system is the same as Equation (1), except the pseudo-time is defined differently as

$$t_{aD} = \frac{\lambda_t}{L^2} \int_0^t \frac{k_f(\bar{p})}{w_{f_D}(\bar{p}) \phi_f(\bar{p}) c_f(\bar{p})} dt, \quad (7)$$

where λ_t is the summation of λ_o and λ_w, which are oil mobility and water mobility defined as $\frac{k_{ro}}{\mu_o}$ and $\frac{k_{rw}}{\mu_w}$. The straight-line relative permeability model is used for fractures in this study i.e.,

$$k_{rw} = S_w, \quad (8)$$

$$k_{ro} = 1 - S_w, \quad (9)$$

where S_w is water saturation. Detailed derivation of governing equations for a two-phase flow without matrix influx is provided in Appendix B.

During the flowback period, initially, water production dominates (region 1) which is gradually replaced by oil production (region 3) after oil breakthrough the fracture as shown schematically in Figure 1. Since the permeability of shale reservoirs is extremely low, no matrix flow from the reservoir is considered for early-time flowback. As a result, the proposed flowback model will be restricted to region 1. Due to the extremely small pore diameter of the shale formation, the large capillary effect is expected. Therefore, breakthrough pressure can be much lower than initial reservoir pressure, which makes the closed-chamber solution works for an extended period of time.

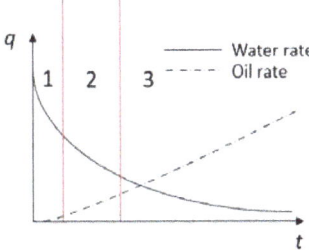

Figure 1. Comparison between water rate and oil rate during flowback.

Even though the solution developed by assuming no matrix influx is accurate in early flowback periods, it cannot predict flowrate accurately for the late flowback period after a significant amount of oil enters the fracture network. To test the applicability of the above-mentioned model without considering matrix flow, we proposed another solution that can consider matrix flow but only limited to wells with bi-wing planar fractures. In the fracture governing equation, the matrix influx is considered as a source term, which is given by

$$\frac{\partial^2 P_p}{\partial x_D^2} = \frac{\partial P_p}{\partial t_{aD}} - q_m, \tag{10}$$

where

$$P_p = \int_P^{P_i} \frac{\lambda_t k_f(\bar{p})}{\phi_f c_f(\bar{p})} dp, \tag{11}$$

$$t_{aD} = \frac{1}{L^2} \int_0^t \frac{\lambda_t}{w_{fD} \phi_f c_f(\bar{p})} dt, \tag{12}$$

$$q_m = \frac{L^2 v_y(t)}{w_f P_i}. \tag{13}$$

The matrix oil velocity $v_y(t)$ is calculated from the matrix governing equation

$$\frac{\partial^2 P_D}{\partial y_D^2} = \frac{\partial P_D}{\partial t_D}, \tag{14}$$

where

$$P_D = \frac{P_m}{P_i}, \tag{15}$$

$$y_D = \frac{y}{D}, \tag{16}$$

$$t_D = \frac{k_m}{\mu_o D^2 \phi_m c_f} t. \tag{17}$$

where P_m represents matrix pressure, D represents the length of stimulated reservoir volume (SRV) or half of the fracture spacing for multi-stage hydraulic fractures in horizontal wells. Although $\frac{\partial S_w}{\partial t}$ the term is canceled out, water saturation inside the fracture should be updated from the volume balance equation to update the relative permeability values.

3. Analytical Solution of Governing Equations

The eigenfunction expansion methods are used to solve the diffusivity equation with time-dependent boundary conditions. The analytical solutions of Equation (1) can be classified into two cases based on the boundary conditions. One case is variable pressure boundary conditions at both fractures' ends and the other case is variable pressure at one end and no flow at the other end. Detailed derivation of the solution is provided in Appendix A. Initial condition of Equation (1) for is given by

$$P_D(x_D, 0) = 1. \tag{18}$$

First, we present the solution for variable pressure boundary conditions at both ends of the fracture without matrix influx, time dependent pressure boundary conditions are prescribed at both ends as

$$P_D(0, t_{aD}) = P_1(t_{aD}), \tag{19}$$

$$P_D(1, t_{aD}) = P_2(t_{aD}). \tag{20}$$

Using an eigenfunction expansion for the solution p as shown in Appendix A, analytical solution of Equation (1) with two Dirichlet boundary conditions are given by

$$P_D(x_D, t_{aD}) = (P_2(t_{aD}) - P_1(t_{aD}))x_D + P_1(t_{aD}) + \sum_{n=1}^{\infty} \left(\int_0^{t_{aD}} e^{-\lambda_n^2(t_{aD}-\tau)} \hat{S}_n(\tau) d\tau + e^{-\lambda_n^2 t_{aD}} c_n \right) \sin(\lambda_n x_D), \quad (21)$$

where c_n is the Fourier series obtained from the initial conditions, $\hat{S}_n(t_{aD})$ is the term that accounts for the time derivative of the two Dirichlet boundary conditions, λ_n is the eigenvalue:

$$c_n = \int_0^1 [1 - P_2(0) + P_1(0)] \sin(\lambda_n x_D) dx_D, \quad (22)$$

$$\hat{S}_n(t_{aD}) = -2 \int_0^1 \sin(\lambda_n x_D) \left(\frac{dP_1}{dt_{aD}} - \frac{dP_2}{dt_{aD}} \right) dx_D, \quad (23)$$

$$\lambda_n = \frac{n\pi}{L}. \quad (24)$$

In the single-phase solution, water rate can be evaluated by taking derivative of Equation (21) with respect to x as

$$q_w|_{x_D=0} = \frac{w_f k_f h P_i}{\mu_w L} \left[P_2(t_{aD}) - P_1(t_{aD}) + \sum_{n=1}^{\infty} \left(\int_0^{t_{aD}} e^{-\lambda_n^2(t_{aD}-\tau)} \hat{S}_n(\tau) d\tau + e^{-\lambda_n^2 t_{aD}} c_n \right) \right], \quad (25)$$

$$q_w|_{x_D=1} = \frac{w_f k_f h P_i}{\mu_w L} \left[P_2(t_{aD}) - P_1(t_{aD}) + \sum_{n=1}^{\infty} \left(\int_0^{t_{aD}} e^{-\lambda_n^2(t_{aD}-\tau)} \hat{S}_n(\tau) d\tau + e^{-\lambda_n^2 t_{aD}} c_n \right) \cos(\lambda_n) \right]. \quad (26)$$

Similarly, in the two-phase solution, the total flow rate at the inlet and outlet are

$$q_t|_{x_D=0} = \frac{\lambda_t w_f k_f h P_i}{L} \left[P_2(t_{aD}) - P_1(t_{aD}) + \sum_{n=1}^{\infty} \left(\int_0^{t_{aD}} e^{-\lambda_n^2(t_{aD}-\tau)} \hat{S}_n(\tau) d\tau + e^{-\lambda_n^2 t_{aD}} c_n \right) \right], \quad (27)$$

$$q_w|_{x_D=1} = \frac{w_f k_f h P_i}{\mu_w L} \left[P_2(t_{aD}) - P_1(t_{aD}) + \sum_{n=1}^{\infty} \left(\int_0^{t_{aD}} e^{-\lambda_n^2(t_{aD}-\tau)} \hat{S}_n(\tau) d\tau + e^{-\lambda_n^2 t_{aD}} c_n \right) \cos(\lambda_n) \right]. \quad (28)$$

For solution with variable pressure boundary condition at one end and no-flow boundary at the other end without matrix influx, time-dependent pressure boundary conditions are prescribed at both ends as

$$P_D(0, t_{aD}) = P_1(t_{aD}), \quad (29)$$

$$\frac{\partial P_D(1, t_{aD})}{\partial x_D} = 0. \quad (30)$$

Using an eigenfunction expansion for the solution p as shown in Appendix A, the analytical solution of Equation (1) with one Dirichlet boundary condition and one Neumann boundary condition is

$$P_D(x_D, t_{aD}) - P_1(t_{aD}) + \sum_{n=1}^{\infty} \left[\int_0^{t_{aD}} e^{-\lambda_n^2(t_{aD}-\tau)} \hat{S}_n(t_{aD}) d\tau + e^{-\lambda_n^2 t_{aD}} c_n \right] \sin(\lambda_n x_D), \quad (31)$$

where

$$c_n = \int_0^1 (1 - P_1(0)) \sin(\lambda_n x_D) dx_D, \quad (32)$$

$$\hat{S}_n(t_{aD}) = -2 \int_0^1 \sin(\lambda_n x_D) \frac{dP_1(t_{aD})}{dt_{aD}} dx_D = -\frac{2}{\lambda_n} \frac{\partial P_{D1}}{\partial t_{aD}}, \quad (33)$$

$$\lambda_n = \frac{(2n-1)\pi}{2}. \quad (34)$$

In the single-phase solution, water rate can be evaluated by taking the derivative of pressure, i.e.,

$$q_w(t_{aD})\Big|_{x_D=0} = \frac{w_f k_f h P_i}{\mu_w L}\left[\sum_{n=1}^{\infty}\left(\int_0^{t_{aD}} e^{-\lambda_n^2(t_{aD}-\tau)}\hat{S}_n(\tau)d\tau + e^{-\lambda_n^2 t_{aD}}c_n\right)\right]. \tag{35}$$

In the two-phase solution, total flow rate can be evaluated as

$$q_t(t_{aD})\Big|_{x_D=0} = \frac{\lambda_t w_f k_f h P_i}{L}\left[\sum_{n=1}^{\infty}\left(\int_0^{t_{aD}} e^{-\lambda_n^2(t_{aD}-\tau)}\hat{S}_n(\tau)d\tau + e^{-\lambda_n^2 t_{aD}}c_n\right)\right]. \tag{36}$$

For two-phase flowback with matrix influx, the concept of pseudo-pressure is used to capture the saturation change inside the fracture. Initial condition of Equation (10) for fracture flow is given by

$$P_p(x_D, 0) = 0. \tag{37}$$

The time-dependent pressure boundary conditions are prescribed at both ends as

$$P_p(0, t_{aD}) = P_{p1}(t_{aD}), \tag{38}$$

$$\frac{\partial P_p(1, t_{aD})}{\partial x_D} = 0. \tag{39}$$

Using the eigenfunction expansion for homogenized pseudo-pressure, the analytical solution of Equation (10) will be

$$P_p(x_D, t_{aD}) = (t_{aD}) + \sum_{n=1}^{\infty}\left[\int_0^{t_{aD}} e^{-\lambda_n^2(t_{aD}-\tau)}\hat{S}_n(t_{aD})d\tau + e^{-\lambda_n^2 t_{aD}}c_n\right]\sin(\lambda_n x_D), \tag{40}$$

where

$$c_n = \int_0^1 (1 - P_1(0))\sin(\lambda_n x_D)dx_D, \tag{41}$$

$$\hat{S}_n(t_{aD}) = -2\int_0^1 \sin(\lambda_n x_D)\left(\frac{dP_{p1}}{dt_{aD}} + q_m\right)dx_D = -\frac{2}{\lambda_n}\left(\frac{dP_{p1}}{dt_{aD}} + q_m\right), \tag{42}$$

$$\lambda_n = \frac{(2n-1)\pi}{2}. \tag{43}$$

Taking the derivative of the pseudo-pressure, the total flow rate at the wellbore can be evaluated as

$$q_t(t_{aD})\Big|_{x_D} = \frac{w_f k_f h P_i}{L}\sum_{n=1}^{\infty}\left[\int_0^{t_{aD}} e^{-\lambda_n^2(t_{aD}-\tau)}\hat{S}_n(t_{aD})d\tau + e^{-\lambda_n^2 t_{aD}}c_n\right]\cos(\lambda_n x_D). \tag{44}$$

Water rate and oil rate can be evaluated as,

$$q_w(t_{aD})\Big|_{x_D} = \frac{\lambda_w}{\lambda_t}q_t(t_{aD}), \tag{45}$$

$$q_o(t_{aD})\Big|_{x_D} = \frac{\lambda_o}{\lambda_t}q_t(t_{aD}). \tag{46}$$

Since the source term in Equation (10) is obtained by solving the matrix flow equation, matrix oil velocity also needs to be updated. The computational cost of the solution is O(M), where M is the number of total segments. To balance the accuracy and computational costs, we divided each half of the fracture into four segments Figure 2 shows hydraulic fractures in a given stimulated reservoir volume (SRV), which is assumed to have a rectangular shape. Matrix oil flux is calculated separately for the four fracture segments. Initial and boundary conditions of the matrix oil flow equation is given by

$$P_D(0, t_D) = \overline{P}_{fD,k} \qquad k = 1, 2, 3, 4 \qquad (47)$$

$$\frac{\partial P_D(1, t_D)}{\partial y_D} = 0, \qquad (48)$$

$$P_D(y_D, 0) = 1, \qquad (49)$$

where $\overline{P}_{fD,k}$ represents the average pressure at segment k. The analytical solution of Equation (14) is given by

$$P_{D,k}(x_D, t_{aD}) = \overline{P}_{fD,k} + \sum_{n=1}^{\infty} \left[\int_0^{t_{aD}} e^{-\lambda_n^2(t_{aD}-\tau)} \hat{S}_n(t_{aD}) d\tau + e^{-\lambda_n^2 t_{aD}} c_n \right] \sin(\lambda_n x_D), \qquad (50)$$

$$v_y(t_{aD})\big|_{x_D=0} = \frac{k_f P_i}{D} \sum_{n=1}^{\infty} \left[\int_0^{t_{aD}} e^{-\lambda_n^2(t_{aD}-\tau)} \hat{S}_n(t_{aD}) d\tau + e^{-\lambda_n^2 t_{aD}} c_n \right]. \qquad (51)$$

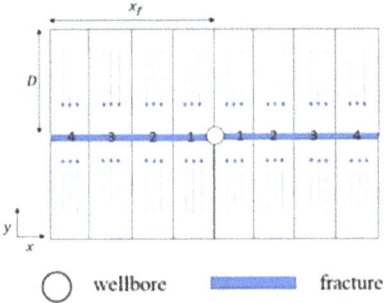

Figure 2. Flowback model with matrix oil influx.

Average water saturation in the fracture segment k is determined using volume balance as

$$S_{w,k} = S_{wi} + \frac{(q_{w,in,k} - q_{w,out,k}) dt}{V_{p,k}}, \qquad (52)$$

where S_{wi} is the initial water saturation, dt is time step and $V_{p,k}$ is the volume of the fracture segment at the current time. $q_{w,in,k}$ and $q_{w,out,k}$ is the inflow and outflow water rates, respectively. The continuous water saturation profile inside the fracture is approximated by a second-order polynomial interpolation between nodal saturations $S_{w,k}$. The two-phase flowback solution with matrix oil influx can be described as following:

1. For the first time step, the average saturation in each segment is equal to initial water saturation ($S_{w,k} = S_{wi}$). Solve Equation (10) by Equations (40) and (44) assuming no oil influx at the initial time ($v_o = 0$) and calculate average fracture pressure $\overline{P}_{f,k}$ for each fracture segment.
2. Based on $\overline{P}_{f,k}$, calculate average oil influx $v_{y,k}$ using Equation (51).
3. For the second time step, update oil rate $v_{y,k}$ calculated from step 2 in Equation (10) and solve it again. Average water saturation for each segment is calculated from the material balance Equation (52). Update pressure and repeat steps 2–3 till reaching the end of the simulation time.

We also show the workflow of the solution in Figure 3.

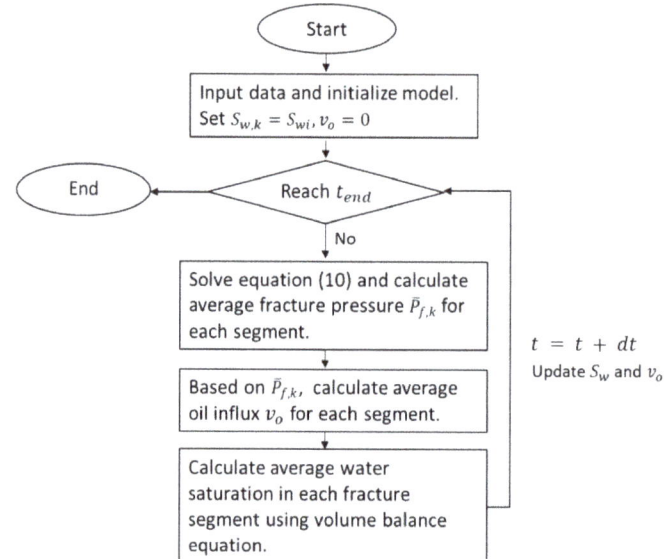

Figure 3. Workflow for two-phase semi-analytical solutions with matrix oil influx.

The analytical solutions presented in this section is only applicable to a single fracture segment. however, fractures may exist in the form of complex networks. In the following sections, we will show how we can discretize the complex networks and solve the systems of equations simultaneously for the whole system.

4. Fractal Fracture Network and Fracture Discretization

Two different fracture network geometries are considered in this study: tree-shape fracture networks and orthogonal fracture networks. Fracture re-initiation at the intersection with natural fractures is a possible branching mechanism during hydraulic fracturing [11]. Tree-shape fracture networks are commonly found as shown in Figure 4b from simulation results. Researchers used a neural network model to process microseismic data to estimate fracture geometry and also believe that, the fracture network of multiple fractured horizontal wells in highly brittle shale formation is usually tree-shaped [12]. An orthogonal fracture network is another common situation in complex fracture networks. In naturally fractured formations, if the joint sets are orthogonal to each other, like the situation in Marcellus Shale [11], induced hydraulic fractures may likely form an orthogonal mosaic network by activating natural fractures as shown in Figure 4b.

Fracture discretization is the first step to simplify calculations. Fracture networks can be divided into multiple segments and nodes as illustrated in Figures 4a and 5a for tree-shape and orthogonal fractures, respectively. The tree-shape network is divided into 18 segments and 8 nodes. The orthogonal fracture network is divided into 24 segments and eight nodes. It is worth noting that in the case of symmetric tree-shape or symmetric orthogonal fracture networks, one may take advantage of the symmetry and reduce the number of nodes and segments to reduce the computational costs.

To generalize fracture networks, tree-shape fracture networks and orthogonal fracture networks can be synthesized using fractal theory as shown in Figure 6. In a tree-shape fracture network (Figure 4a), each fracture segment can be modeled with segment length L_i, width w_{f_i}, and compressibility c_{f_i}, where index i denotes the fracture-segment generation level. The oldest or initial segment is obviously near the well ($i = 0$).

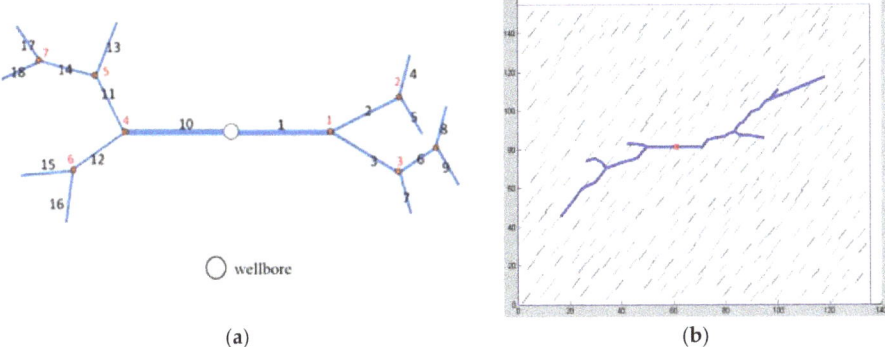

Figure 4. (a) An example of discretizing a tree shape fracture network, with indexed vertices (in red), and indexed fracture panels (in black); (b) Simulation of a hydraulic fracture propagation affected by pre-existing natural fractures [11].

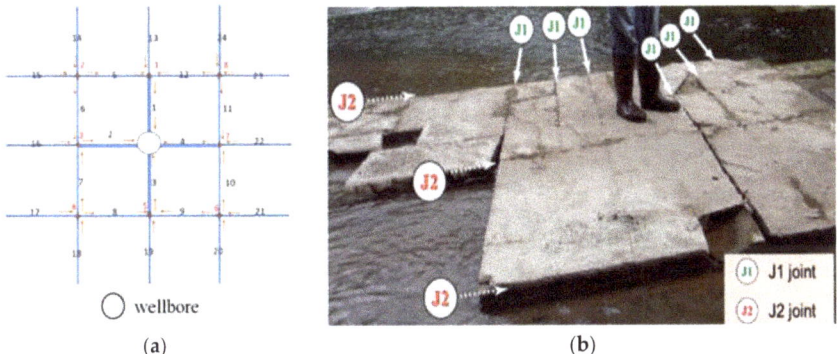

Figure 5. (a) An example of discretization of an orthogonal fracture network, with indexed vertices (in orange), and indexed fracture panels (in black); (b) crosscutting J_1 and J_2 joints in the Marcellus shale exposed in Oatka Creek, Le Roy, New York, USA is a counterpart field example for our model [7].

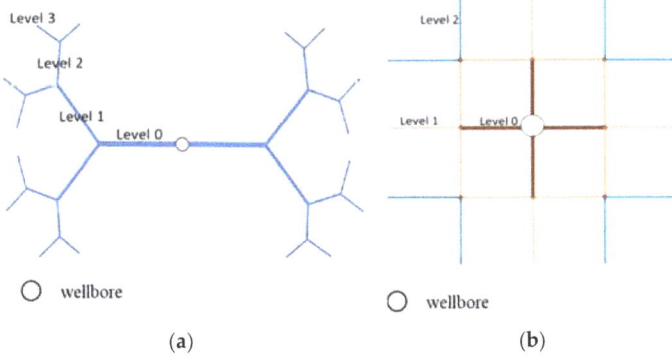

Figure 6. Fracture discretization for fractal fracture networks: (a) tree-shape network; (b) orthogonal fracture network.

The structure and properties of the network can be characterized by fractional changes after each generation based on fractal theory i.e.,

$$L_i = r_L L_0, \tag{53}$$

$$w_{f_i} = r_{w_f} w_{f0}. \tag{54}$$

In orthogonal fracture networks, the width of the fracture segments decreases with the distance from the wellbore, which can be easily verified by stress analysis [13]. The structure and properties of the orthogonal network can also be characterized by fractional reduction after each intersection away from the wellbore as shown in Equations (53) and (54). The compressibility of the activated natural fractures which are labeled with level number larger than 0 in Figure 6 because width reduction also indicates less layers of proppants. The permeability of the activated natural fractures is assumed to be the same.

For the case of long planar fractures, non-uniform fracture width and non-uniform fracture closure can be simulated by discretizing the fracture into several segments as shown in Figure 7. The width of fractures decreases away from the wellbore. For each segment, the width of the segment can be set with an average value based on experience or hydraulic fracturing simulation results [13–15].

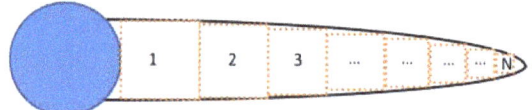

Figure 7. Fracture discretization for one-side of the bi-wing fracture.

To model fracture closure more accurately, understanding of fracture properties is of great importance. Current models for flowback analysis assume that fracture properties, such as fracture permeability and fracture compressibility, do not change over time. In this paper, we will estimate the evolution of these properties with time-based on the mechanical interaction between the formation rock and proppants. For propped fractures, compressibility is contributed by two components: fracture aperture changes and fracture porosity change [16], i.e.,

$$c_f = c_\phi + c_{width}. \tag{55}$$

To incorporate fracture aperture changes during flowback, we adopt the analytical solution for normal compliance of the propped fracture [17]. Assuming there is no slippage between the particles and deformation between the proppant particles is elastic, the compressibility of the fracture due to its width change c_{width} can be obtained by

$$c_{width} = \frac{1}{w_f} \left[\frac{4}{\sqrt[3]{3}} \left(\frac{1 - v_f^2}{E_f} + \frac{1 - v_p^2}{E_p} \right)^{\frac{2}{3}} + (n-1) \left(\frac{1 - v_p^2}{E_p} \right)^{\frac{2}{3}} \right] R \sigma_n^{-\frac{1}{3}}, \tag{56}$$

where n is the number of proppant layers, σ_n is the effective normal stress on proppants, E_f and E_p are the Young's Moduli of the formation and proppants, respectively. v_f and v_p are Poisson's ratios of the formation rock and proppant particles, respectively. We adopt a model to estimate the porosity of the proppants pack [18], that can be simplified as

$$\phi = \phi_i \left[1 - C_0 \left(\frac{\sigma_n}{E_p} \right)^{\frac{2}{3}} \right]^3 \text{ for } \sigma_n \ll E_p, \tag{57}$$

where ϕ_i is the initial porosity of the proppant pack, C_o is a constant depending on the packing structure. Hence, porosity compressibility is estimated from the model by assuming that fracture porosity is equal to the proppant pack as,

$$c_\phi = \frac{2C_o}{\sigma_n} \frac{\frac{\sigma_n}{E_p}}{1 - C_o\left(\frac{\sigma_n}{E_p}\right)^{\frac{2}{3}}}. \tag{58}$$

Permeability can also be modeled as in [18]

$$k = k_i \left[1 - C_o\left(\frac{\sigma_n}{P_o}\right)^{\frac{2}{3}}\right]^4. \tag{59}$$

Thus, total fracture compressibility can be expressed as a function of effective normal stress, proppant properties and mechanical properties of the formation,

$$c_f = \frac{2C_o}{\sigma_n} \frac{\frac{\sigma_n}{E_p}}{1 - C_o\left(\frac{\sigma_n}{E_p}\right)^{\frac{2}{3}}} + \frac{1}{w_f} \frac{4}{\sqrt[3]{3}} \left[\left(\frac{1 - v_f^2}{E_f} + \frac{1 - v_p^2}{E_p}\right)^{\frac{2}{3}} + (n-1)\left(\frac{1 - v_p^2}{E_p}\right)^{\frac{2}{3}}\right] R \sigma_n^{-\frac{1}{3}}. \tag{60}$$

In this study, we compare the flowback response for fracture propped with two kinds of proppants, sand, and ceramics whose properties are shown in Table 1. In comparison to sand, ceramics have a higher strength and is less compressible. Young's modulus and Poisson's ratio of the formation used in this study is assumed to be 25 Gpa and 0.27, respectively. The size of the proppants used in this study is close to 40–70 mesh, we consider an average radius of the proppant is 0.3 mm in the analytical model on fracture compressibility. However, our proposed model is not limited to a specific mesh size and the same procedure can be applied for different proppant mesh sizes.

Table 1. Properties of two types of proppants used in this study.

Properties	Ceramics	Sand
v_p	0.2	0.2
E_p (Gpa)	70	20
R (mm)	0.3	0.3
C_o	2	2

It is worth noticing that in the case with partial proppants coverage, a considerably high fracture compressibility can be used to model the rapid closure of the unpropped part of the fracture.

Upon discretization of the complex fracture networks, the analytical solution for each discretized fracture segment will be solved simultaneously. The next section is dedicated to introducing the numerical technique we utilized for the semi-analytical approach.

5. Workflow of Semi-Analytical Solution for Fracture Networks

Since flowrate is calculated by using the pressure gradient, the pressure boundary condition at each node is solved using the mass balance equation

$$(q_{in})_i = (q_{out})_i \quad i = 1 \dots M. \tag{61}$$

In Equation (35), for the single-phase region, water rate is used while total saturation is used for the two-phase region. The corresponding residual function is defined as

$$R_i = (q_{in})_i - (q_{out})_i \quad i = 1 \dots M. \tag{62}$$

The Newton-Raphson method is utilized to solve the above equations. It should be noted that using this method with a tolerance value of 1×10^{-16}, convergence is achieved by only two to four iterations at each time step. The system of equations is expressed as

$$JdP = -R, \qquad (63)$$

where

$$J = \begin{bmatrix} \frac{\partial R_1}{\partial P_1} & \frac{\partial R_1}{\partial P_2} & \cdots & \frac{\partial R_M}{\partial P_M} \\ \frac{\partial R_2}{\partial P_1} & \frac{\partial R_2}{\partial P_2} & \cdots & \frac{\partial R_2}{\partial P_2} \\ \cdots & \cdots & \cdots & \cdots \\ \frac{\partial R_M}{\partial P_1} & \frac{\partial R_M}{\partial P_2} & \cdots & \frac{\partial R_M}{\partial P_M} \end{bmatrix}, \qquad (64)$$

$$R = \begin{bmatrix} R_1 \\ R_2 \\ \cdots \\ R_M \end{bmatrix}. \qquad (65)$$

Pressure is updated after an arbitrary iteration k as

$$P_{k+1} = P_k + dP_k. \qquad (66)$$

For the two-phase water-oil solution, pseudo-pressure is updated as

$$P_{pk+1} = P_{pk} + dP_{pk}. \qquad (67)$$

A schematic diagram of the semi-analytical workflow is shown in Figure 8.

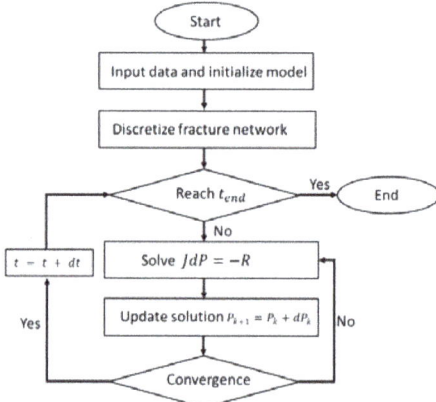

Figure 8. Workflow for single-phase and two-phase semi-analytical solutions.

6. Results and Discussions

To test the accuracy of the proposed semi-analytical model for flowback analysis of the wells completed in shale oil reservoirs, several examples are generated and compared with results from full-fledged numerical simulations with COMSOL Multiphysics and CMG for the two different topologies for fracture networks used in this study. In COMSOL Multiphysics, the diffusivity equation is solved by the built-in two-phase Darcy's flow module using the finite element methods, which allows the generation of fracture networks with non-orthogonal topology. About 3000 quadratic elements is used to solve the two-dimensional, two-phase oil–water flow in both fractal tree-shape and orthogonal fracture networks without matrix influx. Only half of the fracture is modeled due to the symmetry of the problem. A pressure boundary condition is applied at the mouth of the fracture. In CMG, only a quarter of the reservoir is modeled due to the symmetry of the problem with finite difference methods. About 10,000 elements including fracture elements and matrix elements are used to model the two-phase flow with matrix influx. A layer of the elements represents one-half of the fracture, which has higher permeability in comparison to the other elements. The element at the end of the fracture elements is representing the well. The H-adaptive mesh is adopted i.e., elements are coarsening as moving from the fracture to the distant reservoir.

A summary of the reservoir and fracture network properties used in our examples is provided in Table 2. The fracture networks shown in Figure 4 are used for validation purposes. For the orthogonal fracture network, starting from the wellbore, fractures' width and permeability reduce 25% after each intersection away from the wellbore. For the tree-shape fracture network, fracture segments' width reduces 50% after each generation as indicated by earlier geomechanical simulations [13]. In this study, we presumed a smooth pressure drawdown characteristic at the wellbore during early flowback using the below form function,

$$P_{wf}(t) = \frac{1}{1.5}(e^{-\frac{t}{a}} + 0.5)P_i, \tag{68}$$

where a is the drawdown coefficient with the value of 100,000 for our validation example, t is time (in seconds).

Table 2. Input parameters for validation of the semi-analytical solutions.

Properties	Tree-Shape	Orthogonal
Initial fracture pressure (Pa)	3×10^7	3×10^7
Initial Fracture water saturation (fraction)	0.95	0.95
Fracture porosity (fraction)	0.3	0.3
Fracture height (m)	30	30
Fracture width of first level (m)	0.01	0.01
Fracture length of first level (m)	100	50
Width reduction ratio	0.6	0.75
Length reduction ratio	0.5	1
Proppant type	sand	sand
Fracture maximum permeability (m^2)	1×10^{-12}	1×10^{-12}
Water viscosity (Pa-s)	0.001	0.001
Simulation Time (days)	3	3
Number of fracture stages	20	20

A good match for total production between the semi-analytical model and commercial numerical simulations has been obtained to verify the proposed analytical solution as shown in Figure 9. The computational cost of the semi-analytical model in MATLAB (MathWorks, Natick, MA, USA) and the numerical model in COMSOL Multiphysics is compared in Figure 10. In order to examine the influence of the fracture network topology on the flowback behavior, we compare flowback rate from a tree-shape and an orthogonal network fractures sets with the same total length and total fracture volume as shown in Table 2. The reader may notice that the flowrate in the orthogonal fracture network reaches a higher peak in flowrate in comparison to tree-shape fracture but declines faster. The probable reason for this phenomenon is higher conductivity in the orthogonal fracture set in comparison to the tree-shape network with the same volume and length. Different flowback behaviors in the two fracture networks indicate that flowback analysis based on the material balance could be misleading since it cannot capture the effect of the network topology.

For the purpose of doing a sensitivity analysis, we considered the cases described earlier as the base case, and only change a key parameter each time with respect to the base case.

The total fracture volume of the fracture network can be calculated using the length of fracture segments and their corresponding widths. Since it is not realistic to assign a single width to all fractures and on the other hand, it would be computationally very expensive to consider a variable width along all each fracture segment, we use a simplifying hypothesis consistent with geomechanical analyses. We are presuming a value for the width of the fracture segment connected to the well and then assuming a fixed ratio for the fracture width reduction at each intersection away from the wellbore. Therefore, we can investigate the effect of the fracture width distribution by changing the maximum width at the wellbore and the ratio of width changes. In general, increasing fracture width or its length increases the volume inside the fracture network. The influence of the fracture volume on flowback looks similar in both fracture networks.

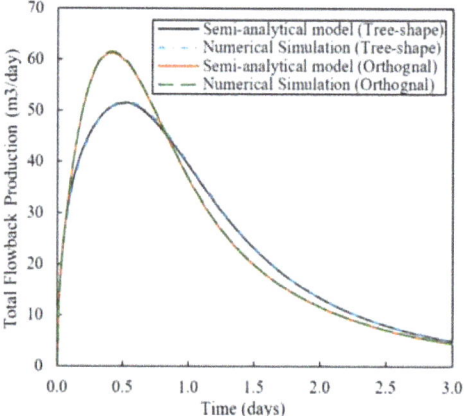

Figure 9. Validation of the two-phase flow model in a tree-shape fracture network and an orthogonal fracture network.

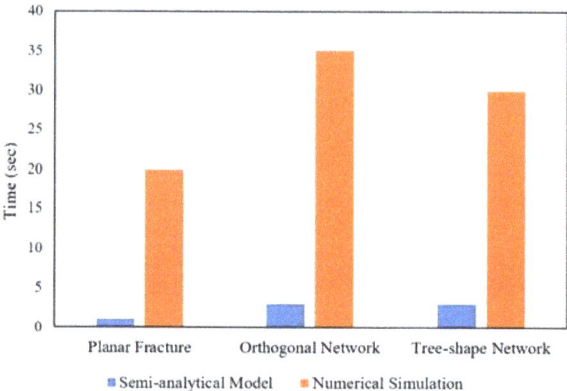

Figure 10. Comparison of computational cost of semi-analytical model and numerical simulation.

6.1. Flowback Responses with Different Fracture Volume

Figure 11 compares the flowback rate for different maximum fracture width (different w_{fi}) in tree-shape networks and orthogonal networks. We can see that in both cases, the flowback rate is nearly proportional to the maximum fracture width (otherwise known as average fracture width). It is also notable that the peak time for the flowrate does not shift with changes in the maximum fracture width as the topology of the fracture network remains the same. In shale gas wells due to nonlaminar effects, this observation may not be necessarily true. Figure 12 compares the flowback rate for different ratios of the fracture width reduction (different r_{w_f}). The result shows that the flowback rate looks similar initially during the transient flow period. Though after a while, the flowrates slightly increase with higher r_{w_f}. We also noticed that changing r_{w_f} shifts the peak time for the flowback rate. The occurrence of peak flowrate delays slightly with higher value of r_{w_f}. A similar phenomenon may also happen with the increase of the total fracture length. Since several parameters affect the flowback behaviors together, the inverse analysis using the proposed model can be non-unique. Hence, constraining one of these values would be helpful to reach a unique fracture network geometry during inverse analyses. For instance, knowledge of micro-seismic data and injection volume can be helpful for estimating the total fracture volume. Fracture width reduction can also be estimated from a coupled analysis of the hydraulic fracturing treatment. Proppant concentration, fracking fluid viscosity, viscosity degradation of the fracking fluids, natural fractures direction with respect to minimum horizontal stress and injection rates are among different factors that may change the fracture width reduction ratio.

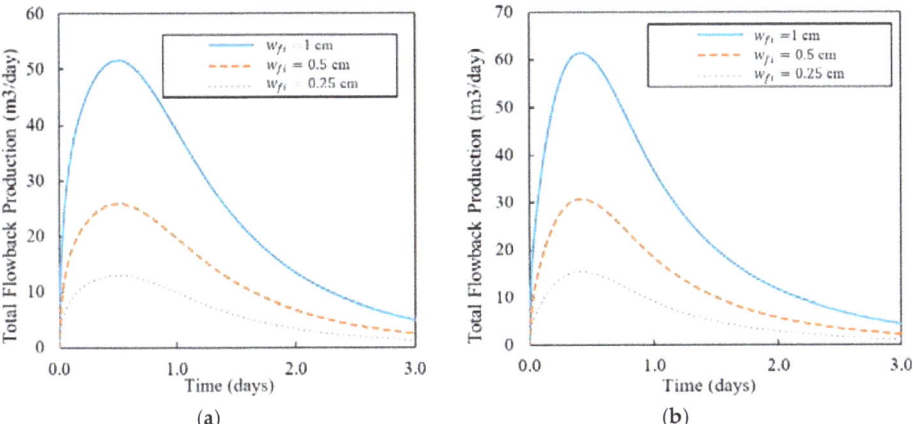

Figure 11. The flowback rate with different fracture widths in the tree-shape fracture network (**a**) and in the orthogonal fracture network (**b**).

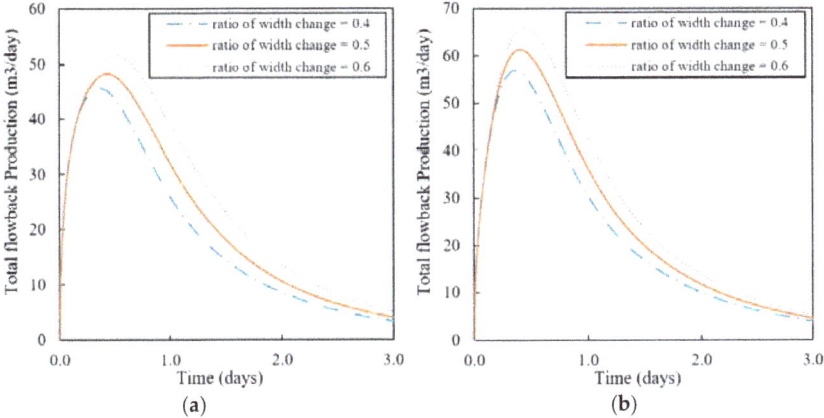

Figure 12. The flowback rate with different r_{w_f} in the tree-shape fracture network (**a**) and in the orthogonal fracture network (**b**).

6.2. Flowback Responses with Different Proppants

Since fracture closure due to in-situ tectonic stresses could be one of the main mechanisms for the flowback of the fracking fluid from induced fracture networks, we investigate the possible effect of the fracture compressibility on the flowback behavior. Figure 13 compares the flowback response in different types of proppants i.e., sand and ceramics. We found that the impact of the proppant type on the flowback is almost the same for the two different fracture networks. Ceramic proppants are less compliant than the regular sand proppants. In both tree-shape and orthogonal networks, flowback rate and recovery can be highly impacted by the proppant type and its corresponding compressibility. As expected, since sand has a lower Young's modulus and a higher compressibility, the water recovery rate increases in comparison to the case with the ceramic proppants. We also noticed that changing proppant type shifts the peak time for the flowback rate. The occurrence of peak flowrate delays slightly with less compressible material. The results suggest that using more compressible proppants may benefit the fracturing fluid recovery, however, it may not provide a high-enough hydraulic conductivity for future production due to the permeability loss driven by compaction.

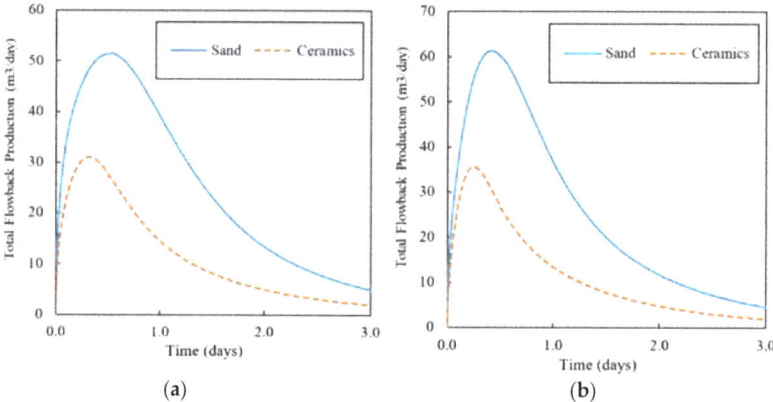

Figure 13. The flowback rate with different proppants in the tree-shape fracture network (**a**) and in the orthogonal network (**b**).

6.3. Flowback Responses with Different Drawdown Strategies

Botthomhole pressure is used as the inner boundary condition. Three different operational schemes are assumed in this study: 1) normal-pressure drawdown, 2) slow pressure drawdown, and 3. Fast pressure drawdown. The drawdown coefficient a in Equation (68) is ranging from 20,0000, 10,0000, to 50,000 to represent slow, normal and fast drawdown cases, respectively. Simulation results for the two fracture networks are shown in Figure 14. The influence of botthomhole pressure drawdown on flowback is similar in the two fracture networks. The initial flowback rate is high, but it also declines faster when the bottomhole pressure drawdown is implemented more aggressively. A faster pressure drawdown results in a higher cleanup rate. However, even though an initial aggressive flowback rate may accelerate the onset of hydrocarbon production, excessively high initial flowback rates may cause proppant flowback and damage to the fracture conductivity as a result of large drag forces driven by the high pressure-gradient forming along the fracture. It can be seen in Figure 14 that using the normal drawdown scheme, we can still achieve a good fracking fluid recovery in the late flowback period in comparison to the aggressive drawdown case, however, a concrete conclusion requires coupling with oil flow from the matrix.

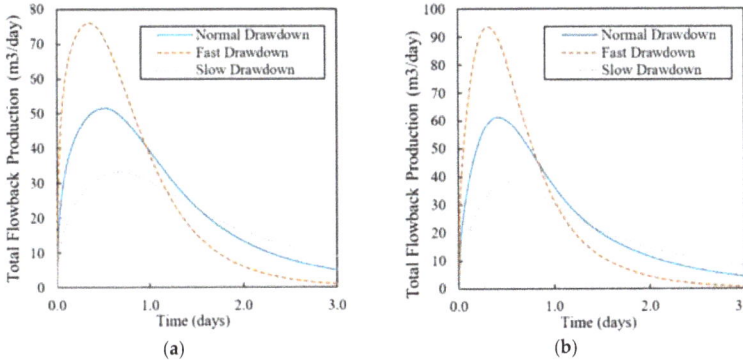

Figure 14. The flowback rate using different botthomhole pressure drawdown schemes in (**a**) the tree-shape fracture network; (**b**) an orthogonal fracture network.

6.4. Flowback Responses with Isotropic Fracture and Anisotropic Fracture Sets

In general, it is expected that the fracture's width decreases with distance from the wellbore. In orthogonal fracture network, while the difference between maximum horizontal stress (S_{Hmax}) and minimum horizontal stress (S_{hmin}) is not significant, fractures' width is expected to remain almost the same in different directions. However, when S_{Hmax} is considerably larger than $S_{H,min}$, it is reasonable to expect that opening for fractures perpendicular to the maximum horizontal stress to be smaller than the one for fractures perpendicular to the minimum horizontal stress. Thus, it makes sense to examine the effect of the difference between maximum and minimum horizontal stresses on the flowback behavior in orthogonal fracture networks (like Figure 5b). The total fracture volume is the same as the volume in the base case, however, the fractures' width is assumed to be different depending on the fracture direction. We assumed that the fracture width in one direction to be 10% higher than the isotropic case while the fracture width in the other direction is 10% lower than the isotropic case. We observe that tectonic stress anisotropy would decrease the flowrate slightly as shown in Figure 15, which is due to the fact that increasing anisotropy can reduce the overall connectivity. However, a concrete conclusion required coupled geomechanical analyses. The difference between the flowrate in the two cases gradually increases till reaching the peak rates and then starts to decrease. The peak of flowback rate decreased about 10% and average flowback rate decreased about 5% in comparison to the case in which the difference between S_{Hmax} and S_{hmin} is not significant. This indicates that the water recovery is slightly higher in the formation which has a smaller difference between S_{Hmax} and S_{Hmin}.

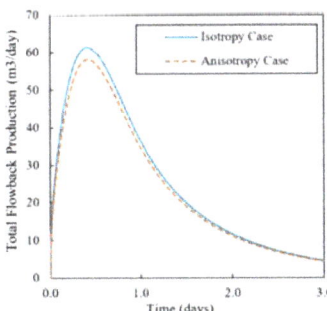

Figure 15. The flowback rate comparison between isotropic fracture and anisotropic fracture sets in an orthogonal network.

6.5. Flowback Responses with Constant Compressibility and Variable Compressibility Models

In current works, flowback analysis is based on the assumption that fracture compressibility remains constant during flowback and production, which indeed according to Equation (60) should be a function of the fracture width and the effective proppant stress. Here, we examine the error that may arise from this assumption. Figure 16 shows that in both fracture networks, the flowback by assuming constant fracture compressibility leads to underestimating flowrate peak and overestimating later production. The result suggests that the flowback analysis using constant compressibility can overestimate the fracture compressibility at the end of the flowback production. By using the proposed semi-analytical model, we can model flowback rate more accurately by dropping the invalid assumption of constant compressibility that can benefit subsequent inverse analyses.

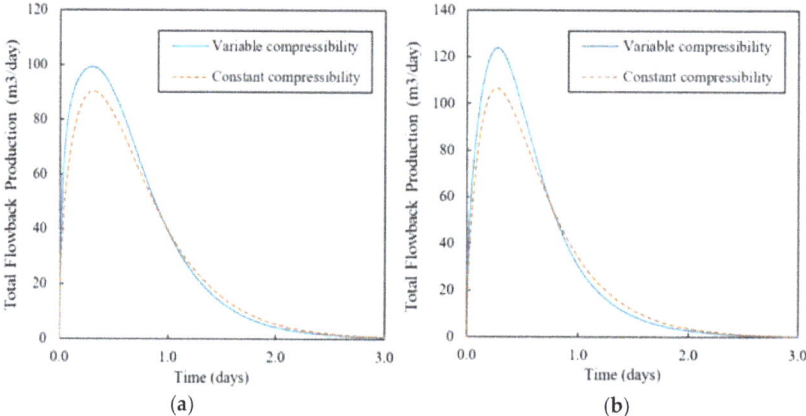

Figure 16. The flowback rate comparison with variable compressibility and constant compressibility in the tree-shape fracture network (**a**) and in the orthogonal network (**b**).

In addition, all the above case studies indicate that with the same fracture volume, an orthogonal fracture network will provide higher flowback production at an early time while the depletion is faster as well. This is because fracture connectivity is higher in orthogonal fracture network.

6.6. Applicability/Limitation of the Proposed Closed-Chamber Solution

The proposed semi-analytical solution presented in this paper only considers fluid flow inside the fracture network. Basically, the matrix flow is neglected in this solution as it takes a while for the reservoir fluid to enter the fracture. In order to gain an insight into the limitation of the proposed closed-chamber solution, we derived a two-phase flowback solution with a matrix influx for bi-wing fractures. The proposed solution with matrix inflow has been validated by CMG as shown in Figure 17 with the input data listed in Table 3. To test the accuracy of the proposed model in the extreme case when oil flux is significant, the oil saturation of the matrix is set to be one. Drawdown coefficient is 50,000 to represent an aggressive choke strategy. To capture the worst-case scenario to limit our flowback solution, the reservoir breakthrough pressure is set to be equal to the initial fracture pressure in the model with a formation influx. However, the true reservoir pressure is much lower than the fracture pressure at the beginning of flowback, additionally, there would be a huge capillary effect which blocks the formation fluid from entering the fracture. Therefore, lower breakthrough pressure should be used in practical analyses. The result indicates that the two-phase flowback solution with matrix oil influx is accurate while its computational cost is much lower than numerical simulation. However, it is worth noting that the model has more error when the oil velocity in matrix is high. This error is rooted in the definition of our two-phase pseudo-pressure, in which the saturation gradient inside the fracture cannot be captured accurately as a continuous function in the case of the large pressure gradient. However, this situation rarely happens during flowback in shale oil reservoirs due to the extremely low permeability of the matrix. This solution is only developed for simple fracture geometries and despite the first solution, this solution cannot be applied to complex fracture networks. But this solution can still serve as a tool to gain insight on the time limitation of the semi-analytical model, i.e., determine when matrix inflow starts to dominate the flowback flow.

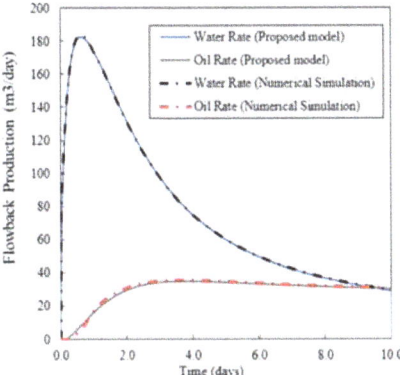

Figure 17. Validation of proposed flowback model with matrix oil influx for a single-stage bi-wing planar fracture.

Table 3. Input parameters of validation case for a two-phase oil–water solution with the matrix influx.

Properties	Value
Initial fracture pressure (Pa)	3×10^7
Initial Fracture water saturation (fraction)	1
Fracture porosity (fraction)	0.6
Fracture height (m)	30
Fracture width (m)	0.02
Fracture length (m)	100
Fracture permeability (m^2)	1×10^{-12}
Matrix permeability (m^2)	1×10^{-18}
Matrix porosity(fraction)	0.05
Total matrix compressibility (1/Pa)	1×10^{-9}
Drawdown coefficient	1×10^6
Proppant Type	Sand
Oil viscosity (Pa·s)	0.001
Water viscosity (Pa·s)	0.001
Number of fracture stages	20

We did three case studies to examine the allowable time interval to use the proposed close-chamber solution. All other inputs are similar to the validation case shown in Table 3. The criteria that we used to define transition time from the fracture linear flow to the matrix linear flow is the time that the flowback rate from the proposed solution deviates more than 20% from the flowback calculated from the solution with matrix oil influx, the model is considered inapplicable hereafter. Figure 18 shows the comparison of the flowrates with and without matrix flow. The results show that the time when the proposed solution without oil matrix influx becomes invalid is 25, 27, and 35 hours for ases 1 to 3, respectively. Case 1 has the highest oil production among all the cases which makes the applicable time relatively short in comparison to the other two cases. Breakthrough pressure is 2.3×10^7 Pa. In case 2, we consider the formation has a smaller pore diameter (i.e., less porosity, less permeability, and higher capillary pressure). Porosity and permeability used is 0.03 and 1×10^{-19} m^2, respectively.

The breakthrough pressure is decreased to 1.9×10^7 Pa. The significant capillary effect postpones the oil breakthrough time. The result indicates the model can be applied for an extended time in shale reservoirs with extremely small pore sizes. In case 3, we consider the formation of oil is more viscous, so oil viscosity is increased to 0.002 Pa·s. When the reservoir fluid is more viscous, matrix oil influx will be slower and the flowback model can be used for a longer time. From the applicability analysis, we show that the proposed semi-analytical solution is working for at least one day or longer times depending to reservoir properties. The allowable time window for this analysis is long enough to make data collection feasible. However, we suggest increasing data collection frequency for more accurate inverse analysis.

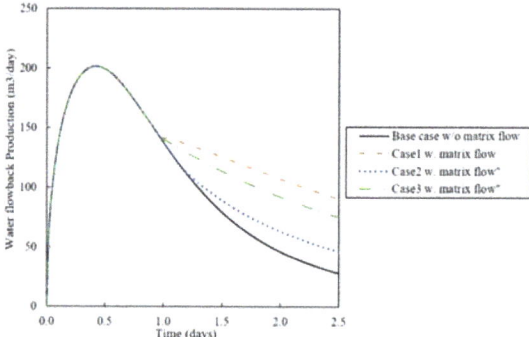

Figure 18. Error of the proposed fast solution and solution with matrix oil influx. Case 1 represents a case with high matrix oil influx. In case 2, a smaller pore radius is used. In case 3, high oil viscosity value is used.

6.7. Field Example

To test the practical applicability of the proposed model, we use the flowback data from a multi-fractured horizontal well completed in a tight oil reservoir in Western Canada [5]. First, using the proposed solution for planar fractures with matrix influx, we showed that a similar production match can be achieved as shown in Figure 19. As we did not include resolved gas in oil in our model, the history matching is based on water production data only.

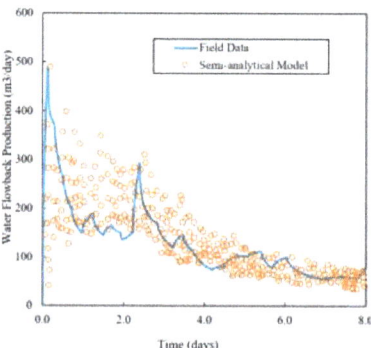

Figure 19. History-match of field case water production using the proposed solution for planar fractures with matrix influx.

As indicated in their study, the authors pointed out that a larger than typical value of fracture width is used in the history matching to account for activated secondary fractures during stimulation.

7. Conclusions

We proposed a semi-analytical solution for two-phase oil-water flowback within the complex fracture network in shale oil wells. The model can capture non-uniform fracture openings within the fracture network and incorporates situations that some oil exists inside the fractures before initiating flow in the well due to imbibition. The solution collapses to the single-phase solution when initial water saturation is one. The results of the semi-analytical solution show a good match with numerical simulations. However, due to the assumption of no matrix influx, the model is only applicable for early-time flowback. The error may arise when a significant amount of oil enters into fractures. As a result, the operators may only apply early data for the purpose of inverse analysis to avoid significant error. We have also proposed a two-phase flowback solution with a matrix oil influx to test the limitation of the closed-chamber solution. Analysis shows that the proposed semi-analytical closed-chamber model is applicable for at least one day for a well with quick oil breakthrough which ensures enough time for the data collection. For wells with bi-wing fractures for formations without a significant amount of natural fractures, the later model with matrix influx can be used to access reservoir properties (e.g., formation permeability, porosity) as well. The semi-analytical model with complex fracture networks can also be further improved by including matrix flow in future works. The proposed models provide a quick tool for inverse analysis which will be addressed in future studies.

Author Contributions: Conceptualization, Y.C., and A.D.T.; methodology, Y.C. and A.D.T.; software, Y.C.; validation, Y.C.; formal analysis, Y.C. and A.D.T.; investigation, Y.C. and A.D.T.; resources, A.D.T.; writing—original draft preparation, Y.C.; writing—review and editing, A.D.T.

Conflicts of Interest: The authors declare no conflict of interest.

Nomenclature

a	Botthomhole pressure drawdown coefficient
c_t	Total compressibility of rock and fluid
c_f	Fracture compressibility
D	Length of stimulated reservoir volume (SRV)
h	Reservoir thickness, m
k_f	Permeability of fracture, md
k_{fi}	Initial permeability of fracture, md
k_{rw}	Water relative permeability
k_{ro}	Oil relative permeability
L_f	Length of fracture segments
M	Number of nodes
N	Number of fracture segments
P	Pressure, Pa
P_i	Pressure, Pa
P_p	Pressure, Pa
q_w	Water rate, m^3/day
q_o	Oil rate, m^3/day
q_t	Total production rate, m^3/day
r_L	Fracture length reduction ratio
r_{wf}	Fracture width reduction ratio
S_w	Water saturation
S_{wi}	Initial water saturation
S_o	Oil saturation
t	time, s
t_{aD}	Dimensionless pseudo-time
w_f	Width of fracture segments
w_{fi}	Initial width of fracture segments
x	Coordinate in the fracture direction
x_D, y_D	Dimensionless coordinates

X_n	Eigen function
ϕ_f	fracture porosity, fraction
ϕ_m	Matrix porosity, fraction
μ_w	Water viscosity, cp
μ_o	Oil viscosity, cp
λ_n	Eigen value
λ_w	Water mobility
λ_o	Water mobility
λ_f	Total mobility

Appendix A. Derivations of Analytical Solution

Eigenfunction expansion is adopted in this study to solve the problem with time-dependent boundary conditions. In case 1, time-dependent Dirichlet boundary conditions are used for both ends. In case 2, time-dependent Dirichlet boundary condition is used only for one boundary and Neumann boundary condition (flux) is used for the other boundary.

Case 1:

$$\frac{\partial P_D}{\partial t_{aD}} = \frac{\partial^2 P_D}{\partial x_D^2} \quad 0 < x_D < 1, t_{aD} > 0 \tag{A1}$$

$$P_D(0, t_{aD}) = P_{D1}(t_{aD}); \quad P_D(1, t_{aD}) = P_{D2}(t_{aD}), \tag{A2}$$

$$P_D(x_D, 0) = 1. \tag{A3}$$

To homogenize the boundary condition, we define

$$P_D = p(x_D, t_{aD}) + w(x_D, t_{aD}). \tag{A4}$$

Let $w(x_D, t_{aD})$ satisfies the inhomogeneous boundary conditions using langrage interpolation

$$w(x_D, t_{aD}) = P_{D1}(t_{aD}) + x_D(P_{D2}(t_{aD}) - P_{D1}(t_{aD})). \tag{A5}$$

Plug (A4) into (A1), we need to solve the following boundary value problem

$$\frac{\partial p}{\partial t_{aD}} = \frac{\partial^2 p}{\partial x_D^2} - \frac{\partial w}{\partial t_D} \quad 0 < x_D < 1, t_{aD} > 0 \tag{A6}$$

$$p(0, t_{aD}) = 0; \quad p(1, t_{aD}) = 0, \tag{A7}$$

$$p(x_D, 0) = 1 - w(x_D, 0). \tag{A8}$$

The eigenfunctions and eigenvalues associated with the Dirichlet boundary conditions (A7) and (A8) are

$$\lambda_n = n\pi \quad n = 1, 2, \ldots N \tag{A9}$$

$$X_n = \sin(\lambda_n x_D). \tag{A10}$$

Define the source term and corresponding series form,

$$S(x_D, t_{aD}) = -\frac{\partial w}{\partial t_{aD}} = -\frac{\partial P_{D1}}{\partial t_{aD}} - x_D\left(\frac{\partial P_{D2}}{\partial t_{aD}} - \frac{\partial P_{D1}}{\partial t_{aD}}\right) = \sum_{n=1}^{\infty} \hat{S}_n(t_{aD}) X_n(x_D) \tag{A11}$$

and

$$\hat{S}_n = -2\int_0^1 \left\{\frac{\partial P_{D1}}{\partial t_{aD}} + x_D\left(\frac{\partial P_{D2}}{\partial t_{aD}} - \frac{\partial P_{D1}}{\partial t_{aD}}\right)\right\} \sin(n\pi x_D) dx_D. \tag{A12}$$

The solution and its derivatives can be expressed in the series form as

$$p(x_D, t_{aD}) = \sum_{n=1}^{\infty} \hat{p}_n(t_{aD}) X_n(x_D) = \sum_{n=1}^{\infty} \hat{p}_n(t_{aD}) \sin(\lambda_n x_D), \tag{A13}$$

$$\frac{\partial p}{\partial t_{aD}} = \sum_{n=1}^{\infty} \frac{\partial \hat{p}_n}{\partial t_{aD}} \sin(\lambda_n x_D); \quad \frac{\partial^2 p}{\partial x_D^2} = \sum_{n=1}^{\infty} -\hat{p}_n(t_{aD}) \lambda_n^2 \sin(\lambda_n x_D). \tag{A14}$$

Substitute (A13) into (A6) we will have

$$\sum_{n=1}^{\infty}\left\{\frac{\partial \hat{p}_n}{\partial t_{aD}} + \lambda_n^2 \hat{p}_n - \hat{S}_n\right\}\sin(\lambda_n x_D) = 0. \quad (A15)$$

Since the eigenfunctions are linearly independent, we have

$$\frac{\partial \hat{p}_n}{\partial t_D} + \lambda_n^2 \hat{p}_n - \hat{S}_n = 0, \quad (A16)$$

which is a first-order linear ordinary differential equation with an integrating factor

$$F = e^{\lambda_n^2 t_{aD}}. \quad (A17)$$

Solution to (A15) is

$$\hat{p}_n = \int_0^{t_{aD}} e^{-\lambda_n^2(t_{aD}-\tau)} \hat{S}_n(t_{aD}) d\tau + e^{-\lambda_n^2 t_{aD}} c_n. \quad (A18)$$

Thus,

$$p(x_D, t_{aD}) = \sum_{n=1}^{\infty}\left\{\int_0^{t_{aD}} e^{-\lambda_n^2(t_{aD}-\tau)} \hat{S}_n(t_{aD}) d\tau + e^{-\lambda_n^2 t_{aD}} c_n\right\} \sin(\lambda_n x_D). \quad (A19)$$

c_n is the term that reflects the initial condition, which is given by

$$c_n = 2\int_0^1 \{1 - (P_{D2}(0) - P_{D1}(0))x_D - P_{D1}(0)\} \sin(\lambda_n x_D) dx_D. \quad (A20)$$

The solution of the original dimensionless governing Equation (A1) is given by

$$P_D(x_D, t_{aD}) = (P_{D2}(t_{aD}) - P_{D1}(t_{aD}))x_D + P_{D1}(t_{aD}) + \sum_{n=1}^{\infty}\left\{\int_0^{t_{aD}} e^{-\lambda_n^2(t_{aD}-\tau)} \hat{S}_n(t_{aD}) d\tau + e^{-\lambda_n^2 t_{aD}} c_n\right\} \sin(\lambda_n x_D). \quad (A21)$$

Case 2:

$$\frac{\partial P_D}{\partial t_{aD}} = \frac{\partial^2 P_D}{\partial x_D^2} \quad 0 < x_D < 1, t_{aD} > 0 \quad (A22)$$

$$P_D(0, t_{aD}) = P_{D1}(t_{aD}); \quad \frac{\partial P_D(1, t_{aD})}{\partial x_D} = 0, \quad (A23)$$

$$P_D(x_D, 0) = 1. \quad (A24)$$

To homogenize the boundary condition, we can define

$$P_D = p(x_D, t_{aD}) + w(x_D, t_{aD}), \quad (A25)$$

Let $w(x_D, t_{aD})$ satisfies the inhomogeneous boundary conditions,

$$w(x_D, t_{aD}) = P_{D1}(t_{aD}) \quad (A26)$$

We need to solve the following boundary value problem,

$$\frac{\partial p}{\partial t_{aD}} = \frac{\partial^2 p}{\partial x_D^2} - \frac{\partial w}{\partial t_D} \quad 0 < x_D < 1, t_{aD} > 0 \quad (A27)$$

$$p(0, t_{aD}) = 0; \quad \frac{\partial p_D(1, t_{aD})}{\partial x_D} = 0, \quad (A28)$$

$$p(x_D, 0) = 1 - w(x_D, 0). \quad (A29)$$

Similarly, the solution of the boundary problem (A27)–(A29) is given by (A19). The eigenfunctions and eigenvalues associated with the boundary conditions (A28) are given by,

$$\lambda_n = \frac{(2n-1)\pi}{2}, \ n = 1, 2, \ldots N \tag{A30}$$

$$X_n = \sin(\lambda_n x_D). \tag{A31}$$

Define the source term and corresponding series form using eigenfunction expansion,

$$S(x_D, t_{aD}) = -\frac{\partial w}{\partial t_{aD}} = -\frac{\partial P_{D1}}{\partial t_{aD}} = \sum_{n=1}^{\infty} \hat{S}_n(t_{aD}) X_n(x_D) \tag{A32}$$

and

$$\hat{S}_n = \int_0^1 -\frac{\partial P_{D1}}{\partial t_{aD}} \sin(\lambda_n x_D) dx_D = -\frac{2}{\lambda_n} \frac{\partial P_{D1}}{\partial t_{aD}}. \tag{A33}$$

c_n is determined from initial condition (A29) as

$$c_n = 2 \int_0^1 \{1 - P_{D1}(t_{aD})\} \sin(\lambda_n x_D) dx_D. \tag{A34}$$

Thus,

$$P_D(x_D, t_{aD}) = P_{D1}(t_{aD}) + \sum_{n=1}^{\infty} \left\{ \int_0^{t_{aD}} e^{-\lambda_n^2(t_{aD}-\tau)} \hat{S}_n(t_{aD}) d\tau + e^{-\lambda_n^2 t_{aD}} c_n \right\} \sin(\lambda_n x_D). \tag{A35}$$

Appendix B. Derivations of Two-Phase Diffusivity Equation without Matrix Flow

1D water and oil flow inside the fracture is given by

$$\frac{\partial}{\partial x}\left(\frac{k_f k_{rw}}{\mu_w}\right)\frac{\partial P}{\partial x} = S_w \phi_f c_\phi \frac{\partial P}{\partial t} + \phi_f \frac{\partial S_w}{\partial t}, \tag{A36}$$

$$\frac{\partial}{\partial x}\left(\frac{k_f k_{ro}}{\mu_o}\right)\frac{\partial P}{\partial x} = (1-S_w)\phi_f c_\phi \frac{\partial P}{\partial t} - \phi_f \frac{\partial S_w}{\partial t}. \tag{A37}$$

Add (A36) and (A37) up,

$$\frac{\partial}{\partial x}\left(k_f(\lambda_w + \lambda_o)\right)\frac{\partial P}{\partial x} = \phi_f c_\phi \frac{\partial P}{\partial t}. \tag{A38}$$

After observation of Equation (A38), we notice that average saturation is independent of time. Dimensionless variables are defined as

$$p_D = \frac{p}{p_i}, \tag{A39}$$

$$x_D = \frac{x}{L}, \tag{A40}$$

$$t_{aD} = \frac{\lambda_w + \lambda_o}{\mu_w x_f^2} \int_0^t \frac{k_f(\overline{p})}{\phi_f c_t(\overline{p})} dt. \tag{A41}$$

Then the linearized diffusivity equation is given by

$$\frac{\partial p_D}{\partial t_{aD}} = \frac{\partial^2 p_D}{\partial x_D^2}. \tag{A42}$$

Appendix C. Derivations of Two-Phase Diffusivity Equation with Matrix Flow

Governing equation of water and oil flow inside the fracture is given by

$$\frac{\partial}{\partial x}\left(\frac{k_f k_{rw}}{\mu_w}\right)\frac{\partial P}{\partial x} = S_w \phi_f c_\phi \frac{\partial P}{\partial t} + \phi_f \frac{\partial S_w}{\partial t}, \tag{A43}$$

$$\frac{\partial}{\partial x}\left(\frac{k_f k_{ro}}{\mu_o}\right)\frac{\partial P}{\partial x} = (1-S_w)\phi_f c_\phi \frac{\partial P}{\partial t} - \phi_f \frac{\partial S_w}{\partial t} + \frac{2v_o}{w_f}. \tag{A44}$$

where v_o represents the matrix oil flow velocity which is solved using oil flow governing equation as presented in Appendix C. Add (A43) and (A44) up.

$$\frac{\partial}{\partial x}\left(k_f(\lambda_w + \lambda_o)\right)\frac{\partial P}{\partial x} = \phi_f c_\phi \frac{\partial P}{\partial t} + \frac{2v_o}{w_f}. \tag{A45}$$

Total transmissibility, pseudo-pressure and pseudo-time are defined as

$$\lambda_t = \frac{k_f k_{rw}}{\mu_w} + \frac{k_f k_{ro}}{\mu_o}, \tag{A46}$$

$$P_p = \int_{P_D}^1 \lambda_t dp = \int_{P_D}^1 \frac{k_f k_{rw}}{\mu_w} + \frac{k_f k_{ro}}{\mu_o} dp, \tag{A47}$$

$$t_{aD} = \frac{1}{x_f^2}\int_0^t \frac{\lambda_t}{\phi_f c_t} dt. \tag{A48}$$

Then (A45) can be transformed into (A46) as

$$\frac{\partial P_p}{\partial t_{aD}} = \frac{\partial^2 P_p}{\partial x_D^2} + \frac{2v_o}{w_f}\frac{x_f^2}{P_i}. \tag{A49}$$

Define the matrix source term as

$$q_m(t_{aD}) = \frac{2v_o}{w_f}\frac{x_f^2}{P_i}. \tag{A50}$$

Thus, the two-phase oil-water flowback governing equation is given by

$$\frac{\partial P_p}{\partial t_{aD}} = \frac{\partial^2 P_p}{\partial x_D^2} + q_m. \tag{A51}$$

Appendix D. Derivations of Fracture Compressibility Formulation

Fracture compressibility is defined as the relative volume change of the fracture as a response to pressure as

$$c_f = \frac{1}{V_f}\frac{dV_f}{dP}, \tag{A52}$$

where V_f is fracture volume and P is pressure. Fracture volume is defined as the product of bulk volume and porosity as

$$V_f = V_b \phi = w_f L H \phi. \tag{A53}$$

Assuming fracture length and fracture height do not recede with pressure, substitute Equation (A53) into (A52),

$$c_f = \frac{1}{w_f}\frac{dw_f}{dP} + \frac{1}{\phi}\frac{d\phi}{dP} = c_{w_f} + c_\phi, \tag{A54}$$

where fracture width compressibility and fracture porosity compressibility are defined as

$$c_{w_f} = \frac{1}{w_f}\frac{dw_f}{dP}, \tag{A55}$$

$$c_\phi = \frac{1}{\phi}\frac{d\phi}{dP}. \tag{A56}$$

References

1. Nolte, K.G. Fracturing-pressure analysis for nonideal behavior. *JPT J. Pet. Technol.* **1991**, *43*, 210–218. [CrossRef]
2. Arevalo-Villagran, J.A.; Wattenbarger, R.A.; Samaniego-Verduzco, F.; Pham, T.T. Production Analysis of Long-Term Linear Flow in Tight Gas Reservoirs: Case Histories. In Proceedings of the SPE Annual Technical Conference and Exhibition, New Orleans, LA, USA, 30 September–3 October 2001.
3. Crafton, J.W. Well Evaluation Using Early Time Post-Stimulation Flowback Data. In Proceedings of the SPE Annual Technical Conference and Exhibition, New Orleans, LA, USA, 27–30 Sepmtember 1998.
4. Abbasi, M.; Dehghanpour, H.; Hawkes, R.V. Flowback Analysis for Fracture Characterization. In Proceedings of the SPE Canadian Unconventional Resources Conference, Calgary, AB, Canada, 30 October–1 November 2012.
5. Clarkson, C.R.; Qanbari, F.; Williams-Kovacs, J.D. Semi-analytical model for matching flowback and early-time production of multi-fractured horizontal tight oil wells. *J. Unconvent. Oil Gas Resour.* **2016**, *15*, 15–145.
6. Dahi Taleghani, A. How Natural Fractures Could Affect Hydraulic-Fracture Geometry. *SPE J.* **2013**, *19*, 161–171. [CrossRef]
7. Engelder, T.; Lash, G.G.; Uzcátegui, R.S. Joint sets that enhance production from Middle and Upper Devonian gas shales of the Appalachian Basin. *Am. Assoc. Pet. Geol. Bull.* **2009**, *93*, 857–889. [CrossRef]
8. Clarkson, C.R.; Qanbari, F.; Williams-Kovacs, J.D.; Zanganeh, B. Fracture Propagation, Leakoff and Flowback Modeling for Tight Oil Wells Using the Dynamic Drainage Area Concept. In Proceedings of the SPE Western Regional Meeting, Bakersfield, CA, USA, 23–27 April 2017.
9. Jiang, Y.; Dahi Taleghani, A. Modified Extended Finite Element Methods for Gas Flow in Fractured Reservoirs: A Pseudo-Pressure Approach. *J. Energy Resour. Technol.* **2018**, *140*, 073101. [CrossRef]
10. Adefidipe, O.A.; Dehghanpour, H.; Virues, C.J. Immediate Gas Production from Shale Gas Wells: A Two-Phase Flowback Model. In Proceedings of the SPE Unconventional Resources Conference, The Woodlands, TX, USA, 1–3 April 2014.
11. Dahi Taleghani, A. Fracture Re-Initiation As a Possible Branching Mechanism During Hydraulic Fracturing. In Proceedings of the 44th U.S. Rock Mechanics Symposium and 5th U.S.-Canada Rock Mechanics Symposium, Salt Lake City, UT, USA, 27–30 June 2010.
12. Shah, K.; Shelley, R.F.; Gusain, D.; Lehman, L.V.; Mohammadnejad, A.; Conway, M.T. Development of the Brittle Shale Fracture Network Model. In Proceedings of the SPE Hydraulic Fracturing Technology Conference, The Woodlands, TX, USA, 4–6 February 2013.
13. Dahi Taleghani, A.; Yu, H.; Lian, Z. Coupled modeling of complex fracture networks induced during hydraulic fracturing treatments. In Proceedings of the 80th EAGE Conference and Exhibition 2018: Opportunities Presented by the Energy Transition, Copenhagen, Denark, 11–14 June 2018.
14. Yu, H.; Dahi Taleghani, A.; Lian, Z. On how pumping hesitations may improve complexity of hydraulic fractures, a simulation study. *Fuel* **2019**, *249*, 294–308. [CrossRef]
15. Dahi Taleghani, A.; Gonzalez Chavez, M.; Yu, H.; Asala, H. Numerical simulation of hydraulic fracture propagation in naturally fractured formations using the cohesive zone model. *J. Petro. Sci. Eng.* **2018**, *165*, 42–57. [CrossRef]
16. Cai, Y.; Dahi Taleghani, A. Pursuing Improved Flowback Recovery after Hydraulic Fracturing. In Proceedings of the SPE Eastern Regional Meeting, Charleston, WV, USA, 15–17 October 2019.
17. Sotelo, E.; Cho, Y.; Gibson, R.L. Compliance estimation and multiscale seismic simulation of hydraulic fractures. *SEG Tech. Program Expand. Abstr.* **2018**, 3236–3240. [CrossRef]
18. Gangi, A.F. Variation of whole and fractured porous rock permeability with confining pressure. *Int. J. Rock Mech. Min. Sci. Geomech. Abstr.* **1978**, *15*, 249–257. [CrossRef]

© 2019 by the authors. Licensee MDPI, Basel, Switzerland. This article is an open access article distributed under the terms and conditions of the Creative Commons Attribution (CC BY) license (http://creativecommons.org/licenses/by/4.0/).

Article

A Novel Porous Media Permeability Model Based on Fractal Theory and Ideal Particle Pore-Space Geometry Assumption

Yongquan Hu [1], Qiang Wang [1,*], Jinzhou Zhao [1], Shouchang Xie [2] and Hong Jiang [2]

1. State Key Laboratory of Oil-Gas Reservoir Geology & Exploitation, Southwest Petroleum University, Chengdu 610500, China; 201621000663@stu.swpu.edu.cn (Y.H.); 201811000142@stu.swpu.edu.cn (J.Z.)
2. Xinjiang Oilfield Company Development Company, Karamay 834000, China; xieshouc@prtrochina.com.cn (S.X.); hongj@prtrochina.com.cn (H.J.)
* Correspondence: 201822000147@stu.swpu.edu.cn

Received: 28 October 2019; Accepted: 17 January 2020; Published: 21 January 2020

Abstract: In this paper, a novel porous media permeability model is established by using particle model, capillary bundle model and fractal theory. The three-dimensional irregular spatial characteristics composed of two ideal particles are considered in the model. Compared with previous models, the results of our model are closer to the experimental data. The results show that the tortuosity fractal dimension is negatively correlated with porosity, while the pore area fractal dimension is positively correlated with porosity; The permeability is negatively correlated with the tortuosity fractal dimension and positively correlated with the integral fractal dimension of pore surface and particle radius. When the tortuosity fractal dimension is close to 1 and the pore area fractal dimension is close to 2, the faster the permeability changes, the greater the impact. Different particle arrangement has great influence on porous media permeability. When the porosity is close to 0 and close to 1, the greater the difference coefficient is, the more the permeability of different arrangement is affected. In addition, the larger the particle radius is, the greater the permeability difference coefficient will be, and the greater the permeability difference will be for different particle arrangements. With the increase of fractal dimension, the permeability difference coefficient first decreases and then increases. When the pore area fractal dimension approaches 2, the permeability difference coefficient changes faster and reaches the minimum value, and when the tortuosity fractal dimension approaches 1, the permeability difference coefficient changes faster and reaches the minimum value. Our research is helpful to further understand the connotation of medium transmission in porous media.

Keywords: porous media; fractal theory; particles model; permeability; tube bundle model

1. Introduction

Fibrous and reservoir rocks are porous media with complex microstructure. It is very important to reasonably characterize the pore structure and predict the permeability of porous media for industrial application and petroleum exploration and development [1–3]. Pore structure plays an important role in the properties of porous media. However, due to the complexity of microstructure and irregularity of pore structure, it is always a challenging task to predict permeability [4–8]. The microstructures of oolitic graintone and dolograinstone can be found in previous study [9,10]. There are plenty of rounded particles that make up the skeleton of the rock and the blue areas represent random pores or micropores between particles. These pores are randomly distributed in space, with sizes spanning several orders of magnitude and connecting with each other through thick channels, forming a complex network of pores. In order to achieve qualitative research, many researchers use this particle model (see Figure 1a) to construct rock space, with the purpose of reconstructing the complex pore structure of the rock, so as to more accurately describe the properties of the rock [11]. Gebart [12] regards the cross section of fibrous porous media as the cross section of circles of equal diameter arranged in a fixed geometry (I and II in Figure 1a), and deduces the permeability model of the fluid flowing along the fiber and perpendicular to the fiber direction. The model establishes the relationship between the permeability and particle size and the volume fraction of the fiber. Since the randomness and complexity of pore space distribution are not considered in this model, the permeability of porous media with low porosity predicted by this model is quite different from the experimental results [13]. Therefore, fractal theory was introduced to study the relationship between pore structure and permeability of porous media from a more realistic perspective.

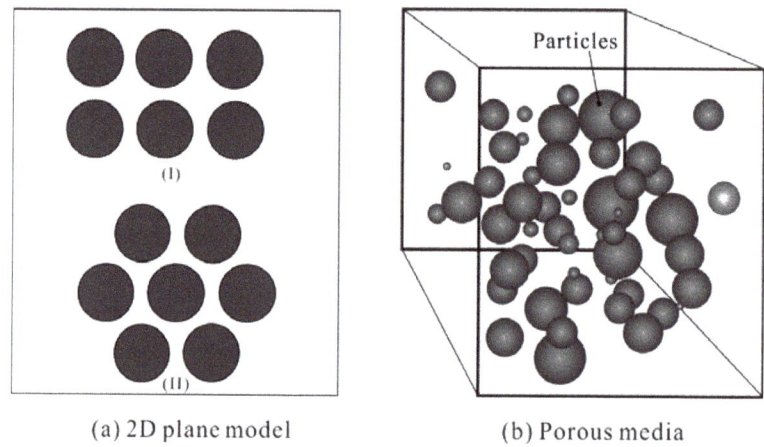

(a) 2D plane model   (b) Porous media

Figure 1. Schematic diagram of porous media model.

Fractal theory describes a natural phenomenon, which allows the self-similarity of objects to represent the properties of objects or scale invariance, and is an effective method to describe the complexity and irregularity of pores in porous media and their related macroscopic transport characteristics [14–17]. The application of fractal theory to the study of porous media is actually to use an appropriate method or model to characterize the structural characteristics of porous media, and then analyze its properties such as transmission and strain [18,19]. At present, there have been many research achievements that use fractal theory and technology to characterize the structure of porous media and analyze its permeability. According to the characterization models of porous media, there are mainly permeability models based on fractal capillary bundle model, fractal improvement for the limitations of the classical Kozeny-Carman (KC) permeability equation, random and deterministic mass fractal porous media permeability model, and fractal effective permeability models based on the Bautista Manero Puig (BMP) model [20–23]. Based on the fractal characteristics of fibrous porous media, Yu et al. [24] used the fractal capillary bundle model to put forward the fractal plane permeability model applicable to various fibrous media earlier. Studies have shown that fiber preforms mainly depends on the fiber bundles of macro pore permeability. Based on Yu's research, Xu and Yu [25] introduced the cross-sectional area of the unit and developed a fractal permeability model for homogeneous porous media. The model is not limited to fiber materials, but also applicable to other fractal porous media. Combining the maximum hydraulic diameter with filament diameter, Xiao et al. [26] expressed the maximum pore diameter as the relationship between porosity and particle size, fractal dimension, and further obtained the dimensionless permeability expression of porous fiber gas diffusion layer. The original fractal permeability model is based on the fractal power law distribution of pore size and Hagen-Poiseuille (H-P) flow equation of circular curved capillary. Xu and Yu, and Xiao et al.'s research is based on the same conditions and assumptions [25–28]. Their models can well fit the permeability test results of the existing high-porosity fiber materials, but the permeability of the low-porosity porous media is not well fitted. Considering the discreteness and discontinuity of fractal geometry, Shou et al. [29] proposed a differential permeability model of fiber porous media, which can fit the permeability results of fiber porous media with a wider porosity range. According to the theory of fluid mechanics, H-P flow is dominant in pore media only when the Knudsen number is less than 0.01. When the Knudsen number is greater than 10, the Knudsen number is dominant. Considering these factors, Zhang et al. developed a new fractal gas permeability model for porous fiber films [30]. Considering electro kinetic phenomena, Zhu et al. studied the flow behavior of porous fibrous media using fractal technique, and derived a fractal permeability model [31]. Using similar methods and theories, Zhu et al. also studied the heat and mass transfer characteristics of fibrous porous media and took capillary force into consideration. However, none of the above fractal studies considered the influence of this factor [32]. Costa, Othman et al. established the porosity permeability model based on the fractal hypothesis of porous media particles and pore area, improved the classic Kozeny Carman permeability equation, and re-verified its validity [13,33]. Considering the porosity connectivity probability, Cihan et al. developed a three-dimensional solid mass fractal porous media permeability model [34]. According to the classic Sierpinski carpet quality fractal model [35–38], Pia and Sanna proposed a new combination of structural units, formed a intermingled fractal units model (IFU), and studied the transmission characteristics of porous media such as permeability and thermal conductivity [39–41]. Subsequent researchers have proposed new pore media models through this method. Similar to the above fractal study on permeability of fibrous porous media, on the basis of Yu's study [24], Turcio et al. studied the effective permeability of non-Newtonian by using the fractal capillary bundle model, and calculated the effective permeability by using the Bautiista Manero Puig (BMP) model [42,43].

With the development and application of fractal theory, more and more factors are considered into the permeability model, and the fit between model results and experimental results is getting higher and higher. However, in researches on porous media permeability based on particle model, most permeability models are still based on two-dimensional particle model, which cannot fully reflect the three-dimensional influence of pore geometry space [44]. In addition, most of the modified KC equations are more applicable to porous media with large porosity, while there are few research results on porous media with small porosity [45–47]. In this paper, the pore space is considered to be randomly distributed and its size spans over two orders of magnitude, which satisfies the scaling law of fractals. In this paper, the matrix structure of porous media is composed of spherical particles (as show in Figure 1b), and then the three-dimensional irregular pore space composed of equal-diameter particles according to ideal geometric model is approximately transformed into capillary bundle model. Finally, the relationship between permeability and pore structure parameters, pore area fractal dimension and capillary tortuosity fractal dimension is established by using fractal principle. In addition, the permeability models of the two particle combination modes were deduced, and finally the permeability models of the loose mode and the compact mode were obtained. Compared with the experimental data and the results of existing analytical formulas, our model is reliable and accurate.

2. Mathematical Model

2.1. Fractal Characteristics of Spherical Particles Matrix

In this paper, the matrix is assumed to be composed of spherical particle clusters, each cluster is composed of particles with the same radius, and the matrix particle radius between clusters is randomly distributed, so there are pores of different sizes in the porous media, and these pore Spaces satisfy the fractal scale theorem [13,48]. As shown in Figure 2, there are two types of clusters: loose mode I cluster (Figure 2a) and compact mode II cluster (Figure 2d). Type I cluster is the structural combination of particles that can form the largest pore space, while type II cluster is the structural combination of particles that can form the smallest pore space. Type I clusters (Figure 2a) consist of eight spherical particles forming a matrix particle cluster. The central point connecting each particle can form a cube. Cutting along the cube surface can form a matrix cube unit as shown in Figure 2b. From Figure 2b,c, eight one eighth of the matrix particles constitute an irregular matrix pore space, which can maximize the matrix pore space. Type II clusters (Figure 2d) consist of four matrix particles, which can form rhombohedrons by connecting the central points of the particles. Cutting along the surface of the rhombohedron can form a unit as shown in Figure 2e,f.

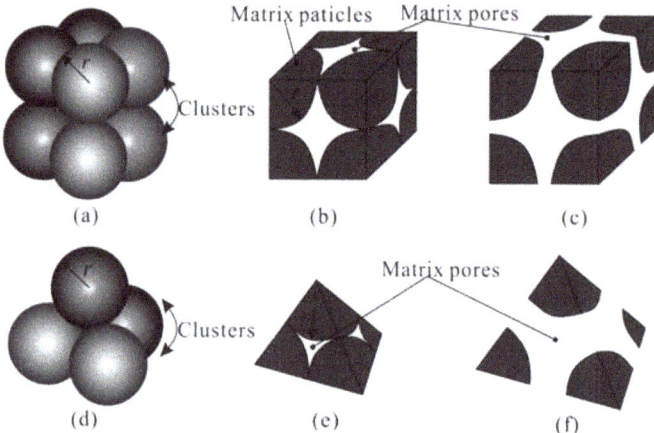

Figure 2. Schematic diagram of structural model of ideal pore space. (**a**) Arrangement of loose particles; (**b**) Ideal loose arrangement; (**c**) Conventional loose arrangement; (**d**) Compact particles arrangement; (**e**) Ideal most compact arrangement; (**f**) Regular compact permutation.

For matrix units composed of type I clusters according to ideal geometry, they are mainly composed of matrix particles and pore space, as shown in Figure 3a. The porosity can be expressed as the relationship between cube volume and matrix particle volume [13]:

$$\varphi = \frac{V_c - \frac{4}{3}\pi r^3}{V_c} \tag{1}$$

where φ is the effective reservoir porosity; V_c represents the volume of the cluster cube; r is the radius of the matrix particle.

Through Equation (1), the cubic unit volume can be deduced as follows:

$$V_c = \frac{4\pi r^3}{3(1-\varphi)} \tag{2}$$

It can be seen from Equation (2) that the volume of cubic unit is related to particle radius and porosity. Since V_c represents the cube space formed by cluster matrix particles, there is:

$$V_c = L_0^3 \tag{3}$$

where l_0 is the side length of a matrix cubic unit.

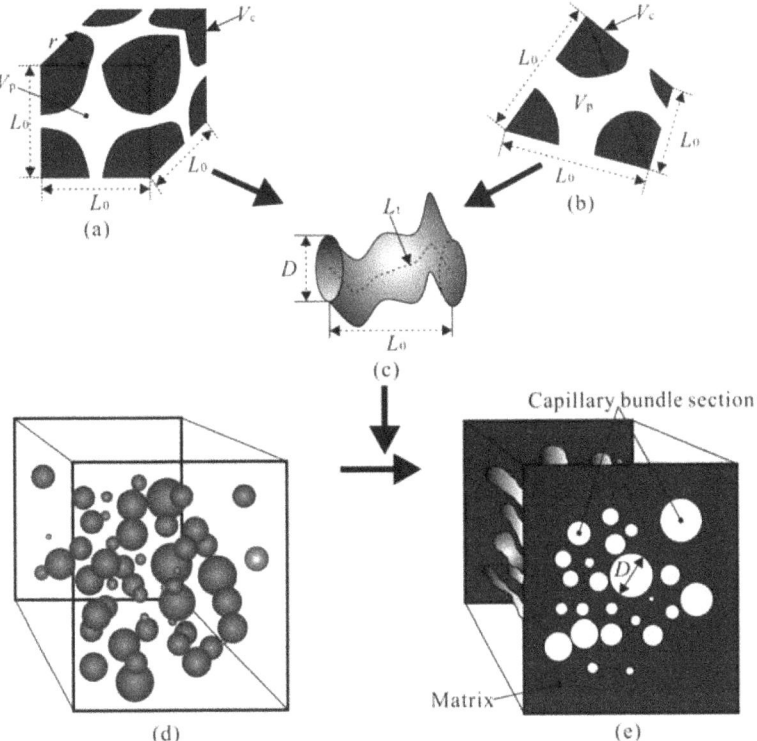

Figure 3. Schematic diagram of irregular pore space transformed into capillary bundle model. (**a**) conventional loose arrangement; (**b**) regular compact permutation; (**c**) capillary channel; (**d**) actual particle model; (**e**) transformed capillary model.

Combining Equations (2) and (3), the expression can be obtained:

$$L_0 = r \times \sqrt[3]{\frac{4\pi}{3(1-\varphi)}} \tag{4}$$

The maximum matrix pore volume $V_{p,max}$ can be expressed as:

$$V_{p,max} = V_c - \frac{4\pi r^3}{3} = \frac{4\pi r^3}{3} \frac{\varphi}{1-\varphi} \tag{5}$$

When the fluid flows through the matrix element, it mainly passes through the three sections shown in Figure 4, which not only passes through the maximum section of the ideal geometric square pore in the middle, but also passes through the minimum section of the ideal geometric irregularity on both sides [13]. The rationality of pore space is considered. Therefore, the irregular matrix pore space composed of solid particles is approximately equivalent to capillary bundle, and the capillary diameter on the equivalent cross section satisfies the fractal scale law, then:

$$V_p = L_t \frac{\pi D^2}{4} \tag{6}$$

where D is the diameter of the capillary bundle section; L_t is the actual length of tortuous bundle, and the ratio of its value to the characteristic length of the external surface can be expressed as:

$$\tau = \frac{L_t}{L_0} \tag{7}$$

Formula (7) is the traditional definition of tortuosity, whose value can be taken as average tortuosity. Yu and Li, Yun et al. and Kou et al. studied the tortuousness model of two-dimensional, three-dimensional porous media and mixed porous media composed of square and circular particles [49–52]. A series of relationships between tortuosity and porosity were established.

Combined with Equations (4)–(7), the maximum tortuous capillary bundle cross-section diameter obtained by cubic element approximation can be derived:

$$D_{max} = 4 \sqrt[3]{\frac{1}{9}} \sqrt[6]{\frac{1}{4\pi}} \sqrt{\frac{\varphi}{\tau}} \sqrt[3]{\frac{1}{(1-\varphi)}} r \tag{8}$$

For type II clusters, the porosity in this structure can also be expressed as the relationship between rhombohedral volume and matrix particle volume:

$$\varphi_e = \frac{V_c - \frac{2}{9}\pi r^3}{V_c} \tag{9}$$

The pore volume expressed in terms of effective porosity, cementing ratio and particle radius can be obtained from Equation (9):

$$V_c = \frac{2\pi r^3}{9(1-\varphi)} \tag{10}$$

Rhombohedron is a regular tetrahedron. According to its geometric characteristics, its volume can be expressed as:

$$V_c = \frac{\sqrt{2}}{12} L_0^3 \tag{11}$$

Side length of rhombohedron can be obtained by combining Equations (11) and (10):

$$L_0 = r \times \sqrt[3]{\frac{24\pi}{9\sqrt{2}(1-\varphi)}} \tag{12}$$

According to Equation (10), the maximum matrix pore volume of rhombohedron with porosity of φ can be obtained as:

$$V_{p,max} = V_c - \frac{2\pi r^3}{9} = \frac{2\pi r^3}{9} \frac{\varphi}{1-\varphi} \tag{13}$$

By combining Equations (6), (7), (12) and (13), the maximum sectional diameter of tortuous bundle approximately obtained by rhombohedron can be obtained as follows:

$$D_{max} = \frac{2\sqrt{2}}{3} \sqrt[6]{\frac{9\sqrt{2}}{24\pi}} \sqrt{\frac{\varphi}{\tau}} \sqrt[3]{\frac{1}{(1-\varphi)}} r \tag{14}$$

It can be seen from Equations (8) and (14) that the two equations have the same form, and the capillary diameter is a function of particle radius, porosity and tortuosity. According to Yu and Li [49], tortuosity is also a function of porosity. Therefore, the key parameters to determine the equivalent capillary diameter are particle radius and porosity.

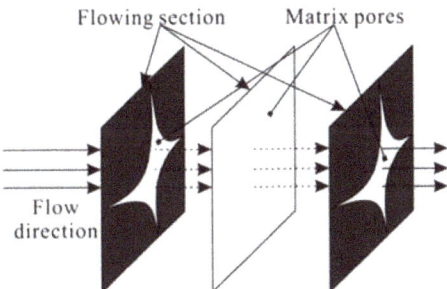

Figure 4. Fluid flow through the pore section diagram.

2.2. Fractal Capillary Bundle Model for Porous Media

The capillary bundle model is often used to simulate the flow and transport characteristics of porous media. The pore size distribution and the fractal scale relationship of curved streamline constitute the basis of fractal capillary bundle model. Most natural rocks have fractal characteristics in a certain range. For example, sandstone, shale and carbonate rocks are self-similar in a range of 3 to 4 orders of magnitude. Fractal dimension can be used to quantitatively describe the characteristics of pore size distribution [53]. In fractal porous media, the cumulative number of pores with diameters larger than the scale N follows the scaling law relationship:

$$N(\geq D) = \left(\frac{D_{max}}{D}\right)^{D_p} \tag{15}$$

where D_p is pore area fractal dimension. When the value is $0 < D_p < 2$, it denotes two-dimensional space, and when $0 < D_p < 3$, it denotes three-dimensional space. According to Yu's research, the fractal dimension of pore area can be expressed as [24,49]:

$$D_p = d - \frac{\ln \varphi}{\ln \frac{D_{min}}{D_{max}}} \tag{16}$$

where D_{min} is the smallest diameter of capillary tube. d is equal to 2 (two-dimension) or 3 (three-dimension).

If the pore diameter ranges from D_{min} to D_{max}, the total number of medium pores under the condition $D > D_{min}$ can be obtained. So we know from Equation (15):

$$N_p(D > D_{min}) = \left(\frac{D_{max}}{D_{min}}\right)^{D_p} \tag{17}$$

Differentiating Equation (15), we can get:

$$-dN = D_p D_{max}^{D_p} D^{-(D_p+1)} dD \tag{18}$$

Equation (18) gives the number of pores in the interval D and $D + dD$, and $-dN > 0$ indicates that the number of pores decreases with the increase of pore diameter. Divide Equations (17) and (18) to get:

$$-\frac{dN}{N_p} = D_p D_{min}^{D_p} D^{-(D_p+1)} dD = f(D) dD \tag{19}$$

where $f(D) = D_p D_{min}^{D_p} D^{-(D_p+1)}$ is the probability density function of pore distribution, which should meet the normalization condition:

$$\int_{-\infty}^{+\infty} f(D)dD = \int_{D_{min}}^{D_{max}} f(D)dD = 1 - \left(\frac{D_{min}}{D_{max}}\right)^{D_p} = 1 \qquad (20)$$

The condition for Equation (20) to be true must be:

$$\left(\frac{D_{min}}{D_{max}}\right)^{D_p} = 0 \qquad (21)$$

Curved capillary or curved streamline of fluid flow also has fractal characteristics. Tortuosity depends on measurement scale and fractal dimension of streamline, which can better reflect the characteristics of curved capillary streamline. The tortuosity fractal dimension is considered to be a more fundamental parameter than permeability. The scaling relationship between the length of curved streamline and the characteristic length of the medium in porous media can be expressed as [48,54]:

$$L_t(D) = L_0^{D_t} D^{1-D_t} \qquad (22)$$

$$D_t = 1 + \frac{\ln \tau_{av}}{\ln \frac{L_0}{D_{av}}} \qquad (23)$$

where D_t is the tortuosity fractal dimension. Average capillary diameter D_{av} and average tortuosity τ_{av} can be expressed as [49,55]:

$$D_{av} = \int_{D_{min}}^{D_{max}} Df(D)dD = \frac{D_p}{D_p - 1} D_{min}\left[1 - \left(\frac{D_{min}}{D_{max}}\right)^{D_p - 1}\right] \qquad (24)$$

$$\tau_{av} = \frac{1}{2}\left[1 + \frac{1}{2}\sqrt{1-\varphi} + \sqrt{1-\varphi}\frac{\sqrt{\left(\frac{1}{\sqrt{1-\varphi}} - 1\right)^2 + \frac{1}{4}}}{1 - \sqrt{1-\varphi}}\right] \qquad (25)$$

2.3. Fractal Permeability Model for Porous Media

The flow of fluid in porous media is regarded as the flow in a curved capillary. The size distribution of the capillary channel satisfies the fractal distribution. According to the modified Hagen-Poiseulle equation, the flow rate q of a fluid passing through a curved capillary can be expressed as [56,57]:

$$q(D) = \frac{\pi}{128} \frac{\Delta p}{L_t(D)} \frac{D^4}{\mu} \qquad (26)$$

Equation (26) is obtained by considering the capillary tube as a circle. Since the size distribution of capillary channels satisfies the fractal distribution, the total flow rate Q can be obtained by integrating the flow rate in a single root canal from the minimum pore diameter D_{min} to the maximum pore diameter D_{max}:

$$Q = -\int_{D_{min}}^{D_{max}} q(D)dN(D) = \frac{\pi}{128\mu} \frac{\Delta p}{LL_0^{D_t-1}} \frac{D_p}{3+D_t-D_p} D_{max}^{3+D_t}\left[1 - \left(\frac{D_{min}}{D_{max}}\right)^{D_p}\left(\frac{D_{min}}{D_{max}}\right)^{3+D_t-2D_p}\right] \qquad (27)$$

Because $1 < D_t < 2$ and $1 < D_p < 2$, the exponent satisfies $3 + D_t - 2D_p > 0$. And there is $0 < \left(\frac{D_{min}}{D_{max}}\right)^{3+D_t-2D_p} < 1$, according to the criterion $\left(\frac{D_{min}}{D_{max}}\right)^{D_p} < 10^{-2}$ of Yu and Li [49], Equation (27) can be simplified into:

$$Q = \frac{\pi}{128\mu} \frac{\Delta p}{LL_0^{D_t-1}} \frac{D_p}{3+D_t-D_p} D_{max}^{3+D_t} \qquad (28)$$

According to Darcy's law, permeability can be expressed as:

$$K = \frac{Q\mu L}{A\Delta p} \tag{29}$$

where A represents the cross section area of capillary bundle passing through the fluid. If the approximately equivalent capillary section is the flow section, and the surface porosity is assumed to be equal to the volume porosity. Then according to the fractal principle, the total pore area and the flow section area on the flow section can be expressed as [6]:

$$\begin{aligned} A_p &= -\int_{D_{min}}^{D_{max}} \tfrac{1}{4}\pi D^2 dN = \int_{D_{min}}^{D_{max}} \tfrac{1}{4}\pi D^2 D_p D_{max}^{D_p} D^{-(D_p+1)} dD \\ &= \frac{\pi D_p D_{max}^2}{4(2-D_p)}\left[1 - \left(\frac{D_{min}}{D_{max}}\right)^{2-D_p}\right] \end{aligned} \tag{30}$$

$$A = \frac{A_p}{\varphi} = \frac{\pi D_p D_{max}^2}{4\varphi(2-D_p)}\left[1 - \left(\frac{D_{min}}{D_{max}}\right)^{2-D_p}\right] \tag{31}$$

Because of $\varphi = \left(\frac{D_{min}}{D_{max}}\right)^{2-D_p}$, the cross-section area can be simplified to [11]:

$$A = \frac{\pi(1-\varphi)D_p D_{max}^2}{4\varphi(2-D_p)} \tag{32}$$

The permeability expression of pore media can be obtained by combining Equations (28), (29) and (32):

$$K = \frac{1}{32}\frac{\varphi}{1-\varphi}\frac{1}{L_0^{D_t-1}}\frac{2-D_p}{3+D_t-D_p}D_{max}^{1+D_t} \tag{33}$$

Substitute Equations (4) and (8) into Equation (33) to get the permeability expression of porous media composed of clusters of type I:

$$K_I = \frac{1}{128}\frac{\varphi}{1-\varphi}\left(\sqrt[3]{\frac{4\pi}{3(1-\varphi)}}\right)^{1-D_t}\left(4\sqrt[3]{\frac{1}{9}}\sqrt[6]{\frac{1}{48}}\sqrt{\frac{\varphi}{\tau_{av}}}\sqrt[3]{\frac{1}{1-\varphi}}\right)^{1+D_p}\frac{2-D_p}{3+D_t-D_p}D_f^2 \tag{34}$$

By substituting Equations (12) and (14) into Equation (33), the relationship between porous media permeability formed by type II clusters and matrix particle size, porosity and fractal dimension can be obtained:

$$K_{II} = \frac{1}{128}\frac{\varphi}{1-\varphi}\left(\sqrt[3]{\frac{24\pi}{9\sqrt{2}(1-\varphi)}}\right)^{1-D_t}\left(\frac{2\sqrt{2}}{3}\sqrt[6]{\frac{9\sqrt{2}}{24\pi}}\sqrt{\frac{\varphi}{\tau_{av}}}\sqrt[3]{\frac{1}{1-\varphi}}\right)^{1+D_p}\frac{2-D_p}{3+D_t-D_p}D_f^2 \tag{35}$$

where D_f is the diameter of particles.

If Equation (25) is substituted into Equations (34) and (35), the permeability under the two ideal modes is a function of reservoir porosity, tortuosity fractal dimension, integral shape dimension of pore surface and radius of solid particles. The dimensionless permeability can be obtained by dividing Equations (34) and (35) by D_f^2.

3. Model Validation

There have been many researches on porous media permeability model based on fractal theory, the most classic one is the KC equation, although this equation is strictly applicable to homogeneous media or actual random filled fiber media. Based on the fractal porous media pore space geometry hypothesis, Costa improved the KC equation and verified its validity [33]. The improved KC equation can be expressed as

$$K = C_{kc} \frac{\varphi^{n+1}}{(1-\varphi)^n} \quad (36)$$

where C_{kc} and n both are the empirical constant, which is related to particle shape and tortuosity. Although appropriate parameters can be set through experience, so that Equation (36) can better match experimental data, the significance of these parameters is not clear, which needs to be determined through experiments. Xiao et al., Xu and Yu continued to improve the classical KC equation based on the fractal bundle model [25,28]. The Xiao's permeability model can be expressed as:

$$K = \left(\frac{4-D_P}{D_P}\right)^{1/2} \frac{[4(2-D_P)]^{(1+D_t)/2} (\pi D_P)^{(1-D_t)/2}}{128(3-D_P+D_t)\ln^2\varphi} \left(\frac{\varphi}{1-\varphi}\right)^{(1+D_t)/2} D_f^2 \quad (37)$$

According to Equations (36) and (37) and experimental data, we verified the model. As can be seen from Figure 5, the permeability of model I and model II are also different due to the different arrangement of particles, and the permeability of model I is relatively larger. In general, the permeability of our model increases with the increase of porosity, and the permeability changes rapidly when the porosity approaches 0 and 1. In order to verify the proposed model accurately, we made artificial cores with an average porosity of 5% to 30% and measured the corresponding permeability. As shown in Figure 5, our experimental permeability is very close to the proposed model permeability when the porosity is between 5% and 30%. In addition, we also compared the Costa's experimental data with the model we proposed [33]. The comparison results show that when the porosity is between 30% and 60%, our model is closer to the experimental data of Costa than Xiao's model. Therefore, the comparison between our experimental data, Costa's experimental data and the proposed model shows that our model is more consistent with the experimental data of medium and low porosity (5%~60%), while Xiao's model is more consistent with the data of high porosity (>60%) [28]. Therefore, our model is more suitable for predicting medium and low porosity porous media permeability. At the same time, the accuracy of the model is proved by comparing the results.

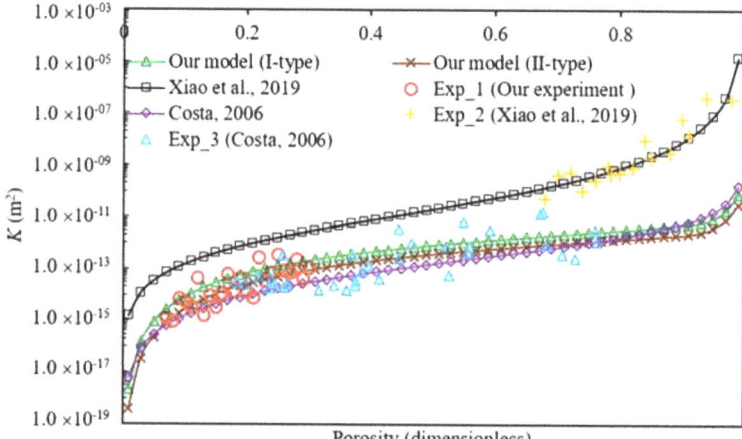

Figure 5. A comparison between the absolute permeability of porous media by the proposed fractal model and existing experimental data (Best parameters are $C_{kc} = 1.77 \times 10^{-12} m^2$; $n = 1.07$; $D_f = 6 \times 10^{-5} m$).

4. Results Discussion and Analysis

Figure 6 shows the relationship between porosity and pore diameter of the largest bundle. It can be seen from Figure 6 that the equivalent capillary bundle pore diameter increases with the increase of porosity. And with the increase of porosity, the diameter difference between the two modes becomes larger and larger. Due to the different arrangement of particles, the equivalent pore diameter is not the same.

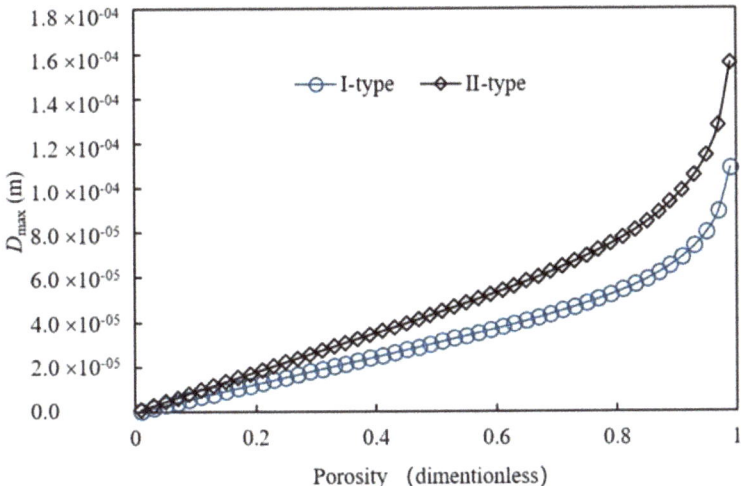

Figure 6. The relationship between pore diameter and porosity of capillary bundle model section.

By substituting Equations (4), (8), (24), and (25) into Equations (22) and (23), it can be known that the tortuosity fractal dimension and the pore area fractal dimension are all functions of porosity. Figure 7 shows the effect of porosity on fractal dimension. With the increase of porosity, the tortuosity fractal dimension decreases, which means that the greater the porosity, the more bent the capillary bundle is, and the longer the capillary bundle is. With the increase of porosity, the pore area fractal dimension also increases, and the maximum pore area on the cross section also increases. In addition, the arrangement of solid particles has little influence on fractal dimension. The pore area fractal dimension and tortuous fractal dimension of type I arrangement are larger than those of type II arrangement under different arrangement modes of solid particles.

Figures 8 and 9 respectively show the influence of the tortuosity fractal dimension on permeability of porous media and the influence of pore area fractal dimension on permeability of porous media. The permeability of porous media has a strong nonlinear relationship with the tortuosity fractal dimension and the integral shape dimension of pore surface. Figure 8 shows that the permeability decreases with the increase of tortuosity fractal dimension, and the permeability changes rapidly when the tortuosity is close to 1. Figure 9 shows that the permeability of porous media increases with the increase of the pore area fractal dimension, and the permeability changes faster when the pore area fractal dimension approaches 2. By comparing the two figures, the tortuosity fractal dimension has the opposite effect on permeability with the pore area fractal dimension. It can also be seen that the larger the radius of particles, the greater the permeability.

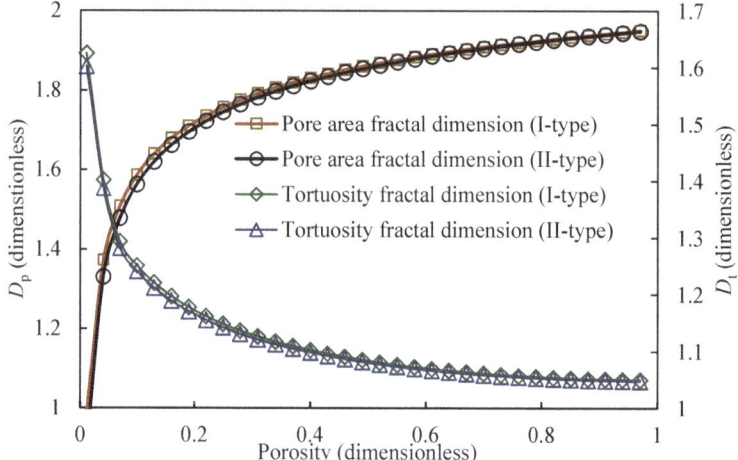

Figure 7. Influence of porosity on the tortuosity fractal dimension (D_t) and pore area fractal dimension (D_p) in reservoir rock porous media.

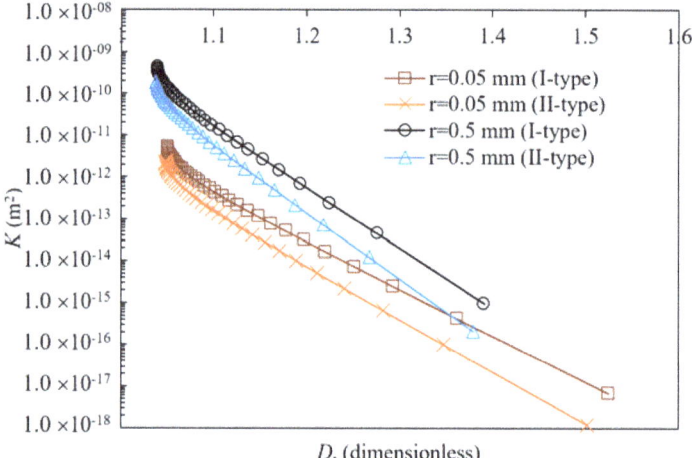

Figure 8. Effect of tortuosity fractal dimension on permeability of porous media.

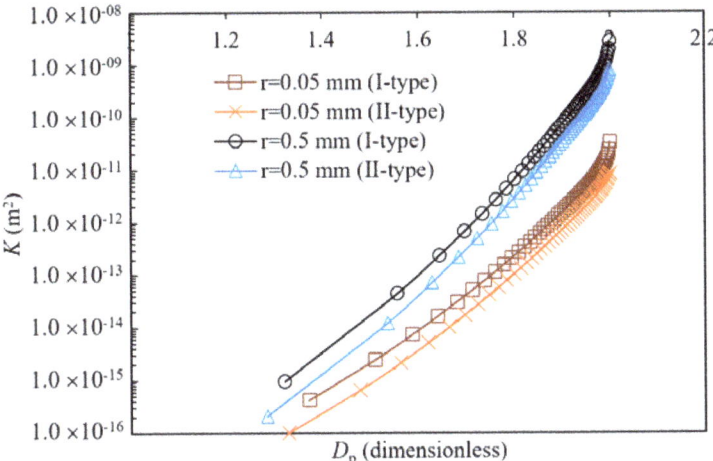

Figure 9. Effect of pore area fractal dimension on permeability of porous media.

In order to analyze the influence of particle arrangement mode on permeability, the permeability difference coefficient R caused by different particle arrangement mode is introduced. The connotation of R is to reflect the influence degree of solid particle arrangement mode on permeability. The larger R value is, the greater the influence degree of permeability is by solid particle arrangement mode. R can be expressed as:

$$R = \frac{K_I - K_{II}}{K_{II}} = \left(\sqrt[3]{\frac{3\sqrt{2}}{6}}\right)^{1-D_t} \left(3\sqrt{2}\sqrt[3]{\frac{1}{9}}\sqrt[6]{\frac{\sqrt{2}\pi}{36}}\right)^{1+D_p} - 1 \tag{38}$$

Figure 10 shows the effect of changes in porosity and solid particle radius on the difference coefficient. As can be seen from the figure, under the same particle radius, with the change of porosity, the difference coefficient is larger when the porosity is close to 0 and 1, and smallest when the porosity is close to 0.6. Therefore, the prediction of permeability of porous media with high porosity and low porosity should pay particular attention to the influence of particle arrangement. In addition, the difference coefficient increases with the increase of the radius of solid particles. This indicates that the arrangement of particles has a great influence on the permeability of porous media when the particle radius is large. It can be seen from Equation (38) that the difference coefficient is a function of the pore area fractal dimension and the tortuosity fractal dimension. Figure 11 shows the relationship between fractal dimension and permeability difference coefficient. The permeability difference coefficient decreases first and then increases with the increase of pore area fractal dimension, reaching the minimum when its value is close to 2. At the same time, as the tortuosity fractal dimension increases, the permeability difference coefficient decreases first and then increases, reaching the minimum when its value is close to 1. The permeability difference coefficient changes rapidly when the pore area fractal dimension approaches 2 and the tortuosity fractal dimension approaches 1, respectively.

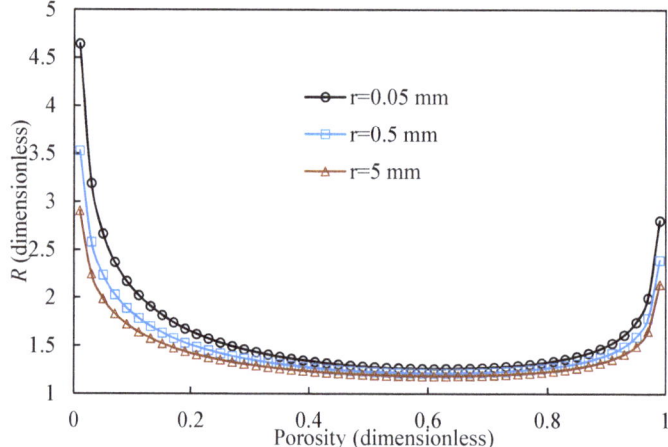

Figure 10. The relationship between porosity and permeability differential coefficient.

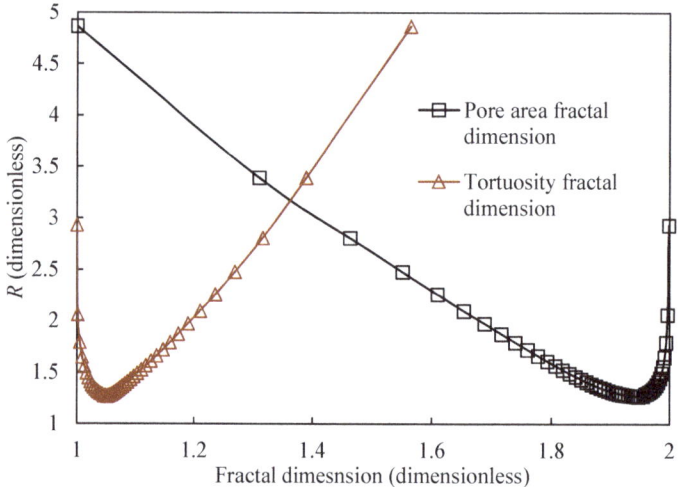

Figure 11. The relationship between the fractal dimension and the permeability difference coefficient (R).

5. Conclusions

In this paper, we derive a novel porous media permeability model considering irregular pore space based on fractal theory and ideal particle space geometry. The results show that our model is more suitable for medium and low porosity porous media. In addition, the following conclusions are obtained through multivariate analysis: (1) the equivalent capillary bundle pore diameter increases with the increase of porosity; (2) tortuosity fractal dimension has a negative correlation with porosity, while the pore area fractal dimension of the pore surface has a positive correlation with porosity; (3) the permeability is negatively correlated with the tortuosity fractal dimension and positively correlated with the pore area fractal dimension and particle radius. When the tortuosity fractal dimension is close to 1 and the pore area fractal dimension is close to 2, the faster the permeability changes, the greater the impact. (4) different particle arrangement has great influence on porous media permeability. When the porosity is close to 0 and close to 1, the permeability is especially affected by the greater the difference coefficient. In addition, the larger the particle radius is, the greater the permeability difference coefficient is, and the greater the permeability difference is for different particle arrangement. (5) With the increase of fractal dimension, the permeability difference coefficient first decreases and then increases. When the pore area fractal dimension approaches 2, the permeability difference coefficient changes quickly and reaches the minimum value, and when the tortuosity fractal dimension approaches 1, the permeability difference coefficient changes quickly and reaches the minimum value.

Author Contributions: Conceptualization, Y.H. and J.Z.; methodology, Q.W.; formal analysis, S.X.; investigation, Q.W.; writing—original draft preparation, Q.W.; writing—review and editing, Y.H. and J.Z.; supervision, H.J. All authors have read and agreed to the published version of the manuscript.

Funding: This research was funded by National Natural Science Foundation of China grant number 51404204 And the APC was funded by National Science and Technology Major Project of the Ministry of Science and Technology of China grant number 2016ZX05060. The authors would like to acknowledge the financial support of the China Scholarship Council.

Conflicts of Interest: The authors declare no conflict of interest.

References

1. Mousavi, M.; Prodanovic, M.; Jacobi, D. New classification of carbonate rocks for process-based pore-scale modeling. *SPE J.* **2013**, *18*, 243–263. [CrossRef]
2. Wu, K.; Chen, Z.X.; Li, X.F. Real gas transport through nanopores of varying cross-section type and shape in shale gas reservoirs. *Chem. Eng. J.* **2015**, *281*, 813–825. [CrossRef]
3. Wu, K.; Chen, Z.X.; Li, X.F.; Guo, C.; Wei, M. A model for multiple transport mechanisms through nanopores of shale gas reservoirs with real gas effect-adsorption-mechanic coupling. *Int. J. Heat Mass Transf.* **2016**, *93*, 408–426. [CrossRef]
4. Hosa, A.; Curtis, A.; Wood, R. Calibrating Lattice Boltzmann flow simulations and estimating uncertainty in the permeability of complex porous media. *Adv. Water Resour.* **2016**, *94*, 60–74. [CrossRef]
5. Chen, Z.L.; Wang, N.T.; Sun, L.; Tan, X.H.; Deng, S. Prediction method for permeability of porous media with tortuosity effect based on an intermingled fractal units model. *Int. J. Eng. Sci.* **2017**, *121*, 83–90. [CrossRef]
6. Pitchumani, R.; Ramakrishnan, B. A fractal geometry model for evaluating permeabilities of porous preforms used in liquid composite molding. *Int. J. Heat Mass Transf.* **1999**, *42*, 2219–2232. [CrossRef]
7. Miguel, A.F.; Rosa, R.; Silva, A.M. Fractal geometry description of the permeability of a natural fissured rock. In *Proceedings of the 9th International Congress on Deterioration and Conservation of Stone*; Elsevier Science B.V.: Amsterdam, The Netherlands, 2000; Volume 1, pp. 595–600.
8. Karacan, Q.; Halleck, P.M. A Fractal Model for Predicting Permeability around Perforation Tunnels Using Size Distribution of Fragmented Grains. *J. Pet. Sci. Eng.* **2003**, *40*, 159–176. [CrossRef]
9. Dravis, J.J. Carbonate applied to hydrocarbon exploration and exploitation, HTC Internal report. Training course note. 2007. Available online: www.dravisinterests.com (accessed on 7 November 2009).
10. Lucia, F.L. Petrophysical Parameters Estimated from Visual Descriptions of Carbonate Rocks: A Field Classification of Carbonate Pore Space. In Proceedings of the 1981 SPE Annual Technical Conference and Exhibition, San Antonio, TX, USA, 5–7 October 1983.
11. Wu, J.S. *Analysis of Resistance for Flow Through Porous Media*; Zhejiang University: Hangzhou, China, 2006.
12. Gebart, B.R. Permeability of unidirectional reinforcements for RTM. *J. Compos. Mater.* **1992**, *26*, 1100–1133. [CrossRef]
13. Othman, M.R.; Helwani, Z.; Martunus. Simulated fractal permeability for porous membranes. *Appl. Math. Model.* **2010**, *34*, 2452–2464. [CrossRef]
14. Zhu, F.; Hu, W.X.; Cao, J.; Sun, F.; Liu, Y.; Sun, Z. Micro/nanoscale pore structure and fractal characteristics of tight gas sandstone: A case study from the Yuanba area, northeast Sichuan Basin, China. *Mar. Pet. Geol.* **2018**, *98*, 116–132. [CrossRef]
15. Qin, X.; Cai, J.C.; Xu, P. A fractal model of effective thermal conductivity for porous media with various liquid saturation. *Int. J. Heat Mass Transf.* **2019**, *128*, 1149–1156. [CrossRef]
16. Wang, Q.; Hu, Y.; Zhao, J.; Ren, L.; Zhao, C.; Zhao, J. Multiscale Apparent Permeability Model of Shale Nanopores Based on Fractal Theory. *Energies* **2019**, *12*, 3381. [CrossRef]
17. Zhao, J.; Wang, Q.; Hu, Y.; Ren, L.; Zhao, C. Numerical investigation of shut-in time on stress evolution and tight oil production. *J. Pet. Sci. Eng.* **2019**, *179*, 716–733. [CrossRef]

18. Maik, N.K.; Sirisha, M.; Inani, A. Permeability characterization of polymer matrix composites by RTM/VARTM. *Prog. Aerosp. Sci.* **2014**, *65*, 22–40.
19. Parnas, R.S.; Salem, A.J.; Sadiq, T.A.; Wang, H.P.; Advani, S.G. The interaction between micro-and macro-scopic flow in RTM preforms. *Compos. Strcut.* **1994**, *27*, 83–107. [CrossRef]
20. Kozeny, J. Ueber kapillare leitung des wassers im boden. *Sitzh. Akad. Wiss. Wein* **1927**, *136*, 271–306.
21. Carman, P.C. Fluid flow through granular beds. *Trans. Inst. Chem. Eng.* **1937**, *15*, 150–167. [CrossRef]
22. Griffin, P.R.; Grove, S.M.; Russell, P.; Short, D.; Summerscales, J.; Guild, F.J.; Taylor, E. The effect of reinforcement architecture on the long-range flow in fibrous reinforcements. *Compos. Manuf.* **1995**, *6*, 221–235. [CrossRef]
23. Khabbazi, A.E. *Numerical and Analytical Characterization of Transport Properties for Single Phase Flows in Granular Porous Media*; University of Toronto: Toronto, ON, Canada, 2015.
24. Yu, B.M.; Lee, L.J.; Cao, H.Q. Fractal characters of pore microstructures of textile fabrics. *Fractals* **2001**, *9*, 155–163. [CrossRef]
25. Xu, P.; Yu, B.M. Development a new form of permeability and Kozeny-Carman constant for homogeneous porous media by means of fractal geometry. *Adv. Water Resour.* **2008**, *31*, 74–81. [CrossRef]
26. Xiao, B.Q.; Fan, J.T.; Ding, F. A fractal analytical model for the permeabilities of fibrous gas diffusion layer in proton exchange membrane fuel cells. *Electrochem. Acta* **2014**, *134*, 222–231. [CrossRef]
27. Xiao, B.Q.; Wang, W.; Zhang, X.; Long, G.; Fan, J.T.; Chen, H.; Deng, L. A novel fractal solution for permeability and Kozeny-Carman constant of fibrous porous media made up of solid particles and porous fibers. *Powder Technol.* **2019**, *349*, 92–98. [CrossRef]
28. Xiao, B.Q.; Fan, J.T.; Ding, F. A novel fractal model for relative permeability of gas diffusion layer in proton exchange membrane fuel cell with capillary pressure effect. *Fractals* **2019**, *27*, 1950012. [CrossRef]
29. Shou, D.H.; Fan, J.; Ding, F. A difference-fractal model for the permeability of fibrous porous media. *Phys. Lett. A* **2010**, *374*, 1201–1204. [CrossRef]
30. Zhang, L.Z. A fractal model for gas permeation through porous membranes. *Int. J. Heat Mass Transf.* **2012**, *55*, 1716–1723. [CrossRef]
31. Zhu, Q.Y.; Xie, M.H.; Yang, J.; Li, Y. A fractal model for the coupled heat and mass transfer in porous fibrous media. *Int. J. Heat Mass Transf.* **2011**, *54*, 1400–1409. [CrossRef]
32. Zhu, Q.Y.; Xie, M.H.; Yang, J.; Li, Y. Analytical determination of permeability of porous fibrous media with consideration of electro kinetic phenomena. *Int. J. Heat Mass Transf.* **2012**, *9*, 365–372.
33. Costa, A. Permeability-porosity relationship: A reexamination of the Kozeny-Carman equation based on a fractal pore-space geometry assumption. *Geophys. Res. Lett.* **2005**, *33*, L02318. [CrossRef]
34. Cihan, A.; Sukop, M.C.; Tyner, J.S.; Perfect, E.; Huang, H. Analytical predictions and lattice Boltzmann simulations of intrinsic permeability for mass fractal porous media. *Vadose Zone J.* **2009**, *8*, 187–196. [CrossRef]
35. Garrison, J.R.; Pearn, W.C.; Vonrosenberg, D.U. The fractal menger sponge and Sierpinski carpet as models for reservoir rock/por systems: I. theory and image analysis of Sierpinski carpets. *In Situ* **1992**, *16*, 351–406.
36. Garza-Lopez, R.A.; Naya, L.; Kozak, J.J. Tortuosity factor for permeant flow through a fractal solid. *J. Chem. Phys.* **2000**, *112*, 9956–9960. [CrossRef]
37. Cihan, A.; Perfect, E.; Tyner, J.S. Water retention models for scale-variant and scale-invariant drainage of mass prefractal porous media. *Vadose Zone J.* **2007**, *6*, 786–792. [CrossRef]
38. Atzeni, C.; Pia, G.; Sanna, U.; Spanu, N. A fractal model of the porous microstructure of earth-based materials. *Constr. Build. Mater.* **2008**, *22*, 1607–1613. [CrossRef]
39. Pia, G.; Sanna, U. An intermingled fractal units model and method to predict permeability in porous rock. *Int. J. Eng. Sci.* **2014**, *75*, 31–39. [CrossRef]
40. Pia, G.; Sanna, U. An intermingled fractal units model to evaluate pore size distribution influence on thermal conductivity values in porous materials. *Appl. Therm. Eng.* **2014**, *65*, 330–336. [CrossRef]
41. Pia, G.; Sanna, U. Intermingled fractal units model and electrical equivalence fractal approach for prediction of thermal conductivity of porous materials. *Appl. Therm. Eng.* **2013**, *61*, 186–192. [CrossRef]
42. Turcio, M.; Reyes, J.; Camacho, R.; Lira-Galeana, C.; Vargas, R.O.; Manero, O. Calculation of effective permeability for the BMP model in fractal porous media. *J. Pet. Sci. Eng.* **2013**, *103*, 51–60. [CrossRef]
43. Bautista, F.; Santos, J.M.; Puig, J.E.; Manero, O. Understanding thixotropic and antithixotropic behavior of viscoelastc micellar solutions and liquid crystalline dispersions. I. the model. *J. Non-Newton. Fluid Mech.* **1999**, *80*, 93–113. [CrossRef]

44. Miao, T.J.; Chen, A.; Zhang, L.W. A novel fractal model for permeability of damaged tree-like branching networks. *Int. J. Heat Mass Transf.* **2018**, *127*, 278–285. [CrossRef]
45. Gostick, J.T.; Fowler, M.W.; Loannidis, M.A.; Pritzker, M.D.; Volfkovich, Y.M.; Sakars, A. Capillary pressure and hydrophilic porosity in gas diffusion layers for polymer electrolyte fuel cells. *J. Power Sources* **2006**, *156*, 375–387. [CrossRef]
46. Shou, D.H.; Fan, J.; Ding, F. Hyfraulic permeability of fibrous porous media. *Int. J. Heat Mass Transf.* **2011**, *54*, 4009–4018. [CrossRef]
47. Ju, Y.; Xi, C.; Zhang, Y.; Mao, L.T.; Gao, F.; Xie, H.P. Laboratory In Situ CT Observation of the Evolution of 3D Fracture Networks in Coal Subjected to Confining Pressures and Axial Compressive Loads: A Novel Approach. *Rock Mech. Rock Eng.* **2018**, *51*, 3361–3375. [CrossRef]
48. Fan, W.P.; Sun, H.; Yao, J.; Fan, D.Y.; Zhang, K. A new fractal transport model of shale gas reservoirs considering multiple gas transport mechanisms, multi-scale and heterogeneity. *Fractals* **2018**, *26*, 1850096. [CrossRef]
49. Yu, B.M.; Li, J.H. Some fractal characters of porous media. *Fractals* **2001**, *9*, 365–372. [CrossRef]
50. Yun, M.J.; Yu, B.M.; Zhang, B.; Huang, M.T. A geometry model for tortuosity of stream tubes in porous media with spherical particles. *Chin. Phys. Lett.* **2005**, *22*, 1464–1467.
51. Yun, M.J.; Yu, B.M.; Xu, P.; Wu, J.S. Geometrical models for tortuosity of streamlines in three-dimensional porous media. *Can. J. Chem. Eng.* **2006**, *84*, 301–309. [CrossRef]
52. Kou, J.L.; Tang, X.M.; Zhang, H.Y.; Lu, H.J.; Wu, F.; Xu, Y.; Dong, Y. Tortuosity for streamlines in porous media. *Chin. Phys. Rev.* **2012**, *21*, 044701. [CrossRef]
53. Daigle, H. Application of critical path analysis for permeability prediction in natural porous media. *Adv. Water Resour.* **2016**, *96*, 43–54. [CrossRef]
54. Wheatcraft, S.W.; Tyler, S.W. An explanation of porous dispersivity in heterogeneous aquifers using concepts of fractal geometry. *Water Resour. Res.* **1988**, *24*, 566–578. [CrossRef]
55. Sun, Z.; Shi, J.T.; Wu, L.; Zhang, T.; Feng, D.; Huang, L.; Shi, Y.; Ramachandean, H.; Li, X. An analytical model for gas transport through elliptical nanopores. *Chem. Eng. Sci.* **2019**, *199*, 199–209. [CrossRef]
56. Cussler, E.L. *Diffusion-Mass Transfer in Fluid Systems*; Cambridge University Press: Cambridge, UK, 2000.
57. Skjeltorp, A.T.; Feder, A.A. (Eds.) *Fractals in Physics*; Elsevier: Amsterdam, The Netherlands, 1990.

© 2020 by the authors. Licensee MDPI, Basel, Switzerland. This article is an open access article distributed under the terms and conditions of the Creative Commons Attribution (CC BY) license (http://creativecommons.org/licenses/by/4.0/).

Review

A Review of Rheological Modeling of Cement Slurry in Oil Well Applications

Chengcheng Tao, Barbara G. Kutchko, Eilis Rosenbaum and Mehrdad Massoudi *

U.S. Department of Energy, National Energy Technology Laboratory (NETL), Pittsburgh, PA 15236, USA; Chengcheng.Tao@netl.doe.gov (C.T.); Barbara.Kutchko@netl.doe.gov (B.G.K.); Eilis.Rosenbaum@netl.doe.gov (E.R.)
* Correspondence: Mehrdad.Massoudi@netl.doe.gov; Tel.: +1-412-386-4975

Received: 25 November 2019; Accepted: 21 January 2020; Published: 24 January 2020

Abstract: The rheological behavior of cement slurries is important in trying to prevent and eliminate gas-migration related problems in oil well applications. In this paper, we review the constitutive modeling of cement slurries/pastes. Cement slurries, in general, behave as complex non-linear fluids with the possibility of exhibiting viscoelasticity, thixotropy, yield stress, shear-thinning effects, etc. The shear viscosity and the yield stress are two of the most important rheological characteristics of cement; these have been studied extensively and a review of these studies is provided in this paper. We discuss the importance of changing the concentration of cement particles, water-to-cement ratio, additives/admixtures, shear rate, temperature and pressure, mixing methods, and the thixotropic behavior of cement on the stress tensor. In the concluding remarks, we propose a new constitutive model for cement slurry, considering the basic non-Newtonian nature of the different models.

Keywords: cement slurries; non-Newtonian fluids; rheology; constitutive relations; viscosity; yield stress; thixotropy

1. Introduction

In oil well cementing applications, cement slurries are placed in the annulus space between the well casing and the geological formations surrounding the wellbore; this is done primarily to provide zonal isolation from the surrounding fluid flow and to prevent the corrosion to the casing for the life of the well [1,2]. Figure 1 shows a typical schematic of an oil well cementing operation. The space must be filled before the cement begins to harden [3]. Failure of the zonal isolation and flow of the formation fluids penetrating the cement can cause disasters, financial loss, and other serious consequences. Thus, understanding the behavior of the cement is one of the key factors in designing a gas resistant cement column [4]. Cement, in general, is mixed on the rig before being pumped into the well [5]. Gas migration caused by the fluid flow in the wellbore requires expensive remedial techniques [6,7]. Studying rheological properties of cement slurry plays an important role in understanding the factors that can prevent or eliminate the gas migration. In the early stages of cement hydration, cement changes from a fluid to a solid-like material and develops compressible strength. Gelation during the cementing process is the buildup of the gel strength, or the premature increase in the cement viscosity. Early gelation is a serious problem in petroleum industry, but it is desired that the cement gel quickly once placed [8]. Static gel strength (SGS) [9] is the yield point showing the phase change; this is related to the safety of cement jobs and the gas migration problem [10].

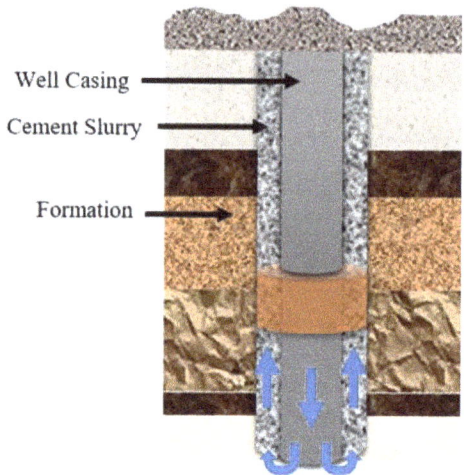

Figure 1. Schematic of an oil well cementing operation [11].

According to the U.S. Department of Transportation, Type I/II ordinary Portland cement provides adequate strength and durability for common applications [12]. However, more specific performance criteria are required for cementing around the steel casing of gas and oil wells. Wells with an average depth of up to 8000 feet below the mud line [13] can require handling cement slurries under high temperature and high pressure conditions (200 °C and 150 MPa in deep wells) [14]. For oil and gas wells, cement should not set too early so that it can be pumped during the placement and the setting time should not be too long once placed to prevent fluids from penetrating the cement barrier. Cement requires a high level of consistency, related to the minimum quantity of water required to initiate the chemical reactions between water and cement. The API Class G and H cements are two common forms of cements used today [15]. Cements ground too fine (Blaine fineness > 9000 cm^2/gm) are not suitable for oil well cementing because of insufficient compressive strength to hold the casing and inadequate sulphate resistance. However, the microfine cements are good for oil well repairing of small cracks [16].

In cementing operations, different cement slurries including "lead" slurries at the upper section and "tail" slurries at the lower section are pumped into the wellbore, as shown in Figure 2. The tail slurries, under higher pressures and temperatures, usually have higher density and higher strength than the lead slurries and are pumped after the lead slurries [17]. Portland cement is widely used in the oil well and construction industries because of its flexibility and widespread availability of its constituent materials [18]. It is produced from the grinding of clinker, which is produced by the calcination of limestone and other raw materials in a cement rotary kiln. After the cement is mixed with water, a series of exothermal chemical reactions occur which results in strength development and cement hardening. Cement slurries are reactive systems, changing continuously chemically and physically [19].

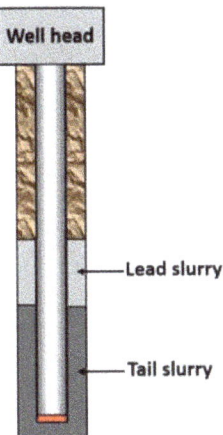

Figure 2. Different types of cement slurry in an oil well.

The first few hours (after mixing the cement and water) is a dormant period for the reactions. The slurry is in a fresh state, after which the setting is initiated, and the cement begins to harden. The fresh cement slurry should be pumpable into the wellbore, where it would harden soon after placement [20]. The design of the cementing operation requires a comprehensive understanding of the mechanical and rheological behavior of cement slurries [21]. Flowability and stability of cement slurries are two important factors for successful oil well cementing [22]. Retarders are usually applied to control the duration of the fresh state, providing a safety time for the pumping operations. Many researchers have put considerable effort to understand the mechanisms for fluid migration during cementing, using experimental and computational models. To monitor the conditions of cement slurry in real-time, wireless sensor network-based monitoring systems can be used [23–26]. In many cases, computational fluid dynamics (CFD) applications can also be used (see Appendix C). Table 1 shows the general cement properties.

Table 1. Cement properties [27–29].

Cement Properties	Value
Cement Powder Density	3.15 g/cm^3
Cement Slurry Density	1.38 g/cm^3–2.28 g/cm^3
Cement Particle Size	0.1 to 100 µm
Compressive Strength	20–40 Mpa
Maximum Solid Concentration Packing	0.65 (spherical particles)
β in the Krieger and Dougherty's relation	1.5–2
Viscosity of the Continuous Phase (Base Fluid)	1 cP for water at 20 °C
Reynolds Number	2716–3971

The objective of this paper is to review the existing constitutive equations for cement slurries, used in oil well applications; specifically, we look at the relationships suggested for the yield stress and the viscosity of the cement as functions of parameters such as: volume fraction, cement surface, shear rate, cement chemistry, mineralogical composition, concentration of additives, temperature and pressure variations, mixing conditions, etc. In the next sections, we review different constitutive models for cement slurries. In Appendix A, a discussion of the chemistry of cement is provided. In Appendix B, we present the governing equations for a single phase approach and the multi-phase approach to the flow of cement slurry. In Appendix C, we provide a short review of the various CFD studies related to cement slurries.

2. Constitutive Modeling of Cement Slurries

Constitutive modeling and mathematical modeling of cement present challenging opportunities in research. Beginning as a (1) powder, (2) becoming a paste (when mixed with water), (3) flowing as a slurry (suspension), and then (4) hardening to become a solid-like material, cement can be modeled and studied as (1) granular materials (powder), (2) visco-plastic fluids with a yield stress (paste), (3) viscoelastic (non-Newtonian fluids, slurry), and (4) a poro-elastic solid. Cement as a paste or as a suspension, can be modeled mathematically using different approaches. For example, it can be considered as a multi-component material composed of water and cement particles, and other additives or it can be considered as a non-homogeneous (single component) fluid that behaves in a non-linear fashion. In this review report, we are basically interested in the latter approach. As such, the cement paste or cement slurry behave as complex fluids exhibiting certain non-linear characteristics such as thixotropy, viscoelasticity, yield stress, presence of normal stress effects, shear-rate dependent viscosity, etc., (see [30–33]).

Fresh cement paste/slurries are suspensions with a high concentration of particles. The size of cement particles is between 0.1 µm and 100 µm. Cement, in general, behaves as a thixotropic material with a yield stress; after the hydration process, it begins to behave as a solid-like material [34]. The rheological studies of cement reveal various phenomena [2]: (1) Rapid formation of gel at rest; (2) failure of gel under a critical value of stress, which is dependent on the interparticle forces; (3) destruction of gel at some shear rate, resulting in a shear-thinning behavior; (4) reconstructions at higher shear rates and jamming, etc.

In the petroleum industry, we need to know the rheological properties of cement in order to evaluate the displacement and the flow rates for optimal mud removal and cement placement before setting [2]. The cement and concrete industries prefer high performance cement with high impermeability, high density, and high strength for structures, which are related to the cement composites [35]. The quality of the concrete structure depends on the behavior of fresh cement during its placement into formwork at the jobsite [36]. The sustained-casing-pressure (SCP) tests in the drilling industry test the solid cement after hydration though gas migration occurs in cement slurries and mud. In general, non-Newtonian fluids models should be taken into consideration when studying cement slurries and muds. Yield stress effects and thixotropy should also be taken into consideration in these studies.

Some of the basic issues in studying cement rheology are (1) lack of reliable experimental data along with inaccurate experimental procedures and difficulties to reproduce the measurements; (2) complex flow behavior of cement affected by various physical and chemical factors; (3) complex cement hydration and setting process. If cement behaves as a viscoplastic fluid, then the yield stress and the plastic viscosity also need to be measured and studied [37]. These properties can depend on different factors [35,38–40] such as:

1. *Physical factors*, such as water-to-cement ratio, geometry of the cement grain (specific surface of cement particles) [38];
2. *Chemical and mineralogical factors*, such as the cement type, its chemical composition, additive types, cement particle concentration in the mix, structural modifications after hydration [41];
3. *Mixing conditions*, such as the type of blender/grinding condition, hydration time, storage/transport condition, curing temperature, stirrer rate, and time [38];
4. *Measurement conditions* (experimental equipment and procedures).

For oil well applications, these properties can also depend on temperature, water-to-cement ratio, and the type of additives used.

The basic governing equations for flow of cement are given in Appendix B. For a purely mechanical case, i.e., where thermo-chemical and electro-magnetic effects are ignored, the constitutive parameter of interest is the stress tensor T. In a sense, we think the rheological response of the cement can be described through constitutive modeling of the stress tensor. As we will discuss, it is well-known that cement, in general, exhibits visco-elastic behavior, along with a yield stress. Some researchers focus on the modeling of the yield stress, T_y, and some on the viscous stress, T_v, and some on the total stress, T. In general, we can assume

$$T = T_v + T_y \qquad (1)$$

In the remainder of this report, we provide reviews related to these three approaches.

The shear viscosity, in general, can be measured using a coaxial cylinder viscometer. Based on the shear rate vs. shear stress curves, it can be seen that cement paste exhibits both thixotropic and rheopectic behavior at different times [35,39,42,43]. The experimental procedure, sometimes, can affect these measurements. For example, an apparent wall slip or sedimentation of some particles in the coaxial viscometers could cause error. The Bingham plastic model and the power-law models are the two most commonly used models for cement slurries. In this section, we review some of the mathematical models used for cement slurries.

2.1. Models for the Total Stress Tensor T

The Bingham fluid model [44] is widely used to describe the relationship between the shear stress and the shear rate for cement slurries at low shear rates. It also is one of the simplest models to describe the visco-plastic nature of some non-Newtonian fluids, where

$$\tau = \tau_y + \eta_p \dot{\gamma} \qquad (2)$$

where τ is the shear stress, τ_y is the constant yield stress, η_p is the plastic viscosity, and $\dot{\gamma}$ is the shear rate. The *total* viscosity of the material can be defined as:

$$\eta = \frac{\tau}{\dot{\gamma}} = \eta_p + \frac{\tau_y}{\dot{\gamma}} \qquad (3)$$

The (one-dimensional) model given in Equation (2) indicates a linear relationship after the initial yield stress is reached. However, this is not an accurate model for the non-linear behavior of many of the cement slurries [45]; in some cases, there is no clear relationship between the shear rate and the volumetric flow rate inside the pipe or annulus [46].

The three-dimensional tensorial form of the Bingham model is given by in the books, for example, Macosko [47] (p. 93), Lootens et al. [48] for cement:

$$T = \left[\tau_y / \left| \left(\frac{1}{2} D : D \right)^{1/2} \right| + \eta_p \right] D \qquad (4)$$

where T is the total stress tensor, $D = \dot{\gamma}$ is the symmetric part of the velocity gradient, τ_y is the yield stress, and η_p is the plastic viscosity, and where ":" indicates the inner (scalar) product of two tensors, $\dot{\gamma} = (2D_{ij}D_{ij})^{1/2} = (\Pi)^2$, $D = \frac{1}{2}(L + L^T)$, $L = \text{grad}v$, $\Pi = 2trD^2$. Later, we will also use the following notation: $A_1 = 2D$. Oldroyd (1947) derived a proper (frame invariant) 3-D form for the Bingham fluid by assuming that the material behaves as a linear elastic solid below the yield stress; he used the von Mises criterion for the yield surface. Thus

$$T = \left[\eta_p + \frac{\tau_y}{\sqrt{\frac{1}{2}\Pi_{A_1}}} \right] A_1 \quad \text{when} \quad \left[\frac{1}{2} T : T \right] \geq \tau_y^2$$
$$T = GE \quad \text{when} \quad \left[\frac{1}{2} T : T \right] < \tau_y^2 \qquad (5)$$

where G is the shear modules, indicating that below the yield stress, the material behaves as a linear elastic solid, obeying the Hooke's Law, and where

$$\begin{gathered} II_{A_1} \equiv A_1 : A_1 \\ A_1 = \operatorname{grad} v + (\operatorname{grad} v)^T \\ \mathbf{E} : \text{strain tensor} \end{gathered} \qquad (6)$$

As Denn [49] indicates, if the material is assumed to be inelastic prior to yielding, then $G \to \infty$, and Equation (5) is replaced by

$$A_1 = 0 \quad \left[\frac{1}{2} T : T\right] < \tau_y^2 \qquad (7)$$

Macosko [31] (p. 96) mentions that for many fluids with a yield stress, there is a lower *Newtonian* regime rather than a *Hookean* one, and thus one can use a two-viscosity (bi-viscous) model, such as

$$\begin{aligned} T &= 2\eta_p \mathbf{D} & \text{for } II_{2D}^{1/2} \leq \dot{\gamma}_c \\ T &= 2\left[\frac{\tau_y}{|II_{2D}|^{1/2}} + K|II_{2D}|^{\frac{n-1}{2}}\right]\mathbf{D} & \text{for } II_{2D}^{1/2} > \dot{\gamma}_c \end{aligned} \qquad (8)$$

where K is the consistency factor and n is the power-law exponent; when $n = 1$, the fluid is Newtonian; when $n > 1$, the fluid is shear-thickening, and when $n < 1$, the fluid is shear-thinning. In this formulation, one uses a critical shear rate instead of a yield criterion and this makes the numerical solution easier. For additional and interesting applications of the flow of a Bingham-type fluid, see White [50], and Lipscomb and Denn [51]. Mendes and Dutra [52] provide further insight into the viscosity of a shear-thinning (yield stress) fluids.

Herschel and Bulkley [53] generalized the Bingham model by introducing a three-parameter model where:

$$\tau = \tau_y + K\dot{\gamma}^n \qquad (9)$$

where τ_y, K, and n are constants. According to Banfill [3], K could be chosen as 2.5 or 0.25 and n as 0.75 or 1.25 for cement. Jones and Taylor [54] and Atzeni [45] have used this model to study cement. This model has been found to describe the rheological behavior for the sealing cement slurries used in drilling technologies [55].

The power-law model, also known as the Ostwald-de Waele model [56] is one of the most popular models in describing the pseudoplastic fluids without yield stress. Here,

$$\tau = K\dot{\gamma}^n \qquad (10)$$

where K is the consistency factor, n is the flow behavior index (the power-law exponent), measuring the degree of non-Newtonian behavior, and $\dot{\gamma}$ is the shear rate.

Williamson's model [57] was used by Lapasin [58] to describe the fresh cement pastes, where

$$\tau = \eta_\infty \dot{\gamma} + \tau_y \frac{\dot{\gamma}}{\dot{\gamma} + \Gamma} \qquad (11)$$

where η_∞ is the viscosity at infinite shear rate and Γ is a parameter indicating the deviation from the Bingham behavior.

The Eyring model [59] was applied to Portland cement pastes by Atzeni et al. [45]; this is suitable for suspensions with dispersed particles:

$$\tau = \sum_{i=1}^{n} a_i \sinh^{-1}(\dot{\gamma} b_i) \qquad (12)$$

where a_i and b_i are constants, and n is the number of flow units. For the simplest case $n = 2$, where there is one Newtonian fluid and one non-Newtonian fluid. This model does not have a yield stress component and is more suitable for non-Newtonian behavior at high shear rates.

Sisko's model [60], applied by Papo [61] to cement pastes, does not include a yield stress term:

$$\tau = a\dot{\gamma} + b\dot{\gamma}^c \tag{13}$$

where a is related to the infinite shear rate viscosity η_∞, and b and c are adjustable parameters with $c < 1$.

The Casson equation [62], applied to cement pastes by Atzeni et al. [45] is given by:

$$\sqrt{\tau} = \sqrt{\tau_y} + \sqrt{\eta_p \dot{\gamma}} \tag{14}$$

where the yield stress τ_y is constant and η_p is the plastic viscosity. According to Kok and Karakaya [63], the Casson model is considered to be one of the most accurate and useful models describing the rheological properties of cement slurries.

Shangraw–Grim–Mattocks model [64] was used by Papo (1988) to study cement, with a constant yield stress:

$$\tau = \tau_y + \eta_\infty \dot{\gamma} + \alpha_1 \left[1 - \exp(-\alpha_2 \dot{\gamma}) \right] \tag{15}$$

where η_∞ is the viscosity at infinite shear rate, α_1 and α_2 are adjustable parameters.

The Ellis's model (1965), applied to cement by Atzeni et al. [45] is given by:

$$\dot{\gamma} = a\tau + b\tau^c \tag{16}$$

where a is a function of the initial viscosity η_0, b is related to the shear stress τ, and c is a constant with the value 2.

Robertson and Stiff [46] proposed a three-parameter model for a yield-pseudoplastic fluid where:

$$\tau = a(\dot{\gamma} + b)^c \tag{17}$$

where a, b, and c are constants. The values for these constants, suggested by Banfill [3] are: 20, 1.5, and 0.35. By adjusting these three parameters, the model can predict the behavior of Newtonian fluids, the Bingham plastic, and the power-law models. However, Beirute and Flumerfelt [65] indicated that the Robertson–Stiff model is not able to describe the yield stress and is limited to the fluids with no yield stress.

According to Batra and Parthasarathy [66], the Robertson–Stiff fluid model can be rewritten as

$$\dot{\gamma}_{ij} = \frac{2A^{1/n}\Pi_T^{1/2}}{\Pi_T^{1/2n} - \tau_y^{1/n}} e_{ij} \tag{18}$$

where $\dot{\gamma}_{ij}$ and e_{ij} are the shear rate and the strain rate tensors, respectively, A and n are constants, Π_T is the second invariant of the stress tensor and τ_y is the yield stress. The equation is satisfied when $\sqrt{\Pi_T} \geq \tau_y$.

The Vom Berg model [67] is given by

$$\tau = \tau_y + a\sinh^{-1}(b\dot{\gamma}) \tag{19}$$

where a and b are constants, and for cement, their values were given to be 26.5 and 0.1, respectively (see Banfill [3]). The model considers yield stress, τ_y, explicitly, describing the flow behavior of cement at shear rates up to 380 s^{-1}, and it also describes the yield behavior at zero shear rate. This model seems to be effective in describing shear-dependent behavior of cement pastes [45,58].

Lapasin et al. [68] suggested the following model:

$$\tau = \tau_\infty + (\tau_0 - \tau_\infty)\exp(-bt_s) \qquad (20)$$

where τ is the shear stress at time t, τ_0 and τ_∞ are the shear stresses at the initial and the equilibrium states, b is the thixotropic exponent and t_s is the shear time. This model neglects the flocculation process. In this model, there is no yield stress.

The initial shear stress is also known as the gel strength at rest, which is a measure of the electrochemical forces within the fluid under static conditions. The relation between the shear stress and the resting time is given by [69]:

$$\tau_0 = \tau_{r0} + ct^m \qquad (21)$$

where τ_{r0} is the initial gel strength, t is the resting time, c is the recovery coefficient, m is the gel exponent.

Quemada's model [70] was reviewed by Banfill [3] and applied to cement slurries, where:

$$\tau = \left(\frac{1+\sqrt{(a\dot\gamma)}}{b+c\sqrt{(a\dot\gamma)}}\right)^2 \dot\gamma \qquad (22)$$

where a, b, and c are material parameters, with the values of 0.14, 10^{-4}, and 0.14 for cement [3]. This model does not include a yield stress term.

Atzeni [45] and Banfill [3] applied a modified form of the Casson's model to study cement pastes:

$$\dot\gamma = a + b\tau^{1/2} + c\tau \qquad (23)$$

where a, b, and c are functions of the viscosity, concentration, and other flow parameters such as the aggregation of solid particles. This model considers interaction forces among the particles; however, it is not accurate when predicting the behavior of suspensions with high particle concentration.

Atzeni et al. [45] reviewed the models proposed by Ellis, Casson, Eyring, and Vom Berg for fresh cement pastes and proposed a new model with a constant yield stress, where:

$$\tau = \tau_y + a\dot\gamma + \sinh^{-1}(\dot\gamma/c) \qquad (24)$$

where a and c are the constants.

Lapasin et al. [58] proposed a 6-parameter equation for fresh cement paste by combining the Vom Berg, the Shangraw–Grim–Mattocks, and the Williamson models, where:

$$\tau = \tau_y + \left[A_0 + (A_1 - A_0)\exp\left\{-\frac{\dot\gamma}{a+b\dot\gamma}t\right\}\right]\sinh^{-1}\frac{\dot\gamma}{c} \qquad (25)$$

where the yield stress τ_y is independent of the structural parameter λ, $\eta_0 = A_0/C$ and $\eta_1 = A_1/C$ are the dynamic viscosities at $\dot\gamma = 0$ when the structural parameter λ is 0 and 1. Equation (25) was checked and compared with experimental data and it was found to be a reasonable model to describe the behavior of different cements. For the Portland cement PTL 425 with water-to-cement ratio 0.40, the following values are suggested: τ_y = 12.1 Pa, A_0 = 43.6 Pa, $A_1 - A_0$ = 87.6 Pa, a = 2100, b = 9.45, and c = 56.1 [58].

Using experimental data, Lapasin et al. [58] proposed an equation describing the effects of specific surface S_{vB} and volume concentration of cement particles ϕ on the dynamicf viscosities η_0 and η_1 and the yield stress τ_y:

$$\eta_0 = \left(1.1\times 10^{-3}e^{13.2\phi}\right)S_{vB}^{4.2} \qquad (26)$$

$$\eta_1 = \left(8.6\times 10^{-5}e^{20.9\phi}\right)S_{vB}^{6.3} \qquad (27)$$

$$\tau_y = \left(2.1 \times 10^{-3} e^{19.2\phi}\right) S_{vB}^{2.5} \qquad (28)$$

De Kee and Chan Man Fong [71] suggested a three-parameter model for structured fluids such as hydrolyzed polyacrylamide solutions; their model was used by Yahia and Khayat to study cement grout [72]:

$$\tau = \tau_y + \eta_p \dot{\gamma} e^{-a\dot{\gamma}} \qquad (29)$$

where η_p is the plastic viscosity, $\dot{\gamma}$ is the shear rate, and a is a time-dependent parameter, with the value 10^{-3} for cement [3]. This model can predict shear-thinning behavior at lower shear rates and shear-thickening behavior at higher shear rates.

Yahia and Khayat [73] suggested a modified Bingham model for cement by adding a second order term:

$$\tau = \tau_y + \eta_p \dot{\gamma} + c\dot{\gamma}^2 \qquad (30)$$

where c is a parameter for capturing the second order effects, with the value of -0.0035 for cement [3].

Additionally, Yahia and Khayat [73] developed a new model for cement grouts with pseudoplastic or shear-thinning behavior, by combining the Casson and the De Kee–Chan Man Fong models:

$$\tau = \tau_y + 2(\sqrt{\tau_y \eta_p}) \sqrt{\dot{\gamma} e^{-a\dot{\gamma}}} \qquad (31)$$

where a is a time-dependent parameter, which allows for the shear-thinning behavior (positive) or shear-thickening behavior (negative) of the cement grouts, which is equal to 10^{-3} according to Banfill [3].

Vipulanandan et al. [74] proposed a new model for cement slurry:

$$\tau = \tau_y + \frac{\dot{\gamma}}{a + b\dot{\gamma}} \qquad (32)$$

where a and b are model parameters fitted with experimental data. For oil well cement slurry with the water-to-cement ratio 0.44 and temperature 85 °C, the following values can be used: $\tau_y = 40$ Pa, $a = 1.38$, $b = 0.004$ [74]. This model has a maximum shear stress limit at relatively high rate of shear strains.

Yuan et al. [75] substituted Equation (21) into (20) for an improved thixotropy model:

$$\tau = \tau_\infty + (\tau_{r0} + ct^m - \tau_\infty) \exp(-bt_s) \qquad (33)$$

where τ_{r0} and τ_∞ are the shear stresses at the initial and the equilibrium states, t is the resting time, c is the recovery coefficient, m is the gel exponent, b is the thixotropic exponent, and t_s is the shear time.

Papo [61] found that the Herschel–Bulkley model, the Sisko model, and the Robertson–Stiff model can all match reasonably well with the experimental data of cement pastes; he recommended the Herschel–Bulkley model mostly because of its ability to describe the yield behavior. Table 2 provides a summary of the existing constitutive relations for cement. The parameters used in various constitutive relations for cement are listed in Table 3. Figure 3 shows the stress vs. shear rate curves for different constitutive relations with the parameters listed in Table 3.

Table 2. Summary of constitutive relations for cement.

Author(s)	Model	Equation No.
Bingham (1922)	$\tau = \tau_y + \eta_p \dot{\gamma}$	(2)
Bingham (tensor form)	$T = \left[\tau_y / \left\| \left(\frac{1}{2} D : D \right)^{1/2} \right\| + \eta_p \right] D$	(4)
Herschel–Bulkley (1926)	$\tau = \tau_y + K \dot{\gamma}^n$	(9)
Power-Law (1929)	$\tau = K \dot{\gamma}^n$	(10)
Williamson (1929)	$\tau = \eta_\infty \dot{\gamma} + \tau_y \frac{\dot{\gamma}}{\dot{\gamma} + \Gamma}$	(11)
Eyring (1936)	$\tau = \sum_{i=1}^{n} a_i \sinh^{-1}(\dot{\gamma} b_i)$	(12)
Sisko (1958)	$\tau = a\dot{\gamma} + b\dot{\gamma}^c$	(13)
Casson (1959)	$\sqrt{\tau} = \sqrt{\tau_y} + \sqrt{\eta \dot{\gamma}}$	(14)
Shangraw–Grim–Mattocks	$\tau = \tau_y + \eta_\infty \dot{\gamma} + \alpha_1 [1 - \exp(-\alpha_2 \dot{\gamma})]$	(15)
Ellis (1965)	$\dot{\gamma} = a\tau + b\tau^c$	(16)
Robertson and Stiff (1976)	$\tau = a(\dot{\gamma} + b)^c$	(17)
Robertson and Stiff (tensor form)	$\dot{\gamma}_{ij} = \frac{2A^{1/n} \Pi_T^{1/2}}{\Pi_T^{1/2n} - \tau_y^{1/n}} e_{ij}$	(18)
Vom Berg (1979)	$\tau = \tau_y + a \sinh^{-1}(b\dot{\gamma})$	(19)
Lapasin (1979)	$\tau = \tau_\infty + (\tau_0 - \tau_\infty) \exp(-bt_s)$	(20)
Quemada (1984)	$\tau = \left(\frac{1 + \sqrt{(a\dot{\gamma})}}{b + c\sqrt{(a\dot{\gamma})}} \right)^2 \dot{\gamma}$	(22)
Modified Casson (1985)	$\dot{\gamma} = a + b\tau^{1/2} + c\tau$	(23)
Atzeni (1985)	$\tau = \tau_y + a\dot{\gamma} + \sinh^{-1}(\dot{\gamma}/c)$	(24)
Lapasin (1983)	$\tau = \tau_y + \left[A_0 + (A_1 - A_0) \exp\left\{ -\frac{\dot{\gamma}}{a+b\dot{\gamma}} t \right\} \right] \sinh^{-1} \frac{\dot{\gamma}}{C}$	(25)
De Kee (1994)	$\tau = \tau_y + \eta_p \dot{\gamma} e^{-a\dot{\gamma}}$	(29)
Modified Bingham (2001)	$\tau = \tau_y + \eta_p \dot{\gamma} + c\dot{\gamma}^2$	(30)
Yahia and Khayat (2001)	$\tau = \tau_y + 2(\sqrt{\tau_y \eta_p}) \sqrt{\dot{\gamma} e^{-a\dot{\gamma}}}$	(31)
Vipulanandan (2014)	$\tau = \tau_y + \frac{\dot{\gamma}}{a+b\dot{\gamma}}$	(32)
Yuan (2015)	$\tau = \tau_\infty + (\tau_{r0} + ct^m - \tau_\infty) \exp(-bt_s)$	(33)

Table 3. Parameters for various models

Equation	τ_y (Pa)	η_p (Pa·s)	K	n	a	b	c	A_0(Pa)	$A_1 - A_0$ (Pa)
Bingham (1922)	20	0.8	-	-	-	-	-	-	-
Herschel–Bulkley (1926)	20	-	2.5 / 0.25	0.75 / 1.25	-	-	-	-	-
Power-Law (1929)	-	-	2.5	0.75	-	-	-	-	-
Casson (1959)	20	0.31	-	-	-	-	-	-	-
Robertson and Stiff (1976)	-	-	-	-	20	1.5	0.35	-	-
Vom Berg (1979)	20	-	-	-	26.5	0.1	-	-	-
Quemada (1984)	-	-	-	-	0.14	10^{-4}	0.14	-	-
Lapasin (1983)	12.1	-	-	-	2100	9.45	56.1	43.6	87.6
De Kee (1994)	20	0.89	-	-	10^{-3}	-	-	-	-
Modified Bingham (2001)	20	1.15	-	-	-	-	−0.0035	-	-
Yahia and Khayat (2001)	20	0.9	-	-	10^{-3}	-	-	-	-
Vipulanandan (2014)	40	-	-	-	1.38	0.004	-	-	-

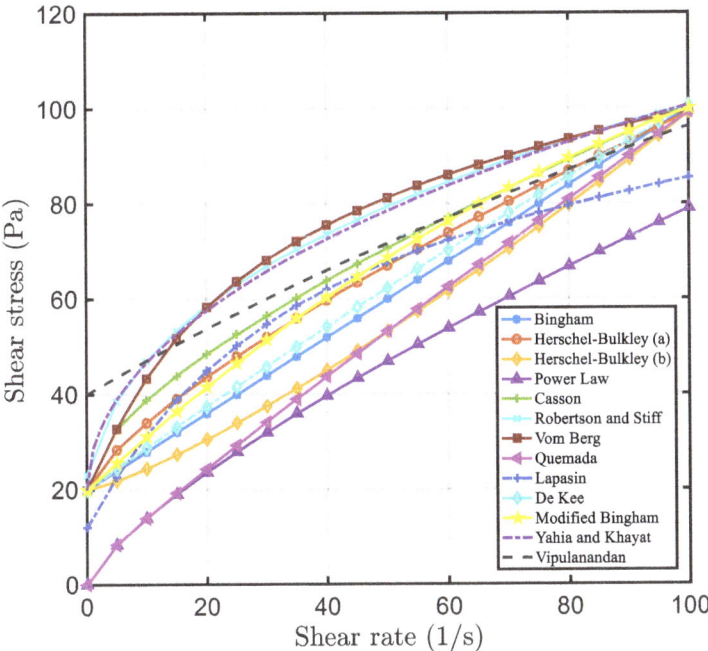

Figure 3. Stress vs. shear rate curves for various models for cement.

A look at Tables 2 and 3 and Figure 3 reveals that almost all the constitutive relationships given for the stress tensor of the cement, other than the Bingham model and the Herschel–Bulkley model, are given in 1-dimensional form where the (shear) stress τ is related to the shear rate $\dot{\gamma}$, a component of the velocity gradient tensor. In general, most problems of interests are 3-dimensional and as a result, constitutive relations are needed for the stress tensor T (with nine components).

2.2. The Importance of Yield Stress and Viscosity

In the previous section, we focused on the constitutive models available for the (total) stress tensor T for cement. Many researchers have looked at the yield stress T_y and the viscous stress T_v separately, and they have also looked at the factors and the parameters that can affect T_y and T_v. In this section we will focus on these issues. For example, it is known that the range of yield stress for a cement paste/slurry/grout is 10–100 N/m^2 and the range of plastic viscosity is 0.01–1 Ns/m^2 [3]. It has been shown that both of these parameters increase when finer cement particles are used [67]; this is mostly due to water and cement interaction. The effect of particle size is related to the surface area in the fine-grained pastes, rather than the volume of the coarse grains. Some of the variables which can affect the rheological properties of cement are: time, shear rate, concentration (volume fraction of the solid particles and the water-to-cement ratio), cement composition (Portland cement), fineness, flyash, silica fume, slags, chemical admixtures, age and temperature, pressure etc., [3,76].

In Section 2.2.1, we review some important yield stress models and in Section 2.2.2, we provide a review of various viscosity correlations, while looking at the different effects.

2.2.1. Yield Stress Models

The idea of fluids with yield stress perhaps can be traced back to Bingham ([44]). However, with all the successes of this model and the subsequent generalizations of it, Barnes and Walter [77] and Barnes [78] have questioned the concept or the reality of fluids with yield stress (see also Barnes [79]).

With the publications of [77], a series of interesting exchanges among different scientists started. Hartnett and Hu [80] responded to [77] with a new paper titled "The Yield stress—An engineering reality," which was followed by other papers Astarita [81], and Evans [82]. For additional and more recent and important discussions on the status of yield stress fluids, we refer the reader to Papanastasiou [83], Bonn and Denn [84], and Denn and Bonn [85]. Moller et al. [86] provide an excellent discussion on the relationship between thixotropic fluids and fluids with yield stress. For a historical survey of the yield stress fluids, see [79], and for a comprehensive review of the flow of visco-plastic materials, see [87].

The yield stress, to a large extent, determines the transition between or the point at which the solid-like behavior changes to a fluid-like behavior. It is one of the most important and difficult properties of the fluid to measure, which in theory, can be obtained at low shear rate tests. Møller et al. [88] discuss some of the difficulties of measuring the yield stress. Dinkgreve et al. [89] talk about the various ways of measuring the yield stress. Nguyen and Boger [90] provide a detailed review of the flow properties of yield stress fluids, and Coussot [91] provides a recent review of the experimental data in yield stress fluids. Different methods to measure the yield stress have been suggested [92]. Unlike ideal yield stress fluids, concentrated colloidal fluids stop flowing abruptly at a critical stress and begin to flow with a high velocity at another critical stress, which increases with the duration of the preliminary rest (a period of rest after pre-shearing over a larger number of applied stresses) [92]. When the shear stress reaches a critical value, the shear rate changes from zero to a critical shear rate abruptly [93]; the critical values of shear rate and the shear stress are considered intrinsic material parameters and independent of flow condition. Below the yield stress, fresh cement behaves as a poro-elastic solid. After the yield stress is reached, the slurry exhibits plastic strains [35].

Roussel et al. [94] identified two different critical strains in fresh cement pastes, while studying the origins of the thixotropy and the mechanism of yielding: (1) the largest critical strain (of the order of a few %), is the strain at the yield point obtained from measurements [95,96]. This strain is related to the network breakage of the colloidal interactions between cement particles (e.g., C-S-H particles) shortly after mixing. It takes only a couple of seconds to form the network; (2) the smallest critical strain (of the order of a few hundredths of %), is the strain when the shear modulus drops significantly [97,98]. This strain is related to the breakage of the early hydrates, which are caused by the contact of flocculated cement grains. Short-term thixotropy is related to colloidal flocculation and long-term thixotropy is related to the ongoing hydrates nucleation. According to the two different critical strains mentioned above, the static yield stress measurement is either determined by the strength of the C-S-H bonds at the rigid critical strain at long times or determined by the C-S-H nucleation at the colloidal critical strain at short times. Perrot et al. [99] analyzed the effects of three parameters on the yield stress and the stability of fresh cement pastes: (1) Brownian motion, depending on temperature; (2) colloidal attractive forces, depending on the average distance between the interacting particles; (3) gravity, depending on the grain size. According to the Perrot et al. [99], cement displays yield stress when the colloidal attractive forces dominate the Brownian motion, while no yield stress is observed if the Brownian motion dominates the colloidal attractive forces. A cement suspension is stable and homogenous without bleeding (cement grains unstably suspended in water) if the colloidal attractive forces dominate gravity, while cement particles settle if gravity dominates the colloidal attractive forces. These forces are time driven and are affected by the interstitial fluid viscosity.

Because of thixotropy, there is more than one state of flocculation for the yield stress measurement: the dynamic yield stress measured through the flow curve at zero shear rate (equilibrium state) and the static yield stress needed to initiate flow before the structure is broken down. The dynamic and the static yield stresses of fresh cement are usually measured in a rotational rheometer with a vane [100,101]. Qian and Kawashima [101] measured the yield stress through shear-rate-controlled and shear-stress-controlled tests. The authors detected a negative slope in the equilibrium flow curve (torque vs. angular velocity) at steady-state condition, shear banding, and stick-slip in the shear-rate-controlled test. They also detected viscosity bifurcation and considered the creep stress as the static yield stress based on the stress-controlled test [101]. Michaux and Defosse [102] reported that the yield stress of cement slurries exhibited a peak value at low dispersant (such as the sodium salt of Polynapthalene Sulphonate) concentration and decreased to zero at high concentration. The dispersant seemed to break the structure through attractive interparticle forces.

In this section, we review the existing models for the yield stress for cement considering different effects such as volume fraction, water-to-cement ratio, additives and damage.

Effect of Concentration on the Yield Stress

From the experimental data by Lapasin [58] it can be noticed that the yield stress τ_y (see Equation (25)) increases with increasing the specific surface area of cement and with decreasing the water-to-cement ratio as well as water film thickness [103]. Larger cement solid concentration and flocculation result in larger yield stress [104].

Legrand [105] proposed a relationship between the yield stress and the concentration

$$\tau_y = A_0 \alpha^{(\phi-0.5)} \tag{34}$$

where A_0 and α are related to the particle size and the shape, and ϕ is the concentration of the cement particles. This relationship is valid when ϕ is in the range of [0.475–0.677], corresponding to the w/c range [0.15–0.35] [106].

Sybertz and Reick [107] suggested an equation showing the influence of concentration of cement particles on the yield stress:

$$\tau_y = P_1 \cdot e^{P_2 \cdot \phi} \tag{35}$$

where P_n (n = 1, 2) are material parameters obtained from experiment, where $P_1 = 2.31 \times 10^{-3}$ and 4.75×10^{-3} and $P_2 = 20.7$ and 19.1 for different types of cement.

Zhou et al. [108] proposed a yield stress model for concentrated flocculated suspensions with different size particles, using the yield stress of individual components in the suspension:

$$\tau_y = \left(\sum \phi_{vi} \tau_{yi}^{1/2}\right)^2 \tag{36}$$

where ϕ_{vi} is the volume fraction of ith component.

Zhou et al. [108] and Flatt and Bowen [109] suggested a yield stress function of the type:

$$\tau_{y,max} = K\left(\frac{\phi}{1-\phi}\right)^c \frac{1}{d^2}$$
$$K = \frac{3.1 A b}{24\pi h_0} \tag{37}$$

where A is the Hamaker constant [110] of colloidal material (5.3 × 10^{-20} J), b and c are fitting parameters ranging from 0.1 to 0.53, h_0 is the distance between the two particles (2.4 nm) and d is the particle diameter.

Flatt and Bowen [111] proposed a yield stress model (YODEL) that depends on the volume fraction:

$$\tau_y = m_1 \frac{\phi^2(\phi - \phi_{perc})}{\phi_m(\phi_m - \phi)} \tag{38}$$

where ϕ_m is the maximum packing volume fraction (equal to 0.57 [111]), ϕ_{perc} is the percolation threshold (equal to 0.026 [111]), and m_1 is given by:

$$m_1 = \frac{1.8}{\pi^4}\left(\frac{G_{max}}{R_{v,50}}\right) F_{\sigma, \Delta} \tag{39}$$

where G_{max} is the maximum attractive force between the particles, $R_{v,50}$ is the median particle radius and $F_{\sigma, \Delta}$ is a function of coordination number related to the contacts between particles. Figure 4 shows the yield stress versus volume fraction for various samples. Equation (38) from YODEL [111] is applied to fit Zhou et al. [108] experimental data.

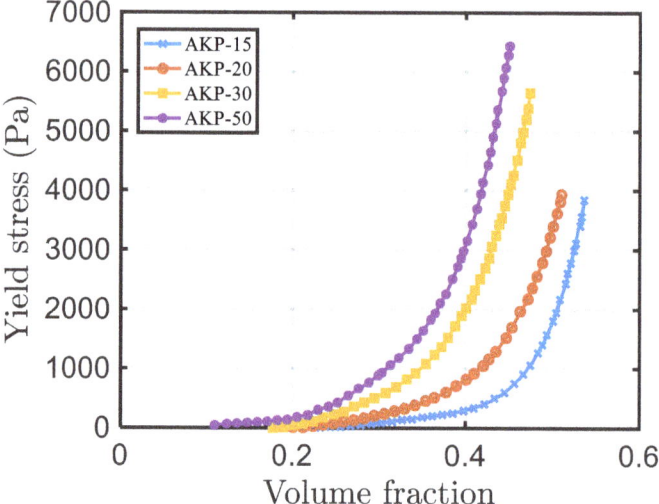

Figure 4. Yield stress vs. volume fraction for concentrated suspensions. YODEL [111] is applied to fit the experimental data of Zhou et al. [108], where AKP-15, 20, 30, and 50 are different alumina particulate samples.

Based on YODEL, Ma and Kawashima [112] related the yield stress to the hydration degree $\alpha(t)$ by assuming $\phi = \phi_0(1 + \chi\alpha(t))$:

$$\tau_y = m_1 \frac{(\phi_0(1+\chi\alpha(t)))^2 \left(\phi_0(1+\chi\alpha(t)) - \phi_{perc}\right)}{\phi_m(\phi_m - \phi_0(1+\chi\alpha(t)))} \tag{40}$$

where χ is an expansion parameter obtained from experiment, related to the density difference between unhydrated cement clinkers and the hydration products. For neat cement pastes (cement without sand or aggregate), $\phi_{perc} = 0.37$ and $\phi_m = 0.59$, $\phi_0 = 0.425$ for cement with water-to-cement ratio of 0.43, $\alpha(t) = 1 - e^{-kt^n}$, where $k = 6.73 \times 10^{-10}$ and $n = 2.85$ [112].

Mahaut et al. [95] used the Chateau–Ovarlez–Trung yield stress model [113] to describe the behavior of a thixotropic cement paste:

$$\frac{\tau_y(\phi)}{\tau_y(0)} = \sqrt{\frac{1-\phi}{(1-\phi/\phi_m)^{2.5\phi_m}}} \tag{41}$$

where $\frac{\tau_y(\phi)}{\tau_y(0)}$ is the dimessionless yield stress of a monodisperse suspension, where $\phi_m = 0.56$. This equation is suitable for yield stress fluids consisting of rigid spherical non-colloidal particles with no interactions between the particles and the paste. Figure 5 shows the fit for the Chateau–Ovarlez–Trung yield stress model for cement.

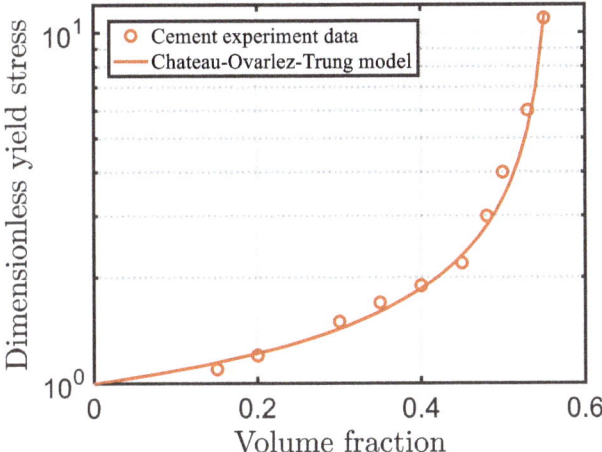

Figure 5. Dimensionless yield stress $\frac{\tau_y(\phi)}{\tau_y(0)}$ vs. volume fraction of cement: comparing the experimental data and the Chateau–Ovarlez–Trung yield stress model [108].

Chougnet et al. [114] also proposed a correlation for the dynamic yield stress, which is the shear stress at the zero limit of shear rate, as a function of the volume fraction

$$\tau_y^{dyn} = \frac{F}{a^2}\left(\frac{\phi}{\phi_m}\right)^{\frac{1}{m(3-f)}} \tag{42}$$

where F is the adhesion force necessary to separate the particles from each other, a is the particle radius, m and f are material parameters.

Based on the experimental data, Lapasin et al. [58] proposed a correlation for the yield stress τ_y describing the effect of cement specific surface S_{vB} and concentration of cement particles ϕ:

$$\tau_y = 2.1 \times 10^{-3} e^{19.2 \cdot \phi} \cdot S_{vB}^{2.5} \tag{43}$$

Effect of Water-to-Cement Ratio on the Yield Stress

Ivanov and Roshavelov [115] found that the yield stress decreases when the water-to-cement ratio increases. Rosquoët et al. [116] suggested a coefficient K in the power-law relation [56] (see Equation (10)), where K is a function of w/c:

$$\tau = K\left(\frac{\dot{\gamma}}{\dot{\gamma}_0}\right)^n = (-175 w/c + 137)\left(\frac{\dot{\gamma}}{\dot{\gamma}_0}\right)^{0.6} \tag{44}$$

where $\dot{\gamma}_0$ is the reference shear rate with a value of 1000 s^{-1} [116]. Figure 6 shows the shear stress vs. the shear rate curves for various values of water-to-cement ratio.

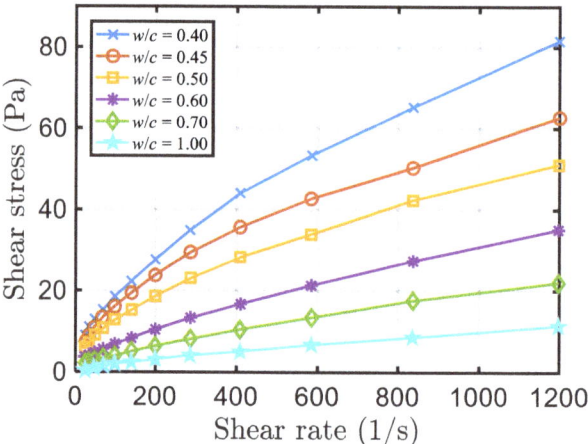

Figure 6. Shear stress vs. shear rate for different w/c [116].

Lapasin et al. [68] found that the yield stress increases linearly when the specific surface bearing (SSB) obtained from Blaine permeability apparatus increases,

$$\tau_y = K(SSB - 2000) \tag{45}$$

where K depends on the water-to-cement ratio (w/c) and it decreases while the w/c increases. Figure 7 shows the effect of water-to-cement ratio on the yield stress, reported by Banfill [3] from various published experimental data of cement. From Figure 7, we can see that there is a reverse log-linear relationship between the yield stress and the water-to-cement ratio.

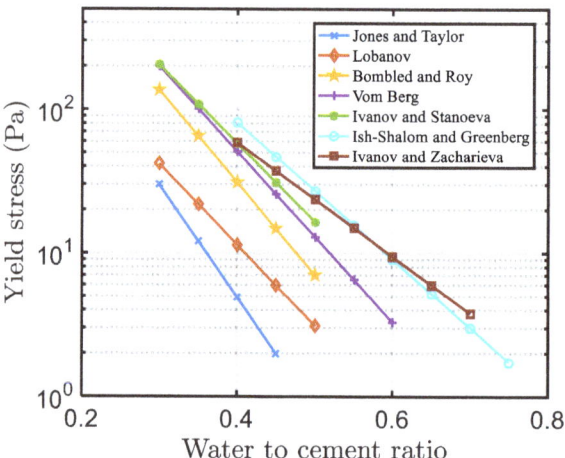

Figure 7. Effect of water-to-cement ratio on the yield stress of cement [3]. Data are from different published experimental data [39,42,54,67,117–121].

Effect of Additives/Admixtures on the Yield Stress

Ivanov and Roshavelov [115] proposed a polynomial equation, based on the experimental data fitted with a regression analysis, describing the effect of each clinker component on the yield stress (with units Pa):

$$\begin{aligned}\tau_y = &-118.5 - 72.3X_1 - 104.8X_2 + 103.6X_3 - 46.1X_4 - 48.7X_5 + 75.4X_1X_2 - \\ &77.4X_1X_3 + 38X_1X_4 + 42.4X_1X_5 - 107X_2X_3 + 39.9X_2X_4 + 41.6X_2X_5 - \\ &54.5X_3X_4 - 50.7X_3X_5 + 21.5X_1^2 + 19.3X_2^2 + 158.7X_3^2 + 38.9X_4^2 + 37.1X_5^2\end{aligned} \quad (46)$$

where X_1, X_2, X_3, X_4, and X_5 are parameters indicating the contributions of the water-to-cement ratio (w/c), concentration of superplasticizer, condensed silica fume (CSF), tricalcium aluminate (C_3A), and SO_3, respectively. Equation (46) indicates that CSF has the strongest influence on the behavior of the cement; as indicated by Ivanov and Roshavelov [115] τ_y initially decreases when CSF increases to a certain value, and then it begins to increase when the CSF is in the range 7.5–15%.

Sybertz and Reick [107] studied the effect of fly ash on the behavior of cement paste and suggested the following equation:

$$\tau_y = P_1 \cdot e^{P_2 \cdot \phi} \cdot (1 - \phi_f) + P_3 \cdot e^{P_4 \cdot \phi} \cdot \phi_f \quad (47)$$

where ϕ_f is the fly ash content and P_n, Q_n (n = 1, 2, 3, 4) are the experimental fitting parameters, where P_1 = 2.31 × 10^{-3} (Fly Ash I) and 4.75 × 10^{-3} (Fly Ash II), P_2 = 20.7 (Fly Ash I) and 19.1 (Fly Ash II), P_3 = 7.96 × 10^{-5} (Fly Ash I) and 1.29 × 10^{-5} (Fly Ash II), P_3 = 17.0 (Fly Ash I) and 18.9 (Fly Ash II). Figure 8 shows the effect of different types of fly ash on the yield stress. Both Fly Ash I and Fly Ash II lower the value of the yield stress.

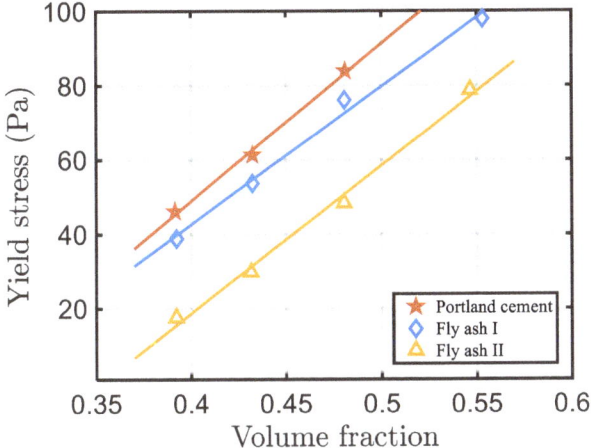

Figure 8. Effect of fly ash additives on yield stress [107].

Effect of Damage on the Yield Stress

Chen et al. [122] suggested a strain rate-dependent constitutive equation for the stress, including the effect of damage because of the dynamical experiment of measuring the mechanical properties of cement-based materials:

$$\tau = \left[1 - \left(D_1 + D_0 \dot{\gamma}_d^\xi\right)\gamma\right] \cdot \left[A_0 + A_1 \left(\frac{\dot{\gamma}_d}{\dot{\gamma}_s}\right)^B\right] \gamma \quad (48)$$

where $\xi = \lambda - 1$, and D_0, D_1, and λ are constants or parameters related to damage; A_0, A_1, and B are constants or parameters related to the shear rate, and $\dot{\gamma}_d$ and $\dot{\gamma}_s$ are the dynamic strain rate and static strain rate, respectively, and γ is the strain. Figure 9 shows the stress–strain curves for cement at various strain rates (104/s and 134/s).

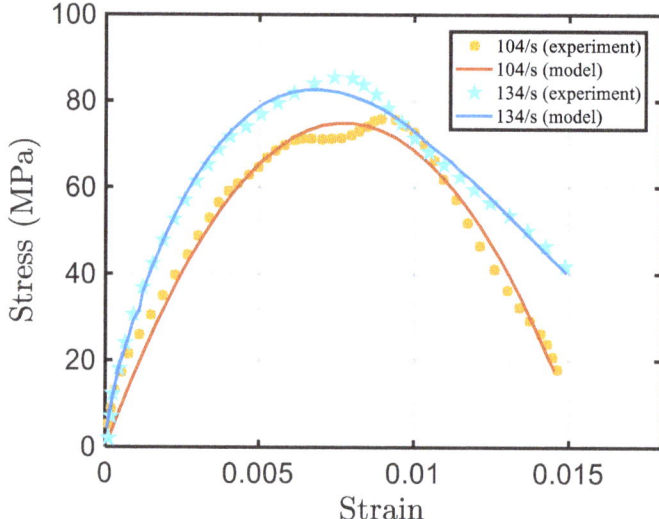

Figure 9. The stress–strain curves for cement using experiments and the damage-based model [122].

We summarize the yield stress models for cement in Table 4:

Table 4. Summary of the yield stress models for cement.

Effect	Author(s)	Model	Equation No.
Effect of concentration	Legrand (1970)	$\tau_y = A_0 \alpha^{(\phi-0.5)}$	(34)
	Sybertz and Reick (1991)	$\tau_y = P_1 \cdot e^{P_2 \cdot \phi}$	(35)
	Zhou et al. (1999)	$\tau_y = \left(\sum \phi_{vi} \tau_{yi}^{1/2}\right)^2$	(36)
	Zhou et al. (1999)	$\tau_{y,max} = K\left(\frac{\phi}{1-\phi}\right)^c \frac{1}{d^2}$	(37)
	Flatt and Bowen (2006)	$\tau_y = m_1 \frac{\phi^2(\phi - \phi_{perc})}{\phi_{max}(\phi_{max}-\phi)}$	(38)
	Ma and Kawashima (2019)	$\tau_y = m_1 \frac{(\phi_0(1+\chi\alpha(t)))^2(\phi_0(1+\chi\alpha(t))-\phi_{perc})}{\phi_{max}(\phi_{max}-\phi_0(1+\chi\alpha(t)))}$	(40)
	Chateau–Ovarlez–Trung (2008)	$\frac{\tau_y(\phi)}{\tau_y(0)} = \sqrt{\frac{1-\phi}{(1-\phi/\phi_m)^{2.5\phi_m}}}$	(41)
	Chougnet (2008)	$\tau_y^{dyn} = \frac{F}{a^2}\left(\frac{\phi}{\phi_{max}}\right)^{\frac{1}{m(3-f)}}$	(42)
	Lapasin et al. (1983)	$\tau_y = 2.1 \times 10^{-3} e^{19.2 \cdot \phi} \cdot S_{vB}^{2.5}$	(43)
Effect of water-to-cement ratio	Rosquoët et al. (2003)	$\tau = (-175w/c + 137)\left(\frac{\gamma}{\gamma_0}\right)^{0.6}$	(44)
	Lapasin et al. (1979)	$\tau_y - K(w/c)(SSB - 2000)$	(45)
Effect of additives/admixtures	Ivanov and Roshavelov (1990)	$\tau_y = -118.5 - 72.3X_1 - 104.8X_2 + 103.6X_3 - 46.1X_4 - 48.7X_5 + 75.4X_1X_2 - 77.4X_1X_3 + 38X_1X_4 + 42.4X_1X_5 - 107X_2X_3 + 39.9X_2X_4 + 41.6X_2X_5 - 54.5X_3X_4 - 50.7X_3X_5 + 21.5X_1^2 + 19.3X_2^2 + 158.7X_3^2 + 38.9X_4^2 + 37.1X_5^2$	(46)
	Sybertz and Reick (1991)	$\tau_y = P_1 \cdot e^{P_2 \cdot \phi} \cdot (1 - \phi_f) + P_3 \cdot e^{P_4 \cdot \phi} \cdot \phi_f$	(47)
Effect of damage	Chen et al. (2013)	$\tau = \left[1 - \left(D_1 + D_0 \dot{\gamma}_d^\zeta\right)\gamma\right] \cdot \left[A_0 + A_1\left(\frac{\dot{\gamma}_d}{\gamma_s}\right)^B\right]\gamma$	(48)

2.2.2. Viscosity Relationships

The non-linear time-dependent response of complex fluids such as cement slurries constitutes an important area of mathematical modeling of non-Newtonian fluids. In general, for many complex fluids such as cement slurries, or drilling fluids, the shear viscosity can be a function of one or all of the following [123]:

- Shear rate $\dot{\gamma}$
- Volume Fraction ϕ
- Temperature θ
- Pressure P
- Thixotropic behavior (structural parameter $\lambda(t)$)
- Water-to-cement ratio w/c
- Additives (Superplasticiser)
- Mixing method
- Electric field
- Magnetic field
- . . . .

For certain materials or under certain conditions, the dependence of viscosity on some of these can be dropped. In this section, we look at the effects of some of these parameters on the viscosity.

Effect of Shear Rate on the Viscosity

Recall that according to the power-law (the Ostwald-de Waele) model [56,124], shown in Section 2.1, the shear viscosity η, can be defined as

$$\eta = \frac{\tau}{\dot{\gamma}} = K\dot{\gamma}^{n-1} \qquad (49)$$

For pseudoplastic (or shear-thinning) fluids $n < 1$; in this case, the viscosity decreases when the shear rate increases. For Newtonian fluids $n = 1$, where the viscosity is independent of the shear rate. For dilatant (or shear-thickening) fluids $n > 1$, the viscosity increases with the shear rate [see Figure 10]. Equation (49) provides an explicit relationship between the shear rate and the shear stress; however, it is not suitable for fluids with yield stress.

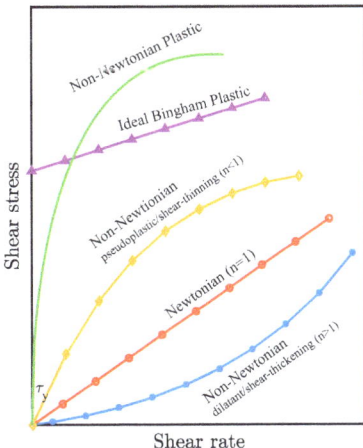

Figure 10. Shear stress vs. shear rate for different types of fluids [125].

Based on the Robertson and Stiff model [46] shown in Section 2.1, the effective viscosity for cement slurries can be written as

$$\eta_{eff} = \frac{A(\dot{\gamma}_R + C)^B}{\dot{\gamma}_R} \qquad (50)$$

where η_{eff} is the effective viscosity and $\dot{\gamma}_R$ is the value of the shear rate at the pipe or the annulus wall.

As shown in Section 2.1, the Casson model [62] is widely used for shear-thinning non-Newtonian fluids. According to this model, the viscosity is:

$$\eta = \frac{1}{\dot{\gamma}}\left[k_0 + k_1 \sqrt{\dot{\gamma}}\right]^2 \qquad (51)$$

where k_0 and k_1 are material parameters obtained from experiment.

From the Carreau–Yasuda model [33], viscosity approaches a lower limit when the shear rate is close to zero and approaches an upper limit when the shear rate is close to infinity. According to this model, the relation between viscosity and the shear rate is:

$$\eta = \eta_\infty + (\eta_0 - \eta_\infty)\frac{1 + \ln(1 + k\dot{\gamma})}{1 + k\dot{\gamma}} \qquad (52)$$

where η_0 and η_∞ are the lower and the upper viscosities when the shear rate is close to zero and infinity, respectively, and k is the shear-thinning parameter. These parameters are obtained from experimental measurements.

Effect of Volume Fraction on the Viscosity

Experiments indicate that the viscosity increases with higher particle concentration and larger cement specific surface [104].

Einstein [126] first presented the simplest mathematical expression for the effects of concentration on the (dimensionless) viscosity, commonly referred as *relative viscosity* of a fluid containing very small number of rigid spheres (low concentration or dilute limit):

$$\eta_r = \frac{\eta}{\eta_0} = 1 + \alpha\phi \qquad (53)$$

where η is the viscosity of the suspension, η_0 is the viscosity of the pure liquid (no particles) (1cP for water at 20 °C), $\alpha = 2.5$ for rigid spheres, and ϕ is the concentration. At higher concentrations, the viscosity reaches an infinite value, and this equation is no longer applicable. The Einstein equation has been applied to cement suspensions by Roussel et al. [127].

For spheres of different sizes, Roscoe [128] suggested:

$$\eta_r = \frac{\eta}{\eta_0} = (1 - \phi)^{-2.5} \qquad (54)$$

Mooney [129] proposed an equation for densely packed particles:

$$\eta_r = \frac{\eta}{\eta_0} = e^{\left[\frac{\alpha\phi}{(1-\frac{\phi}{\phi_m})}\right]} \qquad (55)$$

where η is the apparent viscosity (the applied shear stress divided by the shear rate, i.e., $\eta = \frac{\tau}{\dot{\gamma}}$) of the suspension, η_0 is the apparent viscosity of the continuous/liquid phase without any particles, ϕ is the concentration, α is a parameter depending on particle shape (2.5 for spheres), and ϕ_m is the maximum solid concentration, depending on the particle size distribution and the shape. However, Mooney's equation does not seem to fit the measured data at high concentrations [104].

Roscoe [128] also suggested an expression for the viscosity of concentrated suspensions considering non-uniform particle size:

$$\eta_r = \frac{\eta}{\eta_0} = (1 - 1.35\phi)^{-2.5} \quad (56)$$

Figure 11 shows the dimensionless viscosity versus volume fraction using various relationships.

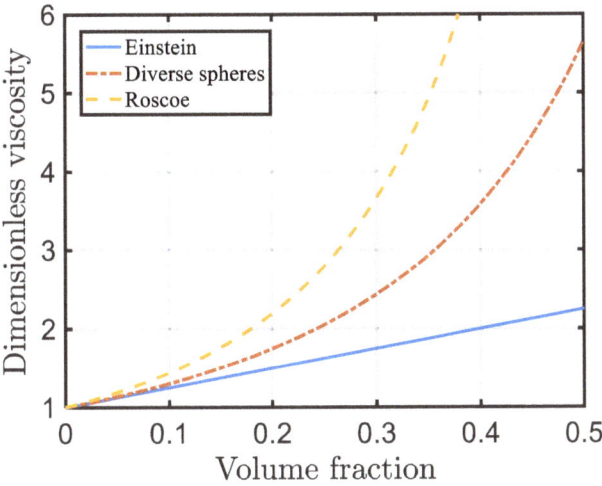

Figure 11. Dimensionless viscosity vs. volume fraction [128].

Krieger and Dougherty [130] provided an equation that is widely applied in cement industry [103, 104,127,131–139]. They suggested,

$$\eta_r = \frac{\eta}{\eta_0} = \left(1 - \frac{\phi}{\phi_m}\right)^{-\alpha\phi_m} \quad (57)$$

For dispersed cement pastes, $\phi_m \cong 0.7$, $\alpha \cong 5$, which increases at higher shear rates. Cement pastes that are not dispersed have higher viscosity and lower ϕ_m [104]. Cement pastes at lower concentrations (higher w/c) exhibit Newtonian behavior, while at higher concentrations (lower w/c) they show pseudo-plastic or plastic behavior. Table 5 shows the Krieger–Dougherty parameters for various types of cements for different strain rates (25 1/s and 500 1/s) [104]. Figure 12 shows the Krieger Dougherty curves for dimensionless viscosity versus volume fraction using the parameters in Table 5.

Table 5. Krieger–Dougherty parameters for cement [104].

Cement Type	Strain Rate (1/s)	ϕ_m	α
Type I, dispersed	25	0.64	5.1
	500	0.76	6.2
Type I, flocculated	500	0.64	6.3
White cement, dispersed	25	0.67	5.7
	500	0.80	6.8
Type V, dispersed	25	0.68	4.5
	500	0.75	5.2

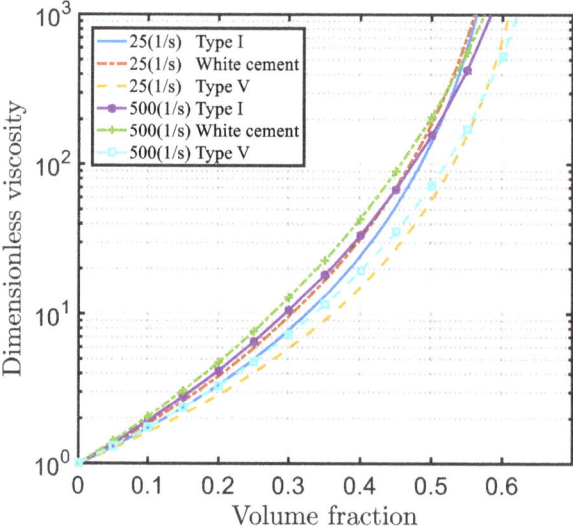

Figure 12. Krieger–Dougherty curves for various types of cements at various strain rates [104].

Chen and Lin [140] combined the Krieger–Dougherty equation with the fitting curves of apparent viscosity measured from a vibrational viscometer and suggested the following equation:

$$\eta_r = \left(1 - \frac{\phi}{\phi_m}\right)^{-2.78} e^{0.073t} \quad (58)$$

where t is the hydration time, $\phi_m = 0.635$ for monodisperse systems. Equation (58) describes the relationship between the viscosity, the volume fraction and the hydration time. It shows that the viscosity increases with the hydration time within 20 min.

Murata and Kikukawa [141] proposed an exponential expression for viscosity; their equation was used by Asaga and Roy to study cement slurry [39]:

$$\eta_r = B_0 e^{(K_1\phi + K_2)} \quad (59)$$

where B_0, K_1, and K_2 are constants obtained from experiments.

Sybertz and Reick [107] suggested an equation showing the effect of concentration of cement particles on the initial viscosity:

$$\eta_0 = Q_1 \cdot e^{Q_2 \cdot \phi} \quad (60)$$

where ϕ is concentration of cement particles and Q_n ($n = 1, 2$) are parameters that can be obtained from experiments, where $Q_1 = 8.97 \times 10^{-5}$ and 4.04×10^{-5} and $Q_2 = 17.1$ and 19.0 for different types of cement.

Chougnet et al. [114] used Mills [142] viscosity correlation to look at the effect of particle aggregation on viscosity.

$$\eta_r = \frac{\eta}{\eta_0} = \frac{(1-\phi)}{\left(1 - \frac{\phi}{\phi_m}\right)^2} \quad (61)$$

where $\phi_m = 4/7$ for randomly packed monodispersed spheres.

Liu [143] proposed a dimensionless viscosity relationship for cement pastes that was reviewed by Bentz et al. [137]:

$$\eta_r = \frac{\eta}{\eta_0} = [a(\phi - \phi_m)]^{-n} \quad (62)$$

where a and n are fitting parameters with values 0.95 and 2.14.

Chong et al. [144] proposed a dimensionless viscosity relationship for cement pastes which was also reviewed by Bentz et al. [137]:

$$\eta_r = \frac{\eta}{\eta_0} = \left(1 + \frac{0.75\frac{\phi}{\phi_m}}{1 - \frac{\phi}{\phi_m}}\right)^2 \tag{63}$$

According to Bentz et al. [137], Liu's model predicts the experimental results better than Chong et al.'s model.

Effects of Temperature and Pressure on the Viscosity

Temperature can play an important role in (oil well) cement operations, especially, in the early stages, when the yield stress increases; its effect on the viscosity is not that obvious because of the decrease of water viscosity (base fluid) at higher temperatures (with increasing rate 0.027 °C/m) [112,145]. With the increase of temperature in deep wells, the viscosity tends to decrease because of thermal thinning [146]. Cement slurries experience different pressure conditions when pumped into the wells. Ma and Kawashima [112] found that the high pressure in deep wellbores (with increasing rate 9.8 kPa/m) increases the yield stress and the viscosity; high pressure seems to accelerate the hydration process without affecting the water viscosity. Kim et al. [147] found that pressure causes the yield stress to decrease by 15% at lower water-to-cement ratios (<0.4), while the effect is not obvious at higher water-to-cement ratios. High pressure also changes the microstructure of the cement, causing deflocculation while increasing the dispersion of cement particles, resulting in a decrease in the yield stress. Figure 13 shows the shear stress versus shear rate curves for different water-to-cement ratios at different temperatures.

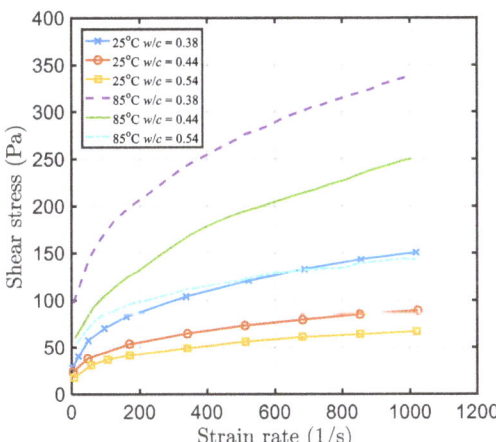

Figure 13. Shear stress vs. shear rate for cement for different water-to-cement ratios at different temperatures (25 °C and 85 °C) [146].

The microscopic and macroscopic properties (e.g., rheometric behavior, mechanical characteristics) of hydration products including calcium silicate hydrate (CSH) and calcium hydroxide (CH) are found to be sensitive to the curing temperature. Vlachou and Piau [19] found that cement particles are more spherical-like at lower temperatures (20 °C) while they are more rod-like at higher temperatures (60 °C), as shown in Figure 14. Temperature seems to accelerate the hydration process. Some studies indicate that both the yield stress and the viscosity decrease when the temperature increases [148]. In some applications, pressure seems to have negligible effects on the flow of cement slurries; at lower water-to-cement ratios, cement slurries become more sensitive to pressure [22].

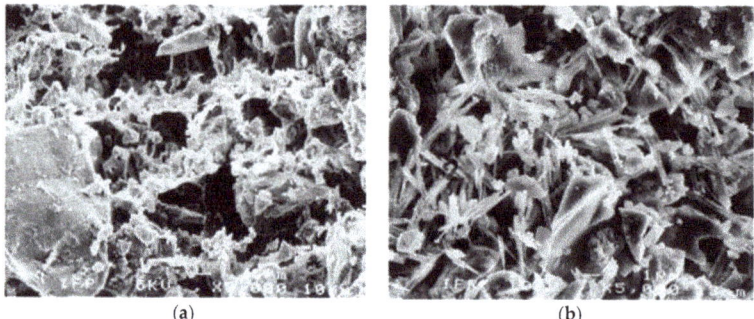

Figure 14. Morphology of cement particle under SEM (**a**) at 20 °C; (**b**) at 60 °C [19].

Sercombe et al. [149] suggested an equation showing the temperature effect on the viscosity for long-term creep behavior of cementitious material:

$$\frac{1}{\eta_\theta} = \frac{1}{\eta_{\theta,0}} e^{[-\frac{U}{R}(\frac{1}{\bar{\theta}} - \frac{1}{\theta})]} \qquad (64)$$

where U is the activation energy of long-term creep, R is the universal gas constant, $U/R = 2700\ K$ [150], $\bar{\theta}$ is the reference temperature (20 °C in their study), $\eta_{\theta,0}$ is the value of η_θ when $\theta = \bar{\theta}$.

The effects of temperature and pressure could also be implemented through the active hydration process, which suggests that the viscosity is related to a fixed hydration reaction rate. Some researchers [149,151] use an Arrhenius-type equation for the hydration kinetics process with some knowledge of chemoplasticity of cement and concrete. The chemical affinity $\widetilde{A}$ is expressed as:

$$\widetilde{A}(\xi) = \dot{\xi} e^{(\frac{E_a}{R\theta})} \qquad (65)$$

where $E_a/R \approx 4000K$, $\dot{\xi}$ is the hydration degree, which is estimated from the evolution of compressive strength depending on the temperature θ.

Scherer et al. [14] assumed that the viscosity of cement slurry is related to the degree of cement hydration and suggested an equation showing the effect of temperature and pressure on the hydration process. From the Avrami–Cahn model [152], the degree of hydration reaction is given by:

$$X \approx \frac{\pi}{3} O_v^B I_B G^3 t^4 \qquad (66)$$

where X is the volume fraction of the transformed reactant, O_v^B is the boundary area, I_B is the nucleation rate on the boundary, G is the linear growth rate of the product and t is the time, and I_B and G are functions of temperature θ and pressure p, where:

$$G(\theta,p) \approx G_0 e^{(-\frac{\Delta E_G + p\Delta V_G}{R\theta})} \qquad (67)$$

$$I_B(\theta,p) \approx I_0 e^{(-\frac{\Delta E_I + p\Delta V_I}{R\theta})} \qquad (68)$$

where ΔE_G and ΔE_I are the activation energy for growth and nucleation, ΔV_G and ΔV_I are activation volumes for growth and nucleation and G_0 and I_0 are constants.

Pang et al. [153] developed a simple scale factor connecting the degree of hydration to temperature and pressure:

$$\dot{\xi}(t) = \dot{\xi}_r(C(\theta, P), t) \tag{69}$$

where $\dot{\xi}_r$ is the degree of hydration at reference temperature θ_r and pressure P_r. The scale factor C is given by:

$$C = e^{[\frac{E_a}{R(\frac{1}{\theta_r} - \frac{1}{\theta})} + \frac{\Delta^{\ddagger}}{R}(\frac{P_r}{\theta} - \frac{P}{\theta})]} \tag{70}$$

where R is the gas constant, $\Delta^{\ddagger}$ is the activation volume and E_a is the activation energy.

Wu et al. [154] studied the effect of temperature and cement hydration on viscosity and suggested two viscosity equations for temperature effect [155] and cement hydration effect [156],

$$\eta_c = \eta_r \cdot e^{[\frac{E}{R\theta_c}]} \tag{71}$$

$$\eta_c = \eta_{c0} + (1000 - \eta_{c0}) \cdot (t/t_v)^n \tag{72}$$

$$\eta_c = \eta_r \cdot e^{[\frac{E}{R\theta_{c0}}]} + \left(1000 - \eta_r \cdot e^{[\frac{E}{R\theta_{c0}}]}\right) \cdot (t/t_v)^n \tag{73}$$

Equation (71) shows the influence of temperature, where θ_c and θ_r are the temperatures of fresh cement slurry and reference temperature, η_c and η_r are viscosities at temperature θ_c and θ_r, respectively. Equation (72) shows the cement hydration effect, where η_{c0} is the initial viscosity of the fresh cement slurry and n is a parameter related to cement hydration kinetics, t_v is the time needed to reach a very high viscosity (e.g., 1000 Pa·s). According to Papo and Caufin [156], t_v is a function of water-to-cement ratio:

$$t_v = t_{v0} + X[w/c - (w/c)_0]^Y \tag{74}$$

where t_{v0} and $(w/c)_0$ are the initial values of t_v and w/c, and X and Y are parameters obtained from experiment. Equation (73) combines Equations (71) and (72), where θ_{c0} is the initial temperature of fresh cement slurry.

Effect of Additives/Admixtures on the Viscosity

Chemical additives/admixtures have the following functions: (1) To disperse the cement particles; (2) to modify the kinetics of the hydration process; (3) to react with the hydration subproducts; (4) to add binders to cement [157]. Additives are applied to either retard or accelerate the curing process of the cement slurries; they could also be used as viscosifiers or dispersant.

Superplasticizing admixtures could make the cementations materials denser, more impermeable and more durable. A superplasticizer provides better dispersibility. As the amount of the superplasticizer increases, the slurry behavior changes from Newtonian to non-Newtonian and finally back to Newtonian, where the mixture is well dispersed [158]. Figure 15 shows the mechanism of how a superplasticizer works. Flocculated cement particles are dispersed because of the negatively charged superplasticizer and the motion of the entrapped water. A superplasticizer generally improves the flow behavior of cement slurries and provides highly amorphous hydrates [159]. It is observed that increasing the amount of a superplasticizer admixture increases the cement gelation threshold [160], while decreasing the yield stress and viscosity and delaying the cement hydration process [161–164]. These effects are not that obvious when the admixture concentration is higher than 0.75% or when the shear rate is high (>128 rpm). In general, the cement slurry seems to behave as a dilatant/shear-thickening fluid with admixture concentrations higher than 0.75% [165,166]. Effects of ultrafine particles on the superplasticizer is to decrease the flow resistance and viscosity [167–169]. Researchers [170,171] have found that the polycarboxylate (PC) superplasticiser admixture reduces the yield stress and the plastic viscosity by 70%; this reduction is thought to be related to the interconnected flocs or the weak cohesive forces.

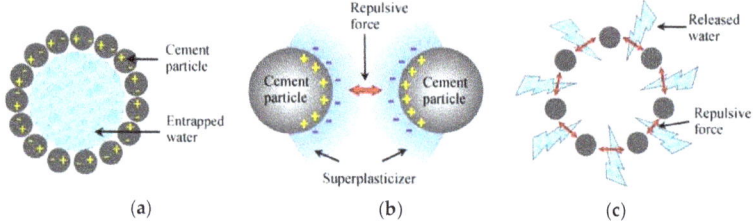

Figure 15. Effect of superplasticizer on cement particles [159]. (**a**) Flocculated cement particles; (**b**) dispersing cement particles by repulsive force; (**c**) release of entrapped water releasing.

Condensed silica fume (CSF), considered as an effective microfiller as superplastizier, has a significant effect on the viscosity and the yield stress of the cement pastes. Ivanov and Roshavelov [115] proposed a polynomial equation fitted with regression analysis using the experimental data, while describing the effect of each clinker component on the apparent viscosity (mPa·s):

$$\begin{aligned}\eta = &-300.8 - 357.6X_1 - 553.4X_2 + 575.4X_3 - 80.8X_4 - 22.4X_5 + \\ &293.7X_1X_2 - 329.4X_1X_3 + 14.7X_1X_4 + 57X_1X_5 - 528.4X_2X_3 + 48.9X_2X_4 + \\ &21.2X_2X_5 - 151.1X_3X_4 - 29.3X_3X_5 + 175.2X_4X_5 + 114.4X_1^2 + 152.8X_2^2 + \\ &664.6X_3^2 + 159.2X_4^2 + 213X_5^2\end{aligned} \quad (75)$$

where X_1, X_2, X_3, X_4, and X_5 are the parameters related to the amount of water-to-cement ratio (w/c), concentration of superplasticizer, condensed silica fume (CSF), tricalcium aluminate (C_3A), and SO_3, respectively. Equation (75) indicates that CSF has the strongest influence on the viscosity of the cement; it is observed that η initially decreases when CSF increases and reaches a certain value, and then it begins to increase when the CSF is in the range 7.5–15%. Wong and Kwan [172] indicated that the addition of CSF to cement causes an increase in the packing density and the flowability in the lower ranges (<15%) and lower w/c while causing a decrease in the packing density at the higher ranges (>15%). They also mention that the addition of pulverized fuel ash (PFA) increases the packing density and flowability of cement paste. The influences of C_3A and SO_3 are much less than other factors.

Glycerin, as a viscosifier, when added to the cement slurry, can increase the viscosity and accelerate the hydration process at about 26% volume content [17]. The effect of glycerin on viscosity is more obvious at larger shear rates. The shear-thinning behavior is reduced, and the slurry begins to be more Bingham-like with the increase of glycerin content. Another viscosifying agent, hydroxypropylmethylcellulose (HPMC) increases the plastic viscosity, while decreasing the fluid loss, increasing the thickening time, and increasing the compressive strength of cement slurries [173]. This is very suitable for oil well cementing under high temperature conditions.

In general, adding these modifying admixtures can increase the yield stress and the viscosity of cement whereas high-range water reducers can decrease the viscosity at low shear rates much more when compared to the viscosity at high shear rates [165]. A combination of these two additives can increase the performance of cement.

Ultra-fine admixtures (UA), including blast furnace slag (BFS), silica fume (SF), fly ash (FA), limestone (LS), and anhydrous gypsum (AG), also can influence the rheological properties of cement paste. It has been observed that the yield stress decreases with an increase in the UA content and the viscosity decreases with addition of ultra-fine LS, SF, FA, and slag while it increases with AG [174–180]. The effect of ultra-fine slag is more obvious when the UA content is more than 15%. The spherical shape of fly ash particles is found to reduce the viscosity and the yield stress of fresh cement pastes [181,182]. Sybertz and Reick [107] studied the effect of fly ash on the rheological behavior of cement paste and suggested an equation showing the influence of fly ash content on the initial viscosity:

$$\eta_0 = Q_1 \cdot e^{Q_2 \cdot \phi} \cdot (1 - \phi_f) + Q_3 \cdot e^{Q_4 \cdot \phi} \cdot \phi_f \quad (76)$$

where ϕ_f is the fly ash content and Q_1, Q_2, Q_3, and Q_4 are experimental fitting parameters, where $Q_1 = 8.97 \times 10^{-5}$ (Fly Ash I) and 4.04×10^{-5} (Fly Ash II), $Q_2 = 17.1$ (Fly Ash I) and 19.0 (Fly Ash II), $Q_3 = 6.25 \times 10^{-5}$ (Fly Ash I) and 0.54×10^{-5} (Fly Ash II), $Q_3 = 18.4$ (Fly Ash I) and 20.9 (Fly Ash II). Figure 16 shows the effect of different types of fly ash on the initial viscosity. Fly Ash I slightly increases the initial viscosity while Fly Ash II decreases the initial viscosity significantly.

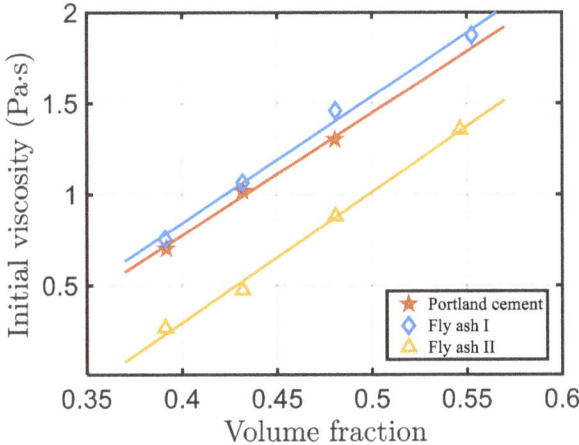

Figure 16. Effect of fly ash additives on the initial viscosity [107].

Hou and Liu [183] found that the addition of synthesizing dispersant extended the thickening time and improved the mobility of the cement slurry. Nehdi [184] noticed that the carbonate filler containing magnesium (MgO) impeded the particle dispersion while accelerating the C_3S hydration, resulting in a rapid increase in the viscosity and the loss of workability for the cement placement.

Sometimes silica flour is added to cement slurries in oil well applications when the temperature exceeds 120 °C, in order to prevent decrease in the compressive strength. This additive increases the slurry viscosity and decreases the thickening time [185].

Addition of ultrafine particles to the cement slurry can decrease the yield stress and the plastic viscosity [186]. Application of high content limestone powder to cement seems to improve the fluidity; this can also reduce the amount of the superplasticizer, and reduce the viscosity and the water-to-cement ratio [187]. The addition of nano-additives including nano-SiO_2 (nS) and nano-TiO_2 (nT) can increase the yield stress, the plastic viscosity, and the torque significantly [188]. The addition of Fe_2O_3 nanoparticles can improve the performance of wellbore cement slurries by increasing the viscosity, the elastic properties after early hydration and the suspending ability of cement particles while decreasing the free water, fluid loss, the compressive strength and the thickening time [189]. The addition of cellulose nanofibers (CNF) increases the yield stress, the degree of hydration, and the flexural strength of the cement slurry [190,191]. According to Colombo et al. [192], the addition of softwood calcium lignosulfonate seems to decrease the yield stress and the viscosity of cement pastes. However, it is reported that the application of densified microsilica or nano-silica may have a negative effect on the rheological properties of cement slurry, e.g., inadequate gel strength and stability, poor zonal isolation [193,194]. For a recent numerical study of the cement sheath and wellbore integrity, see Wang and Taleghani [195].

Phan et al. [196] compared the effect of high range water reducing admixture (HRWRA), (changing the granular phase configuration) and viscosity modifying admixture (VMA) (changing the aqueous solution) by applying the Krieger–Dougherty equation (see Equation (57)). It was found that the HRWRA had a bigger influence on the properties (namely, the viscosity increased) than the VMA. The Krieger–Dougherty equation indicates that the viscosity depends more on the configurational skeleton than on the fluid phase. Emoto and Bier [197] stated that MF 2651 as a plasticizer reduced the viscosity of cement slurry without delaying the hydration process when compared with other plasticizers.

Lu et al. [198] studied the effect of thermo-sensitive viscosity controller (TVC), consisting of inorganic and organic polymeric materials. The cement slurries with TVC become thermally more stable with little thermal thinning between 20 to 120 °C. Velayati et al. [199] investigated the effect of Cassia fistula dry extract on wellbore cement and found that this additive exhibits a retardation property by increasing the thickening time with high efficiency and low costs. Wang et al. [200] studied the effect of chloride additives for cements at low temperatures and found that some chlorides such as LiCl effectively shorten the thickening time and decrease the transition time for static gel strength, while improving the stability of the cement slurry and accelerating the hydration process.

Effect of Water-to-Cement Ratio on the Viscosity

Lapasin et al. [68] showed that the water-to-cement ratio affects the flow behavior of cement slurries. This ratio is defined as the ratio of the weight of water to the weight of cement. From the definition, there is a reverse correlation between volume fraction of cement and water-to-cement ratio. Cements with larger w/c have a smaller concentration. At early stages, w/c has little effect on cement hydration while it has a bigger effect on the physical properties such as strength development, and the setting time [201,202]. The viscosity was found to decrease when the water-to-cement ratio increased [98,115,116,145], shown in Figure 17. It is necessary to increase the w/c of cement slurries for the minimum flow resistances and appropriate injection time when sealing the casing pipes in wellbores [203]. At $w/c = 0.4$, the effect of superplasticizer on rheological properties is not obvious [39]. Massidda and Sanna [204] observed that cement exhibits thixotropic behavior at $w/c = 0.35$ while anti-thixotropic behavior at $w/c > 0.40$. Larger rigidity of slurries (lower w/c and longer hydration time) causes thixotropic behavior while smaller rigidity (higher w/c and shorter hydration time) results in reversible and anti-thixotropic behavior.

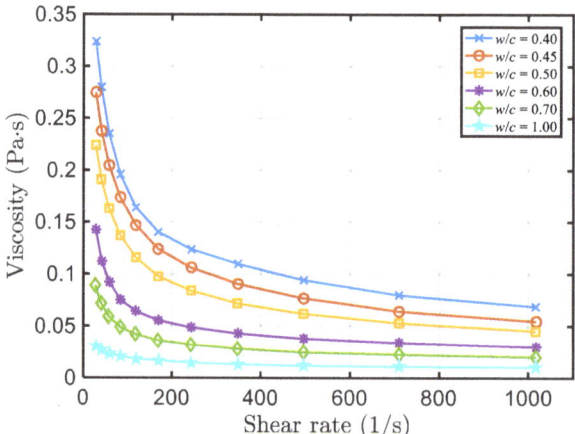

Figure 17. Effect of water-to-cement ratio on the viscosity [116].

Effect of Mixing Method and the Wall Slip on the Viscosity

Yang and Jennings [205] studied the influence of mixing method on the properties of cement paste and found the rheological behavior of cement pastes during the first two hours were significantly affected by the mixing method. Cement pastes have higher peak stresses with hand or paddle mixing than with blender mixing. Williams et al. [206] also noticed that the shear rate during mixing can significantly affect the rheological properties of fresh cement. Orban et al. [207] observed that the plastic viscosity of cement slurries decreases with increasing shear rate. The Yard mixer was found to increase the plastic viscosity and the yield point significantly [208]. ASTM C1738 protocol increases the plastic viscosity by increasing the volume fraction or the content of polycarboxylate superplasticizer (SP) in cement paste [209]. Increasing the mixing rate can also accelerate the cement hydration process and increase the overall heat in this process [210]. High shear rate of mixing can break down the cement agglomerates before adding water; this is called "irreversible structural breakdown" [5,76].

In oil well cementing, cement slurries are pumped through the annular spaces. Particle migration in cement slurries can cause slip at the walls during shearing, similar to the paper pulp flow inside a transparent pulp. Rubio–Hernández [37] found slippage at the walls or the pipe surface can be a problem during the measurement in a rheometer. This can cause an error in the rheological test. This "slip" is usually caused by particle migration. A clear water layer is detected adjacent to the walls, where the fluid has lower viscosity, resulting in a pseudo wall slip. This is caused by a thin film of liquid less than 1 μm, lubricating the walls of the viscometer [211]. Application of vane methods could eliminate the effect of wall slip [212].

Bannister [213] studied the slip effect in a rotational viscometer. The effect of slip on viscosity using the Metzner–Reed power-law model is:

$$\eta_s = C_S \frac{\tau}{\dot{\gamma}} \tag{77}$$

where C_S is the slip coefficient, depending on the consistency index K. The slip coefficient decreases when K (thicker slurry) or the water content (less particles in the slurry) increases. Table 6 shows a summary of the existing viscosity relationships for cement.

2.2.3. Thixotropic Nature of Cement

Cement slurries can also behave as time-dependent non-Newtonian fluids with disperse particles of many different sizes (10 nm to 100 μm) [58,127,213,214]. In general, as discussed earlier, viscosity of cement depends on the shear rate, the application time, particle concentration, temperature and pressure, etc. Thixotropy and yield stress are two important concepts that need to be considered [215]. According to Barnes [216], the term "*thixotropy*", first suggested by Freundlich [217], describes the reversible sol-gel transformation under isothermal conditions. This occurs when some of the chemically formed linkages between the cement particles break under shear (also known as structural breakdown). Thus, the process is shear rate dependent and time dependent. In thixotropic fluids, viscosity decreases with time for constant shear rate and the shear stress gradually reaches a steady value depending on the shear rate. This behavior is reversible when the system is in rest and the stress is removed [218]. "*Pseudoplasticity*" is also sometimes used to represent thixotropy with reference to shear rate dependency. For pseudoplastic fluids, viscosity decreases with the increasing of the shear rate. The opposite of "*thixotropy*" is "*antithixotropy*" [219] or "*rheopexy*", which implies that the structure builds up under shear and breaks down at rest. Antithixotropy (shear-thickening) is the recovery process of thixotropy (shear-thinning), which is different from "*dilatancy*" (shear- thickening) of suspensions with high solid concentration under high shear rates [35]. Figure 18 shows the shear stress vs. shear rate response of different fluids for a constant shear rate test [125].

Table 6. Summary of viscosity relationships for cement.

Effect	Author(s)	Model	Equation No.
Effect of shear rate	Ostwald-de Waele (1929)	$\eta = \frac{\tau}{\dot{\gamma}} = K\dot{\gamma}^{n-1}$	(49)
	Robertson and Stiff (1976)	$\eta_{eff} = \frac{A(\dot{\gamma}_R + C)^B}{\dot{\gamma}_R}$	(50)
	Casson (1959)	$\eta = \frac{1}{\dot{\gamma}}\left[k_0 + k_1\sqrt{\dot{\gamma}}\right]^2$	(51)
	Carreau–Yasuda (1997)	$\eta = \eta_\infty + (\eta_0 - \eta_\infty)\frac{1+\ln(1+k\dot{\gamma})}{1+k\dot{\gamma}}$	(52)
Effect of volume fraction	Einstein (1906)	$\eta_r = \frac{\eta}{\eta_0} = 1 + \alpha\phi$	(53)
	Roscoe (1952)	$\eta_r = \frac{\eta}{\eta_0} = (1-\phi)^{-2.5}$	(54)
	Mooney (1951)	$\eta_r = \frac{\eta}{\eta_0} = e^{\left[\frac{\alpha\phi}{(1-\frac{\phi}{\phi_m})}\right]}$	(55)
	Roscoe (1952)	$\eta_r = \frac{\eta}{\eta_0} = (1-1.35\phi)^{-2.5}$	(56)
	Krieger and Dougherty (1959)	$\eta_r = \frac{\eta}{\eta_0} = \left(1-\frac{\phi}{\phi_m}\right)^{-\alpha\phi_m}$	(57)
	Chen and Lin (2017)	$\eta_r = \left(1-\frac{\phi}{\phi_m}\right)^{-2.78} e^{0.073t}$	(58)
	Murata and Kikukawa (1973)	$\eta_r = \frac{\eta}{\eta_0} = B_0 e^{(K_1\phi + K_2)}$	(59)
	Sybertz and Reick (1991)	$\eta_0 = Q_1 \cdot e^{Q_2 \cdot \phi}$	(60)
	Mills (1985)	$\eta_r = \frac{\eta}{\eta_0} = \frac{(1-\phi)}{\left(1-\frac{\phi}{\phi_m}\right)^2}$	(61)
	Liu (2000)	$\eta_r = \frac{\eta}{\eta_0} = [a(\phi-\phi_m)]^{-n}$	(62)
	Chong et al. (1971)	$\eta_r = \frac{\eta}{\eta_0} = \left(1+\frac{0.75\frac{\phi}{\phi_m}}{1-\frac{\phi}{\phi_m}}\right)^2$	(63)
Effects of temperature and pressure	Sercombe et al. (2000)	$\frac{1}{\eta_\theta} = \frac{1}{\eta_{\theta,0}} e^{[-\frac{U}{R}(\frac{1}{\theta}-\frac{1}{\theta_0})]}$	(64)
		$\tilde{A}(\xi) = \dot{\xi} e^{\left(\frac{E_a}{R\theta}\right)}$	(65)
		$X \approx \frac{\pi}{3} O_v^B I_B G^3 t^4$	(66)
	Scherer et al. (2010)	$G(T,p) \approx G_0 e^{(-\frac{\Delta E_G + p\Delta V_G}{R\theta})}$	(67)
		$I_B(T,p) \approx I_0 e^{(-\frac{\Delta E_I + p\Delta V_I}{R\theta})}$	(68)
	Pang et al. (2013)	$\dot{\xi}(t) = \dot{\xi}_r(C(\theta,P),t)$	(69)
		$C = e^{[\frac{E_a}{R(\frac{1}{\theta_r}-\frac{1}{\theta})} + \frac{\Delta^\ddagger}{R}(\frac{P_r}{\theta} - \frac{P}{\theta})]}$	(70)
	Wu et al. (2014)	$\eta_c = \eta_r \cdot e^{\left[\frac{E}{R\theta_c}\right]}$	(71)
		$\eta_c = \eta_{c0} + (1000 - \eta_{c0}) \cdot (t/t_v)^n$	(72)
		$\eta_c = \eta_r \cdot e^{\left[\frac{E}{R\theta_{c,0}}\right]} + \left(1000 - \eta_r \cdot e^{\left[\frac{E}{R\theta_{c,0}}\right]}\right) \cdot (t/t_v)^n$	(73)
Effect of additives/admixtures	Ivanov and Roshavelov (1990)	$\eta = -300.8 - 357.6X_1 - 553.4X_2 + 575.4X_3 - 80.8X_4 - 22.4X_5 + 293.7X_1X_2 - 329.4X_1X_3 + 14.7X_1X_4 + 57X_1X_5 - 528.4X_2X_3 + 48.9X_2X_4 + 21.2X_2X_5 - 151.1X_3X_4 - 29.3X_3X_5 + 175.2X_4X_5 + 114.4X_1^2 + 152.8X_2^2 + 664.6X_3^2 + 159.2X_4^2 + 213X_5^2$	(75)
	Sybertz and Reick (1991)	$\eta_0 = Q_1 \cdot e^{Q_2 \cdot \phi} \cdot (1-\phi_f) + Q_3 \cdot e^{Q_4 \cdot \phi} \cdot \phi_f$	(76)
Effect of measurements (slip)	Bannister (1980)	$\eta_s = C_s \frac{\tau}{\dot{\gamma}}$	(77)

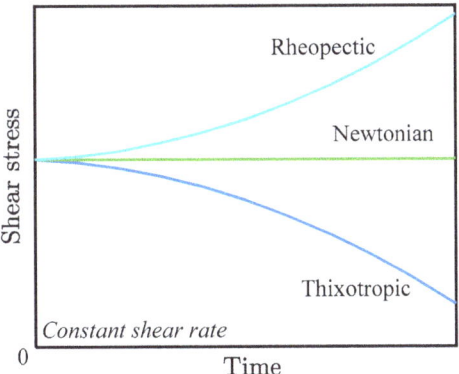

Figure 18. Effect of time on the behavior of various viscous fluids [125].

There are many ways to model the thixotropic effects in fluids. An excellent review is given by Barnes [216]. For a mathematical perspective of modeling thixotropic fluids with yield stress, we refer to [220–222]. An early formulation of suspensions with thixotropy was given by Fredrickson [223]. A more recent mathematical perspective is given by Renardy [224]. There have been other attempts to develop unified models to capture elasto-viscoplastic thixotropic yield stress fluids (see [225,226]).

The thixotropic aspect of cement is an indication that it can become a gel-like structure at rest which is unpumpable while it can be pumped again when stirred [200]. That is, thixotropy results in the development of gel strength and yield stress after the pumping has stopped. A high pressure is needed to restart the pumping. Thus, the thixotropy of cement and the static gel strength along with the yield stress affect the pumping and the restarting difficulties which can be of concern in cement safety (Wang et al. 2017). The thixotropic process (reversible) and the hydration process (irreversible) of cement usually occur simultaneously after the cement clinker is mixed with water. Initially, the thixotropic effect dominates and later the cement hydration process dominates [93]. Yuan et al. [75] assumed that in the intermediate period, only the thixotropic (reversible) process is important and the hydration (irreversible) process can be neglected.

As mentioned before, fresh cement pastes are initially thick colloidal suspensions consisting of cement particles dispersed in water [92]. The structure of suspensions is affected by water-to-cement ratio, particle size distribution, interparticle forces and attraction of the water to solid surfaces [35]. In the initial stages, just after mixing, the dispersed cement particles with high solid concentration coagulate and form a structure within the paste; this structure does not change in the next several hours until setting [97]. Interparticle forces in cement particles are attractive forces overcoming the repulsive forces between the particles, causing very small strains in the cement pastes (0.03%). When the interparticle forces dominate, cement pastes have poor flow properties [227]. Floccules are small clusters of cement particles formed in dilute cement suspensions and combine into large flocculent structures. Aggregations with non-uniformly distributed particles occur in the large flocculent structures, which are composed of high concentration flocs of particles. Flocculated particles form discrete aggregates or gels, as shown in Figure 19.

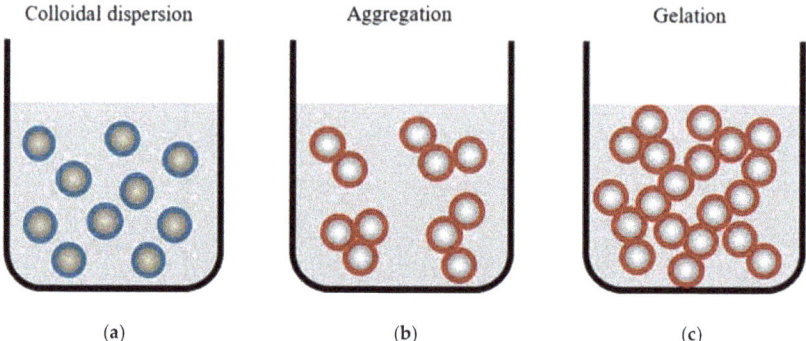

Figure 19. Microstructure of (**a**) dispersed particles; (**b**) aggregated particles; (**c**) gelled particles [228].

Thixotropic behavior is also related to coagulation (particle contacts), dispersion, and re-coagulation of cement particles [36]. The flocculated suspension behaves as viscoelastic solids with storage modulus (the ability to store energy related to deformation of the material) 14–24 kPa in low-strain linear-viscoelastic regions below yield [98]. As the strain increases to a critical value of 10^{-4}, the storage modulus decreases. With the increase of water-to-cement ratio, both the storage modulus and the critical strain decrease. The forces between the flocculated particles are weak and can be broken by shear force. Cement pastes show liquid-like behavior after the flocculated system is broken. During the mixing process, the aggregated cement particles become more uniformly distributed. Flocculent structure keeps breaking down because of the shear during the mixing, causing the thixotropic behavior of cement. At the same time, structural reconstruction occurs, and the flocculent structures rebuild when the shear force is stopped. Thus, cement pastes display two different states: liquid-like under shear conditions and a weak solid-like with a limited yield stress under static conditions. Accordingly, flocculation increases the yield stress and causes pseudoplastic/shear-thinning behavior. It has also been reported that early hydration reaction affects and changes the rheological behavior of fresh cement pastes/slurries from anti-thixotropic to thixotropic [42]. Effects of coagulation, dispersion, and re-coagulation of cement particles have been studied extensively. Cement particles can coagulate in a paste or a slurry; this is primarily caused by the surface attraction forces. Junctions are connections or contacts between the cement particles, consisting of reversible junctions (two particles can be separated) and permanent junctions (two particles cannot be separated).

The cement hydration process consists of a series of chemical reactions between the cement particles and water. From these reactions, the initial fluid-like suspension is transformed into a solid-like material. In general, the effect of cement hydration should be considered when studying rheological behavior of cement.

Hydration reactions result in structural changes, which affect the rheological behavior of cement [229]. The hydration process can cause an exponential increase in the yield stress [131]. Lapasin et al. [58] studied the time-dependent response of cement pastes. They noticed that the shear stresses reached a peak value with increasing age where hydration occurs during the initial stages of the flow (1 min) of the test and decaying to the equilibrium value and then increasing a little [229]. The steady-state flow curves of the cement pastes exhibit shear-thinning behavior and the presence of yield stress. The authors found partially-thixotropic phenomena for the time-dependent behavior of cement pastes. Before the shear rate reaches the initial shear rate value, the shear stress does not increase. At this stage, the kinetics of the structural rebuilding is much less than the kinetics of the structural break-down. After the steady-state conditions has been reached and as the shear rate begins to increase, further structural break-down could be detected because of the different aggregation state of the disperse system at different shear rates. Cement pastes are usually found partially thixotropic in three stages: power-law shear-thinning fluid under low shear rate, a Newtonian plateau at high shear rates, and a shear-thickening paste [196].

Moore [41] introduced a structural parameter $\lambda(t)$ to describe the structural state of a thixotropic fluid, where λ indicates the degree of flocculation or aggregation (also known as the "degree of jamming"), describing the state of the material at a given time and giving the percentage of the particles in potential wells for colloidal fluids [92,214,230–233]. The value of λ is from 0 to 1. The slurry is considered to be dispersed when $\lambda = 0$, and fully flocculated when $\lambda = 1$. According to Feys and Asghari [234] and other references, λ decreases with time and increases with shear rate, that is,

$$\frac{d\lambda}{dt} = k_+(1-\lambda) - k_-\dot{\gamma}\lambda \tag{78}$$

where k_+ and k_- represent the structural buildup and the structural breakdown coefficients. The structural parameter λ displays a characteristic relaxation time of t with the change of shear rate:

$$t = 1/(k_+ + k_-\dot{\gamma}) \tag{79}$$

Cheng and Evans [235] proposed a constitutive equation, including the structural parameter λ:

$$\tau = \eta(\lambda, \dot{\gamma})\dot{\gamma} \tag{80}$$

where the viscosity η is a function of shear rate $\dot{\gamma}$ and λ.

A rate-type equation gives the evolution of λ with time. In general, the rate at which the structure changes can be a function of the shear rate and the structural parameter. A rate equation for the structural parameter can be given [235]:

$$\frac{d\lambda}{dt} = g(\lambda, \dot{\gamma}) = K_1(\dot{\gamma})(1-\lambda)^p - K_2(\dot{\gamma})\lambda^q \tag{81}$$

where $K_1(\dot{\gamma})$ and $K_2(\dot{\gamma})$ are rate constants related to the build-up and the break-down processes. In this equation, the effect of Brownian motion and the shear rate are also considered. The parameters p and q are indications of the orders of the two processes. Lapasin et al. [58] applied Cheng and Evans equation to cement pastes and suggested an alternative equation, where Equation (80) is replaced by:

$$\tau = f_0(\dot{\gamma}) + f_1(\dot{\gamma})\lambda \tag{82}$$

For fresh cement, the structural build-up can be negligible because of the partially thixotropic behavior [236]. Thus, Equation (81) becomes:

$$\frac{d\lambda}{dt} = -K_2(\dot{\gamma})\lambda^q \tag{83}$$

Hattori and Izumi [237] considered the effect of coagulation/junctions on the apparent viscosity at time t and shear rate $\dot{\gamma}$ by suggesting the following relationship for the viscosity:

$$\eta = B_3 \left[\frac{n_3[U_0(\dot{\gamma}Ht^2 + 1) + Ht]}{(Ht + 1)(\dot{\gamma}t + 1)} \right]^{2/3} \tag{84}$$

where B_3 is the friction coefficient between the cement particles, n_3 is the number of uncoagulated particles in unit volume, U_0 is the initial degree of coagulation (percentage of junctions to total particles), H is the coagulation rate, which is related to particle attraction and cement hydration reactions.

De Kee and Chan Man Fong [71] suggested the following relationship for the viscosity:

$$\eta = \frac{\eta_0 k_c}{\alpha_0}\left[1 + (bf_1 - cf_2)e^{-\alpha_0 t}\right] \quad (85)$$

where

$$\alpha_0 = k_c(1 + bf_1 - cf_2) \quad (86)$$

where the parameters η_0, k_c, b, c, f_1 and f_2 are functions of the shear rate. When $(bf_1 - cf_2) > 0$, the fluid behaves as a thixotropic model and when $(bf_1 - cf_2) < 0$, it is anti-thixotropic (or rheopectic). They also showed that if $(bf_1' - cf_2') > 0$, the fluid behaves as a shear-thinning fluid, and if $(bf_1' - cf_2') < 0$, as a shear-thickening fluid, where "'" (the prime) denotes the derivative with respect to $\dot{\gamma}$. For further details about this model, see Carreau et al. [33] (p. 471).

Coussot et al. [92] suggested the following relationship between λ and $\dot{\gamma}$:

$$\frac{d\lambda}{dt} = \frac{1}{t_0} - \alpha\lambda\dot{\gamma} \quad (87)$$

where t_0 is the characteristic time of aging or rejuvenation, and α is a system-dependent constant.

The instantaneous viscosity based on Coussot's model is defined as a function of the flocculation parameter and the shear rate:

$$\eta = \eta_0 f(\lambda, \dot{\gamma}) = \eta_0(1 + \lambda^n) \quad (88)$$

where η_0 is the viscosity when the flow is not affected by the particle interactions ($\lambda = 0$), and n is a parameter indicating the effect of structural breakdown and reconstruction.

Coussot et al. [92] also proposed an equation (based on the experiment) that a stress ramp is applied after different times of rest. The dimensionless form of shear stress–shear rate relation when the cement is at rest is written as:

$$\tau_s = \dot{\gamma}_s \exp\left(\frac{1}{\dot{\gamma}_s}\left[1 + (\lambda_0\dot{\gamma}_s - 1)\exp(-\dot{\gamma}_s\bar{t}_r)\right]\right) \quad (89)$$

where τ_s is the shear stress under steady state condition, $\tau_s = \tau\alpha T_0/\eta_0$, the shear rate is $\dot{\gamma}_s = \alpha T_0\dot{\gamma}$, $\bar{t}_r$ is the dimensionless time during which the cement is at rest for restructuring. This equation is able to describe the rheological behavior of fluids with thixotropy and yield stress.

A general form of a model based on the above ideas was given by Roussel [238,239]:

$$\tau = \tau_y(1 + \lambda) + k\dot{\gamma}^n \quad (90)$$

Roussel [214] suggested a viscosity that depends on the shear rate, as shown in Figure 20a:

$$\eta = \eta_\infty(1 + a^n\dot{\gamma}^{-n})a = \frac{1}{\alpha t_r} \quad (91)$$

where n is an experimental parameter with positive value, η_∞ is the viscosity where the shear rate is infinite, a is a system-dependent constant, and t_r is the time during which the cement is at rest for restructuring, which is a constant. For transient flow, the rheological properties change with time. Roussel [214] also gave an evolution equation for the structural flocculation/jamming parameter λ, shown in Figure 20b.

$$\lambda(t) = \frac{a}{\dot{\gamma}} + \left(\lambda_0 - \frac{a}{\dot{\gamma}}\right)\exp\left(-\frac{\dot{\gamma}t}{\alpha t_r}\right)a = \frac{1}{\alpha t_r} \quad (92)$$

where λ_0 is the initial value of λ, α is a system-dependent constant and the applied strain rate $\dot{\gamma}$ is constant. The characteristic time t_c decreases with an increase in the strain rate, which makes reaching the steady state condition harder,

$$t_c = \frac{at_r}{\dot{\gamma}} = \frac{1}{\alpha\dot{\gamma}} \tag{93}$$

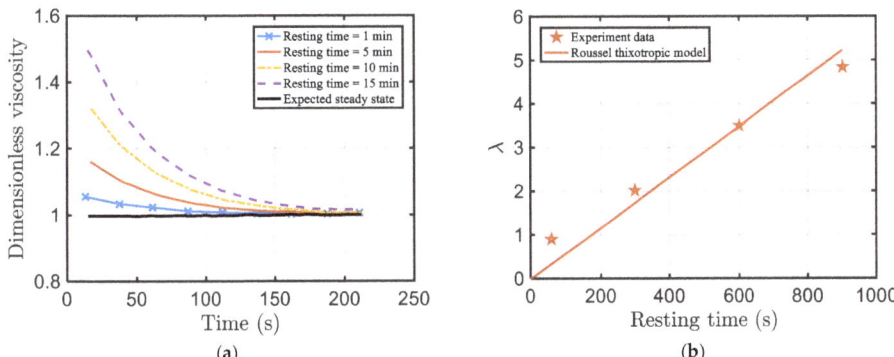

Figure 20. The thixotropic model of Roussel [238]: (**a**) Dimensionless viscosity at constant shear rate for various resting time; (**b**) state of flocculation λ vs. resting time.

Wallevik [36] suggested an equation for the shear viscosity that depends on a temporary shear viscosity η_{tmp}, where

$$\eta_s = \left(\eta_p + \frac{\tau_y}{\dot{\gamma}}\right) + \eta_{tmp} \tag{94}$$

where η_p is the plastic viscosity and τ_y is the yield stress; η_{tmp} is related to the number of reversible junctions J_t:

$$\eta_{tmp} = \tilde{\eta}_p + \frac{\tilde{\tau}_y}{\dot{\gamma}} = a_1 B_3 J_t^{2/3} + a_2 B_3 J_t^{2/3}/\dot{\gamma} \tag{95}$$

where $\tilde{\eta}_p$ and $\tilde{\tau}_y$ are the thixotropic counterparts of η_p and τ_y, B_3 is the friction coefficient between the cement particles, and a_1 and a_2 are the two empirical parameters.

Mewis and Wagner [240] provide a generalized thixotropic model for colloidal suspensions:

$$\eta(\dot{\gamma}, \lambda) = \lambda\tau_y + \lambda K_{st}\dot{\gamma}^n + K_{\infty}\dot{\gamma}^n \tag{96}$$

$$\frac{d\lambda}{dt} = -k_1\dot{\gamma}\lambda + k_2\dot{\gamma}^m(1-\lambda) + k_3(1-\lambda) \tag{97}$$

In Equation (96), $\lambda\tau_y$ shows the effect of λ on the yield stress. No yield stress occurs when there is no flocculation. The term $\lambda K_{st}\dot{\gamma}^n$ indicates that the viscosity increases with λ.

Marchesini et al. [34] studied the irreversible time-dependent rheological behavior of cement slurries and presented a transient constitutive model considering the irreversible effect. The structural parameter is given by:

$$\frac{d\lambda}{dt} = \frac{1}{t_{eq}}\left\{\left[1-\lambda-\left(1-\lambda_{final}\right)\zeta(t)\right]^a - \left(1-\lambda_{eq}\right)^a\left(\frac{\lambda-\lambda_{final}\zeta(t)}{\lambda_{eq}}\right)^b\right\} + \lambda_{final}\dot{\zeta}(t) \tag{98}$$

where t_{eq} and λ_{eq} are the characteristic time and the structural parameter for the thixotropic medium at equilibrium, λ_{final} is the structural parameter of the material at the final equilibrium state, $\zeta(t)$

is a function indicating the evolution of the irreversible processes at time t and a and b are the positive constants.

For an ideal thixotropic behavior, all irreversible processes are neglected, where the irreversible process function $\zeta(t) = 0$, and the structural parameter becomes:

$$\frac{d\lambda_{thixo}}{dt} = \frac{1}{t_{eq}}\left[(1-\lambda_{thixo})^a - (1-\lambda_{eq})^a \left(\frac{\lambda_{thixo}}{\lambda_{eq}}\right)^b\right] \qquad (99)$$

The irreversible process function ζ_i could be written as a function of time and the structural parameter:

$$\zeta(t, \lambda_{eq}) = 1 - \frac{1}{\left[1 + \left(\frac{t}{t_{reac}}\right)^{1/\lambda_{eq}}\right]^{\frac{(1-m\lambda_{eq})\lambda_{eq}}{l}}} \qquad (100)$$

where t_{reac} is the characteristic time for the hydration reactions, l and m are parameters related to the irreversible processes before and after t_{reac} is reached. In this approach, the viscosity function consists of a structural viscosity related to the structural parameter and a pure viscous part in the unstructured state [226,241].

Based on Marchesini et al.'s model, viscosity as a function of the structural parameter is given by:

$$\eta(\lambda) = \eta_\infty \left(\frac{1}{1-\lambda_{thixo}}\right)^\alpha \left(\frac{\lambda_{final} - \lambda_{thixo}}{\lambda_{final} - \lambda}\right)^\epsilon \qquad (101)$$

where η_∞ is the viscosity of the complete unstructured state and α and ϵ are positive constants.

In this section, we have provided a review of the concept of the thixotropy in cements. A structural parameter λ is introduced to represent the degree of flocculation, which is related to the cement hydration process. Thus, the viscosity is a function of the structural parameter λ and a rate equation for λ (e.g., Equations (81), (97), and (98)) should be used along with the governing equations.

2.2.4. A New Model for Cement Slurry

We suggest that the total stress tensor T is defined as

$$T = T_y + T_v \qquad (102)$$

where T_y is yield stress tensor and T_v is the viscous stress tensor. In general, the yield stress can be a function of many parameters, for example, volume fraction, w/c, etc.

$$T_y = T_y\left(\phi, \frac{w}{c}, \ldots\right) \qquad (103)$$

As mentioned earlier, one of the distinct features of some non-Newtonian fluids is the presence of normal stress effects (see [242]), which are manifestations of non-equal normal stresses when the fluid is sheared. In polymeric fluids, this phenomenon is usually observed as rod-climbing or die-swell; in fact, one of the earliest observations is due to Reynolds [243,244] related to the expansion of the voids in wet sands; he called this effect "dilatancy." Massoudi and Mehrabadi [245] discuss this concept along with the Mohr–Coulomb yield criterion. Later Reiner [246] developed a stress tensor model for wet sand; his theory was generalized by Massoudi [247] where the effects of density (volume fraction) gradient were included. Interestingly, in our review, we did not find any direct reference either to attempts at measuring the normal stress effects in cement or modeling them, despite the similarities between cement paste and wet sand. We propose a new constitutive model for the viscous stress tensor of the cement slurry, T_v, by generalizing the traditional second-grade fluid model [248,249].

$$T_v = -pI + \eta A_1 + \alpha_1 A_2 + \alpha_2 A_1^2 \qquad (104)$$

where the viscosity can be a function of various parameters, $\eta = \eta(\dot{\gamma}, \phi, \theta, p, \lambda(t), w/c, \dots)$, α_1 and α_2 are the normal stress coefficients assumed to be constants, $\dot{\gamma}$ is the shear rate, ϕ is the volume fraction, θ is the temperature, p is the pressure, $\lambda(t)$ is the structural parameter describing the degree of flocculation or aggregation and w/c is the water-to-cement ratio. The kinematical tensors A_1 and A_2 are defined as:

$$A_1 = grad v + (grad v)^T \qquad (105)$$

$$A_2 = \frac{dA_1}{dt} + A_1(grad v) + (grad v)^T A_1 \qquad (106)$$

The general idea behind suggesting a new constitutive model for the viscous stress tensor T_v that would include the normal stress effects, is that many non-Newtonian fluids such as polymers, dense suspensions, drilling fluids, blood, granular materials, slags, etc. [123,250–255] have been shown to exhibit these non-linear effects. In our previous paper [11], we showed some preliminary results using the newly proposed viscous stress model where the effects of shear-rate dependent viscosity in cement slurries was studied. The Clausius–Duhem inequality shows [256] that for the classical second grade fluids:

$$\eta \geq 0 \alpha_1 \geq 0 \alpha_1 + \alpha_2 = 0 \qquad (107)$$

As shown in this paper, for cement slurries/pastes, the viscosity depends on the shear-rate. In Section 2.1, we noticed that the power-law model, also known as the generalized Newtonian fluid (GNF) model [33] is one of the simplest models for shear-rate dependency of viscosity. Therefore, we use this idea and suggest a new model for T_v.

$$T_v = -pI + \mu_0 \left(1 - \frac{\phi}{\phi_m}\right)^{-\beta} (1 + \lambda^n) \left[1 + \alpha tr A_1^2\right]^m A_1 + \alpha_1 A_2 + \alpha_2 A_1^2 \qquad (108)$$

$$\frac{d\lambda}{dt} = \frac{1}{t_0} - \kappa \lambda \dot{\gamma} \qquad (109)$$

where we have used Krieger's idea for the volume fraction dependence of viscosity, η_0 is the (reference) coefficient of viscosity, tr is the trace operator, and m is the power-law exponent, a measure of non-linearity of the fluid, related to the shear-thinning effects (when $m < 0$) or shear-thickening effects (when $m > 0$). Equation (108), we believe, is a general expression for the viscous contribution of the stress tensor, T_v; for cement slurries, this model potentially is capable of exhibiting normal stress effect, through the terms α_1 and α_2, thixotropy effects because of the presence of the structural parameter λ, shear-rate dependent effects of the viscosity through the two parameters α and m (showing shear-thinning or shear-thickening effects), and the concentration dependency of viscosity through the two parameters ϕ_m and β. We do plan to use this equation in our future studies, and we do plan to develop/propose a yield stress model, T_y, for cement. A simplified version Equation (108) was used in our earlier study, Tao et al. [11,257].

Alternatively, in the absence of any experimental data related to the normal stress coefficients for the cement, we can suggest a more traditional formulation for the stress tensor T and use Macosko's formulation [31]:

$$\begin{aligned} T &= 2\eta_p D & \text{for } \text{II}_{2D}^{1/2} \leq \dot{\gamma}_c \\ T &= 2\left[\frac{\tau_y}{|\text{II}_{2D}|^{1/2}} + K|\text{II}_{2D}|^{\frac{n-1}{2}}\right]D & \text{for } \text{II}_{2D}^{1/2} > \dot{\gamma}_c \end{aligned} \qquad (110)$$

where all the parameters are defined in Equation (8).

By using one of the correlations given for τ_y in Table 4, namely

$$\tau_y(\phi, w/c) = m_1 \frac{\phi^2(\phi - \phi_{perc})}{\phi_m(\phi_m - \phi)} \times (-175 w/c + 137) \qquad (111)$$

and substituting it in (110), we can obtain a three-dimensional form for the stress tensor considering different effects. Therefore, we assume

$$T_y = 2 \begin{cases} \eta_p \mathbf{D} & \text{for } \Pi_{2D}^{1/2} \leq \dot{\gamma}_c \\ \left[m_1 \frac{\frac{\phi^2(\phi-\phi_{perc})}{\phi_m(\phi_m-\phi)} \times (-175w/c+137)}{|\Pi_{2D}|^{1/2}} + K |\Pi_{2D}|^{\frac{n-1}{2}} \right] \mathbf{D} & \text{for } \Pi_{2D}^{1/2} > \dot{\gamma}_c \end{cases} \quad (112)$$

We can also assume that K could depend on volume fraction, temperature, etc.

3. Concluding Remarks

Rheological behavior of cement is important in cement, concrete, and petroleum engineering industries. In this article, we have provided a comprehensive review of the rheological models used for cement slurries. We have looked at the models describing the total stress tensor for cement as well as the yield stress models and models focusing on the viscous stress tensor. Different effects such as changing the cement concentration, the water-to-cement ratio, the additives/admixtures, the shear rate, the temperature and pressure, the mixing method and the thixotropic nature of cement are taken into consideration. We also propose a new constitutive model where the traditional second-grade fluid model is generalized to include the contributions from variable shear viscosity, concentration, thixotropy, etc.

Author Contributions: M.M. developed the framework of the paper and C.T. did most of the literature review and writing (with help from M.M.). All authors (C.T., B.G.K., E.R. and M.M.) contributed to the editing of the paper. All authorsAll authors have read and agreed to the published version of the manuscript.

Funding: This research was supported by the U.S. Department of Energy, National Energy Technology Laboratory.

Acknowledgments: This paper is supported by an appointment to the U.S. Department of Energy Postgraduate Research Program for C.T. at the National Energy Technology Laboratory, administered by the Oak Ridge Institute for Science and Education.

Conflicts of Interest: The authors declare no conflict of interest.

Disclaimer: This report was prepared as an account of work sponsored by an agency of the United States Government. Neither the United States Government nor any agency thereof, nor any of their employees, makes any warranty, express or implied, or assumes any legal liability or responsibility for the accuracy, completeness, or usefulness of any information, apparatus, product, or process disclosed, or represents that its use would not infringe privately owned rights. Reference herein to any specific commercial product, process, or service by trade name, trademark, manufacturer, or otherwise does not necessarily constitute or imply its endorsement, recommendation, or favoring by the United States Government or any agency thereof. The views and opinions of authors expressed herein do not necessarily state or reflect those of the United States Government or any agency thereof.

Appendix A. Cement Chemistry

Figure A1 shows the microstructure of the anhydrous cement powder for silicate phase, aluminate phase, and gypsum.

Figure A1. Dry cement image under scanning electron microscope (SEM): (**a**) silicate phase; (**b**) aluminate phase; (**c**) gypsum particle [19].

A cement particle contains: alite (C_3S), belite (C_2S), aluminate phase (C_3A), ferrite phase (C_3AF), alkali sulfate, free lime and gypsum [12,258]. Alite, tricalcium silicate (Ca_3SiO_5), accounts for 40–70% mass of a clinker. It displays a hexagonal crystal behavior with the size up to 150 μm. It reacts with water in a short time and to a large extent determines the strengths at the early age (within 28 day). Belite, dicalcium silicate (Ca_2SiO_4), accounts for 15–45% mass of a clinker. It displays a rounded grain habit with size 5–40 μm [259]. It is less reactive with water than alite and determines the strengths at later age (after 28 days). Alite and belite generate well-crystallized calcium hydroxide and poorly-crystallized calcium silicate hydrate (C-S-H). Tricalcium aluminate ($Ca_3Al_2O_6$), accounts for 1–15% mass of a clinker. It displays lath-like or irregular behavior with size 1–60 μm. It is also reactive with water. *American Petroleum Institute* standards limits the tricalcium aluminate content of a class G cement to lower than 3% [2]. Ferrite is tetracalcium aluminoferrite (Ca_2AlO_5, Ca_2FeO_5) that accounts for 0–18% mass of a clinker. It displays crystal behavior. Aluminate and ferrite phases are interstitial/matrix phases to bind the silicate crystals. Periclase (MgO) displays crystal behavior with the size up to 30 μm and free lime (CaO) behavior as rounded crystals is isolated or joined with other phases. Both periclase and free lime have less quantities but affect the performance. Alkali sulfates and calcium sulfates are also found in clinker which affect the hydration rates and strength development [259,260]. Besides cement phase composition, surface area of each phase also affects the cement performance. Surface area is affected by the texture of clinker and the grinding conditions.

Increasing the content of C_3S and C_3A is considered to be an effective way to obtain high-initial-strength cement and concrete. However, in modern cement and concrete application, researchers have found lower water-to-cement ratio to also play an important role. It is the closeness of the cement particles and cement concentration that determine the compressive strength of cement and concrete [157]. It is reasonably easy to control rheological properties such as viscosity and yield stress for cements without too much C_3S and C_3A content. For sustainable purpose, Aïtcin [157] suggests more mineral components to be added with clinker to lower the water-to-cement ratio and to increase the life cycle of cement and concrete structures and extend the effect of hydraulic binders and aggregates. When mixed with water, the reactions generate calcium silicate and aluminate ions in the solution. Cement hydration consists of three stages: (1) an initial reaction after mixing (at least 15 min); (2) a dormant period during which only weak chemical reactions occur; (3) an actual cement hydration [40]. Among the cement hydrates generated, calcium-silicate-hydrate (C-S-H) and calcium hydroxide (portandite) are two of the most important hydrates. The dissolution-diffusion precipitation process changes anhydrous material into hydrates, which decrease the porosity of the cement [261]. Through the hydration process, cement develops strength in two steps: gelation and setting. Gelation occurs almost immediately after cement is mixed with water. The coagulated system develops some strength, which could easily be broken by mixing. It is not possible to measure the yield stress because of the weakness of the network. Setting starts a few hours after the coagulation/flocculation. During the time between coagulation and setting, also called the dormant period, hydrates nucleate and growth occur. Shear modulus is developed to the GPa range until setting.

Appendix B. Governing Equations of Motion and Heat Transfer

Cement slurries could be mathematically described using the method of continuum mechanics. It is possible to model these slurries using either a single-phase (component) approach or a multi-phase (component) approach. We provide a summary of the governing equations for these two approaches.

Appendix B.1. Single Component Approach

The governing equations of motion include the conservation of mass, linear momentum, angular momentum, convection-diffusion equation [262,263], the energy equation and the entropy inequality. When the slurry is assumed to behave as a non-homogeneous and possibly a (non-linear) fluid, then we have:

- Conservation of mass

$$\frac{\partial \rho}{\partial t} + \text{div}(\rho v) = 0 \tag{A1}$$

where $\partial/\partial t$ is the partial derivative with respect to time, div is the divergence operator, and v is the velocity vector, ρ is the density of the slurry.

- Conservation of linear momentum

$$\rho \frac{dv}{dt} = \text{div} T + \rho b \tag{A2}$$

where b is the body force vector, T is the Cauchy stress tensor, and d/dt is the total time derivative, given by $d(.)/dt = \partial(.)/\partial t + [grad(.)]v$.

- Conservation of angular momentum

$$T = T^T \tag{A3}$$

This equation indicates that in the absence of couple stresses the stress tensor is symmetric.

- Convection-diffusion equation for the particles

$$\frac{\partial \phi}{\partial t} + \text{div}(\phi v) = -\text{div} N \tag{A4}$$

where N is the particle flux, and ϕ is the particle volume fraction function or volume distribution (related to concentration), which is a continuous function of position, and time, where $0 \leq \phi(x,t) \leq \phi_{max} < 1$. In reality, $\phi = 1$ at a particle and $\phi = 0$ at a void space (fluid).

- Conservation of energy

$$\rho \frac{d\varepsilon}{dt} = T : L - \text{div } q + \rho r \tag{A5}$$

where ε is the specific internal energy, L is the gradient of velocity, q is the heat flux vector, and r is the specific radiant energy. For complete thermodynamical considerations, the application of the second law of thermodynamics (or the entropy inequality), is also needed [264,265]:

$$\rho \dot{\eta} + \text{div} \varphi - \rho s \geq 0 \tag{A6}$$

where $\eta(x,t)$ is the specific entropy density, $\varphi(x,t)$ is the entropy flux, and s is the entropy supply density from external sources, and $\dot{\eta}$ means the material time derivative of η. If we assume that $\varphi = \frac{q}{\theta}$, and $s = \frac{r}{\theta}$, where θ is the absolute temperature, Equation (A6) reduces to the more familiar form of the Clausius–Duhem inequality:

$$\rho \dot{\eta} + \text{div} \frac{q}{\theta} - \rho \frac{r}{\theta} \geq 0 \tag{A7}$$

In this approach, constitutive relations for T, N, q, ε and r are needed in order to achieve "closure" for the governing equations. In this paper, we have followed this approach and we have focused on modeling T.

Appendix B.2. Multi-Phase (Component) Flow Approach

Many researchers have considered cement slurries to behave as multi-phase fluids; in this case the governing equations should be given for all the components [266,267]. In general, two distinct approaches have been used for multi-component flows: (1) "dilute phase approach," or the Lagrangian approach; (2) "dense phase approach," or the Eulerian approach. It is the latter approach that is often used for cement. Here we present a summary of this approach using the mixture theory, also known as theory of interacting continua systems [268–270]. The details of mixture theory are provided in the books [271,272]. Conservation laws are written for each component taking into account interaction with other constituents. Constitutive relations are required to achieve "closure." The mixture theory equations for a multi-component system are as follows [269,270]:

- Conservation of mass:

$$\frac{D^{(\alpha)}\rho_\alpha}{Dt} + \rho_\alpha \operatorname{div} v^{(\alpha)} = m_\alpha \tag{A8}$$

where $\alpha = 1, 2$.

- Conservation of linear momentum

$$\rho_\alpha \frac{D^{(\alpha)}v}{Dt} = \operatorname{div} T^{(\alpha)} + \pi^{(\alpha)} - m_\alpha\left(v^{(\alpha)} - J^{(\alpha)}\right) + \rho_\alpha F^\alpha \tag{A9}$$

- Conservation of angular momentum

$$T_s^{(\alpha)} = \lambda^{(\alpha)} \tag{A10}$$

- Conservation of energy

$$\rho_\alpha \frac{D^{(\alpha)}U_\alpha}{Dt} = \rho_\alpha r_\alpha - \operatorname{div} q^{(\alpha)} + \psi_\alpha + \operatorname{tr}\left(T_s^{(\alpha)} D^{(\alpha)}\right) - m_\alpha\left[U_\alpha + \left(J^\alpha - \frac{1}{2}v^{(\alpha)}\right)\cdot v^{(\alpha)} - G_\alpha\right] \tag{A11}$$

- Entropy

$$\sum_\alpha \left[\rho_\alpha \frac{D^{(\alpha)}S_\alpha}{Dt} + m_\alpha S_\alpha + \operatorname{div}\left(\frac{q^{(\alpha)}}{\theta_\alpha}\right) - \frac{\rho_\alpha r_\alpha}{\theta_\alpha}\right] \geq 0 \tag{A12}$$

where

$$\rho v - \sum_\alpha \rho_u v^{(\alpha)} \tag{A13}$$

$$\frac{D^{(\alpha)}\beta}{Dt} = \frac{\partial \beta}{\partial t} + v^{(\alpha)} \cdot \operatorname{grad}\beta \tag{A14}$$

for any scalar β, and

$$\frac{D^{(\alpha)}w}{Dt} = \frac{\partial w}{\partial t} + (\operatorname{grad} w)v^{(\alpha)} \tag{A15}$$

for any vector w, where

$$\sum_\alpha m_\alpha = 0 \tag{A16}$$

$$\sum_\alpha \left(\pi^{(\alpha)} + m_\alpha J^{(\alpha)}\right) = 0 \tag{A17}$$

The entropy inequality becomes:

$$\sum_\alpha \eta_\alpha \geq 0 \qquad (A18)$$

where

$$\eta_\alpha = -\rho_\alpha \frac{D^{(\alpha)} A_\alpha}{Dt} - \rho_\alpha S_\alpha \frac{D^{(\alpha)} \theta_\alpha}{Dt} - m_\alpha \left[A_\alpha + \left(J^{(\alpha)} - \tfrac{1}{2} v^{(\alpha)} \right) \cdot v^{(\alpha)} - G_\alpha \right] + \psi_\alpha \\ + \mathrm{tr}\left(T_s^{(\alpha)} D^{(\alpha)} \right) - \frac{q^\alpha \cdot \mathrm{grad} \theta_\alpha}{\theta_\alpha} \qquad (A19)$$

where ρ_α is the density of the α component, v is the velocity vector, D is the symmetric part of the velocity gradient, T is the stress tensor, π represents the interaction forces, θ is the temperature, U is the internal energy, r is the radiant heating, A_α is the Helmholtz free energy $A_\alpha = U_\alpha - \theta_\alpha S_\alpha$, S_α is the entropy per unit mass, q is the heat flux vector, and ψ_α, m_α, and G_α are the supply terms. $T_s^{(\alpha)}$ is the symmetric part of the partial stress tensor $T^{(\alpha)}$, and $\lambda^{(\alpha)}$ is an anti-symmetric second-order tensor. The entropy inequality is used to impose certain restrictions on the types of motions and processes. Constitutive relations are required for A_α, S_α, π, q, and $T^{(\alpha)}$.

In reactive flows, for example those encountered in cement slurries, a series of hydration reactions happen after the components are mixed with water. Some researchers have applied a finite rate model, using the eddy dissipation concept [273]. For cement, the volumetric reaction model and the eddy dissipation model were applied for hydration reactions and interaction between turbulence and chemical reactions [274,275].

Appendix C. Computational Fluid Dynamics Studies

Wallevik [36,276] developed a computational fluid dynamics (CFD) software called viscometric-visco-plastic-flow (VVPF) to study the steady-state and transient flows. Visco-plastic materials with yield stress and thixotropic behavior such as fresh concrete, mortar, and cement paste were studied using the Bingham fluid model. Finite difference method based on alternating direction implicit technique was applied in this software. Moyers-Gonzalez and Frigaard [277] investigated the kinematic instabilities in two-layer eccentric annular flows during oil and gas well cementing process. A non-Newtonian fluid model with shear-thinning and yield stress, namely the Herschel–Bulkley model, was used. Manzar and Shah [278] applied FLUENT for the CFD analysis of slurries used in petroleum industry in different pipes. Lootens et al. [48] used a 3D Bingham fluid model to study the visco-plastic flow of cement pastes under penetrometer test using the Flow3D solver [279]. Malekmohammadi et al. [266] studied the buoyancy-driven flow of non-Newtonian fluids in a horizontal pipe and solved the multi-phase momentum conservation equations using the finite volume method and volume of fluid method (VOF) [280]. Cremonesi et al. [281] simulated the transient flow of non-Newtonian fresh cement suspensions by using the Lagrangian formulation of the Navier–Stokes equations based on the particle finite element approach. The governing equations including the momentum and the mass conservation for the Bingham model are solved using FLUENT software. Aranha et al. [282] developed an in-house software to simulate the annular fluid displacement in vertical and directional offshore wells for Newtonian and non-Newtonian fluids. Wu et al. [155] developed a mathematical model for fresh cement pastes using the numerical software COMSOL ("Heat Transfer in Fluids" model), considering the effects of temperature and cement hydration. Bu et al. [283] modeled the flow characteristics of cement slurry in an eccentric annulus for laminar conditions. The authors used the Herschel–Bulkley model and obtained numerical solution using ANSYS FLUIENT. Zulqarnain and Tyagi [284] studied the fluid displacements in the casing-formation annulus to predict the final cement volume fraction in the annulus for different operating conditions. Zhao et al. [285] applied Lattice Boltzmann method (LBM) to investigate the behavior of shear-thinning fluids in a cementing operation in a horizontal eccentric annulus. A 2D flow model was used along with ANSYS FLUENT solver. Anglade et al. [286] analyzed the flow of fresh cement suspensions using the finite element method, where both shear-thinning and shear-thickening behavior are considered. Wu et al. [287] applied a modified LBM for Herschel–Bulkley fluids for more stable and accurate simulations. This modified Herschel–Bulkley model was used to study the flow between parallel plates and flow in the screw extruder in cement 3D printing. Bao et al. [288] used the Bingham and the power-law fluid models in porous media to study the flow between parallel plates with discrete fracture modeling (DFM) by applying the open-source MATLAB Reservoir Simulation Toolbox (MRST). Tardy and Bittleston [289,290] solved the flow in 2D and 3D axial-azimuthal-radial space and annular displacement of wellbore completion by considering Newtonian fluids with CAFFA and ANSYS FLUENT CFD software. Zhou et al. [291] developed a new model to look at gas migration in non-Newtonian fluids with yield stress and the factors affecting the wellhead pressure change by the analysis of variance (ANOVA). Naccache et al. [292] performed a numerical simulation to study the displacement of two fluids (e.g., cement slurries and drilling fluids) in vertical annular pipe in the cementing operation in the petroleum industry. For the multiphase fluid problem, the governing equations were solved by using the ANSYS FLUENT. A generalized Newtonian fluid (GNF) was used and the regularized Herschel–Bulkley equation was applied for the viscosity of viscoplastic fluids. Foroushan et al. [293] modeled the instability of interface and fluid mixing of cement slurry and drilling mud during cementing operations in oil and gas wells in 3D using ANSYS FLUENT. Skadsem et al. [294] studied the flow of non-Newtonian fluid in an inclined wellbore with concentric and eccentric configurations numerically and experimentally. The 3D flow model for the two fluids was solved with biviscosity approach and finite element simulation in OpenFOAM. Liu et al. [273] modeled the multi-phase pipe flow considering the hydration effects of cemented paste backfill slurry by using a CFD model with ANSYS FLUENT. Murphy et al. [295] simulated the shear flow of two Bingham plastic cement slurries containing Portland cement and fly ash particles by applying fast lubrication dynamics-discrete element model with LAMMPS. Rosenbaum et al. [296] studied the influence of bubbles on foamed cement viscosity using an Extended Stokesian Dynamics Approach with LAMMPS.

Many optimization algorithms are also used in the cement industry. For example, Ohen and Blick [297] used the golden section search method to determine the parameters in the Robertson–Stiff non-Newtonian fluid model. Shahriar and Nehdi applied an artificial neural network (ANN) [298], multiple regression analysis (MRA) [299] and factorial design approach [300] to predict the shear flow and rheological properties (Bingham parameters including yield stress and plastic viscosity) of oil well cement slurries by comparing and validating with the experimental database. The rheological properties with this ANN model were sensitive to the effect of temperature and admixture content of oil well cement. Velayati et al. [4] optimized the cement formulations based on a series of criteria to evaluate the performance of cement slurry under gas migration. The optimization plan was verified by fluid migration analysis (FMA) test. Optimization algorithms including the deterministic simplex method and the stochastic genetic method were applied to determine the parameters in the Herschel–Bulkley model for fresh cement suspensions [9]. Multi-objective optimization methods, such as support vector regression (SVR), model tress (MT), M5 model rules, mix and design approach, particle swarm optimizations (PSO) and response surface method (RSM) are also used in cement chemistry and hydration processes to obtain optimal chemical combination and proportion for better cement performance, workability, and energy consumptions [201,301–304].

References

1. Sabins, F.; Wiggins, M.L. Parametric study of gas entry into cemented wellbores. *Oceanogr. Lit. Rev.* **1998**, *1*, 182.
2. Lootens, D.; Hébraud, P.; Lécolier, E.; Van Damme, H. Gelation, Shear-Thinning and Shear-Thickening in Cement Slurries. *Oil Gas Sci. Technol.* **2004**, *59*, 31–40. [CrossRef]
3. Banfill, P.F. The rheology of fresh cement and concrete-a review. In Proceedings of the 11th International Cement Chemistry Congress, Durban, South Africa, 11–16 May 2003; Volume 1, pp. 50–62.
4. Velayati, A.; Tokhmechi, B.; Soltanian, H.; Kazemzadeh, E. Cement slurry optimization and assessment of additives according to a proposed plan. *J. Nat. Gas Sci. Eng.* **2015**, *23*, 165–170. [CrossRef]

5. Hodne, H.; Saasen, A.; O'Hagan, A.B.; Wick, S.O. Effects of time and shear energy on the rheological behaviour of oilwell cement slurries. *Cem. Concr. Res.* **2000**, *30*, 1759–1766. [CrossRef]
6. Sabins, F.; Childs, J.D. Method of Using Thixotropic Cements for Combating Gas Migration Problems. U.S. Patent 4524828, 25 June 1985.
7. Shahin, G.T., Jr. Method for Monitoring Well Cementing Operations. U.S. Patents 6125935, 3 October 2000.
8. Vuk, T.; Ljubicĭ-Mlakar, T.; Gabrovsĭek, R.; Kaucĭicĭ, V. Tertiary gelation of oilwell cement. *Cem. Concr. Res.* **2000**, *30*, 1709–1713. [CrossRef]
9. Li, Z.; Vandenbossche, J.; Iannacchione, A.; Brigham, J.; Kutchko, B. Theory-Based Review of Limitations with Static Gel Strength in Cement/Matrix Characterization. *SPE Drill. Complet.* **2016**, *31*, 145–158. [CrossRef]
10. Prohaska, M.; Ogbe, D.O.; Economides, M.J. Determining wellbore pressures in cement slurry columns. In Proceedings of the SPE Western Regional Meeting, Anchorage, AK, USA, 26–28 May 1993.
11. Tao, C.; Kutchko, B.G.; Rosenbaum, E.; Wu, W.-T.; Massoudi, M. Steady Flow of a Cement Slurry. *Energies* **2019**, *12*, 2604. [CrossRef]
12. Taylor, H.F. *Cement Chemistry*; Thomas Telford: London, UK, 1997; ISBN 0-7277-2592-0.
13. Brothers, L.E.; Palmer, A.V. Cementing in Deep Water Offshore Wells. U.S. Patents 6835243, 28 December 2004.
14. Scherer, G.W.; Funkhouser, G.P.; Peethamparan, S. Effect of pressure on early hydration of class H and white cement. *Cem. Concr. Res.* **2010**, *40*, 845–850. [CrossRef]
15. American Petroleum Institute (API). *10B Recommended Practice for Testing Oil Well Cements and Cement Additives*; American Petroleum Institute: Washington, DC, USA, 1997.
16. Kumar, S.; Singh, C.J.; Singh, R.P.; Kachari, J. Microfine cement: Special superfine Portland cement. In Proceedings of the 7th NCB International Seminar, New Delhi, India, 30 April 2002.
17. Saasen, A.; Rafoss, E.; Behzadi, A. Experimental investigation of rheology and thickening time of class G oil well cement slurries containing glycerin. *Cem. Concr. Res.* **1991**, *21*, 911–916. [CrossRef]
18. Mindess, S.; Young, J.F. *Concrete*; Prentice Hall: Englewood, NJ, USA, 1981.
19. Vlachou, P.-V.; Piau, J.-M. Physicochemical study of the hydration process of an oil well cement slurry before setting. *Cem. Concr. Res.* **1999**, *29*, 27–36. [CrossRef]
20. Frigaard, I.A.; Paso, K.G.; de Souza Mendes, P.R. Bingham's model in the oil and gas industry. *Rheol. Acta* **2017**, *56*, 259–282. [CrossRef]
21. Banfill, P.F.G. Rheology of fresh cement paste. In *Annual Transactions of The Nordic Rheology Society*; 1997; Available online: https://researchportal.hw.ac.uk/en/publications/rheology-of-fresh-cement-paste (accessed on 23 January 2020).
22. Shahriar, A. Investigation on Rheology of Oil Well Cement Slurries. Ph.D. Dissertation, University of Western Ontario, London, ON, Canada, 2011.
23. Rice, J.A.; Gu, C.; Li, C.; Guan, S. A radar-based sensor network for bridge displacement measurements. In Proceedings of the Sensors and Smart Structures Technologies for Civil, Mechanical, and Aerospace Systems 2012, San Diego, CA, USA, 11–15 March 2012; Volume 8345, p. 83450I.
24. Guan, S.; Rice, J.A.; Li, C.; Gu, C. Automated DC offset calibration strategy for structural health monitoring based on portable CW radar sensor. *IEEE Trans. Instrum. Meas.* **2014**, *63*, 3111–3118. [CrossRef]
25. Guan, S.; Rice, J.A.; Li, C.; Li, Y.; Wang, G. Dynamic and static structural displacement measurement using backscattering DC coupled radar. *Smart Struct. Syst.* **2015**, *16*, 521–535. [CrossRef]
26. Guan, S.; Rice, J.A.; Li, C.; Li, Y.; Wang, G. Structural displacement measurements using DC coupled radar with active transponder. *Struct. Control Health Monit.* **2017**, *24*, e1909. [CrossRef]
27. Helsel, M.A.; Ferraris, C.F.; Bentz, D. Comparative study of methods to measure the density of Cementitious powders. *J. Test. Eval.* **2015**, *44*, 2147–2154. [CrossRef]
28. Bentz, D.P.; Garboczi, E.J.; Haecker, C.J.; Jensen, O.M. Effects of cement particle size distribution on performance properties of Portland cement-based materials. *Cem. Concr. Res.* **1999**, *29*, 1663–1671. [CrossRef]
29. Chindaprasirt, P.; Jaturapitakkul, C.; Sinsiri, T. Effect of fly ash fineness on compressive strength and pore size of blended cement paste. *Cem. Concr. Compos.* **2005**, *27*, 425–428. [CrossRef]
30. Schowalter, W.R. *Mechanics of Non-Newtonian Fluids*; Pergamon Press: Oxford, UK, 1978; ISBN 0-08-021778-8.
31. Macosko, C.W.; Larson, R.G. *Rheology: Principles, Measurements, and Applications*; Wiley: Hoboken, NJ, USA, 1994.
32. Bird, R.B.; Armstrong, R.C.; Hassager, O. Dynamics of Polymeric Liquids. Volume 1: Fluid Mechanics. 1987. Available online: https://inis.iaea.org/search/search.aspx?orig_q=RN:18088690 (accessed on 23 January 2020).

33. Carreau, P.J.; De Kee, D.C.R.; Chhabra, R.P. *Rheology of Polymeric Systems: Principles and Applications*; Hanser Publishers: New York, NY, USA, 1997.
34. Marchesini, F.H.; Oliveira, R.M.; Althoff, H.; de Souza Mendes, P.R. Irreversible time-dependent rheological behavior of cement slurries: Constitutive model and experiments. *J. Rheol.* **2019**, *63*, 247–262. [CrossRef]
35. Shaughnessy, R.; Clark, P.E. The rheological behavior of fresh cement pastes. *Cem. Concr. Res.* **1988**, *18*, 327–341. [CrossRef]
36. Wallevik, J.E. Thixotropic investigation on cement paste: Experimental and numerical approach. *J. Non-Newton. Fluid Mech.* **2005**, *132*, 86–99. [CrossRef]
37. Rubio-Hernández, F.-J. Rheological Behavior of Fresh Cement Pastes. *Fluids* **2018**, *3*, 106. [CrossRef]
38. Lapasin, R.; Papo, A.; Rajgelj, S. Flow behavior of fresh cement pastes. A comparison of different rheological instruments and techniques. *Cem. Concr. Res.* **1983**, *13*, 349–356. [CrossRef]
39. Asaga, K.; Roy, D.M. Rheological properties of cement mixes: IV. Effects of superplasticizers on viscosity and yield stress. *Cem. Concr. Res.* **1980**, *10*, 287–295. [CrossRef]
40. Barbic, L.; Tinta, V.; Lozar, B.; Marinkovic, V. Effect of Storage Time on the Rheological Behavior of Oil Well Cement Slurries. *J. Am. Ceram. Soc.* **1991**, *74*, 945–949. [CrossRef]
41. Moore, F. The rheology of ceramic slip and bodies. *Trans. Brit. Ceram. Soc.* **1959**, *58*, 470–492.
42. Ish-Shalom, M.; Greenberg, S.A. The Rheology of Fresh Portland Cement Pastes. 1960. Available online: https://trid.trb.org/view/102102 (accessed on 23 January 2020).
43. Odler, I. Hydration, setting and hardening of Portland cement. *LEA's Chem. Cem. Concr.* **1998**, *4*, 241–297.
44. Bingham, E.C. *Fluidity and Plasticity*; McGraw-Hill: New York City, NY, USA, 1922; Volume 2.
45. Atzeni, C.; Massidda, L.; Sanna, U. Comparison between rheological models for portland cement pastes. *Cem. Concr. Res.* **1985**, *15*, 511–519. [CrossRef]
46. Robertson, R.E.; Stiff, H.A., Jr. An improved mathematical model for relating shear stress to shear rate in drilling fluids and cement slurries. *Soc. Pet. Eng. J.* **1976**, *16*, 31–36. [CrossRef]
47. Macosko, C.W.; Krieger, I.M. Rheology: Principles, Measurements, and Applications. *J. Colloid Interface Sci.* **1996**, *178*, 382.
48. Lootens, D.; Jousset, P.; Martinie, L.; Roussel, N.; Flatt, R.J. Yield stress during setting of cement pastes from penetration tests. *Cem. Concr. Res.* **2009**, *39*, 401–408. [CrossRef]
49. Denn, M.M. *Polymer Melt Processing: Foundations in Fluid Mechanics and Heat Transfer*; Cambridge University Press: Cambridge, UK, 2008; ISBN 1-316-58314-7.
50. White, J.L. Approximate constitutive equations for slow flow of rigid plastic viscoelastic fluids. *J. Non-Newton. Fluid Mech.* **1981**, *8*, 195–202. [CrossRef]
51. Lipscomb, G.G.; Denn, M.M. Flow of Bingham fluids in complex geometries. *J. Non-Newton. Fluid Mech.* **1984**, *14*, 337–346. [CrossRef]
52. Mendes, P.R.S.; Dutra, E.S. Viscosity function for yield-stress liquids. *Appl. Rheol.* **2004**, *14*, 296–302. [CrossRef]
53. Herschel, W.H.; Bulkley, R. Measurement of consistency as applied to rubber-benzene solutions. *Am. Soc. Test. Mater.* **1926**, *26*, 621–633.
54. Jones, T.E.R.; Taylor, S. A mathematical model relating the flow curve of a cement paste to its water/cement ratio. *Mag. Concr. Res.* **1977**, *29*, 207–212. [CrossRef]
55. Choi, M.; Prud'homme, R.K.; Scherer, G. Rheological evaluation of compatibility in oil well cementing. *Appl. Rheol.* **2017**, *27*, 43354.
56. Ostwald, W. Ueber die rechnerische Darstellung des Strukturgebietes der Viskosität. *Colloid Polym. Sci.* **1929**, *47*, 176–187. [CrossRef]
57. Williamson, R.V. The flow of pseudoplastic materials. *Ind. Eng. Chem.* **1929**, *21*, 1108–1111. [CrossRef]
58. Lapasin, R.; Papo, A.; Rajgelj, S. The phenomenological description of the thixotropic behaviour of fresh cement pastes. *Rheol. Acta* **1983**, *22*, 410–416. [CrossRef]
59. Eyring, H. Viscosity, plasticity, and diffusion as examples of absolute reaction rates. *J. Chem. Phys.* **1936**, *4*, 283–291. [CrossRef]
60. Sisko, A.W. The flow of lubricating greases. *Ind. Eng. Chem.* **1958**, *50*, 1789–1792. [CrossRef]
61. Papo, A. Rheological models for cement pastes. *Mater. Struct.* **1988**, *21*, 41–46. [CrossRef]
62. Casson, N. Rheology of disperse systems. In *Proceedings of a Conference Organized by the British Society of Rheology*; Perga-mon Press: New York, NY, USA, 1959.

63. Kok, M.V.; Karakaya, G. Fluid migration and rheological properties of different origin G-class cements in oil well drilling applications. *Int. J. Oil Gas Coal Technol.* **2018**, *19*, 449–457. [CrossRef]
64. Shangraw, R.; Grim, W.; Mattocks, A.M. An Equation for Non-Newtonian Flow. *Trans. Soc. Rheol.* **1961**, *5*, 247–260. [CrossRef]
65. Beirute, R.M.; Flumerfelt, R.W. An Evaluation of the Robertson-Stiff Model Describing Rheological Properties of Drilling Fluids and Cement Slurries. *Soc. Pet. Eng. J.* **1977**, *17*, 97–100. [CrossRef]
66. Batra, R.L.; Parthasarathy, H. Flow of a Robertson-Stiff Fluid Between Two Approaching Disks. *Polym. Plast. Technol. Eng.* **1996**, *35*, 497–516. [CrossRef]
67. Vom Berg, W. Influence of specific surface and concentration of solids upon the flow behaviour of cement pastes. *Mag. Concr. Res.* **1979**, *31*, 211–216. [CrossRef]
68. Lapasin, R.; Longo, V.; Rajgelj, S. Thixotropic behaviour of cement pastes. *Cem. Concr. Res.* **1979**, *9*, 309–318. [CrossRef]
69. Billberg, P. Development of SCC static yield stress at rest and its effect on thelateral form pressure. In *SCC 2005, Combining the Second North American Conference on the Design and Use of Self-Consolidating Concrete and the Fourth International RILEM Symposium on Self-Compacting Concrete*; 2005; pp. 583–589. Available online: http://www.diva-portal.org/smash/record.jsf?pid=diva2%3A343052&dswid=4012 (accessed on 23 January 2020).
70. Quemada, D. Models for rheological behavior of concentrated disperse media under shear. *Adv. Rheol.* **1984**, *2*, 571–582.
71. De Kee, D.; Chan Man Fong, C.F. Rheological properties of structured fluids. *Polym. Eng. Sci.* **1994**, *34*, 438–445. [CrossRef]
72. Yahia, A.; Khayat, K.H. Applicability of rheological models to high-performance grouts containing supplementary cementitious materials and viscosity enhancing admixture. *Mater. Struct.* **2003**, *36*, 402–412. [CrossRef]
73. Yahia, A.; Khayat, K. Analytical models for estimating yield stress of high-performance pseudoplastic grout. *Cem. Concr.Res.* **2001**, *31*, 731–738. [CrossRef]
74. Vipulanandan, C.; Heidari, M.; Qu, Q.; Farzam, H.; Pappas, J.M. Behavior of piezoresistive smart cement contaminated with oil based drilling mud. In Proceedings of the Offshore Technology Conference, Houston, TX, USA, 5–8 May 2014.
75. Yuan, B.; Yang, Y.; Tang, X.; Xie, Y. A starting pressure prediction of thixotropic cement slurry: Theory, model and example. *J. Pet. Sci. Eng.* **2015**, *133*, 108–113. [CrossRef]
76. Tattersall, G.H.; Banfill, P.F. *The Rheology of Fresh Concrete*; The National Academies of Sciences, Engineering, and Medicine: Washington, DC, USA, 1983; ISBN 0-273-08558-1. Available online: https://trid.trb.org/view/199391 (accessed on 23 January 2020).
77. Barnes, H.A.; Walters, K. The yield stress myth? *Rheol. acta* **1985**, *24*, 323–326. [CrossRef]
78. Barnes, H.A. The 'yield stress myth'? paper–21 years on. *Appl. Rheol.* **2007**, *17*, 43110-1–43110-5. [CrossRef]
79. Barnes, H.A. The yield stress—A review or 'παντα ρει'—Everything flows? *J. Non-Newton. Fluid Mech.* **1999**, *81*, 133–178. [CrossRef]
80. Hartnett, J.P.; Hu, R.Y. The yield stress—An engineering reality. *J. Rheol.* **1989**, *33*, 671–679. [CrossRef]
81. Astarita, G. Letter to the Editor: The engineering reality of the yield stress. *J. Rheol.* **1990**, *34*, 275–277. [CrossRef]
82. Evans, I.D. Letter to the editor: On the nature of the yield stress. *J. Rheol.* **1992**, *36*, 1313–1318. [CrossRef]
83. Papanastasiou, T.C. Flows of materials with yield. *J. Rheol.* **1987**, *31*, 385–404. [CrossRef]
84. Bonn, D.; Denn, M.M. Yield stress fluids slowly yield to analysis. *Science* **2009**, *324*, 1401–1402. [CrossRef]
85. Denn, M.M.; Bonn, D. Issues in the flow of yield-stress liquids. *Rheol. acta* **2011**, *50*, 307–315. [CrossRef]
86. Moller, P.; Fall, A.; Chikkadi, V.; Derks, D.; Bonn, D. An attempt to categorize yield stress fluid behaviour. *Philos. Trans. Royal Soc. A Math. Phys. Eng. Sci.* **2009**, *367*, 5139–5155. [CrossRef] [PubMed]
87. Bird, R.B.; Dai, G.C.; Yarusso, B.J. The rheology and flow of viscoplastic materials. *Rev. Chem. Eng.* **1983**, *1*, 1–70. [CrossRef]
88. Møller, P.C.; Mewis, J.; Bonn, D. Yield stress and thixotropy: On the difficulty of measuring yield stresses in practice. *Soft Matter* **2006**, *2*, 274–283. [CrossRef]
89. Dinkgreve, M.; Paredes, J.; Denn, M.M.; Bonn, D. On different ways of measuring "the" yield stress. *J. Non-Newton. Fluid Mech.* **2016**, *238*, 233–241. [CrossRef]

90. Nguyen, Q.D.; Boger, D.V. Measuring the flow properties of yield stress fluids. *Annu. Rev. Fluid Mech.* **1992**, *24*, 47–88. [CrossRef]
91. Coussot, P. Yield stress fluid flows: A review of experimental data. *J. Non-Newton. Fluid Mech.* **2014**, *211*, 31–49. [CrossRef]
92. Coussot, P.; Nguyen, Q.D.; Huynh, H.T.; Bonn, D. Viscosity bifurcation in thixotropic, yielding fluids. *J. Rheol.* **2002**, *46*, 573–589. [CrossRef]
93. Jarny, S.; Roussel, N.; Rodts, S.; Bertrand, F.; Le Roy, R.; Coussot, P. Rheological behavior of cement pastes from MRI velocimetry. *Cem. Concr. Res.* **2005**, *35*, 1873–1881. [CrossRef]
94. Roussel, N.; Ovarlez, G.; Garrault, S.; Brumaud, C. The origins of thixotropy of fresh cement pastes. *Cem. Concr. Res.* **2012**, *42*, 148–157. [CrossRef]
95. Mahaut, F.; Mokéddem, S.; Chateau, X.; Roussel, N.; Ovarlez, G. Effect of coarse particle volume fraction on the yield stress and thixotropy of cementitious materials. *Cem. Concr.Res.* **2008**, *38*, 1276–1285. [CrossRef]
96. Schmidt, G.; Schlegel, E. Rheological characterization of CSH phases–water suspensions. *Cem. Concr. Res.* **2002**, *32*, 593–599. [CrossRef]
97. Nachbaur, L.; Mutin, J.C.; Nonat, A.; Choplin, L. Dynamic mode rheology of cement and tricalcium silicate pastes from mixing to setting. *Cem. Concr. Res.* **2001**, *31*, 183–192. [CrossRef]
98. Schultz, M.A.; Struble, L.J. Use of oscillatory shear to study flow behavior of fresh cement paste. *Cem. Concr. Res.* **1993**, *23*, 273–282. [CrossRef]
99. Perrot, A.; Lecompte, T.; Khelifi, H.; Brumaud, C.; Hot, J.; Roussel, N. Yield stress and bleeding of fresh cement pastes. *Cem. Concr. Res.* **2012**, *42*, 937–944. [CrossRef]
100. Assaad, J.J.; Harb, J.; Maalouf, Y. Effect of vane configuration on yield stress measurements of cement pastes. *J. Non-Newton. Fluid Mech.* **2016**, *230*, 31–42. [CrossRef]
101. Qian, Y.; Kawashima, S. Distinguishing dynamic and static yield stress of fresh cement mortars through thixotropy. *Cem. Concr. Res.* **2018**, *86*, 288–296. [CrossRef]
102. Michaux, M.; Defosse, C. Oil well cement slurries I. Microstructural approach of their rheology. *Cem. Concr. Res.* **1986**, *16*, 23–30. [CrossRef]
103. Vance, K.; Kumar, A.; Sant, G.; Neithalath, N. The rheological properties of ternary binders containing Portland cement, limestone, and metakaolin or fly ash. *Cem. Concr. Res.* **2013**, *52*, 196–207. [CrossRef]
104. Struble, L.; Sun, G.-K. Viscosity of Portland cement paste as a function of concentration. *Adv. Cem. Based Mater.* **1995**, *2*, 62–69. [CrossRef]
105. Legrand, C. Rheologie Des Melanges De Ciment Ou De Sable et d'eau. *Rev. Mater Constr. Trav. Publics Cim. Beton* **1970**.
106. Kurdowski, W. *Cement and Concrete Chemistry*; Springer Science & Business: Berlin/Heidelberg, Germany, 2014; ISBN 94-007-7945-3.
107. Sybertz, F.; Reick, P. Effect of fly as on the rheological properties of cement paste. *Rheol. Fresh Cem. Concr.* **1991**, 13–22.
108. Zhou, Z.; Solomon, M.J.; Scales, P.J.; Boger, D.V. The yield stress of concentrated flocculated suspensions of size distributed particles. *J. Rheol.* **1999**, *43*, 651–671. [CrossRef]
109. Flatt, R.J.; Bowen, P. Yield Stress of Multimodal Powder Suspensions: An Extension of the YODEL (Yield Stress mODEL). *J. Am. Ceram. Soc.* **2007**, *90*, 1038–1044. [CrossRef]
110. Hamaker, H.C. The London—van der Waals attraction between spherical particles. *Physica* **1937**, *4*, 1058–1072. [CrossRef]
111. Flatt, R.J.; Bowen, P. Yodel: A Yield Stress Model for Suspensions. *J. Am. Ceram. Soc.* **2006**, *89*, 1244–1256. [CrossRef]
112. Ma, S.; Kawashima, S. A rheological approach to study the early-age hydration of oil well cement: Effect of temperature, pressure and nanoclay. *Constr. Build. Mater.* **2019**, *215*, 119–127. [CrossRef]
113. Chateau, X.; Ovarlez, G.; Trung, K.L. Homogenization approach to the behavior of suspensions of noncolloidal particles in yield stress fluids. *J. Rheol.* **2008**, *52*, 489–506. [CrossRef]
114. Chougnet, A.; Palermo, T.; Audibert, A.; Moan, M. Rheological behaviour of cement and silica suspensions: Particle aggregation modelling. *Cem. Concr. Res.* **2008**, *38*, 1297–1301. [CrossRef]
115. Ivanov, Y.P.; Roshavelov, T.T. The effect of condensed silica fume on the rheological behaviour of cement pastes. In *Rheology of Fresh Cement and Concrete: Proceedings of an International Conference, Liverpool, 1990*; CRC Press: Boca Raton, FL, USA, 1990; pp. 23–26.

116. Rosquoët, F.; Alexis, A.; Khelidj, A.; Phelipot, A. Experimental study of cement grout. *Cem. Concr. Res.* **2003**, *33*, 713–722. [CrossRef]
117. Lobanov, V.P. The visco-plastic properties of building mortars. *Kolloidn. Zhurnal* **1950**, *12*, 352–358.
118. Bombled, J.P. Influence of sulphates on the rheological behaviour of cement pastes and their evolution. In *Proceedings of the 7th International Congress on the Chemistry of Cement: Paris*; Cement Research Institute of India: New Delhi, India, 1980; pp. 164–169.
119. Roy, D.M.; Asaga, K. Rheological properties of cement mixes: V. The effects of time on viscometric properties of mixes containing superplasticizers; conclusions. *Cem. Concr. Res.* **1980**, *10*, 387–394. [CrossRef]
120. Ivanov, Y.; Stanoeva, E. Influence of cement composition and other factors on the rheological behavior of cement pastes. *Silic. Ind.* **1979**, *44*, 199–203.
121. Ivanov, Y.; Zacharieva, S. Influence of fly ash on the rheology of cement pastes. In *Proceedings of the 7th International Congress on the Chemistry of Cement: Paris*; Cement Research Institute of India: New Delhi, India, 1980; pp. 103–107.
122. Chen, X.; Wu, S.; Zhou, J. Experimental and modeling study of dynamic mechanical properties of cement paste, mortar and concrete. *Constr. Build. Mater.* **2013**, *47*, 419–430. [CrossRef]
123. Massoudi, M.; Wang, P. Slag behavior in gasifiers. Part II: Constitutive modeling of slag. *Energies* **2013**, *6*, 807–838. [CrossRef]
124. Fredrickson, A.G. *Principles and Applications of Rheology*; Prentice-Hall: Upper Saddle River, NJ, USA, 1964.
125. White, F.M. *Fluid Mechanics fourth edition*; McGraw and Hill: Singapore, 1994.
126. Einstein, A. Eine neue bestimmung der moleküldimensionen. *Ann. Der Phys.* **1906**, *324*, 289–306. [CrossRef]
127. Roussel, N.; Lemaître, A.; Flatt, R.J.; Coussot, P. Steady state flow of cement suspensions: A micromechanical state of the art. *Cem. Concr. Res.* **2010**, *40*, 77–84. [CrossRef]
128. Roscoe, R. The viscosity of suspensions of rigid spheres. *Br. J. Appl. Phys.* **1952**, *3*, 267. [CrossRef]
129. Mooney, M. The viscosity of a concentrated suspension of spherical particles. *J. Colloid Sci.* **1951**, *6*, 162–170. [CrossRef]
130. Krieger, I.M.; Dougherty, T.J. A mechanism for non-Newtonian flow in suspensions of rigid spheres. *Trans. Soc. Rheol.* **1959**, *3*, 137–152. [CrossRef]
131. Struble, L.; Szecsy, R.; Lei, W.-G.; Sun, G.-K. Rheology of cement paste and concrete. *Cem. Concr. Aggreg.* **1998**, *20*, 269–277.
132. Mansoutre, S.; Colombet, P.; Damme, H.V. Water retention and granular rheological behavior of fresh C3S paste as a function of concentration. *Cem. Concr. Res.* **1999**, *29*, 1441–1453.
133. Justnes, H.; Vikan, H. Viscosity of cement slurries as a function of solids content. *Ann. Trans. Nordic Rheol. Soc.* **2005**, *13*, 75–82.
134. Bonen, D.; Shah, S.P. Fresh and hardened properties of self-consolidating concrete. *Prog. Struct. Eng. Mater.* **2005**, *7*, 14–26. [CrossRef]
135. Hendrickx, R.; Rezeau, M.; Van Balen, K.; Van Gemert, D. Mortar and paste rheology: Concentration, polydispersity and air entrapment at high solid fraction. *Appl. Rheol.* **2009**, *19*. [CrossRef]
136. Tregger, N.A.; Pakula, M.E.; Shah, S.P. Influence of clays on the rheology of cement pastes. *Cem. Concr. Res.* **2010**, *40*, 384–391. [CrossRef]
137. Bentz, D.P.; Ferraris, C.F.; Galler, M.A.; Hansen, A.S.; Guynn, J.M. Influence of particle size distributions on yield stress and viscosity of cement–fly ash pastes. *Cem. Concr. Res.* **2012**, *42*, 404–409. [CrossRef]
138. Ouyang, J.; Tan, Y. Rheology of fresh cement asphalt emulsion pastes. *Constr. Build. Mater.* **2015**, *80*, 236–243. [CrossRef]
139. O'Neill, R.; McCarthy, H.O.; Montufar, E.B.; Ginebra, M.-P.; Wilson, D.I.; Lennon, A.; Dunne, N. Critical review: Injectability of calcium phosphate pastes and cements. *Acta Biomater.* **2017**, *50*, 1–19. [CrossRef]
140. Chen, C.-T.; Lin, C.-W. Stiffening Behaviors of Cement Pastes Measured by a Vibrational Viscometer. *Adv. Civ. Eng. Mater.* **2017**, *6*, 20160061. [CrossRef]
141. Murata, J.; Kikukawa, H. Studies on rheological analysis of fresh concrete. In *Proceedings of the RILEM Seminar: Fresh Concrete*; RILEM: Paris, France, 1973.
142. Mills, P. Non-Newtonian behaviour of flocculated suspensions. *J. De Phys. Lett.* **1985**, *46*, 301–309. [CrossRef]
143. Liu, D.-M. Particle packing and rheological property of highly-concentrated ceramic suspensions: φm determination and viscosity prediction. *J. Mater. Sci.* **2000**, *35*, 5503–5507. [CrossRef]

144. Chong, J.S.; Christiansen, E.B.; Baer, A.D. Rheology of concentrated suspensions. *J. Appl. Polym. Sci.* **1971**, *15*, 2007–2021. [CrossRef]
145. Aiad, I.; El-Sabbagh, A.M.; Adawy, A.I.; Shafek, S.H.; Abo-EL-Enein, S.A. Effect of some prepared superplasticizers on the rheological properties of oil well cement slurries. *Egypt. J. Pet.* **2018**, *27*, 1061–1066. [CrossRef]
146. Vipulanandan, C.; Mohammed, A. Smart cement rheological and piezoresistive behavior for oil well applications. *J. Pet. Sci. Eng.* **2015**, *135*, 50–58. [CrossRef]
147. Kim, J.H.; Kwon, S.H.; Kawashima, S.; Yim, H.J. Rheology of cement paste under high pressure. *Cem. Concr. Compos.* **2017**, *77*, 60–67. [CrossRef]
148. Ravi, K.M.; Sutton, D.L. New rheological correlation for cement slurries as a function of temperature. In Proceedings of the SPE Annual Technical Conference and Exhibition, New Orleans, LA, USA, 23–26 September 1990.
149. Sercombe, J.; Hellmich, C.; Ulm, F.-J.; Mang, H. Modeling of Early-Age Creep of Shotcrete. I: Model and Model Parameters. *J. Eng. Mech.* **2000**, *126*, 284–291. [CrossRef]
150. Bazant, Z.P. Creep and damage in concrete. *Mater. Sci. Concr.* IV **1995**, 355–389. Available online: http://cee.northwestern.edu/people/bazant/PDFs/Papers%20-%20Backup%202_20_2013/331.pdf (accessed on 23 January 2020).
151. Ulm, F.-J.; Coussy, O. Strength growth as chemo-plastic hardening in early age concrete. *J. Eng. Mech.* **1996**, *122*, 1123–1132. [CrossRef]
152. Avrami, M. Kinetics of phase change. I General theory. *J. Chem. Phys.* **1939**, *7*, 1103–1112. [CrossRef]
153. Pang, X.; Meyer, C.; Darbe, R.; Funkhouser, G.P. Modeling the effect of curing temperature and pressure on cement hydration kinetics. *ACI Mater. J.* **2013**, *110*, 137.
154. Wu, D.; Cai, S.; Huang, G. Coupled effect of cement hydration and temperature on rheological properties of fresh cemented tailings backfill slurry. *Trans. Nonferrous Metals Soc. China* **2014**, *24*, 2954–2963. [CrossRef]
155. Wu, D.; Fall, M.; Cai, S.J. Coupling temperature, cement hydration and rheological behaviour of fresh cemented paste backfill. *Miner. Eng.* **2013**, *42*, 76–87. [CrossRef]
156. Papo, A.; Caufin, B. A study of the hydration process of cement pastes by means of oscillatory rheological techniques. *Cem. Concr. Res.* **1991**, *21*, 1111–1117. [CrossRef]
157. Aïtcin, P.-C. Cements of yesterday and today: Concrete of tomorrow. *Cem. Concr. Res.* **2000**, *30*, 1349–1359. [CrossRef]
158. Jayasree, C.; Murali Krishnan, J.; Gettu, R. Influence of superplasticizer on the non-Newtonian characteristics of cement paste. *Mater. Struct.* **2011**, *44*, 929–942. [CrossRef]
159. Ma, G.; Wang, L. A critical review of preparation design and workability measurement of concrete material for largescale 3D printing. *Front. Struct. Civ. Eng.* **2018**, *12*, 382–400. [CrossRef]
160. Bellotto, M. Cement paste prior to setting: A rheological approach. *Cem. Concr. Res.* **2013**, *52*, 161–168. [CrossRef]
161. Mikanovic, N.; Jolicoeur, C. Influence of superplasticizers on the rheology and stability of limestone and cement pastes. *Cem. Concr. Res.* **2008**, *38*, 907–919. [CrossRef]
162. Dauksys, M.; Skripkiunas, G.; Janavicius, E. Complex influence of plasticizing admixtures and sodium silicate solution on rheological properties of Portland cement paste. *Mater. Sci.* **2009**, *15*, 349–355.
163. Rehman, S.K.U.; Ibrahim, Z.; Jameel, M.; Memon, S.A.; Javed, M.F.; Aslam, M.; Mehmood, K.; Nazar, S. Assessment of Rheological and Piezoresistive Properties of Graphene based Cement Composites. *Int. J. Concr. Struct. Mater.* **2018**, *12*, 64. [CrossRef]
164. Sanfelix, S.G.; Santacruz, I.; Szczotok, A.M.; Belloc, L.M.O.; De la Torre, A.G.; Kjøniksen, A.-L. Effect of microencapsulated phase change materials on the flow behavior of cement composites. *Constr. Build. Mater.* **2019**, *202*, 353–362. [CrossRef]
165. Khayat, K.H.; Yahia, A. Effect of welan gum-high-range water reducer combinations on rheology of cement grout. *Mater. J.* **1997**, *94*, 365–372.
166. Cyr, M.; Legrand, C.; Mouret, M. Study of the shear thickening effect of superplasticizers on the rheological behaviour of cement pastes containing or not mineral additives. *Cem. Concr. Res.* **2000**, *30*, 1477–1483. [CrossRef]
167. Nehdi, M.; Mindess, S.; Aïtcin, P.-C. Rheology of High-Performance Concrete: Effect of Ultrafine Particles. *Cem. Concr. Res.* **1998**, *28*, 687–697. [CrossRef]

168. Gołaszewski, J.; Szwabowski, J. Influence of superplasticizers on rheological behaviour of fresh cement mortars. *Cem. Concr. Res.* **2004**, *34*, 235–248. [CrossRef]
169. Aiad, I. Influence of time addition of superplasticizers on the rheological properties of fresh cement pastes. *Cem. Concr. Res.* **2003**, *33*, 1229–1234. [CrossRef]
170. Puertas, F.; Santos, H.; Palacios, M.; Martınez-Ramırez, S. Polycarboxylate superplasticiser admixtures: Effect on hydration, microstructure and rheological behaviour in cement pastes. *Adv. Cem. Res.* **2005**, *17*, 77–89. [CrossRef]
171. Zingg, A.; Winnefeld, F.; Holzer, L.; Pakusch, J.; Becker, S.; Gauckler, L. Adsorption of polyelectrolytes and its influence on the rheology, zeta potential, and microstructure of various cement and hydrate phases. *J. Colloid Interface Sci.* **2008**, *323*, 301–312. [CrossRef]
172. Wong, H.H.C.; Kwan, A.K.H. Rheology of Cement Paste: Role of Excess Water to Solid Surface Area Ratio. *J. Mater. Civ. Eng.* **2008**, *20*, 189–197. [CrossRef]
173. Abbas, G.; Irawan, S.; Kumar, S.; Memon, R.K.; Khalwar, S.A. Characteristics of Oil Well Cement Slurry using Hydroxypropylmethylcellulose. *J. Appl. Sci.* **2014**, *14*, 1154–1160. [CrossRef]
174. Zhang, X.; Han, J. The effect of ultra-fine admixture on the rheological property of cement paste. *Cem. Concr. Res.* **2000**, *30*, 827–830. [CrossRef]
175. Park, C.K.; Noh, M.H.; Park, T.H. Rheological properties of cementitious materials containing mineral admixtures. *Cem. Concr. Res.* **2005**, *35*, 842–849. [CrossRef]
176. Vikan, H.; Justnes, H. Rheology of cementitious paste with silica fume or limestone. *Cem. Concr. Res.* **2007**, *37*, 1512–1517. [CrossRef]
177. Stryczek, S.; Brylicki, W.; Małolepszy, J.; Gonet, A.; Wiśniowski, R.; Kotwica, Ł. Potential use of fly ash from fluidal combustion of brown coal in cementing slurries for drilling and geotechnical works. *Arch. Min. Sci.* **2009**, *54*, 775–786.
178. Baldino, N.; Gabriele, D.; Lupi, F.R.; Seta, L.; Zinno, R. Rheological behaviour of fresh cement pastes: Influence of synthetic zeolites, limestone and silica fume. *Cem. Concr. Res.* **2014**, *63*, 38–45. [CrossRef]
179. Li, D.; Wang, D.; Ren, C.; Rui, Y. Investigation of rheological properties of fresh cement paste containing ultrafine circulating fluidized bed fly ash. *Constr. Build. Mater.* **2018**, *188*, 1007–1013. [CrossRef]
180. Burroughs, J.F.; Weiss, J.; Haddock, J.E. Influence of high volumes of silica fume on the rheological behavior of oil well cement pastes. *Constr. Build. Mater.* **2019**, *203*, 401–407. [CrossRef]
181. Laskar, A.I.; Talukdar, S. Rheological behavior of high performance concrete with mineral admixtures and their blending. *Constr. Build. Mater.* **2008**, *22*, 2345–2354. [CrossRef]
182. Provis, J.L.; Duxson, P.; van Deventer, J.S.J. The role of particle technology in developing sustainable construction materials. *Adv. Powder Technol.* **2010**, *21*, 2–7. [CrossRef]
183. Hou, J.; Liu, Z. Synthesizing Dispersant for MTC Design and Its Effect on Slurry Rheology. *SPE Drill. Complet.* **2000**, *15*, 31–36. [CrossRef]
184. Nehdi, M. Why some carbonate fillers cause rapid increases of viscosity in dispersed cement-based materials. *Cem. Concr. Res.* **2000**, *30*, 1663–1669. [CrossRef]
185. Hodne, H.; Saasen, A.; Strand, S. Rheological properties of high temperature oil well cement slurries. *Annu. Trans. Nordi. Rheol. Soc.* **2001**, *8*, 31–38.
186. Kaufmann, J.; Winnefeld, F.; Hesselbarth, D. Effect of the addition of ultrafine cement and short fiber reinforcement on shrinkage, rheological and mechanical properties of Portland cement pastes. *Cem. Concr. Compos.* **2004**, *26*, 541–549. [CrossRef]
187. Sonebi, M.; Svermova, L.; Bartos, P.J.M. Factorial Design of Cement Slurries Containing Limestone Powder for Self-Consolidating Slurry-Infiltrated Fiber Concrete. *ACI Mater. J.* **2004**, *101*, 136–145.
188. Senff, L.; Hotza, D.; Lucas, S.; Ferreira, V.M.; Labrincha, J.A. Effect of nano-SiO_2 and nano-TiO_2 addition on the rheological behavior and the hardened properties of cement mortars. *Mater. Sci. Eng. A* **2012**, *532*, 354–361. [CrossRef]
189. Soltanian, H.; Khalokakaie, R.; Ataei, M.; Kazemzadeh, E. Fe_2O_3 nanoparticles improve the physical properties of heavy-weight wellbore cements: A laboratory study. *J. Nat. Gas Sci. Eng.* **2015**, *26*, 695–701. [CrossRef]
190. Sun, X.; Wu, Q.; Lee, S.; Qing, Y.; Wu, Y. Cellulose Nanofibers as a Modifier for Rheology, Curing and Mechanical Performance of Oil Well Cement. *Sci. Rep.* **2016**, *6*, 31654. [CrossRef]

191. Tang, Z.; Huang, R.; Mei, C.; Sun, X.; Zhou, D.; Zhang, X.; Wu, Q. Influence of Cellulose Nanoparticles on Rheological Behavior of Oil Well Cement-Water Slurries. *Materials* **2019**, *12*, 291. [CrossRef]
192. Colombo, A.; Geiker, M.R.; Justnes, H.; Lauten, R.A.; De Weerdt, K. On the effect of calcium lignosulfonate on the rheology and setting time of cement paste. *Cem. Concr. Res.* **2017**, *100*, 435–444. [CrossRef]
193. Senff, L.; Labrincha, J.A.; Ferreira, V.M.; Hotza, D.; Repette, W.L. Effect of nano-silica on rheology and fresh properties of cement pastes and mortars. *Constr. Build. Mater.* **2009**, *23*, 2487–2491. [CrossRef]
194. Quercia, G.; Brouwers, H.J.H.; Garnier, A.; Luke, K. Influence of olivine nano-silica on hydration and performance of oil-well cement slurries. *Mater. Des.* **2016**, *96*, 162–170. [CrossRef]
195. Wang, W.; Taleghani, A.D. Three-dimensional analysis of cement sheath integrity around Wellbores. *J. Pet. Sci. Eng.* **2014**, *121*, 38–51. [CrossRef]
196. Phan, T.H.; Chaouche, M.; Moranville, M. Influence of organic admixtures on the rheological behaviour of cement pastes. *Cem. Concr. Res.* **2006**, *36*, 1807–1813. [CrossRef]
197. Emoto, T.; Bier, T.A. Rheological behavior as influenced by plasticizers and hydration kinetics. *Cem. Concr. Res.* **2007**, *37*, 647–654. [CrossRef]
198. Lu, H.; Xie, C.; Gao, Y.; Li, L.; Zhu, H. Cement Slurries with Rheological Properties Unaffected by Temperature. *SPE Drill. Complet.* **2016**, *30*, 316–321. [CrossRef]
199. Velayati, A.; Soltanian, H.; Pourmazaheri, Y.; Aghajafari, A.H.; Kazemzadeh, E.; Barati, P. Investigation of Cassia fistula fruit dry extract effect on the oil and gas well cement properties. *J. Nat. Gas Sci. Eng.* **2016**, *36*, 298–304. [CrossRef]
200. Wang, C.; Chen, X.; Wang, R. Do chlorides qualify as accelerators for the cement of deepwater oil wells at low temperature? *Constr. Build. Mater.* **2017**, *133*, 482–494. [CrossRef]
201. Tao, C.; Watts, B.; Ferraro, C.C.; Masters, F.J. A Multivariate Computational Framework to Characterize and Rate Virtual Portland Cements. *Comput. Aided Civ. Infrastruct. Eng.* **2019**, *34*, 266–278. [CrossRef]
202. Tao, C. Optimization of Cement Production and Hydration for Improved Performance, Energy Conservation, and Cost. Ph.D. Dissertation, University of Florida, Gainesville, FL, USA, 2017.
203. Stryczek, S.; Wiśniowski, R.; Gonet, A.; Złotkowski, A.; Ziaja, J. Influence of Polycarboxylate Superplasticizers on Rheological Properties of Cement Slurries Used in Drilling Technologies/Wpływ Superplastyfikatorów Z Grupy Polikarboksylanów Na Właściwości Reologiczne Zaczynów Cementowych Stosowanych W Technologiach Wiertniczych. *Arch. Min. Sci.* **2013**, *58*, 719–728.
204. Massidda, L.; Sanna, U. *Rheological Behaviour of Portland Cement Pastes Containing Fly Ash*; Il Cemento: Milano, Roma, 1982.
205. Yang, M.; Jennings, H.M. Influences of mixing methods on the microstructure and rheological behavior of cement paste. *Adv. Cem. Based Mater.* **1995**, *2*, 70–78. [CrossRef]
206. Williams, D.A.; Saak, A.W.; Jennings, H.M. The influence of mixing on the rheology of fresh cement paste. *Cem. Concr. Res.* **1999**, *29*, 1491–1496. [CrossRef]
207. Orban, J.; Parcevaux, P.; Guillot, D. Influence of shear history on the rheological properties of oil well cement slurries. In Proceedings of the 8th International Congress on the Chemistry of Cement, Rio de Janeiro, Brazi, 22–27 September 1986; Volume 6, pp. 243–247.
208. Saleh, F.K.; Salehi, S.; Teodoriu, C. Experimental investigation of mixing energy of well cements: The gap between laboratory and field mixing. *J. Nat. Gas Sci. Eng.* **2019**, *63*, 47–57. [CrossRef]
209. Han, D.; Ferron, R.D. Effect of mixing method on microstructure and rheology of cement paste. *Constr. Build. Mater.* **2015**, *93*, 278–288. [CrossRef]
210. Saleh, F.K.; Teodoriu, C. The mechanism of mixing and mixing energy for oil and gas wells cement slurries: A literature review and benchmarking of the findings. *J. Nat. Gas Sci. Eng.* **2017**, *38*, 388–401. [CrossRef]
211. Mannheimer, R.J. Laminar and turbulent flow of cement slurries in large diameter pipe: A comparison with laboratory viscometers. *J. Rheol.* **1991**, *35*, 113–133. [CrossRef]
212. Saak, A.W.; Jennings, H.M.; Shah, S.P. The influence of wall slip on yield stress and viscoelastic measurements of cement paste. *Cem. Concr. Res.* **2001**, *31*, 205–212. [CrossRef]
213. Bannister, C.E. Rheological evaluation of cement slurries: Methods and models. In Proceedings of the SPE Annual Technical Conference and Exhibition, Dallas, TX, USA, 21–24 September 1980.
214. Roussel, N. Steady and transient flow behaviour of fresh cement pastes. *Cem. Concr. Res.* **2005**, *35*, 1656–1664. [CrossRef]
215. Mewis, J. Thixotropy-a general review. *J. Non-Newton. Fluid Mech.* **1979**, *6*, 1–20. [CrossRef]

216. Barnes, H.A. Thixotropy—A review. *J. Non-Newton. Fluid Mech.* **1997**, *70*, 1–33. [CrossRef]
217. Freundlich, H. *Thixotropy*; Hermann & Cie: Paris, France, 1935; Volume 1.
218. Barnes, H.A.; Hutton, J.F.; Walters, K. *An Introduction to Rheology*; Elsevier: Edinburgh, London, UK, 1989; ISBN 0-08-093369-6.
219. Hartley, G.S. Negative thixotropy. *Nature* **1938**, *142*, 161. [CrossRef]
220. Renardy, M.; Wang, X. Boundary layers for the upper convected Maxwell fluid. *J. Non-Newton. Fluid Mech.* **2012**, *189*, 14–18. [CrossRef]
221. Renardy, M.; Renardy, Y. Thixotropy in yield stress fluids as a limit of viscoelasticity. *IMA J. Appl. Math.* **2016**, *81*, 522–537. [CrossRef]
222. Larson, R.G. A constitutive equation for polymer melts based on partially extending strand convection. *J. Rheol.* **1984**, *28*, 545–571. [CrossRef]
223. Fredrickson, A.G. A model for the thixotropy of suspensions. *AIChe J.* **1970**, *16*, 436–441. [CrossRef]
224. Renardy, M. The mathematics of myth: Yield stress behavior as a limit of non-monotone constitutive theories. *J. Non-Newton. Fluid Mech.* **2010**, *165*, 519–526. [CrossRef]
225. de Souza Mendes, P.R.; Thompson, R.L. A unified approach to model elasto-viscoplastic thixotropic yield-stress materials and apparent yield-stress fluids. *Rheol. Acta* **2013**, *52*, 673–694. [CrossRef]
226. de Souza Mendes, P.R. Thixotropic elasto-viscoplastic model for structured fluids. *Soft Matter* **2011**, *7*, 2471–2483. [CrossRef]
227. Chappuis, J. Rheological measurements with cement pastes in viscometers: A comprehensive approach. In *Proceedings of Fresh Cement and Concrete*; Taylor & Francis: London, UK, 1990; pp. 3–12.
228. Al Ghanami, R.C.; Saunders, B.R.; Bosquillon, C.; Shakesheff, K.M.; Alexander, C. Responsive particulate dispersions for reversible building and deconstruction of 3D cell environments. *Soft Matter* **2010**, *6*, 5037–5044. [CrossRef]
229. Sant, G.; Ferraris, C.F.; Weiss, J. Rheological properties of cement pastes: A discussion of structure formation and mechanical property development. *Cem. Concr. Res.* **2008**, *38*, 1286–1296. [CrossRef]
230. Tsenoglou, C. Scaling concepts in suspension rheology. *J. Rheol.* **1990**, *34*, 15–24. [CrossRef]
231. Usui, H. A thixotropy model for coal-water mixtures. *J. Non-Newton. Fluid Mech.* **1995**, *60*, 259–275. [CrossRef]
232. Potanin, A.A.; De Rooij, R.; Van den Ende, D.; Mellema, J. Microrheological modeling of weakly aggregated dispersions. *J. Chem. Phys.* **1995**, *102*, 5845–5853. [CrossRef]
233. Quemada, D. Rheological modelling of complex fluids: IV: Thixotropic and "thixoelastic" behaviour. Start-up and stress relaxation, creep tests and hysteresis cycles. *Eur. Phys. J. Appl. Phys.* **1999**, *5*, 191–207. [CrossRef]
234. Feys, D.; Asghari, A. Influence of maximum applied shear rate on the measured rheological properties of flowable cement pastes. *Cem. Concr. Res.* **2019**, *117*, 69–81. [CrossRef]
235. Cheng, D.C.; Evans, F. Phenomenological characterization of the rheological behaviour of inelastic reversible thixotropic and antithixotropic fluids. *B. J. Appl. Phys.* **1965**, *16*, 1599. [CrossRef]
236. Legrand, C. Contribution à l'étude de la rhéologie du béton frais. *Mater. Constr.* **1972**, *5*, 275–295. [CrossRef]
237. Hattori, K.; Izumi, K. A rheological expression of coagulation rate theory. *J. Dispers. Sci. Technol.* **1982**, *3*, 129–145. [CrossRef]
238. Roussel, N. A thixotropy model for fresh fluid concretes: Theory, validation and applications. *Cem. Concr. Res.* **2006**, *36*, 1797–1806. [CrossRef]
239. Roussel, N. Rheology of fresh concrete: From measurements to predictions of casting processes. *Mater. Struct.* **2007**, *40*, 1001–1012. [CrossRef]
240. Mewis, J.; Wagner, N.J. *Colloidal Suspension Rheology*; Cambridge University Press: Cambridge, UK, 2012; ISBN 0-521-51599-8.
241. de Souza Mendes, P.R. Modeling the thixotropic behavior of structured fluids. *J. Non-Newton. Fluid Mech.* **2009**, *164*, 66–75. [CrossRef]
242. Truesdell, C.; Noll, W. *The Non-linear Field Theories of Mechanics*; Springer: Berlin, Germany, 1992.
243. Reynolds, O. LVII. On the dilatancy of media composed of rigid particles in contact. With experimental illustrations. *Lond. Edinb. Dublin Philos. Mag. J. Sci.* **1885**, *20*, 469–481. [CrossRef]
244. Reynolds, O. IV. On the theory of lubrication and its application to Mr. Beauchamp tower's experiments, including an experimental determination of the viscosity of olive oil. *Philos. Trans. Royal Soc. Lond.* **1886**, *1*, 157–234.

245. Massoudi, M.; Mehrabadi, M.M. A continuum model for granular materials: Considering dilatancy and the Mohr-Coulomb criterion. *Acta Mech.* **2001**, *152*, 121–138. [CrossRef]
246. Reiner, M. A mathematical theory of dilatancy. *Am. J. Math.* **1945**, *67*, 350–362. [CrossRef]
247. Massoudi, M. A generalization of Reiner's mathematical model for wet sand. *Mech. Res. Commun.* **2011**, *38*, 378–381. [CrossRef]
248. Man, C.-S. Nonsteady channel flow of ice as a modified second-order fluid with power-law viscosity. *Arch. Ration. Mech. Anal.* **1992**, *119*, 35–57. [CrossRef]
249. Man, C.-S.; Sun, Q.-X. On the significance of normal stress effects in the flow of glaciers. *J. Glaciol.* **1987**, *33*, 268–273. [CrossRef]
250. Gupta, G.; Massoudi, M. Flow of a generalized second grade fluid between heated plates. *Acta Mech.* **1993**, *99*, 21–33. [CrossRef]
251. Massoudi, M. Heat transfer in complex fluids. In *Two Phase Flow, Phase Change and Numerical Modeling*; No. NETL-PUB-234; National Energy Technology Lab (NETL): Pittsburgh, PA, USA; In-house ResearchIn-house Research: Morgantown, WV, USA, 2012.
252. Wu, W.-T.; Aubry, N.; Massoudi, M.; Kim, J.; Antaki, J.F. A numerical study of blood flow using mixture theory. *Int. J. Eng. Sci.* **2014**, *76*, 56–72. [CrossRef] [PubMed]
253. Miao, L.; Massoudi, M. Heat transfer analysis and flow of a slag-type fluid: Effects of variable thermal conductivity and viscosity. *Int. J. Non-Linear Mech.* **2015**, *76*, 8–19. [CrossRef]
254. Wu, W.-T.; Aubry, N.; Antaki, J.F.; Massoudi, M. Normal stress effects in the gravity driven flow of granular materials. *Int. J. Non-Linear Mech.* **2017**, *92*, 84–91. [CrossRef]
255. Wu, W.-T.; Aubry, N.; Antaki, J.; Massoudi, M. Flow of a Dense Suspension Modeled as a Modified Second Grade Fluid. *Fluids* **2018**, *3*, 55. [CrossRef]
256. Dunn, J.E.; Fosdick, R.L. Thermodynamics, stability, and boundedness of fluids of complexity 2 and fluids of second grade. *Arch. Ration. Mech. Anal.* **1974**, *56*, 191–252. [CrossRef]
257. Tao, C.; Rosenbaum, E.; Kutchko, B.G.; Massoudi, M. *Effects of Shear-rate Dependent Viscosity on the Flow of A Cement Slurry*; NETL-PUB-22169; National Energy Technology Lab (NETL): Pittsburgh, PA, USA, 2018.
258. Hanehara, S.; Yamada, K. Interaction between cement and chemical admixture from the point of cement hydration, absorption behaviour of admixture, and paste rheology. *Cem. Concr. Res.* **1999**, *29*, 1159–1165. [CrossRef]
259. Stutzman, P. Scanning electron microscopy imaging of hydraulic cement microstructure. *Cem. Concr. Compos.* **2004**, *26*, 957–966. [CrossRef]
260. Gebauer, J.; Kristmann, M. The influence of the composition of industrial clinker on cement and concrete properties. *World Cem. Technol.* **1979**, *10*, 46–51.
261. Powers, T.C. Structure and physical properties of hardened Portland cement paste. *J. Am. Ceram. Soc.* **1958**, *41*, 1–6. [CrossRef]
262. Slattery, J.C. *Advanced Transport Phenomena*; Cambridge University Press: Cambridge, UK, 1999; ISBN 1-316-58390-2.
263. Probstein, R.F. *Physicochemical Hydrodynamics: An Introduction*; John Wiley & Sons: Hoboken, NJ, USA, 2005; ISBN 0-471-72512-9.
264. Liu, I.-S. *Continuum Mechanics*; Springer Science & Business Media: Berlin/Heidelberg, Germany, 2013; ISBN 3-662-05056-0.
265. Ziegler, H. *An introduction to Thermomechanics*; Elsevier: Amsterdam, The Netherlands, 2012; Volume 21, ISBN 0-444-59893-6.
266. Malekmohammadi, S.; Naccache, M.F.; Frigaard, I.A.; Martinez, D.M. Buoyancy driven slump flows of non-Newtonian fluids in pipes. *J. Pet. Sci. Eng.* **2010**, *72*, 236–243. [CrossRef]
267. Aranha, P.E.; Miranda, C.R.; Magalhães, J.V.M.; Campos, G.; Martins, A.L. Dynamic Aspects Governing Cement-Plug Placement in Deepwater Wells. *SPE Drill. Complet.* **2011**, *26*, 341–351. [CrossRef]
268. Atkin, R.J.; Craine, R.E. Continuum theories of mixtures: Applications. *IMA J. Appl. Math.* **1976**, *17*, 153–207. [CrossRef]
269. Atkin, R.J.; Craine, R.E. Continuum theories of mixtures: Basic theory and historical development. *Q. J. Mech. Appl. Math.* **1976**, *29*, 209–244. [CrossRef]
270. Bowen, R.M. Theory of mixtures in continuum physics. In *Mixtures and Electromagnetic Field Theories*; Academic Press: New York, NY, USA, 1976.

271. Truesdell, C. Thermodynamics of diffusion. In *Rational Thermodynamics*; Springer: Berlin/Heidelberg, Germany, 1984; pp. 219–236.
272. Rajagopal, K.R.; Tao, L. Mechanics of mixtures. *J. Fluid Mech.* **1996**, *323*, 410.
273. Liu, L.; Fang, Z.; Qi, C.; Zhang, B.; Guo, L.; Song, K.I.-I.L. Numerical study on the pipe flow characteristics of the cemented paste backfill slurry considering hydration effects. *Powder Technol.* **2019**, *343*, 454–464. [CrossRef]
274. Rohani, B.; Wahid, M.A.; Sies, M.M.; Saqr, K.M. Comparison of Eddy Dissipation Model and Presumed Probability Density Function Model for Temperature Prediction in a Non-Premixed Turbulent Methane Flame. In *AIP Conference Proceedings*; AIP Publishing LLC: Melville, NY, USA, 2012; Volume 1440, pp. 384–391. Available online: https://doi.org/10.1063/1.4704240 (accessed on 23 January 2020).
275. Wang, P. The model constant A of the eddy dissipation model. *Prog. Comput. Fluid Dyn. Int. J.* **2016**, *16*, 118–125. [CrossRef]
276. Wallevik, J.E. *Rheology of Particle Suspensions: Fresh Concrete, Mortar and Cement Paste with Various Types of Lignosulfonates*; Fakultet for Ingeniørvitenskap og Teknologi: Trondheim, Norway, 2003; ISBN 82-471-5566-4. Available online: http://hdl.handle.net/11250/236410 (accessed on 23 January 2020).
277. Moyers-Gonzalez, M.A.; Frigaard, I.A. Kinematic instabilities in two-layer eccentric annular flows, part 2: Shear-thinning and yield-stress effects. *J. Eng. Math.* **2009**, *65*, 25–52. [CrossRef]
278. Manzar, M.A.; Shah, S.N. Particle Distribution and Erosion During the Flow of Newtonian and Non-Newtonian Slurries in Straight and Coiled Pipes. *Eng. Appl. Comput. Fluid Mech.* **2009**, *3*, 296–320. [CrossRef]
279. Lootens, D. Ciments et Suspensions concentrées Modèles. ÉCoulement, Encombrement et Flocculation. Ph.D. Dissertation, Université Pierre et Marie Curie-Paris VI, Paris, France, 2004. Available online: https://pastel.archives-ouvertes.fr/tel-00007217/ (accessed on 23 January 2020).
280. Patankar, S. *Numerical Heat Transfer and Fluid Flow*; CRC Press: Boca Raton, FL, USA, 1980; ISBN 1-4822-3421-1.
281. Cremonesi, M.; Ferrara, L.; Frangi, A.; Perego, U. Simulation of the flow of fresh cement suspensions by a Lagrangian finite element approach. *J. Non-Newton. Fluid Mech.* **2010**, *165*, 1555–1563. [CrossRef]
282. Aranha, P.E.; Miranda, C.; Cardoso, W.; Campos, G.; Martins, A.; Gomes, F.C.; de Araujo, S.B.; Carvalho, M. A comprehensive theoretical and experimental study on fluid displacement for oilwell-cementing operations. *SPE Drill. Complet.* **2012**, *27*, 596–603. [CrossRef]
283. Bu, Y.; Li, Z.; Wan, C.; Li, H.A. Determination of optimal density difference for improving cement displacement efficiency in deviated wells. *J. Nat. Gas Sci. Eng.* **2016**, *31*, 119–128. [CrossRef]
284. Zulqarnain, M.; Tyagi, M. Development of simulations based correlations to predict the cement volume fraction in annular geometries after fluid displacements during primary cementing. *J. Pet. Sci. Eng.* **2016**, *145*, 1–10. [CrossRef]
285. Zhao, Y.; Wang, Z.; Zeng, Q.; Li, J.; Guo, X. Lattice Boltzmann simulation for steady displacement interface in cementing horizontal wells with eccentric annuli. *J. Pet. Sci. Eng.* **2016**, *145*, 213–221. [CrossRef]
286. Anglade, C.; Papon, A.; Mouret, M. Constitutive parameter identification: An application of inverse analysis to the flow of cement-based suspensions in the fresh state from synthetic data. *J. Non-Newton. Fluid Mech.* **2017**, *241*, 14–25.
287. Wu, W.; Huang, X.; Li, Y.; Fang, C.; Jiang, X. A modified LBM for non-Newtonian effect of cement paste flow in 3D printing. *Rapid Prototyp. J.* **2019**, *25*, 22–29. [CrossRef]
288. Bao, K.; Lavrov, A.; Nilsen, H.M. Numerical modeling of non-Newtonian fluid flow in fractures and porous media. *Comput. Geosci.* **2017**, *21*, 1313–1324. [CrossRef]
289. Tardy, P.M.J. A 3D model for annular displacements of wellbore completion fluids with casing movement. *J. Pet. Sci. Eng.* **2018**, *162*, 114–136. [CrossRef]
290. Tardy, P.M.J.; Bittleston, S.H. A model for annular displacements of wellbore completion fluids involving casing movement. *J. Pet. Sci. Eng.* **2015**, *126*, 105–123. [CrossRef]
291. Zhou, Y.; Wojtanowicz, A.K.; Li, X.; Miao, Y. Analysis of gas migration in Sustained-Casing-Pressure annulus by employing improved numerical model. *J. Pet. Sci. Eng.* **2018**, *169*, 58–68. [CrossRef]
292. Naccache, M.F.; Pinto, H.A.M.; Abdu, A. Flow displacement in eroded regions inside annular ducts. *J. Braz. Soc. Mech. Sci. Eng.* **2018**, *40*, 420. [CrossRef]
293. Foroushan, H.K.; Ozbayoglu, E.M.; Miska, S.Z.; Yu, M.; Gomes, P.J. On the Instability of the Cement/Fluid Interface and Fluid Mixing (includes associated erratum). *SPE Drill. Complet.* **2018**, *33*, 063–076. [CrossRef]

294. Skadsem, H.J.; Kragset, S.; Lund, B.; Ytrehus, J.D.; Taghipour, A. Annular displacement in a highly inclined irregular wellbore: Experimental and three-dimensional numerical simulations. *J. Pet. Sci. Eng.* **2019**, *172*, 998–1013. [CrossRef]
295. Murphy, E.; Lomboy, G.; Wang, K.; Sundararajan, S.; Subramaniam, S. The rheology of slurries of athermal cohesive micro-particles immersed in fluid: A computational and experimental comparison. *Chem. Eng. Sci.* **2019**, *193*, 411–420. [CrossRef]
296. Rosenbaum, E.; Massoudi, M.; Dayal, K. The Influence of Bubbles on Foamed Cement Viscosity Using an Extended Stokesian Dynamics Approach. *Fluids* **2019**, *4*, 166. [CrossRef]
297. Ohen, H.A.; Blick, E.F. Golden section search method for determining parameters in Robertson-Stiff non-Newtonian fluid model. *J. Pet. Sci. Eng.* **1990**, *4*, 309–316. [CrossRef]
298. Shahriar, A.; Nehdi, M.L. Artificial intelligence model for rheological properties of oil well cement slurries incorporating SCMs. *Adv. Cem. Res.* **2012**, *24*, 173–185. [CrossRef]
299. Shahriar, A.; Nehdi, M. Modeling Rheological Properties of Oil Well Cement Slurries Using Multiple Regression Analysis and Artificial Neural Networks. *Int. J. Mater. Sci.* **2013**, *3*, 13.
300. Shahriar, A.; Nehdi, M.L. Optimization of rheological properties of oil well cement slurries using experimental design. *Mater. Struct.* **2012**, *45*, 1403–1423. [CrossRef]
301. Bassioni, G.; Ali, M.M.; Almansoori, A.; Raudaschl-Sieber, G.; Kühn, F.E. Rapid determination of complex oil well cement properties using mathematical models. *RSC Adv.* **2017**, *7*, 5148–5157. [CrossRef]
302. el Mahdi Safhi, A.; Benzerzour, M.; Rivard, P.; Abriak, N.-E. Feasibility of using marine sediments in SCC pastes as supplementary cementitious materials. *Powder Technol.* **2019**, *344*, 730–740. [CrossRef]
303. Watts, B.; Tao, C.; Ferraro, C.; Masters, F. Proficiency analysis of VCCTL results for heat of hydration and mortar cube strength. *Constr. Build. Mater.* **2018**, *161*, 606–617. [CrossRef]
304. Ferraro, C.C.; Watts, B.; Tao, C.; Masters, F. *Advanced Analysis, Validation and Optimization of Virtual Cement and Concrete Testing*; Florida Department of Transportation: Tallahassee, FL, USA, 2017; Available online: https://trid.trb.org/view/1569681 (accessed on 23 January 2020).

© 2020 by the authors. Licensee MDPI, Basel, Switzerland. This article is an open access article distributed under the terms and conditions of the Creative Commons Attribution (CC BY) license (http://creativecommons.org/licenses/by/4.0/).

MDPI
St. Alban-Anlage 66
4052 Basel
Switzerland
Tel. +41 61 683 77 34
Fax +41 61 302 89 18
www.mdpi.com

Energies Editorial Office
E-mail: energies@mdpi.com
www.mdpi.com/journal/energies

www.ingramcontent.com/pod-product-compliance
Lightning Source LLC
LaVergne TN
LVHW070126100526
838202LV00016B/2240

*9 7 8 3 0 3 9 2 8 7 2 0 8 *